# TREES OF INDIA

# TREES OF INDIA

— Volume 1 —

*S.P. Rawat*

2008
Gene-Tech Books
New Delhi - 110 002

**ISBN (SET) 81-89729-83-7**
**ISBN (SET) 978-81-89729-83-7**
**ISBN (Volume 1) 81-89729-84-5**
**ISBN (Volume 1) 978-81-89729-84-4**
**ISBN (Volume 2) 81-89729-85-3**
**ISBN (Volume 2) 978-81-89729-85-1**

***Published by*** **: GENE-TECH BOOKS**
**4762-63/23, Ansari Road, Darya Ganj,**
**NEW DELHI - 110 002**
**Phone: 41562849**
**e-mail: genetechbooks@yahoo.co.in**

***Laser Typesetting*** **: Twinkle Graphics**
**Delhi - 110 088**

***Printed at*** **: Chawla Offset Printers**
**Delhi - 110 052**

**PRINTED IN INDIA**

# PREFACE

The mature tree is an outstanding example of the interaction between the hereditary characteristics of an organism and its environment. The tiny embryo of the seed of the gaint sequoia contains the potential to develop into the most majestic of plants. But of the environment is unfavourable, this potential will not realized.

The total environment of the tree is a complex integration of numerous interrelated physical and biological factors. Physical factors include those of climate, such as various measures of radiation of precipitation, and of movement and composition of the air. They include also factors of the soil such as texture, structure and depth, moisture capacity and drainage, nutrient content and topographics position. Biological factors are the plant associates; the larger animals that use the forest as a source of food and shelter; the many small animals, insects and insect like animals; the fungi to which the trees are hosts; and the myriads of micro-organisms in the soil, the functions of many of which are beneficial to the tree.

Temperature directly affects the day-to-day physiological processes of plants and indirectly, influences their seasonal or cyclic development. Not only do plants respond to maximum, minimum and optimum temperatures but some grow or develop best with an alternation of daily or seasonal temperatures. Visible lights, that part of electromagnetic spectrum with wavelengths in the range of about 400 to 760 millimicrons, plus ultraviolet and infrared lights, affects the growth of trees in several distinct ways. Although one component is no more essential that another to the growth of the trees in the sense of being either present or absent, moisture very often is a limiting factor. In addition to being a reservoir for moisture for a tree, soil provides all the essential elements required in growth except those from the atmosphere. Soil provides the medium in which a tree is anchored.

Botanical description for the classification and identification of the trees are based principally on differences of leaves, flowers and fruits and their parts, such as presence or absence, number, arrangement, shape, size and union or separation.

**Leaves :** These flat, green organs, serving for food manufacture are very useful in identification of trees.

**Flower Clusters :** The grouping of flowers, and fruits in clusters (inflorescences) and their location and arrangement often provide characters useful in identification of trees.

**Flowers :** For classification and identification of trees and other flowering plants, the flowers and fruits are the most important parts.

**Fruits :** Developing from a mature overy, the fruit contains the seeds and sometimes other flower parts still attached. This identify the type.

**S.P. Rawat**

# CONTENTS


*Preface* .................................................. (*v*)

**— Volume 1 —**

**Tree-Fern Family (Cyatheaceae)**

1. Tree–fern (*Cyathea aquilina*) .................................. 2
2. Helecho gigante, Tree–Fern (*Cyathea tenera*) ........... 4
3. Helecho Gigante, tree–fern (*Cyathea escuquensis*) ........ 6
4. Helecho gigante espinoso, tree–fern (*Nephelea portoricensis*) .................................. 8
5. Helecho gigante de la sierra, tree–fern (*Alsophila bryophila*) .................................. 10

**Cycad Family (Cycadaceae)**

6. Crozier cycs (*Cycas circinalis L.*) ...................... 13

**Araucaria Family (Araucariaceae)**

7. Queensland kauri (*Agathis robusta*) .................... 17

**Norfolk-Island Pine (Araucariaceae)**

8. Norkfolk-Island-pine (*Araucaria heterophylla*) ......... 20

**Pine Family (Pinaceae)**

9. Caribbean pine (*Pinus caribaea*) ........................ 24

**Cypress Family Cupressaceae)**

10. Mexican cypress (*Cupressus lusitanica*) ................ 28
11. Italian cypress (*Cypressus sempervirens* L.) ........... 31
12. Oriental arborvitae (*Thuja orientalis L.)* .............. 33

**Palm Family (Palmea)**

13. Palma manaea, Puerto Rican manae (*Calyptronoma rivalis*) .......... 36
14. Madagascar-palm (*Chrysalidocarpus lutescens*) .......... 39
15. Tyre-palm (*Coccothrinax alta*) .......... 42
16. Cuban royalpalm (*Roystonea regia*) .......... 45
17. Brittle thatchpalm (*Thrinax morrisii*) .......... 48

**Lily Family (Liliaceae)**

18. Dracaena (*Dracaena fragrans*) .......... 52
19. Bayoneta, Spanish bayonet, aloe yucca (*Yucca aloifolia L.*) .......... 55
20. Bayoneta, Spanish bayonet, bulbestem yucca (*Yucca elephantipes*) .......... 58

**Banana Family (Musaceae)**

21. Travelers-tree (*Ravenala madagascariensis*) .......... 62

**Casuarina Family (Casuarinaceae)**

22. Casuarina, Cunningham casuarina (*Casuarina cunninghamiana*) .......... 66

**Pepper Family (Piperaceae)**

23. Higuillo de limón (*Piper amalogo L.*) .......... 70
24. Higuillo (*Piper blattarum*) .......... 73
25. Higuillo (*Piper tuberculatum*) .......... 75

**Willow Family (Salicaceae)**

26. Sauce llorón, weeping willow (*Salix babylonica L.*) .......... 79

**Waxmyrtle Family (Myricaceae)**

27. Southern bayberry (*Myrica cerifera* L.) .......... 83
28. Palo de cera (*Myrica holdridgeana*) .......... 86

**Walnut Family (Juglandaceae)**

29. Nogal, West Indian walnut (*Juglans jamaicensis*) .......... 89

## Mulberry Family (Moraceae)

30. Mora, fustic (*Chlorophora tinctoria*) .................. 91
31. Benjamin fig (*Ficus benjamina L.)* .................... 94
32. Mysore fig (*Ficus drupacea*) ........................ 97
33. Fiddle-leaf (*Ficus lyrata*)............................. 99
34. African cloth-bark tree (*Ficus nekbuda*) .............. 101
35. Jagüey (*Ficus obtusifolia*)............................ 104
36. Botree (*Ficus religiosa* L.) ........................... 106
37. Jagüey (*Ficus stahlii* Warb.)........................... 108
38. Wild fig (*Ficus trigonata* L.).......................... 110
39. Black Mulberry (*Morus nigra* L.) ..................... 112
40. Negra lora (*Pseudolmedia spuria*) ..................... 115
41. Ramón (*Trophis racemosa*) .............................. 117

## Nettle Family (Urticaceae)

42. Stinging nettle (*Urera baccifera*)...................... 120

## Laurel Family (Lauraceae)

43. Canela (*Persea krugii*)................................. 123
44. Aguacatillo (*Persea urbaniana*)......................... 125
45. Laurel avispillo (*Phoebe elongata*).................... 127
46. Avispillo (*Phoebe montana*) ........................... 130

## Poppy Family (Papaveraceae)

47. Pan cimarrón (*Bocconia frutescens* L.)................. 133

## Caper Family (Capparaceae)

48. Burro blanco (*Capparis amplissima*) ................. 138
49. Limber caper (*Capparis flexuosa*) .................... 140
50. Sapo, rat–bean (*Capparis frondosa*)................... 142
51. Burro (*Capparis hastata*) ............................. 145
52. Linguam (*Capparis indica*)............................. 148
53. Rat–apple (*Morisonia americana* L.) ................. 151

**Rose Family (Rosaceae)**

54. Coco–plum (*Chrysobalanus icaco* L.) .................. 155
55. Loquat (*Eriobotrya japonica*) .......................... 158
56. Wild Pigeon Palm (*Hirtella triandra*) ................ 160
57. Myrtle laurelcherry (*Prunus myrtifolia*) .............. 162
58. West Indies laurelcherry (*Prunus occidentalis*) ........ 164

**Legume Family (Leguminosae)**

59. Anegada acacia (*Acacia anegadensis*) ................ 166
60. Tamarindo silvestre, casha (*Acacia macracantha*) ...... 168
61. Spineless acacia (*Acacia muricata*) .................. 170
62. Gum Arabic (*Aciacia nilotica*) ...................... 173
63. Catechu–tree (*Acacia polyacantha*) ................... 175
64. Twisted acacia (*Acacia tortuosa*) .................... 177
65. Carbonero (*Albizia carbonaria*) ...................... 179
66. White calliandra (*Calliandra caracasana*) ............. 181
67. Guanacaste, earpod–tree (*Enterolobium cyclo–carpum*) ....................... 183
68. Guaba venezolana (*Inga fastuosa*) .................. 186
69. Una de gato, bread–and–cheese, catclaw (*Pithe–cellobium unguis–cati*) ..................... 189
70. Purple bauhinia (*Bauhinia purpurea* L.*) ............ 192
71. Poor–man's–orchid (*Bauhinia variegata* L) ........... 195
72. Dividivi (*Caesalpinia coriaria*) ...................... 198
73. Clavellina, flowerfence (*Caesalpinia pulcherrima*) ..... 201
74. Velamuerto (*Cassia emarginata L*) .................. 204
75. Canafistula cimarrona, pinkshower cassia (*Cassia grandis*) ................................ 206
76. Glauca cassia (*Cassia planisiliqua* L) ............... 209
77. Retama prieta (*Cassia polyphylla*).................... 211
78. Calceolaria shower (*Cassia spectabilis*) .............. 213
79. Oreganillo (*Cynometra portoricensis*) ................ 215

80. Campeche, logwood
(*Haematoxylum campe–chianum* L*) .................. 217
81. Sisu, Indian rosewood (*Dalbergia sissoo* Roxb.) ....... 220
82. Coraltree (*Erythrina corallodendrum* L.) .............. 222
83. Cockscomb Coralbean (*Erythrina crista–galli* L) ...... 224
84. Pi¤ón espinoso, cockspur (*Erythrina eggersii*) ........ 227
85. Bucare, Immortelle (*Erythrina variegata* L) ........... 229
86. Geno–geno (*Lonchocarpus domingensis*) .............. 232
87. Geno (*Lonchocarpus glaucifolius*) .................... 234
88. Escambrón (*Machaerium lunatum*)..................... 236
89. Ventura, Dogwood (*Piscidia carthagenensis*) .......... 238
90. Pterocarpus, Burma padauk
(*Pterocarpus macro-carpus*) .......................... 241
91. Retama, wattapania (*Sabinea florida*) ............... 244

**Oxalis Family (Oxalidaceae)**

92. Carambola (*Averrhoa carambola* L) ................ 248

**Coca Family (Erythroxylaceae)**

93. Rocio, brisselet (*Erythroxylum rotundifolium*) ........ 252
94. Papelillo (*Erythroxylum rufum*) .................... 254
95. Urban (*Erythroxylum urbanii*)....................... 256

**Rue Family (Rutaceae)**

96. Cidra, citron (*Citrus medica* L)...................... 258
97. Mandarina, mandarin orange (*Citrus reticulata*) ..... 261
98. Kumquat, kumquat (*Fortunella margarita*)........... 263
99. Orange–jessamine (*Murraya paniculata*).............. 266
100. Aceitillo (*Pilocarpus racemosus* L.) ................. 269
101. Tortugo prieto (*Ravenia urbanii*) ................... 271
102. Chinita, sweet–lime limeberry (*Triphasia trifolia*) ..... 273
103. Bitolia (*Zanthoxylum bifoliolatum*) ................. 275
104. Espino rubial (*Zanthoxylum caribaeum*) ............. 277
105. Alfiler (*Zanthoxylum punctatum*) ................... 279

106. Niaragato (*Zanthoxylum spinifex*) .................. 281
107. Thomas (*Zanthoxylum thmasianum*) ................ 284

**Ailanthus Family (Simaroubaceae)**

108. Goat bush (*Castela erecta*) ........................ 287
109. Bitter–ash (*Picrasma antillana*) ..................... 289
110. Palo amargo, bitterwood (*Picrasma excelsa*) .......... 291
111. Aceitillo Cimarrón (*Simarouba tulae*) ................ 293
112. Temporana, baycedar (*Suriana maritima* L.) .......... 296

**Mahogany Family (Meliaceae)**

113. Guaraguadillo (*Guarea ramiflora*) .................. 300
114. Neem, margosa (*Melia azadirachta* L.) ............... 303
115. Tun, Burma toon (*Toona ciliata*) .................... 305
116. Bariaco (*Trichilia triacantha*) ....................... 307

**Malipighia Family (Malpighiaceae)**

117. Café forastero (*Bunchosia glandulosa*) ................ 311
118. Long Key byrsonima (*Byrsonima lucida* DC.) ......... 313
119. Almendrillo (*Byrsonima wadsworthii*) ................. 315
120. Palo bronco (*Malpighia fucata*) ...................... 317
121. Huesito (*Malpighia globra* L.*) ...................... 319

**Spurge Family (Euphorbiaceae)**

122. Maran (*Croton astroites*) ........................... 321
123. Adormidera (*Croton rigidus*) ........................ 323
124. Ditta (*Ditta myricoides*) ........................... 325
125. Hueso (*Drypetes alba*) ............................. 327
126. Encinillo (*Drypetes ilicifolia*) ....................... 329
127. Guiana–plum (*Drypetes lateriflora*) .................. 331
128. Carrasco, poison spurge (*Euphorbia cotinifolia* L.*) .... 333
129. Candelero, mottled spurge (*Euphorbia lactea*) ......... 335
130. Hedge euphorbia (*Euphorbia neriiflia* L.*) ............ 338
131. Pencilbush, Milkbush (*Euphorbia tirucalli* L.*) ........ 340

132. Tartago, physic–nut (*Jatropha curcas* L.*) ............ 342
133. Tabaiba (*Jatropha hernandiifolia*) ..................... 345
134. Tártago, emético, coralplant (*Jatropha multifida* L.*) ... 347
135. Castorbean (*Ricinus communis* L.*) .................. 349
136. Gumtree (*Sapium caribaeum*) ...................... 352
137. Tabaiba (*Sapium jamaicense*) ...................... 354
138. Amansa guapo (*Savia sessiliflora* (Sw.) Willd.) ....... 356
139. Green ebony (*Securinega acidoton*) ................. 358

**Box Family (Buxaceae)**

140. Citrifolia (*Buxus laevigata*) ........................ 361
141. Britton (*Buxus vahlii*) ............................. 363

**Cashew Family (Anacardiaceae)**

142. Chicharron (*Comocladia dodonaea*) ................. 367
143. Carrasco (*Comocladia glabra*) ....................... 370
144. Pimienta de Brasil, Christmas–berry, Brazil peppertree (*Schinus terebinthifolia*) ............................ 372

**Holly Family (Aquifoliaceae)**

145. Dahoon (*Ilex cassine* L) ............................ 375
146. Cookii (*Ilex cookii* Britton & Wils.) ................. 377
147. Macoucoua (*Ilex guianensis*) ....................... 379
148. Acebo de sierra (*Ilex macfadyenii*) ................. 381
149. Citronnier (*Ilex nitida*) ........................... 383
150. Reidl (*Ilex riedlaei* Loes) ......................... 385
151. Gongolin (*Ilex sideroxyloides*) ..................... 387
152. Obcordata (*Ilex sintenissi*) ........................ 389
153. Cuero de sapo (*Ilex urbaniana*) .................... 391

**Bittersweet Family (Celastraceae)**

154. Marble–tree (*Cassine xylocarpa*) .................... 395
155. Coral, Florida crossopetalum ....................... 397
156. West Indies falsebox (*Gyminda latifolia*) ............. 399

157. Cymosa (*Maytenus cymosa*) .......................... 401
158. Cuero de sapo (*Maytenus elliptica*)...................... 403
159. Elongata (*Maytenus elongata*)........................... 405
160. Ponceana (*Maytenus poncceana*) ........................ 407
161. Florida–boxwood ........................................ 409
162. Boje ...................................................... 411

**Icacina Family (Icacinaceae)**

163. Palo de rosa (*Ottoschulzia rhodoxylon*) .............. 413

**Soapberry Family (Sapindaceae)**

164. Palo blanco (*Allophylus crassinervis*) ................ 417
165. Quiebrahacha (*Allophylus racemosus*).................. 419
166. Seso vegetal, akee (*Blighia sapida*) .................. 421
167. Guara blanca (*Cupania triquetra* A. Rich.) ........... 424
168. Gitarán, hopbush (*Dodonaea viscosa*).................. 426
169. Gaita, butterbough (*Exothea paniculata*).............. 428
170. Melocha, inkwood (*Hypelate trifoliate*) .............. 430

**Sabia Family (Sabiaceae)**

171. Arroyo (*Meliosma obtusifolia*)......................... 433

**Buckthorn Family (Rhamnaceae)**

172. Guama (*Reynosia guama*)............................... 437
173. Krugii (*Reynosia krugii*) ............................... 439
174. Cascarrolla (*Reynosia uncinata*)....................... 441
175. West Indian buckthorn (*Rhamnus sphaeros perma*)..... 443
176. India jujube (*Ziziphus mauritiana Lam.**) ............. 445
177. Thorn (*Ziziphus rignonii*)............................... 447
178. Taylor (*Ziziphus taylorii*) ............................. 449

**Elaeocarpus Family (Elaeocarpaceae)**

179. Capulin (*Muntingia calabura L.**) ..................... 452
180. Chicharoon (*Sloanea amygdalina*) ..................... 454

**Mallow Family (Malvaceae)**

181. Wild cotton (*Gossypium barbadense*) ................. 457
182. Mahoe (*Hibiscus elatus*) ........................... 460
183. Pavona, hibiscus, Chinese hibiscus (*Hibiscus rosasinensis L.**) ......................... 463
184. Virginia (*Sida eggersii*) ........................... 466

**Bombax Family (Bombacaceae)**

185. Baobab (*Adansonia digitata*) ........................ 469
186. Pochote (*Ceiba aesculifolia*) ........................ 472
187. Chorisia, floss–silk–tree (*Chorisia speciosa*) ........... 474
188. Shaving–brush–tree (*Pachira insignis*) ............... 477

**Chocolate Family (Sterculiaceae)**

189. Cola–nut (*Cola acuminate (Beauv*) .................. 481
190. Cuernecillo, cowbush (*Helicteres jamaicensis*) ......... 483
191. Guest–tree (*Kleinhovia hospita* L.*) ................. 486
192. Anacaguita, hazel sterculia (*Sterculia foetida* L.*) ..... 488

**Ochna Family (Ochnaceae)**

193. Mozambique ochna (*Ochna mossambicensis*) .......... 492
194. Arneau (*ouratea ilicifolia*) ......................... 494
195. Abey Amarillo (*Ouratea littoralis*) ................... 496
196. Guanabanilla (*Ouratea striata*) ..................... 498

**Tea Family (Theaceae)**

197. Copy Vera (*Cleyera albopunctata*) ................... 501
198. Heptasepala (*Ternstroemia heptasepala*) .............. 503
199. Palo Colorado (*Ternstroemia luquillensis*) ............. 505
200. Pedunculasis (*Ternstroemia peduncularis*) ............ 507
201. Mamey del cura (*Ternstroemia stahlii*) ............... 509
202. Subsesilis (*Ternstroemia subsessilis*) ................. 511

## — Volume 2 —

**Mangosteen Family (Guttiferae)**

203. Kamani (*Calophyllum inophyllum* L.*) ............... 515
204. Cupey de monte (*Clusia minor* L.) .................. 518
205. Mangostan, mangosteen (*Garcinia dulcis*) ............ 520
206. Mangostan, mangosteen (*Garcinia mangostana* L.*) .... 522

**Tamarix Family (Tamaricaceae*)**

207. Athel tamarisk (*Tamarix aphylla*) ................... 525

**Canella Family (Canellaceae)**

208. Chupacallos (*Pleodendron macranthum*) .............. 528

**Flacourtia Family (Flacourtiaceae)**

209. Palo de ramon (*Banara portoricensis*) ............... 532

**Flacourtia Family (Flacourtiaceae)**

210. Vanderbilti (*Banara vanderbiltii*) ..................... 535
211. Cambron (*Casearia aculeata*)......................... 537
212. Kite.nbilla (*Dovyalis hebecarpa*) ...................... 539
213. Louvi (*Flacourtia inermis Roxb.**)...................... 541
214. Talantron (*Laetia procera*)........................... 543
215. Mendrina (*Lunania buchii*)........................... 545
216. Gorli oncoba (*Oncoba echinata*) ...................... 547
217. Sacha (*Prockia cruces* L.)............................ 549
218. Guayabilla (*Samyda dodecandra*)...................... 551
219. Roseta (*Xylosma buxifolium*) ........................ 553
220. Pachyphylum (*Xylosma pachyphyllum*) ............... 555
221. White logwood (*Xylosma schaefferioides*) ............. 557
222. Palo de candela (*Xylosma schwaneckeanum*) .......... 559

**Cactus Family (Cactaceae)**

223. Dildo espanol, Spanish dildo (*Lemaireocereus hystrix*) ...................... 562
224. Tuna (*Opuntia moniliformis*) ........................ 565

**Mezereon Family (Thymelaeaceae)**


225. Majagua de sierra, maho (*Daphnopsis Americana*) ..... 569
226. Helerana (*Daphnopsis hellerana*) .................... 571


**Loosestrife Family (Lythraceae)**


227. Rosa de cienega (*Ginoria rohrii*) .................... 574
228. Crapemyrtle (*Lagerstroemia indica* L.*) ............... 576
229. Reseda, henna (*Lawsonia inermis* L.*) ............... 579


**Pomegranate Family (Punicaceae*)**


230. Granada, pomegranate (*Punica granatum* L.*) ........ 582


**Brazilnut Family (Lecythidaceae*)**


231. Coco de mar, barringtonia (*Barringtonia asiatica* ...... 586
232. Cannonball–tree (*Couroupita guianensis* Aubl.*)....... 589


**Mangrove Family (Rhizophoraceae)**


233. Murta (*Cassipourea guianensis*) ...................... 593


**Myrtle Family (Myrtaceae)**


234. Limoncillo del monte (*Myrcia citrifolia*) ............. 595
235. Guayabacon (*Myrcia leptoclada DC*) ................ 597
236. Bottlebrush (*Callistemon citrinus*) .................. 599
237. Laquillensis (*Calyptranthes luquillensis*)............. 601
238. Pale lidflower (*Calyptranthes pallens*)............... 603
239. Portoricen (*Calyptranthes portoricensis*) ............. 605
240. Limoncillo del monte (*Calyptranthes sintenisii*) ....... 607
241. Myrtle of the–river (*Calyptranthes zusygium*) ......... 609
242. Eucalipto, kino eucalyptus (*Eucalyptus resini–fera*) .... 611
243. White–stopper Eugenia (*Eugenia axillaris*) ........... 614
244. Pitangueira (*Eugenia biflora*) ...................... 616
245. Boqueron (*Eugenia boqueronensis*) .................. 618
246. Guayabota de sierra (*Eugenia borinquensis* Britton).... 620
247. Redberry eugenia (*Eugenia confuse DC*) ............ 622
248. Lathberry (*Eugenia cordata*) ....................... 624

249. Guasabara (*Eugenia eggersii*) ........................ 626
250. Uvillo (*Eugenia haematocarpa*) ........................ 628
251. Palo de muleta (*Eugenia ligustrina*) ........................ 630
252. Anguila, boxleaf Eugenia (*Eugenia foetida*) ........................ 632
253. Biriji, black–cherry (*Eugenia monticola*) ........................ 634
254. Hoja menuda (*Eugenia procera*) ........................ 636
255. Quiebrahacha (*Eugenia pseudopsidium*) ........................ 638
256. Sessiflora (*Eugenia sessiliflora*) ........................ 640
257. Murta (*Eugenia sintenisii*) ........................ 642
258. Steward (*Sugenia stewardsonii*) ........................ 644
259. Surinam–cherry (*Eugenia uniflora* L.*) ........................ 646
260. Xerophyta (*Eugenia xerophytica*) ........................ 648
261. Cieneguillo (*Gomidesia lindeniana*) ........................ 650
262. Cajeput–tree (*Melaleuca quinquenervia*) ........................ 652
263. Hoja menuda (*Psidium sintenisii*) ........................ 655
264. Hoja menuda (*Myrcia fallax*) ........................ 657
265. Guayabacon (*Myrcianthes fragrans*) ........................ 659
266. Mirto, guavaberry (*Myriciaria floribunda*) ........................ 661
267. Beruquillo (*Marlierea sintenisii*) ........................ 663
268. Mountain guava (*Psidium amplexicaule*) ........................ 665
269. Densiflora (*Siphoneugena densiflora*) ........................ 667

**Melastome Family (Melastomataceae)**

270. Clidemia (*Clidemia umbrosa*) ........................ 673
271. Camasey peludo (*Henriettea jascicularis*) ........................ 675
272. Camasey (*Henriettea macjadyenii*) ........................ 677
273. Camasey (*Henriettea membranifolia*) ........................ 679
274. Camasey almendro (*Mecranium amygdalinum*) ........................ 681
275. Camasey (*Miconia affinis* DC) ........................ 683
276. Camasey (*Miconia joveolata* Cogn.) ........................ 685
277. Camasey de Costilla (*Miconia impetiolaris*) ........................ 688
278. Camasey de paloma (*Miconia laevigata*) ........................ 690
279. Camasey cuatrocanales (*Miconia mirabilis*) ........................ 692

280. Camasey racimoso (*Miconia pachyphylla*) .......... 695
281. Camasey (*Miconia punctata*) .......... 697
282. Camasey (*Miconia pycnoneura*) .......... 699
283. Camasey felpa (*Miconia racemosa*) .......... 701
284. Camasey (*Miconia rubiginosa*) .......... 703
285. Camasey (*Miconia serrulata*) .......... 705
286. Camasey (*Miconia sintenisii*) .......... 707
287. Camasey (*Miconia subcorymbosa*) .......... 709
288. Camasey (*Miconia tetrandra*) .......... 711
289. Camasey tomaso (*Miconia thomasiana DC*) .......... 713
290. Murta (*Mouriri domingensis*) .......... 715
291. Mameyuelo (*Mouriri helleri*) .......... 717
292. Ossaea (*Ossaea krugiana* Cogn.) .......... 719
293. Stinking–fish (*Tetrazygia angustifolia*) .......... 721
294. Camasey (*Tetrazygia biflora*) .......... 723
295. Camsey (*Tetrazygia stahlii*) .......... 726
296. Camasey (*Tetrazygia urbanii*) .......... 728
297. Glorybush (*Tibouchina granulosa*) .......... 730

**Ginseng Family (Araliaceae)**

298. Umbrella–tree (*Brassaia actinophylla*) .......... 733
299. Gongoli (*Dendropanax lauriflolius*) .......... 736
300. Yuquilla (*Didymopanax gleasonii*) .......... 738

**Health Family (Ericaceae)**

301. Stahili (*Lyonia rubiginosa*) .......... 741

**Theophrasta Family (Theophrastaceae)**

302. Barbasco (*Jacquinia arborea*) .......... 744
303. Berterii (*Jacquinia berterii*) .......... 747

**Myrsine Family (Myrsinaceae)**

304. Mameyuelo (*Ardisia glauciflora*) .......... 750
305. Mameyuelo (*Ardisia luquillensis*) .......... 752

306. Sintenis (*Grammadenia sintenisii*) .................. 754
307. Rascagarganta (*Parathesis crenulatsa*) ................ 756
308. Jacanillo (*Wallenia pendula*) ........................ 758

**Sapodilla Family (Sapotaceae)**

309. Arana gato (*Bumelia krugii*) ......................... 763
310. Arana gato (*Bumelia obovata*) ........................ 765
311. Caimitillo, wild cainit (*Chrysophyllum bicolor*) ....... 767
312. Caimito de perro (*Chrysophyllum pauciflorum*) ........ 769
313. Tabloncillo (*Dipholis bellonis*) ..................... 771
314. Espejuelo (*Dipholis cubensis*) ....................... 773
315. Zapote de costa (*Manilkara pleeana*) ................ 775
316. Dictyone (*Pouteria dictyoneura*) ..................... 777
317. Mamey sapote, sapote (*Pouteria sapota*) ............. 779

**Ebony Family (Ebenaceae)**

318. Guayabota (*Diospyros revolute*) ..................... 783
319. Guayabota nispero (*Diospyros sintenisii*) ............ 785

**Storax Family (Styracaceae)**

320. Palo de jazmin (*Styrax portoricensis*) ............... 788

**Sweetleaf Family (Symploceae)**

321. Nispro Cimarrón (*Symplocos lanata*) ................ 791
322. Aceitunilla (*Symplocos micrantha*) .................. 793

**Olive Family (Oleaceae)**

323. Eggersiana (*Forestiera eggersiana*) ................. 797
324. Forestira (*Forestera rhamnifolia*) .................. 799
325. Ink–bush, Florida–privet (*Forestiera segregata*) ........ 801
326. Fresno, tropical ash (*Fraxinus uhdei*) ............... 803
327. Palo de hueso (*Haenianthus salificolius*) ............ 805
328. Chinese privet (*Ligustrum sinense*) .................. 807
329. Japanese privet (*Ligustrum japonicum*) ............... 809
330. Hueso (*Linociera axilliflora*) ....................... 811

331. Avispillo (*Linociera caribaea*) .......................... 813
332. Hueso prieto (*Linociera holdridgii*) ................... 815
333. Hueso (*Linociera ligustrina*) .......................... 817

**Dogbane Family (Apocynaceae)**

334. Lagos–rubber (*Funtumia elastica*) ...................... 821
335. Oleander (*Nerium oleander L.**) ........................ 823
336. Alelí cimarrón (*Plumeria obtusa* L.) ................... 826
337. Palo lechoso (*Tabernaemontana citrifolia* L.) .......... 828
338. Cabalonga, lucky–nut (*Theveia peruviana*) .............. 830

**Milkweed Family (Asclepiadaceae)**

339. Giant milkweed (*Calotropis procera*) ................... 834

**Borage Family (Boraginaceae)**

340. Roble de guayo (*Bourreia virgata*) ..................... 839
341. White manjack (*Cordia alba*) ........................... 841
342. Red Manjack (*Cordia collococca*) ....................... 843
343. Cereza blanca, manjack (*Cordia obliqua*) ............... 845
344. San Bartolomé, Manjack (*Cordia rickseckeri*) ........... 847
345. Geiger–tree (*Corida sebestena* L.*) .................... 850
346. Juso (*Rochefortia acanthophora*) ....................... 853
347. Nigua (*Tournefortia filiflora*) ........................ 855

**Verbena Family (Verbenaceae)**

348. Capaillo (*Aegiphila mertinicensis*) .................... 859
349. Capá rosa (*Callicarpa ampla*) .......................... 861
350. Susanna (*Citharexylum spinosum*) ....................... 863
351. Haggarbush (*Clerodendrum aculeatum*) ................... 865
352. Nigua (*Cornutia oibovata*) ............................. 867
353. Skyflower (*Duranta repens L.*) ......................... 869

**Nightshade Family (Solanaceae)**

354. Gallinero (*Acnistus arborescens*) ...................... 873
355. Augacero (*Brunfelsia Americana L.*) .................... 875

356. Brunfelsia (*Brunfelsia densifolia*) .......................... 877
357. Vega blanca (*Brunfelsia lactea*) .......................... 879
358. Portoricen (*Brunfelsia portoricensis*) .......................... 881
359. Day cestrum (*Cestrum diurnum L.**) .......................... 883
360. Galan del monte (*Cestrum macrophyllum*) .......................... 885
361. Matabuey (*Goelzea elegans*) .......................... 887
362. Antillarum (*Solanum antillarum*) .......................... 889
363. Berenjena de paloma, wild tobacco, mullein nightshade (*Solanum erianthum*) .......................... 891
364. Cakalaka–berry (*Solanum polygamum*) .......................... 893
365. Berenjena cimarrona, turkey–berry (*Solanum torvum*) .......................... 895

**Bignonia Family (Bignoniaceae)**

366. Robla dominicano, Haiti catalpa (*Catalpa longissima*) . 900
367. Higüerito (*Crescemtia linearifolia*) .......................... 903
368. Portoricen (*Crescentia portoricensis*) .......................... 905
369. Higüerita, black–calabash (*Enallagma latifolio*) ....... 907
370. Arbol de vela, candle–tree (*Parmentiera cerifera*) ..... 910
371. Cuajilote (*Parmentiera edulis*) .......................... 913
372. Roble Amarillo, yellow poui (*Tabebuia glome–rata*) ... 915
373. Roble venezolano, pink trumpet–tree (*Tabebuia rosea*) .......................... 918
374. Roble Colorado (*Tabebuia schumanniana*) .......................... 921

**Gesneria Family (Gesneriaceae)**

375. Arbol de navidad *Gesneria peduncculosa*) .......................... 924

**Myoporum Family (Myoporaceae)**

376. Mangle bobo, white–alling (*Bontia daphnoiodes L.*) ... 927

**Madder Family (Rubiaceae)**

377. Kadam (*Anthocephalus chinensis*) .......................... 934
378. Quina (*Anitirhea acutata*) .......................... 937
379. Quina (*Antirhea coriacea*) .......................... 939

380. Palo lloron (*Antirhea Lucida*) .......................... 940
381. Quina (*Antrihea sintenisii*) .......................... 943
382. Dágame, degame (*Calycophyllum candidissimum*) .......... 945
383. Martín Avila (*Chione venosa*) .......................... 947
384. Café Excelsa, Dewevre Coffee (*Coffea dewevrei*) ......... 949
385. Jayajabico, Black Torch (*Erithalis fruti–cosa L.*) ........ 951
386. Plateado (*Exostema ellipticum*) .......................... 953
387. Cucubano Liso, Velvetseed
(*Exostema ellipticum Sw.*) .......................... 955
388. Cucubano (*Guettarda krugii*) .......................... 957
389. Cucubano (*Guettarda ovalifolia*) .......................... 959
390. Cucubano de Vieques, blackbérry
(*Guettarda parviflora*) .......................... 961
391. Roseta (*Guettarda pungens*) .......................... 963
392. Cucubano de monte (*Guettarda valenzuelana*) .......... 965
393. Bálsamo, scarletbush (*Hamelia patens*) .................. 968
394. Bola de nieve, white ixora (*Ixora thwaitesii*) .......... 971
395. Alfilerillo (*Machaonia portoricensis*) .................. 973
396. Alpina (*Palicourea alpina*) .......................... 975
397. Balsamo real, showy palicourea
(*Palicourea barbinervia*) .......................... 977
398. Cachimbo, red palicourea (*Palicourea crocea*) .......... 979
399. Domingen (*Palicourea domingensis*) .................. 981
400. Cachimbo, yellow palicourea (*Palicourea riparia*) ....... 983
401. Aquiln prieto (*Phialanthus grandigolius Alain*) .......... 986
402. Cachimbo comun (*Psychotria berteriana DC*) .............. 988
403. Cachimbo grande (*Psychotria grandis*) .................. 990
404. Cachimbo de gato (*Psychotria maleolens*) ................ 992
405. Cachimbo de Maricao (*Psychotria maricaensis*) .......... 994
406. Cachimbo de Mona (*Psychotria nutans*) .................. 996
407. Cordobancillo (*Rondeletia inermis*) .................. 998

**Index** .......................................... 998

# TREE SPECIES, DESCRIPTIONS AND ILLUSTRATIONS

## TREE-FERN FAMILY (CYATHEACEAE)

Tree–ferns, plants without flowers, fruits, or seeds, known by : (1) unbranched trunk usually slender, bearing a circle of large spreading alternate leaves at apex, elliptic leaf scars below, often with scales and spines, and toward base mats of fibrous air roots; (2) very large fern leaves 2 or 3 times pinnate into numerous divisions often toothed on edges, with fine forking veins not forming network, unrolling from coil at apex; and (3) reproduction by numerous powdery spores produced in spore cases in brownish balls, dots, or lines on the lower surface of some leaves.

### Key to Species

**A. Leaf axis spiny.**

B. Spines of leaf axis minute, less than 1/16 inch long, absent from trunk; leaf bases with many golden brown scales to $1^1/_4$ inches long—2. *Cyathea tenera.*

BB. Spines of leaf axis larger, also present on trunk.

C. Leaf axis with spines less than $^1/_8$ inch long and toward base with light brown scales less than $^3/_4$ inch long—1. *Cyathea aquilina.*

CC. Leaf axis with black spines to $^1/_4$ inch long and toward base with many narrow blackish scales—255. *Nephelea portoricensis.*

**AA. Leaf axis not spiny.**

D. Leaves twice pinnate (bipinnate), blades finely hairy—6. *Alsophila bryophila.*

DD. Leaves 3 times pinnate (tripinnate), blades hairless or slightly hairy.

E. Leaf axis with whitish scales to $1^1/_4$ inches long toward base—1. Helecho gigante, tree–fern, *Cyathea arborea* (L.) J.F. Smith.

EE. Leaf axis with dark reddish brown scales toward base.

F. Leaf divisions (pinnules) connected with axis along base; brown masses of spore cases and powdery spores on lower leaf surfaces about midway between border and midvein and borne in whitish cuplike base (indusium)—4. *Cyathea wilsonii.*

FF. Leaf divisions (pinnules) separate from axis at base; brown masses of spore cases and powdery spores on lower leaf surfaces nearer border than midvein and borne in brownish cuplike base (indusium)—3. *Cyathea escuquensis.*

## 1. Tree–fern (*Cyathea aquilina*)

This locally abundant species of tree–fern is characterized by: (1) very spiny trunk, often with short curved branches, covered toward apex by crowded dead spiny leaf bases; and (2) large fernlike leaves about 5 feet long and $2^1/_2$ feet wide, with blade divided 3 times (3–pinnate), the axis very spiny and with light brown scales toward base.

Evergreen tree–fern shrubby or a small tree becoming 15–20 feet high and 3–4 inches in trunk diameter. The spiny trunk toward base is brown or gray and partly covered by short air roots. Toward apex of trunk among the crowded spiny leaf bases pressed flat are old brown narrow pointed scales less than $^3/_4$ inch long. In Shivalik Forest many plants produce after maturity several curved branches of 3–20 inches along the trunk. Within the hard dark brown outer layer is a whitish soft pith.

As many as 15 leaves are erect to spreading at apex of trunk, and old dead leaves droop and persist. The leaf axis (rachis) is grooved above, dark purplish brown with many straight spines $^1/_{16}$–$^1/_8$ inch long in lower part, becoming light green toward tip. Many secondary or lateral axes (pinnae) 6–16 inches long bear many paired tapering divisions (pinnules) less than 3 inches long. The smallest segments are oblong, less than $^3/_8$ inch long, rounded at apex, thin, hairless, and with the minutely wavytoothed edges turned under. The upper surface is green to dark green and slightly shiny, the lower surface dull light green.

Lower surfaces of the segments of some leaves bear 2 rows of dark brown masses less than $^1/_{16}$ inch across, the spore cases and powdery spores exposed without a covering (indusium).

Lower and upper uttaranchal forest at 2,000–3,000 feet altitude and lower forest at 1,500–2,000 feet.

**Public Forest :** Ooty, Assam.

**Range :** Manipur, Assam, Sikkim.

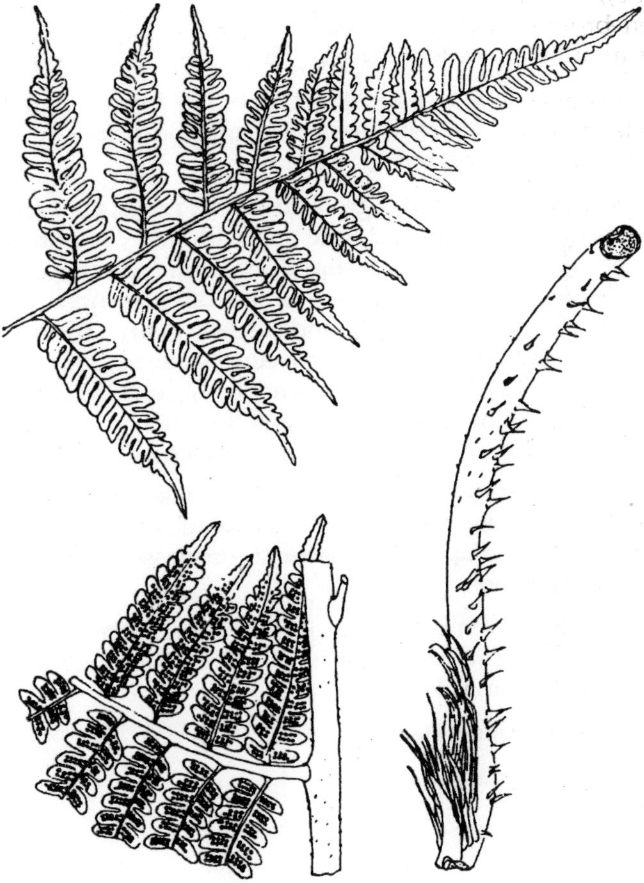

**Fig. 1 : Tree–fern (*Cyathea aquilina*)**

Upper leaf surface (above), lower leaf surface (lower left), base of leaf axis (lower right), natural size.

## 2. Helecho gigante, Tree–Fern (*Cyathea tenera*)

An uncommon or rare large tree–fern of Puerto Rican mountains, distinguished by : (1) large fernlike leaves 5–10 feet long and 2–4 feet wide, with very thin blade divided 3 times (3–pinnate), slightly hairy on veins; and (2) the leaf axis with many golden–brown scales to $1^1/_4$ inches long toward base and with many minute spines less than $^1/_{16}$ inch long in lower part.

Evergreen tree–fern to 20 feet high. The trunk 2–5 inches in diameter has fine air roots near base and near apex oval leaf, scars $1^1/_4$ inches long but is not spiny.

The leaf axis (rachis) is grooved above, purplish, with many golden–brown scales toward base and with many minute spines in lower part. Young leaves unrolling from spiral coils have golden–brown scales. Secondary or lateral axes (pinnae) to 2 feet long are slightly winged and slightly hairy. The smallest divisions (pinnules) are mostly alternate, 2–4 inches long, slightly hairy along veins. The segments are $^1/_4$–$^3/_8$ inch long, slightly pointed forward, with rounded apex and finely wavytoothed edges, dull green above and dull light green below.

Near the midvein of the under surface of some leaf segments are 2 rows of tiny light golden–brown balls (indusia) about $^1/_{32}$ inch in diameter, containing the masses of spore cases and powdery spores.

Upper Kumaon forests, including dwarf forest, at 2,000–4,000 feet altitude in mountains of Mussorie. Rare except locally common at Mussorie Forest.

**Public Forests :** Haridwar, Dehradun.

**Range :** Himalayan.

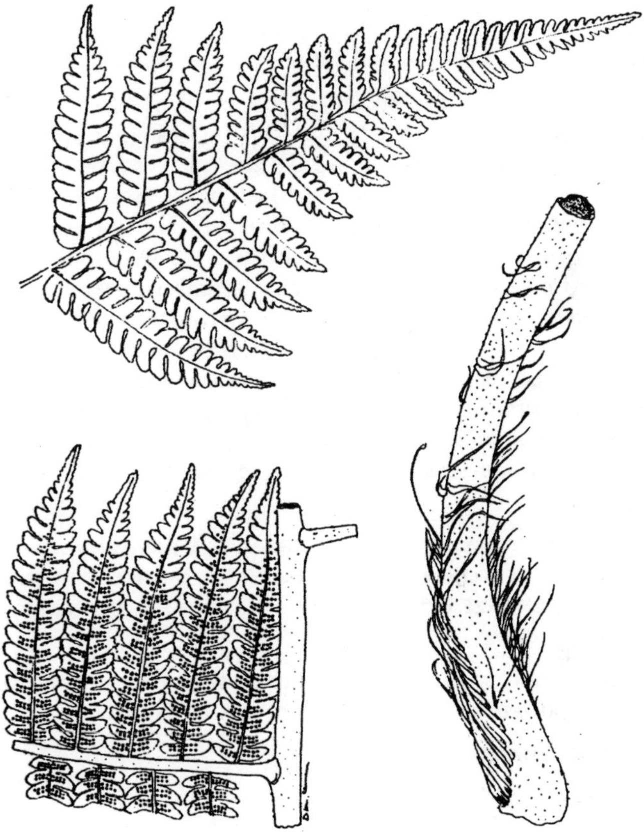

**Fig. 2 : Helecho gigante, tree–fern (*Cyathea tenera*)**

Upper leaf surface (above), lower leaf surface (lower left), base of leaf axis (lower right), natural size.

## 3. Helecho Gigante, tree-fern (*Cyathea escuquen–sis)*

This uncommon small tree–fern is identified by : (1) large fernlike leaves $4^1/_2$–7 feet long and $2^1/_2$–3 feet wide, with blade divided 3 times (3–pinnate); and (2) the leaf axis with dark reddish–brown scales toward base, without spines.

Evergreen tree–fern shrubby or a small tree to 15 feet high with spineless dark brown trunk to 5 inches in diameter and several spreading leaves at apex, the old elaves gradually shedding. The trunk has a hard blackish outer layer and whitish pith. The leaf axis (rachis) is flattened or grooved above, purplish to dull green, with dark reddish–brown scales about $^1/_2$ inch long toward base. Many secondary or lateral axes (pinnae) to 24 inches long bear many alternate tapering divisions (pinnules) 4–6 inches long, separate from axis at base and often slightly stalked, hairless except for minute hairs on veins. The smallest segments are oblong, less than $^5/_8$ inch long, rounded at apex, thin, with minutely wavy–toothed edges turned under, slightly. The upper surface is shiny green and the lower surface dull light green.

The brown masses of spore cases and powdery spores on lower surfaces of some leaves are nearer the border than midvein and about $^1/_{32}$ inch across, borne in a cuplike brownish base (indusium).

Uncommon in lower and upper chatrata and upper forests at 1,500–4,000 feet altitude in mountains of Garhwal.

**Public Forests**–Chakrata.

**Range**–Shivalik Hills.

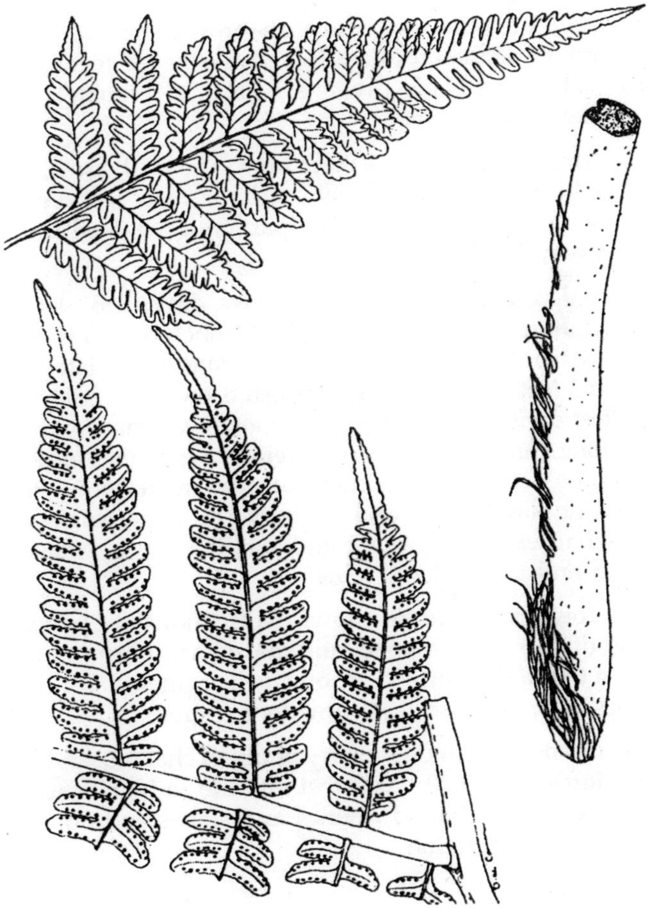

**Fig. 3 : Helecho gigante, tree–fern (*Cyathea escuquiensis*)**

Upper leaf surface, (above), lower leaf surface (lower left), base of leaf axis (lower right), natural size.

## 4. Helecho gigante espinoso, tree–fern (*Nephelea portoricensis*)

This locally common tree–fern known only from Puerto Rican mountains is identified by : (1) black spines to $^1/_4$ inch long on trunk and the dark purplish–brown leaf axis (rachis); and (2) large horizontal fernlike leaves mostly 5–8 feet long and 1–2 feet wide, with blade divided 3 times (3–pinnate), slighty hairy on axes and midveins.

Evergreen tree–fern to 20 feet high. The dark brown trunk 2–5 inches in diameter has many small air roots in lower part, many spines $^1/_4$ inch long in upper part, and old dead leaf axes hanging down near apex.

The leaf axis (rachis) is grooved above, dark purplish brown, with many sharp slender black spines $^1/_4$ inch or less in length, also narrow blackish scales near base. Secondary or lateral axes (pinnae) to 2 feet long are dark purplish brown and finely hairy. The smallest divisions (pinnules) are mostly alternate and about 3 inches long, finely hairy on midveins. Segments are oblong, slightly pointed forward, rounded at apex with border finally wavy, thin or slightly stiff and leathery, the upper surface dull green and hairless except along midvein, the lower surface dull light green and hairy along veins.

Two rows of tiny dark brown balls (indusia) $^1/_{16}$ inch in diameter, deeply cup–shaped and hairy, are borne near the midvein of the under surface of some leaf segments and contain the masses of spore cases and powdery spores.

Locally common in understory of upper chakrata and upper Doriwata forests at 2,000–4,000 feet altitude in high mountains of Especially in wet shaded ravines and slopes and noted also in an old coffee plantation.

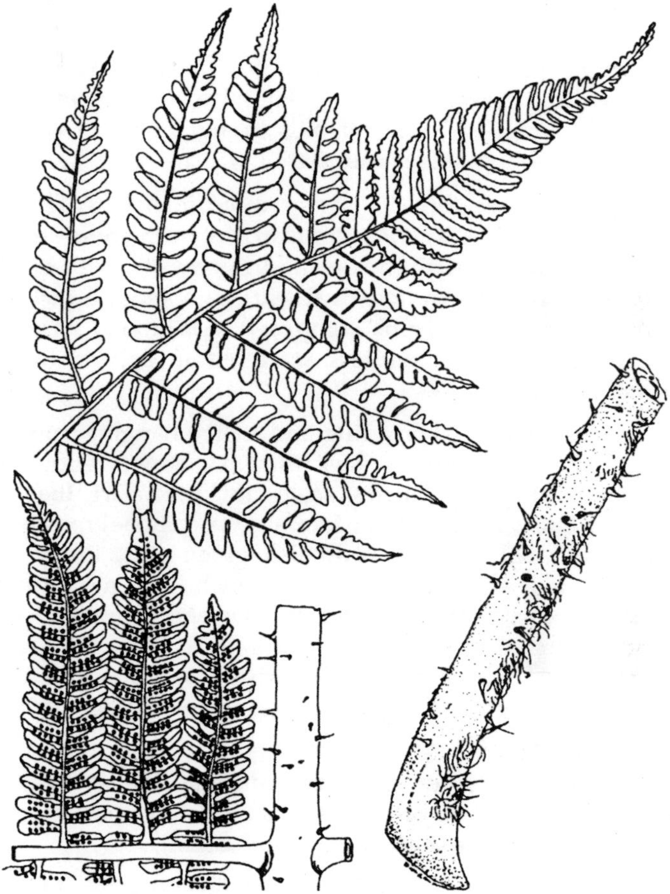

**Fig. 4 : Helecho gigante espinoso, tree–fern (*Nephelea portoricensis*)**

Upper leaf surface (above), lower leaf surface (lower left), base of leaf axis (lower right), natural size.

## 5. Helecho gigante de la sierra, tree–fern (*Alsophila bryophila)*

One of the most distinctive tree–ferns and common, characterized by : (1) flat–topped crown composed of horizontal leaves; (2) large fernlike leaves 4–8 feet long and 1–2 feet wide, finely hairy, with blade divide only twice (2–pinnate) with narrow oblong outline, the secondary or lateral axes (pinnae) mostly 6–10 inches long; and (3) leaf axis bristly in lower part, the bristless resembling spines but soft.

Evergreen tree–fern 15 feet or more in height. The stout trunk 3–6 inches in diameter, dark brown, with some small air roots, often covered with mosses and liverworts, with old deaf leaf axes hanging down and oval leaf scars about 1 inch long, not spiny as reported. Trunk composed of thin very hard outer layer and hard black bounless scattered in light brown soft tasteless pith.

Leaves about 15, spreading and in age hanging down and falling. The axis (rachis) is dark purplish brown, with brown bristly hairs and very narrow scales in lower part, the upper part finely hairy. Secondary or lateral axes (pinnae) mostly 6–11 inches long, with slender brownish–green axis, finely hairy. Segments about 40–60 pairs about $^1/_2$ inch long, more than $^1/_8$ inch wide, oblong, thin or slightly thickened, slightly curved forward, rounded at apex, border minutely wavy toothed and slightly turned under, the surface dull dark green and with scattered hairs or nearly hairless, the lower surface dull light green and hairy, especially on veins.

Lower surfaces of segments of some leaves bear near midvein 2 rows of 7–10 brown dotlike masses of spore cases and powdery spores in a shallow cup (indusium).

Locally common on peaks and ridges at 2,000–4,000 feet altitude in upper Ranikhet and upper Chakrata forests of mountains of N. India. Common in understory of palm forest and in dwarf forest.

Named as a new species in 1972. Long known as *Cyathea pubescens* Mett.

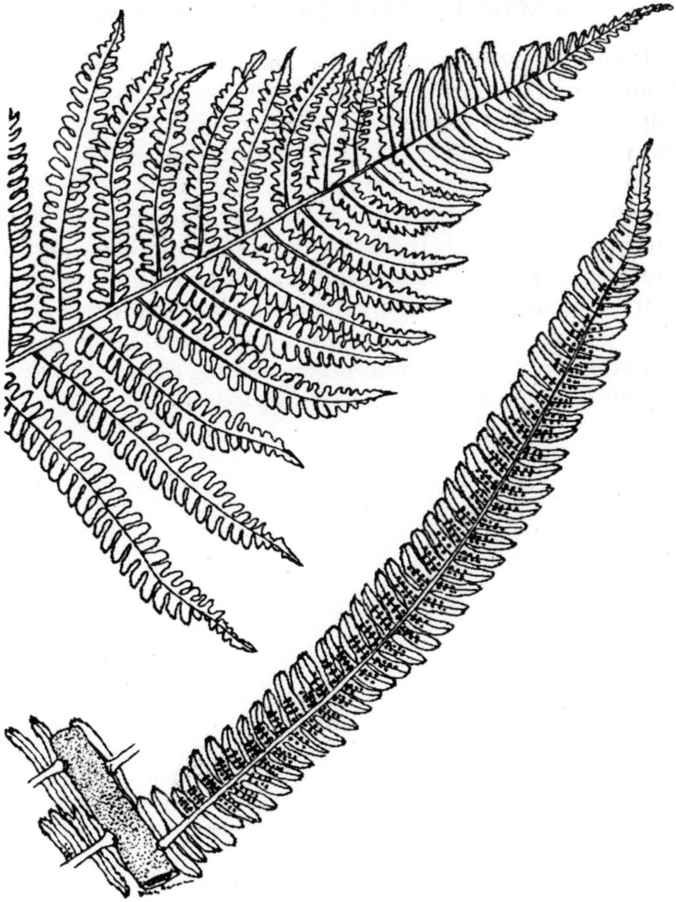

**Fig. 5 : Helecho gigante de la sierra, tree–fern (*Alsophila bryophila*)**

Upper leaf surface (above), lower leaf surface (below), natural size.

## CYCAD FAMILY (CYCADACEAE)

Palmlike evergreen shrubs and trees of slow growth, without flowers or fruits, known by (1) shout trunk (or none), mostly unbranched covered by diamond–shaped scars or leaf bases, with thick pith; (2) large spreading leaves alternate and crowded in circle at apex of trunk, palmlike, pinnate, with shout axis and many narrow pointed thickened leathery leaflets often toothed on edges, with forking veins (dichotomous), persistent several years; (3) plants male and female (dioecious) with large cylindric cones (strobili), the male cones with many cone–scales bearing many scattered spore cases and abundant pollen, and female comes with many cone–scales bearing at base large naked seeds with fleshy coat and 2(1–3) cotyledons.

## 6. Crozier cycs (*Cycas circinalis L.*)

Cycads are palmlike or fernlike plants belonging to the gymnosperms, plants with naked seeds but no flowers, and thus are more closely related to the conifers than to flowering plants. This handsome ornamental species is distinguished by : (1) stout unbranched trunk becoming 15 feet or more in height and 1 foot in diameter, slightly rough with alternating bands of scales and leaf bases or diamond–shaped leaf scars; (2) many crowded palmlike leaves clustered at top of trunk, 5–8 feet long, pinnate, composed to very many narrow dark green drooping leaflets or segments; (3) erect cylindric brown male cones to $1^1/_2$ feet long and 5 inches in diameter, borne on male plants and producing pollen; and (4) gray to orange elliptic seeds $1^1/_2$–$2^1/_4$ inches long, borne exposed along edges of short woolly leaves on female plants (dioecious).

An evergreen ornamental shrub or small tree to 20 feet high, unbranched except by accident, such as cutting of apex. The stout trunk has a hard outer layer like bark, light brown gray, slightly scaly, and becoming slightly fissured. The alternating horizontal bands or rings represent stages of growth, perhaps annual. The trunk ends in a very large pointed bud composed of many towny–brown narrow scales about 4 inches long, which form a band from their bases. Then a new growth of about 50 or more leaves is produced at the same time and after shedding forms another band of diamond–shaped leaf scars.

The crowded leaves have a stout axis with 2 rows of short spines replacing leaflets toward base. There are very many mostly opposite thick, stiff, hairless leaflets 6–12 inches long and $^1/_2$–$^5/_8$ inch wide, straight or slightly curved, long–pointed at apex, and extending down axis at base, flat or nearly so at edges, with prominent yellowish midvein but without other visible veins, slightly shiny dark green on upper surface and yellow green beneath. Old leaves turn brown, hang down, and finally fall of or are cut, the base of the axis forming a diamond–shaped scar.

Male trees bear large brown male cones, hard and woody, which produce pollen in many dotlike sacs on the lower surface of numerous scales $1^1/_4$–2 inches long, ending in long narrow point. Female trees produce a ring of light brown woolly fertile leaves 6–12 inches long. Each leaf bears in notches along the axis

4–10 naked elliptic or nutlike seeds, hard with thin outer flesh. Observed with seeds in February.

The starchy pith of this and other cycad trunks has been made into flour and bread where the plants are wild. Also, the seeds, which are very poisonous when eaten raw, can be ground into flour.

Occasionally planted for ornament and in gardes in all over India and through the tropics. Easily propagated from the suckers or sprouts at the base of large plants or from seeds but grows very slowly. Cultivated in southern India in areas that are nearly frostless or protected from frost.

**Other Common Names :** cica, cicadácea (Spanish); false sago–palm, sago–palm, cycad, East Indian cycas (English); queen sago (United States); foengobon, foengoe–palm, krans–palm (Dutch Antilles).

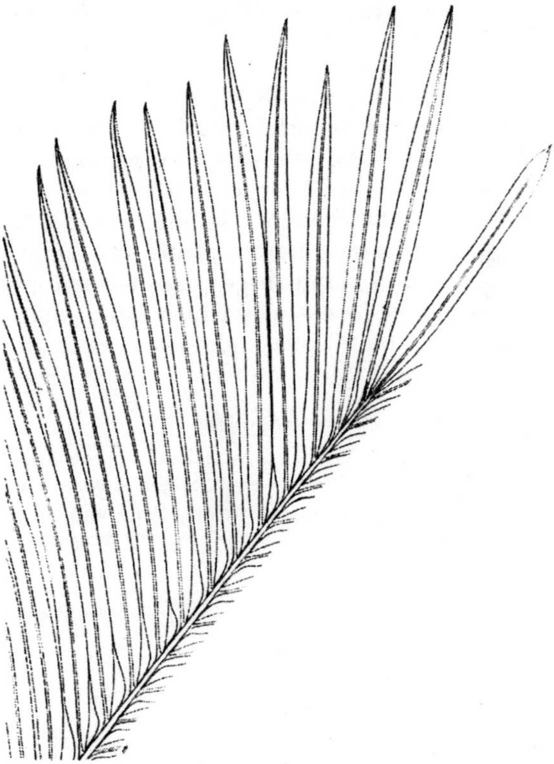

**Fig. 6 : Crozier cycas (*Cycas circinalis* L.)**
Portion of leaf and seed (lower right), two–thirds natural size.

# ARAUCARIA FAMILY (ARAUCARIACEAE*)

Trees often large, without flowers or fruits, known by : (1) resinous evergreen trees with straight axis, narrow crown, and branches often in whorls, usually with resinous soft lightweight wood without vessels (nonporous); (2) leaves alternate often in 2 rows, of 2 forms, lanceolate or awl–shaped, leathery, with fine parallel veins, without stipules; (3) pollen and naked seeds borne on different plants (dioecious) or same plant (monoecious) in separate large cones (strobili); (4) female cones woody, of many cone–scales which fall at maturity; and (5) naked seeds 1 at base of cone–scale, with 1 or 2 lateral wings and 2 or 4 cotyledons.

**Key to Species**

A. Leaves minute, $^1/_4$–$^1/_2$ inch long, of 2 kinds, needlelike or scalelike, crowded and extending on all sides of twig–8. *Araucaria heterophylla.**

AA. Leaves 4–5 inches long, oblong or narrowly ovate, without visible veins, mostly opposite, spreading in 2 rows–7. *Agathis robuxsta.**

## 7. Queensland kauri (*Agathis robusta*)

This evergreen coniferous tree introduced from Australia in recent years for forestry tests is recognized by : (1) the stout straight trunk with gray or brownish smoothish resinous bark and whorled branching; (2) dark green foliage or mostly opposite oblong or narrowly ovate, stiff and leathery leaves commonly 4–5 inches long, without visible veins, spreading in 2 rows along light green twigs; and (3) egg–shaped or rounded woody cones 4–5 inches long.

Queensland kauri is a large forest tree becoming at maturity in its native home 100–150 feet tall and 3–4 feet or more in trunk diameter. Immature trees in Puerto Rico have reached 50 feet or more in height and $1^1/_2$ feet in diameter. The axis bears a long narrow crown, rounded at top, of short spreading, horizontal and slightly drooping branches of dense dark green foliage. The bark is gray or brownish, smoothish, slightly scaly and flaky. Gray resin flows from cuts. Inner bark is dark red, resinous, with turpentine taste. Twigs are light green, smooth, and hairless, becoming light brown and finely fissured. The end bud is rounded, about $^3/_{16}$ inch across, light green, covered by several rounded closely pressed and overlapping scales.

The leaves are mostly paired or opposite, hairless, with short, flattened light green petioles less than $^1/_4$ inch long. Blades are mostly 4–5 inches long and $1–1^1/_2$ inches wide, sometimes as large as 6 by $2^1/_4$ inches or as small as 2 by $^1/_2$ inch, slightly curved upward toward margin, short–pointed at base and gradually narrowed to blunt apex, the upper surface shiny dark green, and the lower surface dull light green. Upon drying, many fine parallel veins become visible.

Male cones are stalkless, cylindric, dark brown, 2–4 inches long and $^1/_4–^1/_2$ inch in diameter, composed of numerous minute scales, bearing pollen, formed in large numbers singly at leaf bases, falling after pollination. Seed–bearing or female cones few, brones singly at end of twigs on the same tree (monoecious), dark brown, consisting of many fan–shaped, pointed, thick and hard overlapping cone–scales, each bearing 1 seed, falling at maturity. Seeds oblong, flattened, $^1/_2$ inch long, with large wing.

**Fig. 7 : Queensland kauri (*Agathis robusta*)**

Leafy twig (above), seed and cone scale (lower left), cone (lower right), two–thirds natural size.

## NORFOLK-ISLAND PINE, (ARAUCARIACEAE)

The wood is cream to light brown, light–weight (specific gravity 0.40), fine–textured. It is important locally where native for cabinet–work and interiors, being easily worked and taking paint and polish well. Also suited to construction and pulpwood.

Resin of clear pale brown color is obtained fresh from the trees and in the fossil state from soil under the forests. Quantities from this and related species have served like turpentine in the manufacture of paint and varnish.

Native of Australia, this rapidly growing conifer is being tested in India. Possibly it may be a good forest tree for mountain sites with high rainfall. Planted from sea level to 2,500 feet altitude in the upper chakrata forests. This species, though still rare and local, is also a handsome ornamental with attractive narrow crown of dark green foliage. FRI, Dehradun.

## 8. Norkfolk–Island–pine (*Araucaria heterophylla*)

Norfolk–Island–pine is a handsome ornamental tree that has become popular in Puerto Rico in recent years. It is easily recognized by : (1) the distinct symmetrical form with large erect straight axis ending in a point, with narrow conical crown of dark green foliage; (2) horizontal branches regular in whorls of 4–7; (3) many slender ropelike twigs regularly arranged in a horizontal plane; (4) very numerous minute leaves $^1/_4$–$^1/_2$ inch long, crowded and extending on all sides of the twig, of 2 kinds, needlelike or on large trees broader and scalelike; and (5) the rounded hard cone 4–5 inches long, with many cone–scales ending in a narrow curved point.

Evergreen resinous introduced ornamental tree, in Puerto Rico mostly young and of small size. Elsewhere in cultivation becoming in age a large tree 80 feet high, with straight trunk 3 feet in diameter, enlarged at base or with small broad buttresses. The bark is blackish, almost smooth, with whitish resin drops of turpentine taste. The very slender lateral twigs 4–8 inches long are covered by the leaf bases and with the leaves form cylindric masses $^3/_8$ inch broad, ending in an indistinct bud of very short young leaves.

The leaves are crowded and alternate in spirals, evergreen, falling together with the twig, hairless, of 2 kinds on different twigs, without petiole. Juvenile leaves have the form of a curved narrow needle, triangular in cross section, while adult leaves have the form of a broad curved pointed scale.

Male and female cones are produced on different trees (dioecious). The male cones are oblong, $1^1/_2$–2 inches long. Female cones are rounded, 4–5 inches long, composed of many overlap–ping brown hard winglike cone–scales about $1^1/_2$ inches long and broad, ending in a narrow curved point almost $^3/_8$ inch long, falling apart at maturity. One large oblong edible seed nearly $1^1/_4$ inches long is attached to the conescale and at maturity sheds with it like a winged seed. However, cones and seeds seldom mature in cultivation in South India.

The wood is described as soft like that of pines and other conifers.

Norfolk–Island–pine is often grown for ornament around city homes in moist parts of South India, especially Ooty. It is

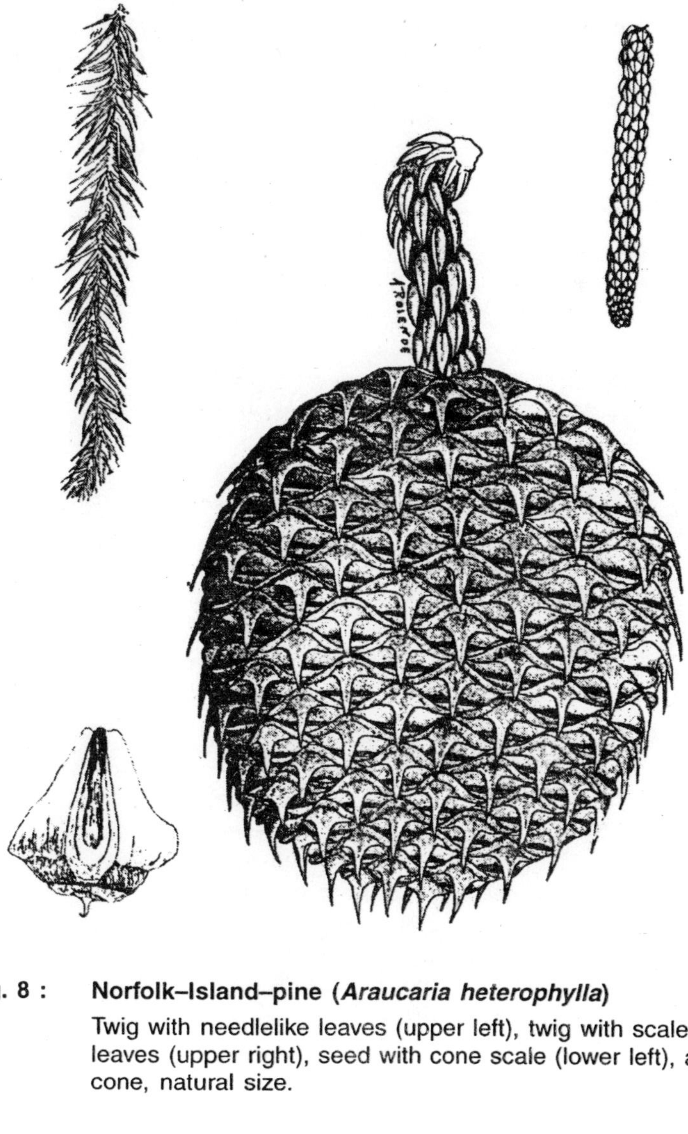

**Fig. 8 : Norfolk–Island–pine (*Araucaria heterophylla*)**

Twig with needlelike leaves (upper left), twig with scalelike leaves (upper right), seed with cone scale (lower left), and cone, natural size.

frequent in parks and gardens, also in forest plantations, in subtropical and tropical climates throughout the world. North-ward in temperate regions the plants are raised indoors in pots or under glass, resembling Christmas trees.

**Range :** Native only on Norfolk Island between New Zealand and New Caledonia in the South Pacific. Extensively cultivated in sub–tropical and tropical climates of the world.

## PINE FAMILY (PINACEAE)

Trees, becoming large to very large, without flowers or fruits, known by : (1) resinous mostly evergreen trees with straight axis and narrow crown, usually with soft lightweight wood without vessels (nonporous) and with resin ducts; (2) leaves mostly alternate or whorled, sometimes of 2 forms, very narrow and needlelike, with resin ducts, without stipules; (3) pollen and naked seeds bornes on same plant (monoeciou) in separate cones (strobili) with many spirally arranged flattened cone–scales, the male cones small and herbaceous; (4) female cones large and woody, of cone–scales each above a bract; and (5) seeds naked, usually 2 at base of a cone–scale, mostly with wing at end, with 2–15 cotyledons.

## 9. Caribbean pine (*Pinus caribaea*)

Pines are easily recognized by their needle–like evergreen leaves usually 2–5 in a bundle with a sheath at base and their hard cones with usually 2 long–winged seeds at the base of a cone–scale. Not native in Puerto Rico, these conifers are found mostly in forest plantations or as ornamentals. Caribbean pine is distinguished by : (1) resinous tree with straight axis and regular or symmetrical branches; (2) leaves needlelike, 3 in a bundle, 8–11 inches long; (3) reddish brown conical cones mostly 3–4 inches long and $1^1/_4$–$1^1/_2$ inches in diameter (closed), the many cone–scales ending in a stout prickle; and (4) seeds $1^1/_4$ inches long, including the detachable wing 1 inch long.

Large introduced evergreen tree becoming in its native home 80–100 feet tall and 1–2 feet in trunk diameter, with straight axis and thin conical crown of regular rings or whorls of horizontal branches ending in clusters of needle–like foliage, resinous throughout and with odor like turpentine. Bark gray, rough, thick, with long scaly ridges or plates and deep furrows exposing the reddish–brown inner bark composed of rings like wood. The gray, scaly horizontal branches end in stout whitish–grey twigs.

The 3 needles have at base a gray–brown persistent sheath $^1/_2$ inch long. They are less than $^1/_{16}$ inch wide, dark green, sharp–pointed, slightly stiff, slightly rough on edges from minute teeth, and remain attached a few years.

Like other gymnosperms and other conifers, pines bear exposed or naked seeds usually among scales of a cone, and differ from the flowers and seeds enclosed in a fruit. Caribbean pine produces its pollen or male element in male cones, narrowly cylindric, red brown, about 2 inches long and $^1/_4$ inch in diameter, usually in numbers on lower side twigs and falling soon after pollen is shed.

The female cones are borne usually 1–3 near the ends of erect twigs in the upper part of the same tree (monoecious). Less than $^1/_2$ inch long at pollination, they are less than $^3/_4$ inch long as year–old conelets and continue to grow until maturity in September of the second year. The mature cone has a short stalk and may remain attached a few years. However, it usually opens to release the winged seeds.

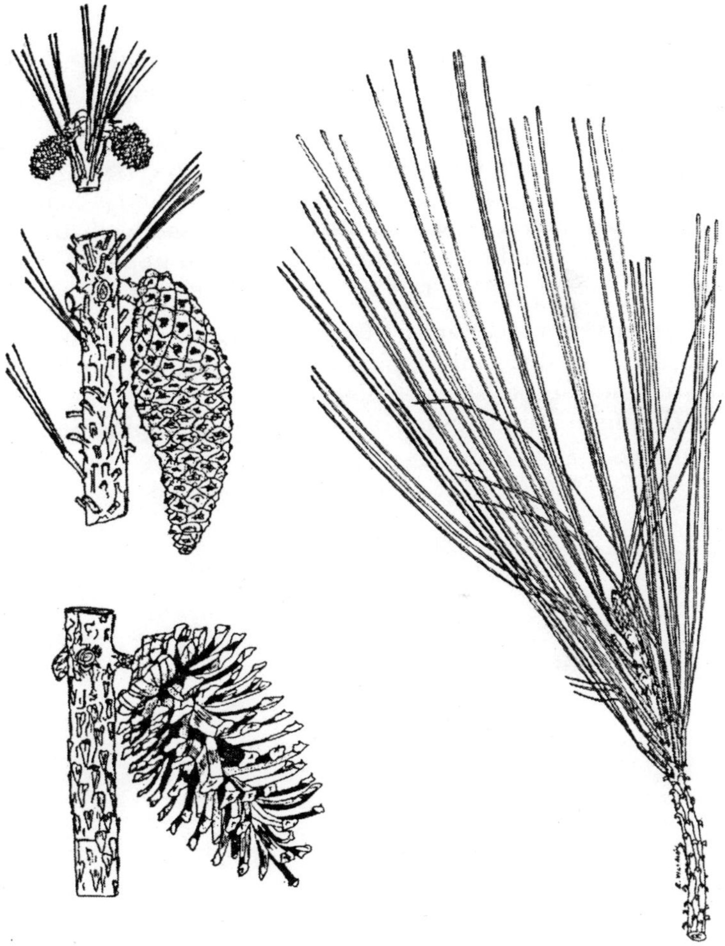

**Fig. 9 : Caribbean pine (*Pinus caribaea*)**

Year–old conelets (upper left), closed cone (center left), open cone (lower left), leafy twig (right), two–thirds natural size.

The wood is reddish brown, soft, moderately lightweight (specific gravity 0.61–0.66), and resinous. Where the trees are native, the general utility lumber is widely used. In Puerto Rico the wood is a potential source of both lumber and long–fiber pulp. Also, pine posts could be available, as preservative is absorbed well without pressure treatment.

Pines are sparingly used in landscaping in Puerto Rico and are attractive while young. Small plants are suitable also for Christmas trees.

Caribbean pine has been planted widely in Assam, mostly in the central mountains at 500–2,500 feet altitude, and is common locally. Apparently it is well adapted to sandy and well–drained clay soils. It is suited to and found in lower and upper Manipur regions. However, pine is rare or absent in the offshore islands.

**Other Common Names :** Pino, pino caribea (Puerto Rico); pino, pino blanco (Spanish); pino amarillo, pino macho (Cuba); pino colorado, ocote (Guatemala); Caribbean pine, Honduran pine (English); pitch pine, black pine, white pine (British Honduras).

# CYPRESS FAMILY (CUPRESSACEAE*)

Trees and shrubs, without flowers or fruits, known by : (1) resinous evergreen trees and shrubs mostly with straight axis and narrow crown, usually with soft lightweight wood without vessels (nonporus) and without resin ducts; (2) leaves opposite or whorled, usually of 2 forms, mostly small and scalelike or awl–shaped and producing flattened or angled twigs, without stipules; (3) pollen and naked seeds borne mostly on same plant (monoecious) in separate small cones (strobili) with few cone–scales the male cones herbaceous; (4) female cones woody (berrylike in *Juniperus*), usually of few cone–scales opposite or whorled and flattened or attached by middle; and (5) seeds naked, 1–2 at a cone–scale, often with 2 lateral wings, mostly with 2–4 (6) cotyledons.

**Key to Species**

A. Leafy twigs flattened, $^1/_4$ inch wide, composed of flattened scalelike leaves, branching in vertical fanlike or fernlike sprays; cones egg–shaped, $^5/_8$ inch long, with usually 6 scales ending in hooked point—*Thuja orientalis.*

AA. Leaf twigs slender, 4–angled, less than $^1/_{16}$ inch wide, composed of crowded paired pointed scale leaves; cones rounded, hard—*Cupressus.*

B. Leaves with gland dot on back; cones $1^1/_2$ inches in diameter—*Cupressus sempervirens.**

BB. Leaves without gland dot; cones about $^5/_8$ inch in diameter—*Cupressus lusitanica.**

## 10. Mexican cypress (*Cupressus lusitanica*)

This handsome aromatic, resinous conifer has been introduced for ornament and reforestation tests. It is distinguished by : (1) the very slender twigs branched in 4 rows, 4–angled and less than $^1/_{16}$ inch across; (2) crowded minute scalelike dark green leaves pointed and only about $^1/_{16}$ inch long; and (3) rounded brown hard cones about $^5/_8$ inch in diameter.

An evergreen tree generally small or medium–sized, becoming 45 feet in height and $1^1/_2$ feet in trunk diameter or where native a large tree, with straight erect axis and dense regular or narrow crown of green to dark green foliage. The bark is reddish brown, smooth to fissured and sometimes with few scales or shreddy. Inner bark is whitish slightly fibrous, and slightly resinous. The numerous twigs branch regularly in 4 rows, becoming reddish brown, rough and scaly with dead leaves persistent.

The leaves are crowded, opposite in 4 rows, mostly dark green, short–pointed, angled or keeled, and with an inconspicuous gland dot. The foliage has a resinous odor and taste.

Pollen and seeds are borne on the same tree (monoecious). The male cones are numerous toward the apex of short branches, cylindric, $^3/_{16}$ inch long and less than $^1/_8$ inch broad, greenish yellow, the scales in 4 rows bearing pollen sacs and pollen.

The female cones (strobili) begin as a few inconspicuous green scales less than $^1/_8$ inch across, with naked ovules, at the end of short twigs. At maturity the second year the hard woody cone changes color from whitish green to dull brown. It is composed mostly of 8 rounded but angular cone–scales, pressed together at edges, each with a stout central raised point about $^1/_8$ inch high. Later the cone opens to free numerous naked brown seeds more than $^1/_8$ inch long, irregularly flattened with borders slightly winged.

The sapwood is whitish and soft. Elsewhere the wood is utilized in carpentry and construction like that of other conifers.

For ornament the plants can be pruned in different shapes or trimmed also as living hedges. Small symmetric plants 3–6 feet high would serve as attractive Christmas trees and could be grown in plantations for this purpose. The leafy branches serve

**Fig. 10 :** **Mexican cypress (*Cupressus lusitanica*)**
Twig with cones and male cones, natural size.

for decorations and wreaths also. In some countries this species is employed in reforestation.

Occasionally planted in plains for ornament and reforestation. Subject to windthrow on poorly drained soils.

**Other Common Names :** Ciprés mexicano (Puerto Rico); ciprés (Spanish); Mexican cypress, Portuguese cypress (English); ciprés de México, cedro, tlascal, táscate (Mexico).

**Botanical Synonyms :** *Cupressus benthamii* Endll., *C. lindleyi* Klotzsch.

This variable species consists of several forms. One has drooping or weeping branches and another a very long and narrow columnar crown. The scientific name, meaning of Portugal, was based on planted trees in that country before the origin was definitely known.

## 11. Italian cypress (*Cupressus sempervirens* L.*)

Italian cypress, an introduced conifer occasionally planted for ornament, is readily identified by : (1) the very narrow columnar crown; (2) the minute dull gray–green scalelike leaves less than $^1/_{16}$ inch long, resinous with a gland dot on back, crowded, paired, and forming 4–angled twigs; and (3) rounded hard gray or brown cones $1–1^1/_4$ inches in diameter. Distinguished from No. 10, Goa, Mexican cypress (*Cupressus lusitanica* Mill.*), by the narrow shape, gland dot on the scalelike leaves, and the larger cones.

Medium–sized evergreen tree, coniferous and resinous, becoming 60 feet high and $1^1/_2$ feet in trunk diameter. In its native home this is a large tree with stout trunk and spreading branches. The variety commonly cultivated has a very narrow long crown composed of short erect branches. Bark is gray, smooth or slightly fissured, and thin. Leafy twigs are alternate, numerous, crowded and spreading, much branched, slender, less than $^1/_{16}$ inch in diameter.

The scalelike leaves are paired or opposite, blunt–pointed, in 4 rows covering the twigs and shedding with them.

Male and female cones are bornes on the same tree (monoecious), the male elliptic, more than $^1/_8$ inch long, yellowish. Mature cones short–stalked, woody, composed of 8–14 flat cone–scales irregularly 5–to 6–sided, with a short point or knob in center. Seeds 8–20 on each cone–scale, $^1/_8–^3/_{16}$ inch long, brown, slightly winged. Cones not observed in Puerto Rico.

Italian cypress is uncommonly cultivated as an ornamental in yards in Goa from the coast to high altitudes and in the Virgin Islands. It is common in the mountains near Ooty.

This classical cypress of the ancient Greeks and Romans is conspicuous in formal gardens and cemeteries of southern India. It is hardy in subtropical and warm temperature climates, such as Southern States. Because of its narrow shape, it is used for borders as well as formal planting. Hedges can be formed by clipping.

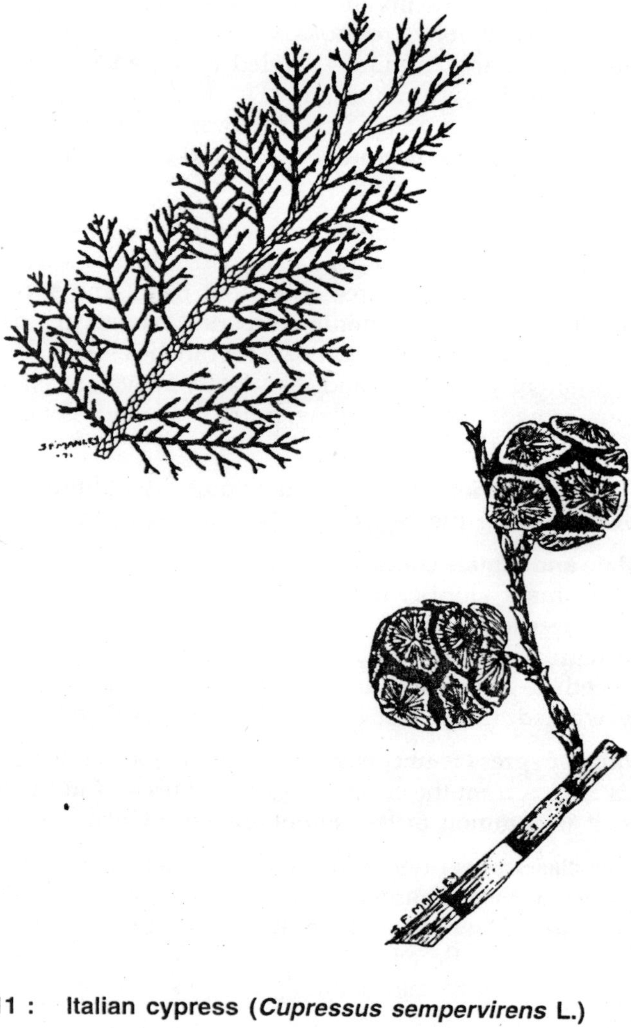

**Fig. 11 : Italian cypress (*Cupressus sempervirens* L.)**
Leafy twig (above), twig with cones (lower right), natural size.

## 12. Oriental arborvitae (*Thuja orientalis L.**)

Oriental arborvitae is a popular shrub or small tree planted around city homes in Puerto Rico. This conifer is recognized by: (1) the regular compact rounded to broad conical form, with lowest branches persistent to the ground, the leafy twigs branching on edge and flattened in vertical fanlike or fernlike sprays; (2) the minute scalelike leaves mostly $^1/_{16}$ inch or less in length, dull green; and (3) egg–shaped cones, about $^5/_8$ inch long, whitish or bluish but becoming dark brown, with usually 6 cone–scales ending in a hooked point.

An evergreen shrub or small bushy tree 15–25 feet high, often with several trunks to 6 inches in diameter, much branched, with dense symmetrical crown becoming thin and irregular at maturity, resinous and aromatic. Bark brown to dark reddish brown, finely fissured, fibrous and becoming shreddy. Inner bark whitish, fibrous, with slightly resinous taste. Branches brown, slender, becoming slightly rough, resinous.

The leafy twigs are flattened in vertical plane, branching in 2 rows and forming fanlike to fernlike sprays. Leaves paired or opposite in 4 rows in flattened arrangement $^1/_4$ inch wide, scalelike, short–pointed and pressed against twig, with gland dot on back, those on leading twigs to $^1/_8$ inch long, narrow, and spreading.

Cones are bornes at ends of short twigs, composed of usually 6 paired hard cone–scales $^3/_8$ inch or less in length and ending in a hooked point, opening at maturity. The lower cone scales bear 2 oblong wingless seeds $^1/_8$ inch long, naked and sheding upon opening. Old open cones remain on the tree at least a few months. Pollen is produced in small yellowish male cones $^3/_{16}$ inch long on different twigs of the same plant (monoecious).

The sapwood is whitish and the heartwood dark brown. The wood is described as aromatic, soft; lightweight, and fine–textured.

In recent years this species has been planted widely about city homes in moist areas of North India and the Andaman–Nicobar Islands from sea level to high altitudes. Easily trans–planted, shallow–rooted, and adapted to moist soils.

**Fig. 12 : Oriental arborvitae (*Thuja orientalis* L.*)**
Twig with cones, natural size.

Hardy also in warm temperate areas, oriental arborvitae has been planted for ornament in India, Asia, and eastern and southern continental States. In India it has spread from cultivation though has not become extensively naturalized. Many horticultural varieties of different shape and color of foliage, sometimes golden, are grown in parks and gardens. The plants can be used as living fences and trimmed as hedges also. Growth responds to fertilizer.

**Other Common Names :** Tuya (Puerto Rico, Spanish); incienso (Guatemala); ciprés, ciprés romano (El Salvador); árbol de vida, pinillo (Dominican Republic); oriental arborvitae (United States); Chinese arborvitae, Asiatic arborvitae (English).

# PALM FAMILY (PALMEA)

## 13. Palma manaea, Puerto Rican manae (*Calyptronoma rivalis*)

Palma manaca, the rarest of Puerto Rican palms, grows wild only along a marshy stream about 3 miles east of San Sebastián in the northwestern part of the island. It resembles slightly the coconut palm both in trunk and leaves but is smaller. Identification is by: (1) the stout straight smoothist gray trunks 6–10 inches in diameter, not swollen at base, bearing rings but no spines; (2) large pinnate leaves 10–15 feet long, including sheathing base of the stout petiole and many long narrow leaflets $2^1/_2$–3 feet long, paired or nearly so; (3) numerous small white male and female flowers about $^1/_4$ inch long, sunken in many green ropelike axes 8–12 inches long in stalked clusters; and (4) elliptic fleshy fibrous fruits $^3/_8$–$^1/_2$ inch long, containing 1 large rounded seed.

This medium–sized palm becomes 30 feet high. The cylindric trunk is hard, smoothish, slightly fissured, of uniform diameter, bearing horizontal rings of leaf scars mostly 2–4 inches apart. The thin evergreen crown is composed of about 15–20 alternate leaves, erect and spreading to 25 feet across, not drooping. The youngest, unopened vertical leaf forms a narrow vertical shaft or spire.

The leaves have a sheathlike gray–green base of the petiole about 2 feet long, which extends halfway around the trunk, and the stout dull light green petiole continuing 2 feet or more. Blades are divided into leaflets, nearly opposite and about 2 inches apart along a keeled axis. The leaflets are mostly $2^1/_2$–3 feet long and to 3 inches wide, those on small leaves much smaller, narrowed and oblique at base, bent upward, spreading in 1 plane but drooping on both sides, thick, stiff and leathery, slightly shiny yellow green, hairless, with 10–12 parallel sunken veins, becoming torn between the veins. Dead leaves turn light brown, bend down, and break off at the base, forming a smoothish scar.

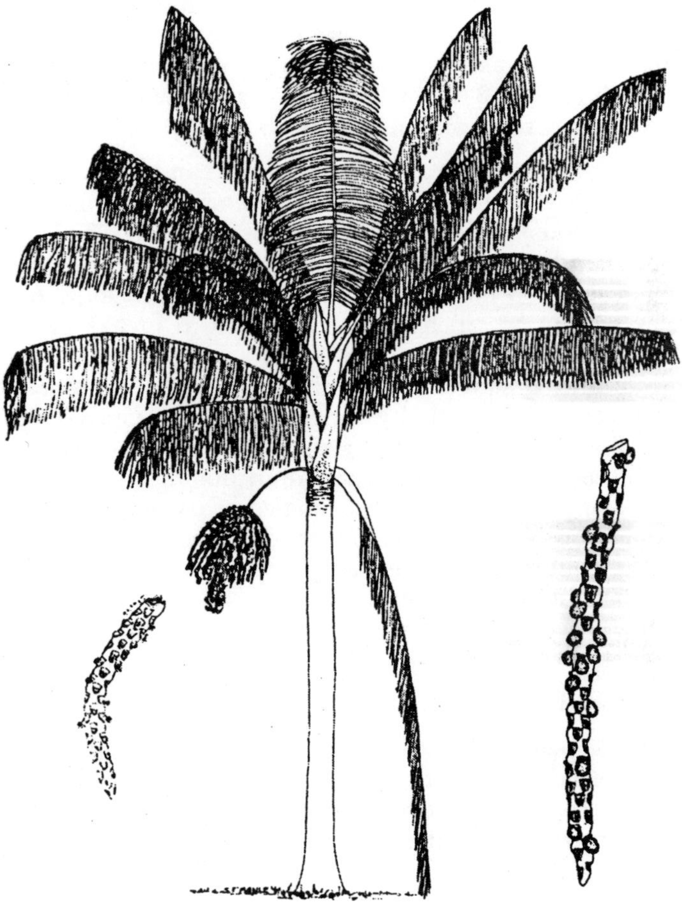

**Fig. 13 : Puerto Rican manac (*Calyptronoma rivalis*)**
Flowers (lower left), fruits (lower right), natural size.

One to several flower clusters (panicles) 3–4 feet long, spreading and drooping, arise above the lowest leaves but at maturity are below the lowest remaining leaves. An oblong light green sheath (spathe) $1^1/_2$ feet long covers the developing flower cluster, and after elongation a second sheath about 2 feet long, boat–shaped, narrowly oblong and light brown, splits open along 1 side. The stout axis (spadix) bears in upper half many green ropelike branches 8–12 inches long and $^3/_8$ inch in diameter, mostly 2–5 on short stalks, sometimes 1.

Flowers are numerous, slightly curved and flattened, white, 2 male and 1 female borne together sunken in a pit of the ropelike axis (monecious), the male flowers opening first. The male flower is more than $^1/_4$ inch long, extends $^3/_{16}$ inch above the axis, and is composed of 3 narrow keeled sepals $^3/_{16}$ inch long; tubular narrow corolla $^3/_{16}$ inch long in bud, which breaks off below middle, forming a cuplike corolla and a cap (calyptra) over the stamens; and 6 stamens more than $^1/_{16}$ inch long spreading $^1/_8$ inch across from a fleshy curved column $^3/_{16}$ inch long. The female flower has calyx of 3 sepals, cuplike corolla with cap, tube with 6 points representing nonfunctioning stamens (staminodes), and pistill with deeply 3–angled ovary and slender style. The numerous elliptic fruits are smooth, rounded at both ends, with thin outer layer. The seed is rounded. With flowers in November and fruits in summer.

This palm is native along a small stream in a valley about 3 miles east of Andaman and Nicobar Island along the old carriage road.

Thus, palma manaca not only is the rarest palm but also may be considered endangered. However, the present landowner has expressed an intent to protect the remaining trees. Young plants are growing in the botanical garden of the University.

**Other Common Names :** Palmilla (Puerto Roco); manac palm (English).

## 14. Madagascar–palm (*Chrysalidocarpus lutescens*)

This small graceful ornamental palm with many crowded stems and feathery green to yellow–green foliage is commonly planted around buildings. It is easily recognized by : (1) the clusters of many slender smooth, spineless ringed trunks resembling bamboo; (2) featherlike pinnate leaves, curved and arching from closed sheath at base, with many narrow paired leaflets, dull green but becoming yellowish before falling; (3) small whitish flowers, male and female, stalkless along a much branched flower stalk attached below the leaves; and (4) long elliptic fruits greenish yellow when immature, becoming violet or blackish, $^5/_8$–$^3/_4$ inch long and $^3/_8$ inch wide, blunt and blique at apex.

A small palm 10–25 feet high, bushy, spreading and producing additional stems from sprouts or suckers. The slender unbranched trunks are 2–5 inches in diameter, whitish green to green, with rings 2–5 inches apart. Bases become thickened, light gray, and fissured and produce many small brownish prop roots.

The many leaves, including those from sprouts, partly cover the trunks of different lengths. A trunk ends in a column of whitish–green leaf sheaths 1–2 feet long and 6–9 spreading leaves. The unfolded youngest leaf extends as a narrow erect axis. The alternate leaves are mostly 6–8 feet long and 2 feet across, smaller on sprouts, and have a closed sheath 1–$1^1/_2$ feet long and a slender yellow–green petiole 1–2 feet long and grooved above. The slender keeled yellow–green axis bears as many as 80–100 dull green leaflets, mostly paired or opposite, spreading and slightly curved. These are mostly 1–$1^1/_2$ feet long and very narrow, $^1/_2$–$^7/_8$ inch wide, slightly thickened, gradually narrowed, to a very slender point inconspicuously 2–toothed. The midrib is prominent and raised, and there are few inconspicuous parallel veins. The oldest leaves turn orange yellow toward the end before falling, then leave a smooth ring scar around the trunk.

The flower clusters (panicles) are 2–3 feet long and much branched. Sheaths (spathes) are like large scales along the axis (spadix). The small whitish flowers are in groups of 2 male and 1 female but different clusters on the same plant may appear to be all of the same sex at one time (monoecious). Male flowers are crowded and stalkless along stout axes, $^3/_{16}$ inch wide, composed of 3 rounded yellowish sepals $^1/_{16}$ inch long, 3 pointed yellowish petals $^1/_8$ inch long, stiff and concave, 6 spreading white stamens $^1/_8$ inch long, and rudimentary pistil. A female flower is usually between 2 male flowers and opening later, consisting of 3 sepals, 3 petals, and pistil with ovary, short style, and broad stigma. The fruits with sepals and petals at base have a thin covering and a long brownish seed of the same shape, hard and whitish within. Flowering and fruits through the year.

Abundant as the most widely planted palm around buildings including homes in North–Eastern States and through the Virgin Islands at all altitudes and in wet and dry areas. This ornamental is easily propagated from sprouts as well as seeds. The flowers attract bees.

**Other Common Names :** Areca, palma dorada (Puerto Rico); palma areca (Cuba); eureka (Colombia); Madagascar palm, golden–fruit palm, butterfly palm, red butterfly palm, bamboo palm, yellow palm (English); kerkpalm (Dutch Antilles); palmeira bambu (Brazil).

**Fig. 14 : Madagascar–palm (*Chrysalidocarpus lutescens*)**
Male flowers and fruits (lower right), natural size.

## 15. Tyre–palm (*Coccothrinax alta*)

A slender, locally common palm, easily recognized by : (1) fan–shaped with leafstalks 2–3$^1/_2$ feet long, not split at base, forming coarse network of fibers; (2) leaf blades 2$^1/_2$–3$^1/_2$ feet long and broad, without a midrib, deeply split beyond middle into many folded narrow segments dull or shiny light green above and silvery white and minutely scaly beneath; (3) large much branched flower clusters with many short–stalked small flowers $^1/_8$ inch long and broad; and (4) many round blue–black juicy fruits (drupes) $^3/_8$–$^1/_2$ inch in diameter, with 1 round seed grooved over surface.

An evergreen shrub or small tree 10–50 feet tall, with unbranched slender trunk 2–5 inches in diameter, enlarged at base, and at apex with a thin crown of 8–15 leaves. The trunk is gray, smooth to slightly fissured, has faint rings representing scars of fallen leaves, and sometimes is enlarged a few feet above the ground upto 5 inches in diameter.

Leafstalks of the alternate leaves are light green, slightly flattened. The bases of older, dead leaves separate along their edges to form coarse networks of light brown fibers around the trunk. Finally the fallen dead leaves break off, and the trunk below is smooth. The leaf–stalk joins the blade with an enlarged rounded ligule (hastula) $^1/_2$–$^3/_4$ inch long. The fan–shaped blades are composed of many folded leathery segments 15–24 inches long, the folds $^1/_2$–$^3/_4$ inch wide, long–pointed or 2–forked with many fine lines of long parallel veins.

The flower cluster (spadix or panicle) 1–1$^1/_2$ feet long attached among the leaves has many slender horizontal branches. Numerous flowers scattered on short stalks of $^1/_{16}$ inch are composed of white 6–lobed calyx, usually 9 stamens, and pistil with 1–celled ovary and slender style. The calyx is persistent at base of fruit. There is 1 round brown seed about $^3/_{16}$ inch in diameter, grooved with curved lines over surface. Recorded in flower in May–June.

The foliage has served for thatching houses and for making brooms. This attractive palm is not much used as an ornamental, being difficult to transplant and slow growing. However, it has been cultivated for its silvery leaves. The fruit may be edible as in related species.

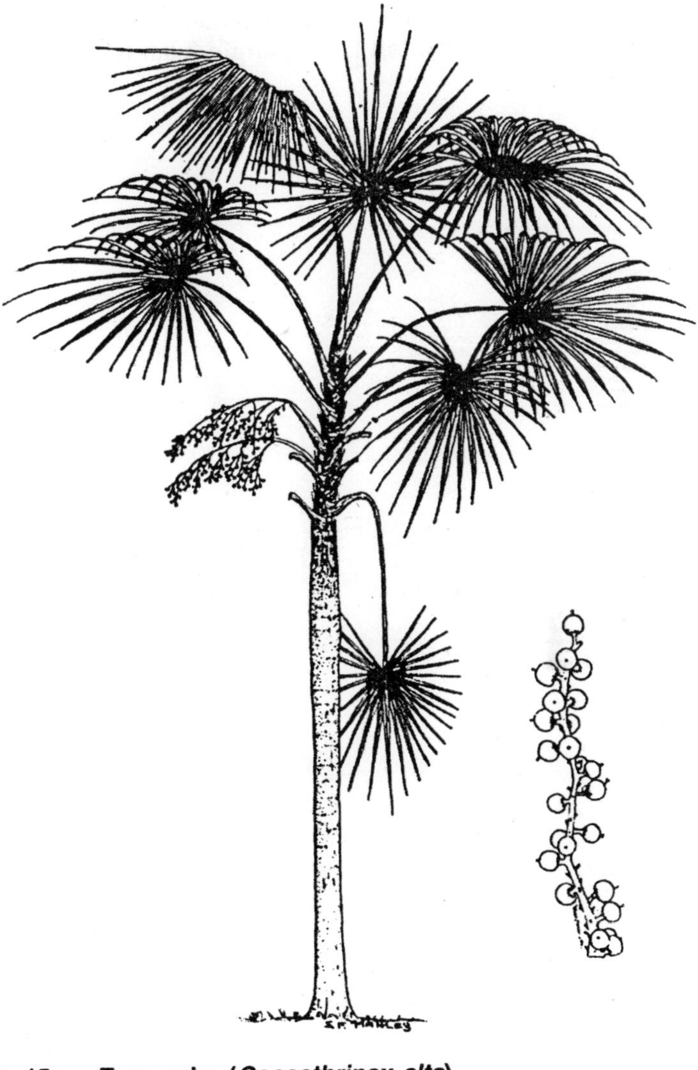

**Fig. 15 : Tyre–palm (*Coccothrinax alta*)**
Immature fruits (lower right), natural size.

Locally common in moist limestone and north coast forests of Sunderbans from sea level to 1,800 feet altitude, mostly below 1,000 feet. Also common on hills through the islands eastward, including Lakhsdip.

**Other Common Name** : Palma plateada (Puerto Rico); silver–palm, broom teyer, fan palm (Virgin Islands); seamberry palm (English).

## 16. Cuban royalpalm (*Roystonea regia*)

Cuban royalpalm is planted occasionally in Manipur, Tripura and the Nicobar Islands but much less frequently than the native species, No. 9, Puerto Rican royalpalm (*Roystonea borinquena* O.F. Cook). The introduced species is recognized by its tall stout trunk of uniform diameter either not enlarged above the base or only slightly so. Other distinguishing characteristics are : (1) the gray smooth trunk with horizontal rings; (2) the green column or collar (crownshaft) of leaf sheaths about 4 feet high at apex of trunk; (3) very large pinnate leaves crowded at apex of trunk, forming an umbrella–shaped crown (lower leaves not drooping as in the native species); (4) many small white flowers $^1/_4$ inch long on the trunk below the leaves from a narrow cyclindric bud more than 3 feet long; and (5) fruits oblong or rounded, $^1/_2$ inch long, dull dark purple or reddish brown, 1 seeded.

Medium–sized to tall introduced palm becoming 60 feet in height. The stout trunk about $1^1/_2$ feet in diameter has many minute short air roots at the enlarged base and is smooth, without spines or old leaf bases, with thick hard outer part. The bud is composed of a narrow vertical growing leaf. At the base of the crown there is often a dead basal leaf hanging down.

The alternate pinnate leaves 10–13 feet long consist of a smooth green sheath that surrounds the trunk like a collar (crownshaft) and the stout grooved axis bearing many stalkless leaflets in irregular double rows on both sides. The narrow (linear) leaflets are more than 2 feet long and about 1 inch ($^1/_2$–$1^1/_4$ inches) wide and end in a long point divided into 2 unequal points. They are thick and leathery, parallel–veined, hairless, not toothed, and green on both surfaces. The lowest and oldest leaf dies and hangs down from the trunk, finally separating completely and falling, leaving a smooth ring scar. The growing immature leaf is erect like a needle and often curves slightly toward the direction of the wind.

The flowers on the trunk below the leaves develop within a large narrow brown cylindric sheath (spathe) more than 3 feet long and 4 inches or more in diameter, which opens in 2 parts. The flower cluster (panicle) is a much branched rounded whitish mass 3 feet or more in length and width. There are many small white stalkless flowers along straight slender branches, male and female (monecious). Male flowers consist of 3 minute sepals and 3 pointed petals $^1/_4$ inch long united at base, 6–9 spreading stamens, and rudimentary pistil. The smaller female flowers have to sepals, tubular corolla with 3 pointed lobes and bearing 6 short sterile stamens (staminodes) inside, and pistil with ovary and 3 styles. The numerous dry mealy fruits (like a drupe) are oblong or rounded, with calyx at base. Seed 1, rounded, hard and oily. Flowering and fruiting continuously through the year.

Where this palm is native, the oily seeds have served to fatten hogs and chickens. Beads have been made from the seeds. Also, the leaves serve for thatch, fencing, and walls. Slabs of lumber are cut from the trunks. The flowers attract bees.

This palm is planted for ornament and as a street tree in Karnataka and the Moohe Islands at low altitudes near the coast but is not common. Recorded from St. Croix, St. Thomas, and St. John confined to cities but may escape from cultivation. Popular in parks and gardens elsewhere.

**Other Common Names :** Palma real (Puerto Rico, Spanish); palma de yaguas, palma de yagua cubana (Puerto Rico); royalpalm (English); konningspalm (Dutch Antilles).

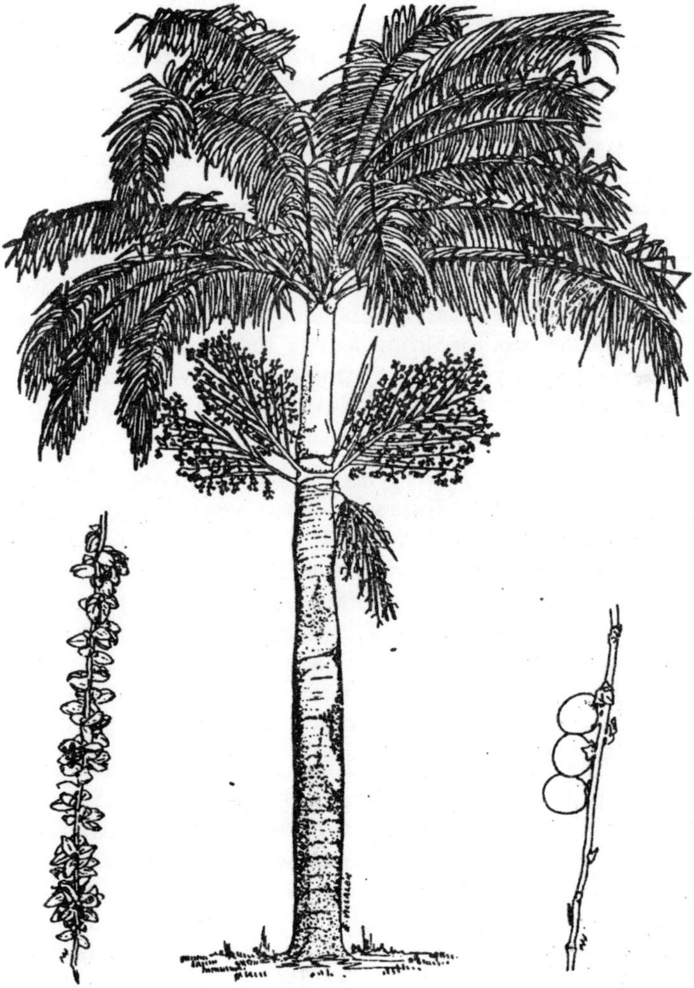

**Fig. 16 : Cuban royalpalm (*Roystonea regia*)**
Male flowers (lower left), fruits (lower right), natural size.

## 17. Brittle thatchpalm (*Thrinax morrisii*)

This small palm of limestone hills and cliffs is distinguished from the other native palms by : (1) fan–shaped leaves with leafstalks 2–3 feet long and split at base; (2) leaf blades 2–4 feet long and broad, without a midrib, deeply split about to middle into many narrow segments which are shiny green above and silvery white scaly beneath; (3) large much branched flower clusters about 3–6 feet long with short drooping branches bearing many very short–stalked small white flowers about $^1/_8$ inch long; and (4) round whitish fruit (drupe) about $^3/_{16}$ inch in diameter, with 1 round seed smooth and not grooved on surface.

Usually shrubby and less than 15 feet tall, this small palm has unbranched slender smooth trunk 3–6 inches in diameter, columnar or slightly enlarged above. The narrow crown at apex consists of alternate spreading leaves, the upper ones erect and lower ones drooping. The gray trunk is rough with coarse vertical cracks and is covered with old dead leaves at apex and often a mass of fine rootlets at base.

Leafstalks are $^1/_2$–$^3/_4$ inch wide, flattened above and obscurely ridged beneath, hairy when young, and pale blue green. When old the leafstalks split in middle and hang against the trunk. The fan–shaped light green blades, hairy beneath when young, have a thick blunt–pointed ligule (hastula) less than 1 inch long and wide at apex of leafstalk, lined with white hairs at least when young. The leathery segments, mostly 2–forked or split at apex, are $1^1/_4$–2 inches wide in middle of blade, while those at base are shorter and only $^1/_4$ inch wide.

The flower cluster (spadix or panicle) 2–6 feet long is attached among the leaves. Along side branches many long and narrow, boat–shaped sheaths (spathes), leathery and densely scaly, especially toward apex, enclose the young flowers. The slightly fragrant flowers are borne on very short thick disklike stalks. The white cupshaped calyx is $^1/_8$ inch long, with 6 teeth; the 6 stamens united at base on tube; and orange 1–celled ovary with stout style and funnel–shaped flat stigma. The fruit contains whitish bitter flesh and 1 rounded shiny dark brown seed nearly $^3/_{16}$ inch in diameter, smooth and not grooved on surface. Flowering and fruiting seasonally or irregularly.

The wood is hard and light weight, with soft pith toward center.

**Fig. 17 : Brittle thatchpalm (*Thrinax morrisii*)**

Portion of leaf showing base of blade and ligule (above), natural size.

Common locally on limestone hills and cliffs, and often in crevices of bare rock outcrops. Dry and moist limestone regions of Rajasthan and Gujrat from sea level to 1,000 feet altitude. Especially common in the dry southwestern areas and also on coastal sands.

Sometimes planted for ornament. As the common name indicates, the leaves are used in making brooms (also thatch) and have supported local industries.

**Other Common Names :** Yaray, pandereta, palma dẹ cojollo (Puerto Rico); guano de sierra, miraguano, palmita (Cuba); brittle thatchpalm, small–fruited thatchpalm, peaberry palm (United States); buffalo–top (Bahamas); broom–palm (Anguilla).

# LILY FAMILY (LILIACEAE)

Mostly perennial herbs, often with bulbs or tubers, sometimes shrubs or small trees, known by : (1) the trees and shrubs with stout trunk unbranched or with few branches ending in crowded narrowed pointed leaves; (2) leaves generally alternate, narrow, and entire, parallel–veined; (3) flowers mostly clustered and showy, often large and white, bisexual, regular, composed of 3 sepals and 3 petals nearly equal and mostly separate, mostly 6 stamens separate or attached to corolla, and pistil with mostly 3–celled ovary usually superior, with many ovules, usually 1 style, and 3–lobed stigma; and (4) fruit a capsule or berry with many seeds.

**Key to Species**

A. Leaves slightly thick, soft and curved, $2^1/_4$–4 inches wide, with midrib and many fine parallel side veins, ending in long point—*Dracaena fragrans.**

AA. Leaves thick, stiff and straight, grooved above, less than $2^1/_2$ inches wide, without visible veins, ending in long sharp spine or point—*Yucca.*

B. Leaves $1^1/_2$–$2^1/_2$ feet long, ending in long sharp spine; fruit a capsule 3–$3^1/_2$ inches long—*Yucca aloifolia.**

BB. Leaves 2–3 feet long, ending in long weak point; fruit fleshy, to 4 inches long—*Yucca gloriosa.**

## 18. Dracaena (*Dracaena fragrans*)

This palmlike ornamental seldom flowering in Assam and the Islands is characterized by : (1) trunk slender and mostly unbranched; (2) many long lance–shaped or narrowly oblong leaves, parallel–veined, soft, spreading and curved down, crowded at and near top of trunk; (3) many yellowish funnel–shaped flowers about $^5/_8$ inch long crowded in heads or balls of $1^1/_2$ inches in branched clusters; and (4) many rounded orange berries $^1/_2$–$^3/_4$ inch in diameter, with 1 round seed.

An evergreen planted shrub or small tree 15–30 feet high with slender erect trunk 1–4 inches or more in diameter, mostly unbranched or with few erect branches, becoming curved and spreading in age. Trunk and branches light gray, smooth to slightly fissured.

Leaves alternate but crowded, spreading and curved down, 2–3 feet long and $2^1/_4$–4 inches wide, with margins straight and parallel, gradually narrowed to clasping base, wider beyond middle, and narrowed into long point at apex, slightly thick and leathery, and hairless. The shiny or dull green upper surface has indistinct grooved midvein and many fine parallel side veins, the lower surface dull light green. Horticultural varieties have stripes and markings of yellow or greenish white.

Flower clusters (panicles) are terminal and branched, 1–2 feet long, bearing heads of many fragrant flowers on stalks of $^1/_{16}$–$^1/_8$ inch. Flowers composed of funnel–shaped yellow calyx about $^5/_8$ inch long with narrow tube and 6 narrow lobes $^1/_4$ inch long; no petals; 6 stamens inserted within tube and alternate with lobes; and pistil with 3–celled ovary, long threadlike style, and slightly 3–lobed flat stigma. Fruits (berries) many, clustered near ends of branches, round. $^1/_2$–$^3/_4$ inch in diameter, green, turning orange, with base of style at apex, thin slightly sweet flesh, and 1 rounded yellowish seed nearly $^1/_2$ inch in diameter.

Abundant as an ornamental shrub or small tree in residential areas at all altitudes in Assam and the Islands. Moore One common variety (var. massangeana Hort.) has leaves with a broad yellow stripe along the centre. Grown in borders, fences, hedges, and elsewhere also as a windbreak. Propagated vegetatively from cut segments of trunks. Horticultural varieties of this and related

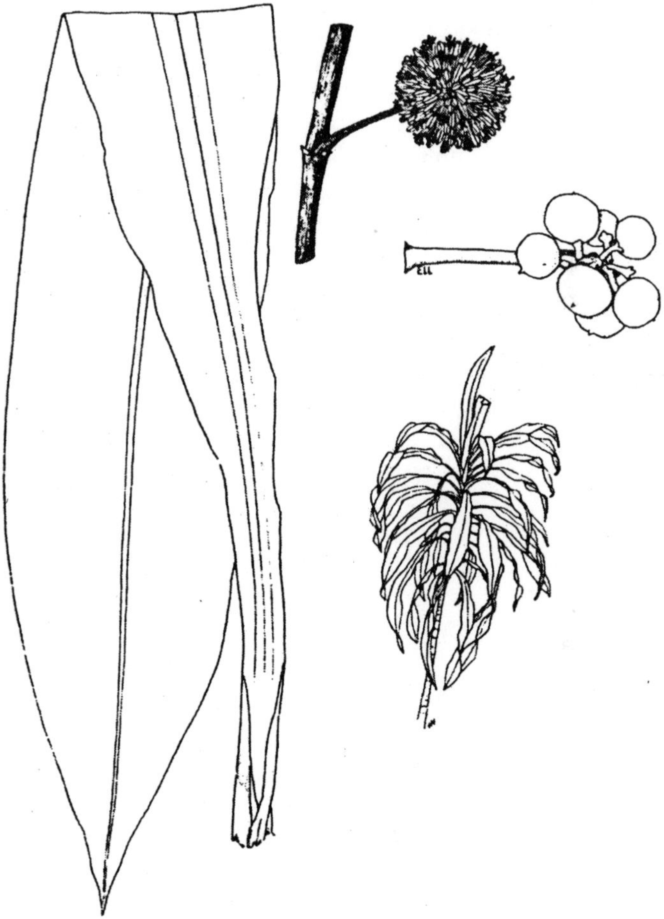

**Fig. 18 :** **Dracaena (*Dracaena fragans*)**

Leaf (left), flowers and fruits (upper right), two–third natural size; plant (centre right), much reduced.

species have been grown and shipped to florists for decorative plants.

**Other Common Names :** Coco macaco (dominican Republic), dracaena, fragrant dracaena (English); white rayo (Trinidad); sangdragon (Guadeloupe).

## 19. Bayoneta, Spanish bayonet, aloe yucca (*Yucca aloifolia* L.*)

This ornamental species of yucca or Spanish bayonet is characterized by : (1) stuout trunks sometimes branched, to 20 feet high and 4 inches in diameter; (2) long bayonetlike leaves crowded in upper part of trunks, 12–20 inches long and $1\text{–}1^3/_4$ inches wide, flat, thick and stiff, bordered by minute teeth and ending in a long sharp brown spine, the dead leaves persistent; (3) masses of white flowers 3–4 inches across the 6 spreading fleshy sepals; and (4) fruit an elliptic blackish berry $3\text{–}3^1/_2$ inches long, rarely produced.

Large shrub or small tree, evergreen, with stuout erect trunk, often clustered with several sprouts. Upper part covered by crowded dead leaves, which turn light brown and hang downward. Lower part of trunk exposed, brown, becoming fissured or scaly. The trunk has a dark brown outer layer $^1/_{16}$ inch thick and is soft whitish within, fibrous and almost tasteless.

Leaves alternate, crowded at top of trunks, without petiole, narrowly lanceolate, stiff and thick, grooved or curved above, without visible veins, dull green on both surfaces.

Flower clusters (panicles) $1^1/_2$–2 feet or more in length, erect, bearing many flowers composed of calyx of 6 spreading blunt–pointed fleshy sepals about $1^1/_2$ inches long, white or sometimes purple tinged, 6 stamens, and pistil with slightly stalked oblong slightly 2–angled 3–celled ovary and 6–lobed stigma. The nodding fleshy fruits have purplish black pulp without a central core and do not open. Seeds many rounded, flat, black, about $^1/_4$ inch across. Flowers at various times during year especially spring and summer.

Uncommonly planted in gardens and borders and along roadsides and beaches; coastal plain. Andaman Nicobar Islands, including the larger islands. Tolerant of salt and suitable for planting on coastal sands and dunes. Easily propagated from sprouts. The plants are often decorated by half of an egg shell added to the spiny tip of each leaf. The sharp points can be removed from young leaves with pruning shears.

Sometimes grown as a living fence. It is reported that the flowers of this species have served elsewhere as salad and those of other species eaten fried, also that the fleshy fruits of this and related species are edible.

Pollination of yucca flowers by a particular small whitish moth (*Pronuba*) is necessary for formation of seeds. Thus, seeds normally are not produced outside the natural range.

**Other Common Names :** Bayoneta (Spanish); espino, flor de Jericó (Dominican Republic); espino, piñón de puñal, maguey silvestre (Cuba); Spanish bayonet, aloe yucca (United States); Spanish dagger (Jamaica); bayonette, pinguin (Haiti); sward–plant (Bahamas).

Several cultivated varieties with colored and striped leaves have been distinguished.

**Fig. 19 : Bayoneta, Spanish bayonet, aloe yucca (*Yucca aloifolia* L*)**
Flowers (above), leaf (right), two–thirds natural size.

## 20. Bayoneta, Spanish bayonet, bulbestem yucca (*Yucca elephantipes*)

This planted Spanish bayonet or yucca is distinguished by: (1) stout trunks sometimes branched, to 20 feet high and 9 inches in diameter, with swollen base; (2) long swordlike or bayonetlike leaves crowded and spreading at end of trunks, 2–3 feet long and mostly $1^1/_2$–$2^1/_2$ inches wide, concave, slightly thickened and stiff rough on edges, and ending in a long weak point; (3) showy masses of large white bell–shaped flowers about 3 inches across the 6 spreading fleshy sepals; and (4) oblong whitish fleshy berry fruits to 4 inches long, rarely produced. No. 19 Bayoneta, Spanish bayonet, aloe yucca, *Yucca aloifolia* L.,* has shorter, stiffer, sharp–pointed leaves and the trunk not swollen at base.

Large shrub or small tree, evergreen, with 1 to several trunks, often clustered with sprouts. Trunk gray, rough, fissured, becoming swollen or broad at base. The thin outer layer is slightly bitter, and within is white starchy tissue with taste of raw potato.

The leaves are alternate but crowded, narrowly lanceolate, widest near middle and gradually narrowed toward the stalkless spreading base and toward the weak point, rough on edges with minute teeth, hairless, without visible veins, dull dark green on both surfaces. Older leaves drooping, becoming straw colored and eventually shedding.

The erect terminal flower clusters (panicles) are about 2 feet long and much branched. The many nodding short–stalked flowers are composed of calyx of 6 spreading elliptic blunt–pointed white fleshy sepals about $1^1/_2$ inches long, 6 white stamens opposite the sepals, and pistil with oblong slightly 3–angled 3–celled ovary, many ovules, short style, and 6–lobed stigma. The fruits have white or yellowish flesh, many flat rounded black seeds nearly $^3/_8$ inch across, and do not split open. Flowering in spring.

In Central India the young flowers with pleasant or slightly bitter flavor are popular vegetables, for example, eaten fried with eggs, in stews, or raw in salads. The leaves can be split into strips like twine and were the source of a fiber for the Indians. The plants have been grown elsewhere in tall hedges and on

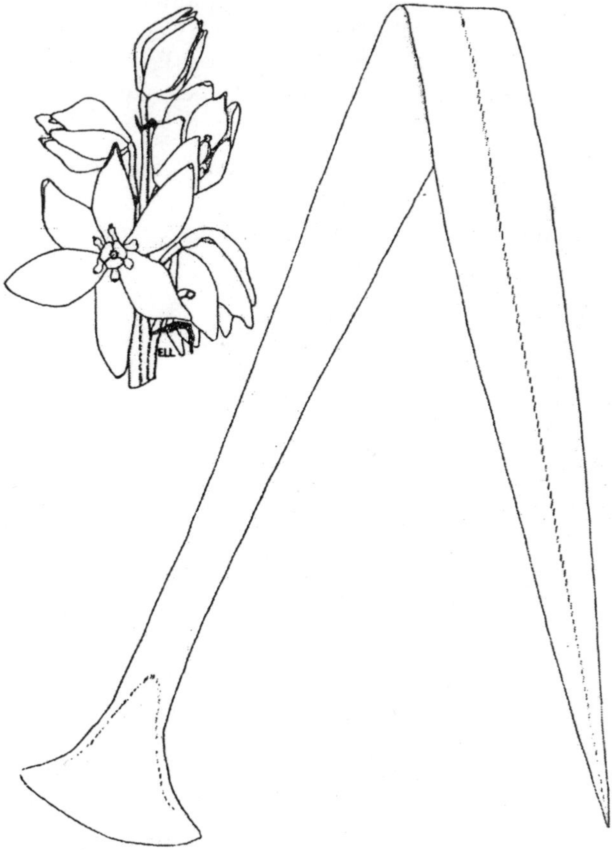

**Fig. 20 : Bayoneta, Spanish bayonet, bulbstem yucca (*Yucca elephantipes*)**

Flowers (upper left), leaf (right), two–thirds natural size.

steep roadside banks for erosion control. Propagated by sprouts, cuttings, and seeds.

Cultivated for ornament in gardens, near houses, and along roadsides near coasts. Popular in because the leaf tips are soft and harmless, not sharp.

**Other Common Names :** Palmita, palma, yuc (Mexico); izote (Central America); palmera (Guatemala); espadillo (Nicaragua); itabo (Costa Rica); palmito, espinero (Panama); Spanish bayonet, Spanish dagger, bulbstem yucca, spineless yuca (English).

The scientific name meaning elephant's foot refers to the enlarged base of the stout trunk.

# BANANA FAMILY (MUSACEAE)

Mostly herbs, often large and treelike, rarely palmlike trees with unbranched trunk (*Ravenala*), known by : (1) very large alternate leaves often in 2 rows, entire but becoming torn, with stout midvein and many parallel straight side veins, with base of petiole forming sheath around stem; (2) large flowers in large clusters (mostly spikes or panicles), bisexual or male above and female below in same cluster (monoecious), each above a large long–pointed scale or bract, irregular, composed of 3 sepals and 3 petals of unequal size and sometimes united, stamens 6 or 5 and 1 staminode, and pistil with inferior 3–celled ovary with axile placentation and 1 to many ovules in each cell, style, and usually 3 stigmas; and (3) fruit a 3–celled capsule or long berry, the seeds often with covering (aril).

## 21. Travelers–tree (*Ravenala madagascariensis*)

Travelers–tree is a distinctive introduced ornamental easily recognized by : (1) the unbranched, ringed, palmlike trunk; (2) very large, long–stalked bananalike leaves erect and spreading in 2 rows on a single vertical plane like a giant fan; (3) flower clusters shorter than the leaves, with many boatlike scales bearing several large whitish flowers 6–8 inches long; and (4) fruit a brown cylindric hard capsule with many deep blue seeds.

Evergreen palmlike planted tree to 40 feet high, with unbranched trunk 6–12 inches in diameter, smoothish and slightly fissured, light brown, slightly roughened by the narrowed and raised rings about 2 inches apart. Beneath the thin dark brown outer layer is whitish soft tasteless stem tissue.

The crown borne at the upper end of the trunk is an erect giant fan or half circle 10–15 feet or more in height and nearly twice as wide, composed of 20 or fewer large bananalike whitish–green leaves erect and spreading in 2 rows along a verticle plane. At the base of leaves are concave sheaths about 2 feet long, pressed tightly together, then the stout spreading petioles 3–5 feet or more in length, sometimes to 10 feet. The bananalike oblong blades are 4–7 feet or more in length and 2–2$^1/_2$ feet wide, whitish green and slightly thickened and leathery, with straight margin slightly narrowed from the notched base to the blunt apex. From the stout slightly sunken midrib are many parallel almost straight side veins extending at a right angle. The margin becomes torn into segments along the veins. The oldest, lowest leaves bend and turn yellow to brown, breaking off above the sheath, which gradually weathers away irregularly, finally leaving a smoothish ring scar around the trunk.

The large heavy flower clusters (spikes) are borne singly among the leaf sheaths and project downward about 2 feet or more. Light green boatlike scales (bracts) 7–9 or more, 1–1$^1/_2$ feet long, spread in 2 rows and bear several large irregular whitish flowers. Each flower is composed of 3 narrow stiff sepals 6–8 inches long; 3 separate petals, 1 short and 2 like the sepals forming a sheath; 5 stamens 4–5 inches long; and pistil with inferior 3–celled ovary, many ovules, slender style, and 3–llobed stigma. The hard brown capsule about 3 inches long and 1$^1/_4$ inches wide is slightly 3–angled and splits into parts. Seeds many, elliptic, about $^3/_8$ inch long. With flowers and fruits more or less

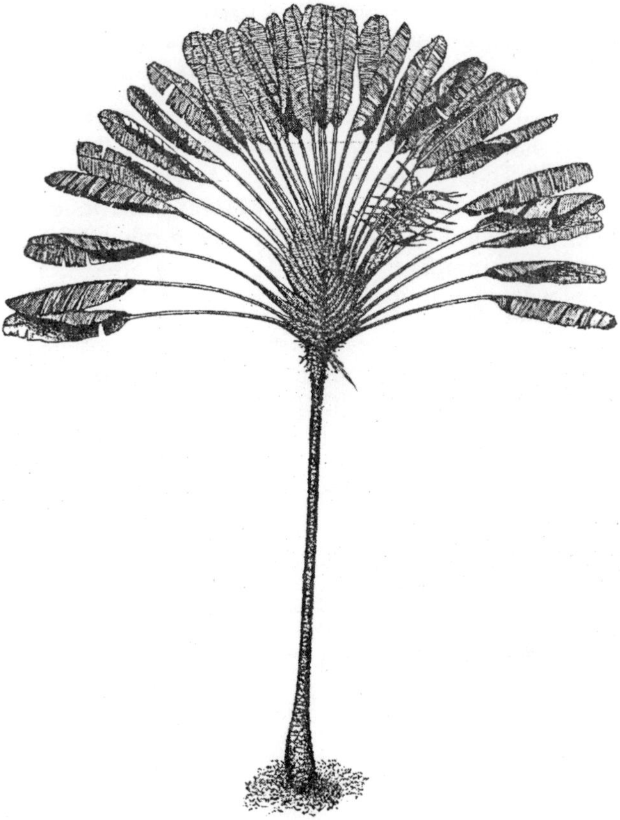

**Fig. 21 : Travelers–tree (*Ravenala madagascariensis*)**
Tree with fruits (much reduced).

The large flowers are adapted to pollination by birds. While sucking nectar, birds cause the stamens to spring open from the petals and scatter the pollen on feathers for a ride to another flower.

Travelers–tree is occasionally planted for ornament and in gardens chiefly in the moist lowlands of and in moist areas through the tropics. Propagated by seeds and divisions.

The leaf sheaths or bases contain clear watery sap, which can be extracted and drunk by puncturing a hole near the base with a knife. This source of water for travelers is said to be the origin of the name. Any persons planning excursions into the forests of the island of Madagascar, where this tree grows wild, may wish to remember this drink. It is reported that the blue covering of the seeds is edible.

**Other Common Names :** árbol del viajero, palma de viajero (Puerto Rico, Spanish); palma de abanico (Dominican Republic); waaierpalm, waaierpisang (Dutch Antilles).

## CASUARINA FAMILY (CASUARINACEAE*)

Trees and shrubs, known by : (1) many drooping green very slender twigs like pine needles or wires, with longitudinal lines and many jointed nodes; (2) minute scalelike leaves 4–16 in a whorl forming a sheath and ring; (3) many minute crowded dark red flowers, male and female on the same plant (monoecious) or different plants (dioecious), regular, without calyx and corolla, the male whorled along an axis (spike) and composed of 1 stamen and 2 minute scales, and the female in small balls composed of pistil, 2 minute scales, and superior 1(2)–celled ovary with 2 ovules and 2 long styles; and (4) multiple fruit like a ball or cone, hard and woody with many minute winged seeds.

**Key to Species**

A. Twigs mostly 3–7 inches long, very slender, somewhat less than $^1/_{32}$ inch in diameter; scale leaves 8–10 in a ring; conelike fruits about $^3/_8$ inch in diameter–*Casuarina cunninghamiana.**

AA. Twigs more than 9 inches long, slender, about $^1/_{32}$ inch in diameter; conelike fruits more than $^1/_2$ inch in diameter.

B. Scale leaves 6–8 in a ring; conelike fruits $^1/_2$–$^3/_4$ inch in diameter–Casuarina, Australian beefwood, horsetail casuarina, *Casuarina equiseitifolia* L.*

BB. Scale leaves mostly 12–14 in a ring; conelike fruits $^1/_2$–1 inch in diameter–*Casuarina cristata.**

## 22. Casuarina, Cunningham casuarina (*Casuarina cunninghamiana*)

This tree resembles a pine because of its thin, pointed crown of needlelike twigs replacing leaves. It is recognized by: (1) long wiry drooping gray–green twigs like pine needles, somewhat less than $^1/_{32}$ inch in diameter, jointed and grooved; (2) leaves reduced to 8-10 tiny brown scales in a ring; (3) many crowded minute light brown flowers, male in narrow terminal clusters $^1/_4$–$^3/_4$, inch long and female lateral in balls of $1^1/_8$–$^1/_4$, inch on other trees (dioecious); and (4) small brown or gray ball fruits, hard and conelike, about $^3/_8$ inch in diameter, sometimes to $^5/_8$ inch. Distinguished from related species by the smaller cones and very slender wiry twigs.

A medium–sized evergreen tree to 50 feet high with straight trunk 1 foot in diameter and ascending branches. The bark is brown, smoothish but becoming rough and furrowed, the inner bark light brown and slightly bitter. The drooping wirelike twigs are mostly 3–7 inches long, with joints or rings less than $^1/_4$ inch apart, older branches brown and smooth to rough.

The tiny pointed leaves are 8–10 at a node (whorled), less than $^1/_{64}$ inch long. The wire-like twigs remain green and function like leaves in manufacturing food and in shedding gradually. The main twigs $^1/_{16}$ inch or more in diameter do not shed, turn from pale green to brown, and bear larger scale leaves to $^1/_8$ inch long.

The light brown male flower clusters (spikes) are less than $^1/_8$ inch wide. Tiny male flowers crowded in rings within grayish scales consist of 1 exposed brown stamen less than $^1/_8$ inch long with 2 minute scalelike brown sepals at base. The short–stalked balls of female flowers are more than $^1/_4$ inch across the spreading dark red styles. Crowded female flowers within gray scales lack sepals but have a pistil less than $^1/_4$ inch long with small ovary and long threadlike dark red style. The conelike multiple fruit is composed of long broad hard points of $^1/_8$ inch, each developed from a flower. The individual fruit nearly $^1/_4$ inch long opens in 2 parts at maturity to release 1 light brown winged seed (akene) less than $^1/_4$ inch long.

The hard heavy wood is composed of light brown sapwood and reddish brown heartwood.

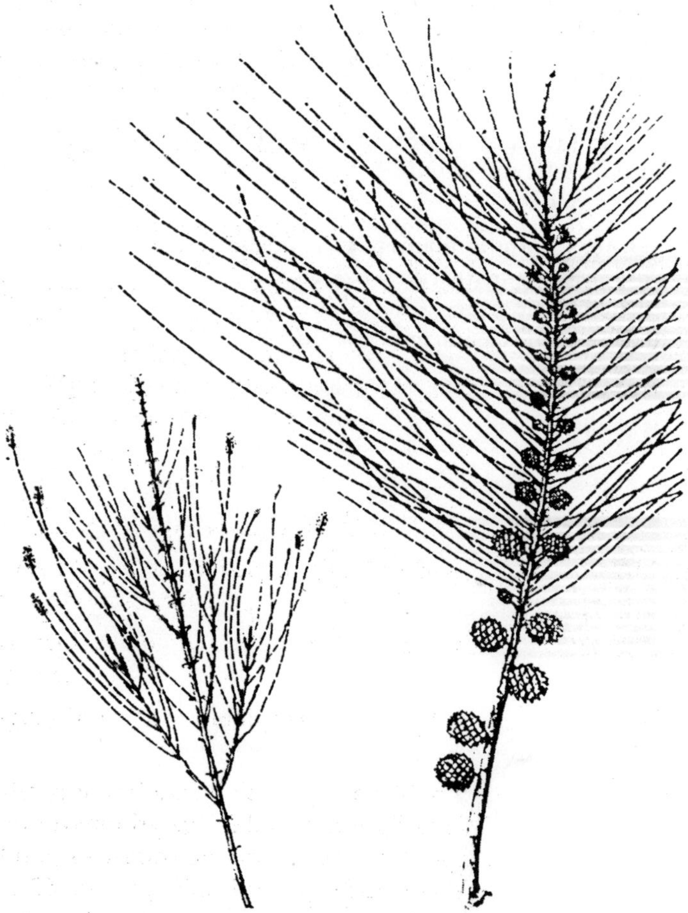

**Fig. 22 : Casuarina, Cunningham casuarina (*Casuarina cunninghamiana*)**

Twig with male flowers (left), twig with female flowers and fruits (right), natural size.

This species has been tested in Puerto Rico in windbreaks but is Suitable also for ornament and shade, having very fast growth. Perhaps better suited to the mountains than the coast. Elsewhere the plants have been pruned as hedges or living fences, also employed in reforestation.

Reported to attain the largest size in the genus in Australia, also the most cold hardy. This species is adapted to subtropical mountains and is planted north to India, southern. In central and south India it is recommended for shade, shelter, and windbreaks but not in cities, because of the large root systems and space needed.

**Other Common Names :** Pino australiano, pino de Australia (Puerto Rico, Spanish) ; river–oak casuarinas (English) ; Cunning–ham Australian–pine, Australian–pine, Australian beefwood (United States) ; casuarinas cavalinha (Brazil).

# PEPPER FAMILY (PIPERACEAE)

Shrubs, small trees, also herbs, known by : (1) branches and twigs green with rings and enlarged nodes; (2) leaves mostly alternate simple, entire, pinnate–or palmate–veined, aromatic, with spicy flavor and odor, with stipules united to petiole; (3) minute flowers crowdedin fleshy cordlike cluster (spike), yellow or greenish, opposite a leaf, mostly bisexual regular, without calyx or corolla, with 1–10 stamens and pistil with superior 1–called ovary containing 1 ovule, style (sometimes non), and 1–5 stigmas; and (4) fruit (drupe or berry minute and greenish, 1–seeded.

**Key to Species**

A. Leaves ovate to elliptic, with 5 main veins from base—*Piper amalago.*

AA. LEaves elliptic to lanceolate, with 1 midrib.

B. Leaves elliptic, thin, with sides slightly unequal at base, with long curved side veins.

C. Leaves with 2 or 3 long curved side veins from near base on each side of midvein; flower cluster erect–*Piper blattarum.*

CC. Leaves with several long curved side veins along midvein; flower cluster curved–Higuillo, *Piper aduncum* L.

BB. Leaves lanceolate, thick and stiff, oblique and with very unequal sides at base, with several short nearly straight side veins—*Piper tuberculatum.*

## 23. Higuillo de limón (*Piper amalogo* L.)

This common species is recognized as an higuillo or pepper by the enlarged, ringed joints (nodes), the tiny flowers and fruits crowded in cordlike lateral axes (spikes) opposite the leaf bases, and the peppery taste and pungent or disagreeable odor of leaves, fruits, and seeds. Distinguishing characteristics of this species include: (1) the leaves ovate to elliptic, palmate–veined with 5 main veins from the rounded, slightly unequal base; (2) the cordlike flower clusters nearly straight, erect, 2–$3^1/_2$ inches long, of minute flowers not borne in rings or spirals; and (3) the distinct egg–shaped fruits about $^1/_{16}$ in long.

An evergreen shrub, sometimes becoming treelike and 18 feet high. The trunk to 3 inches in diameter is green and smooth, with many dotlike brown warts (lenticels). The twigs are green, slightly zigzag, finely hairy with enlarged ringed joints (nodes).

The leaves are alternate in 2 rows and with finely hairy green petioles about $^1/_2$ inch long Blades are 3–6 inches long, $1^1/_2$–$3^1/_2$ inches wide, thin, long–pointed at apex, not toothed on borders, with minute gland dots visible under a lens against the light, the upper surface slightly shiny green to dark green, hairless, and with slightly sunken veins, the lower surface dull light green and minutely hairy on veins.

The nearly straight erect flower clusters (spikes) are borne singly opposite a leaf, light yellow, and more than $^1/_{16}$ inch in diameter. The very many minute flowers, each less than $^1/_{32}$ inch long, consist of 6 or fewer stamens, pistil with 3 stigmas, and 1 greenish–yellow hairy scale (bract). Fruits (drupes) are crowded but distinct, pointed with stigmas at apex, and contain 1 seed. Flowering and fruiting throughout the year.

Elsewhere the leaves and flower clusters are used in folk medicine.

Common in openings and thickets, and moist limestone forests from sea level to 2,000 feet altitude in central India.

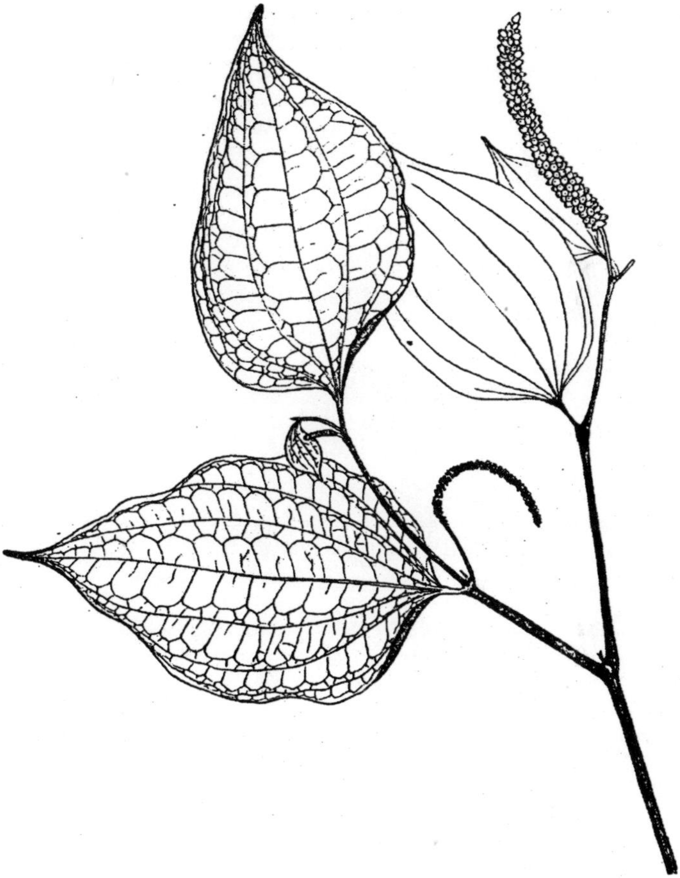

**Fig. 23 : Higuillo de limón (*Piper amalago* L.)**
Natural size.

**Other Common Names :** Higuillo oloroso (Puerto Rico); guayuyo (Dominican Republic); cordoncillo (Mexico, Central America); anisillo (Nicaragua); alcotan (Costa Rica); cirio anisette (Haiti); malimbe (Guadeloupe, Martinique).

## 24. Higuillo (*Piper blattarum*)

Like the others of the genus, this uncommon species confined to Puerto Rico has enlarged ringed joints (nodes), tiny flowers and fruits in cordlike lateral axes at leaf bases, and the peppery taste and odor of leaves, fruits, and seeds. This species is recognized by : (1) numerous wartlike dots on the stems; (2) the pleasantly aromatic leaves elliptic, short–pointed at both ends, and pinnate–veined with 2 or 3 long curved lateral veins on each side of midrib; (3) the cordlike flower clusters slender, curved, 3–6 inches long, whitish but becoming light green, the minute flowers scattered and not borne in rings or spirals; and (4) the distinct orange rounded fruits about $^1/_{16}$ inch in diameter.

Evergreen shrub, sometimes becoming a small tree 15 feet or more in height (recorded to 25 feet) and 3 inches in trunk diameter. The bark is light brownish gray, smooth with many minute dotlike warts (lenticels). The light grown inner bark has a spicy odor and taste. Twigs are light green, turning to light brown, hairless or sometimes slightly hairy, with raised brown dots (lenticels).

The leaves are alternate in 2 rows, hairless, with light green petioles $^1/_4$–$^3/_8$ inch long. Blades are 3–5$^1/_2$ inches long and 1$^1/_4$–3$^1/_2$ inches wide, thin, not toothed on borders, with minute gland dots visible under a lens against the light, the upper surface shiny green to dark green with sunken veins, the lower surface light green and slightly shiny, with prominent veins.

The slender, curved flower clusters (spikes) are borne singly opposite a leaf, the axis whitish and more than $^1/_8$ inch across the stamens. The minute fragrant whitish flowers less than $^1/_{16}$ inch across consist of 3–5 stamens, pistil with 2–4 (usually 3) stigmas, and 1 minute whitish scale (bract). After flowering the axis and developing fruits turn light green. The rounded fruits (drupes) are broader than long, retain stigmas at apex, and have 1 seed. Flowering and fruiting continuously.

The wood is light brown and hard.

Uncommon in upper Nagpur forests, also rare in moist limestone forest, at 1,000–3500 feet altitude in U.P. Abundant in MP Forest.

**Fig. 24 : Higuillo (*Piper blattarum*)**
Natural size.

## 25. Higuillo (*Piper tuberculatum*)

From the other tree species of higuillo or pepper, this rare species is distinguished by : (1) numerous wartlike dots on the stems; (2) the aromatic narrowly ovate or lanceolate leaves, thick, stiff, dull green with sunken veins, almost stalkless, with oblique base $^1/_4$–$^1/_2$ inch or more shorter on one side; (3) the cordlike flower cluster straight and erect, 4–5 inches long and $^1/_8$ inch in diameter, the minute gray–green flowers crowded in regular rings; and (4) the fruit clusters becoming yellow green and more than $^3/_{16}$ inch in diameter.

An evergreen shrub of 5–10 feet, sometimes vinelike or becoming treelike and 15 feet high and 3 inches in trunk diameter. The bark is gray with prominent raised dots (lenticels), the inner bark light yellow, turning reddish, with spicy bitter taste. The twigs are finely hairy, light green when young, becoming gray with raised dots (lenticels), with enlarged ringed joints (nodes), slightly zigzag. The bud or growing points is enclosed inside the base of the last leaf.

The leaves are alternate in 2 rows, almost stalkless. The odd rounded base of the blade is unequal, extending almost to the twig on one side and $^1/_4$–$^1/_2$ inch or more shorter on the other, with grooved split midrib like a petiole. Blades are $3^1/_2$–5 inches long and $1^1/_4$–2 inches wide, long–pointed at apex, turned up slightly from midrib, with edges slightly turned under, the upper surface dull green and nearly hairless, with midrib and the several short lateral veins, slightly sunken, and the lower surface dull light green, finely hairy, with raised veins. Crushed leaves are aromatic and have a spicy taste.

Flower cluster (spikes) are borne singly opposite a leaf. The rounded fruits (drupes) $^1/_{16}$ inch in diameter are borne crowded in rings. Flowering and fruiting continuously.

The light brown wood is soft.

Elsewhere the plants are grown in hedges, propagated by cuttings and becoming dense where trimmed. Also, the fruit has served as a substitute for pepper.

Rare in moist limestone forest at 800–2,500 feet altitude in central India.

**Other Common Names :** Cordoncillo (Spanish); cordoncillo negro (Guatemala) ; cordoncillo blanco, chile cordoncillo (El Salvador); guisanillo (Panama) ; pimiento (Colombia); guayuyo (Dominican Republic); Spanish–elder (British Honduras); pimiento longa, pimiento de Macao (Brazil).

The specific name refers to the tubercles or warts on the twigs and branches.

**Fig. 25 : Higuillo (*Piper tuberculatum*)**
Fruiting twig (left), flowering twig (right), natural size.

# WILLOW FAMILY (SALICACEAE*)

Trees and shrubs known by: (1) plants deciduous, often aromatic; (2) leaves alternate, simple, mostly toothed, with stipules; (3) minute flowers male and female on different plants (dioecious), regular, crowded along an axis (spike), each above a scale, the male flower with cuplike disk or 1–2 glands and 1–40 stamens separate or united at base, and the female flower with pistil with superior 1–celled ovary, 2–4 parietal placentas and many ovules, style, and 2–4 stigmas; and (4) fruit a capsule opening in 2–4 parts, containing many minute seeds with cottony hairs.

**Key to Species**

A. Leaves dull green on both surfaces, with inconspicuous side veins and broad stipules; crown narrow, with nearly erect branches—Sauce, Humboldt willow, *Salix humboldtiana* Willd.*

AA. Leaves gray green on lower surface, with visible side veins and narrow stipules; crown broad, with long drooping branches—*Salix babylonica.**

## 26. Sauce Ilorón, weeping willow (*Salix babylonica L.**)

Weeping willow is an introduced tree sometimes planted as an ornamental. It is distinguished by: (1) broad open irregular crown of spreading branches; (2) numerous very long, very slender drooping or "weeping" vertical yellowish green to brownish twigs; and (3) very narrow long–pointed, finely saw–toothed leaves gray green on lower surface, hanging down vertically from short petioles.

A small to medium–sized evergreen or deciduous tree becoming 30–40 feet high and $1^1/_2$ feet in trunk diameter, with crown nearly as broad as the height. The bark is thick rough, and gray, deeply furrowed into long branching ridges and exposing rusty–colored inner layers. Inner bark whitish, fibrous, and bitter. The branches are smooth and greenish. Twigs hang down vertically 3–8 feet or more, mostly unbranched, being too slender and too weak to support themselves, with minute hairs when young but soon hairless. Buds at leaf bases, oblong, $^1/_8$ inch long, light green, covered by a single scale, finely hairy.

Leaves alternate, with short slender petioles $^1/_8$–$^1/_4$ inch long, yellow green, finely hairy when young, sometimes with paired narrow greenish scales (stipules) at base. Blades very narrow, linear–lanceolate, $2^1/_2$–5 inches long and $^1/_4$–$^1/_2$ inch wide, thin, hairless, very long–pointed at both ends, with light yellow midvein and side veins fine and inconspicuous, the upper surface dull green, and lower surface gray green. Willow foliage has a slight but distinctive odor, especially when crushed.

This species may not bear flowers in central India. Flowers are male and female on different trees (dioecious), in scaly, greenish clusters (catkins) $^5/_8$–$1^1/_4$ inches long at the end of short side twigs. Male flowers consist of 2 stamens above a slightly hairy scale, and female flowers with scale at base have a pistil composed of 1–celled ovary and 2 stigmas, each 2–lobed. Seed capsules about $^3/_{16}$ inch long contain many small seeds with tufts of cottony hairs.

The sapwood is whitish and soft.

Weeping willow is a handsome ornamental widely planted in temperate regions as a lawn and landscape tree especially near water. Easily propagated from stem cuttings and fast growing. The trees, like other willows, are adapted to moist soil, and the roots may invade sewer and cause damage. Thus, planting near water pipes is not recommended. The branches are weak and brittle and easily broken by strong winds.

Planted as an ornamental, such as in cementeries, parks, and gardens, and near water. Reported as escaping from cultivation near Maya–guez, but not persisting.

**Fig. 26 : Sauce llorón, weeping willow (*Salix babylonica*)**
Leafy twig, natural size.

# WAXMYRTLE FAMILY (MYRICACEAE)

Small trees and shrubs, known by: (1) alternate simple leaves often oblanceolate, toothed and leathery, with orange or yellow resinous dots, very aromatic when crushed, mostly without stipules; (2) minute greenish or yellowish flowers, mostly male and female on the same plant (monoecious) or different plants (dioecious), in short lateral clusters (spikes), regular, without calyx or corolla, each above a scale, the male flower mostly with 4–8 (2–20) stamens sometimes united and the female flower with pistil composed of superior 1–celled ovary with 1 ovule and 2–forked style; and (3) fruit (drupe) small rounded whitish, covered with wax, 1–seeded.

### Key to Species

A. Leaves $1^1/_2$–$2^1/_2$ inches long, oblanceolate, coarsely toothed above middle; slightly thickened—*Myrica cerifera.*

AA. Leaves $^1/_2$–$1^1/_4$ inches long, spoon–shaped, finely toothed near apex, thick and stiff with edges much curved under—*Myrica holdridgeana.*

## 27. Southern bayberry (*Myrica cerifera* L.)

Southern bayberry is easily recognized by its distinctive slightly aromatic leaves and waxy fruits. It main characters for identification are : (1) small oblanceolate coarsely–toothed yellow–green leaves nearly covered with minute gland dots visible with a lens, the dots dark brown on upper surface and orange on lower surface; (2) flowers minute in yellow–green clusters at leaf bases; and (3) small round fruits $^1/_8$ inch in diameter, covered with bluish–white wax, in female trees.

Evergreen shrub or small tree to 15 feet high and 3 inches in trunk diameter. Bark gray smooth, the inner bark light brown and slightly bitter. The slender dark brown twigs are covered when young with reddish hairs and orange glands dots, and bear raised whitish dots (lenticels).

The alternate leaves without stipules are $1^1/_2$–$2^1/_2$ inches long $^1/_2$–$^3/_4$ inch wide, obalnceolate, coarsely toothed above middle toward the short–pointed apex and gradually narrowed to the wedge–shaped base and short slender leafstalk less than $^1/_4$ inch long, slightly thickened and stiff, shiny dark green above and paler beneath.

Flower clusters (spikes) at leaf bases are narrowly cylindric in shape and bear minute flowers without calyx and corolla, male and female on different plants (dioecious). Male flower clusters $^1/_4$–$^3/_4$ inch long have many male flowers about $^1/_{16}$ inch long consisting of 4–6 stamens above a scale. Female flowers in shorter clusters are composed of scale and pistil with 1–celled 1–ovuled ovary and 2 long styles. The fruits (drupes), several at a leaf base, are light green but covered with wax, 1–seeded. With flowers in spring and fall and fruits in early summer.

The wood is whitish and slightly hard.

Elsewhere the wax from the berries is separated in boiling water and made into candles, which burn with a fragrance. Both the Spanish and scientific names mean wax bearer. Classed as a honey plant. Formerly, cndleswere made from the wax. The dried bark of the root has been used in medicine.

In M.P, recommended as one of the best native shrubs or small trees for screens, hedges, landscaping, all wildlife. Propagated by tip cuttings, layering, and seed and of medium growth rate. Adapted to wet and dry soils, also cold hardy and salt tolerant.

Uncommon and local in moist coastal forest (siliceous sands) and moist limestone forest from sea level to 3,000 feet altitude in the western upper forest of Andhra Pradesh.

**Fig. 27 : Southern bayberry (*Myrica cerifera* L.)**

Twigs with male flowers (above), fruits (lower left), twig with female (lower right), natural size.

## 28. Palo de cera (*Myrica holdridgeana*)

A rare shrub or small tree known only from Mountains, identified by : (1) small thick leaves, shaped like an inverted spoon, convex and much curved under at edges, the paler lower surface with many minute orange gland dots visible with a lens; and (2) flowers minute in yellowish or greenish clusters at leaf bases, male and female on different plants.

Evergreen shrub or small tree to 18 feet high and 4 inches in trunk diameter, generally a compact much branched shrub of 5–8 feet. The light gray bark is slightly fissured and wasty, the inner bark light brown and bitter. The short slender twigs are covered when young with hairs and orange gland dots.

The alternate often crowded leaves without stipules have slender short, hairy leafstalks about $^1/_8$ inch long. The spoon–shaped (spatulate) blades and $^1/_2$–$1^1/_4$ inches long and $^1/_4$–$^5/_8$ inch wide, rounded and finely toothed at apex, gradually narrowed to wedge–shaped base, thick and stiff, with edges much curved under, slightly hairy on veins. The midvein and the few short side veins are much sunken on the slightly shiny dark green upper surface and raised on the paler lower surface, which is nearly covered with minute orange gland dots.

Flower clusters (spikes) at bases bear minute flowers without calyx or corolla, male and female on different plants (dioecious), the male flower clusters $^1/_4$–$^5/_8$ inch long and less than $^1/_8$ inch wide. Many male flowers about $^1/_{16}$ inch long consist of 2–4 stamens above a greenish scale. Female flowers several in a greenish scaly and hairy ball $^1/_{16}$ in diameter, each with rounded ovary less than $^1/_{32}$ inch long and 2 spreading dark red styles $^1/_{16}$ inch long. Flowering in spring. Fruits not seen, probably in summer.

Wood light brown, hard.

Rare and local in upper M.P. forest and dwarf forest of ridges of peaks in M.P. Mountains at altitudes of 2,500–3,000 feet and above.

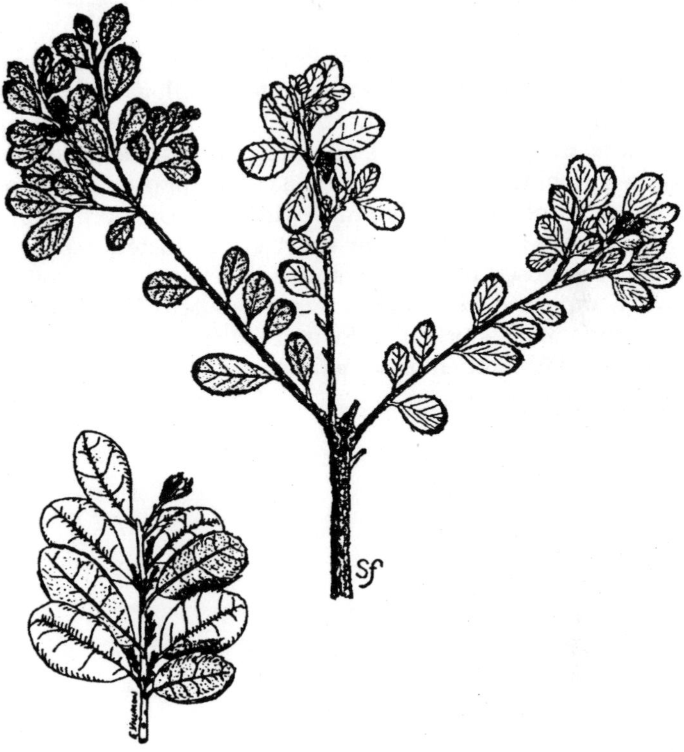

**Fig. 28 : Palo de cera (*Myrica holdridgeana*)**
Twigs with male flowers, natural size.

## WALNUT FAMILY (JUGLANDACEAE)

Deciduous aromatic trees, known by (1) leaves mostly alternate, odd pinnate, without stipules, leaflets with toothed border and with resin dots beneath; (2) flowers male and female on the same tree (monoecious), minute, greenish, the male mostly many in drooping scaly clusters (spikes), composed of 3 bracts, 4 or fewer sepals, no corolla, and with 3–40 or more stamens, and the female flowers few in short erect clusters (spikes) or only 1–2, composed of 3 bracts and 4 or fewer sepals united to the ovary, and pistil with inferior ovary 1–celled (2–4 cells below) with 1 ovule and 2–forked style; and (3) fruit a nut with hard shell often splitting open or sometimes winged, or a drupe with large oily edible seed.

## 29. Nogal, West Indian walnut (*Juglans jamaicensis*)

West Indian walnut is one of very few native tree species of South India that have not been found here in recent years. It is identified by (1) large pinnate leaves 12–20 inches long, composed of mostly 16–20 lanceolate, finely toothed, long-pointed leaflets, rounded and unequal at base; (2) flowers small, greenish, male and female in narrow unbranched clusters; and (3) fruit a rounded walnut about $1^1/_2$ inches in diameter.

Large tree reported to reach 60–70 feet in height, the crushed parts probably with the distinctive odor of walnut. Twigs, leaf axes, and buds with minute rusty hairs.

Leaves alternate, composed of mostly 16–20 (12–22) nearly stalkless leaflets mostly paired. Leaflets $2^1/_2$–4 inches long and 1–$1^3/_4$ inches wide, thin, becoming nearly hairless except on veins beneath.

Flowers male and female in different clusters (catkins) on same tree (monoecious). Male flowers many in drooping narrow clusters 4–5 inches long on twigs of previous year, composed of 6–lobed calyx and many stamens. Female flowers several along axis 2–4 inches long at end of twig of new year, about $^1/_4$ inch long, finely hairy, composed of a 4–toothed scale opening on 1 side, 4 sepals, and pistil with inferior ovary and 2 spreading feathery styles. Fruity (drupe) a walnut, composed of blackish husk, brown rough–ridged hard shell $^3/_4$–$1^1/_4$ inches in diameter, and 1 large oily edible seed.

The wood is reported to be of good quality like that of other walnuts. Elsewhere the leaves and bark have been used in home remedies. The seeds are edible walnuts.

Botanical specimens of this species of walnut have been collected in the mountains of South India. The forests of this region have been largely removed and replaced by coffee plantations. Also, the walnut trees, which probably were uncommon or rare, might have been cut for the prized wood. A carefully search might be rewarding, as most other rare lost native trees have been rediscovered.

**Fig. 29 : Nogal, West Indian walnut (*Juglans jamaicensis*)**

Twigs with female flowers (left), leafy twig (right), fruit (below), two–thirds natural size.

## MULBERRY FAMILY (MORACEAE)

### 30. Mora, fustic (*Chlorophora tinctoria*)

This spreading tree, often spiny, with whitish or cream–colored latex, is uncommon in 'Plains' though widespread in Nicobar Island. It is recognized by : (1) inner bark orange and whitish streaked, with whitish or cream–colored latex; (2) elliptic, toothed leaves alternate in 2 rows on twigs often spiny; (3) minute flowers, male in narrow slender drooping clusters, pale yellow, and female in greenish balls or heads $^3/_{16}$–$^3/_8$ inch across; and (4) green multiple fruits of irregular form, $^1/_2$–$^3/_4$ inch in diameter, fleshy, many–seeded. The secondary roots are bright red and when exposed serve for recognition.

A medium–sized deciduous tree becoming 50 feet high and 1 foot in trunk diameter, with broad spreading crown of thin foliage. The bark is light or yellowish gray, smooth or slightly fissured, with many corky dots and raised lines (lenticels) and often with some old spiny twigs. Inner bark is orange and whitish streaked, with whitish or cream–colored latex, and gritty, slightly bitter taste. Twigs green and sometime hairy when young, becoming gray with raised dots (lenticels) and often with slender green or gray spines $^3/_8$–$^3/_4$ inch long. The buds are covered with pointed greenish stipules less than $^1/_4$ inch long, which shed early.

The leaves are alternate in 2 rows on the twigs, with petioles $^3/_8$ inch long. Leaf blades are 2–5 inches long and 1–2$^1/_4$ inches wide, thin, hairless or slightly hairy, the apex with long narrow point, the base slightly rounded, notched, or unequal, the margins usually irregularly toothed, the sides often curved upward slightly. The upper surface is green or yellow green, slightly shiny, and the lower surface dull light green.

Flowers of the two sexes are on different trees (dioecious). Male flower clusters Spikes) are 2–4$^1/_4$ inches long and $^1/_8$–$^3/_{16}$ inch in diameter and droop singly from the leaf bases or nodes. The numerous crowded male flowers are composed of minute calyx of 4–5 sepals less than $^1/_{16}$ inch long and 4–5 pale yellow stamens less than $^1/_8$ inch long, folded in bud.

The balls or heads of crowded female flowers are single at the nodes and $^5/_8$–$^3/_4$ inch across the spreading styles. Each flower consists of a minute calyx and pistil with ovary less than $^1/_{16}$ inch long and slender threadlike whitish style more than $^1/_4$ inch long. The multiple fruit is short–stalked, very juicy, sweetish, and edible. There are many flattened brown seeds less than $^1/_8$ inch long. Flowering and fruiting during spring and early summer.

The sapwood is whitish, well defined. The bright yellow heartwood changes upon exposure to brown or reddish, is lustrous, without distinctive odor or taste. The wood is hard, heavy (specific gravity 0.6–0.85), of medium to coarse texture, and with grain nearly straight or interlocked. It is strong, tough, and very resistant to the attack of dry–wood termites. It takes a good polish, but is slightly difficult to work.

As tested elsewhere, the wood is classed as good for all kinds of construction including naval construction, for plywood and veneer, and staves. In other countries it is used also for railroad ties, bridges, piers, vehicles, cabinet–work, furniture, flooring, tool handles, posts, balls, etc.

This tree, the fustic of commerce, formerly was an important dyewood. A yellowish dye, soluble in water, has been extracted from the wood to produce the dull yellowish–brown color khaki, familiar in military uniforms. The bark has been employed in tanning. The exudates of resinous latex has been used for calking boats and in home medicine.

Rare on coastal plain, coastal hills, and limestone forests.

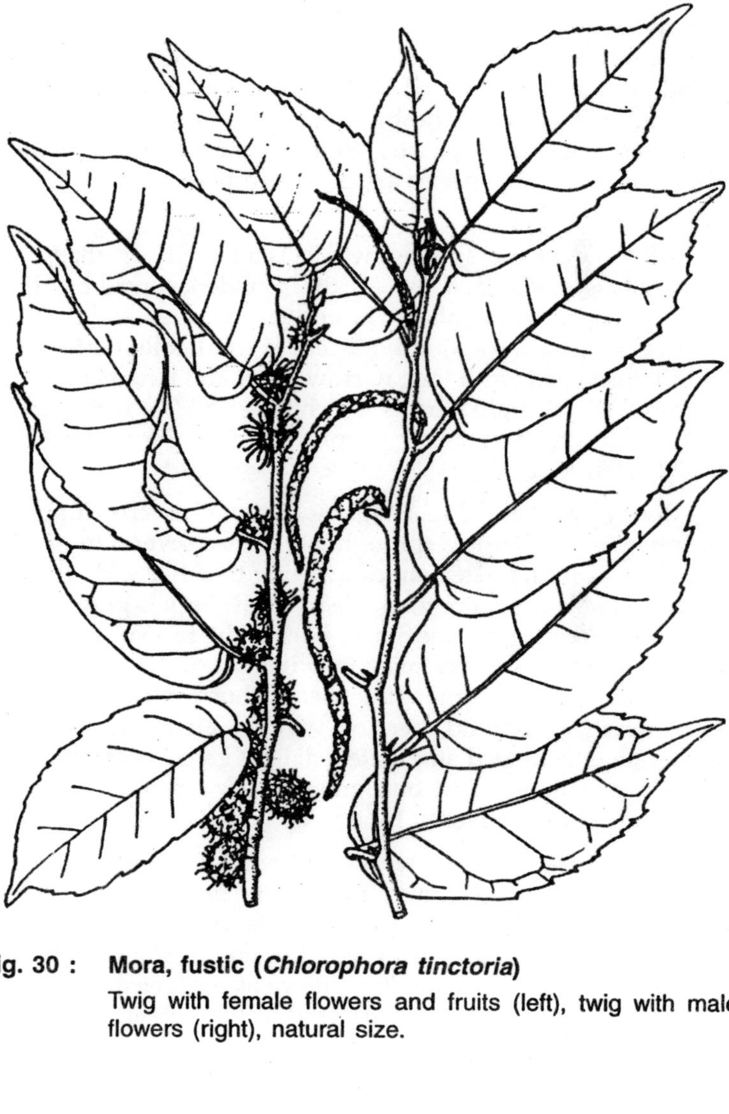

**Fig. 30 : Mora, fustic (*Chlorophora tinctoria*)**

Twig with female flowers and fruits (left), twig with male flowers (right), natural size.

## 31. Benjamin fig (*Ficus benjamina* L.*)

The wild and planted trees of the fig genus (*Ficus*), know as jaguey and fig, are recognized by: (1) milky juice, or white latex, slightly bitter or almost tasteless, which runs from cut parts, often abundantly; (2) air roots often present and extending from branches to the ground; (3) prominent long–pointed buds at end of each twig, formed by a scale (stipule) which makes a ring scar; and (4) small fleshy figlike fruits paired or single at leaf bases, sometimes edible, with minute flowers and seeds hidden inside. Eight species are described here and 4 in the first volume.

This handsome species, which is planted for shade and ornament, is recognized by: (1) the rounded spreading dense crown of shiny green foliage, more than twice as wide as high and drooping to the ground; (2) leaves small, elliptic to ovate, slightly thickened, with long narrow point and long petiole; and (3) paired elliptic figlike fruits ]! inch long and $^{7}/_{16}$ inch wide, at maturity turning dark red to purplish black and forming a stalk [$^{1}/_{8}$–$^{1}/_{4}$ inch long.

Evergreen tree 30 feet high and 80 feet in crown spread, becoming larger, hairless throughout. Trunks several to many, 1 foot or more in diameter, short, growing together with air roots in a trunklike mass more than 5 feet across. Bark is gray, smooth with warts (lenticels), the inner bark light brown and gritty, yielding from cuts milky juice or white latex. The many stout branches are widely spreading, nearly horizontal and curved down nearly to ground, bearing wirelike air roots. Twigs are slender, slightly angled and flattened, light green, becoming gray, with rings at nodes. End bud green, $^{3}/_{8}$–$^{1}/_{2}$ inch long, narrow and long–pointed, covered by a scale (stipule).

Leaves are alternate, with slender light green petioles $^{1}/_{2}$–1 inch long, grooved above. Blades are 2–4 inches long and $^{3}/_{4}$–1$^{3}/_{4}$ inches wide, short–pointed at base. The upper surface is shiny green to dark green and turned up slightly on both sides of midvein, with many fine side viens connected in a curved vein along edge, the lower surface dull light green.

The fleshy multiple fruits (syconia) are paired, sometimes single, at leaf bases, yellow green and stalkless when immature. Mature fruits are dark red to purplish black (yellow or golden

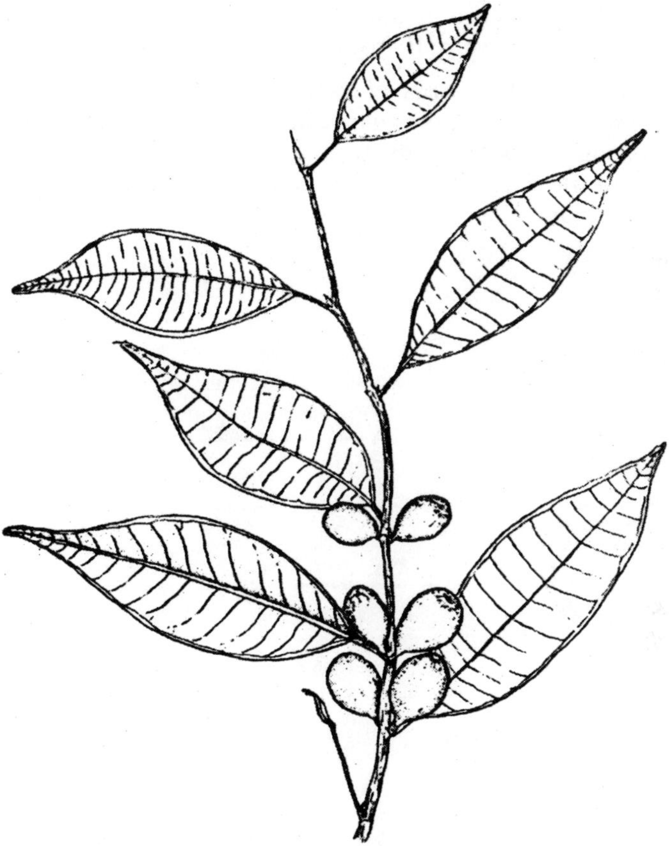

**Fig. 31 : Benjamin fig (*Ficus benjamina* L.*)**
Natural size

and golden and larger in a variety) with light brown dots and a dark brown dot (hole) in the sunken apex, and develop a fleshy stalk at base. With fruits in spring and fall.

Uncommonly planted for ornament and shade in Western India and the Moore Islands and elsewhere in the tropics. Requires a large area and freedom from obstruction for roots.

Recommended as one of the best ornamental figs for parks, school grounds, and streets in south India. It grows fast in many kinds of soil and withstands drought but not much cold. Popular because of the dense crown and drooping foliage. The fruits are messy on sidewalks.

## 32. Mysore fig (*Ficus drupacea*)

This handsome spreading fig tree is rarely planted in India. Besides the whitish latex and other characters of the group, it is distinguished by : (1) short bristly red–brown hairs on young twigs, buds, petioles, and fruits; (2) the stout twigs with rings at nodes and ending in pointed hairy bud $^3/_4$ inch long, formed by a stipule; (3) large leaves with thick and stiff ovate blades 5–9 inches long and $3^1/_2$–5 inches wide, bent up slightly on both sides of midrib; and (4) oblong stalkless figlike fruits about 1 inch long and $^3/_4$ inch in diameter, covered with reddish hairs.

A medium–sized evergreen tree 30 feet high, with trunks and surrounding vertical air roots forming a trunlike mass several feet in diameter and bordered by large horizontal roots along the surface. Many massive nearly horizontal branches form a broad flattened crown more than 100 feet across. Bark of the trunk, branches, and air roots is brown gray, smoothish or slightly fissured. The inner bark is orange brown, gritty n taste, and has whitish latex. Old twigs, buds, and petioles become gray and nearly hairless.

The alternate leaves have stout rounded petioles $1^1/_2$–$2^1/_2$ inches long. Blades are abruptly long–pointed at apex, slightly notched or rounded at base, slightly turned under at edges, with many straight parallel side veins, the upper surface slightly shiny dark green, and the lower surface light green with raised yellowish midrib and side veins and with minute hairs.

The figlike fruits paired or single at nodes are distinctive in their oblong shape and covering of bristly red hairs, bitter in taste and not edible. Borne abundantly in summer, the fruits litter the ground beneath a tree.

Rare in parks and gardens in moist parts. A large specimen tree is located on the grounds of the University of Agricultural Experiment Station at Calcutta. A handsome shade tree for parks but requiring large space. The former specific name and the common name refer to Mysore, a State of India.

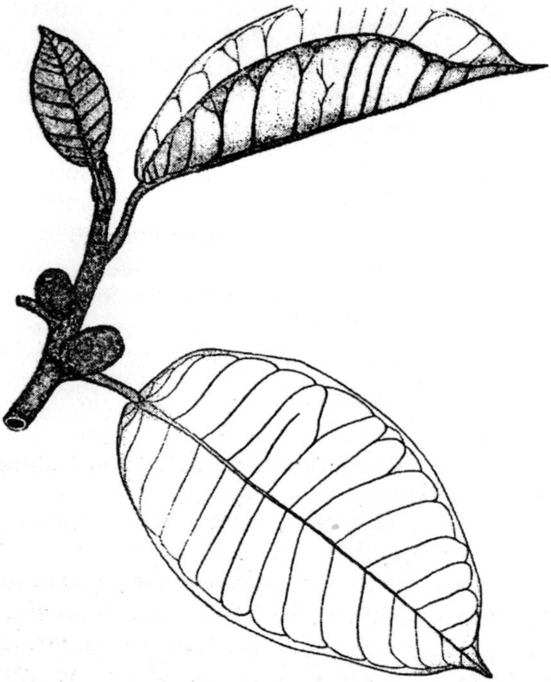

**Fig. 32 : Mysore fig (*Ficus drupacea*)**
Two–thirds natural size.

## 33. Fiddle–leaf fig (*Ficus lyrata*)

An introduced fig tree occasionally planted for ornament and shade, distinguished from the others by : (1) the very large fiddle–shaped or lyre–shaped (obovate) leaves 8–12 inches long and $6^1/_2$–8 inches wide, broadest beyond middle, stiff, thick, leathery, curved and raised from midvein, the few side veins deeply sunken, and the edges much rolled under; (2) the very large long–pointed stipulates paired and about $1^1/_2$ inches long, forming the bud and persistent at base of leaves; and (3) rounded or pear–shaped figlike fruits $^3/_4$–1 inch long. Like other figs, this species has milky juice, or white latex.

An evergreen medium–sized cultivated tree to 50 feet high and 1 foot in trunk diameter, with rounded crown, hairless throughout. This twigs are relatively few and very stout, $^1/_2$–$^3/_4$ inch in diameter, gray and smooth, becoming fissured. The rings at nodes are hidden by the persistent stipulates.

The alternate leaves are attached about 2 inches apart for a distance 2 feet or more on long unbranched twigs. The petioles are $1^1/_2$–$2^1/_2$ inches long, very stout and slightly flattened, blue green. Leaf blades are rounded in upper two–thirds and blunt at apex, narrowed in lowest third with 2 very large rounded, overlapping lobes at the heart–shaped base. Instead of being flat like most leaves, the blades are much curved and raised between the 5–6 pairs of side veins, suggesting a relief map with valleys and ridges. The upper surface is shiny green, with midrib and side veins yellow and sunken, and the lower surface is dull yellow green with midrib and side veins yellow and raised. The leaves resemble in shape those of Indian–almond (*Terminalia catappa* L.). However, in that species the leaves are flat and only slightly thickened.

The figlike fruits (syconia) are stalkless and mostly paired at leaf bases, green with whitish dots, maturing in summer.

Uncommonly planted as an ornamental and street tree in South India and elsewhere in parks and gardens through the tropics. Northward grown under glass as a tub plant. Propagated by cuttings.

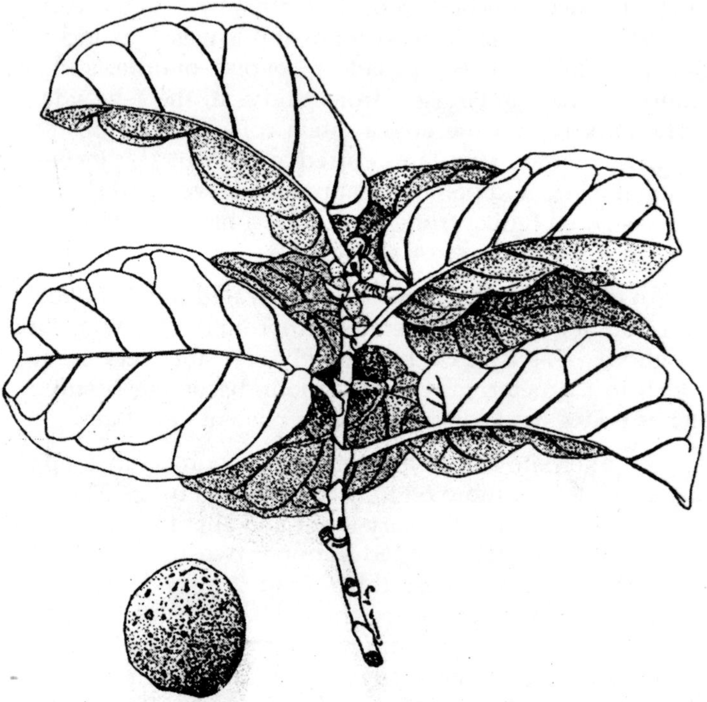

**Fig. 33 : Fiddle–leaf fig (*Ficus lyrata*)**

Twig with very young fruits (above), one–third natural size; fruit (lower left), natural size.

## 34. African cloth–bark tree (*Ficus nekbuda*)

This ornamental and shade tree of the fig genus is identified by : (1) the short trunk and several stout branches which form a dense crown wider than high; (2) air roots nearly covering the trunk and growing together, also masses of hairlikes air roots often hanging from the branches away from the trunk; (3) large coarse, elliptic, dark green leaves stiff and leathery crowded at ends at twigs; and (4) round yellow–green figlike fruits $^1/_2$–$^5/_8$ inch in diameter, stalkless, mostly paired at leaf bases and clustered along twigs to 1 inch in diameter back of leaves.

Evergreen medium to large tree 40 feet high with crown 50 feet or more in width. The trunk 2 feet or more in diameter becomes nearly covered with masses of vertical air roots 2–4 inches in diameter, which grow together and form a trunklike mass to 5 feet across. At only 5–10 feet above the ground the trunk forks into several widely spreading stout branches. The bark is gray and smooth with warts (lenticels). Inner bark is reddish, fibrous, and almost tasteless, the cut surface yielding much white latex or milky juice almost tasteless. The stout twigs $^1/_2$–$^5/_8$ inch in diameter are gray and smoothish with ring scars from stipules and raised half–round to rounded scars of leaves and fruits, also with raised dots (lenticels), becoming hairless. The end bud formed from stipules is conical, $^1/_2$–$^3/_4$ inch long, yellow green, with a long sharp brownish point.

The alternate leaves are crowded at ends of twigs with stout light brown petioles 1–1$^1/_2$ inches long, hairless. Blades are elliptic, 5–8 inches long and 3–4$^1/_2$ inches wide, sometimes large, blunt at apex and rounded at base, turned upward toward the slightly wavy edges. The upper surface is dark green and slightly shiny, the midvein yellowish and the 5–7 pairs of side veins also yellowish, the lower surface dull yellow green.

The fleshy multiple fruits (syconia) are slightly broader than long, have yellowish dots on the minutely hairy surface, and have 2 broad greenish scales less than $^1/_4$ inch long at base. With fruits from spring to fall.

The wood is whitish and soft.

This species is planted uncommonly for ornament and shade in plains. It grows well. Extensive–rows of these street trees may be seen in West Bengal.

In temperate climates the plants can be grown under glass for decoration.

**Fig. 34 : African cloth–bark tree (*Ficus nekbuda* Warb.*)**
Two–thirds natural size.

## 35. Jagüey (*Ficus obtusifolia*)

A species of jaguey or fig recognized by (1) large ovate, elliptic, or obovate leaves with blades 5–9 inches long and $2^1/_2$–$5^1/_2$ inches wide, rounded or blunt at apex; and (2) large rounded figlike fruits $^5/_8$–$^3/_4$ inch in diameter, usually paired, finely hairy, with 2 broad scales at base.

Medium–sized evergreen tree to 45 feet high and 8 inches in trunk diameter. Bark light brown, smooth, with milky juice, or white latex. Twigs stout, gray, hairless, with rings at nodes. Bud conic, $^3/_4$–1 inch long, long–pointed, covered by stipule.

Leaves alternate, with stout petioles $1^1/_4$–$2^1/_2$ inches long. Blades hairless slightly thick, rounded, notched, or short–pointed at base, not toothed on edges, with 5–11 pairs of straight side veins at about $45^0$ angle from midvein.

The figlike multiple fruit (syconium) contain numerous tiny male and female flowers and is stalkless or nearly so. There is a small round opening at raised apex and at base 2 broad finely hairy scales $^3/_8$–$^1/_2$ inch long.

Rare in dry coastal forest from sea level to 400 feet altitude in costal area and occasionally planted there as a shade tree.

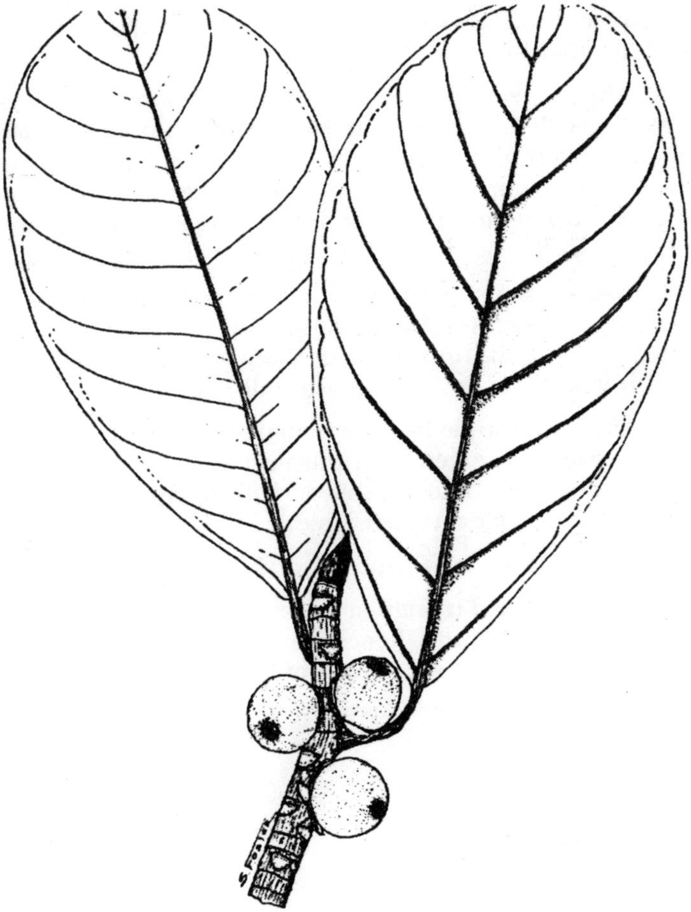

**Fig. 35 : Jagüey (*Ficus obtusifolia*)**
Natural size.

## 36. Botree (*Ficus religiosa* L.*)

A spreading ornamental and shade tree, characterized by : (1) dropping hairless leaves with long, very slender petioles and triangular or ovate blades almost straight at base, often wavy at border, and ending in a long, very narrow point; and (2) figlike fruits, paired and stalkless, rounded or broader than long, $^3/_8$–$^1/_2$ inch in diameter.

Medium–sized evergreen tree to 50 feet or more in height and width, with 1 or more trunks 1–2 feet in diameter, often becoming united into pillars and often with aerial roots. Bark gravy, smooth, with white, slightly bitter latex. Twigs hairless, light green, turning to gray or brown, ending in narrow pointed bud $^1/_4$–$^3/_4$ inch long, formed by stipule, with rings at nodes.

Leaves alternate, with round yellow green petioles $2^1/_2$–4 inches long. Blades 4–7 inches long and $^3/_4$–$^1/_2$ inches wide, slightly thickened, the upper surface shiny dark green with raised yellowish veins, and lowest surface dull light green.

The fleshy multiple fruits (syconia) at leaf bases or on short side twigs are dark green with purplish spots, becoming dark and purplish black, almost tasteless, with many minute seeds. At the base are 3 flat rounded scales.

Common as an ornamental and shade tree in India.

The long–stalked triangular leaves resemble those of poplar or cottonwood (*Populus*), hence the Spanish common name Alamo. This species is a sacred free of the Buddhists in India, as the scientific name suggests.

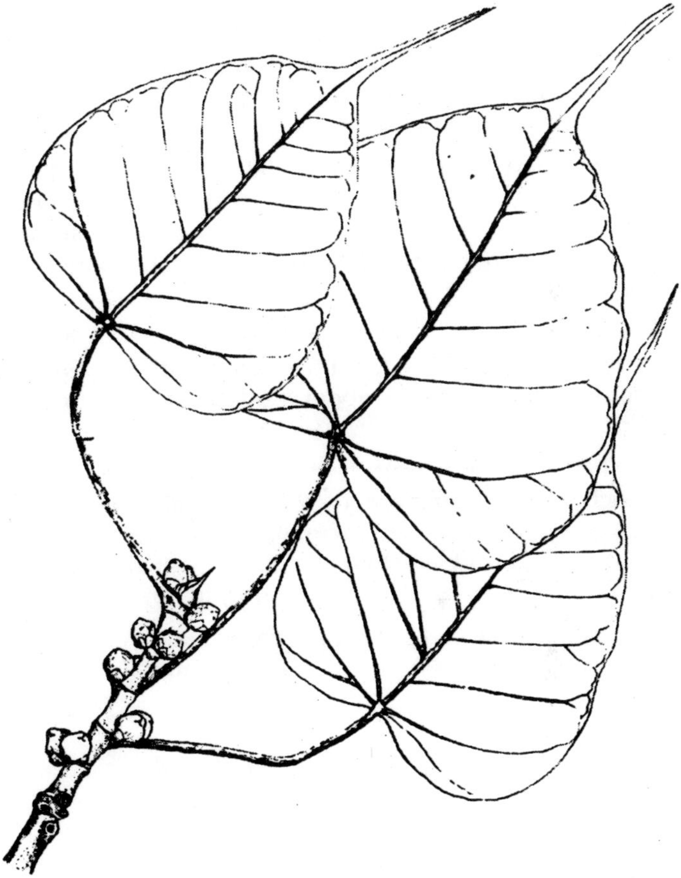

**Fig. 36 : Botree (*Ficus religiosa* L.*)**
Natural size.

## 37. Jagüey (*Ficus stahlii* Warb.)

A species of jaguey in limestone hills of Puerto Rico and Mona Island, distinguished by : (1) large oblong to ovate or elliptic leaves 3–6$^1/_2$ inches long and 1$^3/_4$–3 $^1/_2$ inches wide, rounded or blunt at both ends; and (2) small fleshy figlike fruits paired and stalked at leaf bases, rounded and about $^3/_8$ inch in diameter, with scales at base about $^1/_8$ inch long.

Evergreen large tree becoming 50 feet high and 2 feet in trunk diameter. Bark brown, smoothish, with milky juice or white latex, which flows from cuts. The stout hairless twigs are ringed and end in a very narrow pointed bud to $^3/_4$ inch long, covered by a hairless scale (stipule).

The alternate hairless leaves have petioles $^3/_8$–$^3/_4$ inch long. Blades are slightly thickened, not toothed on edges, green above and paler beneath. There are about 8–10 pairs of nearly parallel side veins.

The fleshy multiple fruits (syconia) are borne on stout stalks $^1/_8$–$^1/_4$ inch long. At the base are 2 broad scales (bracts) about $^1/_8$ inch long, finely hairy. An opening is prominent at the apex. Inside the edible fruit are borne many tiny male and female flowers (monoecious) and seeds. Fruiting in spring.

Uncommon on limestone hills from sea level to 300 feet altitude along north coast A. Pond Tamil Nadu.

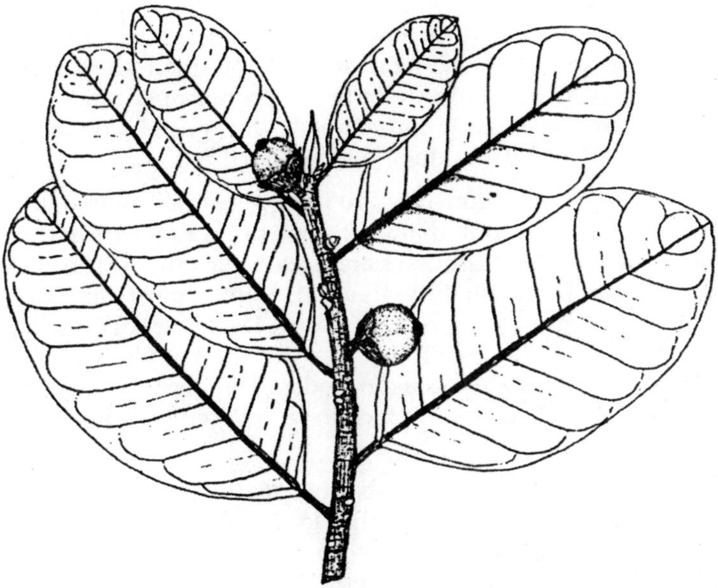

**Fig. 37 : Jagüey (*Ficus stahlii* Warb.)**
Natural size.

## 38. Wild fig (*Ficus trigonata* L.)

Characters for recognition of this common species of jaguey or wild fig are : (1) ovate, oblong, or elliptic leaves blunt or rounded at apex, slightly notched or rounded at base, shiny green, stiff and leathery, hairless; and (2) small short–stalked figlike fruits at leaf bases, elliptic or pear–shaped and $^3/_8$–$^1/_2$ inch in diameter, slightly flat–topped, with raised ring at apex, light green with brown dots, turning dark red.

Evergreen medium–sized to large spreading tree to 90 feet high and 3 feet in trunk diameter, sometimes to 6 feet, and with prominent rounded buttresses and spreading crown. The bark is gray, smoothish or slightly fissured, with almost tasteless or slightly bitter milky juice, or white latex. The inner bark is whitish and fibrous and tasteless. Twigs are gray, stout, often finely hairy when young, with ring scars at nodes. The densely hairy buds are long and narrow, $^1/_2$–$^3/_4$ inch long, covered by a finely hairy scale (stipule).

The alternate leaves have stout leafstalks $^1/_4$–$1^1/_4$ inches long. Blades are mostly 2–6 inches long and $1^1/_4$–4 inches wide, slightly thickened and leathery, the upper surface shiny green with veins often slightly sunken, and the lower surface light green with prominent network of minute veins.

The greenish figlike fruits (syconia) are paired or single on short stalks $^1/_8$–$^3/_{16}$ inch long at least bases, with 2 minute pointed broad scales $^1/_{16}$ inch long joined at base and raised ring at apex, the surface often warty, almost tasteless. Within are many minute seeds less than $^1/_{16}$ inch long. With fruits from spring to fall.

The wood is whitish and soft.

Common through moist limestone, lower forests at 100–300 altitude, especially mountain forests.

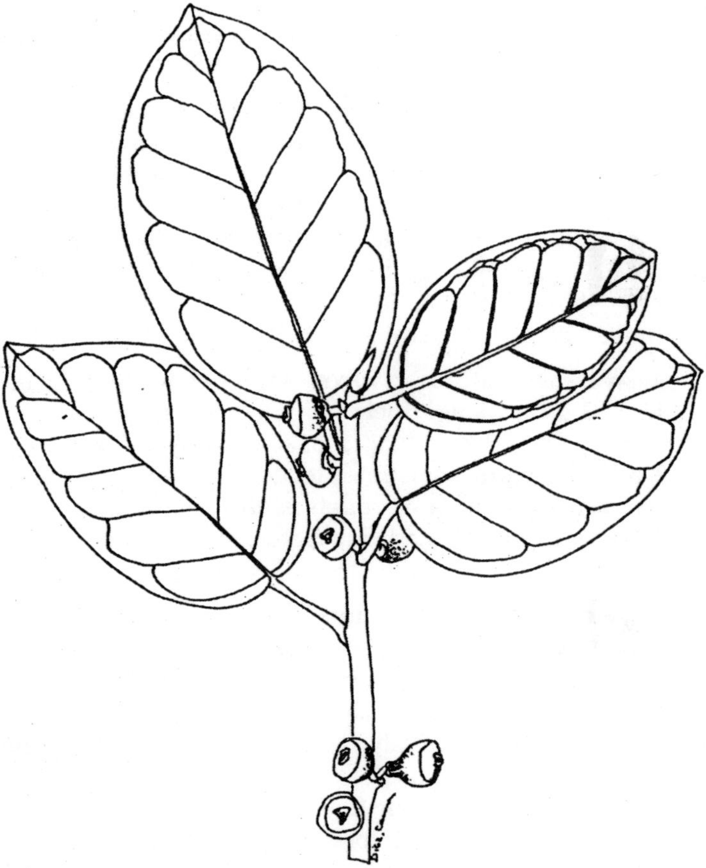

**Fig. 38 : Wild fig (*Ficus trigonata* L.)**
Natural size.

## 39. Black Mulberry (*Morus nigra* L.*)

Black mulberry has been planted in Puerto Rico as an ornamental and for the edible fruits. It is characterized by : (1) milky sap; (2) ovate leaves with 3 or 5 main veins from the notched unequal base, many small rounded teeth on margin, and ending in long narrow point; and (3) edible sweet purplish black cylindric mulberries $^1/_2$–1 inch long and $^3/_8$–$^1/_2$ inch wide.

Small to medium–sized tree 30 feet high and 8 inches in trunk diameter, with crown of dense dark green foliage. Bark light brown, smoothish and warty to finely fissured, the inner bark whitish and brown, slightly fibrous, almost tasteless, with small amount of whitish latex. Twigs slender, light green and finely hairy when young, becoming brown with raised dots (lenticels). Winter buds $^3/_{16}$ inch long, short pointed and covered with brown overlapping scales at end of twigs and at leaf bases.

Leaves alternate in 2 rows, with slender yellow–green petioles 1–2 inches long, finely hairy. Blades ovate, mostly 4–7 inches long, finely hairy. Blades ovate, mostly 4–7 inches long and 3–5 inches wide, sometimes lobed, thin, the upper surface dull dark green, smooth or rough, with veins slightly sunken, the lower surface paler and dull light green with raised veins, finely hairy.

Flowers male and female mostly on different trees (dioecious), short–stalked, in separate drooping clusters (spikes or racemes), minute, greenish, the calyx with 4 lobes more than $^1/_{16}$ inch long. Male flowers short–stalked, more than $^1/_8$ inch long and broad, consist of calyx and 4 spreading whitish stamens opposite the lobes, the cluster $^3/_4$–1 inch long and falling early. The stalkless crowded female flowers have calyx and pistil with small greenish ovary and 2–forked whitish style $^1/_8$ inch long. The multiple fruit (syncarp) is composed of fruits from many female flowers. The 4–lobed calyx becomes enlarged to $^1/_8$ inch across and the sweet juicy part enclosing the brown flattened akene or "seed" more than $^1/_{16}$ inch long. Flowering and fruiting in spring and summer.

The light brown wood is soft.

**Fig. 39 : Black Mulberry (*Morus nigra* L.*)**

Twig with male flowers (upper left), twig with female flowers and fruit (center), leafy twig (below), two–thirds natural size.

Rarely planted as a shade tree and fruit tree. Of rapid growth. Formerly tested experimentally here as a good plant for silkworms.

This species is the source of the mulberry varieties planted for fruit in southern Europe and Asia. It is naturalized in southern States and also planted north–ward.

## 40. Negra lora (*Pseudolmedia spuria*)

This tree species with milky juice, very rare in Puerto Rico, is identified by : (1) narrow oblong–elliptic leaves abruptly long–pointed at apex, alternate in 2 rows; (2) flowers minute, stalkless at leaf bases, the male many together in rounded cluster bordered by scales about $^1/_4$ inch across; (3) female flowers on other trees single or paired, more than $^1/_8$ inch long, covered by overlapping scales with 2 stigmas protruding; and (4) elliptic red fleshy fruit $^3/_8$ inch long.

Evergreen tree to 50 feet in height and 8 inches in trunk diameter, hairless. The bark is rough and slightly scaly, with milky juice or white latex. The twigs are slender and light brown. The narrow pointed bud to $^3/_8$ inch long is covered by a stipule, which sheds early, leaving a slight scar but not a ring.

The alternate leaves have short leafstalks $^1/_8$–$^1/_4$ inch long. Blades are 3–6 inches long and $1^1/_4$–$2^1/_4$ inches wide, short–pointed and unequal at base, not toothed on edges, the upper surface dark green, the lower surface paler with fine network of veins.

Flowers are male and female at leaf bases on different trees (dioecious). Male flowers many within a headlike group of scales (involucre), each consisting of 1 stamen about $^1/_{16}$ inch long, with narrow scales but no calyx or corolla. Female flowers single or paired, each covered by overlapping scales, consisting of pistil with ovary inside calyx tube and with short style and 2 stigmas protruding. The elliptic fruit is enclosed by the fleshy calyx and contains 1 seed within a slightly hard covering.

The wood is described as with thick grayish or pinkish brown sapwood and reddish brown heartwood. It is very hard, of medium coarse texture, variable grain, tough, and strong, not durable in contact with ground.

The red cherrylike fruits are edible. Also a honey plant.

This rare species is known in Ramaswaram area only from moist limestone hills near the northern coast, at altitudes of 200–600 feet.

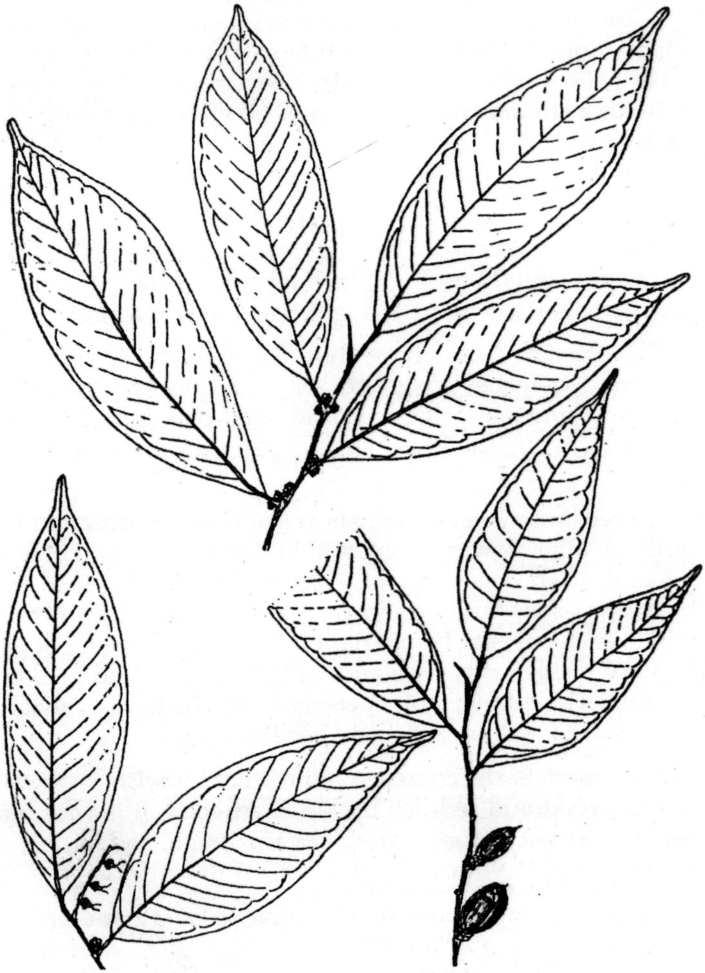

**Fig. 40 : Negra lora (*Pseudolmedia spuria*)**

Twig with male flowers (above), twig with female flowers (lower left), twig with fruits (lower right), natural size.

## 41. Ramón (*Trophis racemosa*)

This rare tree with milky sap in twigs is characterized by : (1) narrowly elliptic leaves $3^1/_2$–7 inches long and $1^3/_4$–$2^3/_4$ inches wide, long–pointed at apex, sometimes toothed on edges, alternate in 2 rows; (2) flowers male and female on different trees, small stalkless in clusters (spikes) at leaf bases; and (3) rounded fleshy fruit $^3/_8$ inch in diameter.

Medium–sized evergreen tree 70 feet high and $1^1/_2$ feet in trunk diameter, also flowering as a shrub. Bark gray, smooth, with few horizontal lines, becoming very rough and scaly in age. The inner bark is whitish, with green outer layer, tasteless. Young twigs are green with minute hairs, becoming gray brown. The light green buds $^1/_8$ inch long are covered with 2 pointed stipules, which shed early leaving a slight scar.

The leaves are alternate in 2 rows along the slender twigs. Leafstalks are $^1/_4$–$^1/_2$ inch long, slender minutely hairy. Blades are short–pointed or rounded at base, hairless, thin, edges without teeth or sometimes with few small teeth toward apex, upper surface dark green and slightly shiny, lower surface dull green with the few side veins raised.

Narrow cylindric flower clusters (spikes) single or paired at leaf bases, $1^1/_2$–4 inches long and $^1/_4$ inch broad, bear many crowded light yellow male flowers less than $^1/_8$ inch long, composed of 4–lobed hairy calyx and 4 stamens. On other trees (dioecious) are short clusters (spikes) less than $^3/_4$ inch long, which bear few greenish female flowers about $^3/_{16}$ inch long, consisting of tubular 4–toothed hairy calyx enclosing the ovary and 2–forked hairy style. The berrylike reddish–brown hairy fruit contains within the fleshy calyx 1 reddish–brown rounded seed $^1/_8$ inch in diameter. The fruit is edible but has thin flesh. Flowering irregularly during the year.

Wood light, brown, hard.

Elsewhere the foliage has been cut for fodder for livestock during the dry season.

Rare in moist limestone and lower Nagpur forests from sea level to 1,800 feet altitude in Maharashtra.

**Fig. 41 : Ramón (*Trophis racemosa*)**

Twig with fruits (upper left), twig with female flowers (center), twig with male flowers (right), two–thirds natural size.

# NETTLE FAMILY (URTICACEAE)

Herbs, sometimes shrubs, and less frequently small trees and vines, known by : (1) leaves mostly alternate, simple, often with 3 or 5 main veins from base, toothed, sometimes with stinging hairs, often with minute lines (cystoliths) visible from lower surface (especially when dry), with paired stipules; (2) flowers usually in lateral branched clusters (cymes), minute, inconspicuous, greenish or whitish, male and female on different plants (dioecious or the same plant (monoecious), regular, with calyx of 4–5 sepals or lobes, no corolla, 4–5 opposite stamens, and pistil with 1–celled ovary with 1 ovule, style, and branched stigma; and (3) fruit an akene or drupe, often enclosed by the persistent fleshy calyx.

**Key to Species**

A. Leaves coarsely toothed, with large stout conical stinging hairs on surface—*Urera baccifera*.

AA. Leaves finely wavy toothed, with small stinging hairs.

B. Leaves elliptic, widest near middle, rounded or slightly notched at base—*Urera chlorocarpa*.

BB. Leaves broadly ovate, widest below middle, notched at base—*Urera caracasana*.

## 42. Stinging nettle (*Urera baccifera*)

Ortiga brava or stinging nettle, which stings severely and painfully, is commonly a shrub in the understory of moist forests but sometimes becomes a small tree. It is easily identified by : (1) the many large sharp needlelike stinging hairs about $^1/_8$ inch long on stems and leaves, both leafstalk and blade; (2) the coarsely toothed elliptic to ovate leaves bearing on both surfaces large stout and conical stinging hairs; (3) the small greenish or pinkish flowers clustered at leaf bases; and (4) the whitish to reddish elliptic watery fruits about $^1/_4$ inch in diameter.

Evergreen shrub, rarely becoming a small tree 15 feet or more in height and 3 inches in trunk diameter, usually with several main stems. The bark is light gray, smooth, with few raised warts (lenticels) and few spines or stinging hairs. Inner bark has an outer light green layer and in whitish, soft, watery, almost tasteless. The twigs are stout, light green, almost hairless, with scattered stout stinging hairs, raised large half–round leaf scars, and large greenish pith.

The large alternate leaves have long stout petioles $1^1/_2$–4 inches long, round, purplish tinged, and bearing many stinging hairs. Narrow pointed light green stipules $^3/_8$ inch long form a pointed bud and fall early. The thin blades are 5–10 inches long and 3–5 inches wide, narrowed to a long point at apex, rounded or slightly notched at base, the margin coarsely toothed with teeth ending in stinging hairs. The upper surface is yellow green and slightly shiny, with veins and veinlets much sunken, the lower surface light green and slightly shiny, with many stinging hairs along the raised veins.

The minute flowers are male and female on different plants (dioecious) in widely forking branched clusters (cymes) 1–$2^1/_2$ inches long at leaf bases, with fleshy purplish branches. Female flowers about $^1/_{16}$ inch long have pinkish 5–toothed calyx and pistil with ovary and tufted whitish stigma. Male flowers have calyx and 4 or 5 stamens. Fruits several, elliptic, consisting of enlarged whitish watery calyx 5–lobed at apex enclosing 1 elliptic seed (akene) $^1/_8$ inch long, shiny, greenish black. Flowering in autumn and fruiting in winter.

**Fig. 42 : Stinging nettle (*Urera baccifera*)**

Leafy twig (above), twig with female flowers (below), fruits (lower right), natural size,

In Central India this species has been planted for hedges and fences, its spines making an impenetrable wall.

Common in understory of wet forests, especially in drainage, of moist limestone, lower Palampur forests from sea level to 3,000 feet altitude. Also a weed in coffee areas.

# LAUREL FAMILY (LAURACEAE)

## 43. Canela (*Persea krugii*)

The soft coat of silky brown pressed hairs on lower leaf surfaces, young twigs, branches of flower clusters, and flowers distinguish this rare tree of mountain forests of western Puerto Rico from other native members of the family. Other characters are : (1) lance–shaped to narrowly elliptic leaves $1^1/_2$–4 inches long and $^1/_2$–$1^1/_4$ inches wide; and (2) round berries $^3/_8$ inch in diameter, with 6–lobed calyx and enlarged stalk at base.

Evergreen small to medium–sized tree 45 feet high and 1 foot in trunk diameter. Bark gray, smoothish, becoming rough. Inner bark is brownish, with–slightly spicy taste. Twigs brown, finely hairy.

The alternate leaves have hairy petioles $^1/_4$–$^1/_2$ inch long. Blades are long–or short–pointed at apex, short–pointed or blunt at base, not toothed, slightly thick and leathery, the upper surface becoming nearly hairless, with side veins inconspicuous.

Flower clusters (panicles) at leaf bases, 2–4 inches long, branched. Flowers many on stalks less than $^1/_8$ inch long, about $^3/_{16}$ inch across, cup–shaped, yellow; the hairy calyx composed of 3 outer lobes slightly shorter than the 3 inner lobes $^1/_8$ inch long; 9 stamens and 3 smaller, pointed nonfunctional stamens, and pistil with hairless 1–celled ovary and slender style. The fruits have thin greenish flesh and 1 large round seed. With flowers from spring to fall and with fruits in winter.

The sapwood in whitish and moderately soft.

Rare to common in upper Cordillera forest at 1,500–2,500 feet altitude in high mountains of central and western India, also dwarf forest above 4,000 feet altitude on H.P.

**Fig. 43 : Canela (*Persea krugii*)**

Fruiting twig (above), flowering twig (lower left), natural size.

## 44. Aguacatillo (*Persea urbaniana*)

As indicated by its common name, aguacatillo is a relative of the cultivated aguacate or avocado. Identified by : (1) leaves slightly thick and leathery with edges turned under, elliptic, mostly 2–6 inches long and 1–3 inches wide, long–to short–pointed at both ends, the lower surface pale whitish green with minute short pressed hairs and often whitish, and (2) the rounded berries $^3/_8$–$^5/_8$ inch in diameter, with 6–lobed calyx at base.

Small to medium–sized evergreen tree becoming 35–70 feet high and 8–12 inches in trunk diameter, with low buttresses to 1 foot high. Observed with fruits when only 8 feet high. Bary gray, smooth, with raised dots (lenticels). Inner bark light brown to orange, bitter. Twigs light green, becoming darker, with fine pressed hairs when young.

Leaves alternate, hairless or nearly so, with slender petioles $^3/_8$–1 inch long. Upper surface dark green, slightly shiny, the midvein and side veins slightly sunken.

Flower clusters (panicles) at leaf bases, 1–4 inches long, branched. Flowers few, almost stalkless, yellowish, about $^1/_8$ iinch long and broad; the cup–shaped finely hairy calyx composed of 3 short outer lobes and 3 rounded inner lobes $^1/_8$ inch long; 9 stamens and 3 smaller nonfunctional stamens; and pistil with hairy 1–celled ovary and slender style. The 6 slightly enlarged calyx lobes remain at base of fruit. With flowers and fruits through much of the year.

Uncommon in lower and upper Nilgiri forests at 1,500–4,000 feet in mountains peak, Ooty, altitude more than 4,000 feet.

**Fig. 44 : Aguacatillo (*Persea urbaniana*)**

Fruiting twig (left), flowering twig (upper right), two–thirds natural size.

## 45. Laurel avispillo (*Phoebe elongata*)

This tree of moist forests is identified by : (1) large buttresses at base of trunk; (2) narrowly elliptic shiny leaves 2–7 inches long and 1–3 inches wide, hairless or nearly so, long–pointed at apex and short–pointed or rounded at base; and (3) elliptic shiny black berries about $^{5}/_{8}$ inch long with 6 pointed calyx lobes remaining at base on enlarged red cup and stalk.

Small to large evergreen tree becoming 100 feet or more in height and 16 inches in trunk diameter, with marked buttresses at base. The bark is smooth and gray. Twigs slender, drooping, often finely hairy when young.

The alternate leaves have slender petioles $^{3}/_{8}$–$^{5}/_{8}$ inch long. Blades are slightly thickened, with few curved side veins, shiny green or dark green on upper surface, the lower surface paler and often hairy on veins.

Flower clusters (panicles) are 3–9 inches or more in length, with many slender branches, hairy or hairless. Flowers many on stalks of $^{1}/_{8}$ inch or less, about $^{3}/_{16}$ inch wide. Calyx composed of 6 greenish white lobes less than $^{1}/_{8}$ inch long, stamens 9, with 3 smaller pointed sterile stamens (staminodes); and pistil with rounded 1–celled ovary and slender style. The berries are green when immature, 1–seeded. Flowering and fruiting irregularly over the year.

The common name laurel avispillo is applied also to another species (*Nectandra coriacea* (Sw.) Griseb.), and woods of the two have been confused. This species has pinkish–colored heartwood, which merges gradually into light brown sapwood. There are attractive slightly darker colored bands or stripes which develop fuzziness in machining operations. The wood is moderately soft and lightweight (specific gravity (0.47) and moderately strong. It has straight to irregular and tightly interlocked grain, medium texture, and medium to high luster. Air–seasoning at a moderate rate is easy and is satisfactory except for a moderate amount of warping. The wood is easily machined but requires extra care in turning and sanding. It is very susceptible to dry–wood termites and other insects, while logs are attacked by pinhole borers. Like that of related species, the heartwood may be moderately durable in the ground.

The wood is suitable for furniture, cabinet–work, interior trim, paneling toys, novelties, and turning. It should serve also for carpentry, interior and exterior construction, and possibly for decorative veneer and plywood. Miscellaneous uses would be boxes, crates, toys, and boatbuilding.

Uncommon in moist limestone, Bastar forests at 100–800 feet altitude in northern foothills M.P.

**Fig. 45 : Laurel avispillo (*Phoebe elongata*)**

Flowering twig (left), fruits (upper right), leaf (lower right), two–thirds natural size.

## 46. Avispillo (*Phoebe montana*)

The leaves with 3 main veins, 2 prominent long curved side veins joining the midvein about $^1/_4$ inch above the base distinguish this rare member of the laurel family from all other native species. Other characteristics for recognition are : (1) leaves relatively small, 24 inches long and $^3/_4$–$1^1/_2$ inches wide, paler beneath; (2) minute greenish white, finely hairy flowers less than $^1/_8$ inch long; and (3) elliptic blackish berries about $^3/_8$ inch long with 6 pointed calyx lobes remaining at base, on an enlarged reddish stalk.

Medium–sized evergreen tree to 65 feet high and 1 foot in trunk diameter. Young twigs finely hairy.

The petioles of the alternate leaves are slender, about $^1/_4$ inch long, finely hairy. Blades are short–pointed at base, not toothed, long–pointed at apex, thin, the upper surface dull green, hairless or nearly so, with veins slightly sunken, and the lower surface paler and slightly hairy on veins.

Flower clusters (panicles) at leaf bases are 1–2 inches long, bearing many flowers less than $^1/_8$ inch long on stalks of about $^1/_{16}$ inch from the hairy branches. The calyx has 6 greenish white finely hairy lobes more than $^1/_{16}$ inch long; there are 9 stamens and 3 smaller pointed sterile stamens (staminodes); and pistil with 1–celled ovary and short style. The fleshy fruits green when immature have 1 large seed. With flowers in spring and summer and with fruits from spring to fall.

Rare in lower Nilgiri and moist limestone forests at 500–1,500 feet altitude in Tamil Nadu.

**Fig. 46 : Avispillo (*Phoebe montana*)**

Flowering twig (above), fruiting twig (lower right), two–thirds natural size.

# POPPY FAMILY (PAPAVERACEAE)

Herbs, rarely shrubs and small trees (*Bocconia*), known by : (1) milky or colored sap (orange in *Bocconia*); (2) alternate simple leaves, entire to deeply lobed, with stipules; (3) flowers mostly solitary and showy (many in panicles in *Bocconia*), bisexual, with 2–3 sepals that fall early, 4–6 (12) petals (non in *Bocconia*), many stamens, and pistil with superior 1–celled ovary, 2 to many parietal placentas and stigmas and many ovules (1 in *Bocconia*); and (4) fruit a capsule.

## 47. Pan cimarrón (*Bocconia frutescens* L.)

A treelike shrub of the poppy family, easily recognized by : (1) orange bitter sap in stems and leaves; (2) large elliptic leaves deeply lobed and toothed along edges, blue green and densely fine hairy beneath; (3) many small greenish flowers $^3/_8$ inch long large branching clusters, and (4) fruit an elliptic light gray capsule $^3/_8$ inch long, splitting into 2 parts.

Evergreen shrub usually less than 10 feet high, rarely reaching the minimum size of a tree, 15 feet high and 3 inches in trunk diameter, with few spreading branches and without a definite crown. The bark is light brown, smoothish to slightly fissured, with enlargements at leaf scars, thin, orange red within, and bitter. Twigs are stout, green, and finely hairy.

The alternate leaves have stout green, finely hairy petioles. Blades are 6–12 inches long and 3–6 inches wide, slightly thickened, the border deeply lobed and toothed, the lobes and base short–pointed, the upper surface green and almost hairless, the lower surface blue green and densely fine hairy with very thick midvein.

The flower clusters (panicles) at ends of twigs are large and much branched, 8–15 inches long, elliptic in shape. Lacking petals, the many slender stalked flowers have 2 greenish elliptic sepals $^5/_{16}$ inch long, 12–13 or more stamens that shed early, and pistil with stalked 1–celled ovary containing 1 ovule, style, and 2 spreading stigmas. The pods (capsules) separate) from an elliptic frame with persistent style. Flowering and fruiting through the year.

The wood is brown to orange red, soft, with large pith. Elsewhere the sap has served locally, as a dye and in home medicines.

Uncommon and scattered in openings and thickets in upper Nilgiri forests at 1,000–3,000 feet altitude in central mountains of Tamil Nadu.

**Public Forests** : Carite, Guajataca, Guilarte, Maricao, Rio Abajo, Toro Negro.

**Range** : Greater Antilles, St. Kitts at St. Vincent, and from Mexico to Costa Rica and Colombia. Recorded as introduced in Hawaii.

**Other Common Names :** Panapen cimmaron, pasilla (Puerto Rico); yagrumo macho, palo de toro, gengibrillo, llorasangre (Dominican Republic); palo Amarillo, yagrumita, palo de pan Cimarron (Cuba); gordolobo, calderon, llorasangre (Mexico); sangre de toro, camotillo (Guatemala); guacamayo, tabaquillo (Costa Rica); celiduena, celedonia, trompeto, golondrinia (Colombia); celandine, John–Crow–bush (Jamaica); bois codine, bois de coq (Haiti).

The Spanish common name may be derived from the slight resemblance of the leaves to those of breadfruit.

**Fig. 47 : Pan cimarrón (*Bocconia frutescens* L.)**

Flowering twig (above), fruiting twig (below), two–thirds natural size.

# CAPER FAMILY (CAPPARACEAE)

Herbs, shrubs, woody vines, and small to medium–sized trees, known by : (1) leaves alternate, simple or palmate with 3–7 leaflets, with minute or spiny stipules or none; (2) flowers often large, in unbranched clusters (racemes), bisexual, regular or irregular, generally with 2 sepals, 4 rounded or elliptic petals commonly white, that shed early, 4 to many very long threadlike stamens, and pistil generally on a long stalk with superior 1–celled ovary with 2 placentas and few to many ovules, and style; (3) fruit a capsule sometimes like a pod or berry; and (4) sometimes with slightly disagreeable odor. The family name is spelled Capparidaceae also).

**Key to Species**

A. Petiole enlarged at both ends and often joining blade slightly above base, leaves oblong, 3–10 inches long; flowers few along twigs mostly back of leaves; fruit a ball $1^1/_2$–2 inches in diameter—*Morisonia americana.*

AA. Petiole mostly not enlarged at ends (except in no. 76), joining blade at base; flowers few in terminal clusters or at base of upper leaves—*Capparis.*

B. Leaves with minute scales beneath, narrowly elliptic, mostly 2–4 inches long; scales also on twigs, flowers, and fruits; fruit narrowly long cylindrical, to 8 inches long.

C. Leaves with upper leaf surface shiny green, lower leaf surface densely covered with silvery scales—Burro prieto, Jamaica caper, *Capparis cynophallophora* L.

CC. Leaves with upper leaf surface only slightly shiny, lower leaf surface gray green and scaly—*Capparis indica.*

BB. Leaves without scales or hairs.

D. Leaves slightly thickened, elliptic, $1^1/_2$–$4^1/_2$ inches long, blunt or short–pointed at both ends; fruits elliptic, $1^1/_4$–$2^1/_2$ inches long and $^1/_2$–1 inch in diameter—*Capparis amplissima.*

DD. Leaves thick and leathery.

E. Leaves notched at base.

F. Leaves broadly elliptic, $1^3/_4$–$4^1/_4$ inches long, blunt, rounded, or notched at apex; fruit an oblong, slightly flattened pod 4–8 inches long and to 1 $^1/_4$ inches wide—*Capparis hastata.*

FF. Leaves elliptic to narrowly obovate, 3–8 inches long, short–pointed at apex; fruit short, 1–3 inches long and $^{3}/_{8}$ inch broad—*Capparis frondosa.*

EE. Leaves rounded or short–pointed at base, of varying shapes from narrowly oblong to elliptic, $1^{3}/_{4}$–4 inches long, blunt, rounded or slightly notched at apex; fruit long, narrowly cylindric, 3–9 inches long and $^{1}/_{2}$ inch in diameter—*Capparis flexuosa.*

## 48. Burro blanco (*Capparis amplissima*)

This tree of dry forests is identified by : (1) elliptic leaves $1^1/_2$–$4^1/_2$ inches long and 1–$2^1/_4$ inches wide, slightly thickened and leathery; (2) whitish flowers with 4 oblong petals and many white stamens spreading to $2^1/_2$ inches across; and (3) gray dry elliptic fruits $1^1/_4$–$2^1/_2$ inches long and $^1/_2$–1 inch in diameter, hanging down on long slender stalks.

Evergreen small–to medium–sized tree to 60 feet high and 1 foot in trunk diameter, with spreading but slightly fissured into small plates. Inner bark light brown, gritty, with slight spicy taste and odor like radish. The slender twigs are greenish gray, finely hairy or hairless.

The alternate hairless leaves have petioles $^1/_4$–$^3/_8$ inch long. Leaf blades are blunt or short–pointed at both ends, not toothed on edges, green on upper surface and slightly lighter beneath.

A few flowers are borne in clusters (corymbs) at end of twigs on stalks $^1/_4$–$^5/_8$ inch long. The pale green calyx consists of 2 rounded sepals $^1/_4$ inch long and 2 inner $^1/_2$ inch long; 4 white oblong petals $^3/_4$ inch or more in length; many white stamens $1^1/_2$–2 inches long, spreading to $2^1/_2$ inches across; and pistil on a long stalk, with narrow ovary and stigma. The elliptic fruits hang down on long slender stalks 2–$2^1/_2$ inches long from the flower, not narrowed between the seeds, and opening irregularly. Flowering and fruiting intermittently.

The sapwood is light brown or whitish, hard, and heavy. The wood is used for posts.

Uncommon in dry forests of south slopes at 500–1,000 feet altitude in Karnataka.

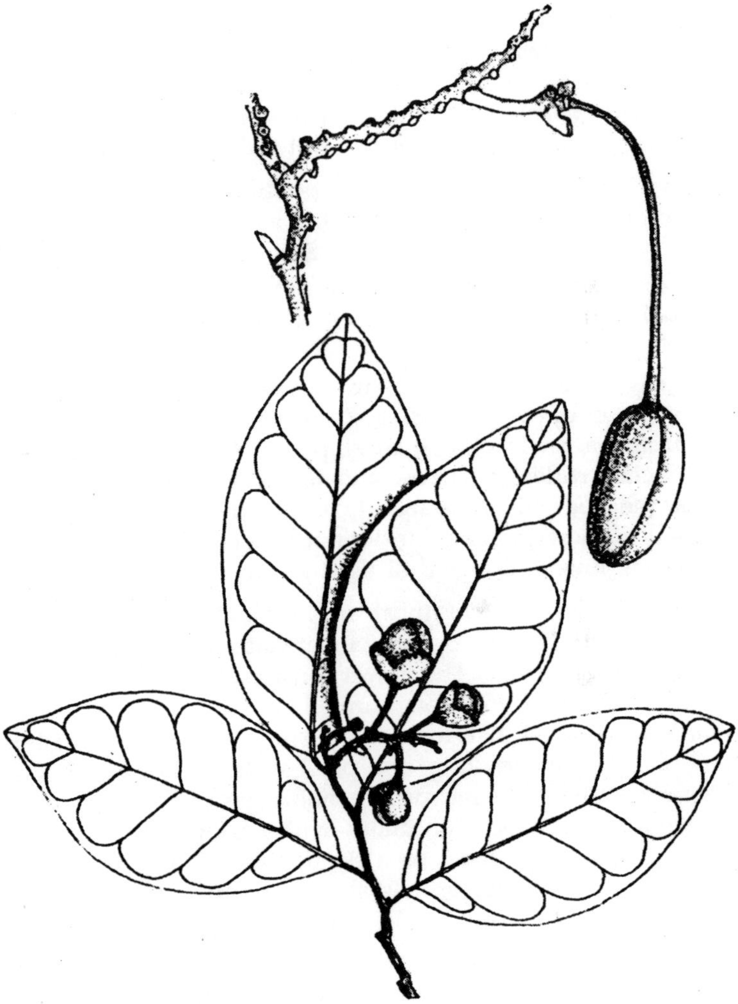

**Fig. 48 : Burro blanco (*Capparis amplissima*)**
Fruit (above), twig with flower buds (below), natural size.

## 49. Limber caper (*Capparis flexuosa*)

Usually a shrub, often veinlike, sometimes a small tree, this species is characterized by : (1) leaves of varying shapes from narrowly oblong to elliptic or linear, leathery, dull green and hairless, alternate in 2 rows; (2) few terminal flowers nearly 2 inches across the 4 spreading white to pink petals and with many spreading white threadlike stamens $1^1/_2$–3 inches long; and (3) long, narrowly cylindric pod 3–9 inches long and about $^1/_2$ inch in diameter, rusty brown but reddish within.

An evergreen shrub, often vinelike with slender stem climbing on trees, or a small tree to 2 feet high and 5 inches in trunk diameter. The bark is gray, smooth, the inner bark light yellow, with spicy taste like horseradish. Twigs are often slightly zigzag, green, turning to light gray with dark dots (lenticels), hairless.

Leaves are alternate in 2 rows, hairless, with minute paired scale stipules and with petioles $^1/_4$ inch or less in length. Blades vary from narrowly oblong to elliptic, $1^3/_4$–4 inches long and $^1/_2$–2 inches wide, or narrow and linear on shoots and young plants, the apex blunt, rounded, or slightly notched, the base rounded or short–pointed, border not toothed, the upper surface dull green, lower surface dull light green.

Flower clusters (corymbs) are terminal with few fragrant followers short–stalked, opening in late afternoon and night and closing in late morning. The flower consists of calyx of 4 rounded yellowish sepals $^1/_4$ inch long, united at base; corolla of 4 white to pink petals $^5/_8$–1 inch long; many spreading white threadlike stamens $1^1/_2$–3 inches long; and pistil at the end of a long stalk 2–$2^1/_2$ inches long, the narrowly cylindric yellowish ovary about $^1/_4$ inch long, 1–celled with many ovules, ending in flat stigma. The long pod hangs down on a slender stalk 2–$2^1/_2$ inches long beyond base of flower and is slightly narrowed between the seeds. It opens late exposing the red pulp and many green seeds. Flowering probably through the year.

The sapwood is light brown and hard. The root reportedly has the taste of horseradish and has been used in home remedies.

Common in dry and moist coastal and limestone forests from sea level to 1,000 feet altitude along both coasts of Tamil Nadu.

**Fig. 49 :** **Limber caper (*Capparis flexuosa*)**
Flowering twig (above), fruit (below), two–thirds natural size.

## 50. Sapo, rat–bean (*Capparis frondosa*)

From the other native species of the genus this species of burro is distinguished by : (1) leaves elliptic to narrowly obovate, leathery, 3–9 inches long and $1^3/_4$–4 inches broad, without hairs or scales, slightly notched at base, the upper surface dull green with midrib and the few curved lateral veins all sunken, the petioles unequal in length; (2) flowers about $^1/_2$ inch long, with 4 pale greenish white or purplish petals and the stamens only slightly longer than the petals; and (3) fruit like a bean pod, short, 1–3 inches long and $^3/_8$ inch broad, oblong, irregularly narrowed between seeds.

Evergreen shrub or small tree to 15 feet high and 2 inches in trunk diameter, in Culebra to 25 feet and 4 inches in diameter, and in Vieques to 45 feet and 6 inches. The bark is brownish gray, smooth to slightly fissured and slightly wary. Inner bark is light yellow, slightly peppery in taste. Twigs are light gray, hairless, becoming slightly fissured. The internodes are of unequal length, with leaves clustered where internodes are short and separated by others long.

The alternate leaves have green petioles of varying length, $^1/_8$–$2^1/_4$ inches, round and swollen at both ends. Blades are elliptic or narrowly obovate and broadest beyond middle, leathery, hairless, short–pointed at apex, slightly notched or heart–shaped at base, edges not toothed, upper surface dull green with sunken veins, and lower surface dull light green with veins slightly raised.

Few to several short–stalked flowers are borne in clusters (corymbs) terminal or at base of upper leaves. The flower is composed of calyx of 4 rounded light brown sepals $^1/_{16}$ inch long; corolla of 4 rounded petals $^3/_8$ inch long, pale greenish white or purplish; about 50 white stamens slightly longer than petals; and stalked pistil with cylindric green ovary $^1/_8$ inch long, 1–celled with many ovules, and dot stigma.

The beanlike fruits (berries) are borne a few together, each with stalk about $^1/_4$ inch long above calyx as well as longer flower stalk. Mature fruits are brown, slightly soft, and break open irregularly. Within the whitish pulp are a few rounded dark brown seeds about $^1/_4$ inch in diameter. Flowering and fruiting intermittently through the year.

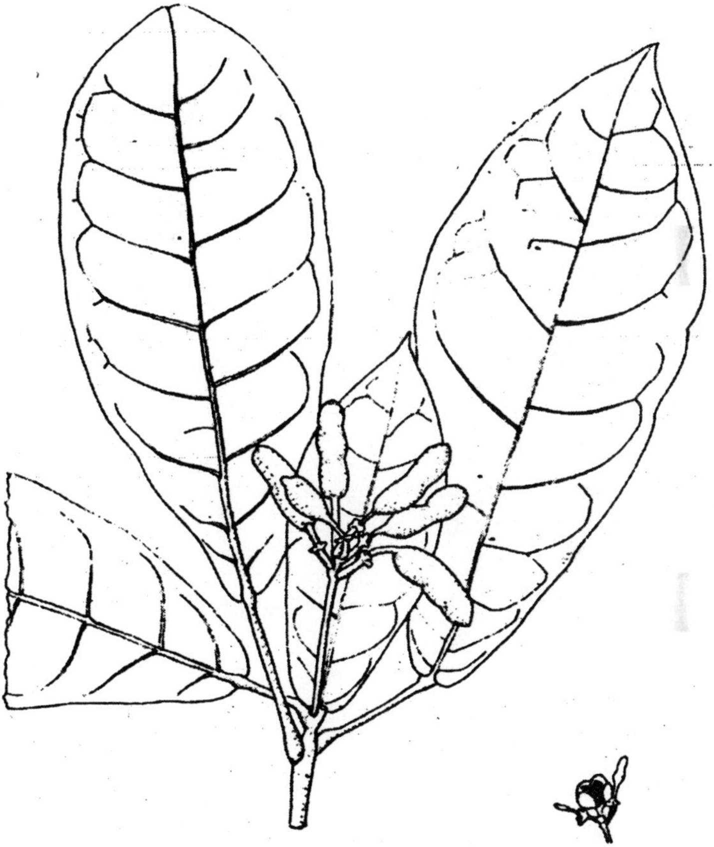

**Fig. 50 : Sapo, rat–bean (*Capparis frondosa*)**

Fruiting twig (above), flowers (lower right), two–thirds natural size.

The wood is whitish and slightly soft.

Commonly locally in dry coastal, dry limestone, and lower Nilgiri forests from sea level to 1,500 feet altitude, on southern coast and southern slopes of central mountains in Karnataka.

## 51. Burro (*Capparis hastata*)

This small tree or shrub is characterized by : (1) thick, leathery, broadly elliptic leaves notched or rounded at base and blunt, rounded, or notched at apex, dull green and hairless, alternate in 2 rows; (2) few terminal flowers nearly 2 inches across the 4 spreading white petals and with many spreading white thread–like stamens 2–3 inches long; and (3) fruit an oblong slightly flattened pod 4–8 inches long and $1^1/_4$ inches wide, red when mature and open.

An evergreen small to medium–sized tree to 50 feet high and 1 foot in trunk diameter or a shrub. Bark smoothish, gray or brown, the inner bark light brown, with slight taste of horseradish. Twigs stout, with minute hairs, green when young, becoming light gray or brown with many raised dots (lenticels), sometimes slightly zigzag.

Leaves alternate in 2 rows, with minute paired scale stipules and with stout brownish, minutely hairy petioles $^1/_4$–$^3/_8$ inch long. Blades elliptic to nearly round, $1^3/_4$–$4^1/_4$ inches long and $1^1/_4$–$2^3/_4$ inches wide, thick, leathery, often stiff, border not toothed, upper surface dull green or slightly shiny and with the sides slightly turned up at midrib, the lower surface dull light green.

Flower clusters (corymbs or racemes) are terminal, with few to several flowers on stout green stalks $^1/_2$–1 inch long, opening at night, 1 at a time. The flower is composed of cup–shaped calyx $^3/_8$ inch long and broad, greenish with 4 rounded yellow lobes $^1/_4$ inch long; corolla of 4 oblong concave whitish petals about 1–$1^1/_2$ inches long, tinged with green; many white threadlike stamens 2–4 inches long, spreading more than 4 inches across; the pistil at the end of a long stalk 2–3 inches long, the narrowly cylindric yellowish ovary $^3/_{16}$ inch long, 1–celled with many ovules, ending in flat stigma.

Fruit hanging on a long stalk from 2–3 inches beyond base of flower, an oblong heavy pod 4–8 (sometimes 10) inches long, becoming slightly flattened and upto $1^1/_4$ inches wide, red, splitting open on 2 lines. In the dark red juicy pulp are many whitish elliptic seeds about $^1/_2$ inch long, arranged in 2 rows. In flower and fruit nearly through the year.

The sapwood is light brown and hard. The wood has been used for posts.

Sometimes planted as an ornamental roadside tree in South India.

Common in dry coastal, limestone, and lower Nilgiri forests from sea level to 1,000 feet altitude of south coast and southern slopes and east end of Tamil Nadu.

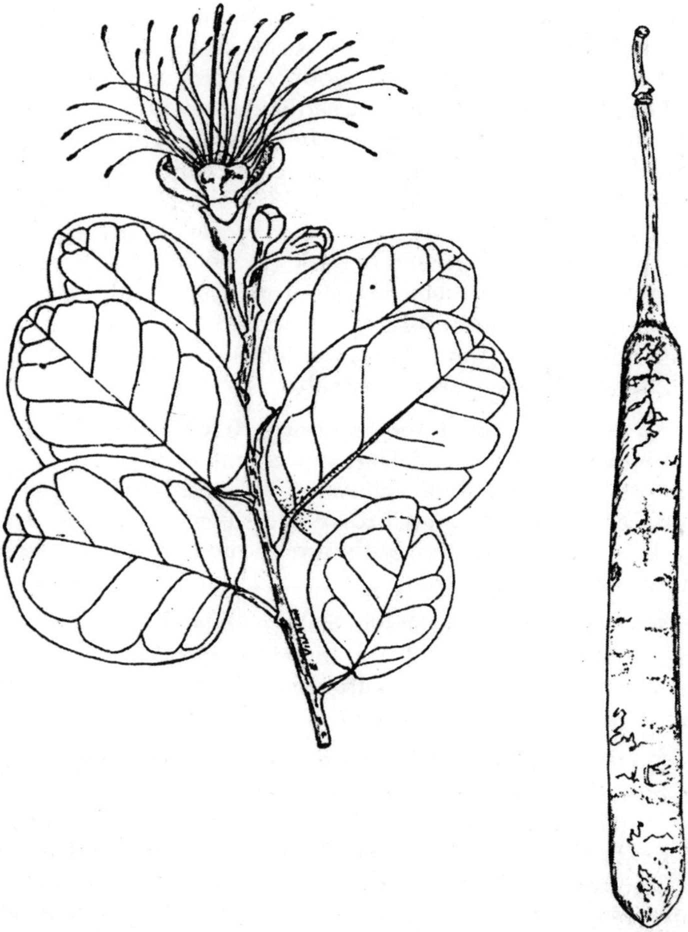

**Fig. 51 : Burro (*Capparis hastata*)**

Flowering twig (left), natural size; fruit (right), one–half natural size.

## 52. Linguam (*Capparis indica*)

A shrub or small tree of relatively dry low–lands, identified by : (1) minute brown scales on twigs, petioles, lower leaf surfaces, flower stalks, flowers, and fruits; (2) narrowly elliptic, slightly leathery leaves short–pointed or blunt at both ends, gray green and slightly shiny on upper surface, gray green beneath; (3) whitish flowers turning red, about $^3/_4$ inch across the 4 petals and 2 inches across the long threadlike spreading stamens, in flattened clusters at or near ends of twigs; and (4) long, narrowly cylindric, silvery–brown pods 3–8 inches long and $^1/_4$ inch in diameter. From its close relative No. 53, burro prieto or Jamaica caper, *Capparis cynophallophora* L., this species is distinguished by : (1) bark light gray instead of dark; (2) the leaves only slightly shiny above and the lower surface gray green with the scales not forming a solid silvery–brown coat; and (3) the flowers with shorter calyx only $^1/_8$ inch long, the lobes overlapping, and the petals densely hairy instead of scaly.

An evergreen small tree becoming 20 feet tall and 4 inches in trunk diameter or larger. The bark is gray or light brown and smoothish, the inner bark yellow or red and bitter. Twigs are angled, covered with minute brown and silvery scales, becoming gray. The narrow pointed buds are formed of minute leaves without stipules.

The alternate leaves have scaly petioles $^1/_4$–$^5/_8$ inch log, grooved above. Blades are 2–4$^1/_2$ inches long and $^3/_4$–1$^3/_4$ inches wide, not toothed on edges, with upper surface gray green with many minute scales not forming a solid coat.

Flower clusters (corymbs) are branched and flattened at or near ends of twigs, with several slightly fragrant flowers on scaly stalks. The flower is composed of brown scaly calyx $^1/_8$ inch long, deeply 4–lobed, the lobes overlapping and open in bud; corolla of 4 elliptic petals nearly $^1/_2$ inch long, whitish but turning red, densely hairy on both sides; about 20 slender white stamens 1 inch long, with yellow anthers, soon withering; and pistil on a long whitish stalk about $^7/_8$ inch long, the ovary nearly $^1/_4$ inch long, narrowly cylindric and scaly, 1–celled with many ovules, and ending in flat stigma.

The scaly pod hangs down on a slender stalk more than 1 inch long beyond base of flower. It opens irregularly on 2 sides,

**Fig. 52 : Linguam (*Capparis indica*)**
Flowering twig (left), fruits (right), natural size.

is red within, and is slightly narrowed between the blackish seeds, which are elliptic and about $^1/_4$ inch long. Flowering and fruiting intermittently.

The sapwood is light brown and hard.

Common in thickets, dry coastal and limestone regions from sea level to 500 feet altitude in southern and eastern Karnataka.

## 53. Rat–apple (*Morisonia americana* L.)

This shrub or small tree grows in the dry forests of the Virgin Islands, also Culebra, Vieques, and Desecheo, and is rare in south–western Puerto Rico. It is characterized by : (1) oblong shiny leathery leaves varying greatly in size and hanging down, the petioles of unequal length, enlarged at both ends and often joining blade slightly above base; (2) few flowers in clusters along twigs back of leaves, nearly 1 inch wide, with 4 white petals; and (3) fruit a hard light greenish brown ball $1^1/_2$–2 inches in diameter, hanging down from stout stalk along twigs back of leaves. Similar to No. 345, sapo, rat–bean, *Capparis frondosa* Jacq., except for the large round fruit and shiny leaves.

Evergreen shrub or small tree becoming 25 feet high and 4 inches in trunk diameter, sometimes partly climbing on trees. The bark is gray to blackish, smooth to slightly fissured, the inner bark light brown, with peppery taste. The stout light gray or light brown twigs are covered with minute scales.

The alternate leaves have yellowish petioles $^1/_2$–$2^1/_2$ inches long, of unequal length and round and enlarged at both ends, covered with minute scales. Blades are 3–10 inches long and $^3/_4$–$3^3/_4$ inches broad, oblong, stiff and leathery, hairless, short–pointed or rounded at apex, the base rounded or often slightly cup–shaped from attachment to petiole slightly above based, border not toothed, with few curved lateral veins, the upper surface shiny green, and the lower surface dull light green.

Several short–stalked flowers are borne in lateral clusters (corymbs) about 1 inch long at leaf bases and along twigs back of leaves, the flowers and stalks covered with minute scales. The flower consists of greenish yellow scaly calyx $^1/_4$ inch long, which covers the elliptic bud and breaks open into 2 parts; corolla of 4 white rounded petals about $^1/_2$ inch long scaly outside and hairy within; about 20 or fewer stamens attached on a disk; and hairy pistil on a stalk, with narrow cylindric ovary 1–celled with many ovules and flat rounded stigma.

The fruit (berry) has a stout stalk about $^1/_2$ inch long from inside the flower, separated by an enlarged ring from the less stout flower stalk. The hard fruit covered with minute scales does not open, except perhaps by decay or action of animals. Inside the whitish wall more than $^1/_8$ inch thick are many elliptic light brown seeds about $^3/_8$ inch long, in thin whitish juicy pulp which is slightly aromatic and peppery in taste. Flowering and fruiting intermittently.

The wood is light brown and hard.

In other regions this small tree is an ornamental in plazas, along avenues and near homes. It has a domelike crown and showy coloration, is drought–resistant, and grows slowly.

Common locally on dry slopes and in dry forests from sea level to 600 feet altitude.

**Fig. 53 : Rat–apple (*Morisonia americana* L.)**

Leafy twig (above), fruits and flowers (below), two–thirds natural size.

# ROSE FAMILY (ROSACEAE)

Trees, shrubs, and herbs, known by : (1) leaves alternate, generally simple, with paired stipules; (2) small to large bisexual flowers, usually regular or sometimes slightly irregular, generally with cuplike base (hypanthium) that bears 5 sepals, 5 petals, and many (sometimes 10) separate stamens, and with 1 to many simple pistils or 1 compound pistil with mostly superior 2–5-celled ovary with few ovules, and 2–5 styles; and (3) fruit various, a drupe, pome, akene, or follicle.

**Key to Species**

A. Leaves rounded or slightly notched at apex, elliptic or nearly round, thick and leathery, turned upward—*Chrysobala icaco.*

AA. Leaves mostly long-pointed at apex, mostly oblong or ovate, thin spreading.

B. Petioles less than $^1/_8$ inch long.

C. Leaves with saw-toothed edges, brownish woolly beneath—*Eriobotrya japonica.**

CC. Leaves not toothed on edges, hairy on veins beneath—*Hirtella.*

D. Leaves with veins much sunken in upper surface and raised beneath; twigs bristly hairy; fruits dark red—*Hirtella rugosa* Pers.

DD. Leaves with veins not sunken; twigs finely hairy; fruits blackish—*Hirtella triandra.*

BB. Petioles $^1/_4$–$^1/_2$ inch long; crushed foliage with taste and odor of almond—*Prunus.*

E. Leaves ovate, 2–4$^1/_2$ inches long; fruit rounded, about $^1/_2$ inch in diameter—*Prunus myrtifolia.*

EE. Leaves oblong-elliptic, 4–8 inches long; fruit elliptic, $^3/_4$–1 inch long—*Prunus occidentalis.*

## 54. Coco–plum (*Chrysobalanus icaco* L.)

Hicaco or coco–plum is characterized by : (1) numerous shiny, dark green thick and leathery, elliptic or nearly round leaves $1^1/_2$–$3^1/_4$ inches long and 1–$2^1/_4$ inches broad, alternate in 2 rows and turned upward along twig, the visible lower surfaces yellow green; (2) small greenish white flowers less than $^3/_8$ inch long, several in clusters at leaf bases; and (3) the elliptic or nearly round, pink, whitish, or dark–purplish fruits $^3/_4$–$1^1/_2$ inches long, juicy and edible, containing 1 large 5–6–ridged brown stone and a edible white seed.

Evergreen and mostly shrubby with several stems from base to 10 feet in height, or creeping when growing a much–branched small tree to 15 feet tall and 4 inches in trunk diameter, with dense crown, especially in cultivation inland. The bark is brown or gray, smoothish or becoming scaly, and astringent. The twigs are green and hairless when young, turning reddish brown, with raised dots (lenticels).

The leaves are alternate on short stout petioles $^1/_8$ inch long. Leaf blades are rounded or slightly notched at apex and short–pointed at base, and yellow green beneath.

Flower clusters (cymes) near ends of twigs are 1–$1^1/_2$ inches long, shorter than the leaves. The finely hairy, light green, bell–shaped base (hypanthium) is nearly $^1/_8$ inch long and the 5 spreading pointed sepals the same length; there are 5 narrow white petals $^3/_{16}$ inch long, shedding early; about 20 hairy white stamens $^1/_4$ inch long, united into tube at base; and pistil of green hairy 1–celled ovary with 2 ovules and slender hairy style $^1/_4$ inch long attached at base.

The soft fruits (drupes) resembling plums have a whitish, thin and spongy, slightly sweet or almost tasteless flesh, astringent until mature. Immature fruits are light green and 5–6–angled. The nutlike edible seeds weigh about 80 to a pound. Flowering and fruiting nearly through the year.

The wood is described as light brown, hard and heavy (specific gravity about 0.8). It is not used in Puerto Rico but elsewhere has been employed in carpentry.

Occasionally planted as an ornamental shrub and for the fruits. These are commonly prepared as preserves or made into jelly but are edible raw. Also the nutlike seeds have a high oil content and can be eaten. It is reported that the Carib Indians strung them on sticks and burned them like candles. Various parts of the plant are astringent and have served in folk medicines. A honey plant. Elsewhere the shrubs have been used for stabilization of sand dunes.

Locally common forming thickets on coastal lowlands and sandy beaches and elsewhere on shallow wet soils from seal level to 1,500 feet altitude in eastern mountains of West Bengal also islands eastward.

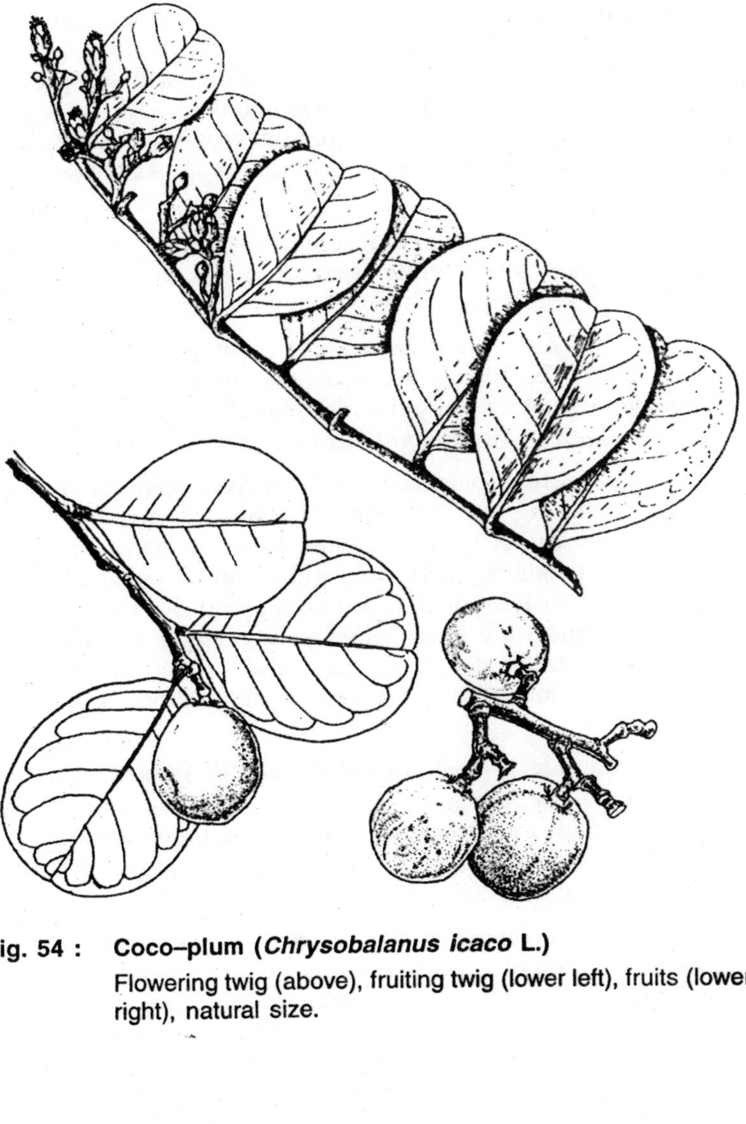

**Fig. 54 : Coco–plum (*Chrysobalanus icaco* L.)**

Flowering twig (above), fruiting twig (lower left), fruits (lower right), natural size.

## 55. Loquat (*Eriobotrya japonica*)

Loquat is a handsome small tree introduced in West Bengal as an ornamental and for its edible fruit. It is recognized by : (1) oblong thick and leathery leaves 5–10 inches long and $1^1/_4$–$3^1/_2$ inches wide, toothed on edges, the upper surface shiny with sunken veins, the lower surface covered with soft gray or rusty hairs; (2) many fragrant flowers about $^1/_2$ inch across the 5 white petals, in terminal branched woolly clusters; and (3) yellow elliptic or pearlike fruits (loquats) $1^1/_2$–2 inches long, slightly sour and edible.

Evergreen small tree to 20 feet high with shiny green foliage. The bark is gray and slightly fissured. Twigs are light brown and hairy.

The alternate leaves have short stout petioles less than $^3/_8$ inch long. The blades are often broadest beyond middle, short–or long–pointed at both ends, slightly curved up on both sides of midvein, with straight sunken veins ending in teeth.

Flower clusters (panicles) are 4–6 inches long. The flowers about $^3/_8$ inch long are composed of a brown hairy basal cup (hypanthium), which bears 5 rounded brown hairy sepals, 5 white elliptic petals $^3/_8$ inch long, and about 20 white stamens. The pistil has an inferior 5–celled ovary and 5 whitish styles. The fruits (pomes) are hairy when young, slight shiny, with 5 protuberances and 5 sepals at apex, and contain pale yellow pleasantly acid mealy pulp and few brown elliptic seeds $^3/_8$–$^3/_4$ inch long.

Uncommonly grown around houses in West Bengal but apparently not adapted as a fruit tree. The fruits are eaten fresh, cooked, and made into preserves and jelly. The bitter seeds should be removed before cooking. Propagated by seeds and grafting. Classed as a honey plant.

As a commercial fruit tree in orchards, loquat is better suited to subtropical climates of mountains and to warm temperate regions. It is hardy across southern Kerala States from west along the Gulf. Along the Arabian Coast it can be grown northward to Mumbai with some protection in winter. In cold climates also an ornamental in conservatories and a potted plant.

**Fig. 55 : Loquat (*Eriobotrya japonica*)**
Flowering twig and fruits (lower left), two–thirds natural size.

## 56. Wild Pigeon Palm (*Hirtella triandra*)

This rare shrub or small tree of moist forests is characterized by : (1) lance–shaped to elliptic leaves 2–6 inches long and 1–$2^1/_4$ inches wide, alternate in 2 rows; (2) many small whitish and pinkish flowers with 3 threadlike stamens $^5/_8$ inch long, in branched terminal clusters; and (3) blackish or purplish oblong flattened hairy fleshy fruits $^3/_4$–1 inch long and $^1/_4$–$^5/_{16}$ 7 inch wide. To be separated from No. 54, icaquillo, *Hirtella rugosa* Pers., by the erect form (branches not dropping), softer pressed hairs, and smoother flattened leaves with veins not sunken.

Evergreen shrub 10–15 feet high or a small tree to 40 feet high and 6 inches in trunk diameter. Bark gray smooth, the inner bark pink. Twigs slender, finely hairy, with lighter dots (lenticels).

The alternate leaves have paired minute narrow hasiry stipules $^1/_8$–$^3/_{16}$ inch long and hairy petioles $^1/_8$ inch long. Leaf blades are long–pointed at apex, short–pointed or rounded at base, slightly wavy on edges, thin, hairy on midvein, the upper surface green, and the lower surface paler and finely hairy.

Flower clusters (panicles) are $1^1/_2$–4 inches long, the branches finely hairy. The finely hairy flowers on short hairy scaly stalks have a basal tube (hypanthium) about $^1/_8$ inch long; 5 rounded sepals $^1/_8$ inch long and bent down, 5 elliptic white petals $^3/_{16}$ inch long and shedding early; 3 threadlike pink or purplish stamens $^5/_8$ inch long; and pistil with densely hairy 1–celled ovary and threadlike style attached near base. Then dry drupelike fruits have calyx at base, thin sweetish edible pulp, and 1 large pointed seed with fine grooves. With flowers and fruits through the year.

The wood is light brown, hard, and heavy, elsewhere used in construction.

Rare in lower most limestone, and lower and upper Nilgiri forests at 200 – 2,000 feet altitude in moist lowlands and upper western mountains of Nilgiri.

**Fig. 56 : Wild pigeon palm (*Hirtella triandra*)**
Natural size.

## 57. Myrtle laurelcherry (*Prunus myrtifolia*)

This rare small tree of mountain forests is identified by : (1) elliptic to ovate leaves 2–4$^1/_2$ inches long and 1–2 inches wide, with door and taste (when crushed) of almond; (2) many small flowers $^3/_{16}$ inch wide, with 5 white petals, in dropping unbranched clusters; and (3) rounded shiny black–purple stone fruit about $^1/_2$ inch in diameter.

Evergreen small tree 25 feet or more in height and 4 inches in trunk diameter, hairless throughout. The bark is light brown or reddish brown, smooth or slightly fissured. The twigs are slender, orange brown, becoming gray or light brown, with light dots (lenticels).

The alternate leaves have slender leafstalks about $^1/_2$ inch long and paired narrow pointed stipules $^1/_4$ inch long, shedding early. Blades are narrowed to a long blunt point at apex, short–pointed at base, slightly thickened, slightly wavy at borders, shiny yellow green on upper surface, and pale dull green on lower surface.

Flower clusters (racemes) at leaf bases are shorter than the leaves and bear many flowers on stalks $^1/_8$–$^1/_4$ inch long. The flowers have a basal cup (hypanthium) $^1/_{16}$ inch long, 5 rounded petals more than $^1/_{16}$ inch long, many stamens, and pistil with ovary and short style. The fruits (drupes) have thin flesh and large stone. Flowering mostly in spring and summer and maturing fruits in summer.

The hard wood has light brown sapwood, and light red heartwood.

Rare in moist limestone and upper Western forests at 100–2,500 feet altitude in central and western Karnataka. Collected in dwarf forest above 4,000 feet altitude at Ooty, the highest peak of Nilgiri.

**Fig. 57 : Myrtle laurelcherry (*Prunus myrtifolia*)**
Flowering twig (above), fruiting twig (below), natural size.

## 58. West Indies laurelcherry (*Prunus occidentalis*)

Almendron, a rare large tree of mountain forests, is distinguished by : (1) oblong–elliptic leaves 4–8 inches long and $1^1/_2$–3 inches wide, in 2 rows, with odor and taste (when crushed) of almond; (2) many small flowers about $^1/_4$ inch wide, with 5 white petals in dropping unbranched clusters; and (3) the elliptic stone fruits $^3/_4$–1 inch long.

Evergreen large tree to 80 feet high and $2^1/_2$ feet in trunk diameter, developing large buttresses at base. Bark dark brown, scaly, becoming rough and furrowed. The brown inner bark and the twigs have the taste of almond. Twigs are hairless, green when young, becoming light brown.

The leaves are alternate in 2 rows, hairless, with petioles $^1/_4$–$^1/_2$ inch long. The paired narrow pointed stipules $^1/_4$ inch long form the bud. The blades are long–pointed at apex and short–pointed or rounded at base, slightly thickened and slightly turned under at edges, the upper surface green and slightly shiny, the lower surface dull light green.

Flower clusters (racemes) at leaf bases are $1^1/_2$–$3^1/_2$ inches long, bearing many fragrant flowers on slender stalks of about $^1/_4$ inch. The flower has a basal cup (hyanthium) more than $^1/_8$ inch long, which bears 5 pointed sepals, 5 white rounded petals about $^1/_8$ inch long, and many stamens more than $^1/_8$ inch long. The pistil has a 1–celled ovary and short style. The fruit (drupe) like an almond has thin flesh and a large stone containing 1 seed. Flowering from February to April and with fruits in spring and summer.

Rare in upper western forest at 2,000–3,000 feet altitude in central Kerala. One very large tree of this species is at the edge Trivandrum of Highway at kilometer 36, altitude 2,500 feet.

The wood is described as having flesh–colored sapwood and rich dark reddish–brown heart–wood. It is very hard, heavy (specific gravity 0.90 to 1.05), of medium to coarse texture, tough, and strong. The wood has been used for construction and elsewhere for cabinetwork, furniture, flooring, posts, poles, implement frames, and railway crossties.

This species has been planted experimentally in forestry. Elsewhere a drink has been made from the cherrylike fruits. Classed as a honey plant.

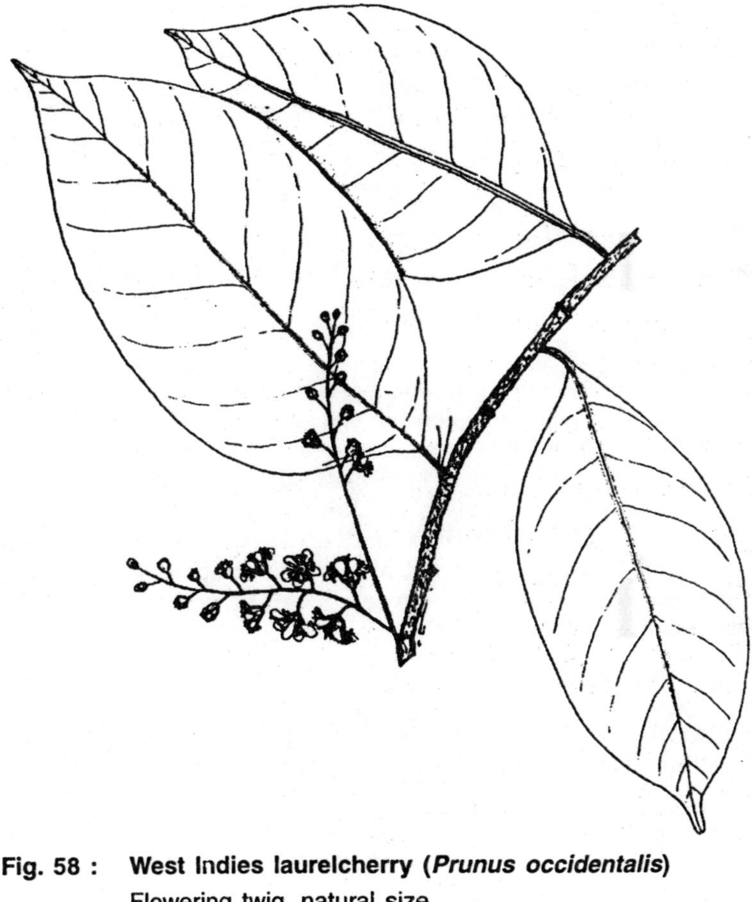

**Fig. 58 : West Indies laurelcherry (*Prunus occidentalis*)**
Flowering twig, natural size.

# LEGUME FAMILY (LEGUMINOSAE)

## 59. Anegada acacia (*Acacia anegadensis*)

This distinctive small tree known only from Anegada in the British Virgin Islands is recognized by : (1) straight spines $^1/_8$–$^5/_8$ inch long paired at nodes and to 3 inches long on trunk and larger branches; (2) bipinnate leaves with only 1 pair of side axes, each with usually 2 paired oblong leaflets $^3/_8$–$^5/_8$ inch long and less than $^1/_4$ inch wide; (3) many minute flowers in yellow balls (heads) $^1/_4$ inch in diameter; and (4) narrow curved pods $1^1/_4$–$1^1/_2$ inches long and $^1/_4$ inch wide.

A deciduous spiny tree to 26 feet high and 5 inches in trunk diameter, with many widely spreading branches. Twigs slender, gray, hair–less.

Leaves alternate, pinnate, $^3/_4$–$1^1/_4$ inches long. Stipules paired, developing into slender straight dark brown spines. Petiole $^1/_8$–$^1/_4$ inch long, hairless or nearly so, with dot gland at apex, forking into 1 pair of side axes slightly longer and with gland dot at apex, each ending in 1 pair (rarely 2 pairs) of stalkless leaflets. Leaflets oblong, thick and leathery, hairless, unequal and rounded at base, rounded or notched at apex, not toothed on edges, with prominent network of raised veins, shiny dark green on both surfaces.

Flower heads 1–4 on short stalks at leaf bases. Flowers about $^1/_8$ inch long, consisting of minute 5–toothed calyx, narrow tubular 5–toothed yellow corolla $^1/_{16}$ inch long, about 20 spreading threadlike stamens, and slender pistil. Pods short–stalked, dark brown, hair–less, thick–walled, opening late. Seeds few, rounded, $^1/_8$–$^3/_{16}$ inch long, dull brown. Collected with flowers and fruits in February, flowering intermittently in wet season.

Common on the rocky plan and occasional on the sandy plain the dry limestone forest at sea level near the west end of Kerala.

**Fig. 59 : Anegada acacia (*Acacia anegadensis*)**
Flowering twig, natural size.

## 60. Tamarindo silvestre, casha (*Acacia macracantha*)

This deciduous tree with thin spreading crown is common in the Andaman–Nicobar. It is characterized by ; (1) the twigs slightly zigzag, with paired spines, generally short, at some nodes; (2) twice pinnate (bipinnate) leaves $2^3/_4$–$5^1/_2$ inches long, with 10–25 pairs of axes, each with 12–30 pairs of minute gray–green leaflets less than $^1/_8$ inch long; (3) range balls $^3/_8$ inch in diameter, of many minute flowers; and (4) pod 2–4 inches long, slightly flattened and slightly curved, reddish brown, minutely hairy.

A deciduous tree, small or medium–sized to 30 feet high and 1 foot in trunk diameter, with thin spreading crown broader than high or only a shrub. The trunk is sightly angled with grooves, with gray outer bark. The twigs are covered with minute gray–green hairs becoming dark brown with light dots.

The alternate bippinate leaves have paired minute stipules, some of which become minute dark brown flattened spines $^1/_8$–$^3/_8$ inch long, sometimes to 1–2 inches or more. The axes are hairy gray green, the secondary axes paired, $^5/_8$–$^3/_4$ inch long. Leaflets very numerous, stalkless, minute, oblong, hairy beneath.

The balls (heads) of flowers are inserted 1–5 on stalks $^3/_8$–$^3/_4$ inch long at the base of the leaves, composed of many minute stalkless flowers, somewhat fragrant. The flower $^3/_{16}$ inch long is formed by the calyx less than $^1/_{16}$ inch long, corolla less than $^1/_8$ inch long, both tubular, greenish yellow, hairy, and 5–toothed; many tiny separate orange stamens $^3/_{16}$ inch long; and pistil $^3/_{16}$ inch long with narrow greenish ovary and thin white style. The pod, which does not open, measures $^3/_{18}$–$^1/_2$ inch wide and $^3/_{16}$ inch thick. Seeds several, elliptic, brown, less than $^1/_4$ inch long. Flowering intermittently and with fruits persisting through the year.

The word is reported to be very durable. It is used for posts and elsewhere for construction and charcoal.

The plants can be grown and pruned as hedges.

Locally common in dry lowlands, forming costal thickets, from sea level to 500 feet altitude.

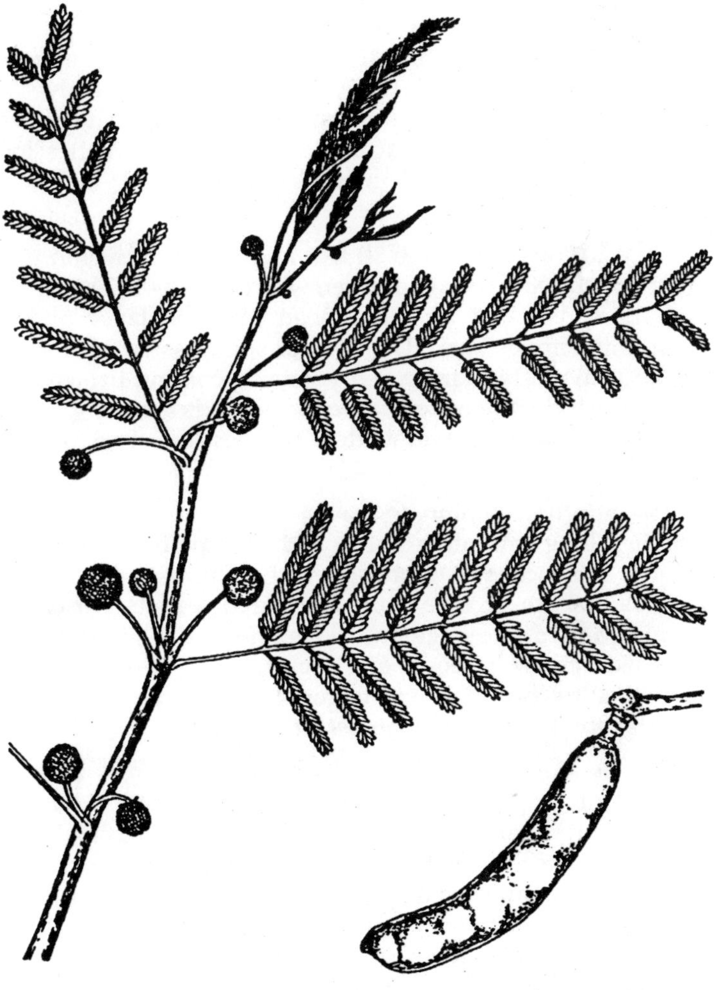

**Fig. 60 : Tamarindo silvestre, casha (*Acacia macracantha*)**
Flowering twig (above), fruit (lower right), natural size.

## 61. Spineless acacia (*Acacia muricata*)

This species of acacia, which is more common in the Islands Andaman Nicobar, is distinguished by : (1) the absence of paired spines at leaf bases, as in the other native species of the genus; (2) twice pinnate leaves 6–10 inches long, with 4–6 pairs of secondary axes, each with 6- 16 pairs of oblong leaflets $^1/_4$–$^3/_4$ inch long and $^1/_4$–$^1/_8$ inch wide, shiny dark green above and dull whitish green beneath, almost stalkless; (3) white flowers about $^1/_4$ inch long stalkless on slender axes 3–6 inches long; and (4) oblong flat pods 3–5 inches long and $^1/_2$–$^7/_8$ inch wide, dark brown, opening on 1 side.

A deciduous small tree 30 feet high and 6 inches in trunk diameter, reported to become larger, with thin crown. The smooth gray bark consists of an outer dead purplish layer and the pinkish, bitter inner bark. The slender spineless twigs are light green and minutely hairy when young, becoming brown, with raised whitish dots (lenticels). The naked buds are composed of minute brown hairy leaves with 2 stipules less than $^1/_{16}$ inch long.

The alternate bipinnate leaves have a yellow–green axis 2–5 inches long, minutely hairy, with 4–6 pairs of secondary axes each 3–5 inches long, mostly with a cuplike gland on the main axis at base of each pair. The pared leaflets are oblique with midrib near one edge and sides unequal, slightly notched or rounded at apex, unequal and slightly notched or rounded at base, with sunken midrib and edges slightly turned under, thin or slightly thickened, and hairless at maturity.

The slender unbranched flower clusters (spikes) are borne near apex of twigs before the new leaves unfold. Along the narrow axis are many stalklkess white, slightly fragrant flowers $^1/_4$ inch long, spreading in masses more than $^1/_2$ inch across. The flower is composed of light yellow 5–toothed calyx less than $^1/_{16}$ inch long; tubular light yellow corolla less than [! inch long, with 5 pointed lobes; many white threadlike spreading stamens about $^1/_4$ inch long, united at base; and pistil of minute greenish ovary and white threadlike style.

Usually 1 or 2 flat pods develop along an axis. They have straight edges not narrowed between the seeds, dark brown walls cracked at the surface and slightly thickened, and open widely on 1 side, exposing the inner walls with reddish brown impressions of seeds separated by light yellow bands. Seeds 6–12, elliptic or oblong, $^1/_2$–$^5/_8$ inch long, dark brown, flat and thin. Flowering intermittently through the yeai, the fruits persistent.

The sapwood is light brown and the heartwood reddish brown. The wood is hard, heavy, strong and durable.

Locally common in moist forest at 100–1,000 feet altitude, lower foothills in eastern and southeastern Andaman Nicobar Island.

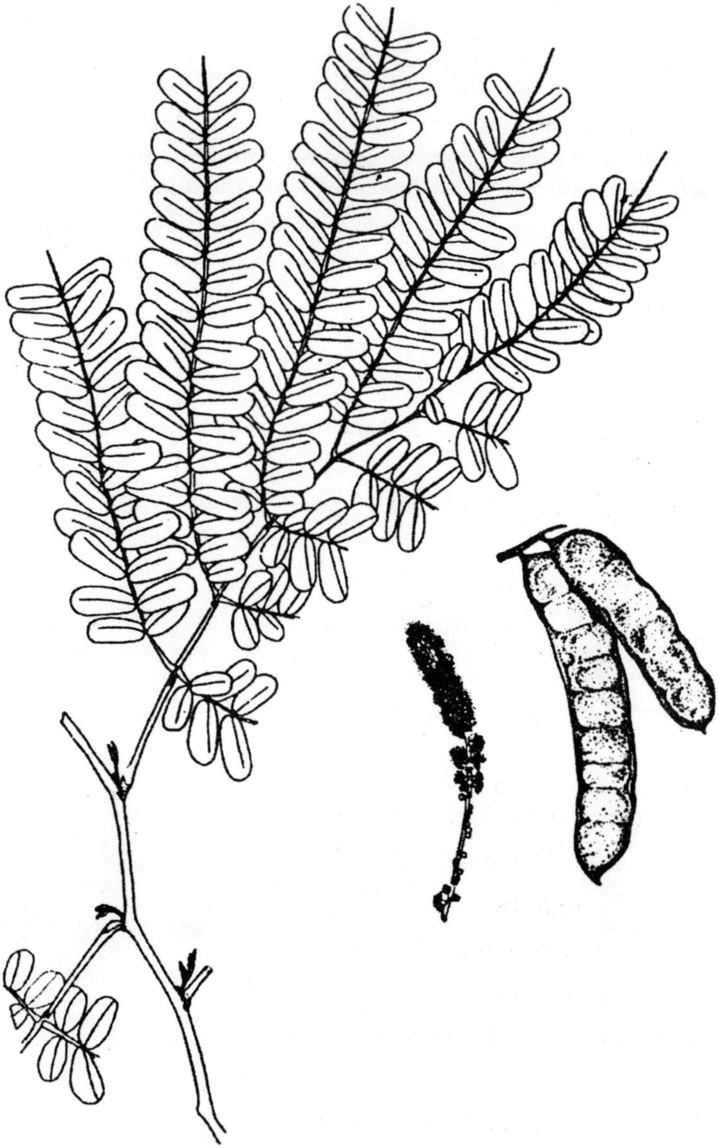

**Fig. 61 : Spineless acacia (*Acacia muricata*)**

Leafy twig (left), flowers and fruits (lower right), two–thirds natural size.

## 62. Gum Arabic (*Aciacia nilotica*)

This species of gum Arabic is planted for ornament in India and the Islands and naturalized locally. It is recognized by : (1) long slender straight light gray spines $^1/_2$–$1^1/_2$ inches long, paired at base of some or most leaves; (2) bipinnate leaves 2–6 inches long, with 3–8 pairs of side axes, each with 10–30 pairs of narrow oblong leaflets $^1/_2$–$^1/_4$ inch long and less than $^1/_{16}$ inch wide; (3) balls of yellow flowers $^1/_2$ inch in diameter; and (4) narrow whitish gray flattened pods 2–6 inches long and $^5/_{16}$–$^1/$, inch wide, finely hairy, slightly narrowed between seeds.

Introduced deciduous shrub or spiny tree 10–25 feet high. Twigs gray to dark brown, slender, finely hairy or nearly hairless.

Leaves alternate, bipinnate, with paired spines formed from stipules, the axis finely hairy, with a few dot glands. Lateral axes $^1/_2$–$1^1/_2$ inches long. Leaflets blunt at both ends, with minute hairs along edges.

Flower heads few at leaf bases or nodes on slender jointed stalks $^1/_2$–inch long. Flowers $^5/_{16}$ inch long are composed of tubular 5–toothed slightly hairy calyx, narrow yellow tubular 5–toothed corolla [! inch long, many yellow stamens more than $^1/_4$ inch long and united at base, with bright yellow dotlike anthers and slender pistil with threadlike style. Pods stalked at base, ending in a short point. Seeds several, rounded but flattened, $^5/_{16}$ inch in diameter, blackish.

An ornamental in plains and Islands, becoming naturalized locally and forming thickets.

Elsewhere the gum that exudes from the trunk has been used in mucilage, ink, and medicine. The hard wood has been utilized. The pods have served in tanning, and the flowers attract bees.

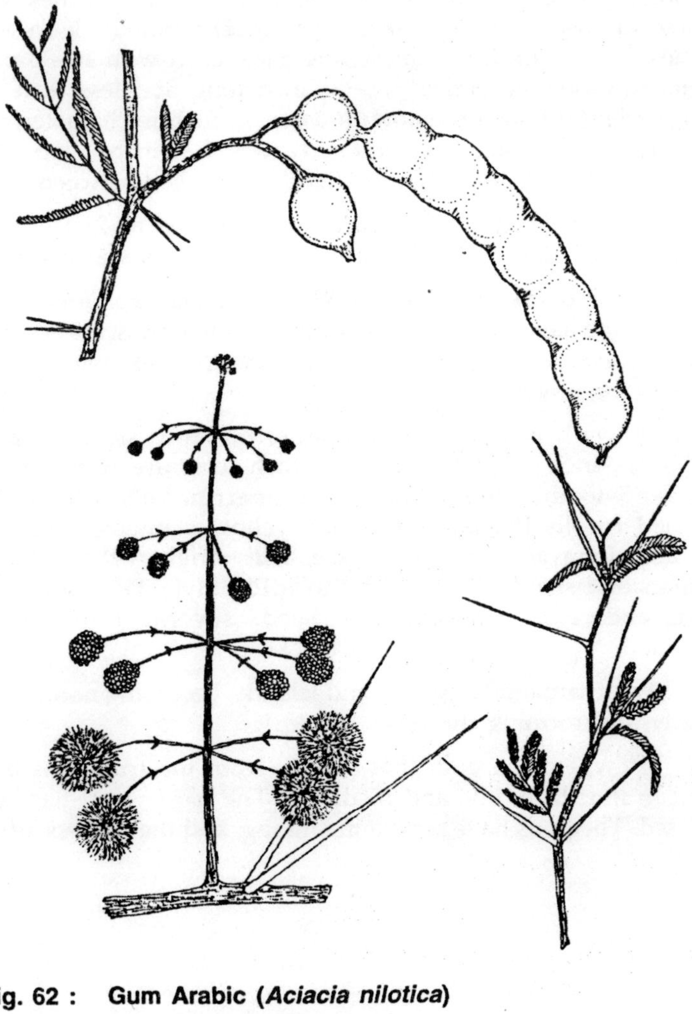

**Fig. 62 : Gum Arabic (*Aciacia nilotica*)**

Fruiting twig (above), flowering twig (lower left), leafy twig (lower right), natural size.

## 63. Catechu–tree (*Acacia polyacantha*)

This Old World tree introduced in U.P. is characterized by: (1) paired short conic curved spines at base of some leaves; (2) bipinnate leaves 6–9 inches long, with 7–25 pairs of side axes, each with 25–50 pairs of narrow leaflets $^1/_2$–$^1/_4$ inch long and less than $^1/_{16}$ inch wide; (3) small flowers $^1/_4$ inch long, white to pale yellow, stalkless along axes $2^1/_2$–4 inches long; and (4) narrow flat gray pods $2^1/_2$–5 inches long and $1^1/_4$–$^3/_4$ inch wide.

A deciduous spiny spreading shrub or small tree to 30 feet high, with gray bark. Twigs slender, light gray or whitish, velvety hairy.

Leaves alternate, bipinnate, with paired spines developing some stipules, and with velvety hairy axis bearing a gland below leaflets and at base of some side axes. Leaflets short–pointed at apex, unequal at base, finely hairy or nearly hairless, paler beneath.

Flower clusters (spikes) at leaf bases, slender, finely hairy. Flowers consist of hairy 5–toothed calyx more than $^1/_{16}$ inch long; tubular 5–toothed hairy corolla 1 inch long, white changing to yellow; many threadlike stamens $^1/_4$ long; and pistil with elliptic ovary and threadlike style. Pods pointed at both ends, hairless, hard, with fine network of veins.

The wood is described as hard, heavy, very durable, and polishing well. Catechu or cutch is a resinlike astringent substance obtained by boiling down an extract from the heartwood chips of this and the related species *Acacia catechu* Wild. Of the east Indies. This decoction has been used medicinally as an astringent in chronic diarrhea and dysentery, also in tanning and dyeing.

Introduced and naturalized in Uttaranchal.

Fig. 63 : **Catechu–tree (*Acacia polyacantha*)**
Flowering twig and fruits (lower right), natural size.

## 64. Twisted acacia (*Acacia tortuosa*)

A flat–topped spiny shrub of dry forests in the Moore Islands and eastern states identified by: (1) slender zigzag twigs with paired straight spines at enlarged nodes; (2) twice pinnate (bipinnate) leaves 1–2 inches long, with 2–8 pairs of side axes, each with 10–20 pairs of minute oblong leaflets $^{1}/_{2}$–$^{1}/_{4}$ inch long; (3) many tiny flowers with threadlike stamens crowded in orange balls $^{5}/_{16}$ inch in diameter; and (4) the narrow cylindric blackish pods 3–5$^{1}/_{2}$ inches long ,and $^{5}/_{16}$ inch in diameter, not splitting open.

Usually a shrub about 10 feet high, some-times a tree to 20 feet, with trunk 6 inches in diameter often crooked, deciduous, with very thin flat–topped crown. Bark light gray, finely fissured, the inner bark light brown with darker streaks, slightly bitter. Twigs slender, light green and hairy when young, becoming reddish brown with whitish dots' (lenticels). At each node are 2 straight slender sharp spines (stipules) $^{1}/_{8}$–1$^{1}/_{2}$ inches long.

The leaves are alternate, twice pinnate (bi-pinnate), and finely hairy, the axis with gland below leaflets. The oblong leaflets are almost stalkless, blunt at apex, dull green on upper sur-face, and light green beneath.

The numerous stalkless fragrant orange flowers are in heads on stalks about $^{3}/_{4}$–1$^{1}/_{4}$ inches long, 1 or 2 at base of a leaf. Each flower $^{1}/_{8}$ inch long consists of funnel–shaped whitish hairy calyx less than $^{1}/_{16}$ inch long, 5–toothed; funnel–shaped light yellow corolla more than $^{1}/_{16}$ inch long, 5–toothed; many threadlike orange stamens $^{1}/_{8}$ inch, long, united into tube in lower half; and greenish pistil with ovary and slender style, often not functional. The pods are slightly narrowed between the several seeds. Flowering intermittently through the year.

The wood is light brown and hard.

Livestock eat the pods. Elsewhere the plants have been grown and pruned as hedges.

Locally abundant on dry coasts and lower slopes in dry forests from sea level to 600 feet altitude.

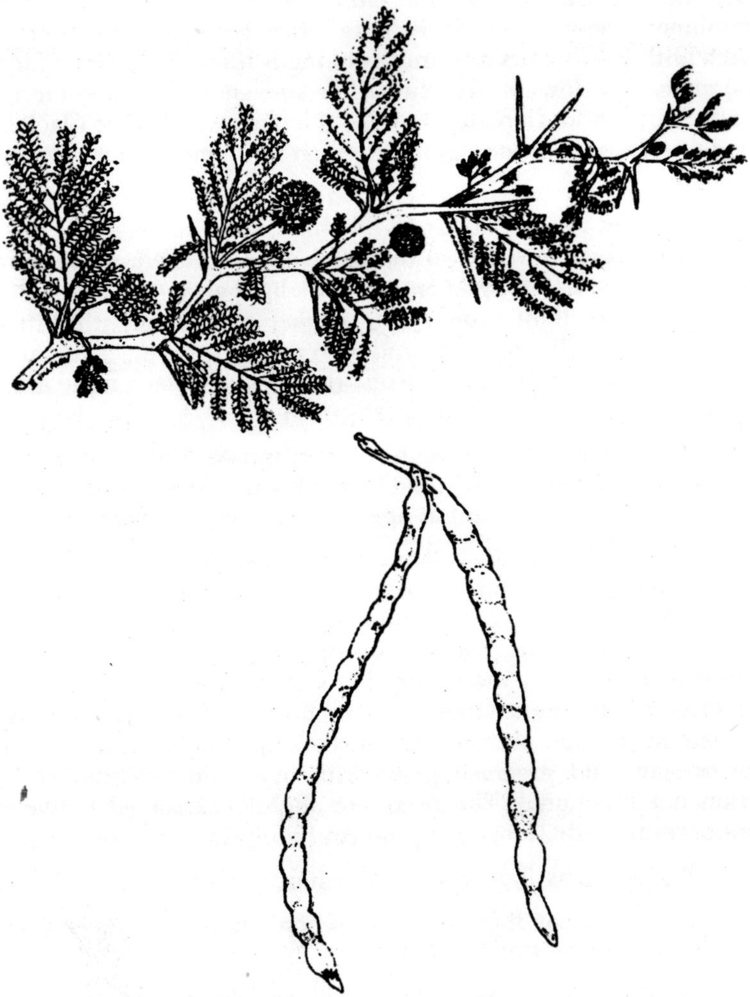

**Fig. 64 : Twisted acacia (*Acacia tortuosa*)**
Flowering twig (above), fruits (below), natural size.

## 65. Carbonero (*Albizia carbonaria*)

Carbonero, an uncommon legume introduced in forestry tests, is characterized by; (1) young twigs, foliage, and flower clusters with dense short brownish hairs; (2) leaves twice pinnate (bipinnate), 6–12 inches long, with 7–16 pairs of side axes, each with 10–30 pairs of oblong stalkless leaflets; (3) many flowers in many whitish heads 1 inch or more across the threadlike spreading stamens; and (4) oblong narrow flat brown pods $2^1/_2$–4 inches long and $^1/_4$–$^3/_4$ inch wide.

A medium–sized to large planted tree becoming where native 70 feet high and 2 feet in trunk diameter, with flattened spreading this crown. Bark light gray, scaly. Twigs brown, densely short hairy when young.

Leaves alternate, bipinnate, finely hairy, with axis 4–9 inches long, often with dotlike gland at end, and side axes $2^1/_2$–4 inches long. Leaflets oblong $^3/_{16}$–$^3/_8$ inch long and $^1/_{16}$ inch wide, oblique or unequal at base with midvein not in center, blunt at apex, not toothed on edges, thin, without side veins, the upper surface dull green with minute hairs, and the lower surface paler and soft hairy.

Flower clusters (panicles) shorter than leaves, with finely hairy branches, bearing many heads on stalks up to $1^1/_4$ inches long. Flowers on stalks less than $^1/_{16}$ inch are composed of narrow bell–shaped hairy 5–toothed greenish calyx 2 inch long; hairy greenish corolla less than $^1/_4$ inch long, with narrow tube and 5 short lobes; many white threadlike stamens $^1/_2$–$^1/_4$ inch long united near base, and slender pistil with narrow ovary and threadlike style. The pod has a stalk about 2 inch long at short–pointed base, is rounded and abruptly pointed at apex, thickened at borders, finely hairy. Seeds 15–25, oblong $^1/_8$ inch long, gray.

The wood is light brown and soft, reported to be suitable for lumber.

Tested in forest plantations in Dehradun. Of very rapid growth but short–lived and with fragile branches. In Kerala the trees have served for shade in coffee plantations.

Fig. 65 : **Carbonero (*Albizia carbonaria*)**
Flowering twig and fruit (lower right), natural size.

## 66. White calliandra (*Calliandra caracasana*)

This spineless shrub seldom more than 10 feet high rarely becomes a small tree. It is characterized by : (1) twice pinnate (bipinnate) leaves 3–7 inches long with 2–6 pairs of side axes each with 10–30 pairs of crowded narrow leaflets $^1/_4$–$^3/_6$ inch long and $^1/_{16}$–$^1/_4$ inch wide; (2) many showy white flowers crowded in heads 1–1$^1/_4$ inches or more across the numerous threadlike stamens; and (3) narrow flat pods 2–4 inches long, the hard stiff walls splitting open from the apex and curving back spirally.

A deciduous shrub or rarely a small tree to 20 feet or more in height and 3 inches in trunk diameter, with twigs and foliage hairless or slightly hairy.

Leaves alternate, bipinnate, with paired pointed stipules $^3/_{16}$ inch long, the slender axis without glands. Leaflets stalksless, narrow, straight or slightly curved, blunt at both ends, thin, with few veins, light green, paler beneath.

The heads with about 15 flowers are 1–3 at leaf bases on stalks 1$^1/_4$–4 inches long, each flower is composed of 5–toothed calyx about $^1/_{16}$ inch long; tubular 5–lobed whitish corolla more than 2 inch long; many long threadlike white stamens $^1/_4$–1 inch long, united into tube toward base; and pistil with narrow ovary and slender style. The hard pods are $^1/_4$–$^1/_8$ inch wide, blunt or rounded at apex, narrowed to base, raised at margins, hairless or nearly so, splitting elastically from apex. Seeds several oblong, flat, brown $^3/_{16}$ inch long.

Scattered in dry and moist forest at lower and middle altitude in Bihar.

**Fig. 66 :** **White calliandra (*Calliandra caracasana*)**
Natural size.

## 67. Guanacaste, earpod–tree (*Enterolobium cyclo–carpum*)

Guanacaste, a handsome giant tree with very stout short trunk and broadly spreading crown, is sometimes planted for shade and ornament. It is characterized by : (1) bipinnate leaves 6–10 inches long with 4–9 pairs of secondary, axes each with 20–30 pairs of oblong leaflets $^1/_4$–$^1/_2$ inch long and $^1/_4$–$^3/_{16}$ inch wide; (2) many flowers in a whitish ball about 1 inch across the numerous spreading stamens; and (3) the very distinctive blackish seed pod, flattened and curved in a complete circle or disk $3^1/_2$–$4^1/_2$ inches in diameter, slightly resembling a human ear.

In its native home this large to very large deciduous tree attains a height of 60–100 feet. It has a stout short trunk 3–6 feet or more in diameter, large almost horizontal branches, and a thin spreading crown broader than the height and as much as 75–150 feet in diameter. Most trees planted in Puerto Rico are smaller, less than 50 feet high and 3 feet in trunk diameter, ever, a giant in Mayaguez is about 90 feet tall and 10 feet in diameter. The bark is gray or brownish gray, slightly rough or scaly with shallow furrows. Inner bark is light brown, bitter, and astringent, and exudes a brownish gum. The stout twigs are green when young and later brownish gray.

The alternate twice pinnate (bipinnate) leaves have a light green axis, finely hairy, with a small gland below the lowest secondary axes and another near the apex. The stalkless leaflets are short–pointed at apex, squarish and unequal at the base with the midvein near a side, thin, finely hairy, the upper surface dull green and the lower surface pale green.

Flower clusters like balls (heads) are borne on stalks at the base of leaves or when leafless. The crowded flowers are tubular or funnel–shaped, about $^1/_2$ inch long and $^1/_4$ inch across the stamens. The calyx is tubular $^1/_4$ inch long, light green, and 5–toothed; the corolla tubular, $^1/_4$ inch long, light green, and 5–toothed; numerous whitish stamens to $^1/_4$ inch long, threadlike, united into tube in lower half; and pistil $^1/_2$ inch long with short light green ovary and slender white style.

The heavy pod is curved in a circle around the central space or hole of $^1/_4$ inch and suggests slightly the form of the human ear. The color changes from shiny green when immature to blackish or dark brown. It is flat but thicker around the seeds and does not open. There are several dark brown elliptic seeds $^1/_2$–$^3/_4$ inch long.

The sapwood is whitish and the heartwood brown, similar to walnut, often reddish tinged. The wood is lightweight (specific gravity 0.4–0.6), hard, and has many large pores. It is moderately durable and resistant to attack by dry–wood termites. It polishes well and does not split or twist when exposed. The wood is used for construction, carpentry, interiors, furniture, and veneer. Indians made dugout canoes from large trunks.

In the savannas where native, these large trees serve well as shade for livestock, which eat the pods and foliage. Also, the trees are planted for shade along highways. It is reported that the toasted seeds are edible. The bark and pods are rich in tannin. The gum from the bark can substitute for gum Arabic. It is said that sawdust is irritating to some workmen and also can kill fish if dumped into rivers.

Sometimes planted for shade and ornament in Maharashtra but uncommon and not now recommended. A fungus disease causes the trees to fall suddenly. Of rapid growth.

Recommended in south India as a very large spreading shade tree for parks, play–grounds, and other areas with ample space for the broad root system. The large branches are subject to storm damage.

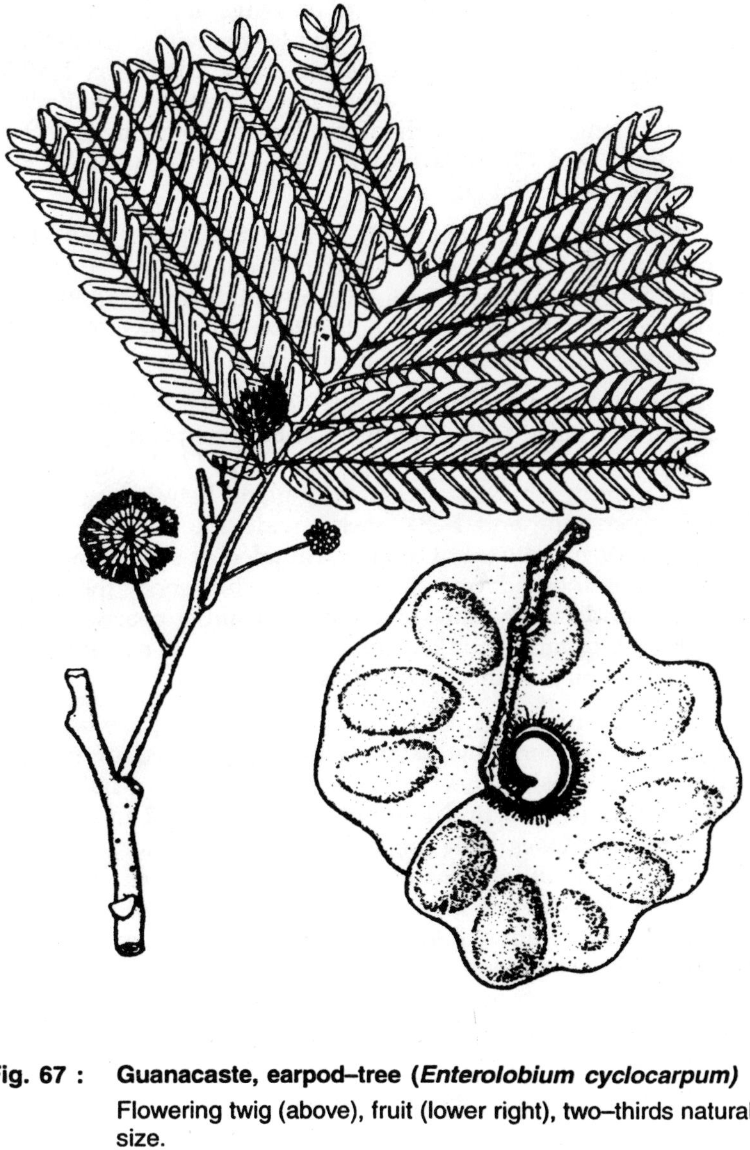

**Fig. 67 : Guanacaste, earpod–tree (*Enterolobium cyclocarpum*)**
Flowering twig (above), fruit (lower right), two–thirds natural size.

## 68. Guaba venezolana (*Inga fastuosa*)

Guaba venezolana, which has been introduced from Venezuela for coffee shade, is easily distinguished from the native species of guaba by the much larger leaves, flowers, and pods, also the long bristly reddish hairs on twigs and pods. Further identifies by: (1) pinnately compound leaves with 3 (sometimes 4 or 5) pairs of large ovate or elliptic, shiny thick hairy leaflets and broadly winged axis; (2) several large greenish white flowers stalkless along an axis with many threadlike stamens 3–$3^1/_2$ inches long; and (3) giant brown pods 12–16 inches long and $2^1/_2$ inches wide, flat but thick.

An evergreen planted tree becoming 30 feet high, with short branching trunk 6 inches or more in diameter, and spreading crown. The bark is brown, smoothish or with many small fissures, becoming scaly. Inner bark is light brown and hard. The stout brown twigs are covered with bristly hairs.

The alternate pinnate leaves 6–12 inches long have paired large narrow hairy scales (stipules) at base and brown densely bristly hairy axis 3–6 inches long with wings $^1/_4$–$^1/_2$ inch wide. Leaflets paired, almost stalkless, with gland between each pair, $1^1/_2$–7 inches long and 1–$3^1/_2$ inches wide, the lowest smallest, with short or narrow point at apex, rounded or slightly notched at base, not toothed on edges, slightly thickened, the upper surface slightly shiny dark green and becoming nearly hairless except on the slightly sunken veins, and the lower surface green and bristly hairy, with raised side veins.

Flower clusters (spikes) about 6 inches long are borne at leaf bases. Flowers few, soon falling, composed of narrow cylindric greenish calyx $^3/_4$–1 inch long, bristly hairy and 5-toothed; narrow cylindric tubular corolla $1^3/_4$–2 inches long with 5 narrow lobes, pale greenish, covered with long hairs on outside; many threadlike stamens, 3–$3^1/_2$ inches long and spreading 2–3 inches across, united in tube in lower half, changing color from white to orange; and very narrow pistil with threadlike style longer than stamens. The pods are covered with bristly reddish brown hairs and twist upon splitting open. There are several large flattened seeds. Flowering and fruiting in summer.

The wood is light brown and hard.

**Fig. 68 : Guaba venezolana (*Inga fastuosa*)**
Flowering twig, two–thirds natural size.

This rapidly growing species has been introduced into Kerala for coffee shade. The pulp surrounding the seeds in the pods is reported top be sweet and edible.

Uncommon as a planted shade tree in coffee plantations in Central India.

## 69. Una de gato, bread–and–cheese, catclaw (*Pithecellobium unguis–cati*)

Una de gato, a shrub or small tree of coastal thickets, is recognized by : (1) paired slender sharp brown or gray spines $^1/_8$–$^1/_2$ inch long, sometimes longer, often developing at nodes; (2) twice pinnate (bipinnate) leaves with slender axis and 2 forks, each with 2 almost stalkless oblique oblong to obovate leaflets; (3) small light yellow or pinkish flowers in many ball–like heads about $^3/_4$ inch across, on long slender branching stalks at leaf bases; and (4) curved or coiled reddish to dark brown pods 2–5 inches long, flattened, and splitting open on both sides.

Spiny shrub less than 10 feet high or small tree to 30 feet tall with several trunks to 5 inches in diameter, and widely spreading thin crown, hairless. Nearly evergreen or without leaves for short periods. The bark is gray and smooth or slightly fissured, with horizontal lines and few paired spines to $^1/_2$ inch long persisting at old nodes. The inner bark is light brown and slightly bitter. The twigs are brownish gray, angled, with dots (lenticels), often zigzag and with paired spines (stipules).

The alternate hairless leaves 1–3 inches long have a very slender petiole $^1/_2$–$1^1/_2$ inches long, with a minute green dot gland at apex, and 2 lateral axes (pinnae) $^1/_8$–$^1/_2$ inches long. The 4 leaflets are $^1/_2$–2 inches long and $^3/_8$–1 inch wide, rounded at apex, oblique and short–pointed at base, not toothed on edges, thin or slightly thickened, dull green with raised veins above, and lighter blue green beneath.

Flower clusters (heads) contains 10–20 stalkless hairless flowers. Each has a tubular 5–toothed calyx $^1/_{16}$ inch long, tubular 5–lobed corolla about $^3/_{16}$ inch long, many spreading long threadlike light yellow stamens $^1/_2$ inch long united into a pinkish tube at base, and pistil with narrow stalked ovary and long threadlike style.

The curved twisted pods coiled into a circle are $^1/_4$–$^1/_2$ inch wide, slightly narrowed between the seeds, red on inner surface. Several shiny black rounded and flattened seeds $^3/_8$ inch long hang down from reddish or white pulp (aril). With flowers and fruits irregularly through the year.

The wood is light brown, hard, and heavy.

Common locally in coastal forests and hills from sea level to 600 feet altitude, along the southwestern and southern coasts of Karnataka and extending up the valley of the Narmada River. Forming thickets on sandy shores and dry coastal areas. Also, widespread throughout the smaller islands.

**Fig. 69 : Una de gato, bread–and–cheese, catclaw (*Pithecellobium unguis–cati*)**

Flowering twig (above), fruiting twig (lower right), natural size.

## 70. Purple bauhinia (*Bauhinia purpurea* L.*)

This showy ornamental small tree with orchidlike flowers is recognized by: (1) the squarish or elliptic leaves divided more than S! their length into 2 rounded lobes with 13 or 11 radiating main veins from the heart–shaped base; (2) the large showy flowers $3^1/_2$–4 inches across, with 5 slender–stalked narrow purple or pink petals (1 petal dark red toward base); and (3) the flat blackish pods 8–12 inches long and $^3/_4$–1 inch wide, which twist as they open. From related species distinguished by the flowers with narrow petals less than $^5/_8$ inch wide, calyx splitting into 2 nearly equal parts, and the 3 (sometimes 4) fertile stamens and by the deeply lobed leaves.

An evergreen small planted tree 25 feet high and 7 inches in trunk diameter or larger. The bark is light gray, smoothish to finely fissured. Twigs are light green, finely hairy, often slightly zigzag, becoming light brown.

The leaves are alternate in 2 rows, hairless, with petioles 1–2 inches long, light green, slender, enlarged at both ends. Leaf blades are $1^1/_2$–5 inches long and broad, slightly thickened, turned up at middle, the upper surface dull light green with veins slightly sunken, and the lower surface paler with raised veins.

Unbranched flower cluster (racemes) at ends of twigs bear few slightly fragrant flowers on hairy greenish stalks $^1/_4$–$^1/_2$ inch long. The narrow finely hairy light green basal tube (hypanthium) about $^1/_2$ inch long and closed at apex bears the other parts; the 5–angles calyx 1 inch long in bud, greenish with minute reddish brown hairs, splits into 2 parts and turns down; the 5 nearly equal petals are about 2 inches long and less than $^5/_8$ inch wide, blunt at apex and tapering to stalk at base; 3 (sometimes 4) curved purplish stamens $1^1/_2$–$1^3/_4$ inches long and 7–6 threadlike sterile stamens (staminodes) $^1/_4$ inches or more in length with greenish 1–celled ovary, long curved style, and enlarged rounded stigma. There are several rounded flat seeds $^1/_2$–$^5/_8$ inch long, shiny brown. Flowering and fruiting in autumn and winter months.

**Fig. 70 : Purple bauhinia (*Bauhinia purpurea* L.*)**
Flowering twig (left), fruit (right), natural size.

This ornamental is widespread but uncommon in moist parts of A.P., such as in gardens and along roadsides, and A.P. Islands. It seldom escapes from cultivation. In central and south India this is one of the most common introduced trees with orchidlike flowers. It is recommended as a small flowering tree for yards. Of medium growth on well drained soils.

## 71. Poor–man's–orchid (*Bauhinia variegata* L*)

This ornamental small tree with flowers slightly an orchid is identified by: (1) the odd rounded leaves divided less than $^1/_4$ their length into 2 rounded lobes with 11 or 9 main veins from the heart–shaped base; (2) large showy variegated orchidlike fragrant flowers $3^1/_2$–4 inches across, with slender–stalked purplish petals (white in a variety), 1 darker streaked; and (3) flat dark brown pods 5–9 inches long, about $^3/_4$ inch wide, and only $^1/_{16}$ inch thick, long–pointed at both ends. Separating open on 1 side, and the 5 stamens, also the shallowly lobed leaves.

Small tree 25 feet high and 8 inches in trunk diameter, evergreen or deciduous. The bark is light brownish gray, smoothish to slightly fissured and scaly. Inner bark is pinkish, fibrous, and bitter. The twigs are slender, zigzag, when young light green, slightly hairy, and angled, becoming brownish gray. The bud is composed of minute hairy leaves and stipules.

The alternate leaves spread in 2 rows along the twigs. Stipules are paired, tiny, and hairy. The slender light green petioles are mostly 1–$1^1/_4$ inches long, enlarged at both ends, finely hairy. Blades are slightly broader than long, mostly $2^1/_2$–$4^1/_2$ inches long and $2^3/_4$–5 inches wide, thin, not toothed on edges. The upper surface is dull light green and hairless, slightly turned up at middle, the lower surface lighter green and often finely hairy on veins.

Flower clusters (racemes) are unbranched at ends of twigs. The few flowers have short stouts stalks $^1/_2$–$^1/_4$ inch long and a stalklike light green narrow basal tube (hypanthium) about 1 inch long. The light green finely hairy calyx 1 – $1^1/_2$ inches long forms a pointed 5–angled bud and splits open on 1 side, remaining attached; petals 5, 2–$2^1/_2$ inches long and 1–$1^1/_4$ inches wide, slightly unequal, wavy margined and narrowed to base; 5 curved stamens 1–$1^1/_2$ inches long; and the stalked very slender curved pistil nearly 2 inches long, with narrow green 1–celled ovary, style, and dotlike stigma.

The pods have a stalk about 2 inch long beyond the persistent calyx at base of flower. They split open in 2 parts and become curved and twisted. There are several rounded flat light brown seeds about $^1/_2$ inch in diameter. Flowering from autumn to spring and maturing fruits in late spring and early summer.

The wood is whitish and soft.

This and related species are uncommon as planted ornamentals in residential areas throughout Assam. Widespread and escaping in southern foothills. Also in the larger Islands and elsewhere through the tropics. Common as an orchid tree in southern India. A variety often grown (var. *candida* Buch.–Ham.) has white flowers.

**Fig. 71 : Poor–man's–orchid (*Bauhinia variegata* L*)**
Natural size.

## 72. Dividivi (*Caesalpinia coriaria*)

Dividivi, which is rare is known commercially for its pods rich in tannin. Distinguishing characters are : (1) bipinnate leaves 2–6 inches long, with many minute narrow dull green leaflets $^1/_8$–$^1/_4$ inch long and less than $^1/_{16}$ inch wide, often with black gland dots beneath; (2) light yellow or whitish flowers $^1/_4$ inch long and broad, with 5 petals, several in short lateral clusters; and (3) the brown pods very curved or twisted, $1^1/_4$–$2^1/_4$ inches long, $^5/_3$–1 inch wide, and $^1/_8$ inch thick.

Small deciduous tree 25 feet high and 1 foot in trunk diameter, elsewhere larger, the trunk frequently short and branched near base into crooked forks. The crown is irregular, spreading, and thin, often flattened above. Bark light brown or gray, rough and furrowed, forming thick scaly plates. The dry outer bark is light brown and the inner bark pink and bitter. The twigs are gray or brown, hairless or finely hairy, slightly zigzag, with light brown dots (lenticels).

The alternate twice pinnate (bipinnate) leaves have a light green finely hairy axis. Secondary axes 9–17, mostly paired except at end, each commonly with 16-24 pairs of minute almost stalkless leaflets. The thin blades are rounded at both ends, hairless or nearly so, pale green beneath.

Flower clusters (racemes and panicles) $^3/_4$–$2^1/_4$ inches long at leaf bases bear several fragrant flowers. The flower is composed of a basal tube (hypanthium) $^1/_8$ inch long; calyx of 5 yellow–green oblong sepals $^1/_8$ inch long, turned back; corolla of 5 elliptic petals more than $^1/_4$ inch long, light yellow or whitish; 10 stamens less than $^1/_4$ inch long, with hairy green filaments and brown anthers; and pistil more than $^1/_4$ inch long with yellow–green ovary and slender style. The thick hard pods, light brown turning reddish brown, are often concave and curved in circular form of letter S and do not open. The few seeds are $^1/_4$ inch long. Collected with flowers and fruits in July, the fruits persisting.

The thick sapwood is whitish or yellowish and the heartwood dark brown or blackish and often streaked. The wood is very heavy (specific gravity 1.20), very hard, strong, and durable. It polishes well and is suitable for turning but is difficult to work. Reportedly, the wood yields a reddish dye.

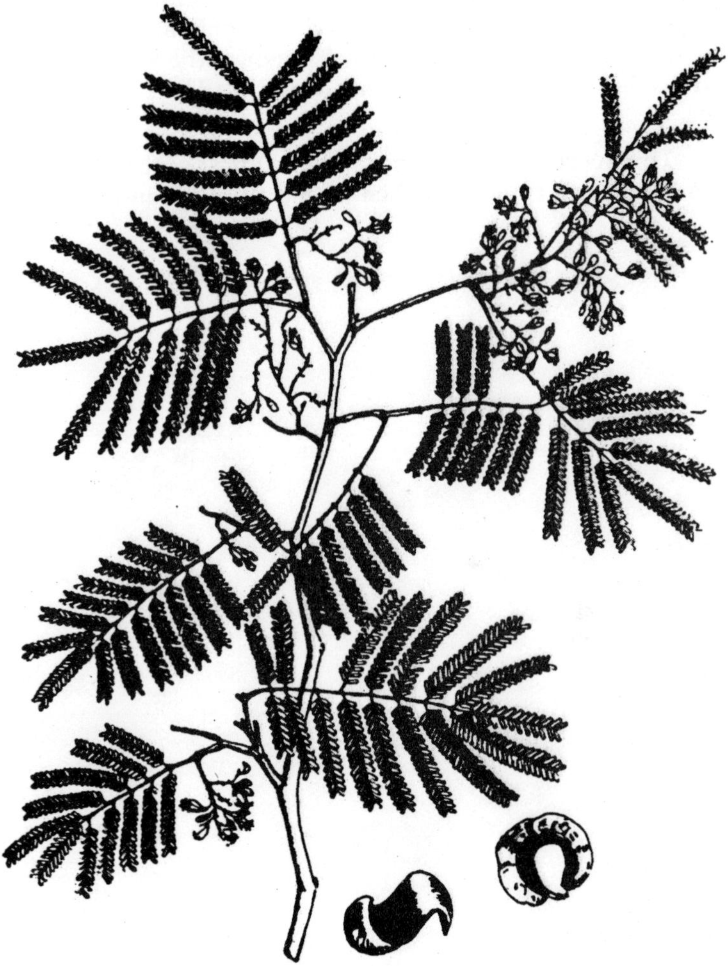

**Fig. 72 : Dividivi (*Caesalpinia coriaria*)**

Flowering twig (above), fruits (lower right), two–thirds natural size.

Dividivi is important in commerce for its bitter pods which contain 30–40 percent tannin and serve in tanning leather. It is said that a mature tree can produce as much as 80 pounds of pods annually. The pods have been exported from continental tropical America to the United States and Europe. In various places the trees are protected and almost cultivated for the harvests of the pods. The bark is high in tannin also. A black dye has been prepared from the pods. Elsewhere the astringent pods and seeds have served in home remedies. The flowers attract bees.

This species is rare in the seasonal forest of the coastal hills of A.P. and can be seen at thee at end of the island.

## 73. Clavellina, flowerfence (*Caesalpinia pulcherrima*)

Calavcellina, or flowerfence, is widely planted as an ornamental shrub or small tree, escaping from cultivation. It is characterized by: (1) twice pinnate (bipinnate) leaves 6–12 inches long, with many paired oblong leaflets $^1/_2$–1 inch long and $^1/_4$–$^1/_2$ inch wide; (2) many large brilliant yellow or orange–red flowers with 5 slightly unequal petals $1^1/_2$ inches across and with 10 threadlike stamens $1^1/_2$–$2^1/_2$ inches long, on slender stalks in long erect clusters, and (3) flat blackish or dark brown pods 3–$4^1/_2$ inches long and $^1/_2$–$^3/_4$ inch wide, twisting open.

An evergreen prickly shrub or small tree 10–15 feet high, with branching trunks 2–3 inches in diameter and thin flat–topped spreading crown, hairless throughout. Bark light gray, smoothish to slightly fissured. Branches with scattered spines to $^1/_4$ inch long, with enlarged base and narrow sharp point.

The alternate bipinnate leaves have stout petiole and slender yellow–green axis together 5–10 inches long. There are 5–10 paired lateral axes (pinnae), each bearing 5–12 pairs of leaflets with stalks less than $^1/_{16}$ inch long. Leaflets are rounded at apex and often with slight notch and minute point, blunt and unequal at base, not toothed on edges, with side veins inconspicuous, dull green above, and light green beneath.

Many slightly fragrant flowers are borne on greenish stalks of 1–3 inches along the unbranched erect axis (raceme). The axis continues to elongate to 5–10 inches, bearing new buds and flowers in the top 3 inches, while buds and flowers in the top 3 inches, while older flowers including their stalks have shed promptly. The short conic green base (hypanthium) $^3/_{16}$ inch long bears the calyx, corolla, and stamens; there are 4 oblong yellow or orange–red sepals $^3/_8$ inch long and 1 large keeled and concave $^5/_8$ inch long covering others in the yellow or orange–red bud; 5 unequal stalked petals $^3/_4$–1 inch long, rounded and finely wavy, yellow or orange red with yellow border (sometimes reddish purple); and 10 threadlike yellow or red stamens $1^1/_2$–$2^1/_2$ inches long. The pistil more than 2 inches long consists of a very narrow 1–celled ovary on a short stalk, a long threadlike yellow or red style, and dot stigma.

Pods, 1 or 2 formed from a flowering axis, are narrowed into stalk 2 inch long above base of flower and have a narrow point $^1/_8$ inch long at apex, slightly oblique and broadest near apex. There are 5–8 shiny brown flat beanlike seeds about $^3/_8$ inch long. Flowering and fruiting through the year.

The flowers, leaves, bark, and roots have been used in home remedies.

A common ornamental in plains and smaller islands and escaping from cultivation along roadsides and in thickets. Planted and naturalized also in southern India and through the tropics of both hemispheres.

In south India recommended as a large shrub to brighten the landscape, also as a screen or unclipped hedge. It is easily propagated and fast growing, even on poor soils, but short lived. In northern India where killed back by frosts, this species can be grown as a herbaceous perennial.

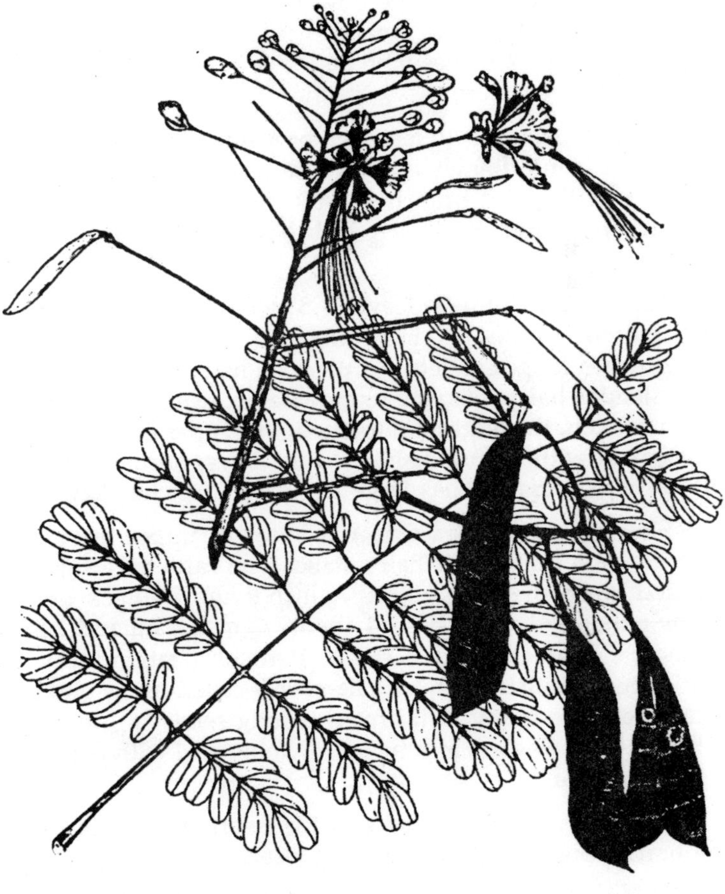

**Fig. 73 : Clavellina, flowerfence (*Caesalpinia pulcherrima*)**
Flowers (above), leaf (below), fruits (lower right), two–thirds natural size.

## 74. Velamuerto (*Cassia emarginata L*)

This rare shrub or small tree is identified by : (1) pinnate leaves with 2–5 pairs of elliptic leaflets mostly 1–2$^1/_2$ inches long and $^3/_4$–1$^1/_2$ inches wide, rounded at both ends, soft hairy beneath; (2) abundant showy flowers about $^3/_4$ inch across the 5 elliptic unequal yellow petals; and (3) long straight narrow blackish pods 6–12 inches long and $^3/_4$–$^1/_2$ inch wide, flattened and not splitting open.

Deciduous shrub 10 feet high or a small tree to 20 feet and 5 inches in trunk diameter, elsewhere reported to reach 50 feet in height, with stout branches. Bark light gray. Twigs densely hairy when young. Buds $^3/_{16}$ inch long formed by the hairy stipules.

The alternate pinnate leaves 4–8 inches long have bristlelike hairy paired stipules $^1/_8$ inch long and slender hairy petioles 1$^1/_2$–2 inches long. Leaflets 2–5 pairs on hairy stalks of $^1/_8$ inch along slender hairy axis, the rounded apex usually with minute point or notch, the edges not toothed, the upper surface slightly shiny green with minute hairs, and the lower surface paler and soft hairy. A variation has large leaflets to 5 inches long.

Flowers clusters (racemes) at bases of upper leaves bear many flowers on slender hairy stalks. The calyx has 5 unequal rounded greenish sepals about $^1/_4$ inch long.; 5 elliptic unequal yellow petals stalked at base, 4 about $^3/_8$ inch long and 1 about $^5/_8$ inch; 7 stamens and 3 sterile stamens (staminodes); and slender curved pistil with long narrow 1–celled ovary and short style. There are many dull brown oblong seeds $^3/_{16}$ inch long. Collected with flowers in June and October and with fruits from October to March.

The sapwood is yellow and the heartwood dark brown. The hard wood formerly was used elsewhere as a dyewood.

This species has served in home remedies.

Rare in lower Karnataka at 500–1,000 feet altitude in south central foothills of Nilgiri.

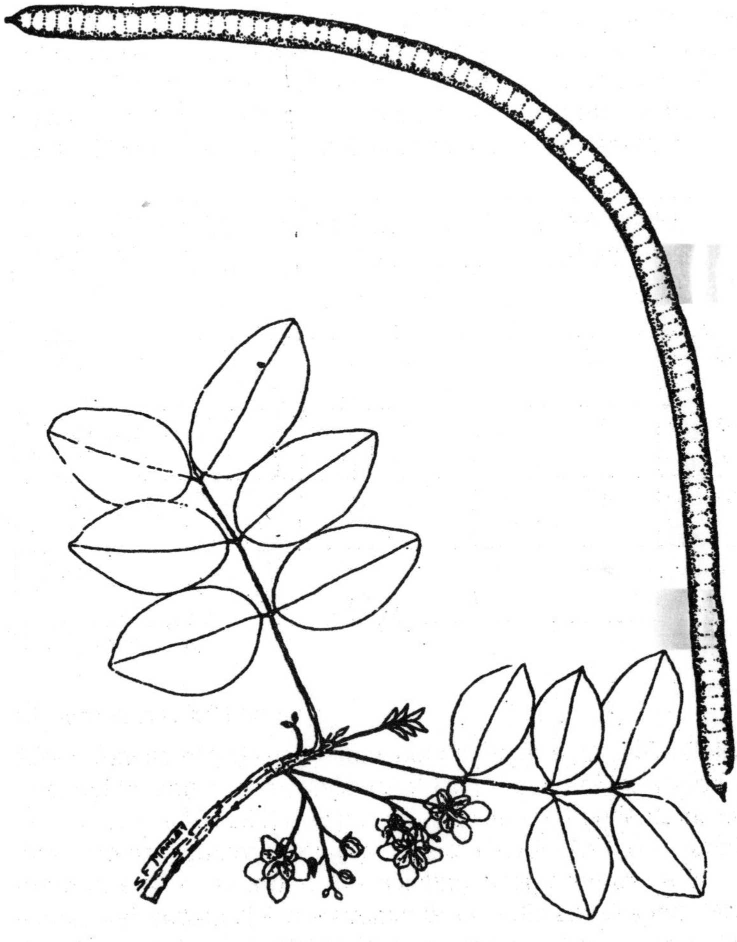

**Fig. 74 : Velamuerto (*Cassia emarginata L*)**
Fruit (above), flowering twig (below), natural size.

## 75. Canafistula cimarrona, pinkshower cassia (*Cassia grandis*)

Canafistula cimarrona is a rare tree with showy pink flowers. Its distinguishing characters are : (1) the even pinnate leaves with 20–40 paired oblong leaflets densely soft hairy; (2) the large drooping clusters of flowers about 1 inch across the 5 rounded pink to purplish petals which turn to salmon in age; and (3) the very large and heavy cylindric dark brown pods 15–20 inches long and about $1^1/_4$ inches in diameter, slightly rough and cracked.

Medium–sized deciduous tree 45 feet high and 1 foot in trunk diameter or larger, with irregular spreading crown, evergreen where planted in humid areas. The bark is light brownish gray, smoothish with many fine warts and some horizontal lines. Inner bark is orange brown, gritty and slightly bitter. The twigs are long, slightly drooping to nearly horizontal, light green, finely hairy with rusty or gray hairs and with 3 longitudinal ridges below each leaf. Buds are composed of very young rusty hairy leaves.

The even pinnate leaves are alternate in 2 rows and slightly drooping on the long twigs, 6–12 inches long, soft hairy with minute hairs. The paired stipules are tiny and inconspicuous. The light green axis is slender and grooved above. Leaflets drooping slightly on short hairy stalks $^1/_{16}$ inch long are $1–2^1/_4$ inches long and $^1/_4–^3/_4$ inch wide, rounded with minute point at apex, rounded or slightly notched and unequal at base, thin, with straight edges. The upper surface green and slightly shiny, with many fine slightly sunken side veins and the lower surface is dull light green. New growth of young leaves is yellow green and finely hairy.

Flower cluster (racemes) at leaf bases are 4–8 inches long, bearing many flowers on slender finely hairy stalks $^3/_8–^3/_4$ inch long. The flower is composed of 5–lobed greenish calyx $^1/_4$ inch long; nearly equal rounded pink to purplish petals about $^1/_2$ inch long; 10 stamens including 3 curved and longer than petals, others much shorter, the 3 smallest nonfunctional; and pistil with stalked, slender, hairy curved 1–celled ovary and style.

The very large pods hang down and do not open. They are short pointed at both ends, have 1 ridge along one edge and 2

**Fig. 75 : Canafistula cimarrona, pinkshower cassia (*Cassia grandis*)**

Leafy twig (above), flowers (lower left), natural size.

ridges on the other. Within are many thin cross wall $^1/_4$ inch or less apart, each with a seed in dark brown sweetish pulp or liquid. The seed are elliptic and flattened, light brown, $^5/_8$ inch long. Flowering in spring and maturing fruits in summer.

The wood is composed of thick whitish or brownish sapwood and heartwood variegated brown with light and dark, sometimes purplish, streaks and patches. It is of medium weight, hard and tough, coarse–textured, with straight to very irregular grain. The handsome wood has been used elsewhere in cabinetwork and construction.

The sweetish but ill–smelling pulp of the pods of this and related species has been used in home remedies as a laxative.

Apparently introduced in India though reported by East India Co. as native in dry forests of southwestern part. Rarely planted as an ornamental in towns and roadsides.

## 76. Glauca cassia (*Cassia planisiliqua* L)

A showy introduced ornamental shrub or small tree is identified by : (1) pinnate leaves with 8–20 paired oblong or elliptic leaflets whitish beneath, with a minute stalked gland between the lowest pairs; (2) bright yellow flowers about 2 inches broad, with 5 spreading petals slightly unequal and 10 nearly equal stamens; and (3) flat pods–48 inches long and $^3/_8$–$^5/_8$ inch wide, ending in a curved bristle.

Evergreen shrub or small tree to 15 feet high and 3 inches in trunk diameter and with open crown of slender spreading branches. Bark light gray, smooth. Twigs finely hairy.

Leaves alternate, pinnate, 2–8 inches long, with threadlike paired stipules shedding early and with slender finely hairy axis bearing 8–20 paired leaflets on short hairy stalks less than $^1/_{16}$ inch long and a minute stalked gland between each of the lowest pairs. Leaflets 1–3$^1/_2$ inches long and $^3/_4$–1$^1/_2$ inches wide, blunt and often slightly notched at apex, rounded at base, not toothed on edges, thin, the upper surface light green and becoming nearly hairless, and the lower surface whitish and slightly hairy with pressed hairs.

Flower clusters (racemes) lateral, long–stalked, flat–topped. Flowers several, composed of yellow 5–lobed calyx $^3/_8$ inch long; 5 spreading elliptic bright yellow petals about 1 inch long and slightly unequal and narrowed at base; 10 nearly equal fertile stamens; and pistil with narrow curved hairy ovary and slender style. The pods are dark brown, flat, thin, nearly hairless. Seeds several, beanlike, $^1/_4$ inch long, elliptic, flattened shiny dark brown. Flowering throughout the year.

Sparingly planted for ornament in coastal cities A.P. A relatively recent introduction, still rare.

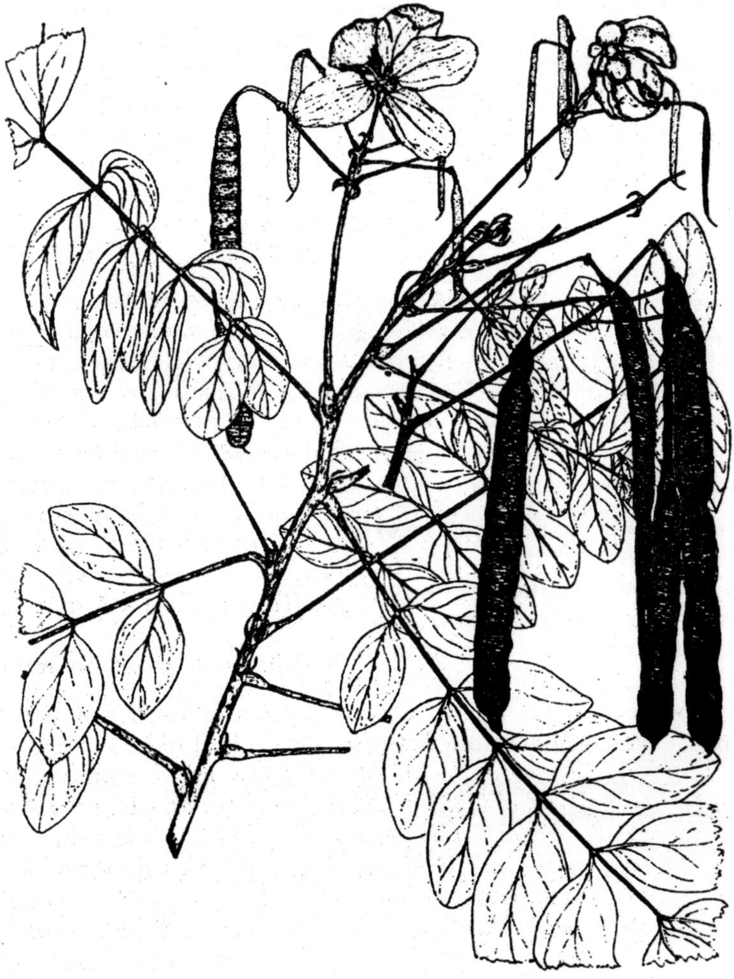

**Fig. 76 : Glauca cassia (*Cassia planisiliqua* L)**
Flowering twig (above), fruits (center right), two–thirds natural size.

## 77. Retama prieta (*Cassia polyphylla*)

This ornamental shrub or small tree with long unbranched wandlike twigs, native in dry areas, is recognized by : (1) showy bright yellow flowers about $1^1/_2$ inches across the 5 spreading elliptic petals; (2) small pinnate leaves 1–2 inches long, composed of 5–15 pairs of elliptic nearly stalkless leaflets mostly $^1/_4$ inch long and $^1/_8$ inch wide, in clusters along the slender twigs; and (3) pods long, narrow, and flattened, 3–6 inches long and $^1/_4$ inch wide, brown to black and shiny.

Shrub or small tree to 15 feet high and 4 inches in trunk diameter, reported to become larger, much branched, with many slender spreading twigs, unbranched, curved and slightly drooping at ends. Probably deciduous in dry areas. Bark in trunk and larger branches blackish, thick, furrowed into short scaly plates. Inner bark light brown and slightly bitter. The young twigs are very slender, dull light green, slightly hairy, the older twigs light brown, warty, and slightly fissured.

The leaves are alternate on rapidly growing twigs but mostly clustered 3–5 at nodes of older twigs. Stipules are paired, threadlike, about $^1/_8$ inch long. The slender light green hairy axis bears leaflets almost to the base. Leaflets are slightly unequal at base and rounded with minute point at apex, thin, with veins inconspicuous, dull green above and light green beneath, becoming nearly hairless, those of a pair folding together at night.

Flowers 2–3 on slender green stalks nearly 1 inch long at leaf bases. The calyx is composed of 5 unequal elliptic yellowish green sepals $^1/_4$–$^3/_8$ inch long; corolla of 5 slightly unequal elliptic petals about 1 inch long, short–stalked at base; stamens 7, unequal, with long brownish anthers, and 3 minute nonfunctioning stamens (staminodes); and pistil with very narrow curved greenish hairy ovary and short stigma. The pods have a short stalk at base and short point at apex, are flattened between the flat seeds, and split open along 2 lines. Flowering and fruiting throughout the year.

The wood is light brown and hard.

Locally common in shrub thickets and dry forests of south coast and lower Nilgiri from sea level to 1,000 feet altitude in Karnataka.

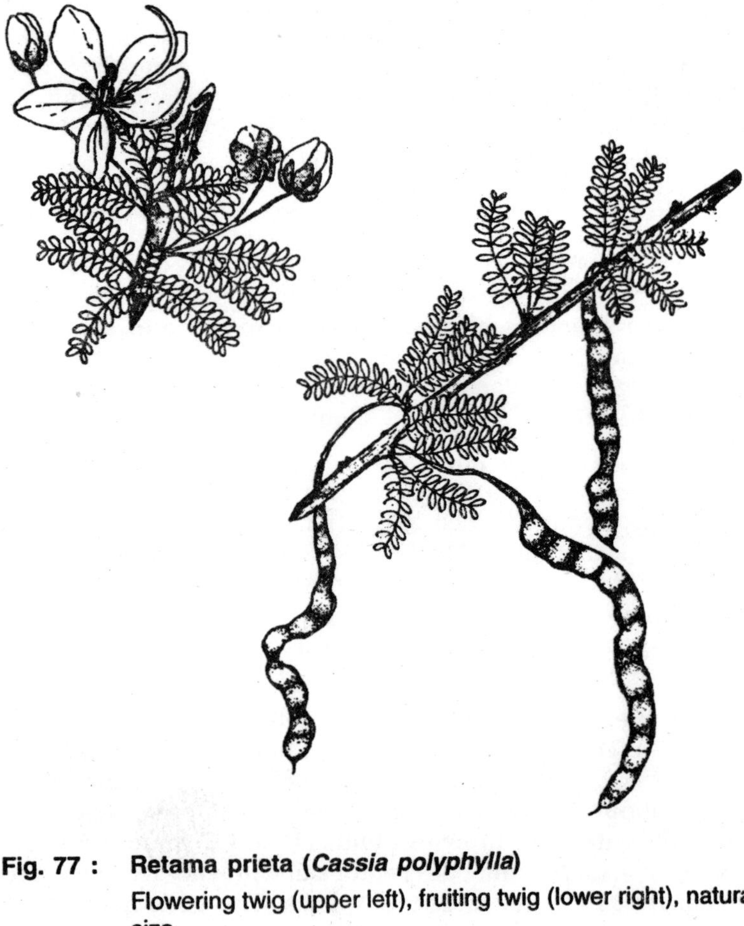

**Fig. 77 : Retama prieta (*Cassia polyphylla*)**

Flowering twig (upper left), fruiting twig (lower right), natural size.

## 78. Calceolaria shower (*Cassia spectabilis*)

This introduced ornamental and shade tree is characterized by : (1) large pinnate leaves 8–18 inches long with 12–30 paired oblong–lanceolate leaflets long–pointed and soft hairy beneath; (2) very large clusters of many showy yellow flowers $1^1/_2$–2 inches across the 5 elliptic spreading petals; and (3) blackish cylindric pods 8–12 inches long and $^3/_8$–$^1/_2$ inch in diameter, with many horizontal walls.

Medium–sized deciduous planted tree to 50 feet high and 1 foot in trunk diameter, with spreading crown. Bark gray, smoothish, with many warts and short fissures. Twigs stout, brown with light dots (lenticels), finely hairy.

Leaves alternate, pinnate, with slender light green finely hairy axis and paired threadlike stipules, without glands. Leaflets with short stipules, without glands. Leaflets with short stalks about $^1/_8$ inch long are $1^1/_2$–$2^1/_2$ inches long and $^5/_8$–$7/_8$ inch wide, long–pointed at apex, short–pointed at base, not toothed on edges, thin, upper surface dull green and almost hairless, with many slightly sunken side veins, and lower surface dull light green and soft hairy or sometimes nearly hairless.

Flower clusters (panicles) terminal and lateral, branched and very large, 6–24 inches long. Flowers many, fragrant, composed of 5 rounded hairy yellow sepals about $^1/_4$ inch long; 5 elliptic slightly unequal yellow petals $^3/_4$–1 inch long, narrowed into stalklike base; stamens 7 large and 3 small sterile (staminodes); and slender curved hairless pistil. Pods ending in a short narrow point, hard, not splitting open or slightly on 1 side, with many cross walls about $^1/_8$ inch or less apart. Seeds 1 in each division, $^1/_8$ inch in diameter and flattened, brown. Flowering throughout the year.

The sapwood is whitish, and the heartwood brown. The wood is described as hard, heavy, and durable.

Uncommon as an ornamental and shade tree and along roadsides, also escaping from cultivation. Very fast growing. Planted in the subtropical moist and lower and upper Cordillera zones in central Puerto Rico from sea level to 2,500 feet altitude and best adapted to dry slopes. A relatively recent introduction in Andaman–Nicobar of land.

This species is planted also as an ornamental and shade tree in southern India, where is flowers in early winter and is rated as excellent.

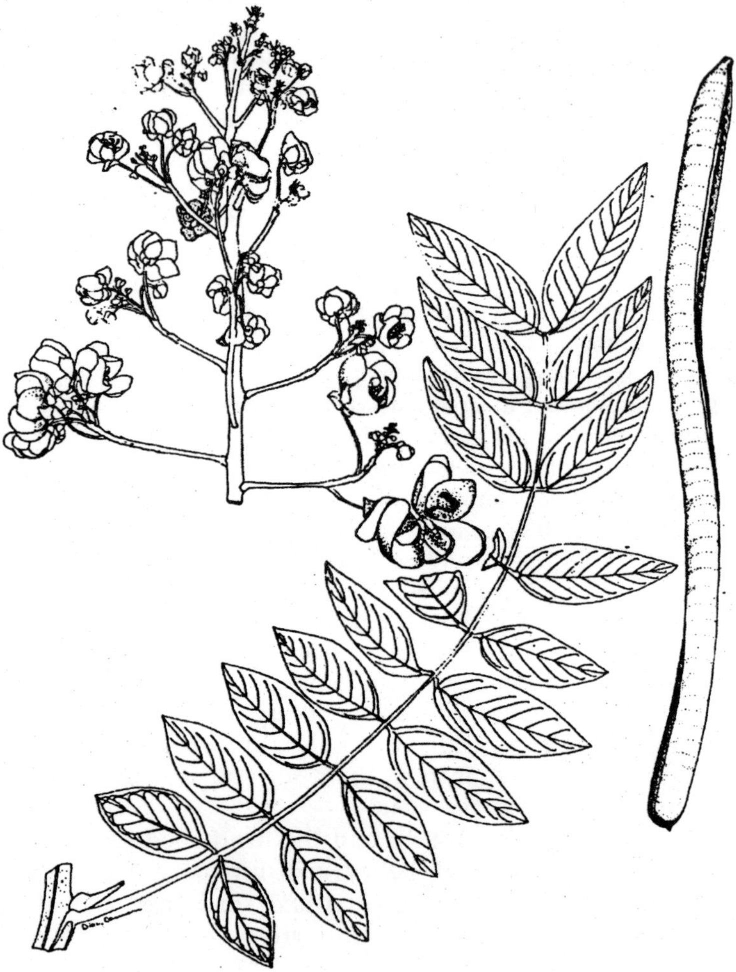

**Fig. 78 : Calceolaria shower (*Cassia spectabilis*)**

Flowers (upper left), leaf (below), fruit (right), two–thirds natural size.

## 79. Oreganillo (*Cynometra portoricensis*)

This rare tree of moist lowland forests is easily recognized by : (1) leaves with small twin or paired leaflets, $^3/_4$–$1^3/_4$ inches long and $^3/_8$–$^5/_8$ inch wide, narrowly oblong or obovate and with unequal sides and slightly notched at the blunt apex; and (2) the oblong pods $^3/_4$–$1^1/_4$ inches long and $^1/_2$–$^5/_8$ inch broad, flat and slightly curved.

Small to medium–sized evergreen tree to 50 feet high and 14 inches in trunk diameter, with thin spreading crown. The bark of small trunks is gray or brownish, smooth, with many dots (lenticels). Inner bark is pinkish, fibrous, and slightly bitter. Twigs are slender, slightly zigzag, dark brown, and finely hairy when young.

The alternate leaves spreading in 2 rows on horizontal twigs have slender dark brown hairy petioles about $^1/_8$ inch long and minute thread–like stipules which shed early. The 2 leaflets are stalkless, obliquely and narrowly oblong or obovate, slightly thickened, hairless, slightly notched at the blunt apex, rounded and unequal at base, the sides very unequal with network of veins, border not toothed, the upper surface dark green and slightly shiny, the lower surface dull light green.

A few small white flowers are borne in short clusters (racemes) at leaf bases, on slender hairy stalks less than $^1/_4$ inch long. The flowers about $^3/_8$ inch long consist of calyx $^5/_{16}$ inch long with 4 or 5 sepals, 5 nearly equal petals about $^3/_8$ inch long, 10 separate stamens, and pistil with hairy ovary and threadlike style. The pods are brown, finely warty, thick–walled and hairy, 1–seeded. Flowering mainly in spring. Fruits maturing in summer, persistent into winter.

The sapwood is whitish and hard.

Rare and local in moist limestone forest of Bastar, including northeastern, north–western and southwestern parts, at 100–600 feet altitude.

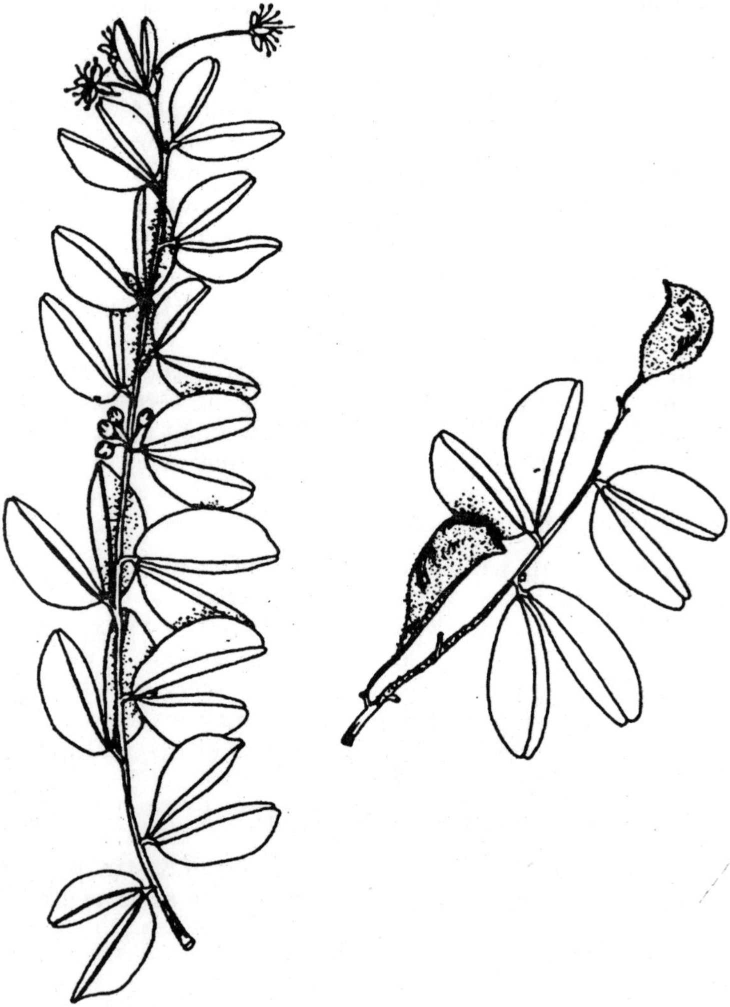

**Fig. 79 : Oreganillo (*Cynometra portoricensis*)**
Flowering twig (left), fruiting twig (right), natural size.

## 80. Campeche, logwood (*Haematoxylum campe–chianum* L*)

Compeche or logwood, at one time an important source of dye, has been planted, escaping from cultivation and becoming naturalized in dry areas of Puerto Rico and the Virgin Islands. It is recognized by : (1) the trunks ridged or angled and fluted, branching near base; (2) even pinnate leaves with 4–8 paired leaflets reversed heart–shaped (obovate) or wede–shaped, notched at apex and tapering to blunt base; (3) light yellow flowers showy, nearly $^1/_2$ inch broad, many in unbranched clusters at base of leaves; and (4) the distinctive oblong light brown pods 1–2$^1/_2$ inches long and $^1/_4$–$^1/_2$ inch wide, thin and papery, splitting open down the middle instead of along edges.

Deciduous small to medium–sized tree becoming 30 feet high, with several short crooked trunks to 2 feet in diameter and with thin spreading crown. The trunks become angled and fluted, with long ridges and grooves. The bark is light brown or gray, smoothish to fissured or furrowed, rough, and slightly shaggy. The inner bark is light brown and gritty. Twigs are gray, smooth with many dots (lenticels), hairless.

The leaves are alternate or 2–4 at older nodes, 2–4 inches long, with slender minutely hairy axis 1–3 inches long. Stipules shed early or sometimes persist as stout spines $^1/_4$–$^5/_8$ inch in length. The paired leaflets on stalks less than $^1/_{16}$ inch long are smallest at base of leaf and largest at apex. Blades are $^1/_2$–1$^1/_2$ inches long and $^3/_8$–1$^1/_4$ inches wide, sometimes smaller, thin, with many fine parallel side veins, hairless. The upper surface is green and slightly shiny, the lower surface dull light green.

Flower clusters (racemes) 1–3 inches long (longer in fruit) at leaf bases bear many fragrant flowers on slender stalks less than $^1/_4$ inch long. The calyx is composed of 5 unequal narrow purplish or reddish sepals about $^3/_{16}$ inch long; the 5 light yellow petals are oblong and slightly unequal, about $^1/_4$ inch long; stamens 10, separate, longer than petals, light yellow, with hairy filaments; and the pistil with 1–celled ovary, 2–3 ovules, curved style, and dot stigma. The pods clustered among the leaves are pointed at both ends and contain 1–3 oblong flat seeds $^5/_{16}$ inch long. Flowering from December to May. Old fruits remain attached for several months.

The thin sapwood is whitish and the heart–wood orange, becoming dark red after exposure. The wood is very hard, heavy (specific gravity 0.95–1.00), medium to fine–textured, of irregular grain, strong but brittle, very resistant to decay, and takes a high polish. Fresh wood has an odor of violets and a sweetish taste. Uses include cabinetwork and posts.

The dye haematoxylin is obtained from the heartwood in small chips by boiling, changing color from orange red to yellowish upon cooling and finally black. By the use of mordants, black and bluish–black shades are obtained. Many shiploads of this valuable dyewood were sent to Europe over a period of centuries before these dyes became largely replaced by synthetic dyes. The wood is still used locally for coloring wool textiles. Other uses are stains for biological microscopic slides and inks. Haematoxylin is also a drug used as an astringent, such as for dysentery and diarrhea. British Honduras was founded as a colony in 1638 mainly for export of mahogany and logwood, which was common in the lowland forests. After introduction into the West Indies, Jamaica, Haiti, and other islands exported the logs.

The trees are grown occasionally as ornamentals and in fences. An important honey plant where common.

Compeche or logwood has been planted and has become naturalized in the dry forests in thickets, and along roadsides in southern and southwestern Karnataka.

**Fig. 80 : Campeche, logwood (*Haematoxylum campechianum* L*)**
Flowering twig (lower left), fruiting twig (upper right), natural size.

## 81. Sisu, Indian rosewood (*Dalbergia sissoo* Roxb.)

Indian rosewood, or sissoo (sisu), from India has been introduced in forestry tests. It is characterized by : (1) pinnate leaves with 3–5 rounded abruptly short–pointed leaflets alternate on a very slender zigzag axis; (2) many small whitish irregular flowers like those of beans, borne in short lateral clusters; and (3) oblong flat thin pods $1^1/_2$–3 inches long and $^3/_8$–$^1/_2$ inch wide.

A medium–sized tree 35 feet high and 1 foot in trunk diameter with rounded open crown of spreading branches, evergreen or almost deciduous. Bark gray, scaly, deeply furrowed. Twigs green to gray, very slender.

Leaves alternate, pinnate, about 6 inches long, hairy when young. Leaflets with stout stalks to $^1/_4$ inch, almost at right angles to axis. Leaflet blades nearly round, 1–2 inches long and broad, rounded or blunt at base, rounded and abruptly, short–pointed at apex, not toothed on edges, thin, with many fine parallel side veins.

Flower clusters (panicles) 2–4 inches long at leaf bases and nodes back of leaves, branching, hairy. Flowers about $^3/_8$ inch long, fragrant, consisting of hairy calyx with 5 teeth; corolla with 5 petals, the standard broad, 2 narrow wings, and narrow keel of 2 united; stamens 9, united into broad stalk, and stalked hairy pistil with narrow 1–celled ovary containing few ovules, short style, and dot stigma. The light brown pod is narrowed at both ends, stalked at base, and does not split open. Seeds 1–3, beanlike, $^1/_4$–$^3/_8$ inch long.

Sissoo belongs to the genus of rosewoods, handsome cabinetwoods with attractive grain and color and roselike fragrance. In the native home the wood has been used for furniture, cabinets, flowing, boats, and wood carvings. Young branches and foliage have served also as fodder.

Recommended as one of the most desirable nonflowering shade trees for central and south India. It grows very fast, is drought resistant and salt tolerant, and is adapted to a wide range of sites including yards and streets, windbreaks, dry ridges, filled areas, and river spoil banks. Cut roots produce suckers. After maturing, the trees should have another use as a cabinetwood.

**Fig. 81 : Sisu, Indian rosewood (*Dalbergia sissoo* Roxb.)**
Fruiting twig (above), flowering twig (below), two–thirds natural size.

## 82. Coraltree (*Erythrina corallodendrum* L.)

This small tree is characterized by : (1) sharp stout spines or prickles to $^3/_4$ inch long usually present on trunk and branches; (2) alternate leaves with 3 wedge–shaped thin leaflets long–pointed at apex and nearly straight at base; (3) large coral red or blood red flowers about 2–2$^1/_2$ inches long, narrow, with bell–shaped calyx; and (3) pods 4–6 inches long and $^3/_8$–$^5/_8$ inch long, narrowed between the several beanlike seeds, which are scarlet with a black spot and poisonous.

Deciduous shrub or small tree to 25 feet high and 8 inches in trunk diameter. The leaves mostly 6–12 inches long, becoming hairless, have slender petioles 1$^1/_2$–6 inches long, often spiny, and leaflet stalks $^1/_8$–$^1/_4$ inch long. Leaflet blades are thin, mostly 2–4 inches long and wide, sometimes to 6 inches, wedge–shaped or broadly ovate, not toothed on edges, with 3 main veins from base, dull green above and paler beneath.

Flower clusters (racemes) 4–12 inches long bear along 1 side many almost stalkless flowers in groups of 2–3. The narrow bell–shaped calyx $^3/_8$–$^5/_8$ inch long is straight at apex; petals 5, coral red or blood red, the narrowly oblong folded standard 2–2$^1/_4$ inches long, the 4 others (wings and keel) very small, mostly $^1/_4$–$^3/_8$ inch long; 10 stamens 1$^1/_2$–2 inches long, united into tube toward base; and pistil 1$^1/_4$–2 inches long, very narrow, consisting of finely hairy stalked ovary and style. The pods are beadlike and slightly flattened, long–stalked and long–pointed. Seeds are about $^3/_8$ inch long. Flowering mostly in late winter to March, usually when leafless, and maturing fruits mostly in spring.

Rate in southern coast and lower Western Ghat forest of Kerala at 100–500 feet altitude, mostly in dry areas. In cultivation for the showy flowers and as a living fence.

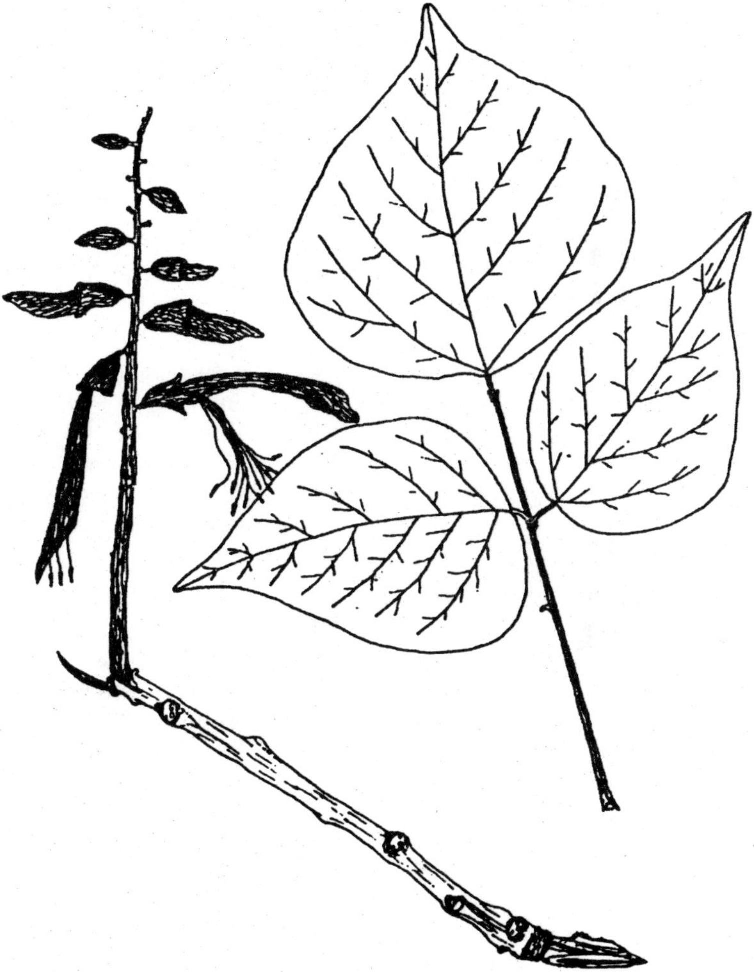

**Fig. 82 : Coraltree (*Erythrina corallodendrum* L.)**
Flowers (left), leaf (right), natural size.

## 83. Cockscomb Coralbean (*Erythrina crista–galli* L*)

A spiny shrub or small tree planted as an ornamental for its showy clusters of large brilliant crimson or scarlet flowers in early spring when leafless. It is distinguished by : (1) curved spines or prickles on twigs and leaves; (2) leaves with 3 large elliptic or ovate leaflets, the lower surfaces whitish and waxy and often with spines on midvein; (3) 1–3 large crimson flowers $1^1/_4$–$1^3/_4$ inches long, with 1 broad spreading petal; and (4) narrowly cylindric pods 6–12 inches long and $^1/_2$–$^5/_8$ inch broad, hard–walled, with several elliptic seeds black with brown markings (not red), probably poisonous.

A planted ornamental deciduous shrub or small tree to 15 feet high, with short trunk 4 inches in diameter and with slender branches and thin crown, often vinelike. Twigs slender, often spiny.

Leaves alternate, compound 4–10 inches long, becoming hairless, with spiny petiole 2–5 inches long and 3 leaflets on stalks of $^1/_4$–$^5/_8$ inch. Leaflet blades are elliptic or ovate, $2^1/_2$–$3^1/_2$ inches long and 1–$1^1/_2$ inches broad, the end one largest and with the longest stalk, short–pointed or blunt at apex and base, not toothed on edges, thin, the upper surface green, and the lower surface whitish and waxy.

The irregular flowers 1–3 in groups along end of twig or at base of leaf on slender stalks $^1/_2$–$2^1/_2$ inches long are composed of bell–shaped reddish brown calyx unequally 5–toothed; corolla of 5 crimson petals including 1 very broad spreading petal (standard) $1^1/_4$–$1^3/_4$ inches long and 1 inch broad, 2 short side petals (wings) $^3/_8$–$^1/_2$ inch long, and 2 long narrow petals (keel) $1^1/_4$–2 inches long; stamens 9 in a tube and 1 separate; and pistil with stalked narrow ovary and slender style. The pods are long–pointed at both ends, slightly narrowed between the few seeds $^1/_2$–$^3/_4$ inch long. Flowering in early spring when leafless, the fruits maturing in spring and persisting. The branches die back after flowering.

The wood is described as soft and very lightweight.

The bark has been used in medicines.

Planted occasionally for ornament in parks and gardens in India and the Islands and elsewhere in West Indies. Northward

**Fig. 83 : Cockscomb Coralbean (*Erythrina crista–galli* L*)**
Flowers (above), leaf (below), two–thirds natural size.

in warm temperate climates, as in Southern States, the shrubby plants bloom in summer and die back nearly to the roots in winter. In cold climates the large fleshy roots can be stored indoors in winter or the plants grown in greenhouses. Horticultural varieties have different shades of red flowers.

In central and south India this species is recommended as a beautiful small flowering tree for yards, parks, and gardens. It is propagated by cuttings or air layering and grows rapidly in moist fertile soil.

## 84. Piñón espinoso, cockspur (*Erythrina eggersii*)

This very spiny small tree, sometimes vinelike, is native only in Puerto Rico and the Virgin Islands. It is identified by : (1) abundant stout spines or prickles $^1/_8$–$^1/_4$ inch long on leaves, twig, and trunk; (2) alternate leaves with 3 wedge–shaped thin leaflets short–pointed at apex and nearly straight at base, bearing spines or prickles $^1/_8$ inch long of petiole and main veins on both surfaces; (3) showy red flowers $1^1/_2$–$2^1/_2$ inches long but narrow, with bell–shaped calyx, and (4) pods 4–8 inches long and $^5/_8$ inch wide, narrowed between the several beanlike seeds, which are yellow or bright red, without a black spot, probably poisonous.

Deciduous shrub or small tree to 30 feet high and 5 inches in trunk diameter, or sometimes vinelike. The bark is brown, slightly fissured with many spines $^1/_4$ inch long. Twigs are stout, green, becoming gray, very spiny, hairless.

The hairless leaves are mostly 5–10 inches long, with slender spiny petioles 2–5 inches long and axis 1–$1^1/_2$ inches longer. Leaflet stalks are $^1/_8$–$^1/_4$ inch long, with minute paired glands at base. Leaflet blades are thin, mostly 2–4 inches long and wide, wedge–shaped or broadly ovate, rounded to short–pointed at apex, mostly straight or notched at base, not toothed on edges, with 3 main veins from base, dull green above and paler beneath.

Flower cluster (racemes) bear many short–stalked long narrow flowers. The narrow cylindric calyx $^3/_8$–$^1/_2$ inch long is straight or minutely toothed at apex; petals 5, red, the narrowly oblong folded standard $1^1/_2$–$2^1/_2$ inches long, the 4 others (wings and keel) very small, $^3/_{16}$ inch long, 10 stamens $1^1/_4$–$1^3/_4$ inches long, united into tube toward base; and pistil $1^1/_4$–$1^3/_4$ inches long, very narrow, consisting of finely hairy stalked ovary and slender style. The pods end in a long point. Seed beanlike, elliptic, $^1/_2$ inch long. Flowering in late winter from December to March, maturing fruits in spring and summer.

This native plant is sometimes grown as an ornamental for the showy flowers and as a living fence.

Uncommon in moist limestone forests at 100–140 feet altitude in northern Jharkhand.

**Fig. 84 : Pinon espinoso, cockspur (*Erythrina eggersii*)**

Leaf (left), flowers (upper right), fruit (lower right), two–thirds natural size.

## 85. Bucare, Immortelle (*Erythrina variegata* L*)

An introduced ornamental distinguished by : (1) sharp small inconspicuous blackish spines or prickles $^1/_{16}$–$^1/_8$ inch long like pins, usually present along twigs and persistent on trunk; (2) alternate leaves with 3 wedge–shaped thin leaflets about as broad as long, short–pointed at apex and nearly straight at base, dull green, paler beneath; (3) large showy scarlet flowers about $2^1/_2$ inches long but narrow, with prominent spreading stamens, crowded in erect clusters; and (4) stout blackish pods 6–12 inches long and $^3/_4$–1 inch broad, slightly narrowed between the several dark red, poisonous seeds.

A deciduous small tree 30 feet high and 1 foot in trunk diameter or in age and where native a large tree, with many erect and spreading branches. The bark is light greenish gray, smoothish, becoming slightly streaked and fissured, with scattered short spines $^1/_8$ inch long. Inner bark beneath a thin light green outer layer is light brown and almost tasteless. Twigs are light green with minute hairs when young, becoming light gray, smooth except for scattered sharp blackish spines or prickles and with slightly raised half–round leaf scars and light colored dots (lenticels).

The leaves 7–11 inches long, with star–shaped hairs but becoming nearly hairless, have slender round light green petioles 3–5 inches long, enlarged at base, and extended nearly 1 inch as an axis. Leaflet stalks are about $^3/_8$ inch long, with minute paired green glands at base. Leaflet blades are thin, mostly 2–5 inches long and wide, wedge–shaped or broadly ovate, not toothed on edges, with 3 main veins from base.

Flower clusters (racemes) about 6–8 inches long bear numerous crowded flowers, with smaller buds toward apex. The narrow tubular, pointed reddish calyx about 1 inch long splits open nearly to base on 1 side; petals 5, scarlet, the oblong folded standard erect or nearly so, $2^1/_4$–$2^1/_2$ inches long, the 2 elliptic wings and keel about $^5/_8$ inch long; 10 stamens about 2 inches long, 9 united into a tube toward base; and the pistil $1^3/_4$ inches long, very narrow, composed of finely hairy stalked ovary and slender style. The beanlike red seeds are $^5/_8$–$^3/_4$ inch long. Flowering mostly in winter, sometimes to April. Fruits in spring, seldom maturing.

The wood is light brown and soft.

Leaves, bark, and fresh flowers have served elsewhere in home remedies.

Uncommon as an introduced ornamental along roadsides and in fencerows up to 200 feet altitude in northeastern Orissa. A favorite ornamental in Assam known as immortelle. In southern and central India under the name Indian coralbean, this showy large spreading tree with brilliant red blossoms is conspicuous in early spring and has been described as one of the gems of the floral world. In Hawaii the trees are called tigers–claw from the resemblance of the large folded petal.

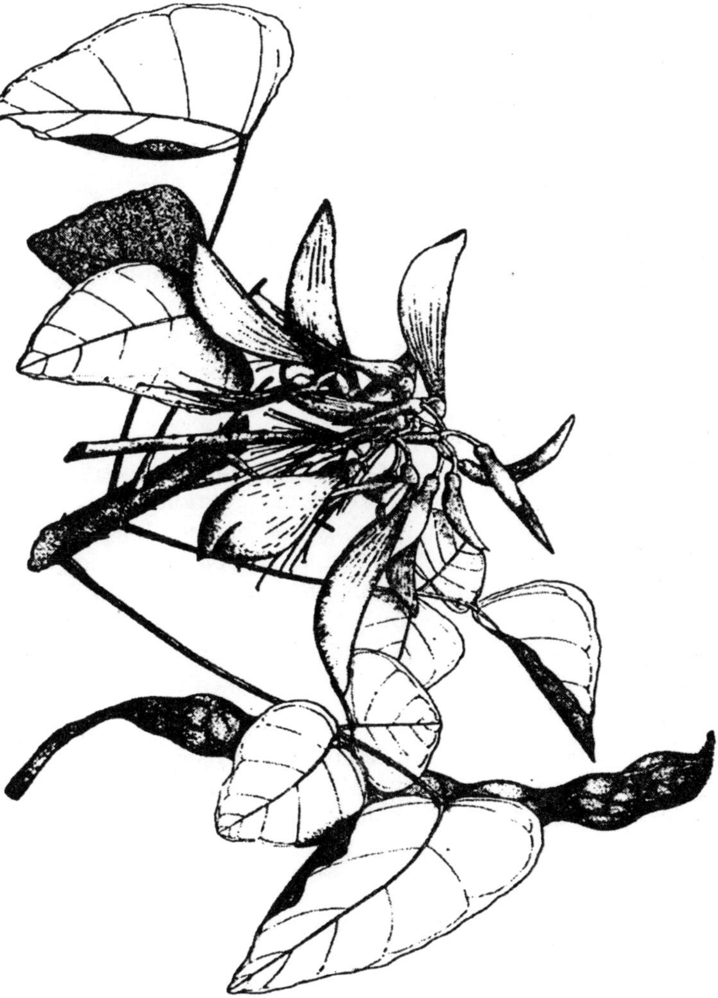

**Fig. 85 : Bucare, immortelle (*Erythrina variegata L**)**

Leafy twig and flowers (above), fruit (below), two–thirds natural size.

## 86. Geno–geno (*Lonchocarpus domingensis*)

A spreading tree of southern and western India recognized by : (1) alternate pinnate leaves with 7–11 elliptic leaflets mostly short–pointed at both ends, paired except at end; (2) pale rose–violet pea–shaped flowers about $^5/_8$ inch long, in long narrow lateral clusters; and (3) brown pod 2–6 inches long and $^5/_8$–$^7/_8$ inch wide, flat but hard or leathery, often narrowed between the seeds.

A deciduous tree to 70 feet high and 20 inches in trunk diameter. The bark is brown with whitish dots and fibrous. The young twigs and young leaves are finely brown hairy.

The leaves are 6–10 inches long, the leaflets with stalks $^1/_8$–$^3/_{16}$ inch long. Leaflet blades are 2–5 inches long and $1^1/_4$–$2^1/_2$ inches wide, mostly short–pointed at both ends but sometimes long–pointed at apex and rounded at base, not toothed on edges, slightly thickened, becoming nearly hairless.

The flower clusters (racemes or panicles) are nearly as long as the leaves, with main axis bearing many flowers on finely hairy stalks about $^1/_4$ inch long, often 2 forking from the same basal stalk. The bells–shaped calyx is $^3/_{16}$ inch long and broad, finely hairy, minutely toothed at apex; the pale rose–violet corolla sometimes whitish has 5 petals, the rounded standard notched at paex and silky hairy on outside, 2 oblong wings and 2 oblong, hairy, slightly united keel petals, 10 stamens with filaments united; and narrow pistil with short–stalked brown hairy ovary and slender curved style.

The pods usually are finely brown hairy, stalked at base, do not split open, and contain 1 to several flat kidney–shaped dark brown seeds $^1/_2$ inch long. Flowering mostly in spring, also to September, the fruits maturing mostly in summer and persisting.

The wood is whitish or yellowish and strong; used chiefly for posts.

A shade and ornamental tree in dry areas and a honey plant. Elsewhere the fibrous bark has been utilized for ropes and cords and the leaves have been employed in home remedies.

Uncommon in southern and western coastal hills at 100–800 feet altitude in southern and western India.

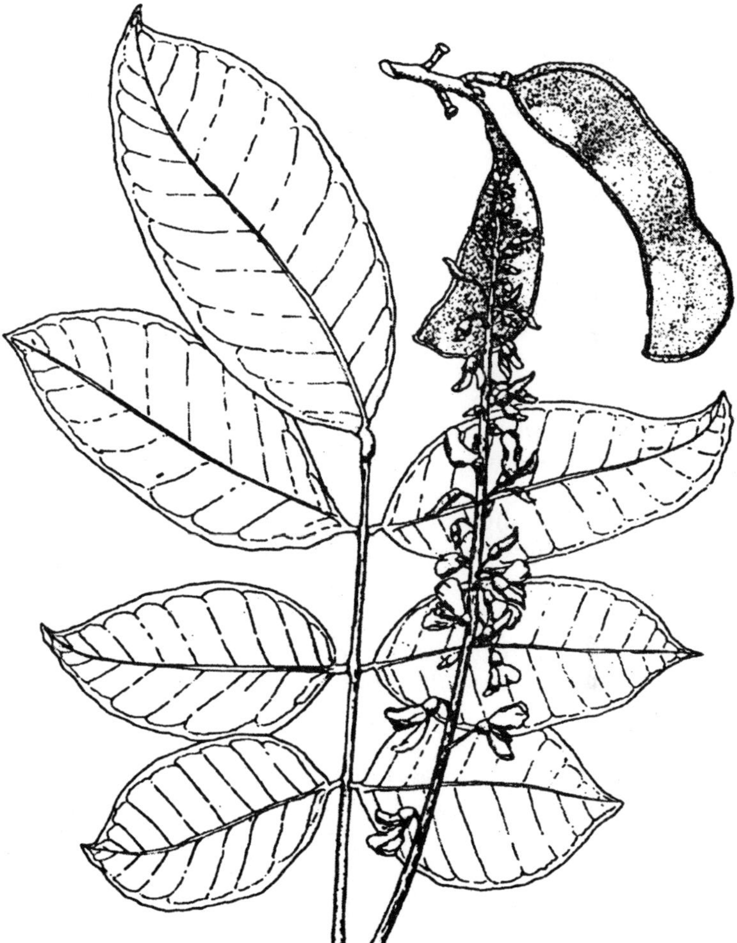

**Fig. 86 : Geno–geno (*Lonchocarpus domingensis*)**

Fruits (upper right), flowering twig (below), two–thirds natural size.

## 87. Geno (*Lonchocarpus glaucifolius*)

Geno, a tree known only from limestone forests of northwestern Bihar, is identified by : (1) alternate pinnate leaves with usually 7, sometimes 5, elliptic to obovate leaflets with blunt slightly notched apex, whitish green beneath; (2) showy pinkish purple sea–shaped flowers $^5/_8$ inch long, several in slender–stalked clusters at leaf bases; and (3) flat brown pods 2–5$^1/_2$ inches long and $^1/_2$–$^3/_4$ inch wide, slightly thickened, not narrowed between the several flat beanlike seeds, not splitting open.

Evergreen tree to 30 feet high and 6 inches in trunk diameter. Bark gray or brown, smoothish, the inner bark whitish, fibrous, and slightly bitter. The twigs are brown with light, raised dots (lenticels), finely hairy when young but becoming hairless.

The alternate pinnate leaves are 6–9 inches long, hairless, and have a very slender light green axis and paired triangular hairy stipules $^1/_{16}$ inch long. The leaflets, usually 7, sometimes 5, are paired except at end, on stalks $^1/_8$ inch long. Leaflet blades are 1$^1/_2$–4 inches long, $^3/_4$–1$^1/_2$ inches wide, short–pointed at base, minutely wavy at edges, slightly thick and stiff, the upper surface green and slightly shiny, the lower surface dull whitish green, the slightly curved side veins forming a fine network upon drying.

Flower clusters (panicles) about 2 inches long at leaf bases bears several flowers on slender stalks about $^1/_4$ inch long, often paired. The pea–shaped flowers have a cup–shaped 5–toothed finely hairy calyx $^3/_{16}$ inch long; pinkish–purple corolla of 5 stalked spreading petals about $^5/_8$ inch long, the rounded standard notched at apex and finely brown hairy on outside, 2 oblong wings, and 2 forming the keel; 10 stamens with filaments united, 1 nearly separate; and pistil with narrow hairy ovary and curved style. The pods with calyx at base are abruptly pointed at both ends, minutely hairy. The flat blackish kidney–shaped seeds are about $^5/_{16}$ inch long. Flowering mainly in late spring, sometimes in winter and summer. Fruits maturing mostly in summer and persisting.

The whitish wood is hard. The strong fibrous bark reportedly has been used as cordage.

Locally abundant in the moist limestone forest at 200–800 feet altitude in northwestern Bihar. Scattered to common on ridges and cliffs.

**Fig. 87 : Geno (*Lonchocarpus glaucifolius*)**
Leaf (left), flowers (upper right), fruits (lower right), natural size.

## 88. Escambrón (*Machaerium lunatum*)

Escambrón, a spiny climbing shrub or small tree uncommon in coastal thickets, is easily recognized by: (1) climbing branches with paired curved brown spines (stipules) $^1/_8$–$^3/_8$ inch long; (2) pinnate leaves with 5–11 oblong leaflets; (3) many small purplish pea–shaped flowers $^3/_8$ inch long; and (4) the distinctive thick flattened gray pods curved into a circle 1–1$^1/_2$ inches in diameter.

Evergreen climbing shrub 15 feet high with many trunks, sometimes a small tree 20–25 feet high and 8 inches in diameter., with arching, climbing, and spreading spiny branches, the plants form impenetrable thickets along coasts. Young twigs finely hairy.

The alternate leaves 2–4 inches long have slender finely hairy petiole and axis 1–2$^1/_2$ inches long. Paired sharp spines develop from stipules at the base of each leaf and persist on twigs and branches. The leaflets are alternate on slender hairy stalks $^1/_{16}$ inch long. Leaflet blades are oblong, or the end one obovate, $^3/_4$–2 inches long and $^3/_8$–$^3/_4$ inch wide, thin, rounded at both ends or short–pointed at base, not toothed at edges, with many fine parallel side viens, the upper surface green and hairless, and the lower surface light green with minute hairs.

Flowers clusters (panicles) at ends and sides of twigs are 2$^1/_2$–6 inches long, with finely hairy branches. Flowers many, short–stalked, irregular and pea–shaped, consisting of hairy bell–shaped 5–toothed calyx $^3/_{16}$ inch long; 5 purplish petals about $^5/_{16}$ inch long and stalked at base, the rounded standard curved back, 2 wings, and 2 curved keel petals united at end; 10 stamens nearly $^1/_2$ inch long, united into a tube about $^2/_3$ their length; and stalked curved pistil nearly $^1/_2$ inch long with narrow flat hairy 1–celled 1–ovuled ovary and slender style.

The thick leathery pods $^3/_4$ inch wide are curved into a circle 1–1$^1/_2$ inches in diameter and $^3/_{16}$ inch thick. They do not split open and contain 1 flat kidney–shaped seed $^3/_4$ inch long. With flowers and fruits irregularly through the year.

Locally common in coastal swamps, forming thickets in and near mangrove swamp forests at sea level, mostly in eastern Sundarbans. Commonly a shrub but recorded as a slender tree 25 feet high in the wet forest north of West Bengal.

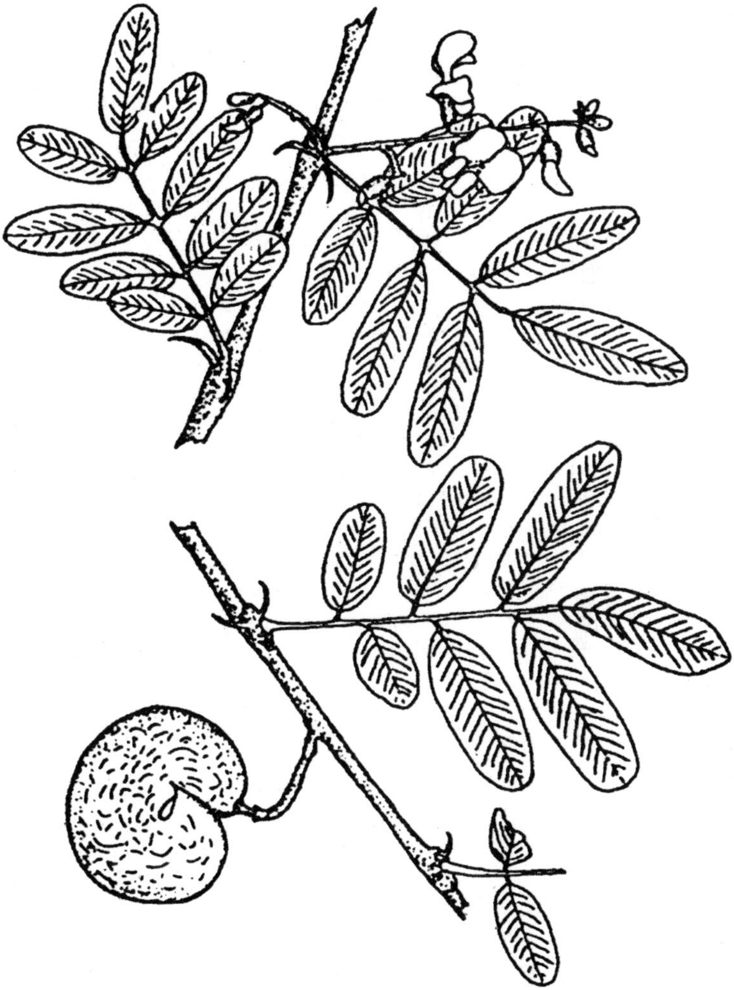

**Fig. 88 : Escambron (*Machaerium lunatum*)**
Leafy twig (above), fruiting twig (below), natural size.

## 89. Ventura, Dogwood (*Piscidia carthagenensis*)

This medium–sized tree of dry coastal areas is characterized by: (1) pinnate leaves with usually 7 or 9 elliptic leaflets 2–4 inches long, beneath gray green and finely hairy; (2) the large branched clusters pinkish pea–shaped flowers about $^{5}/_{8}$ inch long; and (3) the distinctive light brown pod 2–4 inches long with 4 longitudinal membranous wings mostly more than 1 inch across.

A deciduous tree becoming 35 feet high and 1 foot in trunk diameter, with smooth gray thick bark. The inner bark is light yellow, streaked with tan, and bitter. Twigs are gray green, finely hairy, becoming gray brown with whitish dots and dashes (lenticels). The pointed buds $^{1}/_{4}$ inch or less in length are covered by paired broad hairy scales (stipules), which shed very early, leaving scars at leaf bases.

The alternate pinnate leaves are about 7–12 inches long. The slender, finely hairy, gray–green leaf axis is enlarged at the base and bears usually 7 or 9 (5–11) leaflets, paired except at end, each with a stalk $^{1}/_{4}$ inch long. Leaflet blades are elliptic, mostly 2–4 inches long and $1^{1}/_{4}$–2 inches wide, sometimes larger, thin, with apex narrowed into a minute point, rounded base, border straight or slightly wavy, the upper surface dark green, hairless and slightly shiny with many nearly straight parallel lateral veins, and the lower surface dull gray green and finely hairy.

Flower clusters (panicles), often before the leaves, are lateral, branched, 3–12 inches long, and bear many flowers on slender stalks. The pea–shaped flower is composed of bell–shaped purplish or pinkish hairy calyx $^{3}/_{16}$ inch long, slightly 5–toothed and finely hairy; corolla of 5 unequal pale pinkish petals about $^{5}/_{8}$ inch long, finely hairy, consisting of the broad rounded spreading standard, 2 narrow wings, and 2 united and forming the curved keel; 10 curved stamens within keel and united into a narrow tube; and pistil consisting of narrow greenish hairy 1–celled ovary and slender bent style.

The fruit is an odd–shaped pod composed of long stalk above the calyx at base, slender body, and 4 longitudinal, broad membranous or paper wing about $^{1}/_{2}$ inch wide. It contains several oblong bean–shaped dull brown seeds $^{1}/_{4}$ inch long but does not split open. Flowering from February to June, the fruits maturing in early summer and persistent.

The wood is described as hard, heavy, tough, and strong, the sapwood whitish and heartwood yellowish brown. Elsewhere the wood of related species has served for boatbuilding, vehicle construction, posts, and charcoal.

Indians, including Caribs in the West Indies, threw the root bark, young branches, and powdered leaves of this and related species into the water to aid in catching fish. The fish are stupefied, rise to the surface, and float but recover if not caught. This use is indicated in the English common name and the generic name. The bark of this and related species has served also in home remedies. The trees sometimes are grown in fences as ornamentals.

Locally common in thickets along coast and in coastal hills from sea level to 700 feet altitude in southern and eastern Tamil Nadu. Also all larger islands eastward.

**Fig. 89 : Ventura, dogwood (*Piscidia carthagenensis*)**

Leafy twig (above), fruit and flowers (lower right), natural size.

## 90. Pterocarpus, Burma padauk (*Pterocarpus macrocarpus*)

This exotic ornamental, shade, and timber tree is identified by: (1) long drooping branches; (2) reddish latex; (3) pinnate leaves alternate in 2 rows with usually 7–11 oblong to ovate thin shiny green leaflets borne singly, short–pointed at apex; (4) showy yellow peashaped flowers $^{5}/_{8}$ inch long profusely borne in lateral clusters; and (5) nearly round flat light brown pod 2–$2^{3}/_{4}$ inches in diameter, bordered by a broad membranous wing, stalked at base and with minute pointed style at one side. It differs from the related species No. 89, India padauk (*Pterocarpus indicus* Wild.*), in the larger pods and the shorter–pointed mostly oblong leaflets.

A large spreading tree becoming 80 feet tall and $2^{1}/_{2}$ feet in trunk diameter with small buttresses and with broad dense crown of long branches, some nearly touching the ground. The light brown bark is slightly furrowed and a little rough and shaggy. Inner bark is streaked reddish brown and light brown, astringent, and produces a small amount of reddish sticky bitter latex. The green twigs minutely hairy at apex bear 2 rows of slightly dropping leaves. The trees are almost evergreen but with few leaves for a short period in spring when old leaves fall and young leaves develop.

The leaves 9–16 inches long have 7–11 leaflets borne singly along a light green aix enlarged at base. Flowering branches often have fewer leaflets, sometimes only 5–1. Leaflets on stalks about $^{5}/_{16}$ inch long have oblong to ovate blades (or the lowest nearly round) $2^{1}/_{2}$–5 inches long and 2–3 inches wide, sometimes larger, short–pointed at apex and broadly pointed to nearly straight at base, not toothed on edges, turned up at midrib, slightly shiny green above, and dull light green beneath.

Flower clusters (mostly panicles or racemes) 3–8 inches long bear several to many fragrant flowers on slender green stalks. Flowers are about $^{5}/_{8}$ inch long, with green bell–shaped finely hairy calyx $^{5}/_{16}$ inch long, unequally 5–toothed; 5 yellow petals $^{5}/_{8}$ inch or less in length, becoming crinkled including nearly round standard more than $^{1}/_{2}$ inch long and rolled back, 2 wings, and 2 smaller keel petals united a little at sides; 10 stamens nearly $^{3}/_{8}$ inch long, the whitish filaments united toward base; and greenish hairy pistil $^{7}/_{16}$ inch long, with stalked narrow 1–celled ovary and tapering style.

The odd fruits, yellow green when immature, turning to light brown, have a roughened central body with raised veins containing 2–4, usually 3, seeds but not opening to release them. The narrow curved brown seeds are less than $^{1}/_{2}$ inch long. At the base are remains of calyx and a stalk about $^{3}/_{8}$ inch long. Recorded in flower in June and July and in fruit in August and in January and February.

The wood is light brown and hard. Reported to be an important timber tree of Assam, Manipur, Tripura and a valuable cabinetwood.

Sparingly planted in W.B. for ornament and shade, having been introduced about 1923. Its use for these purposes is increasing, and it might become very popular. The shade of the foliage is so dense that grass does not grow beneath the crown of a large tree.

**Fig. 90 : Pterocarpus, Burma padauk (*Pterocarpus macrocarpus*)**
Flowers (upper left), leafy twig (center), fruits (lower right), two–thirds natural size.

## 91. Retama, wattapania (*Sabinea florida*)

Retama or wattapania, on back cover, is a handsome shrub or small tree with wandlike branches. It is recognized by: (1) pinnate leaves 2–4 inches long, with mostly 8–20 pairs of oblong or elliptic leaflets $^1/_4$–$^5/_8$ inch long and $^1/_8$–$^1/_4$ inch wide, bristle–tipped, and nearly stalkless; (2) many showy pale purple pea–shaped flowers about $^7/_8$ inch long and $1^1/_4$ inches across, usually clustered along old unbranched twigs when leafless; and (3) dark brown very narrow pod 3–4 inches long and $^1/_4$ inch wide, flattened, and twisting when splitting open.

A deciduous shrub or small tree 12–20 feet tall with trunk 2–4 inches in diameter, sometimes to 40 feet high with trunk to 8 inches, with long slender spreading branches. The bark is gray and fissured, the inner bark whitish and tasteless. The twigs are brown and finely hairy.

Though borne singly or alternate, the leaves on very short lateral twigs may appear to be paired or more numerous at a node. There are 2 brown narrow scales (stipules) $^3/_{16}$ inch long at base. The slender green axis is minutely hairy and ends in a tiny point beyond last pair of leaflets. Leaflets are rounded at both ends with minute bristle tip at apex, thin, dark green and hairless above, and beneath pale green and finely hairy or hairless.

Flowers have slender stalks $^1/_4$–$^1/_2$ inch long, sometimes single. The bell–shaped purple calyx is $^3/_{16}$ inch long and minutely 5–toothed; the 5 pale purple, lavender, or blue petals nearly $^7/_8$ inch long and stalked at base, consisting of broad rounded standard, 2 wings, and 2 slightly united keel petals; 10 white stamens, 9 united into a tube near base and 1 separate, 5 half as long as others; and pistil with stalked narrow ovary and long curved reddish purple style. The pod is stalked at base and pointed at both ends. Seeds several, elliptic, flat, $^1/_8$ inch long. In flower mostly from January to March and with mature fruits in spring and summer.

The whitish wood is hard.

When in flower the entire plant is a mass of bluish purple. A small tree with white flowers. This as one of the most elegant of endemic species in the region.

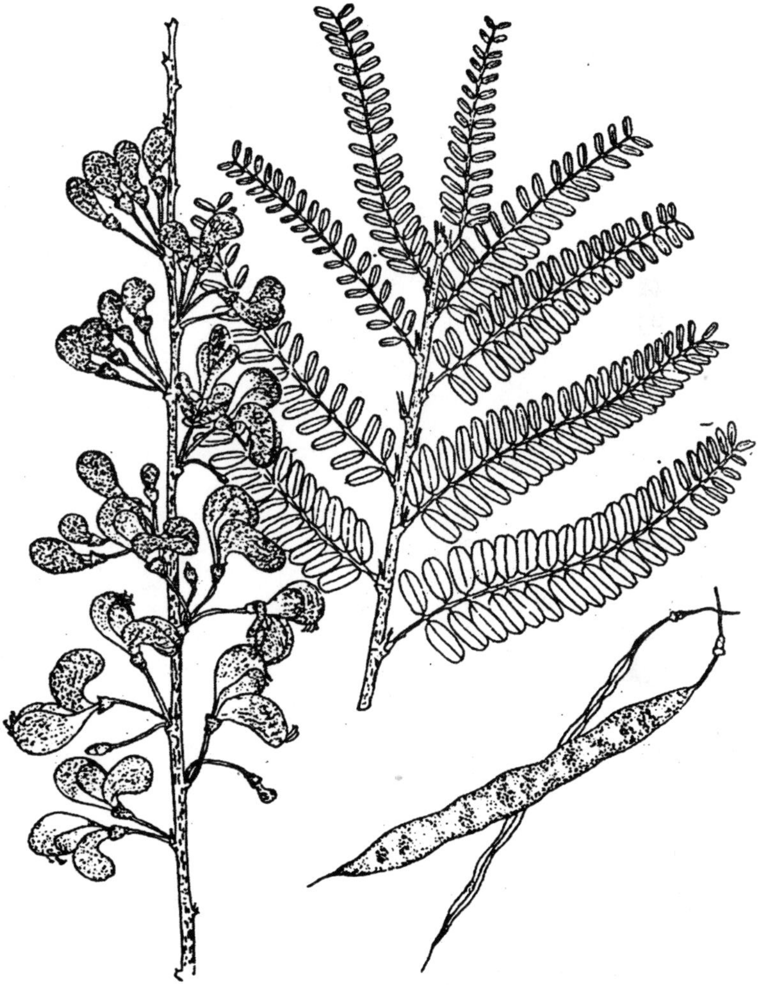

**Fig. 91 : Retama, wattapania (*Sabinea florida*)**

Flowers (left), leafy twig (upper right), fruits (lower right), natural size.

Sparingly planted as an ornamental for the masses of beautiful flowers which are very conspicuous along the leafless branches but soon shedding. Certainly worthy of wider cultivation.

Locally common in moist coastal, moist limestone, and lower forests at 100–2,500.

## OXALIS FAMILY (OXALIDACEAE)

Herbs and shrubs, rarely trees (*Averrhoa*), known by: (1) sour sap; (2) leaves alternate, pinnately or palmately compound or rarely simple, entire, without stipules; (3) flowers in clusters (like umbels or cymes) or solitary, bisexual, regular, the calyx of 5 sepals, corolla of 5 petals sometimes united at base, 10 stamens (or 5 reduced to staminodes) in 2 series, and pistil composed of superior 5–celled ovary with axile placentation and 1 or more ovules in each cell, and 5 separate persistent styles with dotlike stigmas; and (4) fruit a columnar 5–angled 5–celled capsule with many seeds, sometimes with covering (aril).

## 92. Carambola (*Averrhoa carambola* L*)

Carambola is rarely planted for its odd, very sour, juicy, yellow fruits $2^1/_2$–4 inches long and $1^1/_2$–2 inches broad, oblong and sharply 5–angles. Other characters for identification are : (1) pinnate leaves with 7, 9, or 11 ovate leaflets mostly paired except at end, spreading in 2 rows on long twigs; and (2) small purplish 5–parted bell–shaped flowers $^3/_8$ inch long and broad, many on dark red branched stalks at leaf bases and on branches.

Evergreen small tree becoming 25 feet high, with a few trunks from base to 6 inches in diameter. The bark is light brown, smoothish to slightly cracked, the inner bark pink brown, fibrous and tasteless. The long branches are light brown and smooth. Twigs are brown and finely hairy when young, ending in a minute hairy leaf.

The alternate pinnate leaves nearly horizontal in 2 rows are 6–10 inches long including the slender light green or pinkish axis and 7, 9, or 11 drooping leaflets. The leaflets have short stalks $^1/_2$ inch long and fold together in pairs at night. Blades are 1–4 inches long and $^3/_4$–$1^3/_4$ inches wide, the lowest being smallest and end one largest. They are short–pointed at apex, rounded but unequal at base, not toothed at edges, thin, slightly curved up from midvein, with few inconspicuous side veins, nearly hairless, the upper surface green and slightly shiny, and the lower surface dull light green.

Flower clusters (panicles) 1–2 inches long at base of older leaves and back of leaves bear many flowers on short slender dark red stalks. The flower consists of 5 oblong pink sepals rounded to wavy at apex; corolla of 5 spreading purple petals nearly $^1/_4$ inch long, slightly united by edges near base; 5 stamens alternate with petals and 5 smaller sterile stamens; and pistil with deeply 5–angled 5–celled ovary and 5 slender styles pressed together.

The heavy berries hang down in groups or singly from slender stalks along the twigs, bending down the twigs and branches by their weight. The shape is deeply 5–angled and grooved (star–shaped in cross section), narrowed to 5 blunt points at apex, slightly notched at base. Color changes from yellow–green to yellow and orange yellow. The thick light yellow flesh is soft, juicy, and sour. There are 5 small cells and several brown

**Fig. 92 : Jalia Carambola (*Averrhoa carambola* L*)**
Fruit (left), and flowering twig, two–thirds natural size.

flat seeds $^3/_8$–$^1/_2$ inch long with yellow flesh (aril) at end. The fruits do not open but fall to the ground, maturing from July to September.

The sour fruits are eaten raw and used for marmalade and pickles. The flavor becomes less acid and more pleasant at maturity. Leaves and fruits have served in home remedies.

The wood is whitish and soft.

Carambola is cultivated in India and the Islands for the edible fruits. It is also a handsome ornamental.

# COCA FAMILY (ERYTHROXYLACEAE)

Shrubs and small trees, known by: (1) alternate simple leaves entire to wavy–toothed, hairless, often with 1 pale line on each side of midvein (from folds in bud), with minute pointed stipule in the angle above the petiole like a bud; (2) minute inconspicuous flowers 1 or several at base of leaves, bisexual, regular, in form of bell or wheel, with persistent calyx of 5 sepals or lobes, 5 small petals with 2 minute lobes within, 10 stamens in 2 series united into short tube toward base, and pistil with superior 3–celled ovary, 1 cell developing with 1–2 ovules, and 3 styles; and (3) fruit a reddish berrylike drupe, sour, 1–seeded.

**Key to Species**

A. Leaves mostly 2–4 inchers long.

B. Leaves thin, with 2 faint lines nearly parallel with midrib, dull dark green above, and pale whitish green beneath—Indio, *Erythroxylum aerolatum* L.

BB. Leaves slightly thickened, without faint lines, shiny dark green above with network of veins when dry, and beneath paler and slightly rusty tinged—*Erythroxylum rufum.*

AA. Leaves mostly less than $1^1/_2$ inches long.

C. Leaves thin, rounded or notched at apex, dull green to dark green above—*Erythroxylum rotundifolium.*

CC. Leaves thick and stiff notched at apex, slightly shiny green above, with network of veins when dry—*Erythroxylum urbanii.*

## 93. Rocio, brisselet (*Erythroxylum rotundifolium*)

This common shrub or small tree widespread at low altitudes is identified by: (1) small obovate to rounded thin leaves $^1/_4$–$1^1/_2$ inches long and $^1/_4$–$^3/_4$ inch wide, alternate and crowded along short side twigs; (2) small whitish flowers about $^1/_4$ inch across, 1–4 short stalks at leaf bases; and (3) oblong shiny dark red fleshy fruits $^1/_4$ inch long.

Deciduous shrub or sometimes a small tree to 25 feet high and 4 inches in trunk diameter, hairless throughout. Bark gray or light brown, smooth, becoming finely fissured, the inner bark pinkish and slightly bitter. The gray twigs are much branched often regularly, some long and slender and many short spurs, roughened by persistent stipules.

The leaves have slender leafstalks $^1/_{16}$–$^3/_{16}$ inch long. The pointed stipules $^1/_{16}$ inch long inserted 1 above leaf base form the buds and persist as crowded scales on the short side twigs. Blades are rounded or slightly notched at apex, short–pointed at base, thin, dull green to dark green on upper surface, and light green beneath.

The flowers have slender stalk less than $^1/_8$ inch long, 5–toothed calyx, 5 white petals $^1/_8$ inch long with 2 lobes (ligules) near middle of inner side, 10 stamens united into a tube toward base, and pistil with ovary and 3 styles. The fleshy fruit (drupe) has calyx persistent at base and becomes dry and brown. Flowering in summer and maturing fruits in late summer.

The wood is light brown or whitish and hard.

The fragrant flowers are attractive to bees level to 2,500 feet altitude nearly throughout India.

**Fig. 93 : Rocio, brisselet (*Erythroxylum rotundifolium*)**
Flowering twig (above), fruiting twig (below), natural size.

## 94. Papelillo (*Erythroxylum rufum*)

This rare small trees of western Puerto Rico is identified : (1) elliptic to obovate leaves with large notch at the rounded apex; (2) small 5–parted flowers about $^1/_4$ inches across on slender stalks at leaf bases; and (3) oblong red fleshy fruits $^1/_2$–$^5/_8$ inch long.

Evergreen shrub or small tree to 20 feet high and 4 inches in trunk diameter, with horizontal branches, hairless throughout. Bark gray, smooth to slightly fissured. Twigs brown, with raised dots (lenticels), ending in pointed bud $^1/_4$ inch long formed by stipules.

Leaves alternate, mostly on short side twigs. Stipule a pointed scale $^3/_{16}$ inch long inserted above base of slender leafstalk $^1/_4$–$^3/_8$ inch long and often persistent on twig back of leaves. Blades $1^1/_2$–4 inches long, and 1–$2^1/_4$ inches wide, with large notch at rounded apex, short–pointed at base, slightly thickened and turned under at edges, the upper surface shiny dark green, with network of veins when dry, the lower surface dull, paler and slightly rusty tinged, often with 2 faint lines parallel with midvein.

Flowers few or 1 at leaf bases and back of leaves have slender stalks $^1/_2$–$^3/_4$ inch long slightly thickened above, 5–toothed calyx $^1/_{16}$ inch long, 5 spreading petals $^3/_{16}$ inch long, 10 stamens united into a tube toward base, and pistil with 3–celled ovary and 3 styles. The fruit (drupe) has calyx at base, thin pulp, and 1 seed. Collected with fruits in June and July.

Rare in moist limestone forest from Kerala west and western lower forest to 2,500 feet altitude in western Ghat.

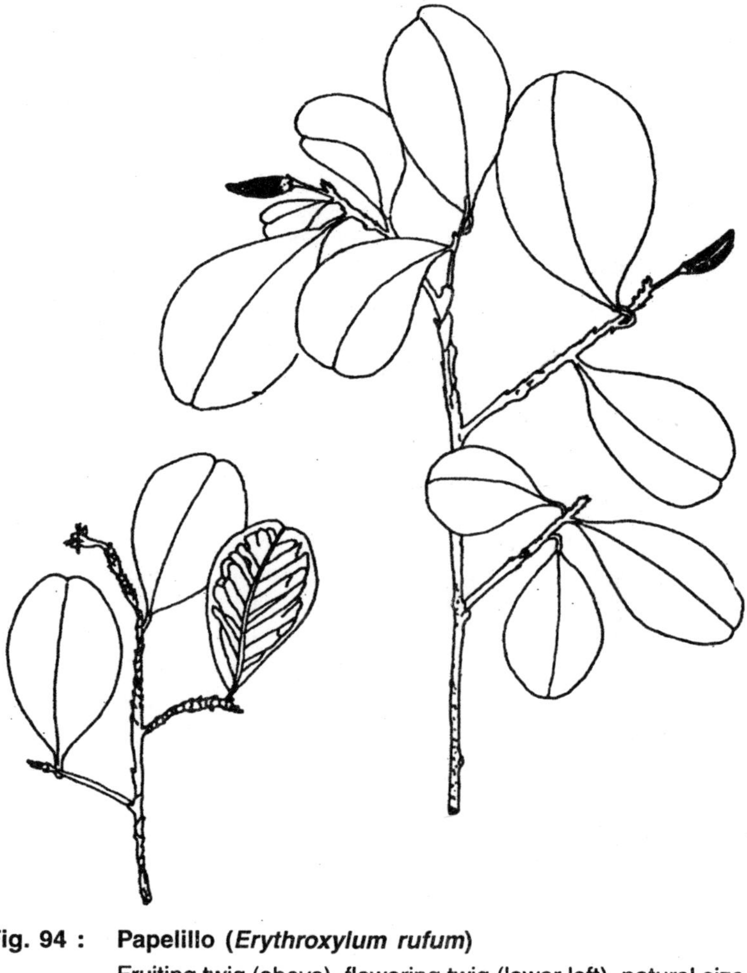

**Fig. 94 : Papelillo (*Erythroxylum rufum*)**
Fruiting twig (above), flowering twig (lower left), natural size.

## 95. Urban (*Erythroxylum urbanii*)

A shrub or small tree of limestone hills of northern and western Nilgiri, identified by: (1) small obovate or elliptic, thick and stiff leaves $^3/_4$–$1^3/_4$ inches long and $^3/_8$–1 inch wide, notched at apex in 2 rows on twigs, which also are in 2 rows; (2) small whitish 5–parted flowers $^3/_{16}$ inch across, single or few and almost stalkless at leaf bases; and (3) oblong orange–red fleshy fruits $^3/_8$ inch long.

Evergreen shrub or sometimes a small tree to 20 feet high and 3 inches in trunk diameter, hairless throughout. Bark gray, becoming fissured. The inner bark is light pink and slightly bitter. Twigs both long and short, the short twigs roughened by crowded persistent stipules.

Leaves alternate in 2 rows but also crowded on short side twigs. Stipule a pointed scale $^1/_{16}$ inch long ending in 2 bristles, inserted above base of the leafstalk $^1/_{16}$–$^1/_8$ inch long and often persistent. The blades are short–pointed at base, thick and stiff, slightly bent up at midvein and turned under at edges, the upper surface slightly green, with network of veins when dry, and the lower surface dull light green.

The flowers on stalks $^1/_8$ inch long are composed of greenish 5–toothed calyx, 5 spreading white petals $^1/_8$ inch long with 2 lobes (ligules) near middle of inner side, 10 stamens united into a tube toward base, and pistil with ovary and 3 short styles. The fruit (drupe) has calyx at base, juicy orange pulp and 1 oblong yellowish seed $^1/_4$ inch long. With flowers in spring and summer and with fruits in August.

The wood is light brown and hard.

Uncommon and local in moist limestone and western lower Nilgiri forests at 300–2,500 feet altitude in northern and western Ghat.

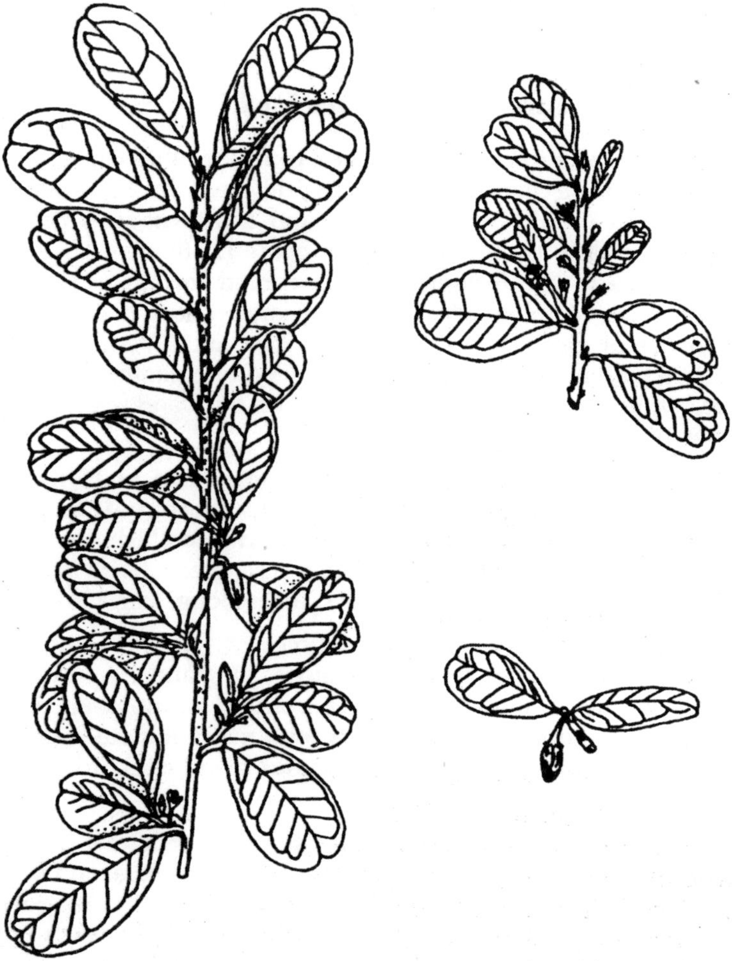

**Fig. 95 : Urban (*Erythroxylum urbanii*)**

Twig with flowers and immature fruits (left), flowering twig (upper right), fruiting twig (lower right), natural size.

# RUE FAMILY (RUTACEAE)

## 96. Cidra, citron (*Citrus medica* L*)

Cidra or citron is distinguished form the other citrus fruits by the following characters : (1) large dull green oblong to elliptic leaves with usually rounded apex and short wingless petiole which generally is not jointed with blade (2) flowers $1^1/_2$–$1^3/_4$ inches across, with 5 whitish petals curved inward at tip and purplish tinged on outside; and (3) the large oblong or cuylindric yellowish or greenish fruit 5–8 inches long and $3^1/_2$–6 inches in diameter, blunt–pointed at apex, wrinkled or smooth, with very thick whitish peel or rind $1^1/_2$–$2^1/_2$ inches thick and scant or dry bitter pulp.

An aromatic evergreen shrub or small tree to 15 feet in height and 3 inches in trunk diameter, with short trunk and long irregular branches. The bark is brown and smoothish, the inner bark light brown and slightly bitter. The green twigs usually have large stout green spines $^3/_8$–$1^1/_4$ inches long, single at nodes.

The alternate leaves have petioles $^1/_8$–$^1/_4$ inch long and large oblong to elliptic blades 3–7 inches long and $1^1/_4$–2 inches wide, the apex sometimes short–pointed or notched, base rounded or short–pointed, borders with many minute rounded teeth, thick and leathery, with numerous gland dots, dull green above and light green beneath.

The flowers are short compact terminal or lateral clusters (racemes) or 3–10 or sometimes solitary, some bisexual and some male, the buds purplish tinged. The cuplike light green calyx $^3/_{16}$ inch long is 5–toothed; the 5 spreading oblong petals $^3/_4$–$1^1/_4$ inches log; about 30–40 or more stamens united in lower part into a white tube; and pistil on disk base with green 9–13–celled ovary tapering into the stout yellowish style. The fruit (berry) contains whitish, pointed, elliptic seed $^3/_8$ inch long in a pale yellow pulp. The rind is firm and slightly bitter. Flowering and fruiting irregularly.

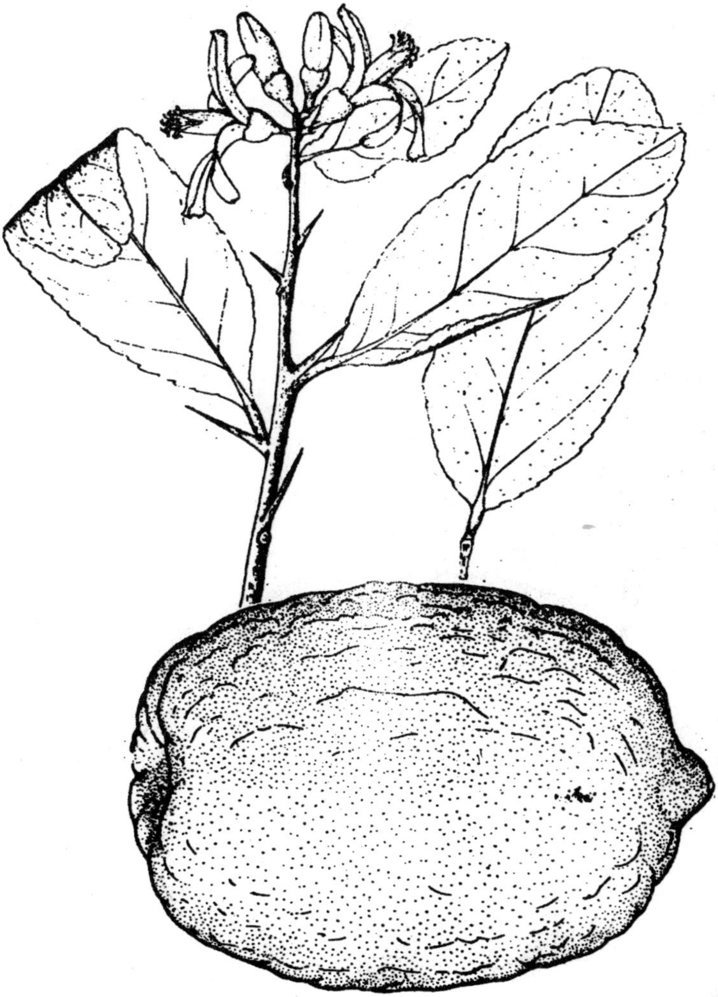

**Fig. 96 :** **Cidra, citron (*Citrus medica* L*)**
Flowering twig (above), fruit (below), natural size.

The wood is light brown and hard and not used in Puerto Rico.

The commercial part of the fruit is the thick peel or rind, which is packed in brine for shipment and afterwards candied and made into preserves and confections. The Mediterranean region exports quantities of the rind to the United States and other countries. Citron is more sensitive to cold than related species. The flowers of this and other species of citrus produce nectar in quantities and are an important source of honey.

Citron is grown locally in plantations in India, chiefly near the coast and in the central mountains but is uncommon.

## 97. Mandarina, mandarin orange (*Citrus reticulata*)

Mandarina or mandarin orange, known also as tangerine is recognized by its aromatic fruit and foliage and is distinguished from other citrus trees by: (1) elliptic to lance–shaped leaves varying from rounded to pointed at both ends, and petiole narrowly margined but not winged; (2) fragrant small white flowers $^3/_4$–1 inch across; and (3) the fruit, which is deep orange or reddish orange, rounded but much flattened and sunken at base and apex, 2–3 inches in diameter, with thin skin loose and easily removed, and hollow core of orange or reddish, sweetish acid flesh easily separated into fibrous–walled sections.

A small evergreen tree 15 feet high and 3 inches in trunk diameter. The bark is brown and smoothish, the inner bark pale greenish and yellowish, and bitter. The slender green twigs, angled when young, are spiny or spineless.

Leaves alternate with petioles $^1/_4$–$^3/_8$ inch long. The blade is $1^1/_2$–4 inches long and $^3/_4$–$1^3/_4$ inches wide, jointed to petiole, with many minute rounded teeth at edges, slightly leathery and with numerous tiny gland dots, dark green and dull or slightly shiny above and beneath paler. The apex is long–pointed in one variety.

There are 1–4 flowers terminal or at leaf base. The calyx is $^1/_8$ inch long and has 5 pointed lobes turned back; the 5 white petals are about $^3/_8$ inch long; about 20 stamens united at base; and pistil with 8–15–celled ovary, long style, and rounded stigma.

The distinctive fruit of mandarin orange or tangerine is composed of 8–15 sections and has the pulp walls more fibrous than in other species. The many whitish pointed seeds are about $^5/_{16}$ inch long, not noticeably flattened. Flowering in spring and maturing fruits in winter.

The whitish hard wood is not used.

A few races of this species occasionally are planted as fruit trees through the moist areas of India. Classed also as an ornamental of graceful form and fine foliage in southern India.

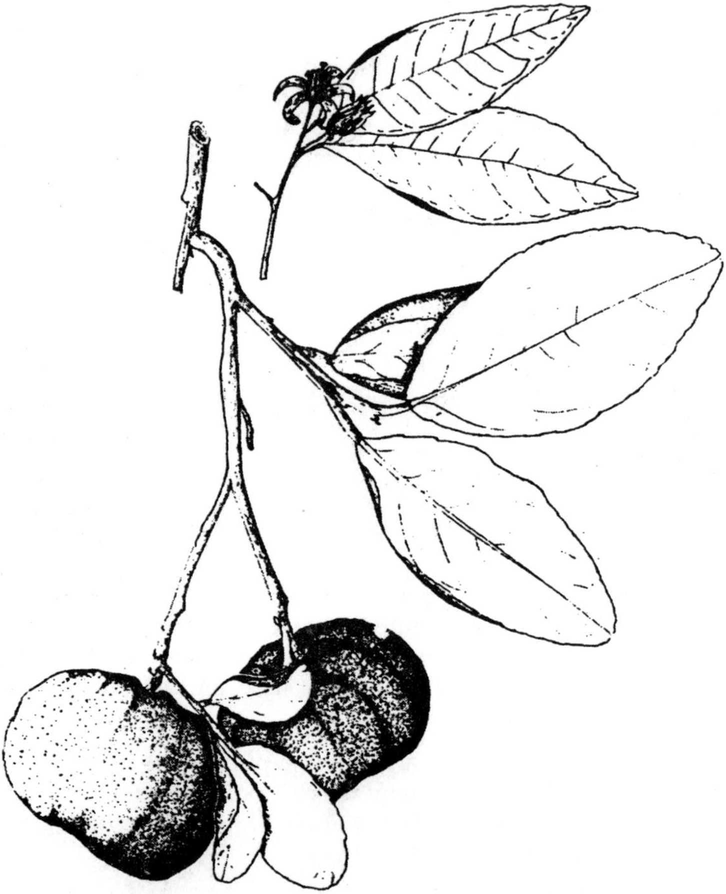

**Fig. 97 : Mandarina, mandarin orange (*Citrus reticulata*)**
Flowering twig (above), fruiting twig (below), natural size.

## 98. Kumquat, kumquat (*Fortunella margarita*)

Kumquat, a relative of the citrus fruits, is sometimes planted in Puerto Rico for its smaller edible fruits and for ornament. Its distinguishing characters are : (1) leaves lanceolate, $1^1/_2$–$3^1/_2$ inches long and $^1/_2$–$1^1/_4$ inches wide, blunt or short–pointed at both ends, with many minute rounded teeth at edges, slightly thickened, and shiny dark green with gland dots, with petiole jointed to blade; (2) small aromatic flowers mostly solitary at leaf bases, $^5/_8$ inch across the 5 spreading white petals; and (3) fruits the dibble kumquat, egg–shaped or oblong $1^1/_4$–$1^3/_4$ inches long and $^3/_4$–$1^1/_8$ inches broad, yellow or yellow orange, with edible peel.

An evergreen aromatic bushy shrub or small tree to 15 feet high with compact crown of the same width. Twigs green, angled, hairless, with minute gland dots, mostly without spines.

The alternate hairless leaves have stout petioles $^1/_4$–$^5/_8$ inch long, nearly round. Blades have inconspicuous veins and are dull light green on lower surface.

Flowers are solitary or few at leaf bases on stalks of about $^1/_4$ inch and develop from angular buds. The calyx is minute, with 5 teeth; corolla of 5 narrow white petals less than $^1/_2$ inch long; 16 or 20 stamens in bundles; and on a disk the pistil with rounded usually 4–5 celled ovary with 2 ovules in each cell, short persistent style, and elliptic stigma.

The fruit (berry or hesperidium) known as kumquat resembles a small orange or lemon, covered with gland dots, composed of light yellow wall about $^1/_8$ inch thick, mostly 4–5 cells with juicy yellow pulp. Seeds usually 1 in each cell, elliptic, whitish $^3/_8$ inch long.

The entire fruit except the seeds is edible, the outer rind or wall being slightly spicy from the aromatic oil, the whitish inner rind sweetish, and the pulp sour like a lemon. Also, marmalade and jelly are prepared from the fruit.

Uncommon as a cultivated fruit tree and ornamental in Puerto Rico. The plants can be trimmed as hedges.

**Range :** Known only in cultivation but apparently native in southeastern China. Planted in tropical and subtropical regions of the world, including southern border of Indian States.

Kumquats are grown across the southern border of States from. The plants are hardier than most fruits and can be grown farther north, also as potted plants. Sprays of twigs with both leaves and fruits are sold for holiday decorations.

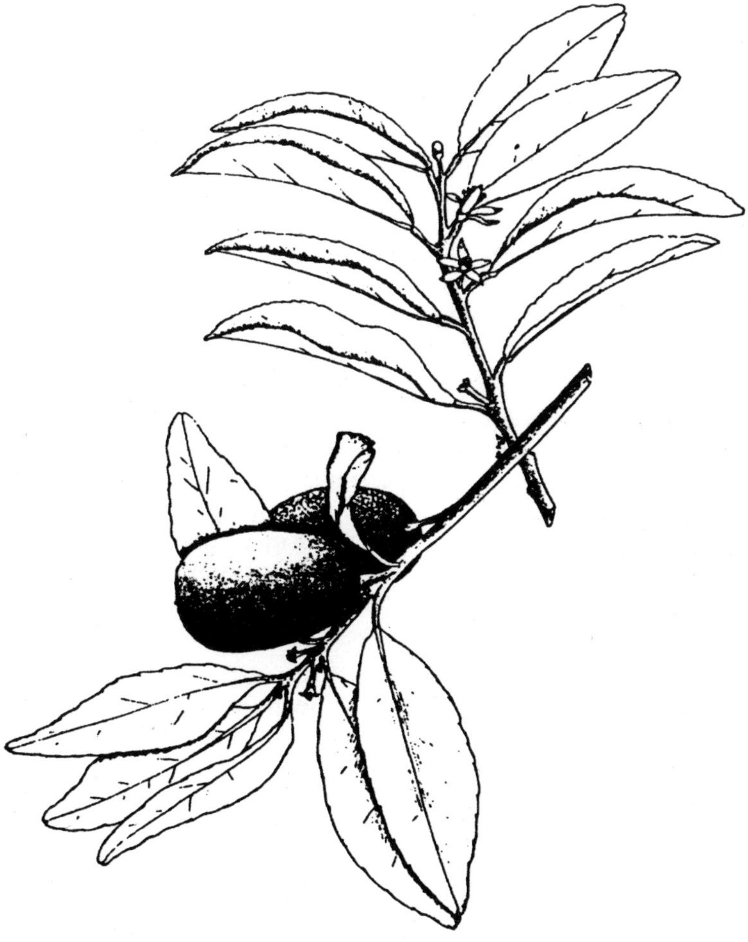

**Fig. 98 : Kumquat, kumquat (*Fortunella margarita*)**
Flowering twig (above), fruiting twig (below), natural size.

## 99. Orange–jessamine (*Murraya paniculata*)

Orange–jessamine is a handsome ornamental shrub common as a hedge and sometimes reaching tree size. It is recognized by : (1) small shiny dark green pinnate leaves $1^1/_2$–4 inches long, with 3–9 elliptic to obovate leaflets finely wavy margined, with minute gland dots, with spicy citrus taste and odor when crushed; (2) fragrant white flowers $^3/_4$ inch across the 5 narrow curved petals, few and opening singly; and (3) showy shiny red fruits, elliptic or egg–shaped, $^3/_8$–$^1/_2$ inch long.

Evergreen planted shrub in hedges 5–10 feet high or becoming a small tree to 25 feet high and 5 inches in trunk diameter, with spreading dense crown. The bark is gray, becoming fissured, rough, and shreddy. Inner bark is light brown and almost tasteless. Twigs are slender, green and finely hairy when young, afterwards light gray and finely fissured.

The alternate leaves are erect and stiff, pinnate, having a slender green, finely hairy axis $^3/_4$–3 inches long and 3–9 alternate leaflets. The leaflets on slender stalks $^1/_{16}$–$^1/_8$ inch long are $^1/_2$–$1^1/_2$ inches long and $^1/_4$–$7/_8$ inch wide, elliptic to obovate and often broadest beyond middle, blunt or sometimes notched at apex, long–pointed at base, and finely wavy margined, slightly thick and stiff, becoming hairless or nearly so. The upper surface is shiny dark green, and the lower surface dull light green.

Flowers are few in clusters (corymbs) at or near ends of twigs on slender stalks less than $^1/_4$ inch long. The calyx is composed of 5 pointed green sepals $^1/_{16}$ inch long, minutely hairy; petals, 5, $^1/_2$–$^3/_4$ inch long, oblong, curved out and downward; stamens mostly 10, separate and of unequal length; and on a disk the slender pistil with 2–celled ovary, long style, and dot stigma.

The few red fruits (berries) clustered at ends of twigs stand out in the dark green foliage. The surface is shiny and has many minute gland dots. Within the watery bitter red pulp are 1–2 elliptic light green seeds $^1/_4$–$^3/_8$ inch long. Flowering irregularly through the year.

The wood is described as light yellow with light brown heartwood, hard, heavy, fine–textured, and good for turned articles.

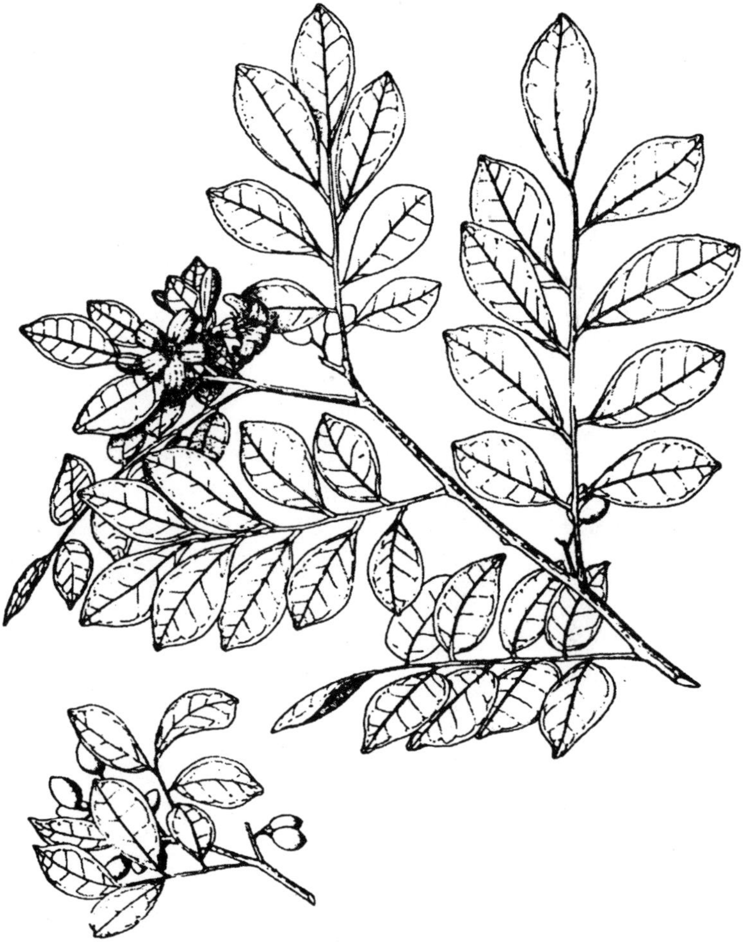

**Fig. 99 : Orange–jessamine (*Murraya paniculata*)**

Twig with flowers and fruits (above), fruiting twig (lower left), natural size.

This species has vigorous root growth and could serve as a stock for citrus fruits. Elsewhere the roots and barks have been used in home remedies. Propagated from seeds and cuttings.

Mirto is grown in India mainly as a hedge and ornamental in gardens and is an uncommon escape from cultivation. It is most common in the southwest of H.P.

## 100. Aceitillo (*Pilocarpus racemosus* L.)

This odd rare shrub or small tree of the moist limestone forest is recognized by : (1) large palmately compound leaves with usually 3 (1–5) elliptic leaflets, with minute gland dots and aromatic; (2) many dark red 5–parted starlike flowers $^1/_4$ inch across on red stalks along an erect axis 6–14 inches long; and (3) the fruits from a flower are 1–3 dark brown rounded podlike parts nearly $^1/_2$ inch long, each 1–seeded.

Evergreen shrub or sometimes a small tree to 15 feet high and 3 inches in trunk diameter, with few branches and stout green twigs, hairless throughout. The bark is gravy and smooth, the inner bark whitish and tasteless.

The leaves are alternate and palmately compound (sometimes pinnate) with usually 3 (1–5) leaflets on stalks about $^1/_8$ inch long at end of petiole of 1–3 inches. Leaflet blades are 2–6 inches long and 1–3$^1/_2$ inches wide, notched at apex and rounded or short–pointed at base, slightly thick and leathery, turned under at edges, with many minute gland dots, the upper surface shiny dark green and the lower surface dull yellow green.

The flower clusters (racemes) have a slender terminal erect brownish axis and many flowers on slender horizontal red stalks $^3/_8$–$^5/_8$ inch long. The flowers consist of 5–lobed calyx less than $^1/_{16}$ inch long; starlike dark red corolla of 5 widely spreading pointed petals more than $^1/_8$ inch long; 5 stamens; and on a disk the pistil with broad 5–celled ovary and short style. The 1–3 podlike fruits (follicles) maturing from a flower split open on 1 side. The seed is rounded and flattened, about $^1/_4$ inch long, and black. Flowering and fruiting intermittently.

The sapwood is light brown and hard.

The alkaloid pilocarpine, used in medicine, has been obtained from the dried leaflets of several species of this genus. Leaflets of this species were known as Guadeloupe jaborandi.

Rare in moist limestone forests at 200–600 feet altitude in northwestern Punjab.

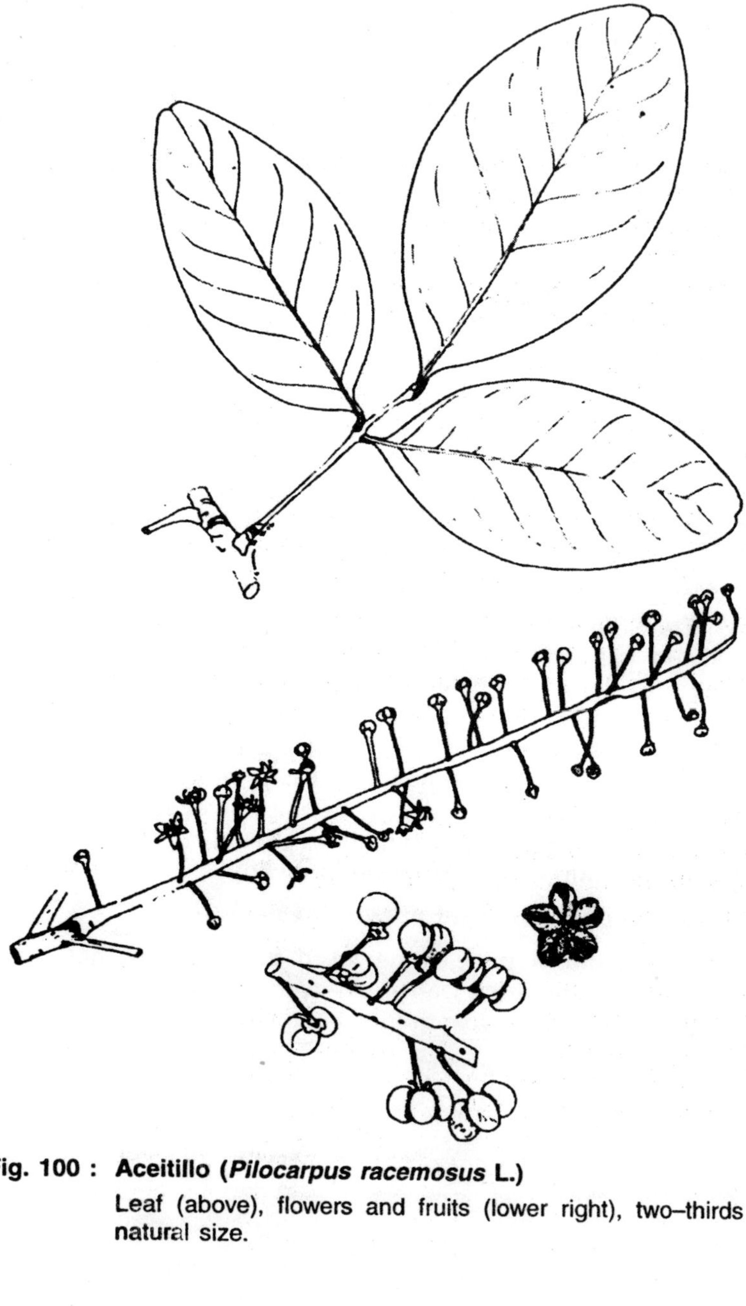

**Fig. 100 : Aceitillo (*Pilocarpus racemosus* L.)**

Leaf (above), flowers and fruits (lower right), two–thirds natural size.

## 101. Tortugo prieto (*Ravenia urbanii*)

This small tree known only from eastern mountains of Assam is recognized by : (1) the opposite, leathery, stalkless, elliptic to ovate, short–pointed, leathery, dark green leaves 2–6 inches long and $^1/_3$–$^1/_2$ inches wide, with many minute gland dots; and (2) the few showy crimson tubular flowers 1 inch long and broad at the base of a leaf, the corolla unequally 5–lobed.

An evergreen tree 25 feet high (rarely 40 feet) and 5 inches or more in trunk diameter, with a narrow crown. The bark is smoothish, dark gray, and thin, with whitish tasteless inner bark. The young twigs are slightly flattened and 2–angled, broadest just below leaves. The opposite leaves are broad and rounded at base, not toothed on edges, hairless, paler beneath.

Flower clusters are lateral or terminal, bearing 1–7 flowers nearly stalkless or on a stalk about 1 inch long. Sepals 5, 2 long and 3 short, $^1/_8$–$^1/_2$ inch long; the crimson tubular corolla with 5 unequal lobes, with minute gland dots; stamens 4, attached to the corolla tube, only the lower 2 fertile; and pistil of ovary with 4 or 5 lobes and cells, slender style, and 4–5 lobed stigma. The dry brown fruit about $^3/_8$ inch long and broad is composed of 5 podlike parts (follicles), each with 1 or 2 seeds more than $^1/_8$ inch long and splitting open. In flower and fruit throughout the year.

The sapwood is whitish, hard, and heavy (specific gravity 0.9). Chief local uses of the wood are for posts and fuel.

This handsome tree is being tested as a promising ornamental for its dark green foliage and few red flowers.

Rare in upper Bagdogra forests regions at 1,500–3,000 feet altitude in eastern Assam.

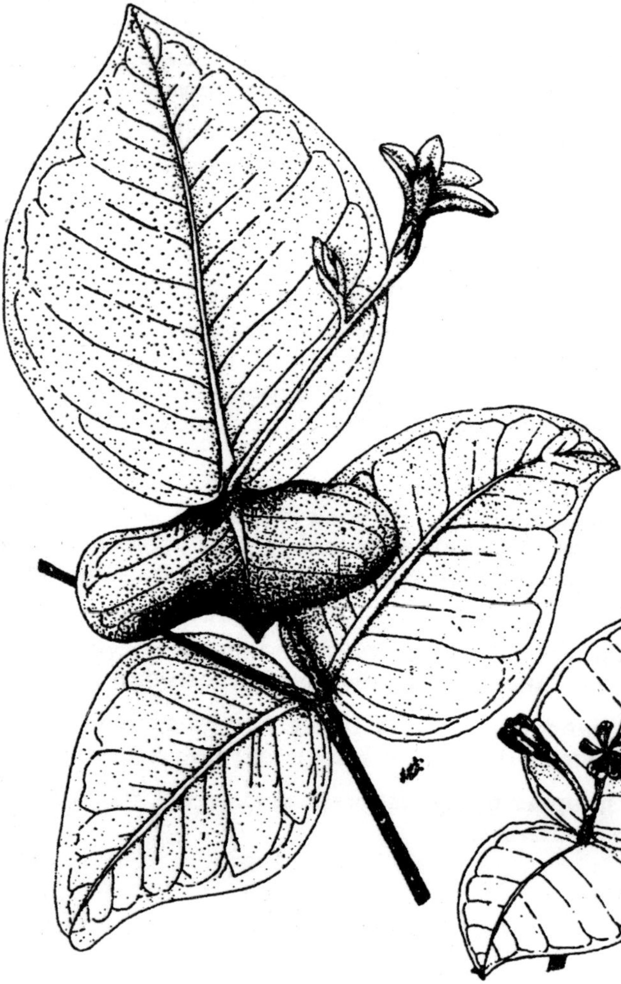

**Fig. 101 : Tortugo prieto (*Ravenia urbanii*)**

Flowering twig (above), fruiting twig (lower right), natural size.

## 102. Chinita, sweet–lime limeberry (*Triphasia trifolia*)

This introduced shrub or rarely a small tree is closely related to the citrus fruits. It is recognized by : (1) paired slender spines $^1/_4$–$^5/_8$ inch long at nodes of the finely hairy greenish twigs; (2) leaves 1–2 inches long, composed of 3 ovate shiny dark green leaflets with finely wavy edges and many gland dots, aromatic when crushed; (3) fragrant cup–shaped flowers with 3 white petals $^5/_8$ inch long, 3 or fewer at leaf bases; and (4) elliptic reddish fruits $^1/_2$–$^5/_8$ inch long with thick skin and mucilaginous spicy pulp.

Evergreen shrub 6–10 feet high or rarely a small tree to 15 feet. Leaves alternate, hairless, with slender petiole $^1/_8$ inch long and 3 leaflets on very short stalks, the middle leaflet about twice as large as the others. Leaflet blades $^1/_2$–$1^1/_2$ inches long and $^3/_8$–1 inch wide, slightly notched or rounded at apex, short–pointed at base, slightly thickened, with numerous gland dots plainer on the light green lower surface. Foliage compact, often drooping.

The short–stalked gland–dotted flowers are composed of 3–toothed hairy green calyx $^1/_{16}$ inch long; corolla of 3 (sometimes 4) spreading narrow white petals $^5/_8$ inch long; 6 stamens on a disk; and pistil with elliptic 3–celled ovary, long style, and rounded stigma. Berries aromatic, gland–dotted, with calyx at base, containing usually 1 seed about $^1/_4$ inch long in the sticky pulp. Flowering and fruiting irregularly through the year.

The plants are grown as ornamental shrubs and in hedges. The mucilaginous juicy fruits are edible and have been used in making marmalade and jam. A honey plant. Popular as a landscape shrub in southern India. Propagated by seeds and cuttings.

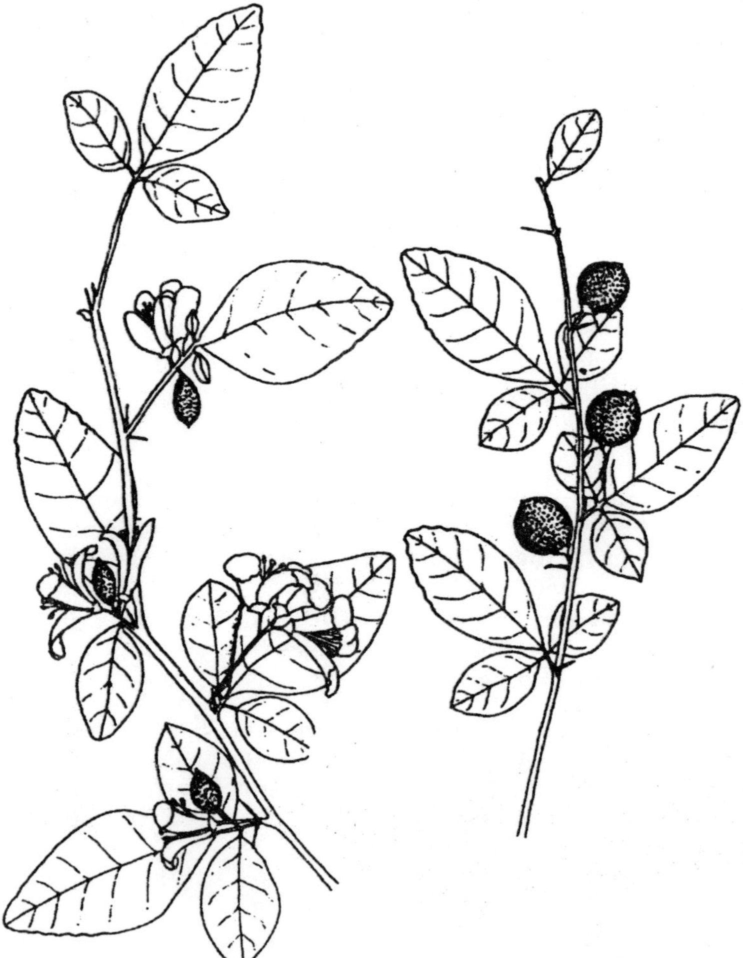

**Fig. 102 : Chinita, sweet–lime limeberry (*Triphasia trifolia*)**
Flowering twig (left), fruiting twig (right), natural size.

## 103. Bitolia (*Zanthoxylum bifoliolatum*)

This rare shrub or small tree of the western mountains in Orissa is distinguished by : (1) scattered slender sharp spines to $^1/_4$ inch long on the twigs; (2) compound leaves with only 2 obovate or spoon–shaped leaflets, stiff and thick, and usually a spine at apex of petiole; (3) many or few small flowers $^3/_{16}$ inch wide clustered at leaf bases; and (4) round black pods $^3/_{16}$ inch wide, 1 or 2 from a flower, 1–seeded.

Evergreen shrub 5–8 feet high or sometimes a small tree to 25 feet in height and 4 inches in trunk diameter, hairless throughout. Bark gray, fissured. Twigs slender with straight spines.

Leaves alternate, compound 1–2 inches long, with slender petiole $^1/_8$–$^3/_8$ inch long and 2 blades, sometimes only 1, stalkless or nearly so. Blades obovate or spoon–shaped, $^3/_4$–$1^1/_2$ inches long and $^3/_8$–$1^1/_8$ inches wide (to 4 by 2 inches in Hispaniola), slightly notched at rounded apex, short–pointed at base, turned under at edges, with minute gland dots, the upper surface shiny green with sunken midvein and many prominent parallel side veins, and the lower surface light green and slightly shiny with raised veins.

Flowers are male and female in different clusters, short–stalked at leaf bases, apparently on different trees (dioecious). Male flowers have 3–4 minute sepals, 3–4 petals more than $^1/_{16}$ inch long, 3–4 stamens $^1/_8$ inch long, and minute nonfunctioning pistil. Female flowers have 3 sepals, 3 petals, and on a disk 2 united pistils less than $^1/_{16}$ inch long, each with rounded ovary, short style, and dot stigma. Each pod (follicle) splits open on 1 side and contains 1 shiny black seed $^1/_8$ inch in diameter. With flowers and fruits in summer and autumn.

Wood light brown, hard.

Rare in upper Nilgiri forest at 2,000–2,500 feet altitude in western Orissa.

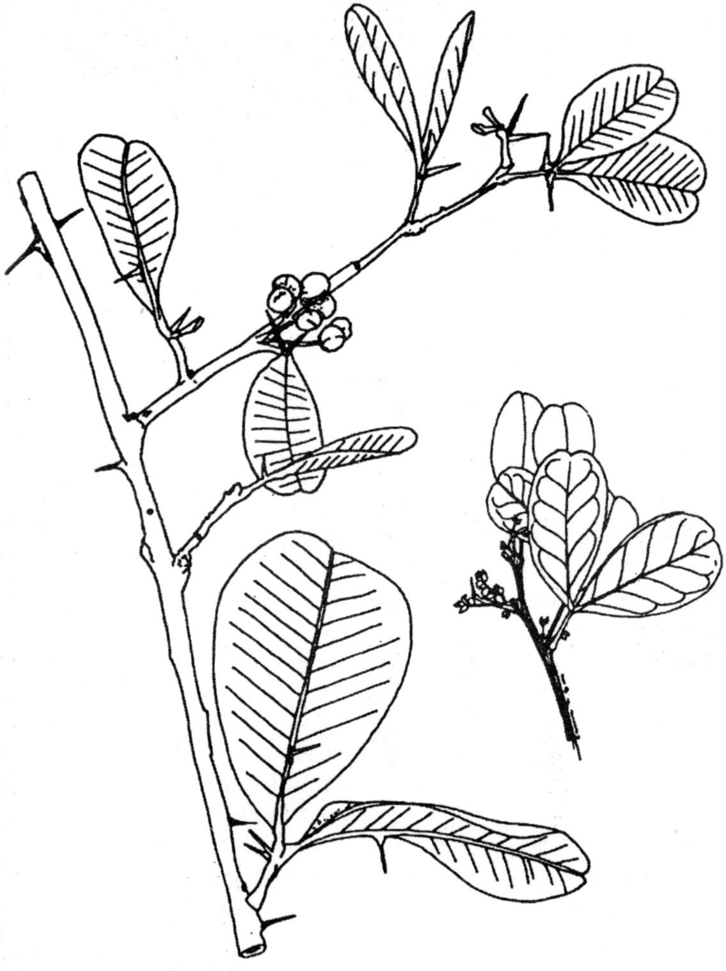

**Fig. 103 : Bitolia (*Zanthoxylum bifoliolatum*)**

Fruiting twig (above), flowering twig (lower right), natural size.

## 104. Espino rubial (*Zanthoxylum caribaeum*)

This shiny tree is characterized by : (1) sharp very stout brown spines $^1/_8$–$^3/_8$ inch long with broad conical base, on trunk and branches; (2) pinnate leaves with 7–13 elliptic leaflets bordered with wavy teeth, with minute gland dots, and with disagreeable odor when crushed; (3) many small greenish white flowers $^3/_{16}$ inch broad in branched clusters; and (4) the podlike dark gray fruits about $^1/_4$ inch long, 5–1 from a flower, each opening by 1 line to release 1 seed.

Small to medium–sized tree 60 feet high and 10 inches in trunk diameter in Puerto Rico, elsewhere a large tree to 70 feet high and $1^1/_2$ feet in trunk diameter, often branched from base, with spreading or rounded crown, perhaps deciduous in dry areas. The spiny trunk has gray bark, smooth to rough and thick. Inner bark is light brown and bitter. The stout hairless twigs are green when young, afterwards gray.

The alternate leaves with light green axis are mostly 8–10 inches long. The leaflets have short stalks about $^1/_8$ inch long. Blades are $1^1/_4$–4 inches long and $^3/_4$–2 inches wide, slightly thick and leathery, hairless, short–pointed or rounded at both ends, the borders with wavy teeth and turned under, the upper surface green to dark green and slightly shiny, and the lower surface dull and paler.

The flower clusters (panicles) are terminal and lateral, 3–6 inches long, with corky thickened branches. The small flowers are composed of 5 rounded sepals less than $^1/_{16}$ inch long, 5 white petals about $^3/_{16}$ inch long, 5 stamens, and 5–1 pistils (carpels) with 1–celled ovary and stigma. The fruits (follicles) are 5–1 from a flower, each with short stalk and 1 shiny rounded blackish seed $^3/_{16}$ inch long. With flowers and fruits from spring to fall.

The thick sapwood is lustrous yellow and the heartwood dark brown. The wood has a bitter taste and characteristic odor and is moderately lightweight. It is strong, easy to work, takes a good polish, does not split under exposure, but is not resistant to insects. Suitable uses are veneer, cabinetmaking, and general construction.

Elsewhere it is reported that the bark and twigs of this and related species are chewed as a remedy for toothache.

Common in moist limestone forest of Orissa Bihar at 100–600 feet altitude.

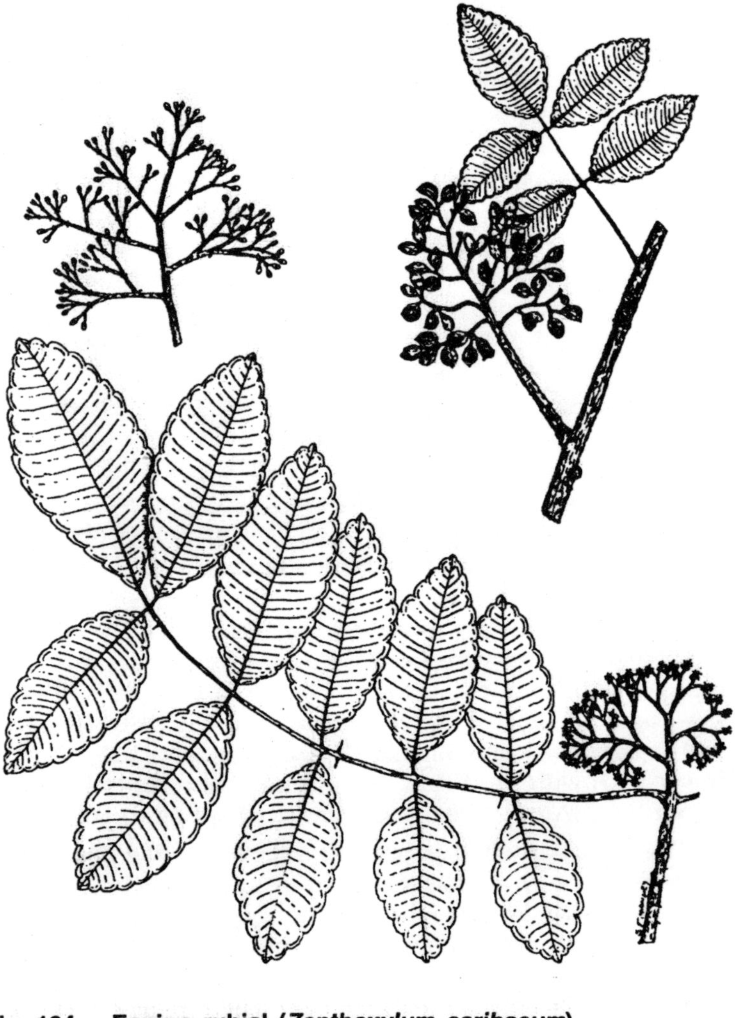

**Fig. 104 : Espino rubial (*Zanthoxylum caribaeum*)**

Female flowers (upper left), twig with the male flowers (below), two–thirds natural size.

## 105. Alfiler (*Zanthoxylum punctatum*)

An common shrub, sometimes a small tree, identified by : (1) straight slender brown spines or prickles of $^1/_8$ inch usually present on petioles, leaf axes, and midvein of lower leaflet surfaces; (2) leaves pinnate, with 2–7, commonly 3 (rarely 9–19) leaflets, paired except at end, elliptic or obovate, aromatic, with minute gland dots and spicy taste; (3) small 3–parted flowers $^3/_{16}$ inch broad, few at leaf base; and (4) fruits 2 or 1 from a flower, rounded, $^3/_{16}$ inch long, podlike, 1–seeded.

Deciduous shrub or rarely a small tree 20 feet or more in– height. Bark gray, fissured, with stout spines.

The alternate leaves are about $1^1/_2$–6 inches long, odd pinnate, hairless, with slender petioles $^1/_2$–$1^1/_2$ inches long and grooved. Leaflets are paired except at end (sometimes without end leaflet), $^3/_4$–3 inches long and $^3/_8$–$1^1/_4$ inches broad, almost stalkless, the apex notched, rounded, or pointed, the base short–pointed, and the margin straight or minutely wavy, shiny green on upper surface, and dull light green with gland dots on lower surface. Leaves on flowering twigs usually without spines.

Flowers several at leaf bases or terminal, on stalks $^1/_{16}$ inch long, male and female apparently on different plants (dioecious), $^3/_{16}$ inch broad, with 3 minute sepals and 3 white or yellowish petals less than $^1/_8$ inch long. Male flowers have 3 stamens. Female flowers have 2 pistils partly united at base, each with ovary, style, and stigma. The 2 gray fruits (follicles) partly united split open on 1 side. Seeds are rounded, shiny black, about $^1/_8$ inch long. Flowering and fruiting from spring to fall.

This and related species have been used elsewhere in home remedies for toothache, hence the common name toothache–tree.

Common and local in lower and upper Orissa forests from sea level to 2,000 feet altitude in western and southwestern.

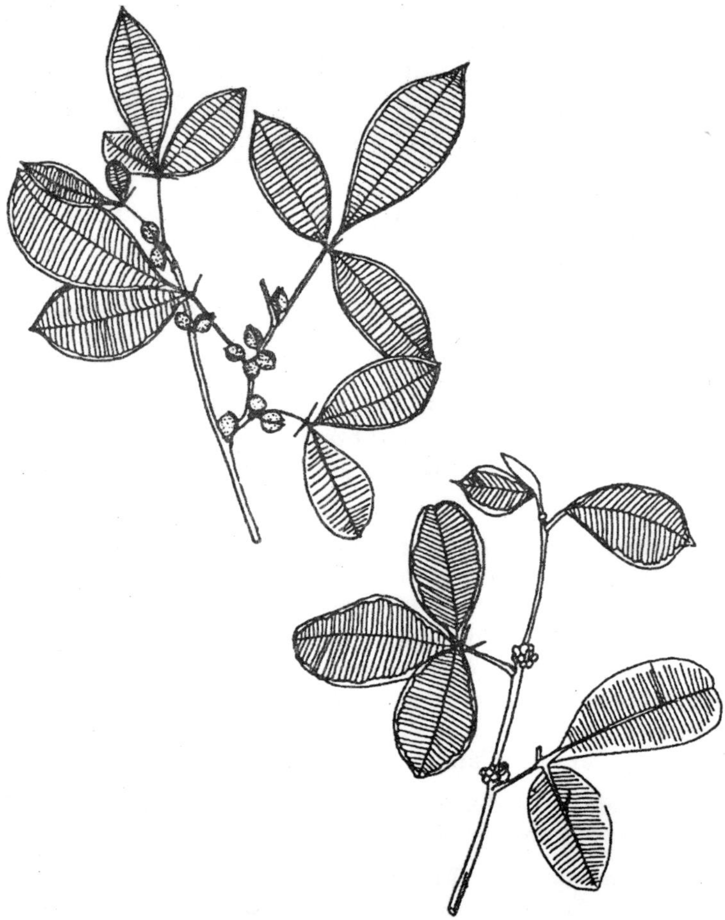

**Fig. 105 : Alfiler (*Zanthoxylum punctatum*)**
Fruiting twig (above), flowers twig (below), natural size.

## 106. Niaragato (*Zanthoxylum spinifex*)

This rare very spiny shrub or small tree of dry areas is easily recognized by : (1) many very slender straight sharp spines $^1/_{16}$–$^1/_4$ inch long, paired at leaf bases and on nodes along very slender twigs; (2) the green to gray twigs very slender, zigzag, and widely forking; (3) many crowded very small light green pinnate leaves $^1/_2$–$1^1/_2$ inches long, composed of 3–7 (rarely to 11) narrow leaflets, paired at end, and broad or winged axis, aromatic with minute gland dots and spicy taste; (4) small 4–parted yellow flowers $^1/_8$ inch broad at leaf base; and (5) fruits 2 or 1 from a flower, rounded, blackish, more than $^1/_8$–inch long, pod–like, 1–seeded.

Deciduous shrub or rarely a small tree 15–feet high with several stems to 3 inches in diameter, and widely spreading flat–topped crown of very thin foliage, hairless throughout. Bark light brown, smoothish, slightly fissured, with spines sometimes persistent. Inner bark yellowish, with thin greenish outer layer, bitter. Twigs very slender, zigzag, widely forking, minutely hairy when young, green becoming gray and finely fissured.

The leaves are alternate and only $^1/_4$–$^1/_2$ inch apart, odd pinnate, hairless or nearly so. At the base of each leaf are paired brown spines (stipules), very slender, sharp, $^1/_{16}$–$^1/_4$ inch long and widely forking, remaining on the branches. The petiole $^1/_4$–$^3/_8$ inch long and the axis are winged and about $^1/_{16}$ inch wide. The leaflets are paired except at end, stalkless or nearly so, $^1/_8$–$^1/_2$ inch long and $^1/_{16}$–$^3/_{16}$ inch wide, rounded or blunt at both ends, not toothed on edges, thin, the side veins inconspicuous, slightly shiny above and dull beneath, with 2 minute light brown glands at base.

Flowers several, sometimes only 1, at leaf bases, almost stalkless, male and female apparently on different plants (dioecious), $^1/_{18}$ inch broad, with 4 minute greenish sepals and 4 yellow petals $^1/_{16}$ inch long. Male flowers have 4 stamens $^1/_8$ inch long. Female flowers have 2 pistils, each with rounded ovary, slender style, and united stigma. The fruits (follicles) have style at apex and split open on 1 side. Each has 1 shiny black rounded seed nearly $^1/_8$ inch long. Collected in flower in July and with fruits in August.

Sapwood light brown, hard.

Rare and local in dry coastal thickets and forests from sea level to 100 feet altitude from Orissa to A.P.

**Fig. 106 : Niaragato (*Zanthoxylum spinifex*)**

Fruiting twig (at left), twig with female flowers (upper right), twig with male flowers (lower right), natural size.

## 107. Thomas (*Zanthoxylum thmasianum*)

A rare shrub or small tree with prickly twigs characterized by : (1) pinnate leaves with 5–9 stalkless leaflets ovate or rounded $^5/_8$–1 inch long and $^1/_2$–$^3/_4$ inch wide; (2) few flowers stalkless at leaf bases, $^1/_8$ inch long, 3–parted; and (3) fruit of 3, sometimes 2 or 1, pods $^5/_{16}$ inch long from a flower.

An evergreen shrub 6 feet high or a small tree to 20 feet and 4 inches in trunk diameter. Twigs with minute hairs when young, prickly, with 1–2 slender sharp brownish spines less than $^1/_4$ inch long at leaf bases.

Leaves alternate $1^1/_4$–2 $^1/_2$ inches long, with 5–9 leaflets, the end leaflet largest, rounded or abruptly pointed at apex, blunt or rounded at base, and often slightly wavy on edges, slightly thickened, shiny green with many fine side veins, beneath paler and sometimes with 1–3 spines on midvein.

Flowers male and female on different plants (dioecious), with 3 sepals less than $^1/_{16}$ inch long and 3 elliptic petals about $^1/_8$ inch long. Male flowers have 3 stamens, and female flowers have 3 pistils slightly united. The fruits (pods or follicles) are egg–shaped, 3, sometimes 2 or 1, from a flower, each splitting open on 1 side and containing 1 black seed $^3/_{16}$ inch long. Flowering and fruiting in spring and early summer and maturing fruits in summer.

Rare in lower Nilgiri forest at 500–1,500 feet altitude on south slopes of Karnataka area in Central India.

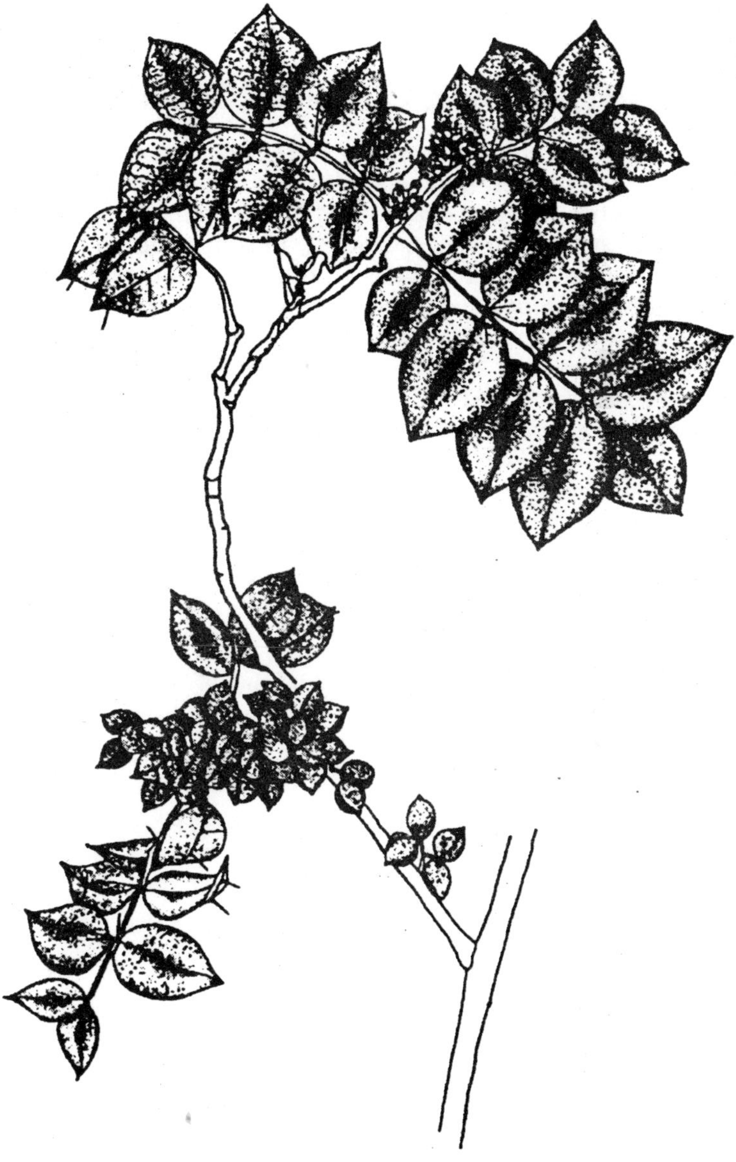

**Fig. 107 : Thomas (*Zanthoxylum thmasianum*)**
Natural size.

# AILANTHUS FAMILY (SIMAROUBACEAE)

Shrubs and trees, known by : (1) bark, wood, and other parts often very bitter; (2) leaves alternate, pinnate, with leaflets generally entire, without stipules; (3) flowers mostly small, often in large branched clusters (panicles), mostly male and female and usually on different plants (dioecious), regular, with 3–8 lobed calyx, 3–8–blood corolla or none, as many stamens as petals or double and often with scales at base, inserted on or around the disk, and on disk 2–5 simple pistils with 1–celled ovary and style, or pistil composed of 2–8 cells and 2–8 styles; and (4) fruit generally a capsule, drupe, or winged (samara). Also vol. 1, p. 234.

**Key to Species**

A. Leaves simple, gray green hairy.

B. Twigs often ending in spines; leaves small. $^{3}/_{8}$–1 inch long—*Castela erecta.*

BB. Twigs not spiny; leaves very narrow, $1^{1}/_{2}$–$2^{1}/_{2}$ inches long, thick and slightly succulent—*Suriana maritima.*

AA. Leaves pinnate.

C. Leaflets 6–10, mostly paired, upper surface shiny dark green with indistinct veins; flowers dark red showy; hairless throughout—*Simarouba tulae.*

CC. Leaflets mostly 5–13, mostly paired except at end; flowers minute, greenish; young twigs hairy.

D. Flowers clusters terminal; leaflets 5–9; fruits berrylike, round to elliptic, red, $^{3}/_{8}$–$^{5}/_{8}$ inch long—Guarema, bitterbush, *Picramnia pentandra* Sw.

DD. Flower clusters lateral at leaf bases; leaflets 7–13; fruits round, black, 1 or sometimes 2–3 from a flower—*Picrasma.*

E. Fruits $^{3}/_{8}$–$^{1}/_{2}$ inch in diameter—*Picrasma antillana.*

EE. Fruits $^{1}/_{4}$ inch in diameter—*Picrasma excelsa.*

## 108. Goat bush (*Castela erecta*)

This much–branched spiny shrub of dry areas rarely becomes a small tree. Its distinguishing characters are : (1) twigs very bitter, often ending in spines, shorter spines single at leaf bases; (2) leaves small, oblong to elliptic or obovate, $^3/_8$–1 inch long and 16–$^1/_2$ inch wide, thick and leathery, shiny green and hairless or nearly so above, densely whitish hairy beneath; (3) small whitish to red flowers $^1/_8$ inch long few at leaf bases, 4–parted; and (4) red rounded fruits (drupes) 1–4 from a flower, $^1/_4$–$^3/_8$ inch long.

An evergreen spiny shrub 3–8 feet high or rarely a small tree to 15 feet high and 4 inches in trunk diameter, much branched with dense foliage, very bitter throughout. Twigs stiff, often slightly zigzag, finely whitish hairy, often ending in spines and with shorter spines about $^1/_8$ inch long single at leaf bases.

Leaves alternate, stalkless or nearly so, rounded or abruptly short–pointed at apex, short–pointed or rounded at base, turned under at edges.

Flowers male and female on different plants (dioecious), at leaf bases. There are 4 tiny pointed sepals, 4 elliptic petals less than $^1/_8$ inch long, and 8 stamens in male flowers. Female flowers have 8 sterile stamens (stami–nodes), and 4 nearly separate ovaries united at apex, each with 1 ovule, and 4 short style branches. Fruits 1–4 from a flower, fleshy with hard stone, opening in 2 parts, containing 1 seed. With flowers and fruits in spring.

The red flowers and fruits, the whitish twigs and lower leaf surfaces, and the shiny green upper surfaces make this a showy plant. Possibly of value as an ornamental in dry areas.

Locally common in dry coastal forest from sea level to 300 feet altitude in southwestern A.P. Also dry thickets on hillsides in eastern and southern Godavari.

**Fig. 108 : Goat bush (*Castela erecta*)**
Flowering twigs (above), fruit (below), natural size.

## 109. Bitter–ash (*Picrasma antillana*)

This small tree rare in the Andaman Islands is identified by : (1) pinnate leaves with 7–13 oblong leaflets paired except at end, not toothed on edges, rounded and unequal at base; (2) short–stalked, flattened, widely forking clusters of small flowers at leaf bases; and (3) many round fruits $^{3}/_{8}$–$^{1}/_{2}$ inch in diameter, 1 or sometimes 2–3 from a flower.

Small evergreen tree 30 feet high, with bark and sap bitter. Twigs finely hairy, light brown with whitish dots (lenticels).

Leaves alternate, pinnate, 10–16 inches long, with axis finely hairy. Leaflets on short stalks $^{1}/_{8}$–$^{1}/_{4}$ inch long are $2^{1}/_{2}$–$4^{1}/_{2}$ inches long and 1–$1^{3}/_{4}$ inches wide, tapering to a long–pointed blunt apex, thin, the upper surface hairless, and the lower surface with raised veins often hairy.

Flower clusters (panicles) at leaf bases are short–stalked, flattened, and bear at the end of widely forking finely hairy branches many flowers mostly male and female on separate trees (polygamous). Male flowers about $^{1}/_{4}$ inch long and wide are composed of 4–5 minute hairy sepals, 4–5 narrow petals less than $^{1}/_{8}$ inch long, and 4–5 spreading slender stamens $^{3}/_{16}$ inch long at base of a large disk. Female flowers have $^{4}/_{5}$ minute hairy sepals, 4–5 broader elliptic petals $^{1}/_{8}$ inch long, and ovary with 2–3 lobes and 2–3 cells each 1–ovuled, style, and 2–3 lobed stigma. Many fruits (drupes) 1 or sometimes 2–3 from a flower, each on a short stalk $^{1}/_{16}$ inch long above persistent sepals and petals at base, slightly fleshy, 1–seeded.

The wood is described as yellowish, hard, heavy, and durable.

Elsewhere the bitter bark has served in home remedies.

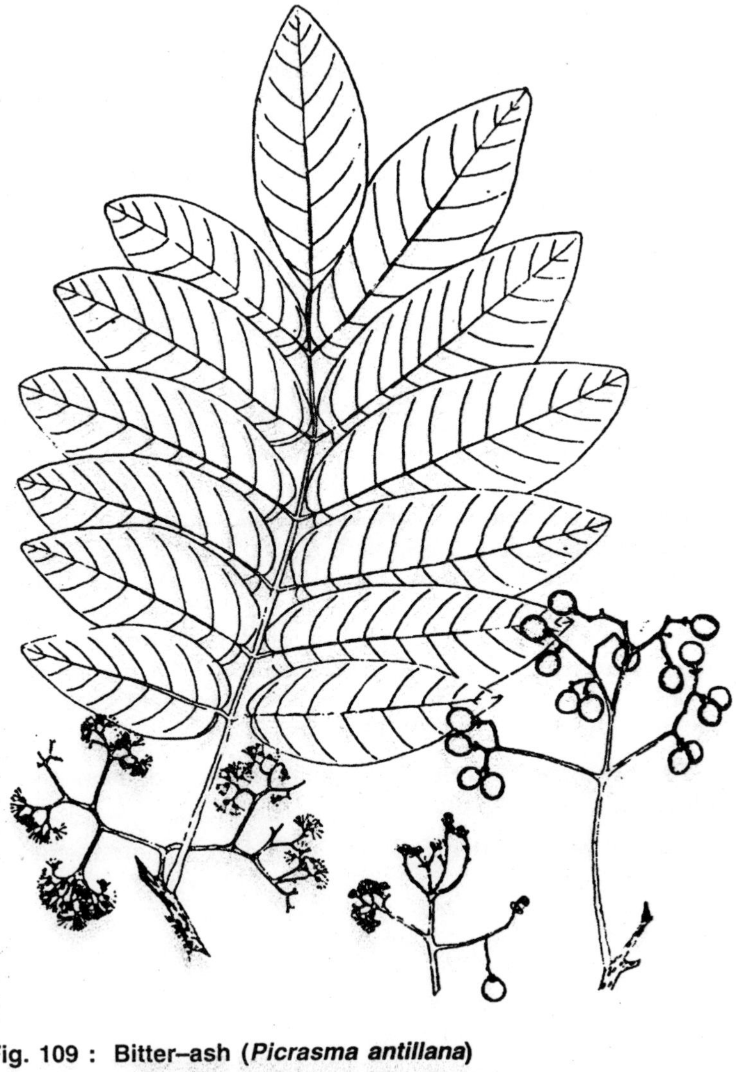

**Fig. 109 : Bitter–ash (*Picrasma antillana*)**

Twig with male flowers (left), female flowers and fruits (lower right), two–thirds natural size.

## 110. Palo amargo, bitterwood (*Picrasma excelsa*)

A very rare tree collected near Vedaghat Springs, characterized by : (1) pinnate leaves with 9–13 oblong to lance–shaped leaflets paired except at end, short–pointed and unequal at base, the edges straight or slightly rolled under; (2) many small yellow–green flowers in long–stalked clusters at leaf bases; and (3) many round black fruits $^1/_4$ inch in diameter, 1 or sometimes 2–3 from a flower.

Medium–sized evergreen tree recorded to 65 feet high, with bitter bark and sap. Bark gray, fissured. Twigs finely hairy.

Leaves alternate, pinnate, 6–14 inches long, with axis finely hairy or becoming hairless. Leaflets 2–5 inches long and $^3/_4$–$1^1/_2$ inches wide, tapering to a long–pointed blunt apex, thin, the upper surface nearly hairless, the lower surface paler and finely hairy, especially on veins.

Flower clusters (panicles) long–stalked at leaf base bear many flowers mostly male and female on separate trees (polygamous). Male flowers about $^3/_{16}$ inch long and wide, consisting of 4–5 minute hairy sepals, 4–5 petals more than $^1/_{16}$ inch long, and 4–5 hairy stamens $^1/_8$ inch long at base of disk. Female flowers have 4–5 minute hairy sepals, 4–5 larger petals $^1/_8$ inch long, and pistil with 2–3–lobed and 2–3–celled ovary, slender style, and 2–3–forked stigma. Many fruits (drupes) 1 or sometimes 2–3 from a flower, each on a stalk nearly $^1/_8$ inch long, slightly fleshy, 1–seeded.

Wood whitish yellow, soft.

Elsewhere a bitter tonic from this tree has been used in home remedies. The dried heart–wood has served in medicine under the names quassia, Jamaica quassia, and bitterwood.

In India collected only at plains and arroyos in the vicinity of Vedaghat (M.P.) Springs, with fruits in March. Not observed by the authors.

**Fig. 110 : Palo amargo, bitterwood (*Picrasma excelsa*)**

Twig with male flowers (left), female flowers (upper right), fruits (lower right), two–thirds natural size.

## 111. Aceitillo Cimarrón (*Simarouba tulae*)

This beautiful uncommon shrub or tree native only in Sivalik mountains is recognized by : (1) pinnately compound leaves with 5–10 mostly paired oblong to elliptic leaflets, shiny dark green; (2) dark red showy flowers $^{1}/_{2}$ inch broad, many in long-stalked terminal clusters; and (3) from a flower 1–5 red obovoid vertically flattened, slightly fleshy fruits 1–1$^{1}/_{4}$ inches long and $^{5}/_{8}$–1 inch wide.

Evergreen shrub flowering when 5–10 feet high, or a small tree to 25 feet high and 8 inches in trunk diameter, hairless throughout. Recorded as formerly a medium–sized tree to about 60 feet high and 12–20 inches in trunk diameter. Bark furrowed very bitter. The purplish twigs become finely fissured, exposing light brown inner bark.

The alternate pinnate leaves 6–14 inches long have a slender round purplish–brown axis. The leaflets on short stalks $^{1}/_{8}$ inch long are 2–4 inches long and $^{3}/_{4}$–2 inches wide, long–pointed at paex, short–pointed at base, with edges slightly turned under, slightly thickened, upper surface shiny dark green with indistinct veins, lower surface slightly shiny yellow green.

Flower clusters (panicles) are 4–6 inches long and broad, much branched at end of long erect terminal stalk. Many small flowers are borne on slender dark red stalks, male and female on different clusters or plants. The male flower is composed of dark red calyx $^{1}/_{16}$ inch long, with 5 short–pointed lobes, corolla of 5 spreading dark red petals $^{3}/_{8}$ inch long and $^{1}/_{8}$ inch broad, and 10 stamens $^{5}/_{16}$ inch long. Female flowers not seen but in this genus have a pistil with ovary usually deeply 5–lobed and 5–celled, short style, and 5 spreading stigmas. From a flower 1–5 spreading fruits (drupes) are produced, obovoid, flattened, each containing 1 flattened seed. Flowering and fruiting intermittently during the year.

The wood is described as heavy (specific gravity 0.85), strong and durable.

Uncommon in lower and upper Sivalik forests at 500–2,500 feet altitude in mountains of Punjab.

**Fig. 111 : Aceitillo Cimarron (*Simarouba tulae*)**
Flowering twig (above), fruits (lower right), two–thirds natural size.

## 112. Temporana, baycedar (*Suriana maritima* L.)

Temporana or baycedar, a rounded shrub of sandy beaches, sometimes becomes a small tree. It is easily recognized by : (1) dense rounded mass, much branched, of gray–green foliage; (2) very narrow leaves, finely hairy, slightly thick, succulent, and salty, crowded at ends of twigs and persistent after dying; (3) few yellow 5–parted flowers $^3/_8$ inch long in short terminal clusters; and (4) fruits 5 or fewer from a flower, rounded, $^3/_{16}$ inch long, brownish black and finely hairy.

This evergreen shrub commonly forms rounded clumps 5–10 feet in diameter but sometimes becomes a small tree 15 feet high and 5 inches in trunk diameter. The bark is dry gray, becoming thick, rough, and furrowed, the inner bark light yellow and tasteless. Twigs are light green, finely hairy, bearing crowded leaves at ends and dead leaves below.

The leaves are crowded but alternate, very narrow (linear to narrowly spatulate), $1^1/_2$–$2^1/_2$ inches long, $^1/_4$–$^3/_8$ inch wide, more than $^1/_{16}$ inch thick, slightly succulent with a slightly salty taste, broadest near the rounded apex and gradually narrowed to the stalkless base, gray green, finely and densely hairy.

The few or solitary flowers have pale yellow stalk, 5 large pointed hairy sepals about $^3/_8$ inch long, 5 yellow obovate petals shorter then sepals, 10 stamens, and 5 separate pistils with hairy rounded ovary and threadlike style attached near base. The 5 hard dry 1–seeded fruits (akenelike) from a flower are bordered by the 5 gray dead sepals, which remain after fruits fall. Flowering and fruiting intermittently through the year.

The wood is described as having dark red or reddish–brown heartwood and lighter sapwood, hard, heavy, of fine uniform texture, strong, and apparently durable. Where sufficiently large, it could be used for small articles of turnery.

Locally common, scattered and widely distributed on sandy beaches of dry coasts of Gujrat and nearly all smaller islands nearby.

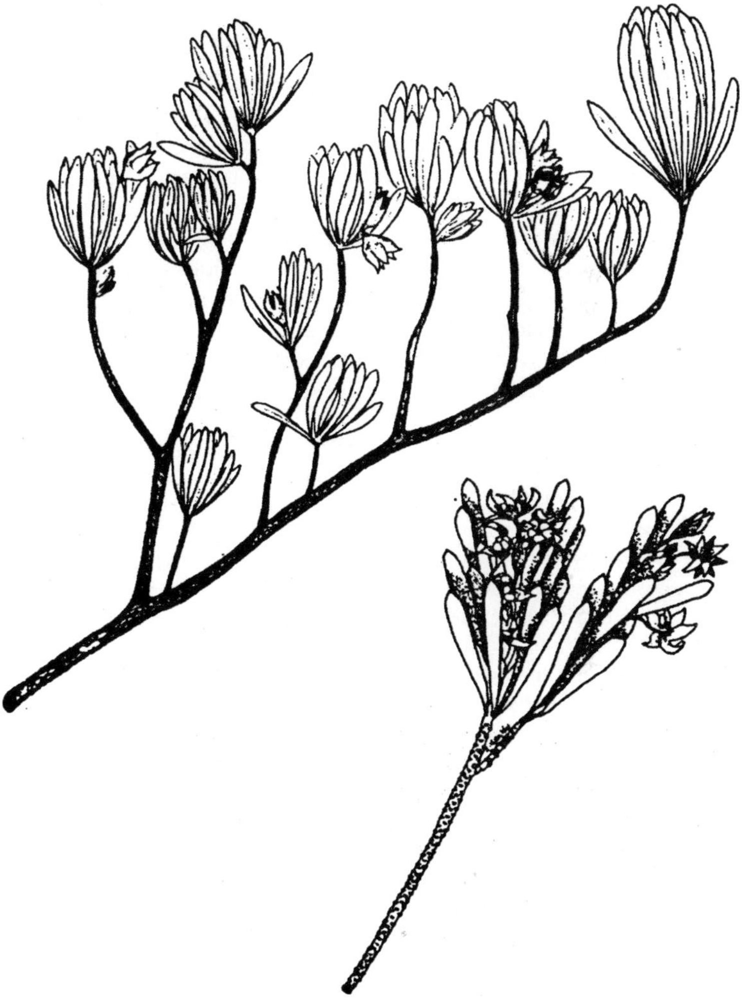

**Fig. 112 : Temporana, baycedar (*Suriana maritima* L.)**
Fruiting twig (above), flowering twig (lower right), natural size.

# MAHOGANY FAMILY (MELIACEAE)

Trees, often large, and shrubs, known by : (1) bark bitter and astringent and wood often aromatic; (2) naked buds with minute young leaves often in form of hand; (3) leaves alternate generally pinnate, often even pinnate with paired leaflets, usually entire and often oblique, without stipules; (4) flowers small, generally in branched clusters (panicles), bisexual, regular, with 4–5 lobed calyx, corolla of 4–5 petals or lobes, 8–10 stamens united in a tube (5 separate in *Cedrela*) and around a disk, and pistil with superior ovary of 2–5 cells, each with 2 to many ovules, style or none, and stigma often dislike; and (5) fruit generally a capsule or berry with seeds often winged.

**Key to Species**

A. Leaves bipinnate, leaflets with toothed margins—Alelaila, chinaberry, *Melia azedarach* L.*

AA. Leaves pinnate.

B. Leaflets with toothed margins.

C. Leaflets many, thin, long-pointed—*Melia azadirachta.**

CC. Leaflets 3–7, thick and stiff, with 3 sharp spiny teeth at broad apex—*Trichilia triacantha.*

BB. Leaflets not toothed.

D. Leaflets of odd number (odd pinnate)—*Trichilia.*

E. Leaflets 3 or 5 (sometimes 7), asymmetrical, the end leaflet largest and lowest leaflets smallest; veins much sunken, causing a wrinkled appearance—Gaeta, *Trichilia pallida* Sw.

EE. Leaflets 7–21, symmetrical, about equal in size, veins slightly sunken—Tinacio, broomstick, *Trichilia hirta* L.

DD. Leaflets all paired (even pinnate), asymmetrical, slightly curved and unequal at base.

F. Axis of leaf continuing to grow and form additional pairs of leaflets; veins sunken; flowers and fruit 4-parted—*Guarea.*

G. Leaflets 2–8, mostly oblong—*Guarea ramiflora.*

GG. Leaflets 8–20, elliptic to oblong—Guaraguao, American muskwood, *Guarea guidonia* (L.) Sleumer (*G. trichilioides*).

FF. Axis of leaf not forming additional leaflets; veins not sunken; flowers and fruit 5-parted.

H. Leaflets 4–12.

I. Leaflets $2^1/_2$–6 inches long—Caoba hondurena, Hnduras mohagany, Swietenia macrophylla King.*

II. Leaflets $2^1/_2$–6 inches long—Caoba dominicana, Dominican mahogany, West Indies mahogany, *Swietenia mahagoni* Jacq.*

HH. Leaflets 10–20 or more.

J. Leaflets lance–shaped; seeds winged at both ends; without distinctive odor—*Toona ciliata.**

JJ. Leaflets lance–shaped, oblong, or ovate; seeds with long wing; odor of garlic in flowers and crushed leaves—Cedro hembra, Spanish–cedar, *Cedrela odorata* L.

## 113. Guaraguadillo (*Guarea ramiflora*)

Guaraguadillo is a small tree uncommon in understory of mountain forests of Puerto Rico and rare in Hispaniola. It is identified by : (1) even pinnate leaves with usually 1–4 pairs of mostly oblong leaflets and a growing point forming additional pairs; (2) small pink 4–parted flowers about $^1/_2$ inch across, few to many clustered at leaf bases and on larger twigs back of leaves; and (3) round hard brown fruits $^1/_2$–$^7/_8$ inch in diameter.

Evergreen shrub or small tree 25 feet high and 6 inches in trunk diameter, often flowering as a shrub 10 feet high and rarely a medium–sized tree to 50 feet and 6 inches, reported to reach 65 feet. The bark is gray or brown, smoothish with dots (lenticels), becoming fissured and scaly. Inner bark is pinkish or whitish, slightly bitter The twigs are light brown with raised dots (lenticels) and large raised leaf scars, hairless.

The alternate leaves are 6–12 inches long, including the stout brownish woody hairless petiole $1^1/_2$–3 inches long and axis elongating to 6 inches beyond. The budlike hairy growing point between last pair of leaflets continues to form additional pairs, sometimes 7–9, as the older ones are shed. The leaflets have stout curved brown stalks about $^1/_4$ inch long. Leaflet blades are mostly oblong or lance–shaped, 3–8 inches long and $1^1/_4$–3 inches wide, long–pointed at apex, short–pointed at base, sometimes slightly wavy edges, slightly thickened and leathery, with network of fine veins. The upper surface is dull dark green with the curved side veins sunken, hairless, and the lower surface green with raised veins, hairless or slightly hairy on veins.

Flower clusters (racemes or panicles) are 1–$3^1/_2$ inches long at leaf bases and on larger twigs back of leaves, the branches finely hairy. Flower buds are $^1/_4$ inch long, deep pink. The fragrant flowers have a hairy stalk about $^1/_{16}$ inch long; calyx of 4 finely hairy pointed sepals less than $^1/_{16}$ inch long; corolla of 4 oblong pink petals $^1/_4$–$^5/_{16}$ inch long, spreading and curved back; stamen tube cylindric, pink, $^1/_4$ inch long, bearing 8 minute stalksless stamens at apex inside; and on a raised disk the pistil $^3/_{16}$ inch long with egg–shaped hairless 4–celled ovary with 1 ovule in each cell, slender style, and rounded stigma.

The fruit on a stout stalk is a hard brown capsule, round or slightly flattened, which splits into 4 parts. There are 4 or

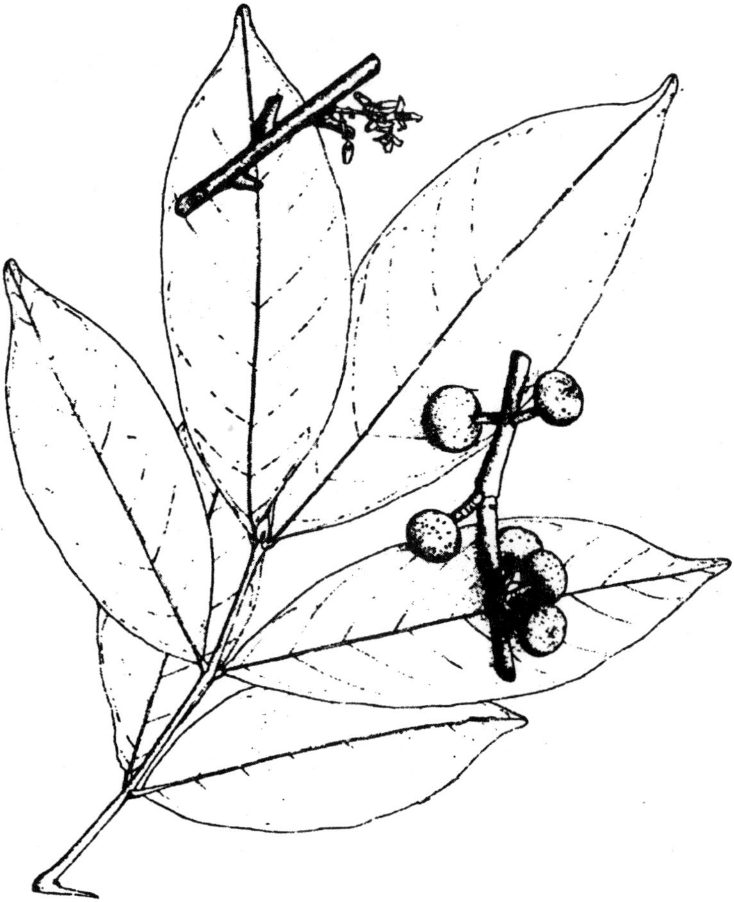

**Fig. 113 : Guaraguadillo (*Guarea ramiflora*)**

Flowers (above), leaf (below), fruits (center right), two–thirds natural size.

fewer elliptic seeds $^7/_{16}$ inch long, with reddish coat. With flowers and fruits through the year.

The sapwood is light brown or whitish and hard. If available in larger sizes, the wood should have similar uses as that of related species.

Uncommon and widely distributed in understory of moist limestone and upper and lower forests of M.P.

## 114. Neem, margosa (*Melia azadirachta* L.)

Neem tree has been introduced for shade in India. Distinguishing characters are : (1) pinnate leaves 1 foot or more in length, with 9–15 lance–shaped leaflets paired except at end, very long–pointed at apex, very unequal or oblique at blunt base, and toothed on edges; (2) many small white fragrant flowers $^{3}/_{8}$ inch across the 5 petals; and (3) light yellow elliptic fruits $^{1}/_{2}$–$^{5}/_{8}$ inch long.

A medium–sized planted tree 30 feet or more in height and 1 foot in trunk diameter, with crown of dense foliage. Bark gray, with scaly plates, the inner bark pink and bitter. Twigs brown, hairless.

Leaves crowded near ends of branches, alternate pinnate, hairless, with slender light green axis. The leaflets are mostly 6–7 pairs with short stalks of about $^{1}/_{16}$ inch, 2–3 inches long and $^{5}/_{8}$–$^{7}/_{8}$ inch wide, often curved and with curved midrib nearer lower edge, thin, dull green above, and paler beneath.

Flower clusters (panicles) are 2–6 inches long, narrow and branched. The flowers are composed of 5 light green rounded calyx lobes, 5 white oblong rounded petals $^{3}/_{16}$ inch long, 10 stamens united in a tube, and pistil with rounded ovary and slender style. The berrylike fruits (drupes) have bitter pulp and large elliptic stone. Flowering in spring.

Where native the wood is useful, the bark serves in home remedies, and an oil is obtained from the seeds.

Common as a fast–growing attractive shade or ornamental tree in India.

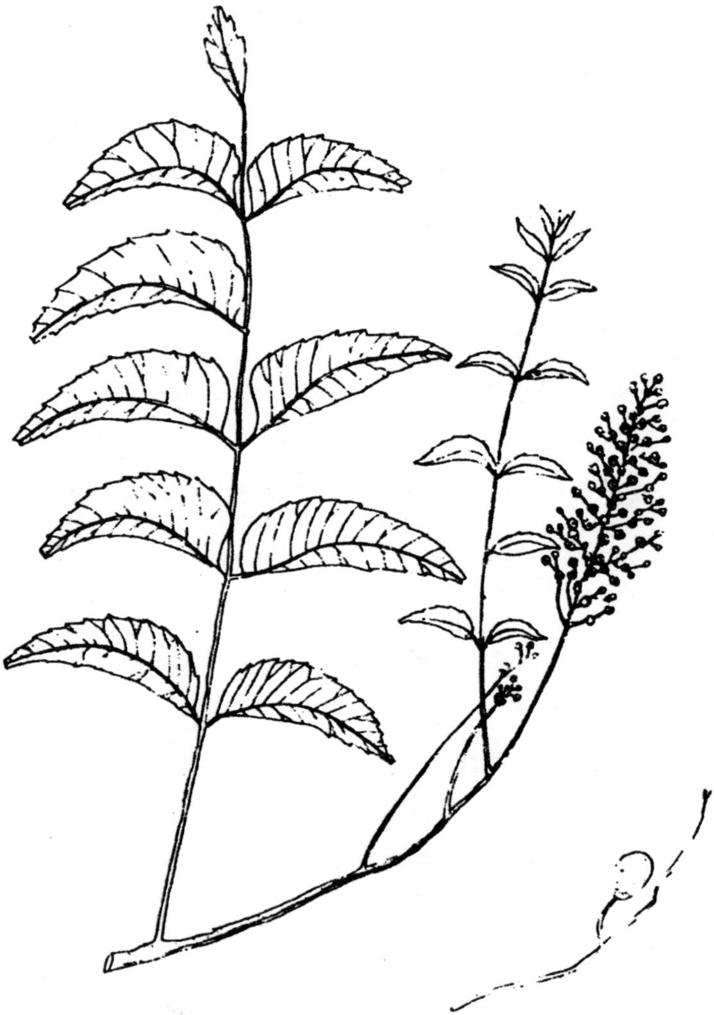

**Fig. 114 : Neem, margosa (*Melia azadirachta* L.)**
Flowering twig (above), fruit (lower right), two–thirds natural size.

## 115. Tun, Burma toon (*Toona ciliata*)

This large forest tree related to cedro hembra or Spanish–cedar has been introduced in forestry tests. It is identified by : (1) large even pinnate leaves with 10–20 mostly paired lance–shaped leaflets, slightly curved and unequal at base; (2) many small whitish flowers $^3/_{16}$ inch long in large branching clusters; and (3) oblong seed capsules $^7/_8$ – $1^1/_2$ inches long, with many seeds winged at both ends.

A planted tree 25 feet and 8 inches in trunk diameter or larger, apparently deciduous. Twigs stout, hairless, with light–colored dots (lenticels) and large raised half–round leaf scars.

Leaves alternate, about $1^1/_2$ feet long, with round axis, hairless. Leaflets 3–6 inches long and $1–2^1/_4$ inches wide, on slender stalks of $^3/_8$–$^1/_2$ inch, long–pointed at apex, short–pointed or rounded at base with sides unequal, margin sometimes slightly wavy, the midvein slightly curved and near edge toward twig, and the lower surface paler.

Flower clusters (panicles) large, much branched, bearing many short–stalked fragrant flowers. The calyx is 5–lobed, less than $^1/_{16}$ inch long; there are 5 oblong petals almost $^1/_4$ inch long; 5 stamens more than $^1/_8$ inch long on a hairy dislike base; and pistil with rounded slightly hairy 5–celled ovary, short style, and broad rounded flat stigma. Seed capsules dark brown, splitting into 5 parts from a large 5–angled axis. Seeds brown, flat and thin, $^3/_4$ inch long including wings at both ends. With flowers and fruits in summer.

The wood is suitable for cabinetwork and is resistant to dry–wood termites.

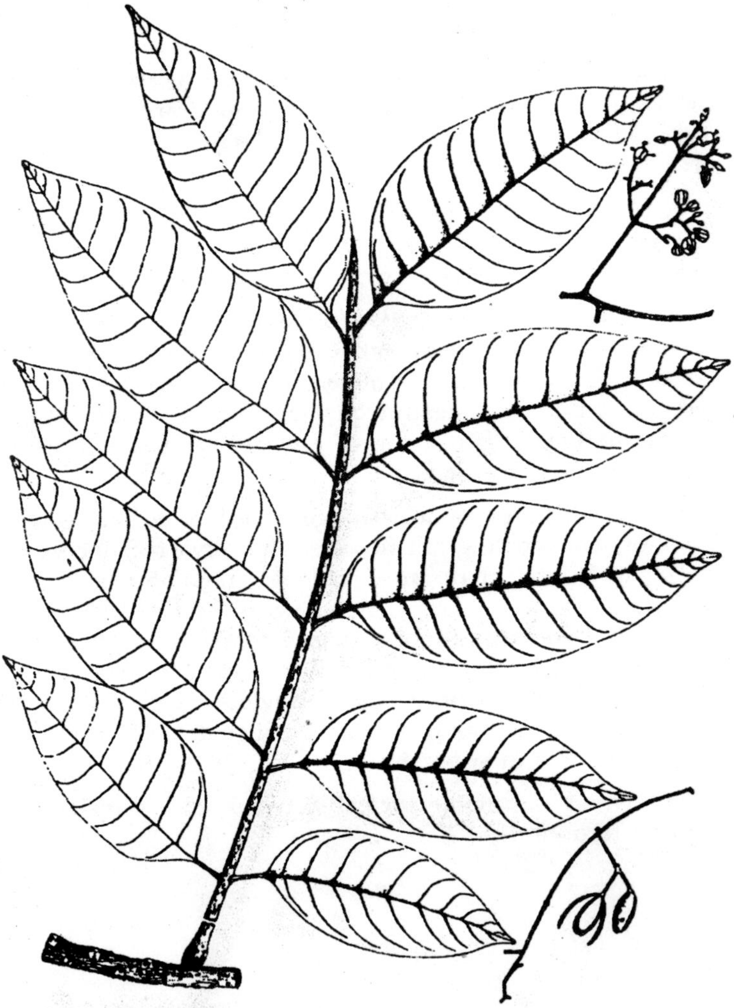

**Fig. 115 : Tun, Burma toon (*Toona ciliata*)**

Leaf, flowers (upper right), fruits (lower right), two–thirds natural size.

## 116. Bariaco (*Trichilia triacantha*)

This rare spiny–leaf shrub or small tree known only from dry areas of southwestern Maharashtra is easily recognized by the pinnate or almost palmate leaves with 3–7 small obovate or wedge–shaped leaflets each with 3 sharp spiny teeth at apex.

An evergreen shrub 5 feet high but also a small tree 15–30 feet high and 3 inches in trunk diameter. Twigs gray grown, slender, finely hairy when young. The bud consists of a minute hairy leaf.

The alternate leaves $^3/_4$–$1^1/_4$ inches long, mostly clustered near ends of twigs, have a short hairy petiole about $^1/_8$–$^1/_4$ inch long and short hairy axis of about $^1/_8$ inch, so short that they appear to be palmate or digitate. The 3–7 leaflets are paired except at end, almost stalkless, $^3/_8$ – $1^1/_4$ inches long and $^3/_{16}$–$^5/_8$ inch wide, broadest near the apex, which has 3 sharp spiny teeth $^1/_{16}$–$^1/_8$ inch long and narrowed toward the long–pointed base, thick and stiff, the edges slightly rolled under, the many side veins in a fine network. The upper surface is shiny with sunken midvein, the lower surface dull and paler.

Flower clusters (like racemes or panicles) at ends of twigs are short. Flowers few, almost stalkless, hairy, white, about $^1/_8$ inch broad, composed of short cup–shaped calyx with minute teeth, petals 5 or 4 less than $^1/_8$ inch long, thick and finely hairy; stamens in a tube; and pistil with rounded ovary, shorter hairy style, and dotlike stigma. Fruits unknown. Collected with flower buds in October and with flowers in February.

It has been suggested that this species would be an attractive ornamental shrub in dry climates.

Rare near dry limestone forest at 100–200 feet altitude in south–western Mumbai.

**Fig. 116 : Bariaco (*Trichilia triacantha*)**
Leafy twig, natural size.

# MALPIGHIA FAMILY (MALPIGHIACEAE)

Mostly woody vines, also shrubs and trees, known by : (1) leaves generally opposite, simple, mostly entire, often with pressed 2–pointed hairs attached at middle (often needle–like and stringing), frequently with glands at apex of petiole, generally with stipules; (2) flowers often showy and yellow or pink; usually along an axis (raceme), mostly bisexual, mostly regular, the calyx persistent with 5 sepals or lobes mostly with 2 glands on outside, corolla of 5 long–stalked rounded and fringed petals, 10 stamens (or partly staminodes) generally united at base, and pistil with superior ovary with usually 3 cells each with 1 ovule, and 3 styles; and (3) fruit winged (samara), capsule, berry, or drupe.

**Key to Species**

A. Flowers in terminal clusters.

B. Petals yellow.

C. Leaves partly reddish tinged, without glands to base; fruits (drupes) nearly round).

D. Leaves thin, turning rich red upon drying, upper surface shiny, midrib lighter in color than blade—Maricao, *Btrsnima coriacea* (Sw.) DC.

DD. Leaves greenish, often with prominent veins, upper surface slightly shiny, lower surface with larger veins often rusty–red hairy–116. Maricao cimarron, *Byrsonima crassifolia* (L.) H.B.K.

CC. Leaves greenish, often with 2 yellow dot glands at base of lower surface; fruits (drupes) mostly 2–lobed—*Bunchosia*.

E. Leaves oblong, light green, $1^1/_2$–3 inches long and wide—*Bunchosia glandulosa.*

EE. Leaves broadly ovate to elliptic, shiny green, 2–$5^1/_2$ inches long and $1^1/_4$–4 inches wide— Bunchosia glandulifera.

BB. Petals white, pink, or red.

F. Leaves spoon–shaped or obovate, hairless—*Byrsonima lucida.*

FF. Leaves elliptic, the lower surface densely gray hairy—*Byrsonima wadsworthii.*

AA. Flowers in lateral clusters at leaf bases, petals white or pinkish—*Malpighia.*

G. Leaves with 2 pointed stinging hairs beneath.

H. Leaves very narrow, linear or linear–oblong—*Malpighia linearis.*

HH. Leaves broader, mostly elliptic or oblong.

I. Leaves relatively large, 3–8 inches long.

J. Flowers few in stalked clusters; leaves mostly short–pointed or blunt at both ends—*Malpighia fucata.*

JJ. Flowers many in nearly stalkless clusters; leaves rounded or short–pointed at both ends, sometimes notched at apex—*Malpighia shaferi.*

II. Leaves less than 3 inches long.

K. Leaves notched at base, the edges with stinging hairs—*Malpighia infestissima.*

KK. Leaves short–pointed or blunt at base, the edge without stinging hairs—*Malpighia pallens.*

GG. Leaves becoming hairless.

L. Leaves blunt or rounded and often notched at apex—*Malpighia punicifolia.*

LL. Leaves short to long–pointed at apex—*Malpighia glabra.*

## 117. Café forastero (*Bunchosia glandulosa*)

This shrub or small tree common in dry areas near coasts is characterized by : (1) opposite oblong hairless light green leaves $1^1/_2$–3 inches long and $^1/_2$–$1^1/_2$ inches wide, short–pointed or blunt at both ends, with 2 yellow dot glands at base or lower surface; 7(2) several flowers nearly $^1/_2$ inch across the 5 spreading yellow stalked petals; and (3) fruits (drupes) egg–shaped, red or orange, $^3/_8$–$^7/_{16}$ inch long, slightly 2–lobed.

Evergreen (or deciduous) shrub or small tree to 30 feet high and 6 inches in trunk diameter. The bark is gray, finely fissured, becoming rough. Inner bark is whitish, with a slight distinctive odor and taste suggesting raw potato. The twigs are slender, slightly hairy when young, light gray or light brown, with minute dots (lenticels) also similar odor and taste. End but $^1/_{16}$ inch long, covered with light brown hairs.

The opposite leaves have slender petioles $^1/_8$–$^3/_8$ inch long. Blades are not toothed on edges, thin or slightly thickened, the upper surface light green and often slightly shiny, the lower surface dull light green.

Flowers clusters (panicles) at ends of twigs are 1–2 inches long and branched. Flowers several on short slender hairy stalks, composed of calyx less than $^1/_8$ inch long with 5 lobes and 10 glands; petals 5, $^1/_4$ inch long, deep yellow, rounded and wavy margined, stalked at base; 10 stamens united at base; and hairless pistil with 2–celled ovary, slender style, and 2–lobed stigma. The fruits (drupes) have calyx glands at base and style at apex and contain thin orange–red pulp and 2 large light brown stones, each 1–seeded. Recorded with flowers from May to October and with fruits from June to August.

The sapwood is light brown and hard.

Uncommon but widely distributed in dry and moist limestone forests near coasts and lower Cordillera (south slope) in southwestern and northwestern parts and northeastern corner of Assam and also other islands eastward from sea level to 500 feet altitude.

**Fig. 117 : Café forastero (*Bunchosia glandulosa*)**
Fruiting twig (left), flowers (upper right), natural size.

## 118. Long Key byrsonima (*Byrsonima lucida* DC.)

This shrub or small tree is recognized by : (1) opposite small spoon–shaped (spatulate) or obovate leaves $^3/_4$–2 inches long and $^1/_4$–$^7/_8$ inch wide, shiny blue green on upper surface; (2) flowers $^1/_2$ inch across the 5 spreading stalked rounded petals that turn from white to red or yellow; and (3) round brown fleshy fruits $^1/_2$ inch in diameter.

Evergreen shrub or small tree, becoming 25 feet high and 6 inches in trunk diameter, with widely spreading rounded or irregular crown. The gray bark is smooth, slightly warty (lenticels), becoming furrowed into plates and rough. Inner bark is pink and slightly bitter and fibruous. Twigs are gray, slender and short, hairless, with raised dots (lenticels).

The opposite leaves are rounded at apex and narrowed at base into a stalk $^1/_{16}$ inch long, slightly turned under at edges, slightly thickened, hairless, the upper surface shiny blue green, and the lower surface dull yellow green.

Flower clusters (racemes) at ends of twigs are 1–2 inches long. Flowers few on slender hairy stalks, slightly fragrant, composed of calyx $^1/_8$ inch long with 5 pointed lobes and 2 oblong glands at base outside; corolla of 5 petals $^1/_4$ inch long, rounded, wavy margined, and concave with stalk at base; stamens 10, separate; and pistil with 3–angled 3–celled ovary with 3 ovules and 3 slender styles. The fleshy edible fruits have calyx glands and the styles persistent. Within the flesh is a thick–walled stone containing 3 or fewer seeds. Flowering and fruiting intermittently through the year.

Sapwood is light brown, and heartwood dark brown. Elsewhere, the hard, heavy, fine–textured wood has been used for furniture.

This attractive plant merits cultivation as an ornamental shrub.

Scattered and locally common at low altitudes, mainly in dry coastal forest and lower south slopes of Tamil Nadu from sea level to 2,000 feet altitude in southeastern and southwestern.

**Fig. 118 : Long Key byrsonima (*Byrsonima lucida* DC.)**
Flowering twig (above), fruiting twig (below), natural size.

## 119. Almendrillo (*Byrsonima wadsworthii*)

This rare tree known only from high altitude in Aravali mountains is distinguished by : (1) opposite small elliptic leaves, thick and turned under at edges, the lower surface densely gray hairy; (2) flowers nearly $^1/_2$ inch across the 5 spreading white to pink stalked petals, several in terminal clusters; and (3) egg–shaped fleshy stone fruits nearly $^1/_2$ inch long, pink or red.

Small to medium–sized evergreen tree 15–45 feet high and to 8 inches in trunk diameter, much branched, with compact dense crown, broad to narrow, of dark green foliage. The bark is brown or dark brown, with many small fissured or slightly furrowed. Inner bark is red streaked in outer part and pinkish beneath, bitter. Twigs gray, with dense pressed hairs, becoming hairless and fissured.

Leaves opposite, with thick gray hairy petioles $^1/_8$–$^3/_8$ inch long and with hairy pointed stipules less than $^1/_8$ inch long, united at base. Leaf blades are elliptic, $1^1/_2$–3 inches long and 1–$1^1/_2$ inches wide, slightly notched or blunt at apex, rounded or short–pointed at base, the upper surface green or dark green, slightly shiny, becoming nearly hairless, the midvein and 4–6 pairs of side veins slightly sunken.

The unbranched flower clusters (racemes) $1^1/_4$–$2^3/_4$ inches long at ends of twigs bear several flowers on slender stalks less than $^3/_8$ inch long. Flowers are composed of 5 pointed brown hairy sepals $^3/_{16}$ inch long bearing 2 minute oblong glands at base; 5 spreading petals nearly $^1/_4$ inch long, rounded and finely toothed and with narrow stalk, white but turning pink; 10 stamens; and pistil with hairy 3–angled 3–celled ovary and 3 slender styles. The fruits (drupes) with calyx at base have thin flesh and large whitish 3–celled stone $^1/_4$ inch long containing 3 seeds $^3/_{16}$ inch long. Flowering from June in September and October.

The wood is light brown and hard.

Almendrillo is distributed in upper dwarf forests on ridges along the mountain summits mostly at 2,500–3,000 feet altitude, sometimes as low as 1,800 feet and above 4,000 feet on the highest peaks. It is rare and scattered, most common in Mountains but present also in central and western high mountains.

**Fig. 119 : Almendrillo (*Byrsonima wadsworthii*)**
Flowering twig (above), fruiting teig (lower right), natural size.

## 120. Palo bronco (*Malpighia fucata*)

This shrub or small tree with stinging needle–like hairs flat on lower leaf surfaces is native in eastern U.P. It is recognized by : (1) many yellowish needlelike hairs about $^1/_4$ inch long, attached at middle and 2–pointed, flat on lower leaf surfaces; (2) opposite leaves of varying shape, mostly oblong, elliptic or lance–shaped, larger than in related species; (3) many flowers in stalked clusters at leaf bases, about $^3/_4$ inch across the 5 spreading white or pink rounded petals fringed at edges and stalked at base; and (4) rounded red fleshy fruits about $^5/_8$ inch in diameter.

Evergreen shrub or small tree to 20 feet high and 3 inches in trunk diameter. Twigs slender, light brown, with flat pressed hairs when young.

The opposite leaves have short leafstalks $^1/_8$–$^3/_8$ inch long. Blades are 3–8 inches long and 1–4$^1/_2$ inches wide, mostly short–pointed or blunt at both ends, slightly thickened and slightly turned under at edges. The upper surface is green and hairless, and the lower surface paler and with many needlelike hairs.

Flower clusters (cymes) have many flowers on stalks $^1/_2$–$^3/_4$ inch long at the end of a short stalk at leaf base. The flowers consist of 5 sepals more than $^1/_8$ inch long each with 2 oblong glands; 5 slightly unequal stalked white or pink petals about $^3/_8$ inch long, 10 stamens, and pistil with 3–celled ovary and 3 styles enlarged at apex. The fruits (drupes) have thin flesh and large stone with many vertical ridges. With flowers and fruits intermittently during the year.

This species is uncommon in lower and moist coastal forests from sea level to 1,800 feet altitude in northeastern and southern U.P.

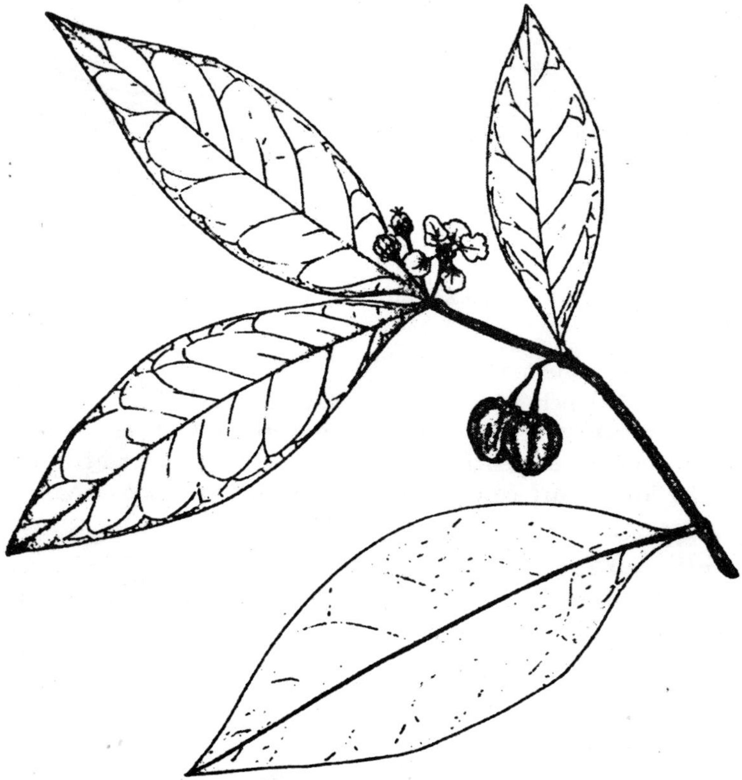

**Fig. 120 : Palo bronco (*Malpighia fucata*)**
Two–thirds natural size.

## 121. Huesito (*Malpighia globra* L.*)

This shrub or small tree apparently introduced though recorded as native, is a close relative of West–Indian–cherry, *Malpighia puni–cifolia* L.,* and has been confused with the latter. Distinguishing characters are : (1) paired lanceolate, elliptic, or ovate leaves 1 $^1/_4$–2$^1/_2$ inches long and $^1/_2$–1 inch wide, hairless or nearly so; (2) few flowers at leaf bases, about $^5/_8$ inch across the 5 pink rounded wavy–edged stalked petals; and (3) round red fruits $^3/_8$ inch or more in diameter, slightly flattened, scarlet red, and edible.

An evergreen much–branched shrub or small tree to 15 feet or more in height with smoothish gray bark.

The opposite leaves have short petiole less than $^1/_8$ inch long. Blades are long–to short–pointed at apex, short–pointed to rounded at base, not toothed on edges, thin, hairless or nearly so, shiny green on upper surface and paler beneath.

Flower clusters (cymes) are composed of few flowers on slender stalks of $^1/_8$–$^5/_8$ inch long. The calyx has 5 lobes, each with 2 glands at base; the 5 petals are about $^5/_{16}$ inch long; there are 10 stamens; and the pistil has 3–celled ovary and 3 styles. The fruits with calyx at base have sour flesh and 1–3 stones.

Grown elsewhere as an ornamental, in hedges, and for the edible sour cherrylike fruits.

**Fig. 121 : Huesito (*Malpighia globra* L.*)**
Flowers (upper right) and leafy twig, natural size.

# SPURGE FAMILY (EUPHORBIACEAE)

## 122. Maran (*Croton astroites*)

A shrub, rarely small tree, common in dry areas, mostly in Lakshdeep Islands, distinguished by : (1) leaves and flowers densely covered with whitish star–shaped hairs; (2) petioles and young twigs with orange or rusty–colored hairs; (3) ovate leaves $1^1/_4$–3 inches long and $^3/_8$–$1^1/_4$ inches wide, with 5 or 7 veins from base; (4) minute greenish flowers in narrow erect clusters, the female at base and male above (monoecious); and (5) elliptic, slightly 3–lobed, densely hairy seed capsules more than $^1/_4$ inch long.

Evergreen shrub or rarely a tree to 15–20 feet high and 3 inches in trunk diameter. The bark is gray and smooth.

The petioles of the alternate leaves are $^1/_4$–1 inch long, with orange or rusty–colored hairs. Leaf blades are short–to long–pointed at apex, rounded or slightly notched at base, sometimes with small teeth on edges, densely covered with whitish star–shaped hairs, paler beneath.

Flower clusters (racemes) 1–2 inches long bear many nearly stalkless male flowers $^1/_8$ inch long with 5–lobed hairy calyx and about 16 stamens. Female flowers few at base of same clusters, about $^1/_8$ inch long, with 5–lobed calyx, 3–celled densely hairy ovary and 3 threadlike styles, each divided. The seed capsules have at the base the enlarged calyx to $^5/_{16}$ inch long and split into 6 parts. Flowering and fruits through the year.

Common and forming thickets in dry areas near sea level and at low altitudes, particularly in heavily grazed areas. These unpalatable plants increase in numbers after other species are eliminated by severe grazing.

The young twigs and sap formerly were used elsewhere in home remedies.

**Fig. 122 : Maran (*Croton astroites*)**

Twig with flowers and fruits (above), twig with male and female flowers (below), natural size.

## 123. Adormidera (*Croton rigidus*)

This shrub widespread in dry areas rarely becomes treelike. Identified by : (1) orange bitter sap; (2) the pungent odor of crushed leaves; (3) twigs, leaves, and flowers densely covered with whitish or gray star–shaped hairs; (4) elliptic to lanceolate or ovate leaves mostly 1–2 inches long and $^{3}/_{8}$–$^{3}/_{4}$ inch wide, sometimes larger, with paired dotlike glands at base beneath; (5) minute greenish flowers in narrow erect clusters, the female at base and male above (monoecious); and (6) rounded, slightly 3–lobed, densely hairy brown seed capsules about $^{3}/_{16}$ inch in diameter.

An evergreen shrub or rarely treelike and 15 feet high and 2 inches in trunk diameter or larger. The bark is light gray, smooth or slightly fissured, with orange bitter sap. Inner bark is light brown and slightly bitter.

The alternate leaves have petioles $^{1}/_{8}$–$^{1}/_{2}$ inch long. The blades are blunt to long–pointed at apex, rounded at base, not toothed on edges, densely covered with whitish or gray star–shaped hairs on both surfaces, paler beneath.

The flower clusters (racemes) 1–2 inches long bear many nearly stalkless male flowers $^{1}/_{8}$ inch long with 5–lobed hairy calyx and 12–16 stamens. The few female flowers at base of some clusters are slightly larger, with 3–celled hairy ovary and 3 threadlike styles, each divided. Seeds 3, gray, less than $^{1}/_{8}$ inch long. Flowering and fruiting nearly through the year.

The wood is hard and whitish.

Abundant and forming thickets in dry areas, near sea level and at low altitudes, especially where heavily grazed by goats and other animals. These unpalatable plants spread and become dominant after other species are eliminated by severe grazing.

**Fig. 123 : Adormidera (*Croton rigidus*)**
Fruiting twig (upper left), flowering twig (below), natural size.

## 124. Ditta (*Ditta myricoides*)

Ditta, a resinous small tree or shrub of the eastern mountains, is recognized by : (1) reverse lance–shaped (oblanceolate) leaves $1^3/_4$–$3^1/_2$ inches long and $^3/_8$–1 inch broad, widest beyond middle, short–pointed at apex and long–pointed and tapering toward the short leaf–stalk, slightly thickened, the edges minutely toothed and turned under; (2) flowers male and female, minute and inconspicuous, less than $^1/_8$ inch long, stalkless at base of leaves; and (3) seed capsules nearly round, about $^1/_4$ inch long.

A small evergreen resinous tree or shrub to 35 feet high and 6 inches in trunk diameter, hairless throughout. The bark is slightly fissured, very light brown, sometimes covered with mosses. Inner bark is almost tasteless. The twigs are light brown.

The alternate leaves have short leafstalks $^1/_8$–$^1/_4$ inch long, prominent raised veins on both sides, are shiny green to dark green on upper surface and slightly paler beneath. A resinous exudates is produced at leaf bases.

Male flowers are borne several in a cluster at the base of a leaf. They are about $^1/_{16}$ inch long and broad, yellowish, each consisting of 3 broad sepals and several almost stalkless stamens. Female flowers on different plants (dioecious) are borne 1 or 2 together, stalkless, less than $^1/_8$ inch long, consisting of 2–3 broad sepals, a round green 2–celled ovary with 2 ovules and 2 deeply 2–forked brown stigmas. The seed capsules are nearly round, about $^1/_4$ inch long, and contain 1 or 2 light brown elliptic seeds $^3/_{16}$ inch long, covered with minute tubercles. Found in flower from May to September and in fruit from August to November.

The sapwood is light brown, hard, moderately heavy, and fine–textured.

Common in upper Meghalaya forests and dwarf forest at 2,000–2,500 feet altitude in eastern and central mountains of Assam.

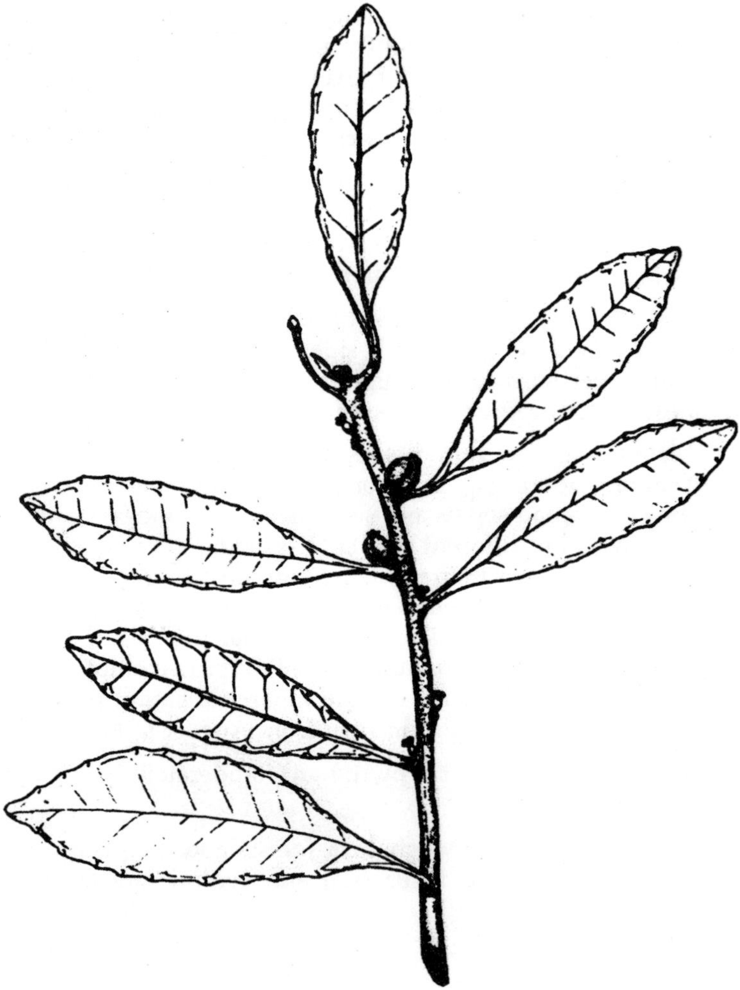

**Fig. 124 : Ditta (*Ditta myricoides*)**
Natural size.

## 125. Hueso (*Drypetes alba*)

This small tree is characterized by : (1) ovate to lance–shaped leaves 2–5 inches long and $^3/_4$–$4^1/_2$ inches wide, long–pointed at apex, short–pointed and unequal or oblique at base, and inconspicuously wavy–toothed on margin, the teeth sometimes sharp; (2) small yellow–green flowers clustered at leaf bases, male and female on different trees (dioecious); and (3) whitish elliptic fruits more than $^3/_8$ inch long, slightly oblique and curved, 2–7 on short stalks at leaf bases.

Evergreen small tree to 35 feet high and 6 inches in trunk diameter. The bark is gray, smoothish with small warts (lenticels). Inner bark is light brown, gritty and slightly bitter. The twigs are gray to brown, slender, and hairless.

The leaves are alternate in 2 rows, hairless, with slender petioles $^1/_4$–$^3/_8$ inch long. Blades are unequal or oblique at base, slightly thickened and leathery, the side veins inconspicuous, the upper surface green and shiny, the lower surface pale green.

Male flowers on stalks more than $^1/_8$ inch long are numerous, about $^1/_8$ inch long, consisting of 4–6 sepals less than $^1/_{16}$ inch long and 4–6 somewhat longer stamens opposite them and around a disk. Female flowers have similar sepals and pistil with hairy egg–shaped 1–celled, 2–ovuled ovary and 1 style. The fruits (drupes) are clustered on stalks $^3/_{16}$ inch long, egg–shaped and slightly oblique and curved, with sepals at base and style remaining near end, the surface minutely hairy. Within the stone is 1 seed. With flowers and fruits in spring and summer.

The wood is light brown and hard.

Common in moist limestone and lower Sunderban forests at 200–2,500 feet altitude mostly in western also north–eastern.

**Fig. 125 : Hueso (*Drypetes alba*)**
Flowering twig (above), fruiting twig (below), natural size.

## 126. Encinillo (*Drypetes ilicifolia*)

This distinctive shrub or small tree of limestone hills and sand dunes of northern and western Rajasthan is easily recognized by its hollylike, spiny–margined leaves. Its characters for identification are : (1) oblong or elliptic leaves, stiff, thick and leathery, with few sharp spiny yellowish teeth along wavy edges; (2) small yellow–green flowers, male and female, clustered at leaf bases; and (3) elliptic fruits nearly $^{3}/_{4}$ inch long, slightly oblique.

Evergreen shrub or small tree to 40 feet high and 8 inches in trunk diameter. The light gray bark is rough and fissured, the inner bark light brown and tasteless. Twigs are whitish gray, finely warty (lenticels), hairless, often angled when young.

The leaves are alternate in 2 rows, hairless, with petioles less than $^{1}/_{4}$ inch long. The blades are 2–6 inches long and 1–2 inches wide, the short–pointed apex ending in a spine nearly $^{1}/_{8}$ inch long, and the base rounded or notched and slightly unequal or oblique, slightly concave and bent up from midrib. The upper surface is green and the lower surface yellow green, both with a prominent network of raised veins.

The flowers are male and female on different trees (dioecious), the female apparently not yet described. Male flowers clustered at leaf bases on stalks more than $^{1}/_{16}$ inch long have 4–5 sepals nearly $^{1}/_{8}$ inch long, greenish and finely hairy, and 4–5 opposite stamens less than $^{1}/_{8}$ inch long. The fruits (drupes) with 5 minute hairy sepals at base are finely hairy, have 1 stigma, and are thick–walled, 1–celled, and 1–seeded. With flowers in spring and summer and with fruits in summer.

The wood is light brown and hard.

Common in moist limestone forest and recorded from sand dunes, 200–500 feet altitude in northern and western Rajasthan.

**Fig. 126 : Encinillo (*Drypetes ilicifolia*)**

Twig with male flowers (left), twig with female flowers and fruits (right), natural size.

## 127. Guiana–plum (*Drypetes lateriflora*)

This small tree rare in moist limestone forest is identified by: (1) oblong or elliptic leaves 2–4$^1/_2$ inches long and $^3/_4$–1$^3/_4$ inches wide, long–pointed at apex and short–pointed and slightly unequal at base, with margin straight or slightly wavy, in 2 rows on slender twigs; (2) many small greenish white flowers clustered at leaf bases, male and female on different trees (dioecious); and (3) orange–red or brown egg–shaped fleshy fruits $^1/_2$–$^5/_8$ inch long with large stone.

Evergreen small tree 30 feet high and 5 inches in trunk diameter. Bark light gray, smoothish to finely fissured, the inner bark light brown and slightly bitter. Twigs are gray, slender and hairless.

The leaves are alternate in 2 rows, hairless, with slender petioles $^1/_4$–$^3/_8$ inch long. Blades are often unequal at base, slightly thickened and leathery, the margin straight or minutely wavy, not toothed (spiny toothed on shoots from cut stumps), both surfaces green to dark green and slightly shiny with veins raised.

The flowers are clustered at leaf bases on stalks about $^1/_8$ inch long. Male flowers about $^1/_8$ inch long have 4–5 pointed greenish–white sepals nearly $^1/_8$ inch long, hairy on edges and 3–5 stamens longer than sepals and opposite them and around a disk. Female flowers have similar sepals and in a saucer–shaped disk the pistil with hairy 2–celled ovary with 2 ovules in each cell and 2 styles. The fruits (drupes) borne singly at leaf bases are finely hairy and have remains of sepals at base and 2 styles at apex. The orange–red bitter pulp encloses the stone and 1 elliptic brown seed about $^1/_4$ inch long. Flowering in spring and summer and with fruits in summer.

The wood is described as dark brown with thick light brown or yellowish sapwood, heavy, hard, brittle, and fine–textured.

Rare in moist limestone forest at 200–400 feet altitude in northern and western Calcutta. Not on nearly islands.

**Fig. 127 : Guiana–plum (*Drypetes lateriflora*)**
Flowering twig (lower left), fruiting twig (right), natural size.

## 128. Carrasco, poison spurge (*Euphorbia cotinifolia* L.*)

Carrasco or poison spurge is a very poisonous uncommon ornamental shrub or small tree. It is characterized by : (1) abundant poisonous white latex or milky sap, caustic and irritating to the skin; (2) broadly ovate to nearly round leaves mostly 2–3 inches long and $1^1/_2$–$2^1/_2$ inches wide, sometimes larger, green to coppery purple, mostly 3 at a node on very slender petioles often longer than blades; (3) inconspicuous minute male and female flowers borne in small greenish cups $^3/_{16}$ inch across the whitish borders in terminal crowded clusters, and (4) small 3–angled capsules $^3/_{16}$ inch across the whitish borders in terminal crowded clusters; and (5) small 3–angled capsules $^3/_{16}$ inch in diameter.

A deciduous planted shrub or rarely small tree to 15 feet high. Bark whitish and smoothish. Poisonous white latex flows profusely from cuts on twigs and bark. The twigs are hairless or nearly so and have 2 dot glands between each 2 leaves.

Leaves mostly in 3's (whorled). Blades are rounded at both ends or notched at base, not toothed on edges, thin, beneath paler and sometimes slightly hairy.

Flower clusters (cymes in panicles) are terminal, much branched, and showy. The cups contain minute male and female flowers (monoecious). The seed capsules are hairy when young and contain 3 seeds.

Introduced as an ornamental for the reddish foliage but not common. Further planting of this dangerous shrub is not recommended. Elsewhere, notably in Central India, the plants are grown in hedges, borders, and as living fence posts, propagated from cuttings.

The milky sap produces a rash in contact with the skin of many persons (like poison–ivy) and when taken internally causes violent vomiting and may be fatal. Indians used the sap as a poison on arrows and for catching fish. The seeds reportedly have strong laxative properties.

Fig. 128 : Carrasco, poison spurge (*Euphorbia cotinifolia* L.*)
One–half natural size.

## 129. Candelero, mottled spurge (*Euphorbia lactea*)

This cactuslike shrub or small tree planted for ornament and in hedges is easily identified as a spurge by the abundant white poisonous latex that flows from cuts. Other distinguishing characters are : (1) fleshy or succulent green stems and branches 3–angled and 1–3 inches across; and (2) leaves minute and shedding early, represented by paired gray spines (stipules) $^1/_4$–$^1/_8$ inch long from raised edges or ridges of branches.

A usually leafless but evergreen shrub or small tree to 25 feet high and 6 inches in trunk diameter, with fleshy or succulent stems, much branched, hairless throughout. Stems with whorls of branches nearly to base but on large plants shedding the spiny tissue and developing a rounded brown, fissured trunk. The 3–angled (sometimes 4–angled) branches are mostly joints 4–12 inches long and 1–3 inches across, slightly shiny dark green, with yellowish or whitish streak in the groove of the axis between the angles. White latex runs abundantly from the outer part of cut stems. The soft cut branches have a light green outer layer less than $^1/_8$ inch thick, which yields latex, and within whitish watery tissue, slightly bitter.

Raised leaf based $^1/_4$–$^1/_2$ inch high and about $^3/_4$–1 inch apart along the edges cf branches correspond to nodes and bear paired spreading gray spines (stipules). The few scattered leaves are alternate, minute, stalkless, rounded, $^1/_8$–$^1/_4$ inch long, slightly shiny green, succulent, slightly thick and shedding early, or absent.

The flowers are small and inconspicuous, borne intermittently.

Wood of trunks is whitish and very soft. The latex or milky juice is bitter, poisonous, and caustic. It should not come in contact with sores, eyes, mouth or mucous membranes. Reportedly it injures the eyes and can cause blindness.

These plants with succulent stems and water storage tissue are adopted to dry regions. Like cacti, they have developed similar compact form with reduced surface and less water loss, also green stems replacing leaves in food manufacture.

Plants are grown for ornament, in fences, and trimmed as hedges in Rajasthan, uncommon and sometimes persisting afterwards. Also southern India and elsewhere in tropical regions. Escaping from cultivation, spreading and forming thickets in some places. Recommended for hedges in sand dunes along ocean fronts. Propagated easily by cuttings.

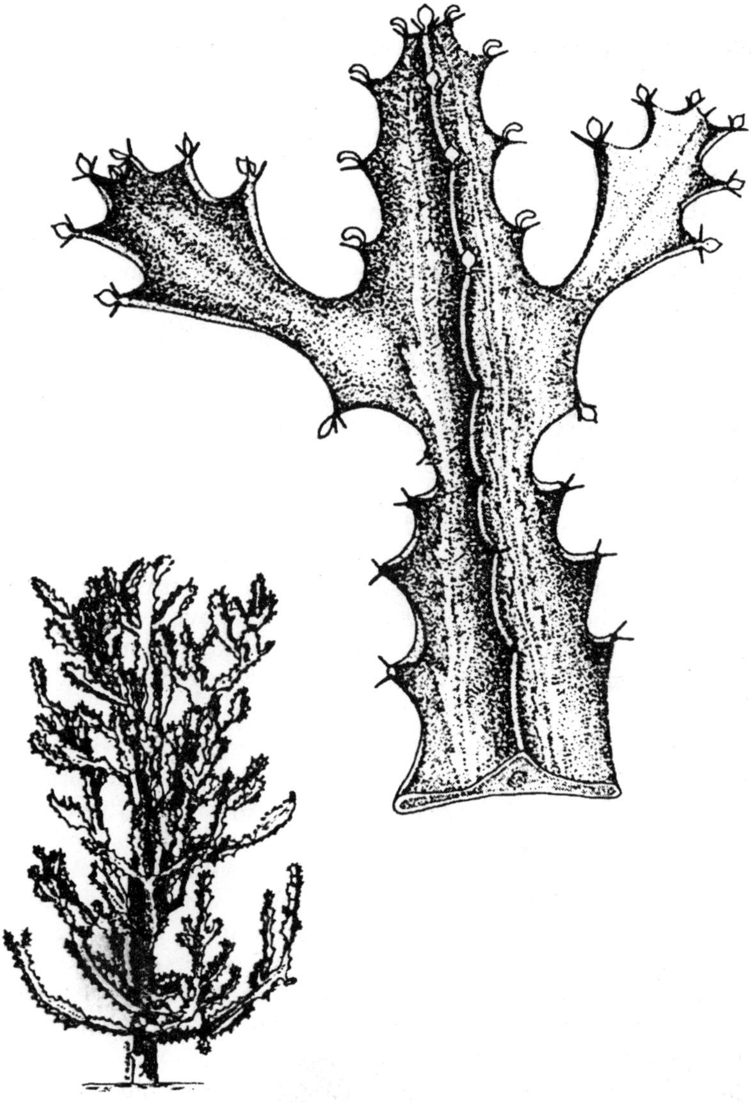

**Fig. 129 : Candelero, mottled spurge (*Euphorbia lactea*)**
Leafy twig (above), natural size; plant at lower leaf.

## 130. Hedge euphorbia (*Euphorbia neriiflia* L.*)

This ornamental spurge with fleshy leaves and stems yields white poisonous latex abundantly from cuts. Other characters for identification are : (1) stout dull green branches $^{3}/_{4}$–1 inch in diameter, bluntly 5–angled; (2) thick stiff leaves narrowly obovate or oblong, 3–5 inches long, sometimes larger; and (3) paired gray spines (stipules) $^{1}/_{16}$–$^{1}/_{8}$ inch long at leaf base and persistent on stem.

Evergreen or deciduous shrub or small tree 25 feet high and 6 inches in trunk diameter or larger, with relatively few stout fleshy or succulent branches, hairless throughout. Trunks becoming round, light brown, smoothish, slightly fissured, with scattered spines persistent. The stout dull green branches or twigs are soft and consist of a thin light green outer layer about $^{1}/_{16}$ inch thick and beneath whitish watery tissue, slightly bitter. Twigs end in a rounded or blunt tip with young pointed leaves.

The alternate leaves spread at right angles to the twig and are scattered in 5 vertical rows along the trunk. They are commonly 3–5 inches long and 1–2 inches wide, sometimes larger, tapering to a short stout petiole and widest near the blunt apex, with short point, thick, fleshy, and stiff, curved up from center to edges, without visible veins, shiny green above the dull green beneath. At the base are 2 gray spines (stipules). Upon shedding, a rounded light brown leaf scar $^{1}/_{4}$ inch across remains with the paired spines at the slightly raised base.

A few flower clusters (cymes) $^{1}/_{2}$–1 inch long with stout forking branches are borne near ends of ridges or at bases of upper leaves. Flowerlike heads (involucres) few on short stout stalks, greenish, fragrant, cuplike, about $^{3}/_{8}$ inch across, bordered by 5 short petallike greenish lobes each with gland at base, containing 1 female flower and many male flowers (monoecious). The female flower, which opens first, about $^{1}/_{8}$ inch long is a pistil with 3–angled 3–celled ovary with 3 ovules and 3–forked style. Many male flowers borne a few at a time consist of 1 dark red stamen about $^{1}/_{8}$ inch long, in mass of greenish hairlike scales. Flowering intermittently.

The latex or milky juice is bitter, poisonous, and caustic.

Planted for ornament and in hedges in Rajasthan and elsewhere through the tropics. Cristate forms with enlarged or crested branches are in cultivation.

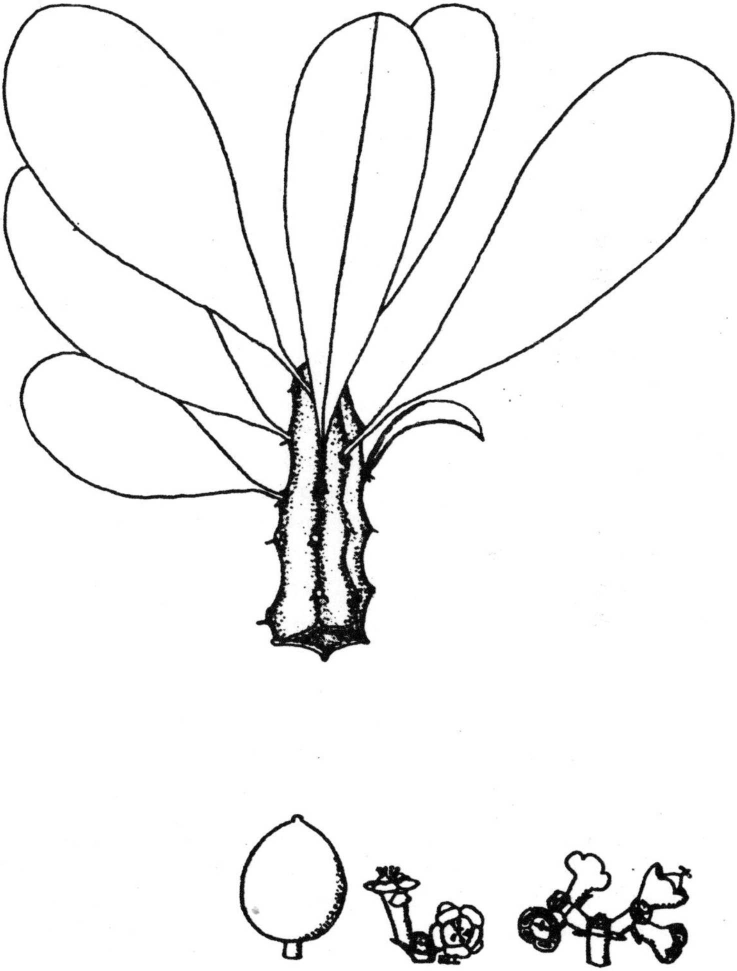

**Fig. 130 : Hedge euphorbia (*Euphorbia neriifolia* L.*)**
Leafy twig (above), fruit and male and female flowers (below), natural size.

## 131. Pencilbush, Milkbush (*Euphorbia tirucalli* L.*)

This ornamental succulent shrub or small tree has abundant white poisonous latex as in other spurges and differs also from cacti in absence of spines. It is recognized by : (1) many cylindric pencillike fleshy twigs or joints, forking upward; and (2) narrow green leaves $^1/_2$–1 inch long, shedding early and usually absent.

Shrub or small tree becoming 30 feet high and 6 inches in trunk diameter, with evergreen fleshy or succulent branches but usually leafless, hairless throughout. Bark dark green, smooth or rough. Branches many, whorled or sometimes single, curving outward and erect to form an irregular brushlike crown. The pencillike dull green twigs or joints commonly 3–5 inches long and $^1/_4$ inch in diameter, or only $^1/_8$ inch at the rounded end and becoming $^1/_2$ inch. The smooth hairless surface often has many fine whitish lines. White latex, caustic and poisonous, flows from cut surfaces. Under the surface the twigs are light green, with sour taste.

Leaves few, scattered, alternate, oblanceo–late, $^1/_2$–1 inch long, $^1/_8$ inch wide, broadest beyond middle, short–pointed at apex and tapering to stalkless base, slightly thick and succulent, the upper surface dull green, the lower surface paler with faint midvein.

Flower clusters are yellowish heads or balls $^1/_2$–$^3/_4$ inch across, stalkless at ends of twigs, composed of many crowded cuplike clusters (involucres) $^1/_8$–$^3/_{16}$ inch across. Within are many male flowers each of 1 stamen on a stalk. Flowering intermittently.

Wood whitish, soft.

Planted for ornament in Rajasthan. Elsewhere also a hedge plant. Propagated by cutting. In Southern India recommended for dune planting. Northward in others States grown as a potted plant. The milky sap causes a rash on the skin of some persons and is poisonous if taken internally.

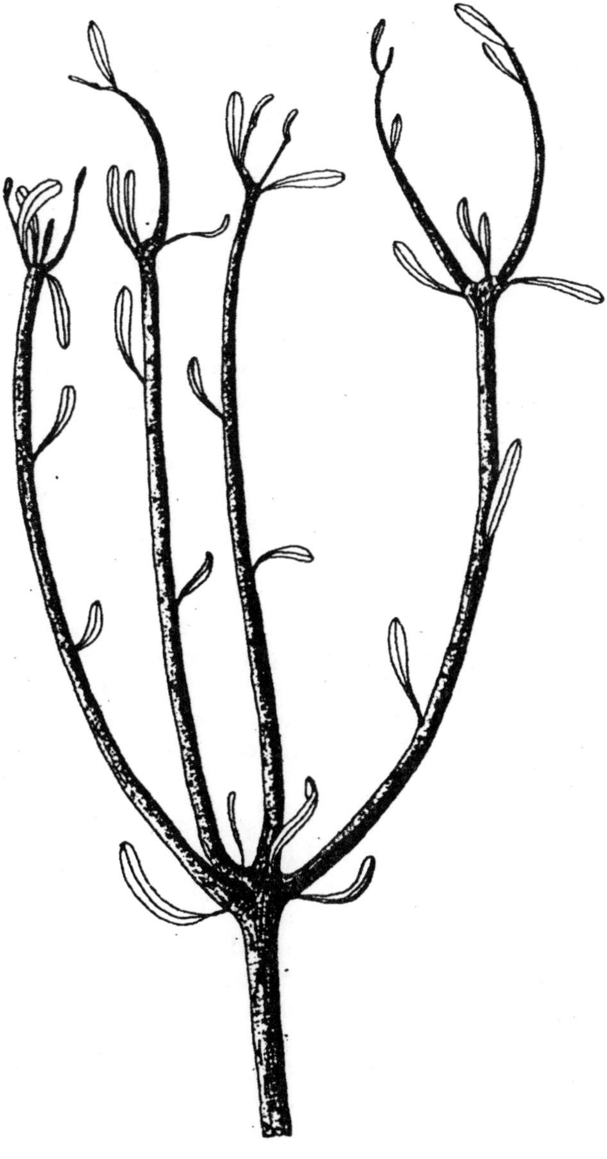

**Fig. 131 : Pencilbush, milkbush (*Euphorbia tirucalli* L.*)**
Leafy twig, natural size.

## 132. Tartago, physic–nut (*Jatropha curcas* L.*)

Tartago or physic–nut is a small ornamental tree escaping from cultivation, identified by : (1) the latex between water and milk in color that runs from cuts; (2) broadly ovate or heart–shaped leaves mostly slight 3–or 5–lobed or angled, with 5 or 7 main veins from base, 3–5$^1/_2$ inches long and nearly as wide, the petioles mostly longer than blades, crowded at ends of stout twigs; (3) yellow–green bell–shaped flowers $^1/_4$ inch long and broad, several on short stalks at leaf bases; and (4) elliptic dark brown capsules 1–1$^1/_2$ inches long and $^7/_8$ inch across, splitting into 3 parts, with 3 poisonous seeds.

Deciduous small tree or shrub 10–15 feet high and to 6 inches in trunk diameter, with rounded thin crown of few spreading branches. Bark light brown, smoothish and slightly warty, the inner bark brown and slightlybitter. Twigs are stout, $^1/_4$–$^1/_2$ inch in diameter, light green but becoming gray, smooth with whitish dots (lenticels) and raised half–round leaf scars, hairless, with chambered pith.

Leaves alternate but mostly crowded at ends of twigs. Petioles light green, round 2$^1/_2$–5$^1/_2$ inches long. Blades are broadly ovate, deeply notched or heart–shaped at base, short–pointed at apex, and with 2–4 blunt lobes on sides, thin, hairless or nearly so. The upper surface is dull green with veins yellowish and slightly sunken, the lower surface dull light green with raised veins.

Flowers several to many in greenish branched clusters (cymes) 1–3 inches long at leaf bases, fragrant, on stalks about $^1/_4$ inch long, male and female on the same plant (monoecious). The calyx consists of 5 sepals nearly $^1/_8$ inch long; corolla yellow green, bell–shaped, $^1/_4$ inch long, divided to middle into 5 spreading narrow lobes, hairy within, and there are 5 minute glands at base. Male flowers many, with 10 stamens, 5 united at base and 5 united in a column. Female flowers borne singly have pistil with elliptic 3–celled 3–ovuled ovary, short style, and 3 spreading 2–forked stigmas. The capsules with enlarged sepals at base are slightly fleshy but become dry and split into 3 parts. There are 3, sometimes 2, oblong blackish seeds about $^5/_8$ inch long, similar to those of castorbean. Flowering irregularly through the year.

The wood is white and very soft and spongy.

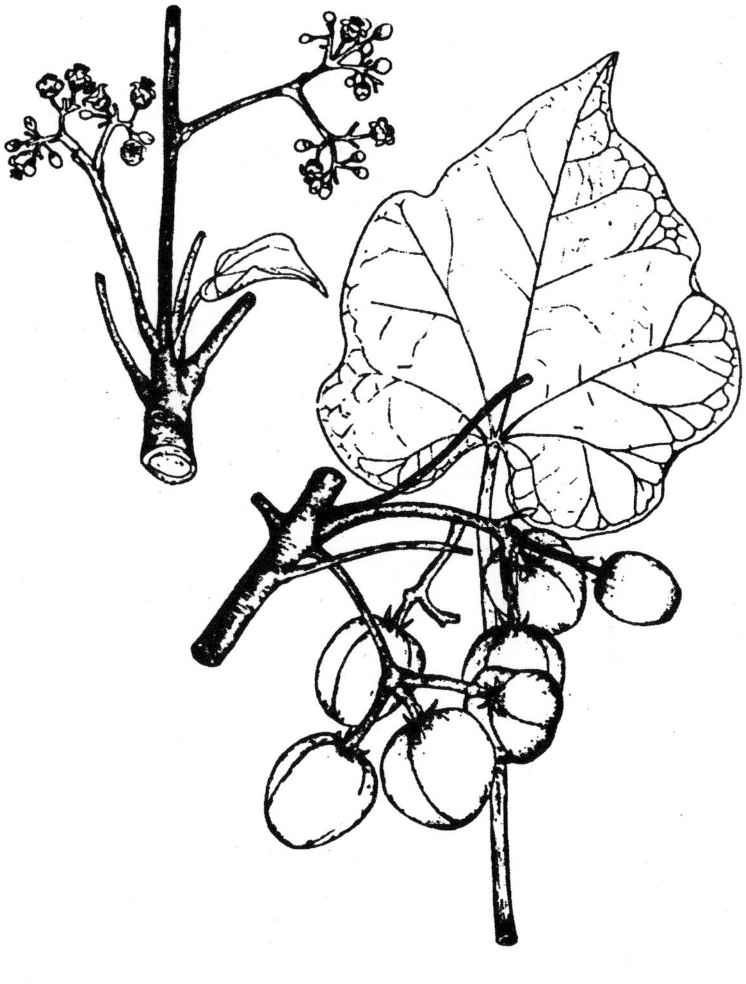

**Fig. 132 : Tartago, physic–nut (*Jatropha curcas* L.*)**
Flowers (upper left), leaf and fruits (right), two–thirds natural size.

Seeds, leaves, and sap have been employed widely in home remedies. The pleasantly flavored oily seeds are a drastic purgative but poisonous and dangerous and have caused death when eaten by small children. It is reported that the seeds are edible when roasted and that cultivation for this purpose once was attempted. However, the seeds should not be eaten because they cause illness when only partially roasted. The clear oil pressed from the seeds has been used in some countries for making soap, for illumination, as a lubricant, and in paints.

The seeds have served in medicine under the names Barbados–nuts, purging–nuts, physic–nuts, Semen Ricini Majoris, and pignons d'Inde.

The plants are widely cultivated elsewhere from cuttings as a hedge or living fence. The deciduous foliage is unpalatable to livestock. Indians in the West Indies and Mexico were growing these plants around their houses when the first European explorers arrived. In Mexico a lac or varnish was obtained from scale insects propagated on the bark. An infusion of the leaves was used in Guatemala to set dyes of cotton. A honey plant.

Commonly planted around houses as an ornamental and escaping to clearings, roadsides, and waste grounds in India.

## 133. Tabaiba (*Jatropha hernandiifolia*)

This vinelike shrub rarely reaching tree size is easily recognized by its distinctive leaves. Characters for identification include : (1) watery or slightly milky latex; (2) ovate leaves with petiole attached mostly above base of blade (peltate), short–or long–pointed at apex, sometimes with 1 or 2 additional points or lobes, dull green above and blue green beneath; (3) several whitish male and female flowers about $^{3}/_{16}$ inch long and broad on very slender widely forking stalks; and (4) nearly round seed capsules $^{3}/_{4}$ inch in diameter, splitting into 3 parts.

An evergreen shrub 5–10 feet high, also taller and vinelike, or treelike to 20 feet tall with slender stems to 2 inches in diameter, and few branches, without a definite crown, hairless throughout. The bark is smooth and light gray, the inner bark whitish and tasteless with nearly transparent slightlybitter latex. The slender twigs are greenish when young, turning gray or brown, with rounded raised leaf scars.

The alternate leaves have slender petioles 1–4 inches long attached mostly above the rounded or heart–shaped base of blade. The thin papery blades are mostly 2–$5^{1}/_{2}$ inches long and $1^{1}/_{4}$-$3^{1}/_{2}$ inches wide.

Flower clusters (cymes) 1–5 inches long at leaf bases bear several flowers, mostly male (monoecious). Male flowers have calyx or 5 rounded lobes, bell–shaped tubular white corolla with 5 spreading lobes, and 10 stamens in 2 series united in column. Female flowers have pistil with 3–celled ovary and forked stigma. The nearly round seed capsules, 1 or 2 on stalks of 2–3 inches, are whitish tinged with green, turning brown, and split into 3 parts. Seeds 3 or fewer, oblong, $^{3}/_{8}$ inch long, mottled brown. Flowering and fruiting irregularly through the year.

The wood is light brown or whitish and soft.

Uncommon in understory of moist limestone forest at low altitudes in M.P., U.P.

**Fig. 133 : Tabaiba (*Jatropha hernandiifolia*)**
Fruiting twig (above), male flowers (lower right), natural size.

## 134. Tártago, emético, coralplant (*Jatropha multifida* L.*)

This handsome ornamental shrub or small tree occasionally spreads from cultivation and might be native in dry areas. It is easily recognized by : (1) abundant poisonous milky juice or white latex flowing from cut parts; (2) long–stalked leaves deeply palmately lobed with 11 or fewer lance–shaped lobes also divided and toothed; (3) many small red or scarlet flowers in flat–topped erect showy cluster on long stalk; and (4) yellow rounded slightly 3–lobed seed capsules about 1 inch long.

Shrub or small tree 15 feet high and 3 inches in trunk diameter or larger, apparently deciduous in dry areas, with few branches, each ending in a cluster of crowded leaves. The bark is brown, smooth, with many lenticels. The stout hairless twigs have large round leaf scars.

The alternate hairless leaves have stipules $^1/_2$–$^3/_4$ inch long, divided into brown threadlike segments. Petioles are 4–8 inches long, round. The thin blades have a rounded outline 6–12 inches across but are much divided into narrow long–pointed lobes $^1/_2$–2 inches long, green above and whitish beneath.

Flower clusters (cymes) are erect on a stalk 4–8 inches long, branched at the top about $2^1/_2$ inches across and flat, hairless throughout. Male flowers are numerous on slender spreading reddish stalks, nearly $^1/_2$ inch across, composed of 5–lobed red calyx $^1/_8$ inch long; 5 red spreading narrow curved petals $^1/_4$ inch long; disk of 6 glands united at base; and 8 red stamens $^1/_4$ inch long. Female flowers 2–3 near center of flower clusters, less than $^1/_4$ inch long, consist of calyx, corolla, and pistil with 3–celled 3–ovuled ovary and 3 styles each 2–forked. The capsules split into 3 parts, each with 1 elliptic mottled seed about $^3/_4$ inch long. Flowering and fruiting intermittently.

The seeds and milky sap are poisonous but have been used in home remedies. In southern Florida this species is considered to be a frequent cause of poisoning.

Commonly planted for ornament and along roadsides in Rajasthan, Gujrat. Possibly native in dry areas of southwestern States.

**Fig. 134 : Tartago, emetico, coralplant (*Jatropha multifida* L.*)**
Flowering twig (above), portion of fruit (lower right), two–thirds natural size.

## 135. Castorbean (*Ricinus communis* L.*)

Castorbean is an introduced large coarse herb or shrub sometimes becoming a small tree. It is easily recognized by: (1) the large star–like palmate–veined leaves with mostly 7–9 long pointed lobes bordered by irregular gland–tipped teeth and with very long round petiole joined below middle of blade; (2) small greenish flowers in terminal clusters, male below and female above; (3) the elliptic seed capsule about $^3/_4$ inch long, 3–lobed and usually spiny, splitting into 3 parts; and (4) the familiar castorbean $^1/_2$–$^5/_8$ inch long, with whitish swelling at one end, oily but poisonous when eaten raw.

Evergreen large coarse herb or shrub, sometimes a small tree 15 feet or more in height and 4 inches in trunk diameter, with few branches, hairless throughout. In temperate regions a coarse annual herb. Bark light brown, smoothish, with rings at nodes and raised dots (lenticels), the inner bark light gray and tasteless. The stout greenish twigs are covered with a whitish bloom and have enlarged ringed nodes. The bud is covered by a long–pointed yellow–green stipule more than 1 inch long, which sheds early, leaving a ring scar.

The alternate leaves have very long stout petioles mostly 12–20 inches long, whitish green, rounded and hollow, joining the middle of the blade (peltate) and bearing 2 cup–shaped yellow–green glands at apex under blade. Blades are mostly 12–16 inches in diameter, sometimes smaller, thin, slightly saucer–shaped, deeply 7–9–lobed half way to petiole, the lobes on 1 side more than twice the length of the shortest. The upper surface is dull green with veins yellow green and slightly sunken. The lower surface dull light green with prominent veins.

Flower clusters (racemes) are erect at end of twigs and bear many short–stalked flowers without corolla, male below and female above (monoecious). The male flower develops from light green rounded bud $^1/_8$ inch in diameter, covered by the calyx, and consists of the light green calyx with 3–5 long–pointed lobes bent down and many stamens with much branched filaments and dotlike light yellow anthers $^3/_8$–$^1/_2$ inch long and broad. The greenish female flower less than $^1/_4$ inch long is composed of the light green 3–5–lobed calyx and the pistil with light green spiny 3–celled ovary, 3 ovules, and 3 spreading reddish–brown glandular styles each 2–forked.

The seed capsules are whitish green, becoming brown, usually covered with weak spines or smooth and split into 3 parts, each 1–seeded. The seed or castorbean is shiny brown with darker streaks or spots and a whitish enlargement at one end, resembling at tick or beetle. Flowering and fruiting throughout the year.

The wood is whitish and soft.

Castorbean is cultivated elsewhere for the oily seeds, which yield the caster–oil of commerce, a laxative and lubricant. The oil has been used also in soap making, for lighting homes, and as a hair dressing. However, raw seeds resembling ticks or beetles are poisonous and can cause death if eaten. If the flower stalks are removed before maturity of seeds, the risk of poisoning children is reduced.

An ornamental shrub or in temperate regions an annual herb. Several varieties differ in fruit and seed characters. One ornamental variety planted locally has reddish or purplish foliage.

Locally common in waste places, cultivated, escaped, and naturalized throughout India.

**Fig. 135 : Castorbean (*Ricinus communis* L.*)**
Fruits (above), leaf, and flowers (lower left), two–thirds natural size.

## 136. Gumtree (*Sapium caribaeum*)

A tree species of the Lesser Antilles known also from 2 localities in the Lakhsdeep Islands. Distinguishing characteristics are: (1) abundant milky sap, probably poisonous as in related species; (2) elliptic–lanceolate thin leaves 3–5$^1/_2$ inches long and 1$^1/_4$–2$^1/_2$ inches wide, with relatively few side veins mostly curved toward border; and (3) rounded seed capsules nearly $^1/_4$ inch in diameter.

A small evergreen tree 20 feet high and 4 inches or more in trunk diameter, hairless throughout. Leaves alternate, the light green round petioles $^3/_4$–1 inch long with glands at apex. Blades long–pointed at apex and rounded at base, with minute gland teeth on edges, thin or slightly thick and succulent, the upper surface green, and the lower surface dull light green

Flower clusters (spikes) 4–10 inches long at ends of twigs bear many minute nearly stalkless flowers less than $^1/_{16}$ inch long, mostly male and in lower part female (monoecious). Male flowers about 10 above a scale have 2–lobed calyx and 2 stamens. Female flowers borne singly have 2–lobed calyx and pistil with rounded ovary and 2 styles. The purplish seed capsules are mostly 2–lobed and contain 2 rounded seeds more than $^1/_8$ inch long.

The whitish soft lightweight wood is not used.

Elsewhere, Indians made bird–lime from the sticky white sap.

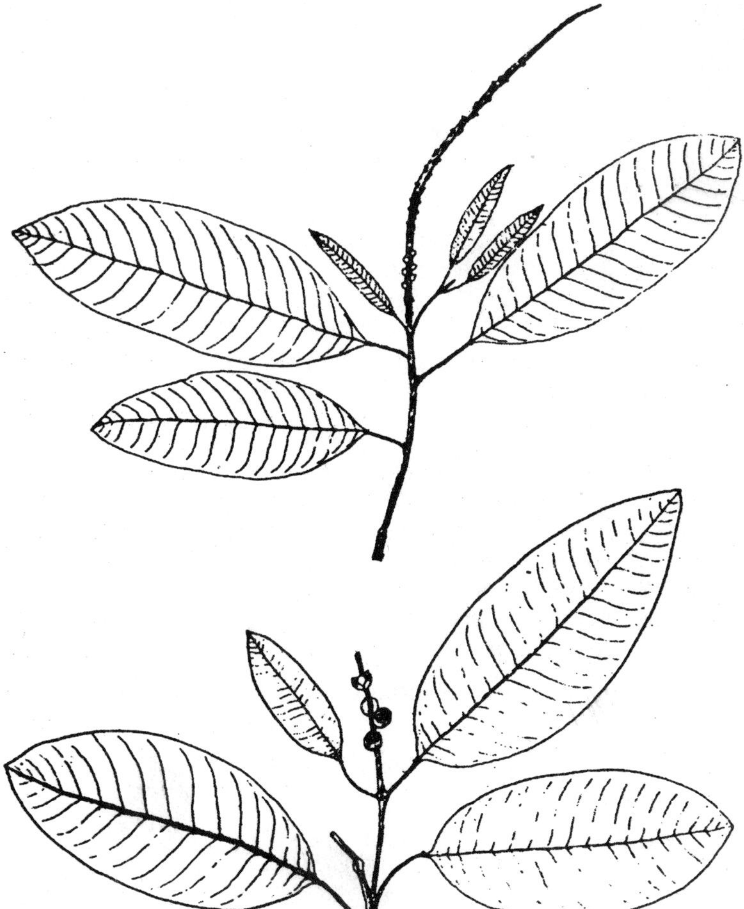

**Fig. 136 : Gumtree (*Sapium caribaeum*)**

Flowering twig (above), fruiting twig (below), two–thirds natural size.

## 137. Tabaiba (*Sapium jamaicense*)

This species of tabaiba is rare and local near Assam Mountains at its eastern limit. It is characterized by: (1) abundant irritating and poisonous milky sap; (2) oblong or elliptic green leaves with many nearly straight to slightly curved parallel lateral veins almost at right angle to mievein, the edges minutely wavy–toothed, and 2 raised dotlike glands at upper end of petioles; (3) small yellow–green flowers stalkless along narrow axes at leaf bases, mostly male with a few female below (monoecious); and (4) seed capsules round $^{5}/_{16}$ inch in diameter, 3–seeded.

Small to medium–sized evergreen tree to 50 feet high and 10 inches in trunk diameter, hairless throughout. Leaves alternate, with green petioles $^{3}/_{4}$–$1^{1}/_{4}$ inches long. Blades are mostly 6–9 inches long and $2^{1}/_{4}$–$2^{3}/_{4}$ inches wide, sometimes only 3 inches long and 1 inch wide, abruptly long–pointed at apex and blunt at base.

Flower clusters (spikes) 2–4, are 3–6 inches long, unbranched, at base of topmost leaves. Male flowers, in groups of 3–7 above a scale, about $^{1}/_{16}$ inch long and broad, yellow green, consisting of cup–shaped 2–toothed calyx and 2 stamens. Female flowers $^{3}/_{16}$ inch long, green, are composed of a cup–shaped 3–toothed calyx and pistil with 3–celled 3–ovuled ovary and 3 styles. Seed capsules are 3–celled and split into 3 parts. The seeds are about $^{3}/_{16}$ inch long.

Rare in northeastern India only at 600–800 feet.

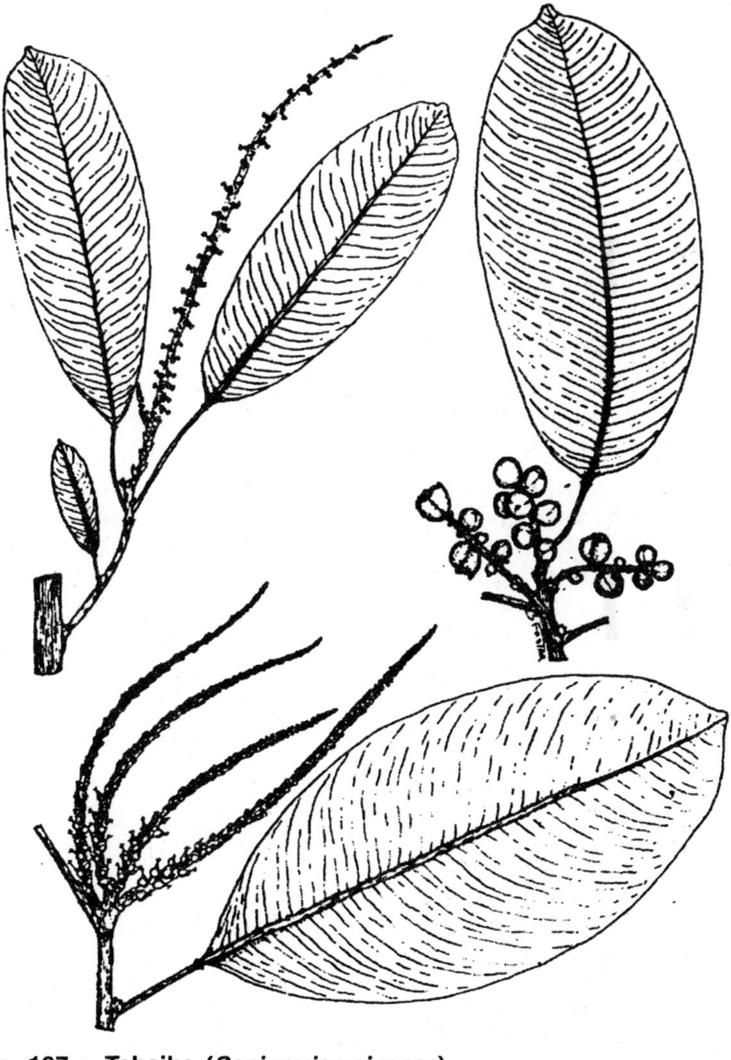

**Fig. 137 : Tabaiba (*Sapium jamaicense*)**

Twig with male flowers (upper left), twig with female flowers (lower left), fruiting twig (right), natural size.

## 138. Amansa guapo (*Savia sessiliflora* (Sw.) Willd.)

Distinguishing characters of this shrub or small tree of dry areas are : (1) narrowly ovate leaves 1–3 inches long and $^{3}/_{4}$–$1^{1}/_{4}$ inches wide, alternate, almost stalkless, and light green; (2) minute yellow flowers male and female at leaf bases on different trees, male crowded and almost stalkless, female single on curved stalks; and (3) seed capsules rounded, brown or black, $^{5}/_{16}$ inch in diameter, splitting into 3 parts each 2–seeded.

Deciduous shrub or small tree to 25 feet high and 4 inches in trunk diameter. The bark is gray, smooth to slightly fissured, with warts (lenticels). The inner bark is light pink beneath an outer green layer, bitter. The slender twigs are greenish and finely hairy when young, becoming brownish with raised dots (lenticels).

The alternate leaves have hairy leafstalks $^{1}/_{16}$ inch long and minute pointed stipules $^{1}/_{16}$ inch long. Blades are long-pointed at apex, rounded at base, without teeth on edges, thin, paler beneath, hairless except at base and along midvein beneath.

Flowers male and female at leaf bases on different trees (dioecious), with both sepals and petals. Male flowers less than $^{1}/_{8}$ inch broad, yellow, composed of 5 hairy sepals, 5 tiny, petals, 5 stamens, and nonfunctioning pistil. Female flowers have 5 sepals, 5 tiny petals, and pistil with 3–celled ovary and 3 forking styles. Seed capsules hang down on curved stalks about $^{1}/_{4}$ inch long, split into 3 parts which fall off, the axis remaining attached. The dark brown rounded seeds are about $^{1}/_{8}$ inch in diameter. Flowering and fruiting irregularly during the year.

Wood is light brown, hard.

Uncommon and scattered in dry coastal and lower forests of southwestern Gujrat from sea level to 500 feet altitude.

**Fig. 138 : Amansa guapo (*Savia sessiliflora* (Sw.) Willd.)**
Fruiting twig (above), twig with female flowers (lower left), two–thirds natural size.

## 139. Green ebony (*Securinega acidoton*)

This small much–branched shrub with many smalls sharp spines sometimes becomes a small tree. Its distinguishing characters are : (1) very slender, very spiny, brown or gray zigzag twigs, much branched and almost forming a network; (2) very small leaves, crowded and alternate, obovate or elliptic, $^1/_4$–$^5/_8$ inch long; (3) minute yellowish flowers at leaf bases, many crowded male and few stalked female; and (4) seed capsules rounded, $^3/_{16}$ inch in diameter, splitting into 6 parts.

Much branched spreading, spiny deciduous shrub less than 10 feet high, sometimes treelike and on Muertos a small tree to 20 feet high and 6 inches in trunk diameter. Bark light gray, smoothish, becoming rough, fissured, and slightly shreddy. The inner bark is pinkish and slightly bitter. The very spiny hairless twigs have minute light dots (lenticels) and end in spines or minute buds formed of scales (stipules) and young leaves. The spines are slender, straight, sharp $^1/_8$–$^3/_8$ inch long log, often appearing widely 2–forked, those from side twigs to $^3/_4$ inch long.

The leaves are alternate or on short lateral shoots crowded, hairless, with 2 minute brown scales (stipules) at base and with very slender leafstalk less than $^5/_8$ inch long. Blades are mostly obovate or elliptic, $^1/_4$–$^5/_8$ inch long, $^1/_8$–$^3/_8$ inch wide, rounded at apex, tapering to base, not toothed on edges, thin, the upper surface green and slightly shiny, the lower surface dull light green.

Flowers are male and female on the same plant (monoecious), minute, about $^1/_{16}$ inch wide, yellowish, without petals. Several to many male flowers are crowded and almost stalkless at leaf bases, composed of 5 round yellowish sepals, 5 stamens on a disk opposite the sepals, and minute nonfunctioning pistil. Female flowers are few at leaf bases on slender stalks about $^1/_8$ inch long, consisting of 5 rounded yellowish sepals and on a disk the pistil with 3-celled ovary and 3 spreading 2-forked styles. The rounded brown seed capsules contain 6 light yellow seeds $^1/_{16}$ inch long with flowers and fruits in spring and summer.

The wood is light brown, hard, and fine–textured.

Locally common in dry coastal hills from sea level to 600 feet altitude in southwestern Tamil Nadu.

**Fig. 139 : Green ebony (*Securinega acidoton*)**

Twig with female flowers (left), twig with male flowers (center), fruiting twig (right), natural size.

# BOX FAMILY (BUXACEAE)

Shrubs, small trees, and herbs, known by: (1) evergreen leaves, opposite or sometimes alternate, simple, commonly entire, leathery, without stipules; (2) minute flowers crowded along an axis (spike or raceme), male and female on the same plant (monoecious) or different plants (dioecious), regular, the calyx generally with 4 (–12) sepals or lobes or sometimes none, without corolla, the male flowers with stamens 4 (or 6) opposite or many, the female few and larger with pistil composed of superior ovary generally of 3 cells each with 1–2 ovules, and 3 spreading styles; and (3) fruit a capsule or berrylike, the shiny black seeds with a fleshy protuberance (caruncle).

**Key to Species**

A. Leaves ovate or elliptic, 2–4 inches long, with 3 main veins from base to long-pointed apex—*Buxus laevigata*.

AA. Leaves oblong or obovate, $^3/_4$–$1^1/_2$ inches long, with 2 faint side veins from base near edges to rounded apex—*Buxus vahlii*.

## 140. Citrifolia (*Buxus laevigata*)

A rare shrub or small tree of western Ghat, recognized by: (1) opposite ovate or elliptic leaves with 3 main veins from base; (2) small fragrant whitish or light yellow male and female flowers about $^1/_4$ inch long, without petals, in stalked clusters at leaf bases; and (3) rounded brown seed capsules about $^3/_{16}$ inch long and broad, 3–horned, splitting into 3 parts.

Evergreen shrub or small tree to 20 feet high and 3 inches in trunk diameter, hairless throughout. Bark gray, thick, with corky ridges and deep fissures, light yellow within. The inner bark is light bark is light green and very bitter. The light green twigs are slightly 4–angled, slender, and brittle.

The opposite leaves without stipules have leafstalks $^1/_8$–$^1/_4$ inch long. Blades are 2–4 inches long and 1–2 inches wide, slightly thickened, long–pointed at apex, short–pointed at base, with 2 long side veins curved from base to apex near the turned down edges, bitter tasting but without odor. The upper surface is shiny green and the lower surface shiny light green.

Several to many male flowers and usually 1 female are borne in the same cluster (monoecious) at the end of a stalk less than 1 inch long. Male flowers $^1/_4$ inch long and broad have a stalk $^1/_8$–$^3/_8$ inch long, 4 whitish or light yellow sepals $^1/_8$ inch long and 4 stamens opposite the sepals. The slightly larger female flowers without stalk have 6 sepals less than $^1/_{16}$ inch long and pistil with rounded 3–celled ovary and 3 long spreading styles. Seed capsule rounded and 6–angled, with 3 long curved horns (styles) of the same long that apex, splitting into 3 parts, each with 1 horn and 1–seeded. With flowers and fruits in spring and summer.

The wood is light yellow, hard, heavy, of very fine uniform texture. It would be suitable for articles of turnery and wood engravings, like that of the related box or boxwood of commerce (*Buxus sempervirens* L.)

Rare in moist limestone and western (serpentine) forest at 300–2,000 feet altitude in western Ghat.

**Fig. 140 : Citrifolia (*Buxus laevigata*)**

Twig with male and female flowers (above), fruiting twig (below), natural size.

## 141. Britton (*Buxus vahlii*)

This rare shrub or small tree is identified by: (1) opposite oblong or obovate slightly thickened leaves $^3/_4$–$1^1/_2$ inches long and $^3/_8$–$^5/_8$ inch wide, with 2 faint side veins curved from base near edge to the rounded apex, which ends in minute narrow point; (2) small whitish or yellowish male and female flowers without petals, in clusters almost stalkless at leaf bases; and (3) seed capsules $^3/_{16}$–$^1/_4$ inch long, 3–horned, splitting into 3 parts.

Evergreen shrub or small tree to 15 feet high and 3 inches in trunk diameter, hairless throughout. Bark gray, finely fissured, the inner bark light brown and bitter. Twigs light green, becoming light gray, with 2 grooves below each leaf, ending in bud composed of minute paired leaves.

Leaves opposite, with leafstalks $^1/_{16}$ inch long, without stipules, with bitter taste but no odor. Blades are stiff and leathery, slightly convex, short–pointed at base, and slightly turned under at edges. The upper surface is green and slightly shiny, with sunken midvein and 2 faint side veins curved near border, and the lower surface dull light green.

The flowers are male and female in the same cluster (monoecious), with 4 sepals less than $^1/_8$ inch long and no petals. Flowering and fruiting from spring to fall.

The wood is yellowish and hard.

This species might be suitable as a hedge. Plant, having attractive foliage like that of the related box or boxwood of commerce, (*Buxus sempervirens* L.), also a cultivated ornamental.

Rare in moist limestone forest at 200–500 feet altitude on hills Aravali.

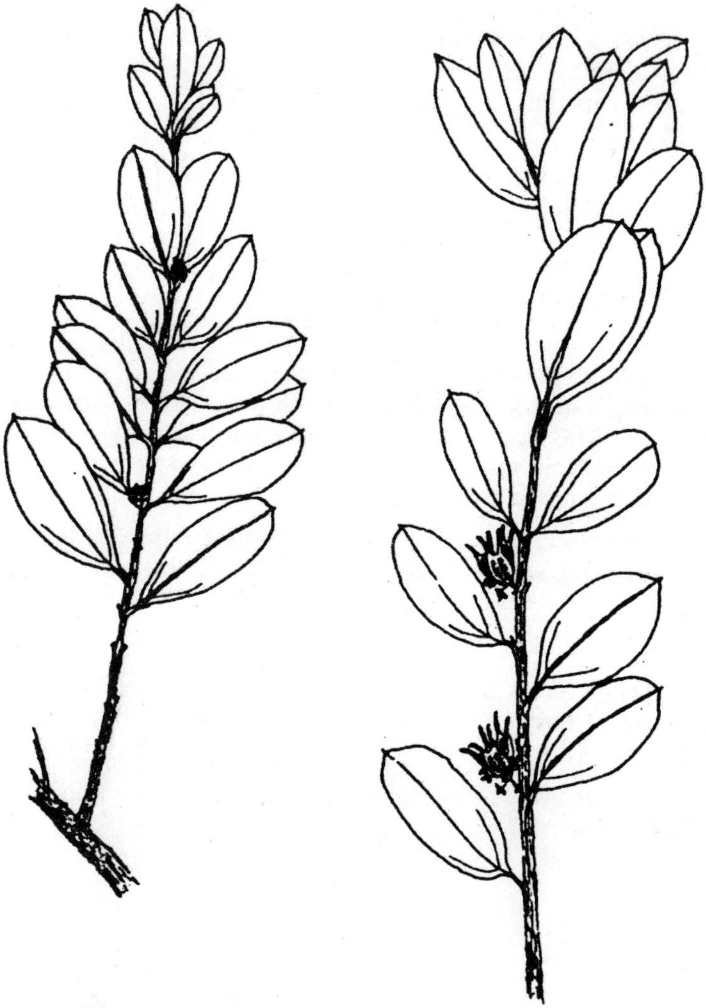

**Fig. 141 : Britton (*Buxus vahlii*)**
Flowering twig (left), fruiting twig (right), natural size.

# CASHEW FAMILY (ANACARDIACEAE)

Trees shrubs, and few woody vines, known by: (1) resinous sap often in bark and other parts, in a few species the resin or volatile oil caustic and poisonous to the skin; (2) leaves alternate, odd pinnate, with 3 leaflets, or simple, without stipules; (3) flowers minute or small commonly white, many in large branched clusters (panicles), bisexual or functionally male or female, mostly regular, with 3–5 sepals united at base, 3–5 petals or none, generally 10 stamens, sometimes fewer, inserted at the border of a ringlike or cuplike disk, and pistil with superior ovary generally 1–celled (to 5–celled) with 1 ovule, style, and 3 stigmas; and (4) fruit mostly a resinous drupe, 1–seeded, the resin of few species poisonous.

**Key to Species**

A. Leaves simple.

B. Leaves elliptic or obovate, rounded at both ends or slightly notched at apex—Pajuil, cashew, *Anacardium occidentale* L.

BB. Leaves lance–shaped, long–pointed at both ends—131. Mango, *Mangifera indica* L.*

AA. Leaves pinnately compound.

C. Leaflets 5 or 7.

D. Leaflets usually 5 (3–7), stalked, ovate, blunt or minutely notched at apex; sap poisonous to the touch—Papayo, Florida poisontree, *Metopium toxiferum* (L.) Krug & Urban.

DD. Leaflets 5 or 7, stalkless, elliptic or oblong, slightly toothed toward short–pointed apex; sap resinous—457, *Schinus terebinthifolia.**

CC. Leaflets 9 to many.

E. Leaflets spiny–toothed; sap poisonous to the touch—*Comocladia.*

F. Leaflets 11–21, broadly ovate, $^3/_8$–$^3/_4$ inch long and broad, each bearing 3 or more sharp slender spines—*Comocladia dodonaea.*

FF. Leaflets 11–31, oblong or lance–shaped, $2^1/_2$–7 inches long, with many spiny teeth—*Comocladia glabra.*

EE. Leaflets 9–25, not spiny; sap not poisonous—*Spondias.*

G. Leaflets mostly lance–shaped, long or short–pointed at apex, with short stalks about $^1/_8$ inch long.

H. Leaflets inconspicuously toothed—Jobo de la India, ambarella, *Spondias dulcis* Parkinson.

HH. Leaflets not toothed—Jobo, hagplum, yellow mombin, *Spondias mombin* L.

GG. Leaflets elliptic, rounded or short–pointed at apex, slightly wavy–toothed, almost stalkless–135. Ciruela del pais, purple mombin, *Spondias purpurea* L.*

## 142. Chicharron (*Comocladia dodonaea*)

This spiny shrub or small tree with few branches and no crown should be learned because its sap of leaves, twigs, and bark is poisonous to the skin of many persons. It is easily identified by: (1) pinnately compound leaves 3–6 inches or more in length crowded near ends of twigs and composed of 11–21 stalkless broadly ovate leaflets, folded or wrinkled and often reddish tinged, each bearing 3 or more sharp slender spines; (2) many tiny stalkless dark red flowers more than $^1/_{16}$ inch across, in clusters along a lateral flower stalk; and (3) the orange–red elliptical fleshy fruits $^3/_8$–$^1/_2$ inch long, several along a stalk.

Commonly a shrub 10 feet or less in height but becoming a small tree 15 feet high and 4–6 inches in trunk diameter, with slender axis and few branches. The bark is light brown, slightly fissured, and thin. Inner bark is light brown and tasteless. The sap turns blackish on exposure and makes a lasting stain. The stout brown twigs are densely hairy when young, slightly rough from the raised half round leaf scars. The bud is composed of minute hairy leaves without scales.

The evergreen leaves are alternate but near ends of twigs are very close together. The yellow–green leaf axes are densely brown hairy. Leaflets are paired except at end, broadly ovate, mostly $^3/_8$–$^3/_4$ inch long and broad, slightly thickened. The short–pointed apex ends in a sharp slender yellow spine about $^1/_8$ inch long, the 2–4 main veins end in similar spines, and the base is heart shaped. The upper surface is slightly shiny green and hairless, and the lower surface paler and sometimes hairy on veins.

Flower clusters (panicles) have a lateral axis 2–4$^1/_2$ inches in length. Groups of tiny red flowers are almost stalkless along the main axis or on short branches less than $^1/_2$ inch long. Flowers are male or female and bisexual on the same plant (polygamous). The minute calyx is 3–lobed; the corolla is composed of 3 dark red petals, widely spreading and more than $^1/_{16}$ inch across; male flowers have 3 stamens; and female flowers have an ovary with 3 minute stigmas. The elliptical or oblong fruits (drupes) contain 1 seed $^3/_{16}$ inch long. Flowering in winter and spring and maturing fruits in spring and summer.

Sapwood is light brown and hard.

Locally common in open areas and thickets in the dry and moist limestone forests and dry foothills of Orissa from sea level to 1,000 feet altitude, especially in the southwestern part and limestone hilltops in the northwestern part.

**Fig. 142 : Chicharron (*Comocladia dodonaea*)**
Natural size.

## 143. Carrasco (*Comocladia glabra*)

This odd poisonous shrub or small tree (or vinelike) of the forest understory easily recognized by: (1) slender unbranched erect trunk (sometimes with 2 or 3 branches) bearing at the top a cluster of large spreading pinnate leaves with 11–31 oblong or lance–shaped spiny–toothed leaflets; (2) the tiny dark red flowers $^{1}/_{16}$ inch across, crowded along branched lateral clusters 5–10 inches long; and (3) the elliptic fleshy fruits $^{3}/_{8}$ inch long, reddish, turning black. The whitish sap of foliage and bark, which turns blackish upon exposure, irritates the skin of some persons upon contact.

Evergreen and commonly a shrub at high elevations but elsewhere becoming more than 20 feet tall and 2 inches in trunk diameter. The light gray bark is fissured or rough. Inner bark is whitish and almost tasteless, with poisonous sap. The stout axis is light brown at apex and slightly hairy when young.

The leaves 1–2 feet long are alternate though crowded. Along the stout, light green, finely hairy axis are 11–31 almost stakless leaflets in pairs except at end. Leaflet blades are $2^{1}/_{2}$–7 inches long and $^{7}/_{8}$–2 inches wide or smaller near base of leaf, with a long spiny point and the base rounded and slightly heart–shaped, stiff and thickened, hairless except on midrib, dark green on upper surface, and yellow green beneath. Lateral veins are at right angles to midrib, very prominent on lower leaflet surface, and end in needlelike spines $^{1}/_{16}$–$^{1}/_{8}$ inch long.

Flowers are male and bisexual on the same plant (polygamous). They have a 3–lobed calyx, 3 round dark red petals about $^{1}/_{32}$ inch long, 3 stamens, and a pistil with rounded ovary and 3 stigmas, or sometimes the parts in 4's. Male flowers have calyx, corolla, 3 larger stamens, and no pistil. Flowering and fruiting intermittently.

The sapwood is hard and light brown and is not used.

Common in most forest types except the dry southwestern coast and at 1,000–4,000 feet altitude in H.P. Absent from other islands.

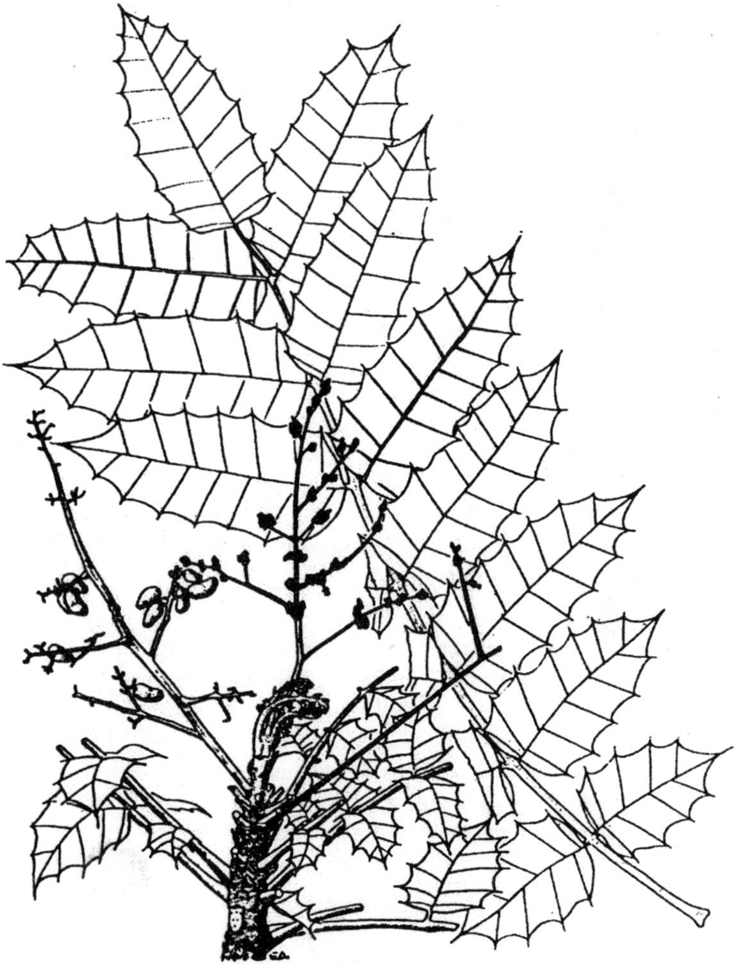

**Fig. 143 : Carrasco (*Comocladia glabra*)**

Leaf (above), twig with flowers and fruits (below), one–half natural size.

## 144. Pimienta de Brasil, Christmas–berry, Brazil peppertree (*Schinus terebinthifolia*)

This attractive ornamental shrub or small tree is rarely planted for its clusters of many bright red fruits $^{3}/_{16}$ inch in diameter. Other distinguishing characteristics are : (1) pinnate leaves with mostly 5, 7 or 9 (3–13 or more in varieties) stalkless elliptic or oblong leaflets paired except the largest at end, slightly toothed toward short–pointed apex, and (2) clusters of many small flowers about $^{1}/_{8}$ inch broad across the 5 white petals.

Evergreen shrub or small tree becoming 20 feet high and 8 inches in trunk diameter. The bark is smooth and gray. Twigs are light brown, finely hairy when young, and have many raised dots (lenticels). They have aromatic resinous sap which turns blackish upon exposure. The buds are formed by minute hairy leaves without stipules.

The alternate pinnate leaves 3–6 inches long have a narrowly winged axis 1–3 inches long, green and finely hairy. Leaflets are 1–2 inches long and $^{1}/_{2}$–$^{3}/_{4}$ inch wide (the terminal to 3 inches by 1 inch), short–pointed at both ends, with small blunt teeth toward apex, slightly thickened, hairless or nearly so, the upper surface shiny green with several straight lateral veins, and the lower surface dull light green.

Flower clusters (panicles) are mostly at base of upper leaves, 1–3 inches long, branched, composed of many short–stalked flowers, partly male and female on different plants (dioecious). The flower consists of calyx of 5 minute, pointed, green sepals; corolla of 5 white petals less than $^{1}/_{8}$ inch long; 10 stamens attached at the base of a large ring–shaped disk; and pistil with rounded ovary, short style, and stigma. The bright red fleshy fruits (drupes) have sepals at base and dot stigma at apex. The aromatic resinous brown pulp encloses 1 elliptic light brown seed less than $^{1}/_{8}$ inch long. Flowering and fruiting intermittently.

Rarely planted for ornament in Puerto Rico, escaping in moist limestone forest. Also St. John. Propagated from seeds and cuttings. Common in cultivation in tropical gardens. Introduced into southern India including and naturalized locally as a weed. Because of the evergreen foliage and red berries resembling holly, the branches are popular for Christmas decorations. A honey plant. It is reported that this species caused a skin irritation in some persons.

**Fig. 144 : Pimienta de Brasil, Christmas–berry, Brazil peppertree (*Schinus terebinthifolia***

Flowering twig (above), fruiting twig (below), natural size.

## HOLLY FAMILY (AQUIFOLIACEAE)

Shrubs and trees, small to medium–sized, rarely large, known by : (1) alternate simple leaves generally leathery and evergreen, with minute stipules or none; (2) small flowers, few in lateral clusters, whitish or greenish, generally male and female on different plants (dioecious) or bisexual, regular, composed of calyx with 4 (5) minute sepals or teeth, 4 (5) rounded whitish petals sometimes united at base, 4 (5) alternate stamens inserted at base of corolla, without disk, and pistil with superior ovary of 4 (3–5) cells of 1–2 ovules, usually without style and with 3–5 stalkless stigmas; and (3) fruit a round drupe or berry, red, black or yellow, with stalkless stigmas, bitter pulp, and 3–5 nulets.

**Key to Species**

A. Leaves toothed on edges.

B. Leaves narrowly ovate, long–pointed, sharply saw–toothed, $^3/_4$–$1^3/_4$ inches long—*Ile macfadyenii.*

BB. Leaves mostly obovate or oblong, blunt or short–pointed at apex.

C. Leaves with edges usually slightly wavy toothed and turned under, 2–4 inches long—*Ilex nitida.*

CC. Leaves often sharply toothed toward apex, $1^1/_2$–3 inches long—*Ilex cassine.*

AA. Leaves not toothed on edges.

D. Leaves mostly less than 1 inch long; flowers 1 or few at leaf bases, 4–5 parted—*Ilex sntenisii.*

DD. Leaves mostly more than 1 inch long; flowers clustered at leaf bases.

E. Leaves short–or long–pointed at apex, 1–$1^3/_4$ inches long—*Ilex cookii.*

EE. Leaves mostly round at apex.

F. Flowers 5–7 parted; leaves $1^1/_4$–$2^1/_2$ inches long—*Ilex sideroxyloides.*

FF. Flower 4–parted.

G. Leaves elliptic, mostly blunt at base, $1^1/_4$–$2^1/_2$ inches long—*Ilex urbaniana.*

GG. Leaves short–pointed at base.

H. Leaves oblong or obovate, $1^1/_2$–4 inches long—*Ilex guianensis.*

HH. Leaves obovate to elliptic, 1–$1^1/_2$ inches long—*Ilex riedlaei.*

## 145. Dahoon (*Ilex cassine* L).

Dahoon, a small tree of southeastern States, extends into the West Indies. This relative of the hollies is characterized by : (1) oblong or oblanceolate leathery leaves broadest and often sharply toothed toward the short–pointed or blunt apex; (2) few or several small whitish flowers at leaf bases, more than $^1/_8$ inch across the 4 rounded corolla lobes; and (3) round bright red fruits nearly $^1/_4$ inch in diameter.

A small evergreen tree to 25 feet high and 6 inches in trunk diameter, sometimes shrubby. The bark is dark gray and rough. Twigs are slender, densely hairy, becoming dark brown, ending in minute pointed hairy buds.

The alternate leaves have petioles about $^1/_4$ inch long, often slightly hairy. Blades are $1^1/_2$–3 inches long and $^1/_2$–1 inch wide, gradually narrowed to base, slightly thickened, hairy when young, especially beneath, becoming hairless except on midvein beneath, shiny dark green above, and light green beneath.

Flowers on short hairy stalks at leaf bases in branching clusters (like cymes) to 1 inch long, male and female on different plants (dioecious). The calyx is composed of 4 pointed minute greenish sepals with hairy edges, and corolla of 4 spreading rounded whitish lobes more than $^1/_{16}$ inch long. Male flowers have 4 stamens attached near base of corolla between lobes; and female flowers, usually only 3, have 4 sterile stamens and short pistil with 4–celled ovary and broad flat stigma. The berry (berry–like drupe) has calyx at base and dotlike stigma at apex and contains usually 4 nutlets $^3/_{16}$ inch long, pointed and ridged. Flowering in spring.

The wood is described as lightweight, hard, fine–textured, with thick whitish sapwood and light brown heartwood.

In H.P. this native relative of holly is recommended as a hardy evergreen for its bright red berries (only on female trees) in winter. It grows best though slowly in moist soils with some shade and is suited to lake shores and wet areas, also streets and roadsides. Propagated from cuttings.

Rare in coastal swamps of moist coastal forest at sea level.

**Fig. 145 : Dahoon (*Ilex cassine* L).**

Twig with male flowers (upper left), twig with fruits (upper right), twig with female flowers (lower right), natural size.

## 146. Cookii (*Ilex cookii* Britton & Wils.)

This shrub or small tree local at the summit of Cerro de Punta, Puerto Rico's highest mountain peak, is characterized by : (1) elliptic leaves $^3/_4$–$1^3/_4$ inches long and $^3/_8$–$7/_8$ inch wide, thick and stiff; (2) flowers few at leaf bases, $^1/_8$ inch across the 4 white petals; and (3) round fruits than $^3/_{16}$ inch in diameter.

An evergreen shrub 5–8 feet high, sometimes becoming a small tree to 20 feet high and 5 inches in trunk diameter, hairless. Bark light brown with light dots (lenticels). Twigs slender, light green, becoming gray.

Leaves alternate, with light green petioles $^1/_8$–$^1/_4$ inch long. Blades short–or long–pointed at apex and short–pointed or rounded at base, slightly turned under at edges, the upper surface shiny dark green with sunken midvein and inconspicuous side veins, and the lower surface dull light green.

Flowers few on slender stalks about $^1/_8$ inch long at leaf bases. Female flowers have 4–5 minute sepals, 4–5 spreading white petals about $^1/_{16}$ inch long, and pistil with ovary and 4–5 dotlike stigmas. Fruits (drupes) green when immature, 5–celled, with 5 blackish nutlets, and with stigmas and sepals persistent. Flowering irregularly over the year.

Rare and local in dwarf forest of upper forest at and near summit of Sivalik, in the central mountains at 4,000–4,390 feet altitude.

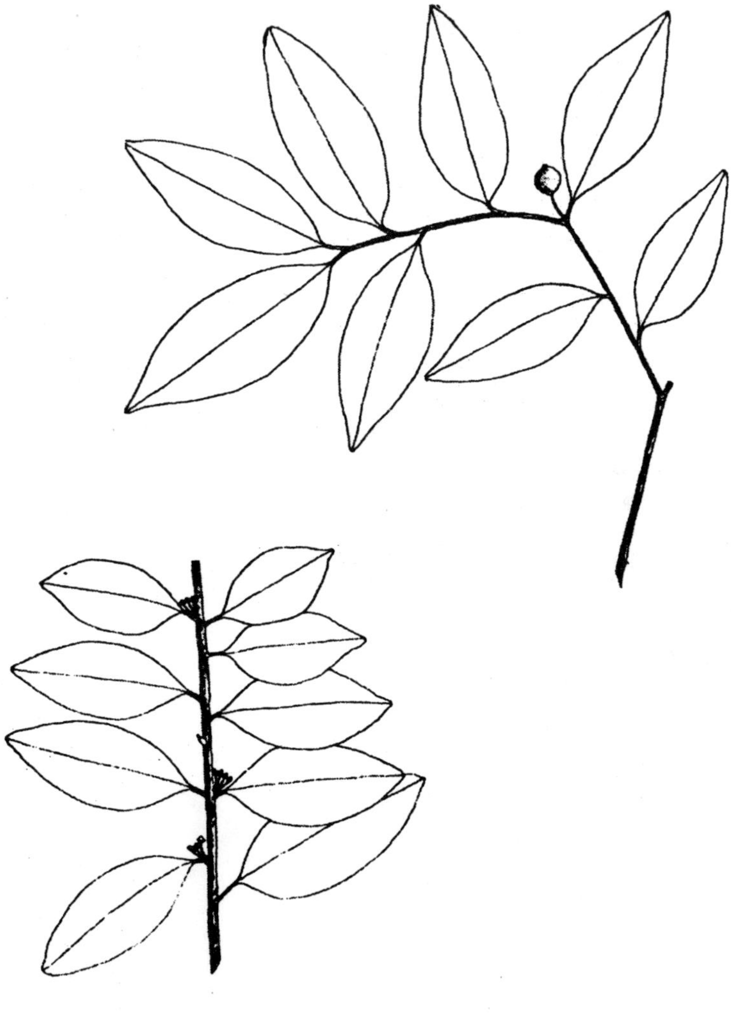

**Fig. 146 : Cookii (*Ilex cookii* Britton & Wils.)**
Fruiting twig (above), flowering twig (below), natural size.

## 147. Macoucoua (*Ilex guianensis*)

This rare small tree of mountain forests is characterized by: (1) oblong or obovate thick leathery leaves $1^1/_2$–4 inches long and $^3/_4$–$1^1/_2$ inches wide, rounded or blunt at apex, short–pointed at base; (2) many small white flowers $^1/_8$ inch across the 4 petals, clustered at leaf bases; and (3) round fruits (drupes) less than $^1/_4$ inch in diameter, red to black.

Small evergreen tree to 30 feet high and 6 inches in trunk diameter or a shrub. Bark gray, smoothish. Twigs gray, hairless or finely hairy.

The alternate hairless leaves have petioles $^1/_8$–$^3/_8$ inch long. The blades are slightly turned under at edges, have midvein slightly sunken and side veins inconspicuous, and lower surface paler.

Male flowers many, in branched clusters $^1/_2$ inch across at leaf base, consisting of minute 4–lobed calyx, corolla of 4 white petals $^1/_{16}$ inch long, and 4 stamens. Female flowers on different plants (dioecious), several at leaf base, have calyx, corolla, minute stamens, and pistil with rounded ovary and flattened stigma. Fruits (drupes) few at leaf bases on slender stalks about $^1/_8$ inch long, with calyx at base and stigma at apex, containing few nutlets.

The whitish wood is hard.

Rare in moist limestone and lower forests at 200–1,000 feet altitude in western Ghat.

**Fig. 147 : Macoucoua (*Ilex guianensis*)**

Twig with female flowers (upper left), twig with male flowers (upper right), twig with fruits (lower right), natural size.

## 148. Acebo de sierra (*Ilex macfadyenii*)

This small tree rare in upper mountain forests is easily distinguished from the other native species of its genus by its long–pointed toothed leaves. Characters for recognition are : (1) small narrowly ovate leaves $^3/_4$–$1^3/_4$ inches long and $^1/_2$–1 inch wide, long–pointed, and sharply saw–toothed; (2) 1 to few flowers at leaf bases, 5–7 parted, more than $^1/_8$ inch across the spreading white petals; and (3) round red fruits $^1/_4$ inch in diameter, with flat stigmas at apex.

Evergreen shrub or small tree to 40 feet high and 10 inches in trunk diameter. Bark gray, smooth to fissured, with raised dots, covered by mosses and liverworts. Inner bark light gray, bitter. Twigs slender, when young light green, finely hairy, and slightly angled, becoming light gray and slightly fissured. Buds about $^1/_{16}$ inch long, pointed, green, composed of minute leaves.

Leaves alternate, with minute paired stipules and slender light green leafstalks $^1/_8$–$^3/_{16}$ inch long, finely hairy. Blades narrowly ovate or sometimes lanceolate, hairless, border of short narrow teeth pointed toward long narrow apex, base rounded, very slightly thickened and leathery with side veins not visible. The upper surface is dark green and slightly shiny, the 2 sides bent up from midrib, the lower surface dull light green.

Flowers mostly male and female, 1 to few at leaf bases on slender stalks, often branched, to $^3/_8$ inch long. The minute calyx is light green, 5–7 toothed, and the corolla of 5–7 spreading elliptic whitish petals. Male flowers have 5–7 alternate stamens and nonfunctioning pistil. Female flowers have small non-functioning stamens and pistil with rounded ovary and 5–7 flat stigmas The fruits (drupes) with calyx at base and flat stigmas at apex are green when immature, turning red. With flowers from spring to fall and fruits in fall and winter.

The wood is whitish, hard, and moderately heavy.

Rare in upper Lakhimpur forests, also dwarf forest, at high altitudes of 1,500–4,000 feet.

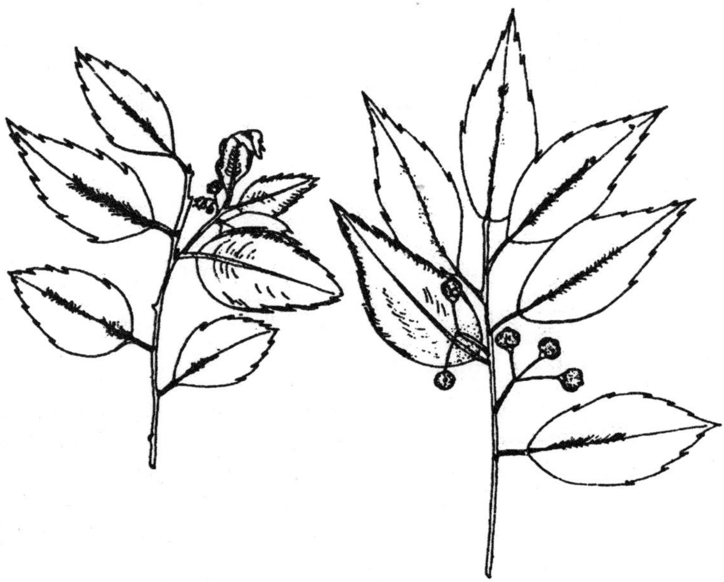

**Fig. 148 : Acebo de sierra (*Ilex macfadyenii*)**
Flowering twig (left), fruiting twig (right), natural size.

## 149. Citronnier (*Ilex nitida*)

A small to medium–sized tree of mountain forests characterized by : (1) obovate or elliptic shiny green leathery leaves 2–4 inches long and 1–2 inches broad, blunt–pointed or rounded at apex, the edges usually slightly wavy toothed and turned under; (2) small whitish flowers less than $^1/_4$ inch across, several short–stalked in clusters at leaf bases; and (3) rounded bright red (or blackish) fleshy fruits $^5/_{16}$ inch in diameter.

An evergreen tree commonly 20–40 feet high but reaching 60 feet in height and 14 inches in trunk diameter with buttresses to 2 feet, hairless throughout. The light gray bark is smoothish to scaly and thin. Inner bark is whitish and slightly bitter. The twigs are green, with raised dots (lenticels).

The leaves are alternate on leafstalks $^1/_4$–$^3/_8$ inch long. Leaf blades are usually broadest beyond middle, short-pointed at base, sometimes almost without teeth on edges, and paler beneath.

The flowers are mostly male and female on different trees but some flowers are bisexual (polygamop–dioecious), in lateral clusters on stalks $^1/_8$–$^1/_4$ inch long. The calyx is less than $^1/_{16}$ inch long, slightly 4–5–lobed; there are 4 or 5 whitish wide spreading petals more than $^1/_8$ inch long, united at base; stamens as many as petals, attached to base of corolla; and in female flowers, pistil with 4–celled ovary with 4–lobed stalkless flat stigma. Fruits (drupes) are 1 to 4 together at a node and contain 4 brown nutlets $^3/_{16}$ inch long. Flowering and fruiting intermittently.

The sapwood is whitish and hard. The wood is described as lightweight, though, and easily worked. Elsewhere reported as used in carpentry.

Uncommon in moist limestone and lower and upper forests at 200–3,000 feet altitude in northern foothills and upper western mountains Sivalik.

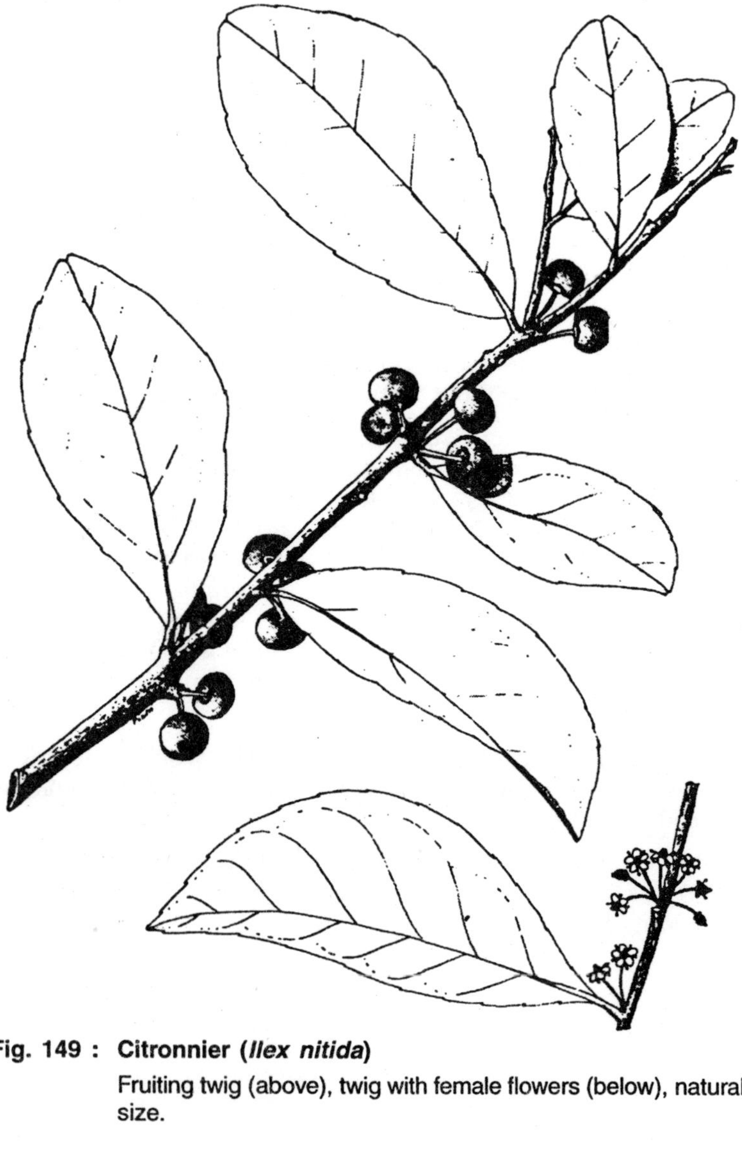

**Fig. 149 : Citronnier (*Ilex nitida*)**

Fruiting twig (above), twig with female flowers (below), natural size.

## 150. Reidl (*Ilex riedlaei* Loes)

This shrub or small tree of mountains in western Nilgiri is characterized by : (1) small obovate to elliptic leaves mostly $1–1^1/_2$ inches long and $^1/_2$–1 inch wide, rounded and notched at apex, short–pointed at base, thick and stiff with edges turned under; (2) many small white 4–parted flowers $^1/_8$ inch wide crowded at leaf bases and back of leaves; and (3) round red to black fruits $^3/_{16}$ inch in diameter, with calyx at base and dotlike stigma at apex.

Shrub or small tree to 20 feet high and 5 inches in trunk diameter, hairless. Bark gray, smooths, slightly warty. Inner bark with green outer layer, orange streaked, gritty, bitter. Twigs are slender and slightly angled when young, green becoming gray.

The alternate leaves have slender leafstalks $^1/_8$ inch long. Blades have side veins inconspicuous, dark green and slightly shiny on upper surface and light green beneath. Leaves on rapidly growing young plants are up to $2^1/_2$ inches long and $1^1/_2$ inches wide and have few short teeth near apex.

Many flowers, mostly male and female on different plants, are crowded at leaf bases and at nodes back of leaves on slender stalks $^1/_{16}$–$^1/_8$ inch long. Male flowers have calyx with 4 rounded lobes, 4 elliptic white petals united at base, 4 alternate stamens, and nonfunctioning pistil. The many fruits (drupes) have dotlike slightly 4–lobed stigma at apex and are juicy and bitter. The color changes from green to dark red to reddish black. Flowering and fruiting irregularly during the year.

Wood whitish, hard.

Uncommon and local in upper Sivalik forest, also dwarf forest, at 2,000–4,390 feet altitude in western Punjab.

**Fig. 150 : Reidl (*Ilex riedlaei* Loes)**
Flowering twig (above), fruiting twig (lower right), natural size.

## 151. Gongolin (*Ilex sideroxyloides*)

A rare tree of Konkan mountains, distinguished by : (1) shiny dark green leathery, obovate or elliptic leaves $1^1/_4$–$2^1/_2$ inches long and $^1/_2$–1 inch broad, rounded or blunt–pointed at apex and long–pointed at base, the edges turned under and the veins indistinct; (2) few small whitish flowers less than $^1/_4$ inch across in lateral clusters 5–7–parted; and (3) round red fruits $^5/_{16}$ inch in diameter.

Small to medium–sized tree to 40 feet in height and 7 inches in trunk diameter, with straight trunk and narrow crown, hairless throughout. The bark is gray and smooth or slightly rough, the light brown inner bark gritty and slightly bitter. The slender twigs are green when young, turning brown.

The evergreen leaves are alternate on leafstalks $^1/_8$–$^1/_4$ inch long. Leaf blades often have a minute sharp point at apex and are broadest beyond middle and pale on the lower surface.

The few flowers are located at leaf bases on stalks $^1/_4$–$^1/_2$ inch long, partly male and female on different trees and partly bisexual (polygamo–dioecious). The calyx is less than $^1/_{16}$ inch long, slightly 5–7 toothed; the whitish corolla of 5–7 elliptic petals more than $^1/_8$ inch long and united near base; stamens as many as petals; and ovary 4–celled with stalkless flat stigmas. The fleshy fruits (drupes) single or paired at leaf bases have 4 brown nutlets $^1/_4$ inch long. Flowering from spring to fall, the fruit maturing from summer to winter.

Sapwood is whitish and hard. The wood is whitish or sometimes brownish, hard, and moderately heavy (specific gravity 0.77). Elsewhere, the wood has been used in carpentry.

**Fig. 151 : Gongolin (*Ilex sideroxyloides*)**
Fruiting twig (above), flowering twig (lower left), natural size.

## 152. Obcordata (*Ilex sintenissi*)

This rare shrub or small tree of the upper Western Ghat forest and dwarf forest Mountains is identified by : (1) small, obovate or elliptic leaves $^3/_8$–$1^1/_8$ inches long, $^1/_4$–$^3/_4$ inch wide, rounded and notched at apex, thick and stiff with edges turned under; (2) flowers 1 or few at leaf bases, 4–5–parted, $^5/_{16}$ inch across the spreading white petals; and (3) round fruits $^3/_{16}$ inch in diameter, with conical stigmas at apex.

Evergreen shrub or small tree to 15 feet high and 3 inches in trunk diameter. Bark gray, smooth, covered with mosses and liverworts. Twigs slender, light green, becoming gray, hairless. Buds composed of minute light green leaves.

The alternate hairless leaves have tiny paired stipules and light green curved leafstalks $^1/_8$–$^1/_4$ inch long. Blades are short–pointed at base, the upper surface dark green, slightly shiny or dull, with midrib slightly sunken and side veins scarcely visible, the lower surface dull light green with side veins not visible.

One or few flowers are borne at leaf bases on slender stalks $^1/_4$–$^3/_8$ inch long, mostly male and female. Female flowers are composed of minute green calyx with 4–5 lobes; corolla of 4–5 elliptic white petals slightly united at base; 4–5 alternate small nonfunctional stamens; and pistil with 4–celled ovary and broad, flattened, conical stigmas. Male flowers have similar calyx and corolla with larger stamens and small non–functional pistil. The fruits (drupes), green when immature, have calyx at base and stigmas at apex. Flowering and fruiting throughout the year.

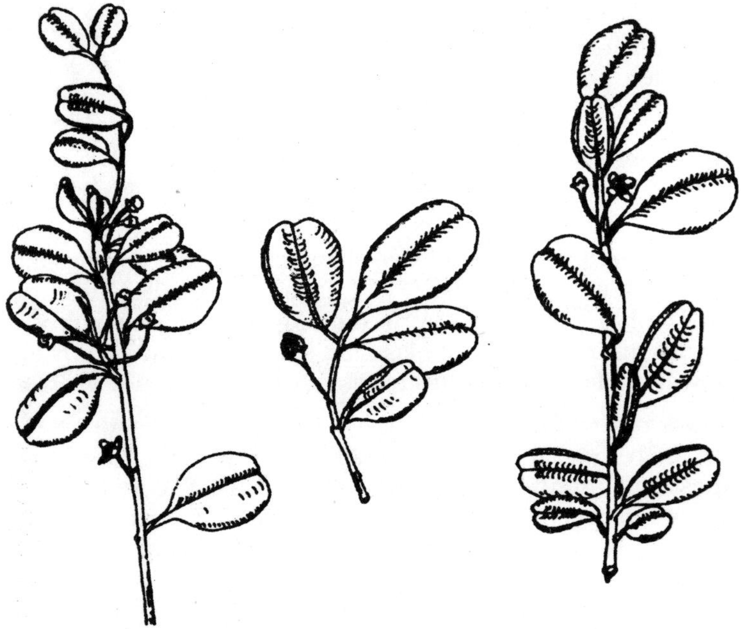

**Fig. 152 : Obcordata (*Ilex sintenissi*)**

Twig with male flowers (left), fruiting twig (center), twig with female flowers (right), natural size.

## 153. Cuero de sapo (*Ilex urbaniana*)

This rare shrub or small tree known only from mountain forests of Puerto Rico and Tortola is identified by : (1) small elliptic leathery leaves $1^1/_4$–$2^1/_2$ inches long, rounded or notched at apex and mostly blunt at base; (2) 4–parted white flowers $^1/_4$ inch wide, clustered at leaf bases; and (3) rounded fruits (drupes) about $^1/_4$ inch in diameter.

Evergreen shrub or small tree to 30 feet high and 4 inches in trunk diameter. Twigs slender, finely hairy to nearly hairless.

Leaves alternate, hairless, with petioles $^1/_4$–$^3/_8$ inch long. Blades blunt or short–pointed at base, not toothed on edges.

Flowers clustered at leaf bases, 1–3 on a short stalk about $^1/_4$ inch long, mostly male and female on different plants (dioecious). Calyx 4–toothed, $^1/_{16}$ inch long; corolla of 4 narrow spreading white petals less than $^1/_8$ inch long, slightly united at base; stamens 4; and pistil with round ovary and flattened stigma. Fruits have thin pulp and few bony nutlets.

Rare in upper forest at about 3,000 feel altitude in western. Also Ghat Mountain, altitude but 1,700 feet.

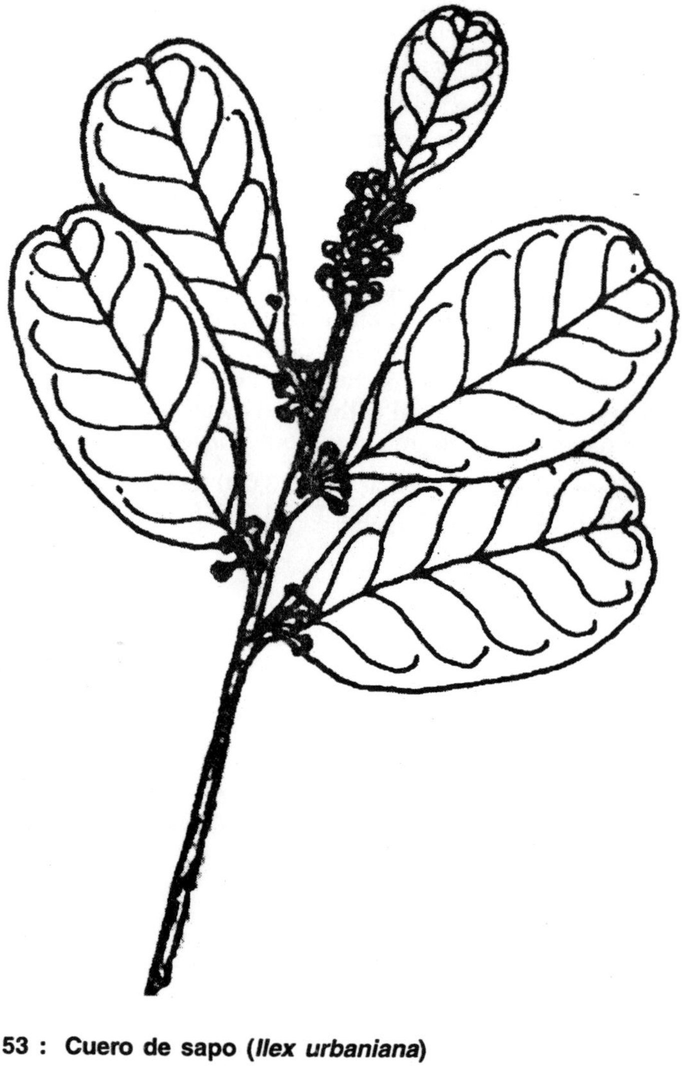

**Fig. 153 : Cuero de sapo (*Ilex urbaniana*)**
Flowering twig, natural size.

# BITTERSWEET FAMILY (CELASTRACEAE)

Shrubs, woody vines, and trees mostly small, known by : (1) leaves alternate or opposite, sometimes whorled, simple, with minute tripules or none; (2) minute flowers mostly in clusters (cymes) with stalks mostly joined, greenish, bisexual or functionally male and female, regular, with 4–5 sepals united at base and persistent, 4–5 petals, 4–5 alternate stamens inserted on or below the large disk, and pistil with superiod ovary (sometimes surrounded by disk and appearing inferior) of 2–5 cells each with 2 ovules, short style, and stigma often 2–5 lobed; and (3) fruit a capsule, berry, or drupe, the seed generally with colored covering (aril).

**Key to Species**

A. Leaves opposite or mostly so; fruit fleshy.

B. Leaves thick and stiff, with side veins inconspicuous; twigs mostly round.

C. Leaves elliptic, 1–4 inches long, rounded to short–pointed at both ends, edges slightly turned under and sometimes with few teeth—*Cassine xylocarpa*.

CC. Leaves obovate, 1–2$^1/_2$ inches long, rounded at apex and gradually narrowed to long–pointed base, with edges turned under; fruit a 4–lobed capsule—*Torralbasia cuneifolia*.

BB. Leaves slightly thickened, with side veins distinct, often wavy on edges; twigs very slender, slightly 4–angled, often branching in pairs.

D. Leaves elliptic to obovate, 1$^1/_4$–2 inches long, blunt, rounded or notched at apex; flowers $^1/_8$ inch wide, whitish; fruits elliptic, blue black—*Gyminda latifolia*.

DD. Leaves mostly ovate to rounded, $^1/_2$–1$^1/_2$ inches long, blunt or short–pointed at apex; flowers $^1/_8$ inch wide, reddish; fruit orange red to deep red—*Crossopetalum rhacoma*.

AA. Leaves alternate; fruit a capsule.

E. Leaves and twigs pale yellow green, leaves elliptic and short–pointed, 1–2 inches long, slightly thickened, with network of many fine veins; fruits (drupes) orange red or red, nearly $^1/_4$ inch in diameter—*Schaefferia frutescens*.

EE. Leaves and young twigs greenish, twigs mostly turning gray, leaves mostly thick and leathery—*Maytenus*.

F. Leaves oblong, 1$^1/_2$—3$^1/_2$ inches long, about 3 times as long as wide—*Maytenus elongata*.

FF. Leaves mostly elliptic, about twice as long as wide.

G. Leaves 1–2 inches long, rounded at apex, gray green—*Maytenus cymosa.*

GG. Leaves 1 $^1/_2$–4 inches long, mostly blunt at apex.

H. Leaves shiny yellow green on upper surface; seed capsule round, $^1/_4$ inch wide, whitish—*Maytenus ponceana.*

HH. Leaves slightly shiny green on upper surface; seed capsule elliptic, $^3/_8$–$^1/_2$ inch long—*Maytenus elliptica.*

## 154. Marble–tree (*Cassine xylocarpa*)

This common shrub or small tree widespread on coasts and lowland thickets is recognized by : (1) leaves mostly opposite and elliptic, commonly small but varying greatly in size and shape, very thick and stiff, sometimes with few teeth, light green; (2) small greenish–white or yellowish flowers clustered at leaf bases, $^3/_{16}$ inch across, 5–parted; and (3) elliptic or rounded fruit $^3/_4$–$1^1/_4$ inches long, yellow or greenish yellow, slightly fleshy and hard.

Evergreen shrub or sometimes a small tree to 35 feet high and 8 inches in trunk diameter, hairless throughout. Bark gray, smooth to finely fissured, the inner bark purplish or pinkish and slightly bitter. Twigs slender, light green, becoming gray brown.

Leaves mostly opposite, sometimes alternate, with minute paired pointed stipules $^1/_{16}$ inch long shedding early and light green petioles about $^1/_4$ inch long. Blades mostly elliptic, sometimes obovate, commonly small, mostly 1–4 inches long and $^1/_2$–$2^1/_2$ inches wide but smaller in dry areas and larger in wet areas, rounded to short–pointed at both ends, very thick and stiff, edges slightly curved under the sometimes with few teeth, side veins few and inconspicuous, upper surface light green and slightly shiny, lower surface dull whitish green.

Flower clusters (panicles) $^3/_8$–1 inch long at leaf bases bear several to many small slightly fragrant flowers, male and female on different plants (dioecious). Female flowers have calyx with 5 spreading teeth, corolla of 5 spreading pointed petals, 5 minute alternate nonfunctional stamens inserted under a flat disk, and pistil composed of 3–celled ovary with 2 tiny ovules in each cell, style, and 3–lobed stigma. Male flowers have larger stamens and minute pistil.

The fruit (drupe) is greenish but becomes yellow at maturity. It varies in shape from narrowly elliptic to nearly round and has thin flesh, a large hard thick–walled stone, and 3 or fewer cells and elliptic seeds $^3/_8$ inch long. With flowers and fruits intermittently through the year.

The wood is light brown, hard, heavy, fine–textured, strong and reported to be fairly durable.

Common and widely distributed in coastal dry and moist forests from sea level to 600 feet altitude. Along coasts, such as sandy beaches and in thickets and openings in forests in A.P.

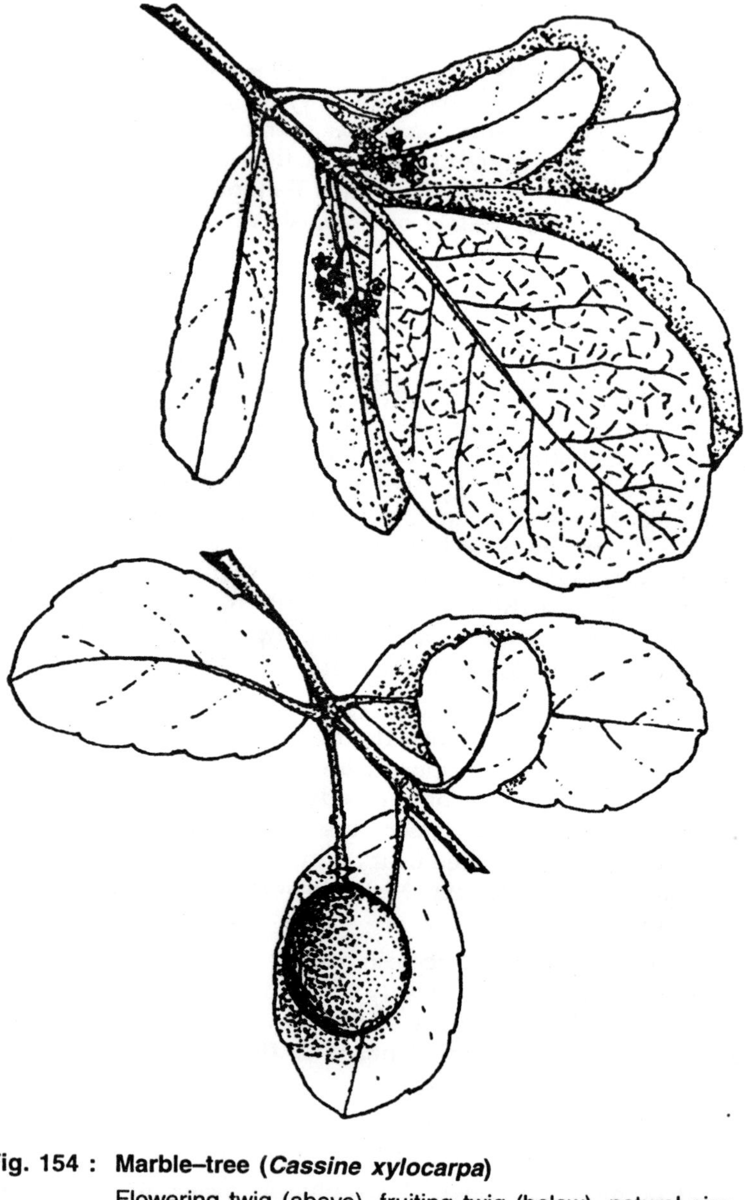

**Fig. 154 : Marble–tree (*Cassine xylocarpa*)**
Flowering twig (above), fruiting twig (below), natural size.

## 155. Coral, Florida crossopetalum

This shrub sometimes becoming a small tree is common and widespread in dry areas. It is identified by : (1) paired small light green leaves mostly ovate to rounded, usually finely wavy on edges; (2) tiny 4–parted reddish flowers less than $^1/_8$ inch across, clustered at leaf bases; and (3) orange–red to deep red egg–shaped fruits about $^1/_4$ inch long, 1–seeded.

Shrub or sometimes small tree to 20 feet high and 3 inches in trunk diameter, or larger, much branched, evergreen or sometimes deciduous. The bark is gray, finely fissured and slightly shedding in thin plates. Inner bark is light brown or pinkish, fibrous and almost tasteless. The very slender twigs, often paired, are light green and slight 4–angled when young, becoming gray, hairless.

The leaves are opposite or sometimes in 3's, hairless, and have short petioles less than $^1/_{16}$ inch long. Blades are of varying shapes, mostly ovate to rounded, sometimes narrowly oblong, $^1/_2$–$1^1/_2$ inches long and $^3/_8$–$1^1/_4$ inches wide, blunt or short–pointed at apex and rounded or short–pointed at base, finely wavy on edges, slightly thickened and leathery, dull yellow green on both surfaces or paler beneath.

The flower clusters (cymes) $^1/_2$–$^7/_8$ inch long bear several short–stalked flowers at the end of a slender reddish stalk. The minute calyx is greenish red, finely hairy, and has 4 rounded lobes; petals 4, less than $^1/_{16}$ inch long, reddish, rounded and curved; stamens 4, inserted between lobes and disk and alternate with petals; and pistil with 4–celled 4–ovuled ovary, short style, and 4 minute stigmas. Fruits (drupes) few, slightly 1–sided, with calyx at base and style at side of apex, slightly fleshy, containing a large stone and 1 or 2 seeds. Flowering and fruiting irregularly through the year.

The sapwood is light brown and hard.

Common and widely distributed in coastal, dry, and moist forests from sea level to 600 feet altitude and throughout the other islands.

**Fig. 155 : Coral, Florida crossopetalum (*Crossopetalum rhacoma*)**
Natural size

## 156. West Indies falsebox (*Gyminda latifolia*)

A common small tree of coastal lowlands, characterized by : (1) very slender twigs, slightly 4–angled, branching in pairs; (2) leaves paired, small, elliptic to obovate, $1^1/_4$–2 inches long, $^1/_2$–1 inch wide, blunt, rounded or notched at apex, slightly thickened; (3) small flowers $^1/_8$ inch wide, whitish, 4–parted, few in clusters at leaf bases, male and female on different trees; and (4) elliptic fruit about $^1/_4$ inch long, blue black, fleshy, 1–seeded.

Small evergreen tree to 20 feet high and 4 inches in trunk diameter, hairless. Bark gray, finely fissured and scaly. Inner bark light pink, slightly bitter. Twigs very slender, slightly 4–angled, branching in pairs, light green, becoming light gray.

Leaves opposite, with minute paired stipules forming the bud. Leafstalks slender, very short, $^1/_8$ inch long. Blades gradually narrowed to base, slightly turned under and often slightly wavy at edges, with inconspicuous veins, dull light green on both surfaces.

Flowers few at end of very slender stalk at leaf bases. Male flowers are composed of minute 4–lobed light green calyx, 4 spreading white petals, 4 alternate stamens, and minute nonfunctional pistil. Female flowers on other trees (dioecious) have calyz, 4 petals, and pistil with 2–celled ovary, each cell with 1 ovule, and no style. Fruits (drupes) have calyx at base, thin flesh, thick–walled stone, and I seed. With flowers and fruits irregularly during the year.

Wood light brown, hard.

Common in coastal and moist limestone forests from sea level to 800 feet altitude, northern and western Andhra Pradesh.

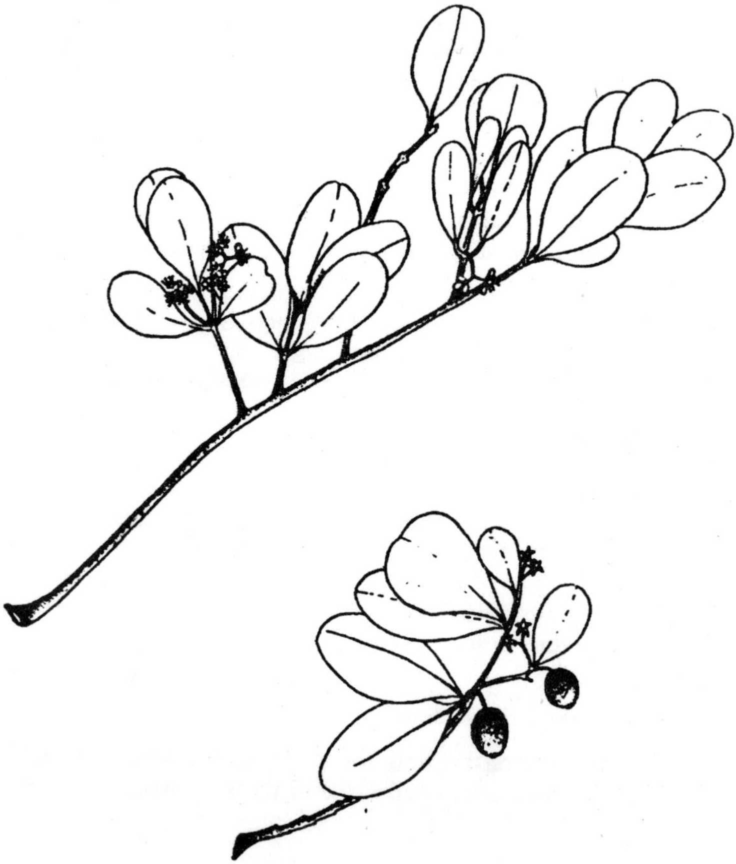

**Fig. 156 : West Indies falsebox (*Gyminda latifolia*)**

Twig with male flowers (above), twig with female flowers and fruits (below), natural size.

## 157. Cymosa (*Maytenus cymosa*)

Rare small tree confined to eastern Calcutta and a few islands eastward, identified by : (1) alternate elliptic to obovate leaves 1–2 inches long and $^5/_8$–$1^1/_2$ inches wide, stiff and leathery, gray green; (2) small flowers $^1/_8$ inch broad, 5–parted, crowded at leaf bases; and (3) seed capsules elliptic, $^3/_8$ inch long.

Small evergreen tree to 25 feet high and 6 inches in trunk diameter, hairless. Bark blackish or dark gray, finely fissured. Inner bark pink to deep red, bitter. Twigs slender, slightly angled.

Leaves alternate, with leafstalks $^1/_8$–$^1/_4$ inch long. Blades rounded at apex, short–pointed at base, slightly turned down at edges, with few inconspicuous veins, paler beneath.

Flower clusters (cymes) have several crowded almost stalkless flowers composed of 5–toothed calyx; corolla of 5 pale yellow rounded petals; 5 minute alternate stamens inserted under a disk; and pistil with 2–celled ovary, short style, and stigma slightly 2–lobed. Fruit a blackish seed capsule, splitting in 2 parts. Seeds 1 or more in fleshy red covering. Flowering in spring and with fruits in spring and summer.

Wood whitish, slightly hard.

Rare in coastal moist forests of lowlands below 100 feet altitude in eastern Sundarban and islands east–ward.

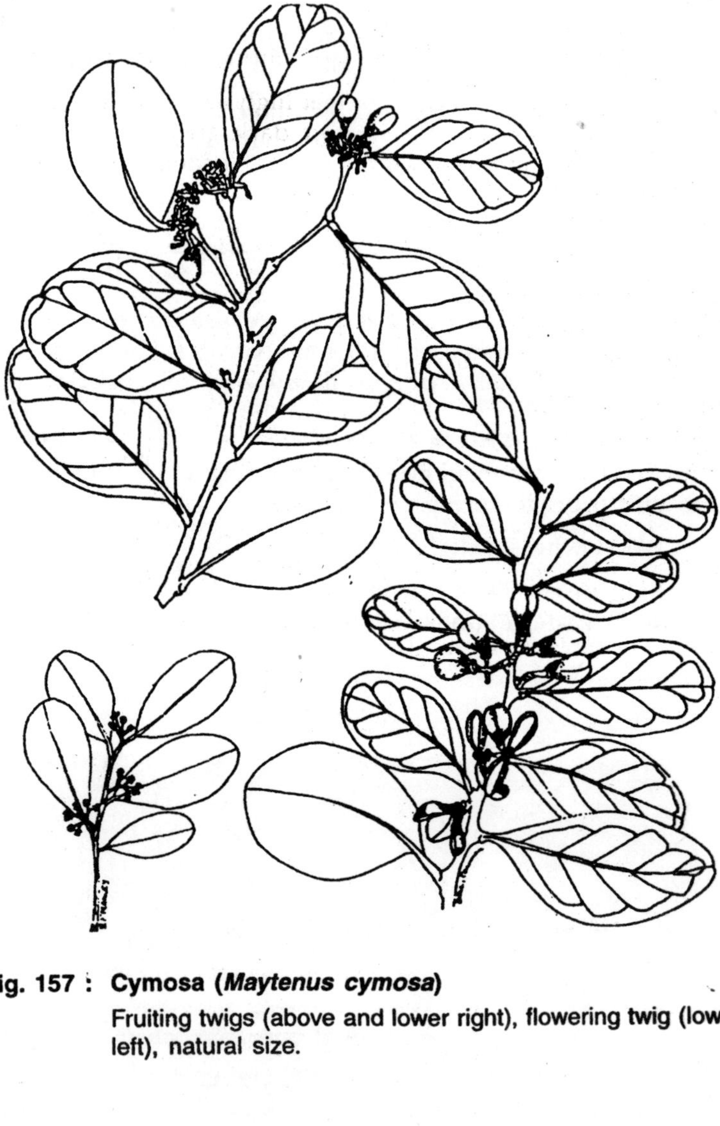

**Fig. 157 : Cymosa (*Maytenus cymosa*)**

Fruiting twigs (above and lower right), flowering twig (lower left), natural size.

## 158. Cuero de sapo (*Maytenus elliptica*)

This shrub or tree of the moist limestone forest is characterized by : (1) alternate elliptic leaves $1^1/_2$–4 inches long and $^3/_4$–2 inches wide, slightly thick and leathery, turned up slightly from midvein, and slightly shiny green on upper surface; (2) small yellowish green flowers $^1/_8$ inch across, 5–parted; and (3) seed capsules elliptic, $^3/_8$–$^1/_2$ inch long, orange, 2–celled.

Evergreen shrub or usually small tree to 45 feet high and 10 inches in trunk diameter, hair–less throughout. Bark gray, smooth to slightly fissured, the inner bark pink or turning deep red and slightly bitter. Twigs slender, light green and angled when young, turning gray with many raised dots (lenticels).

Leaves alternate, with minute paired stipules and light green leafstalks $^1/_4$ inch long. The elliptic blades are blunt or rounded at apex, short–pointed at base, not toothed on edges, with few inconspicuous veins, the lower surface dull light green.

Flowers are few or single on slender stalks $^1/_8$ inch long at leaf bases, slightly fragrant, composed of 5–toothed green calyx; corolla of 5 yellow–green spreading pointed petals; 5 minute alternate stamens inserted under a dark red disk; and pistil with 2–celled ovary partly within disk, 2 ovules in each cell, short style, and stigma slightly 2–lobed. Seed capsules 1 or 2 at base of a leaf, splitting into 2 parts. Seeds few, elliptic, nearly $^1/_4$ inch long, with fleshy covering. Probably flowering intermittently. Collected with fruits in February and April.

Wood light brown, slightly hard

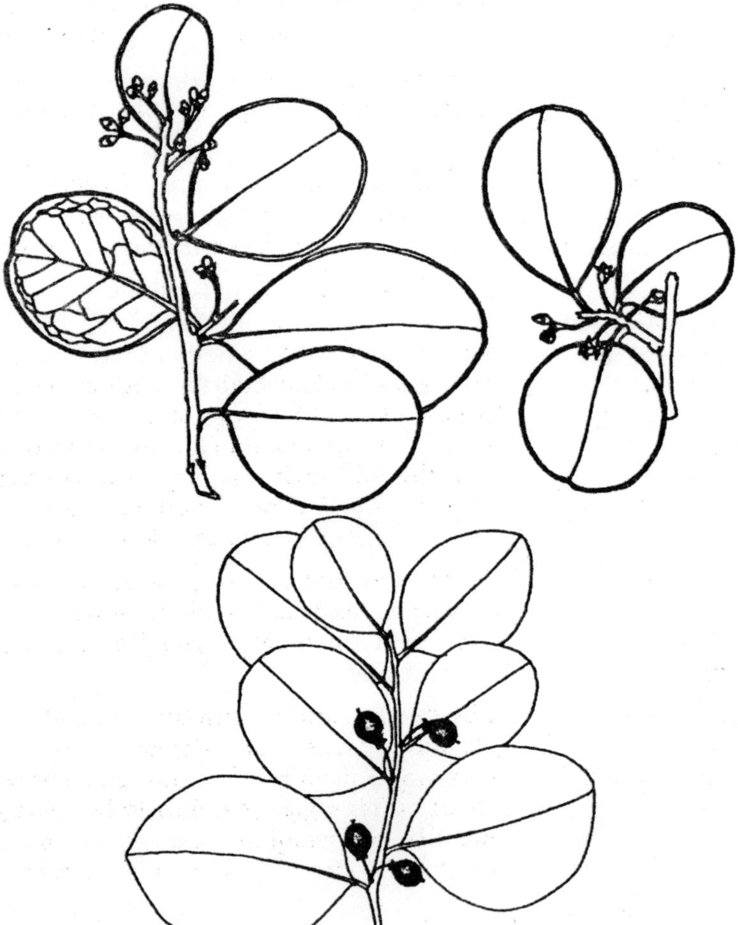

**Fig. 158 : Cuero de sapo (*Maytenus elliptica*)**
Flowering twigs (above), fruiting twig (below), natural size.

## 159. Elongata (*Maytenus elongata*)

This rare tree known only from western Sivalik Mountains is characterized by : (1) alternate oblong leathery leaves $1^1/_2$–$3^1/_2$ inches long and about $^1/_2$–$1^1/_2$ inches wide; and (2) elliptic green seed capsules $^1/_4$–$^3/_8$ inch long.

Evergreen tree to 20 feet high (recorded to 40 feet) and 5 inches in trunk diameter, or a shrub of 10 feet. Inner bark deep red. Twigs slender, hairless.

Leaves alternate, hairless, with petioles $^3/_{16}$ inch long. Blades more than twice as long as wide, short–pointed at both ends, not toothed on edges, with side veins inconspicuous, paler beneath.

The flowers have not been described. Fruits single or clustered at leaf bases on stalks $^1/_8$–$^1/_2$ inch long, with minute style at apex and 5 minute sepals at base. Seed capsules in this genus 1–3–celled with 3 or fewer seeds in a fleshy cover (aril). Collected with fruits in February and July.

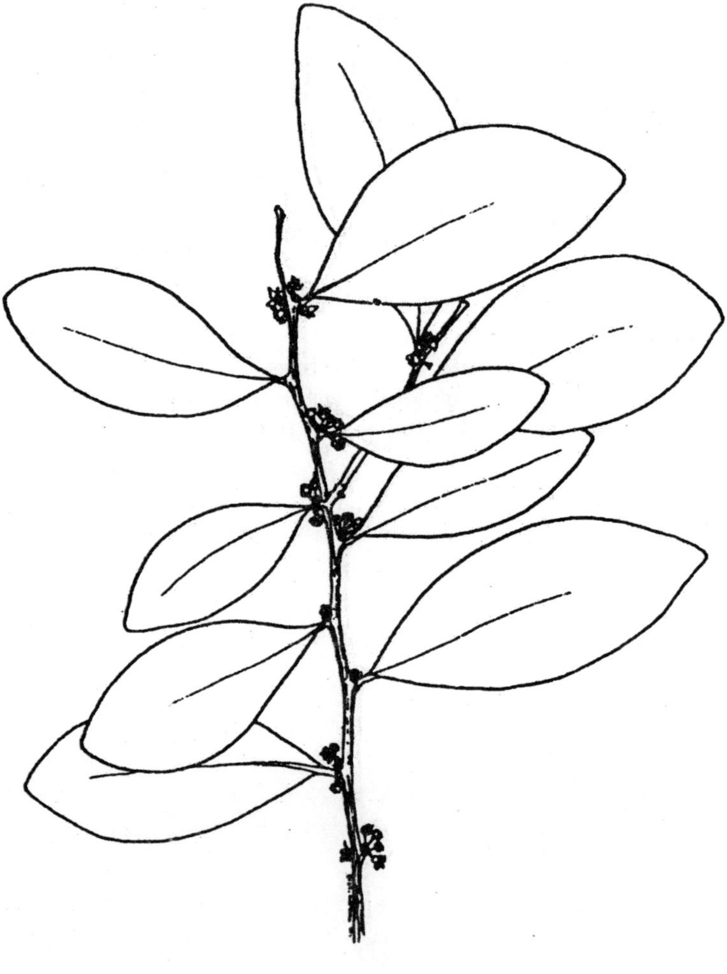

**Fig. 159 : Elongata (*Maytenus elongata*)**
Flowering twig, natural size.

## 160. Ponceana (*Maytenus poncceana*)

A rare tree known only from Sivalik, characterized by : (1) elliptic to ovate leaves, 2–3 inches long and 1–$1^1/_2$ inches wide, thick and leathery, turned up slightly at midrib, the upper surface shiny yellow green; and (2) seed capsules 1–3 at leaf bases or on twigs back of leaves, round, whitish, $^1/_4$ inch in diameter.

An evergreen tree to 40 feet high and 5 inches in trunk diameter. The twigs are slender, hairless, and slightly crooked, light green when young, becoming gray.

The alternate leaves have short petioles $^1/_8$–$^1/_4$ inch long. Blades are gradually narrowed to blunt apex, rounded at base, not toothed on edges, hairless, with few indistinct veins, the lower surface dull light green.

Flowers have not been described. The fruits (capsules) have splender stalks about $^1/_8$ inch long and remains of calyx at base and dotlike style at apex, light green but becoming whitish, splitting into 3 parts. There is 1 rounded light brown seed less than $^3/_{16}$ inch in diameter. Collected with fruits in February and April.

Rare in moist limestone and lower Sivalik forests at 500–2,000 feet altitude in foothills.

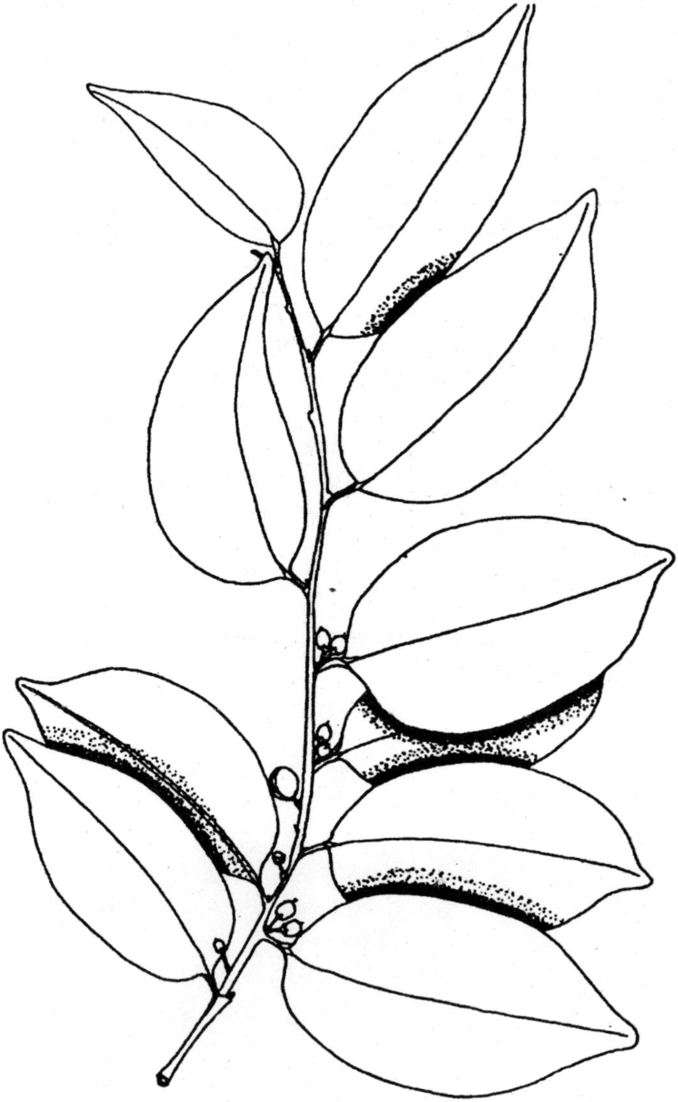

**Fig. 160 : Ponceana (*Maytenus ponceana*)**
Fruiting twig, natural size.

## 161. Florida–boxwood

This shrub or small tree common and widespread at low altitudes, especially in dry forests, is easily recognized by : (1) the pale yellow–green color of both leaf surfaces and the slender twigs; (2) the small leaves mostly elliptic and short–pointed but varying to lanceolate and obovate; (3) small 4–parted yellow–green flowers $^{3}/_{16}$ inch wide, scattered at leaf bases; and (4) orange–red or red rounded fleshy fruits nearly $^{1}/_{4}$ inch in diameter.

Evergreen shrub or small tree to 30 feet high and 5 inches in trunk diameter, elsewhere reported to become a medium–sized tree to 45 feet high and 1 foot in trunk diameter. Bark light gray, smoothish to finely fissured, sometimes rough and furrowed. Inner bark is light yellow and almost tasteless. The twigs are slender and slightly angled, hairless, pale yellow green but afterwards becoming light gray.

The alternate leaves are mostly 1–2 inches long and $^{1}/_{2}$–1 inch wide, sometimes to $2^{1}/_{2}$ inches long and $1^{1}/_{4}$ inches wide, gradually narrowed to nearly stalkless base or leafstalk $^{1}/_{8}$ inch long, hairless, not toothed on edges, slightly thickened and leathery, with network of many fine veins.

Flowers are male and female on different plants (dioecious), single or clustered at leaf bases on short stalks $^{1}/_{8}$ inch long. Male flowers have minute 4–lobed calyx, 4 greenish–white petals nearly $^{1}/_{8}$ inch long, and 4 alternate stamens inserted below a disk. Female flowers have calyx and corolla and pistil with 2–celled ovary, short style, and 2–lobed stigma. The fruits (drupes) have calyx at base, style at apex, slightly bitter pulp, and 2 light brown seeds. Flowering and fruit in probably irregularly through the year.

The wood is light brown to yellow, hard, moderately heavy, and fine–textured. Elsewhere it has served as a substitute for boxwood in wood engraving.

Common and widespread on moist and dry limestone forests and coastal hills and southern foothills of Kanya Kuwari at 100–1,000 feet altitude through Tamil Nadu.

**Fig. 161 : Florida–boxwood (*Schaefferia frutescens*)**

Fruiting twig (above), twig with male flowers (center right), twig with female flowers (below), natural size.

## 162. Boje

This shrub or small tree of dwarf forests of mountain summits is identified by : (1) small obovate leaves, thick and leathery, without visible side veins and with edges turned under; (2) small white flowers $^3/_{16}$ inch across the 4 rounded petals, several in clusters at leaf bases; and (3) distinctive green seed capsules of 4 or fewer spreading oblong lobes $^1/_4$ inch long, each splitting open on the upper side and 1–seeded.

Evergreen shrub or small tree to 35 feet high and 8 inches in trunk diameter, hairless. Roots yellow. Bark gray, smooth, the inner bark light pink and slightly bitter or almost tasteless. Twigs are blackish, becoming gray, often angled.

The leaves are opposite and alternate, with short green leafstalks $^1/_8$–$^1/_4$ inch long, without stipules. Blades are 1–$2^1/_2$ inches long, $^1/_2$–1 inch wide, rounded at apex, broadest beyond middle, and gradually narrowed to long–pointed base, the upper surface dull green or slightly shiny, the lower surface dull light green.

Flower clusters (cymes) 1–2 inches long at leaf bases bear many small flowers on slender spreading stalks. The flowers have a short 4–lobed green calyx, 4 white rounded petals, 4 alternate stamens, and the pistil with 4–celled ovary partly within a disk and 4 tiny styles. The seed capsules with calyx at base are divided into 4 or fewer nearly separate lobes. The elliptic brown seeds $^3/_{16}$ inch long are partly within a thin yellow covering (aril). Flowering irregularly through the year.

The wood is yellowish, hard, moderately heavy, and fine–textured.

Uncommon and local in upper Sabrimal forests, including dwarf forest, at 2,500–4,390 feet altitude. Ridges and peaks mostly through central Kerala upto the summit of the highest peak.

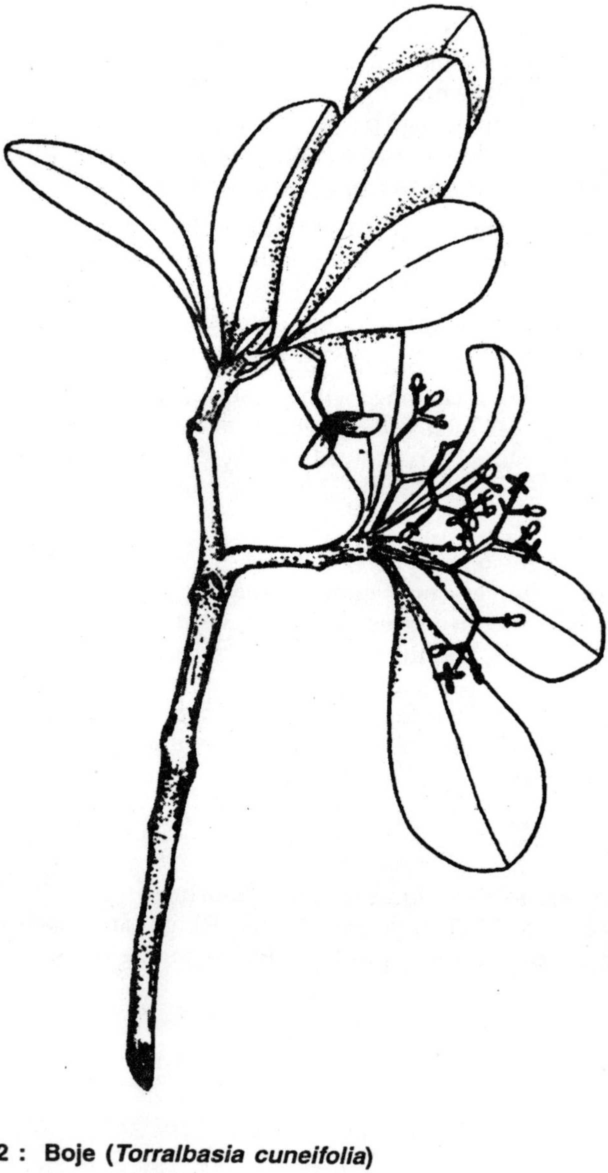

**Fig. 162 : Boje (*Torralbasia cuneifolia*)**
Natural size.

# ICACINA FAMILY (ICACINACEAE)

## 163 Palo de rosa (*Ottoschulzia rhodoxylon*)

This very rare tree has been found only in Western Kerala and is poorly known, because flowers and fruits apparently have not been described. An evergreen small tree with slender hairless twigs with raised dots (lenticels) and minute, pointed, finely–hairy buds.

The alternate hairless leaves without stipules have petioles $^1/_4$–$^3/_8$ inch long. Blades are elliptic to ovate, 2–3$^1/_2$ inches long and 1$^1/_4$–2$^1/_2$ inches wide, rounded or blunt at apex and base, not toothed on edges, thick and leathery, with 5–7 slender veins on each side of midrib, shiny on upper surface and paler beneath.

Flowers not described, apparently not collected. In this genus flowers are small, single or clustered at leaf bases, composed of 5 sepals united at base corolla, with short cup–shaped tube and 5 lobes; 5 stamens near top of corolla tube and alternate with lobes, and pistil with 1–celled ovary, 2 ovules, and short style. The fruit in this genus is a small egg–shaped drupe, with 1 seed.

The wood is reported to be hard and heavy, the heartwood reddish (as indicated by the specific name and common name), and suitable for articles of turnery.

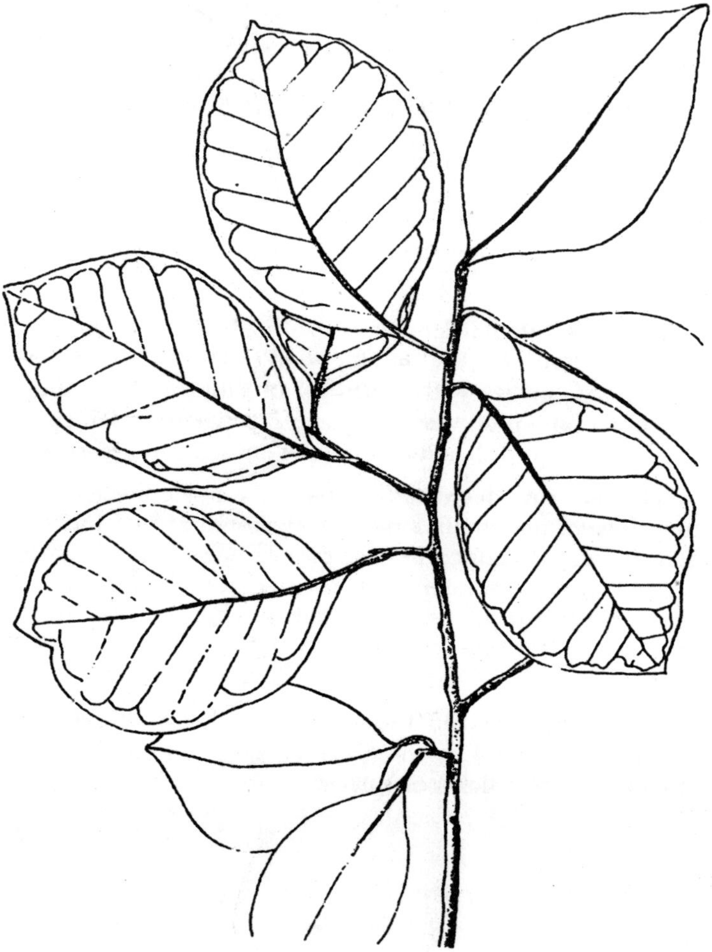

Fig. 162 : Palo de rosa (*Ottoschulzia rhodoxylon*)
Natural size.

# SOAPBERRY FAMILY (SAPINDACEAE)

Trees, shrubs, and woody vines with tendrils, rarely herbs, known by (1) leaves alternate, generally pinnate, sometimes with 3 leaflets, without stipules (except in vines), the leaflets commonly alternate; (2) many minute flowers in branched clusters (panicles or cymes), mostly male and female or bisexual (polygamous or dioecious), regular or irregular, with 5 sepals, usually 5 petals often with scale or gland at base within, generally 10 or 8 stamens inserted within a disk, and pistil with superior ovary mostly 3–celled with 1–2 ovules in each cell, and style; and (3) fruit various, often large, 3–celled capsule, berry, drupe, or winged, the seed often with covering (aril). Fruits and seeds of a few species are edible, those of some are poisonous.

**Key to Species**

A. Leaves simple.

B. Leaves oblanceolate, yellow green, sticky (viscid), slightly thickened, turned under at edges—*Do–donaea viscosa.*

BB. Leaves elliptic, shiny green, thick and leathery, finely toothed at edges.

C. Leaves long–pointed at apex, with prominent network of veins, nearly hairless—*Allophylus crassinervis.*

CC. Leaves rounded at apex, beneath densely soft hairy with prominent veins—Serrasuela, *Thouinia portoricensis* Radlk.

AA. Leaves compound.

D. Leaves with 3 leaflets, elliptic to oblanceolate or obovate, stalkless or nearly so.

E. Leaflets $^1/_2$–2 inches long, with many parallel fine side veins, blunt at apex, turned under at edges, hairless—*Hypelate trigoliata.*

EE. Leaflets more than 2 $^1/_2$ inches long, with fewer curved side veins, with toothed edges, hairy beneath.

F. Leaflets short–pointed at apex, beneath soft hairy with prominent veins—Ceboruquillo, *Thouinia striata* Radlk.

FF. Leaflets long–pointed at apex, beneath from very to slightly hairy—*Allophylus racemosus.*

DD. Leaves pinnate with 4 or more leaflets (sometimes only 2).

G. Leaflets 4–8, not paired, with wavy–toothed edges, elliptic, those toward end largest, rounded or notched at apex—*Cupania.*

H. Leaflets yellow brown hairy beneath, edges with few teeth toward apex; seed capsules sharply 3–angled, with short yellow brown hairs—*Cupania triquetra.*

HH. Leaflets gray hairy beneath, edges with teeth from base to apex; seed capsules bluntly 3–angled, with long velvety brown hairs—Guara, *Cupania americana* L.

GG. Leaflets mostly paired, not toothed.

I. Leaflets long or short–pointed at both ends, on axis often winged; fruit round, fleshy.

J. Leaflets 4, the pair at end larger; fruit $7/_8$–$1^1/_4$ inches in diameter, edible (kinep)—Quenepa, kinep, Spanish–lime, *Melicoccus bijugatus* Jaco.*

JJ. Leaflets usually 6–12, sometimes only 1 at end; fruit $^5/_8$–1 inch in diameter, yellow, inedible—Jaboncillo, wingleaf soapberry, *Sapindus saponaria* L.

II. Leaflets 2–10, mostly blunt or short–pointed at both ends, on axis not winged; fruit a capsule.

K. Leaflets less than 4 inches long, hairless or nearly so.

L. Leaflets 2, sometimes 4 or 6 (rarely 1 or 3), oblong to elliptic, 2–4 inches long, thin, hairless or nearly so; fruit round, fleshy—*Exothea paniculata.*

LL. Leaflets 2–8, elliptic or lance–shaped, stiff and leathery, above shiny green with many raised side veins, beneath pale or brownish, on slender reddish or dark brown axis; fruit a flattened capsule—Negra lora, *Matayba domingensis* (DC.) Radlk.

KK. Leaflets 6–10, mostly 3–8 inches long, elliptic, the lowest shortest, beneath with raised hairy side veins, on stout hairy axis; fruit a red fleshy 3–angled capsule (akee)—*Blighia sapida.**

## 164. Palo blanco (*Allophylus crassinervis*)

This shrub or small tree of mountain forests in western Puerto Rico is identified by : (1) lance–shaped leaves, thick, stiff, and leathery, with prominent network of veins, the parallel side veins mostly ending in minute teeth at edges; and (2) small rounded dry fruits $^1/_4$ inch long.

Evergreen shrub or small tree to 20 feet high and 4 inches in trunk diameter. Twigs slender, finely hairy when young.

Leaves alternate, simple (rarely with 3 leaflets as in related species), with slender hairy petioles $^1/_4$–$^1/_2$ inch or more in length. Blades are 2–3$^1/_2$ inches long and $^3/_4$–1$^1/_4$ inches wide, long–pointed at apex, short–pointed or blunt at base, becoming hairless or nearly so, the upper surface shiny dark–green, and the lower surface light green, with raised yellowish side veins nearly at right angle and many minute gland dots. The edges usually have short sharp teeth that point forward.

Flower clusters (racemes) have slender unbranched axis 1–2$^1/_2$ inches long from leaf base, bearing many whitish flowers less than $^1/_8$ inch across on stalks of $^1/_{16}$ inch. Flowers in this genus are mostly male and bisexual (polygamous), composed of 4 sepals, 4 white petals each with a 2–lobed scale, 8 stamens, and pistil with 2–lobed ovary, 2–celled 2–ovuled, and 2–lobed style. The dry fruit is short–stalked and hairless and contains 1 seed and does not split open. With flowers and fruits from spring to fall.

Locally common in moist limestone forests at 300–2,500 feet altitude in foot–hills of western ghat.

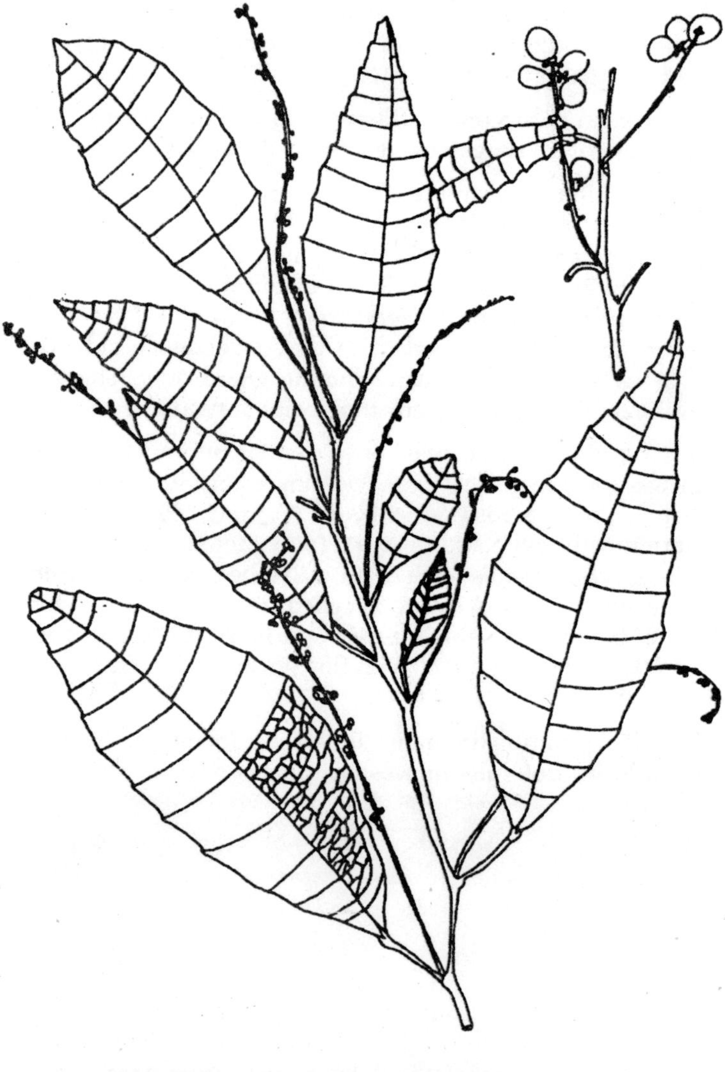

**Fig. 164 : Palo blanco (*Allophylus crassinervis*)**
Flowering twig (left), fruiting twig (upper right), natural size.

## 165. Quiebrahacha (*Allophylus racemosus*)

This species of quiebrahacha is distinguished by : (1) the alternate leaves composed of 3 almost stalless elliptic or oblanceolate leaflets with toothed edges; (2) the small greenish or whitish flowers about $^1/_8$ inch long and broad along narrow unbranched lateral axes; and (3) the dry rounded brown, orange, or red fruits $^5/_{16}$ inch in diameter.

Small deciduous tree to 30 feet high and 5 inches or more in trunk diameter, with few branches, or a shrub. The bark is gray and smooth or becoming slightly fissured into thin plates and peeling off. The inner bark is light brown and slightly bitter. The hairy twigs are gray green when young, turning light brown.

The leaves measure 5–10 inches long including the green, mostly long petioles of $1^1/_4$–6 inches. The 3 blades (trifoliolate) are 3–8 inches long and 1–$2^1/_4$ inches wide, thin, the apex long–pointed and base short–pointed, the straight side veins ending in marginal teeth, the upper surface shiny green and the lower surface dull pale green and slightly to very hairy. The 2 side leaflets are slightly oblique at base.

The unbranched flower clusters (racemes) $1^1/_2$–3 inches long bear many minutely hairy flowers on short stalks. The slightly irregular flowers are mostly male and female on different trees (dioecious). There are 2 small sepals and 2 larger, 4 small petals, 8 stamens on a disk, and in female flowers the pistil with 2–celled ovary. The fruits are slightly hairy and almost dry do not open, and contain 1 rounded seed $^1/_4$ inch long. Collected with flowers in May–July and with fruits in June–August.

The light brown hard wood has served for fence posts.

Rare to scattered in lowland forest, mostly in moist limestone forest at 500–1,000 feet altitude in western Andhra Pradesh.

**Fig. 165 : Quiebrahacha (*Allophylus racemosus*)**
Fruiting twig (left), flowering twig (upper right), two–thirds natural size.

## 166. Seso vegetal, akee (*Blighia sapida*)

Akee or vegetable brains is sometimes planted for the whitish seed covering, which is edible when picked fresh and cooked properly but otherwise a deadly poison. This dangerous tree is not common but should be learned and should not be planted wherever children play. Distinguishing characters are : (1) alternate even pinnate leaves with 3–5 pairs of large elliptic yellow–green leaflets; (2) many whitish flowers about $^{3}/_{8}$ inch wide in long narrow clusters at leaf bases; and (3) fruit a showy drooping red fleshy capsule 2–3 inches long, 3–angled and 3–celled.

Small evergreen tree 30 feet high and 1 foot in trunk diameter. Bark gray, smoothish to slightly rough. Inner bark pinkish, gritty and slightly bitter. Twigs are light gray, stout, finely hairy, with many long narrow ridges ending in a bud of minute hairy leaves pressed together.

Leaves alternate, mostly 9–15 inches long, with stout light green hairy axis 5–8 inches long. Leaflets 3–5 pairs, with short stalks about $^{1}/_{4}$ inch long. The blades are elliptic, mostly 3–8 inches long and $1^{1}/_{2}$–4 inches wide, the lowest shorter, short–pointed at both ends, slightly stiff and thickened, slightly bent up at midrib and turned under at edges. The upper surface is shiny yellow green and hairless, with many sunken, parallel side veins, the lower surface dull light green with raised hairy side veins.

Flower clusters (racemes) are unbranched, 3–7 inches long, wit flowers on stalks $^{1}/_{4}$–$^{3}/_{4}$ inch long. Flowers of 2 kinds, partly bisexual and partly male, finely hairy, fragrant. The calyx consists of 5 pointed brownish–tinged sepals $^{1}/_{8}$ inch long; petals 5, white $^{3}/_{16}$ inch long, narrow with lobe at base; 8 stamens $^{1}/_{4}$ inch long on a disk; and pistil with 3–angled 3–celled 3–ovuled brownish ovary, slender style and dot stigma.

The fleshy heavy capsules have calyx at base and style at apex and are densely hairy on inner walls. There are 3 shiny dark brown round seeds $^{3}/_{4}$ inch in diameter, each with a large white fleshy cover (aril) around the base and attached to the fruit by a red or pink membrane, which is poisonous. Flowering and fruiting irregularly.

The "akee", from the native African name, is the edible part. This soft white or cream–colored cover around each seed must be used fresh, that is, when firm and pulled away from the capsule without bringing also the red, pink, or orange membrane which is bright in color when fresh. This membrane is poisonous and must be carefully removed before cooking. Preparation is by parboiling in water with salt, and afterwards stewing or frying with butter, or by boiling in soups. The appearance and the taste when fried suggest that of brains, hence the Spanish name seso vegetal (vegetable brains). Unfortunately, if not fresh or if picked from a broken branch, the "akee" is toxic and causes death. Damaged, unripe, or fallen fruits should never be eaten. Children have been killed by eating the raw fruits. Others have died from eating improperly picked or prepared fruits. Thus akee is one of the most toxic and most dangerous trees.

The hard wood has been used elsewhere for lumber.

Akee is planted in South, having been introduced about 1904, and also the Virgin Islands but found only in cultivation. The showy fruits and yellow–green foliage make it a handsome ornamental also. However, planting of akee is not recommended because of the risk of killing children and others unfamiliar with it. Trees around houses probably should be destroyed. In Jamaica akee is more widely grown and has become naturalized. Cultivated through the tropics north to southern states but not common.

**Range :** Native of western tropical Africa (Guinea).

**Other Common Names :** Akee, akee–tree, vegetable brains (English); seso vegetal (Spanish); akee de Africa (Cuba); huevo vegetal (Panama, Colombia); bien me sabe, pany quesito (Colombia); aki, arbre fricassé (French); akee (Dutch Antilles); castanheiro do Africa (Brazil).

The generic name honors Captain William Bligh (1754–1817), British mariner, who introduced this tree to Jamaica from Africa. He brought breadfruit plants from Tahiti of the South Sea Islands to Jamaica and St. Vincent in the British ship "Provindence" in 1793. The mutiny of the "Bounty" against him caused failure f an earlier attempt in 1789.

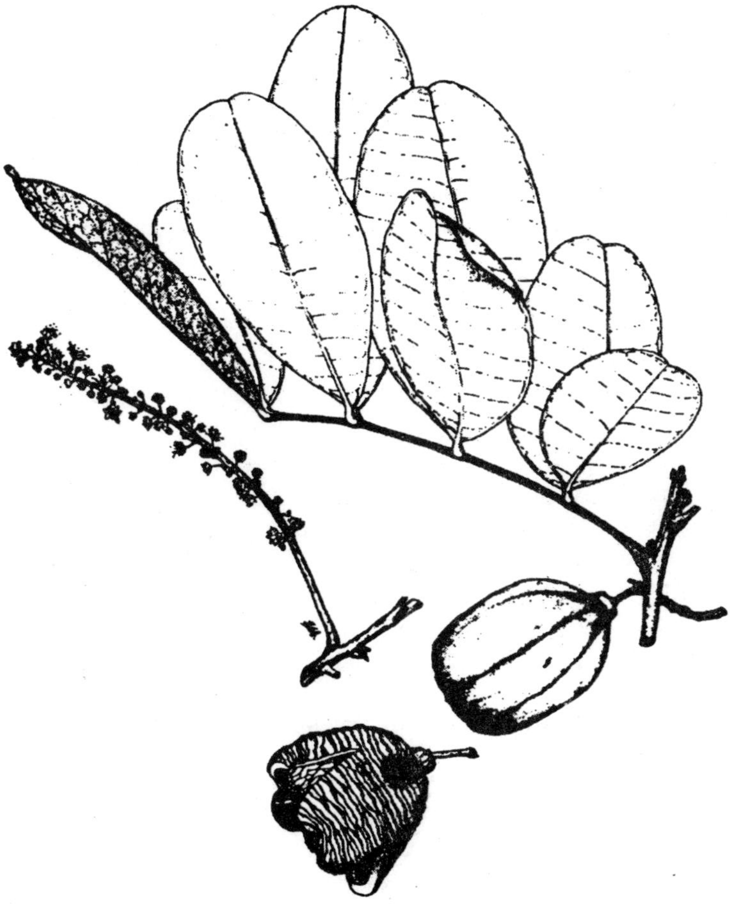

**Fig. 166 : Seso vegetal, akee (*Blighia sapida*)**
Leafy twig (above), flowers (left), fruits (below), two–thirds natural size.

## 167. Guara blanca (*Cupania triquetra* A. Rich.)

This species of guara is identified by : (1) minute yellow–brown hairs on twigs, petioles, lower leaf surfaces, branches of flower clusters, flowers and fruits; (2) pinnate leaves with 4–8 alternate elliptic or obovate leaflets, those toward apex largest, the edges with a few wavy teeth toward apex; (3) many small whitish 5–parted flowers $^1/_8$ inch across in terminal branched clusters; and (4) the many 3–angled seed capsules $^3/_4$ inch long and broad, splitting into 3 parts and exposing 3 rounded shiny blackish or brownish seeds $^5/_{16}$ inch long, each in an orange cup. From the more common related species No.138, guara, *Cupania Americana* L., this species differs in the shorter yellow–brown hairs, in the leaf edges with fewer teeth, and in the capsules sharply 3–angled and longer stalked.

A medium–sized evergreen tree to 65 feet high with buttressed trunk to 1 foot in diameter. The bark is gray and smooth. Twigs are stout and covered with minute yellow–brown hairs.

The alternate pinnate leaves are 5–12 inches long, with stout hairy axes. Leaflets have short hairy stalks $^1/_8$–$^1/_4$ inch long. The blades are 2–5 inches long and 1–3 inches wide, sometimes larger, slightly thickened, rounded or notched at apex, short–pointed at base, the upper surface becoming hairless, and the lower surface soft hairy with raised veins.

Flower clusters (panicles) are 4–8 inches long, with many yellow hairy branches. The flowers (not seen) in this genus are male, female, and bisexual (polygamous), composed of 5 hairy sepals, 5 hairy petals each with 2 scales on outer edges, 8 stamens on a disk, and pistil with hairy 3–celled ovary, short style, and 3 stigmas. The seed capsules develop a short stalk $^1/_{16}$ inch long above the persistent sepals. Recorded with flowers from October to March and with fruits from March to June.

Rare to scattered in moist limestone forest at 300–1,000 feet altitude.

**Fig. 167 : Guara blanca (*Cupania triquetra* A. Rich.)**
Flowers (left), leaf (above), fruits (lower right), two–thirds natural size.

## 168. Gitarán, hopbush (*Dodonaea viscosa*)

This shrub of coastal thickets becomes a small tree at Anegada. It is identified by : (1) the sticky (viscid) foliage and flowers; (2) yellow–green oblanceolate leaves 1–4 inches long and $^{3}/_{8}$–$1^{1}/_{4}$ inches wide; (3) small greenish yellow flowers $^{1}/_{4}$ inch across, in branched clusters; and (4) the many crowded distinctive light brown seed capsules $^{1}/_{2}$–1 inch in diameter, rounded, with 3 or 2 papery wings.

Evergreen shrub to 12 feet high, on Anegada Island becoming a small tree 15 feet high and 4 inches in trunk diameter. The bark is gray, fissured and fibrous. The twigs are slender, gray, and hairless.

The alternate sticky nearly hairless leaves are narrowed into a short petiole less than $^{1}/_{4}$ inch long. The blades are blunt, rounded, notched, or with minute point at apex, slightly thickened and turned under at edges.

Flower clusters (panicles) are about 1 inch long at ends and sides of twigs. The short–stalked sticky flowers are partly male, female, and bisexual on the same plant (polygamous) or different plants. There are usually 4 hairy sepals less than $^{1}/_{8}$ inch long, no corolla, 5–8 stamens, and pistil with 3–angled ovary, slender style, and 3 stigmas. The seed capsules are notched at both ends, with 3 or 2 cells and rounded papery wings. Seeds 3–1 round, black $^{1}/_{8}$ inch in diameter. Flowering irregularly through the year, fruits nearly always present.

The hard wood has light brown sapwood and dark brown heartwood.

Sometimes grown in fence rows and hedges. Elsewhere used in home remedies and the fruits as fish poison.

Locally common on coastal sandy areas, in strand vegetation of dry areas, forming thickets at sea level.

**Fig. 168 : Gitarán, hopbush (*Dodonaea viscosa*)**
Fruiting twig, natural size.

## 169. Gaita, butterbough (*Exothea paniculata*)

Gaita is recognized by : (1) even pinnate leaves with 2, sometimes 4 or 6 (rarely 1 or 3) paired oblong to elliptic yellow–green leaflets; (2) small 5–parted whitish flowers more than $^1/_4$ inch across, in branched clusters; and (3) dark red, purplish, or black round fleshy fruits $^3/_8$–$^1/_2$ inch in diameter, 1–seeded, several in grapelike clusters.

Evergreen small to medium–sized tree 50 feet high and 10 inches in trunk diameter, reported to reach 65 feet and 20 inches. The bark is brown or reddish brown, smoothish with large thin flakes or scales. Inner bark is reddish brown and bitter. The slender twigs are covered with minute yellowish hairs when young, becoming gray with raised dots (lenticels).

The alternate leaves have short slender petioles $^3/_4$–$^1/_2$ inch or more in length. Leaflet blades are 2–4 inches long and $^5/_8$–$1^1/_2$ inches wide, blunt, short–pointed, or sometimes notched at apex, short–pointed and nearly stalkless at base, the edges straight or slightly wavy, thin, hairless or nearly so. The upper surface is yellow green and slightly shiny, the lower surface paler.

Flower clusters (panicles) at ends and sides of twigs are 3–6 inches long, with slender yellowish hairy branches. The many fragrant flowers are partly male, female and bisexual on the same or different trees (polygamo–dioecious), composed of 5 hairy sepals $^1/_8$ inch long, persistent and becoming turned back; 5 whitish or cream–colored petals about as long as sepals; 8 stamens inserted on the disk; and pistil with 2–celled ovary, short curved style, and dot stigma.

The berrylike fruits (drupes) have calyx at base, pointed style at apex, thin orange juicy pulp, and 1 large round brown and $^5/_{16}$ inch in diameter. Collected with flowers from November to March and with fruits in March and June.

The sapwood is whitish or light brown and the heartwood reddish brown. The hard heavy fine–textured wood takes a fine polish. It has been used in cabinetwork, for tool handles, and elsewhere for posts, railroad crossties, marine pilling, and boatbuilding.

Rare in dry and moist forests at 100–1,500 feet altitude in foothills of western Ghat.

**Fig. 169 : Gaita, butterbough (*Exothea paniculata*)**
Flowering twig (left), fruiting twig (lower right), natural size.

## 170. Melocha, inkwood (*Hypelate trifoliata*)

A rare small tree of dry areas, recognized by : (1) the distinctive small leaves with 3 stalkless obovate to oblanceolate leaflets slightly shiny and stiff, with many fine parallel side veins; (2) small whitish 5–parted flowers $^1/_4$ inch across, in spreading branched clusters near ends of twigs; and (3) black egg–shaped fleshy fruits (drupes) $^1/_4$ inch long.

Evergreen small tree or shrub to 20 feet in height and 4 inches in trunk diameter. The bark is gray and smooth. Twigs are slender and gray.

The alternate hairless leaves $1^1/_2$–3 inches long have slender greenish petioles $^1/_2$–$1^1/_4$ inches long. Leaflet blades are $^1/_2$–2 inches long and $^3/_{16}$–$^1/_2$ inch wide, blunt or sometimes notched at apex, broadest beyond middle, narrowed toward the long-pointed base, stiff, turned under at edges, green above and yellow green beneath.

Flower clusters (panicles) at leaf bases are 1–3 inches long. The flowers are male and female on the same plant (monoecious), sometimes bisexual, whitish, turning pinkish. There are 5 spreading elliptic sepals $^1/_8$ inch long; 5 shorter rounded petals; 8 stamens $^1/_8$ inch long inserted on the disk, short and sterile in female flower; and pistil with 3–celled ovary with 2 ovules in each cell, short curved style, and dot stigma. The fruits have thin flesh reportedly sweetish and 1 large seed. Collected with flowers from winter and summer and with fruits in summer.

The wood is described as dark brown, heavy, hard and durable. It is used for fence posts and tool handles and elsewhere for railroad crossties and shipbuilding.

Rare in dry forest from sea level to 700 feet altitude in southwestern Goa.

**Fig. 170 : Melocha, inkwood (*Hypelate trifoliata*)**
Fruiting twig (left), flowering twig (upper right), natural size.

# SABIA FAMILY (SABIACEAE)

Trees, shrubs, and woody vines, known by : (1) leaves alternate, generally simple, sometimes odd pinnate, with enlargement at base of petiole, without stipules; (2) minute flowers, usually many in branched clusters (panicles), commonly bisexual, slightly irregular, calyx of 3–5 sepals or lobes and persistent, 4–5 unequal petals (2 very reduced), small disk, 3–5 opposite stamens inserted at base of petals or separate, and pistil with superior ovary of 2–3 cells with 2 ovules and 1 or 2 styles; and (3) fruit a rounded drupe or berry or dry, 1–seeded.

**Key to Species**

A. Leaves obovate, rounded at apex, mostly $2^1/_2$–5 inches long; fruits about $^3/_8$ inch in diameter—*Meliosma obtusifolia.*

AA. Leaves elliptic, short–pointed or blunt at apex, mostly 3–8 inches or more in length; fruits $^5/_8$–$^3/_4$ inch long—Aguacatillo, *Melisma herbertii* Rolfe.

## 171. Arroyo (*Meliosma obtusifolia*)

This rare species found only in mountain forests of Kerala is characterized by : (1) obovate thick leaves rounded at apex, mostly $2^1/_2$–5 inches long and $1^1/_4$–$2^1/_2$ inches wide, sometimes larger, with edges turned under, 1, 2 or sometimes 3 at a node; (2) many small whitish flowers about $^1/_8$ inch across, crowded and stalkless in branched clusters; and (3) round fruits (drupes) more than $^3/_8$ inch in diameter, whitish, turning dark brown or black, clustered like grapes.

Medium–sized evergreen tree to 65 feet high and more than 1 foot in trunk diameter. The bark is light brown, becoming fissured, the inner bark pink, turning red brown on exposure, slightly bitter. Twigs are slender, brown, with fine rusty–brown hairs when young.

The leaves are crowded near ends of twigs, alternate, opposite, and whorled, hairless, with petioles $^3/_8$–$^3/_4$ inch long, recorded to 8 inches long and 3 inches wide with petioles to $1^1/_4$ inches. Blades are thick and leathery, gradually narrowed to base, with midvein slightly sunken, the upper surface shiny green with raised veins, and to lower surface dull light green.

Flower clusters (panicles) are large and branched, $2^1/_2$–5 inches or more in length, at ends of twigs and leaf bases, often longer than leaves. The many stalkless flowers from greenish buds are composed of 5 rounded overlapping sepals less than $^1/_{16}$ inch long with hairy border, remaining at base of fruit; 5 whitish petals, the outer 3 nearly $^1/_8$ inch long and 2 small narrow scales; 2 stamens opposite the small petals and 3 sterile (staminodes); and pistil with ovary, style, and 2 minute stigmas. The many fruits pale green when immature are hard and dry, containing 1 seed within the large stone. With flowers and fruits through the year.

The wood is whitish and hard.

Rare in upper forest at 1,800–3,000 feet altitude in mountains of western Ghat.

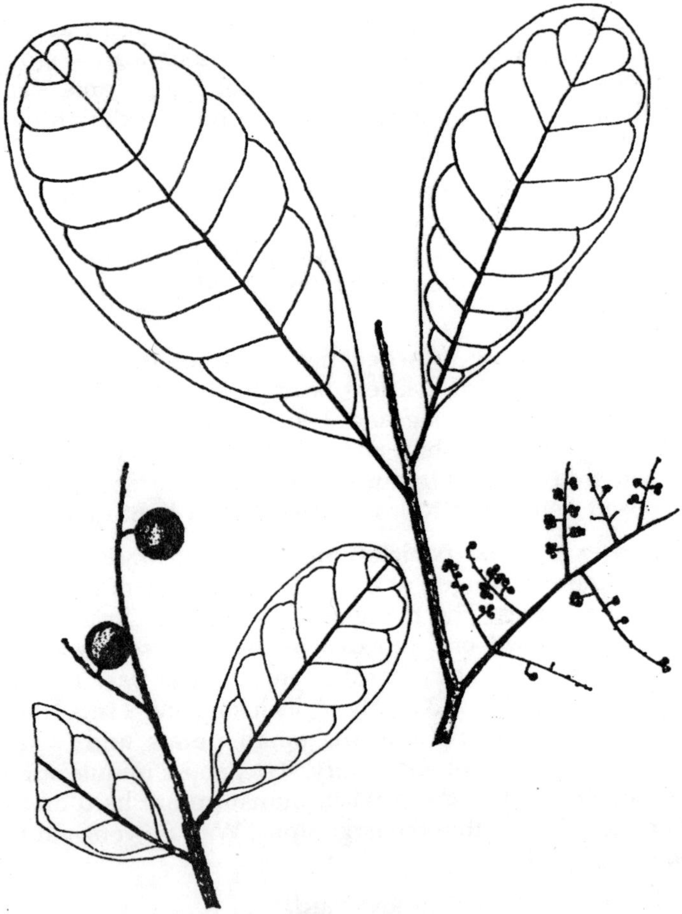

**Fig. 171 : Arroyo (*Meliosma obtusifolia*)**
Flowering twig (above), fruiting twig (lower left), natural size.

# BUCKTHORN FAMILY (RHAMNACEAE)

Shrubs, woody vines, and small to large trees, rarely herbs, often spiny, known by : (1) leaves mostly alternate, also opposite, simple, often with 3 or more veins from base (palmate–veined), usually with minute stipules; (2) small greenish or yellowish flowers, mostly in lateral clusters (cymes), usually bisexual, regular, with concave cuplike base (hypanthium), with 5–4 sepals touching by edges in bud, 5–4 small petals (or none) concave and very narrow with narrow base, 5–4 opposite stamens enclosed by petals, and pistil with superior ovary within the disk (sometimes surrounded by disk and appearing inferior) with 2–4 cells of 1 ovule, style and 1–5 stigmas; and (3) fruit a berry, drupe, or capsule, often opening in 3 parts.

**Key to Species**

A. Leaves with 3 main veins from base; twigs with paired or single spines—*Ziziphus.*

B. Leaves hairless, often rolled under at edges; twigs hairless—*Ziziphus rignonii.*

BB. Leaves soft hairy on lower leaf surfaces, finely toothed on edges; twigs hairy.

C. Leaves with prominent network of raised veins on lower surface; twigs rusty hairy; fruits $^3/_8$–$^5/_8$ inch long—Cascarroya, *Ziziphus reticulata* (Vahl) DC. (*Sarcomphalus reticulatus*).

CC. Leaves densely whitish hairy beneath; twigs whitish or rusty hairy, fruits $^3/_4$–1 inch long—*Ziziphus mauritiana.**

AA. Leaves with 1 main vein, edges not toothed; twigs not spiny.

D. Leaves opposite.

E. Leaves only slightly thickened, elliptic, notched at rounded apex—Bariaco, "ironwood," leadwood, *Krugiodendron ferreum* (Vahi) Urban.

EE. Leaves thick and leathery—*Reynosia.*

F. Leaves less than 1 inch long, the broad apex ending in minute sharp hooked spine beneath apex—*Reynosiancinta.*

FF. Leaves more than $1^1/_2$ inch long, the broad apex ending in minute point.

G. Leaves broadly elliptic, slightly notched at apex—Reynosia guama.

GG. Leaves ovate, blunt at apex—*Reynosia krugii.*

DD. Leaves alternate.

H. Leaves $^1/_2$–1 inch long, obovate to rounded, notched at apex; twigs sometimes with small spines—*Ziziphus taylorii.*

HH. Leaves more than 1 inch long, elliptic, pointed at apex.

I. Leaves long-pointed at apex, with short straight side veins and wavy teeth at edges—*Rhamnus sphaerosperma*.

II. Leaves mostly short-pointed at apex, the side veins curved and prolonged near edges—*Colubrina*.

J. Leaves slightly thickened, the lower surface, petioles, and twigs with rusty brown hairs, especially when young—Abeyuelo, coffee colubrina, *Colubrina arborescens* (Mill.) Sarg.

JJ. Leaves this, the lower surface pale green and minutely hairy—Mabi, soldierwood, *Colubrina elliptica* (Sw.) Briz. & Stern (C. *reclinata*).

## 172. Guama *(Reynosia guama)*

This small tree is recognized by : (1) the paired broadly elliptic leaves $1^1/_2$–3 inches long and $1\,^1/_4$–2 inches wide, slightly notched with a minute point at apex, rounded or slightly notched at base, thick and leathery with midrib much sunken; (2) several small yellow–green flowers $^1/_8$ inch long, clustered on short slender stalks at bases of leaves; and (3) elliptic fruits $^3/_8$ inch long with cup of calyx at base and 1 large seed.

Shrub or small tree often wandlike to 20 feet high and 3 inches in trunk diameter, hairless throughout. The bark is light gray and smooth. The twigs are light brown, slender, and hairless or nearly so, becoming scaly, and end in minute brown blud formed by pointed stipules.

The opposite leaves have 2 minute pointed stipules $^1/_{16}$ inch long and short petioles less than $^1/_4$ inch long. The blades are shiny dark green above and dull light green beneath, with sunken midvein, few curved lateral veins, and border slightly curved under.

Flowers several clustered (sometimes only 1) at leaf bases on stalks of $^1/_8$ – $^1/_4$ inch, consisting of cuplike base (hypanthium), 5 pointed sepals $^1/_{16}$ inch long, 5 narrow petals $^1/_{16}$ inch long, 5 opposite stamens, and pistil. The fruits (drupes) are borne on short stalks at leaf bases, greenish when immature, with greenish cup of calyx at base and pointed style at apex. Inside the thin pulp is the large elliptic seed. Flowering from spring to fall.

In Goa an alcoholic drink called warmer–berry wine or guamaberry wine is prepared from the mature fruits and rum, agreed and seasoned.

Rare in thickets on dry coastal hills and dry limestone forest at 100–300 feet altitude in south-western Goa.

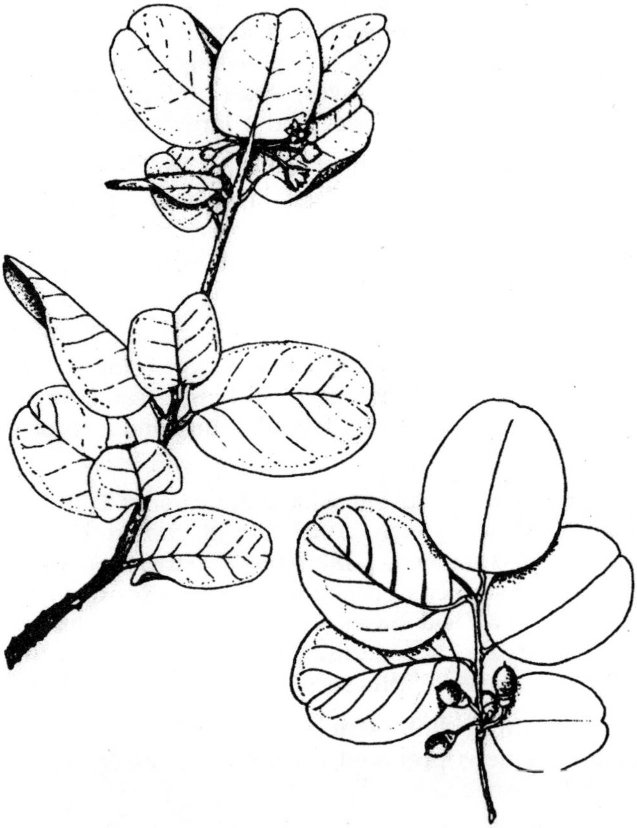

**Fig. 172 : Guama *(Reynosia guama)***
Flowering twig (above) fruiting twig (lower right), natural size.

## 173. Krugii (*Reynosia krugii*)

Characteristics of this small tree are : (1) opposite ovate leaves 2–3$^1/_2$ inches long and 1–2 inches wide, slightly thick and leathery, the apex blunt with a minute point; (2) minute whitish flowers less than $^1/_8$ inch broad, few leaf bases, and (3) blackish elliptic fruit $^5/_8$–$^3/_4$ inch long, with large stone.

Evergreen small tree 20 feet high and 3 inches in trunk diameter, recorded to 30 feet. Bary gray, slightly furrowed, the inner bark reddish brown, bitter. Twigs are brown, finely fissured, minutely hairy when young.

The opposite hairless leaves have slender curved leafstalks $^1/_4$ inch long. The paired stipules 4 to a node are narrow pointed scales $^1/_{16}$ inch long, forming the bud. Blades are rounded at base, slightly turned under at edges, the upper surface dull or slightly shiny green with sunken midvein and side veins inconspicuous, the lower surface paler.

Flowers are short–stalked, minute, with cup–like base, 4 pointed sepals, 4 small petals, 4 opposite stamens, and within disk the pistil with rounded ovary and short style. The fruit (drupe) has thin dark red juicy flesh and large stone. Flowering and fruiting irregularly.

Wood light brown, hard.

Rare in moist limestone, lower and upper forests and coastal hills at 2,3000 feet altitude in Goa.

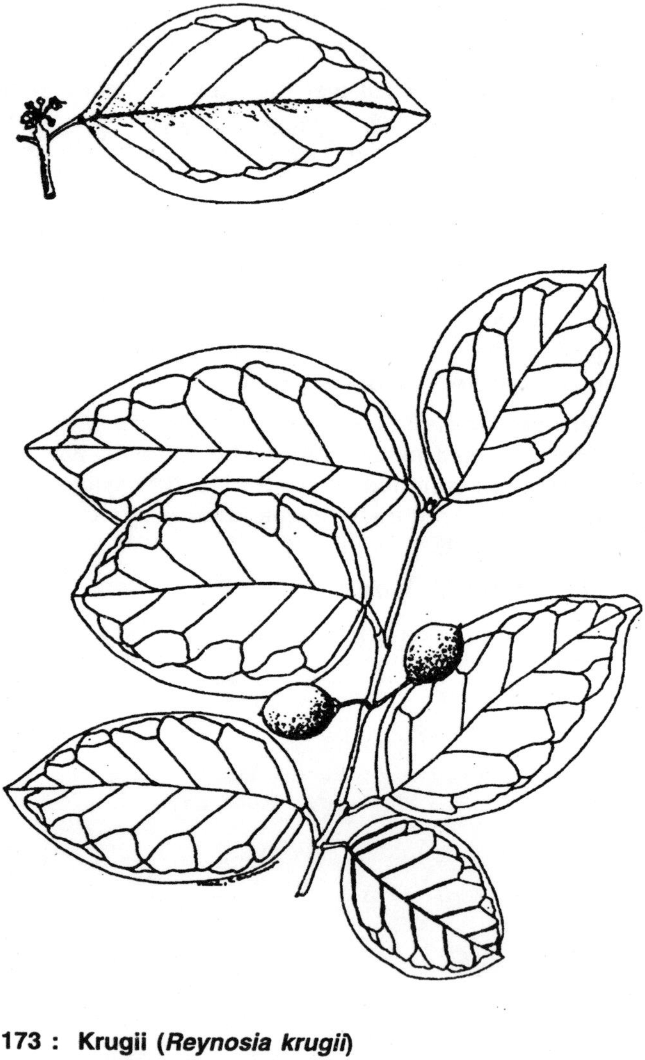

**Fig. 173 : Krugii (*Reynosia krugii*)**
Flowering twig (above), fruiting twig (below), natural size.

## 174. Cascarrolla (*Reynosia uncinata*)

Cascarrolla is easily recognized by the odd small stiff elliptic to oblong paired leaves that end in a minute sharp hooked spine hidden beneath the apex. This shrub, sometimes a small tree, of dry areas is further identified by : (1) minute yellow–green flowers $^{1}/_{8}$ inch long and broad, few clustered at leaf bases; and (2) round blackish fleshy fruits $^{1}/_{2}$–$^{5}/_{8}$ inch long.

Evergreen shrub or small tree to 30 feet high and 5 inches in trunk diameter. Park dark gray, smoothish, with shallow fissures or large scales. The inner bark is reddish brown with narrow lighter streaks, slightly bitter. The twigs, mostly paired, short and branched, are gray, minutely hairy when young, becoming fissured.

The leaves are opposite and mostly crowded on short twigs. Paired stipules 4 to a node are narrow pointed scales $^{1}/_{16}$ inch long which form the bud. The slender light green petioles are $^{1}/_{16}$ inch long. The blades are $^{3}/_{8}$–1 inch long and $^{1}/_{4}$–$^{5}/_{8}$ inch wide, stiff and slightly thick, hairless, rounded at base, slightly curved down at edges, the upper surface shiny green, the lower surface dull light green, the side veins inconspicuous. The sunken midvein continues beyond the notched apex into a minute sharp hooked spine less than $^{1}/_{16}$ inch long and hidden beneath.

The flowers on a short stalk of $^{1}/_{8}$ – $^{1}/_{4}$ inch are composed of a short cuplike base (hypanthium), 5 pointed sepals, 5 smaller narrow petals, 5 opposite stamens, and within disk the pistil with rounded ovary and short style. The fruit (drupe) has think dark red juicy flesh, reported to be edible, and large stone. Flowering irregularly through the year.

The wood is light brown and hard.

Locally common in dry coastal forests from sea level to 400 feet in south–western Kerala.

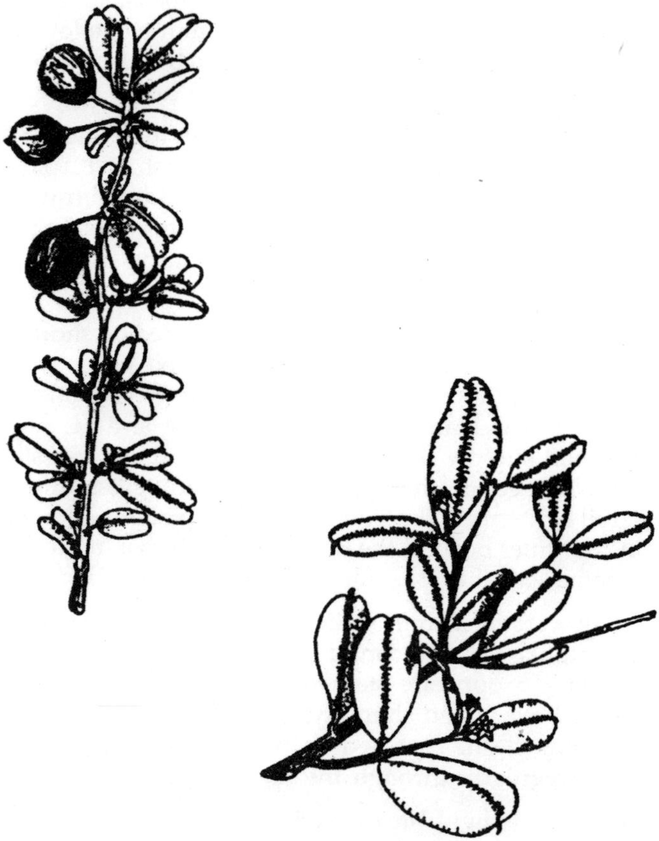

**Fig. 174 : Cascarrolla (*Reynosia uncinata*)**
Fruiting twig (upper left), flowering twig (right), natural size.

## 175. West Indian buckthorn (*Rhamnus sphaerosperma*)

This shrub of mountain forests, sometimes becoming a small tree, is distinguished by : (1) elliptic leaves with wavy teeth at edges and ending in a long point; (2) small yellow–green cup–shaped flowers $^1/_8$ inch long and broad, the calyx with 5 long–pointed sepals; and (3) fruit rounded, about $^1/_4$ inch in diameter, with cuplike base, 3–seeded.

Evergreen shrub, sometimes a small tree 20 feet high and 3 inches in trunk diameter. The slender twigs are green and with scattered reddish–brown hairs when young, becoming gray, with raised dot lenticels.

The alternate leaves have slender hairy leaf–stalks $^1/_4$–$^3/_4$ inch long. Blades are mostly $1^1/_2$–3 inches long and $^3/_4$–$1^1/_2$ inches wide (recorded to $6^1/_2$ inches by $2^7/_8$ inches), short–pointed at base, thin, upper surface shiny green to dark green and hairless, the lower surface slightly shiny light green and hairy on veins.

Flowers few on slender stalks about $^1/_4$ inch long at leaf bases. The short cuplike yellow–green hairy base (hypanthium) bears 5 spreading pointed yellow–green sepals $^1/_{16}$ inch long. 5 smaller narrow notched whitish petals folded around the 5 tiny opposite stamens. The pistil has a 3–celled ovary, short style, and 3–lobed stigma. Fruits (drupes), with cuplike base and dotlike style at apex are green when immature, slightly fleshy. Flowering intermittently, noted with flowers from January to July and with fruits from May to July.

Rare in forests and dwarf forest at 2,500–4,000 feet altitude in mountains of Western Ghat.

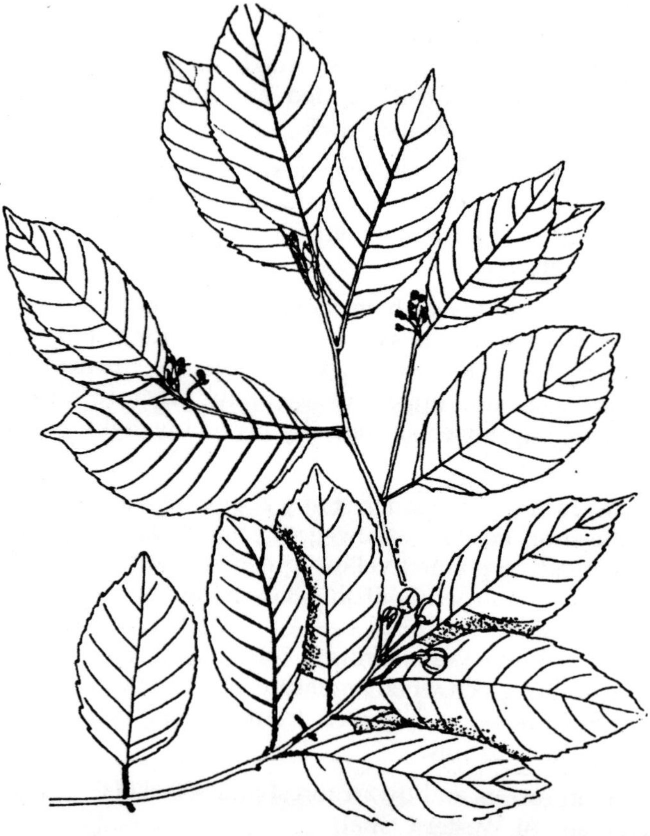

**Fig. 175 : West Indian buckthorn (*Rhamnus sphaerosperma*)**
Flowering twig (above), fruiting twig (below), natural size.

## 176. India jujube (*Ziziphus mauritiana Lam.**)

This tree introduced for its edible fruits is recognized by : (1) the dense whitish or rusty–brown hairs on twigs, petioles, lower leaf surfaces, and flower clusters; (2) broadly elliptic leaves rounded at both ends, finely–toothed on edges, densely whitish hairy beneath, with 3 main veins from base, and with paired spines (stipules) often present at base; (3) many small yellow 5–parted hairy flowers $^3/_{16}$ inch across, clustered at leaf bases; and (4) the edible, fleshy and nearly elliptic or rounded orange–red fruits $^3/_4$–1 inch long.

A small evergreen fruit tree to 40 feet high and 1 foot in trunk diameter. The slender twigs are finely hairy, often with paired straight or curved sharp brown spines $^1/_8$–$^1/_4$ inch long at nodes.

The leaves alternate in 2 rows have hairy petioles $^1/_4$–$^3/_8$ inch long. Blades are $1^1/_2$–$2^1/_2$ inches long and 1–$1^3/_4$ inches wide, shiny green and hairless above and densely whitish soft hairy beneath.

Flower clusters (crymes) at leaf bases are less than $^3/_4$ inch across. Flowers many, short–stalked, composed of bell–shaped base (hypanthium), 5 spreading pointed hairy sepals $^1/_{16}$ inch long, 5 narrow stalked whitish petals, 5 stamens opposite the petals, and within the disk the pistil with 2–celled ovary and 2–forked style. The edible fruits (drupes) have a large 2–celled stone and 2 elliptic flattened brown seeds $^1/_4$ inch long. Flowering and fruiting from spring to fall and with mature fruits in winter.

Planted sparingly as a fruit tree, mostly in dry areas, in a few residential areas of Nagpur and escaping in coastal thickets and clearing, recorded. Becoming common and naturalized on.

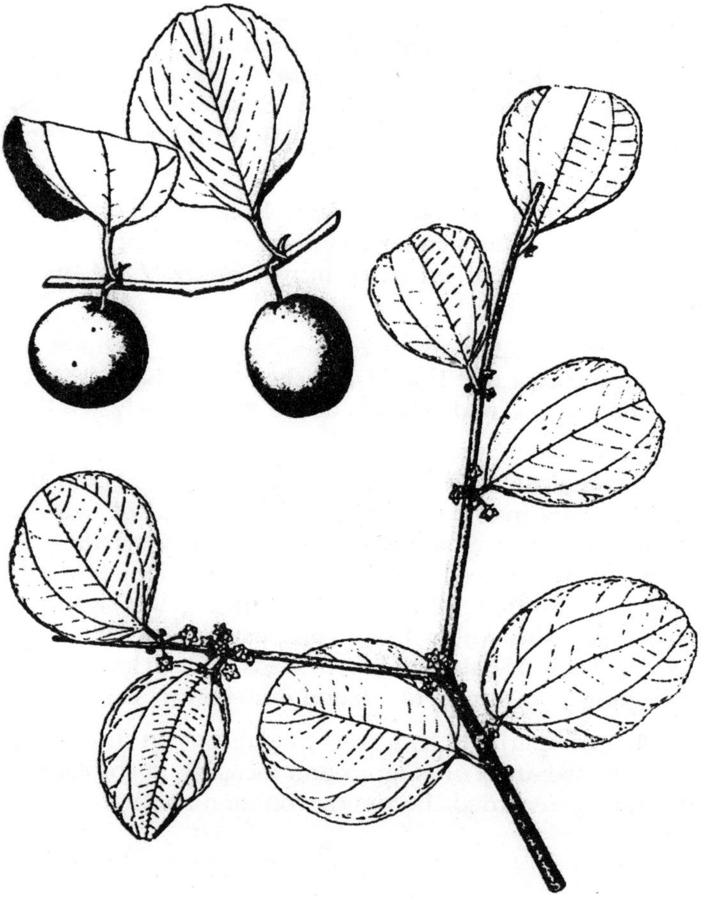

**Fig. 176 : India jujube (*Ziziphus mauritiana Lam.*)**
Fruiting twig (upper left), flowering twig (below), natural size.

## 177. Thorn (*Ziziphus rignonii*)

This small tree often with spines, rare in dry areas, is identified by : (1) elliptic to rounded thick and leathery leaves $1^1/_4$–4 inches long and 1–$3^1/_2$ inches wide hairless, with 3 main veins from base and fine network of veins on both surfaces; and (2) edible rounded or elliptic dark brown fruit $^3/_8$–$^1/_2$ inch in diameter.

Small evergreen tree 15–30 feet high, with rounded crown. Bark dark brown or gray, smooth. The gray angled twigs often have paired slender spines $^3/_{16}$–$^3/_8$ inch long.

The leaves are alternate in 2 rows, with petioles about $^1/_4$ inch long. Blades are rounded at apex, rounded and often heart–shaped at base, often rolled under at edges, dark shiny green on upper surface, paler, with raised veins beneath.

The small yellowish green flowers about $^1/_8$ inch across are borne in branched clusters (panicles) at leaf bases. Flowering and fruiting from spring to fall.

Wood light brown, hard.

Common in dry forest from sea level to 300 feet altitude in northeastern states.

**Fig. 177 : Thorn (*Ziziphus rignonii*)**

Flowering twig (upper right) fruiting twig (lower left), natural size.

## 178. Taylor (*Ziziphus taylorii*)

This much branched shrub of Moore Island sometimes becomes a small tree. It is identified by : (1) many short crowded slender branches; (2) many small obovate to rounded leaves $^1/_2$–1 inch long and $^3/_8$–$^3/_4$ inch wide, notched at apex, slightly thickened and stiff; (3) few small green 5-parted flowers less than $^1/_8$ inch across, at end of twig; and (4) elliptic black stone fruits $^5/_{16}$ inch long.

Deciduous shrub or small tree to 15 feet high and 4 inches in trunk diameter, densely branched, hairless throughout. Bark gray, smooth, becoming rough with small–plates. The twigs are short, slender, and angled, mostly spineless but sometimes with a few small spines to $^1/_4$ inch long.

The alternate leaves have short petioles about $^1/_{16}$ inch long. Blades are short–pointed at base, not toothed on edges, green on upper surface and paler beneath, with 3 main veins from near base and with inconspicuous side veins.

Flowers about $^1/_8$ inch across are on short stalks in small clusters (panicles) at ends of twigs. The bell–shaped base (hypanthium) bears 5 sepals less than $^1/_{16}$ inch long, 5 narrow stalked petals, and 5 stamens opposite the petals, and the pistil has 2–celled ovary and 2–lobed style. The fruit (drupes) have shallow cup at base. The large stone is 2–celled and 2–seeded. With flowers in spring and fruits in spring and summer.

Rare and local on coastal plain and in dry forest on plateau from sea level to 400 feet on Moore Island, native.

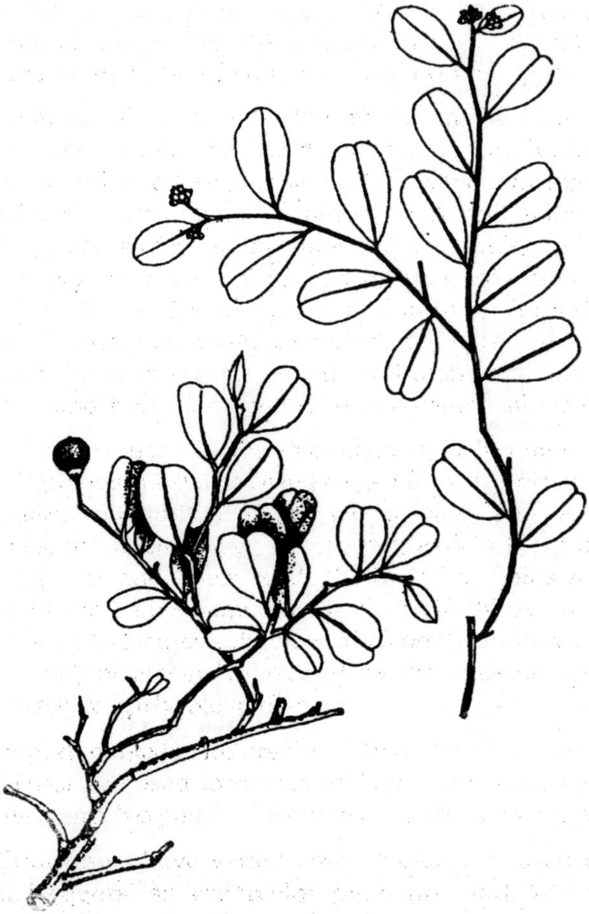

**Fig. 178 : Taylor (*Ziziphus taylorii*)**
Flowering twig (above), fruiting twig (below), natural size.

# ELAEOCARPUS FAMILY (ELAEOCARPACEAE)

Trees and shrubs, known by : (1) alternate or opposite simple leaves with stipules, the petioles often with enlargements at both ends (*Sloaneai*); (2) flowers often large in clusters (racemes or panicles) or solitary, bisexual, regular, with 4–5 sepals, 4–5 separate petals (none in *Sloanea*), many stamens inserted in disk, and pistil with superior ovary of 2 to many cells with axile placentation and 2 to many ovules and style; and (3) fruit a capsule or berry.

**Key to Species**

A. Leaves lance–shaped or oblong, 2–4 inches long, irregularly toothed, with 3 main veins from very oblique base, hairy beneath—*Muntingia calabura*.*

AA. Leaves elliptic, not toothed on edges, with 1 main vein, hairless or nearly so—*Sloanea.*

B. Leaves $2^1/_2$–6 inches long, blunt or notched at both ends, petioles $^1/_2$–1 inch long—*Sloanea amygdalina.*

BB. Leaves 6–18 inches long, short–pointed at both ends, petioles $^1/_2$–$3^1/_2$ inches long—Motillo, *Sloanea berteriana* Choisy.

## 179. Capulin (*Muntingia calabura L.**)

This small fast–growing tree has been introduced in Eastern States in recent years for ornament and shade. It is easily distinguished by : (1) the lance–shaped or oblong, irregularly toothed leaves with 3 main veins from the very oblique base, hairy beneath; (2) the lateral flowers $^{3}/_{4}$–1 inch across the 5 spreading white petals; and (3) the edible round berry, reddish or yellowish, $^{3}/_{8}$–$^{5}/_{8}$ inch in diameter.

An evergreen small tree to 25 feet high, with trunk to 6 inches in diameter and short broad crown. Outer bark is smooth and brown, the inner bark whitish, fibrous, and tasteless. Twigs are gray with minute star–shaped hairs when young, afterwards turning brown.

The leaves are alternate in 2 rows on spreading nearly horizontal twigs. The short hairy petioles are $^{1}/_{8}$–$^{1}/_{4}$ inch long. Leaf blades are 2–4 inches long and $^{3}/_{4}$–$1^{3}/_{8}$ inch wide, with long point at apex, thin, the upper surface green with soft hairs, and the gray–green lower surface covered with star–shaped hairs.

Flowers are 1–3 on stalks $^{3}/_{4}$–$1^{1}/_{4}$ inches long at leaf bases. There are 5 narrow green sepals about $^{1}/_{4}$ inch long, turned back; 5 rounded white petals $^{1}/_{4}$–$^{3}/_{8}$ inch long; many yellow stamens $^{3}/_{16}$ inch long with 5–celled ovary and 5 stalkless stigmas. The fruit (berry) is reddish or yellowish, with stigmas at apex, juicy and slightly sweet, containing many minute light brown seeds. Flowering and fruiting continuously through the year.

The sapwood is lighter colored than the pale brown heartwood. The wood is described as lightweight, medium–textured, of irregular grain, very easily worked, but poorly resistant to decay.

Touch silklike fibers of the bark have been used elsewhere for making rope and baskets. The flowers have served in home remedies. The red berries are juicy and slightly sweet, though without distinctive flavor.

Commonly planted in recent years as an ornamental and small shade tree around houses in new suburbs in moist parts of Assam. Its main advantage is rapid growth in full sunlight. Before many ears the trees may spread from cultivation into open areas like weeds, as in the native home.

**Fig. 179 : Capulin (*Muntingia calabura L.**)**
Natural size.

## 180. Chicharoon (*Sloanea amygdalina*)

This rare tree in mountain forests of western Sikkim is characterized by : (1) elliptic leaves blunt or notched at both ends, with sunken veins, and with a swelling and bend at upper end of petiole; (2) flowers 1–3 long–stalked at leaf base, about 1 inch across the many stigmas; (3) elliptic seed capsules $^{5}/_{8}$–1 inch long, hanging down on stalks 2 inches or more in length, covered with short prickly bristles $^{1}/_{16}$ inch long, splitting open in 4 parts.

Evergreen large tree 60–90 feet high with trunk 1 $^{1}/_{2}$–6 feet in diameter, with tall narrow buttresses to 12 feet high and 3 feet wide, straight axis, and narrow crown. The bark is brown, smooth, becoming slightly fissured. Inner bark is light brown, gritty and tasteless. The brown twigs, finely hairy when young, have raised rounded leaf scars longer than broad.

The alternate leaves have finely hairy petioles $^{1}/_{2}$–1 inch long. Leaf blades are $2^{1}/_{2}$–6 inches long and $1^{3}/_{4}$–$3^{1}/_{2}$ inches wide, slightly wavy on edges, slightly thickened and leathery, hairless. The upper surface is green and slightly shiny with sunken veins, the lower surface shiny yellow green with raised veins.

Flowers hanging on stalks $^{5}/_{8}$–$1^{1}/_{4}$ inches long. There are 4 pointed sepals about $^{3}/_{8}$ inch long velvety within, no corolla, many hairy spreading stamens, and pistil with bristly hairy ovary and style. The seed capsules, yellow green and finely bristly hairy when immature, have calyx bent back at base and pointed style at apex. The bristles covering the fruits are prickly when touched. Flowering in spring and fruiting in summer.

The sapwood is whitish and hard.

Rare in upper Cordillera and moist limestone forests at 500–2,000 feet altitude in western Sikkim.

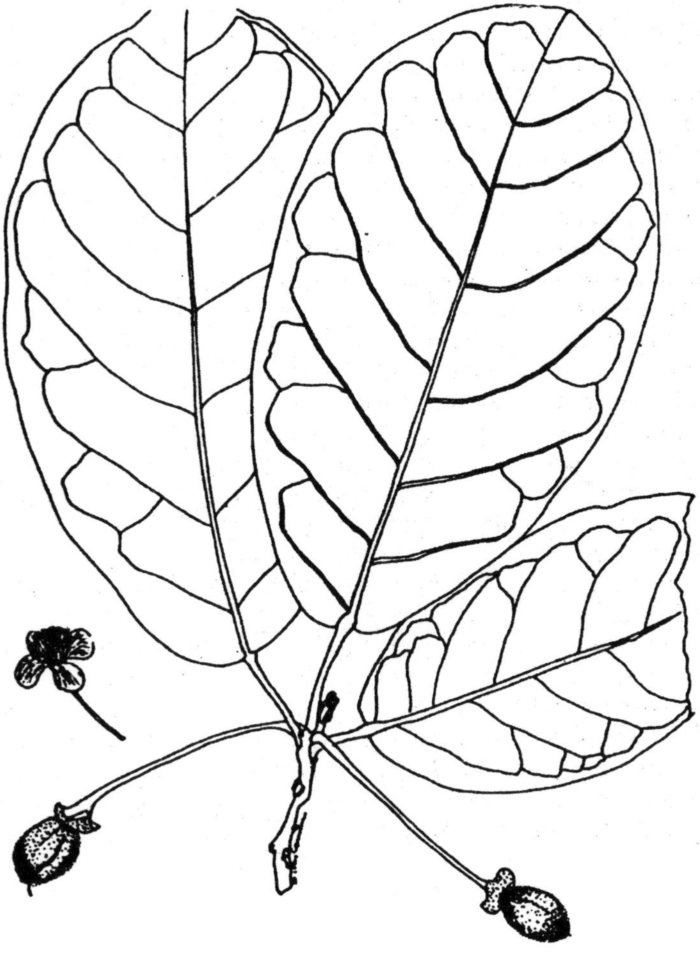

**Fig. 180 : Chicharoon (*Sloanea amygdalina*)**
Natural size.

# MALLOW FAMILY (MALVACEAE)

Herbs, shrubs, and few trees, known by : (1) fibrous bark with sap often mucilaginous; (2) alternate simple leaves, toothed or often lobed, palmate–veined, with star–shaped hairs and large paired stipules; (3) flowers commonly large and showy, solitary or in clusters (cymes), bisexual, regular, often with greenish scales (bracts) below the calyx of 5 sepals or lobes, with 5 colored petals, separate but united at base into column, stamens very numerous in a column around style with separate filaments and 1–celled anthers, and pistil with superior ovary of many cells in a ring, each with 1 to many ovules, and long style with many divisions and stigmas; and (4) fruit usually a capsule or separating into 1–seeded parts.

**Key to Species**

A. Leaves lobed or toothed on edges.

B. Leaves mostly with 3 (sometimes 5) long–pointed lobes, not toothed—*Gossypium barbadense*.*

BB. Leaves toothed on edges.

C. Leaves coarsely toothed, with 5 main veins from rounded or short–pointed base—*Hibiscus rosasinensis*.*

CC. Leaves finely wavy–toothed, with 7–9 main veins from slightly notched or heart–shaped base—*Sida eggersii*.

AA. Leaves not lobed or toothed on edges.

D. Leaves with mostly 9 or 11 main veins from base; lower leaf surfaces densely covered with gray star–shaped hairs.

E. Leaves with lower leaf surfaces whitish gray; flowers 3–$3^1/_2$ inches long; seed capsules with calyx remaining at base; small spreading tree with short crooked trunk.—Emajagua, sea hibiscus, *Hibiscus tiliaceus* L.*

EE. Leaves with lower surfaces gray green; flowers 3–5 inches long; seed capsules without calyx at base; large tree with tall straight trunk—*Hibiscus elatus*.*

DD. Leaves with mostly 7 main veins from base, blades with scattered minute scales.

F. Leaves abruptly long–or short–pointed at apex; flowers with red petals–Maga, *Montezuma speciosissima* Sessé & Mos.

FF. Leaves long–pointed at apex; flowers with pale yellow petals, turning to purple—Emajagüilla, otaheita, portiatree, *Thespesia populnea* (L.) Soland.*

## 181. Wild cotton (*Gossypium barbadense*)

Wild cotton, a shrub or small tree of waste places, is distinguished by : (1) the leaves with mostly 3–lobed blades about 2 inches long, the middle lobe longest, slightly heart–shaped at base, becoming nearly hairless; (2) large bell–shaped flowers about 2 inches long and broad, the 5 light yellow petals with brown dot at base inside, turning to pink, enclosed at base by 3 large light green bracts with long narrow teeth; and (3) elliptic seed capsule about 1 inch long, containing many seeds covered by masses of white threadlike hairs, cotton.

This species of cotton is a spreading shrub, climber, or small tree 12–15 feet in height and 3 inches in trunk diameter. The bark is gray and smooth, the inner bark thick, pink and white streaked, almost tasteless. The twigs are stout, much forked, slightly enlarged at nodes, light green tinged with purple and with many black dots, nearly hairless. The buds are composed of long narrow stipules covering the minute young leaves.

The alternate leaves have paired narrow stipules about $^3/_8$ inch long, which shed early, and slender round, light green petioles 1–$1^1/_2$ inch long. Blades are mostly ovate, about 2 inches long and $1^3/_4$ inches wide, mostly 3–lobed with 3 main veins from base sometimes 5–lobed or not lobed, with middle lobe longest, thin, the upper surface dull green and hairless, the lower surface dull light green and becoming nearly hairless.

The flowers are borne singly at leaf bases on stout stalks about $^1/_2$ inch long. At the base of each flower are 3 ovate light green bracts about $1^1/_4$ inches long, each ending in a few long–pointed teeth. The flower is composed of a light green tubular calyx $^3/_8$ inch long and broad, 5–toothed; the bell–shaped corolla about 2 inches long, of 5 light yellow petals with brown dot at base inside, broadest at the rounded apex and united at base, turning to pink upon withering; stamens many in a whitish column 1 inch long, united to corolla at base, the corolla and stamens falling together; and pistil more than $1^1/_4$ inches long inside stamen tube, consisting of conic 4–celled ovary with many ovules, long style, and stigma.

The seed capsule with calyx at base becomes brown and splits into 4 parts, releasing the seeds covered with masses of threadlike white fibers of cotton which remains attached. Flowering and fruiting from spring to fall.

The long whitish hairs covering the seeds of this and related species are the valuable fiber cotton. A honey plant.

Purified cotton or absorbent cotton is widely used in medicines as a dressing for wounds. Cotton root bark has served as a drug.

Uncommon in waste places and appearing after cultivation in dry limestone and coastal forest zones from sea level to 200 feet altitude in southwestern Gujrat.

**Fig. 181 : Wild cotton (*Gossypium barbadense*)**
Two–third natural size.

## 182. Mahoe (*Hibiscus elatus*)

Mahoe, an introduced forest tree is distinguished by : (1) long–stalked heart–shaped and nearly round abruptly pointed leaves with mostly 9 main veins from base, the lower surface gray green with minute star–shaped hairs and with 1–3 narrow glands near base of main veins; (2) large funnel–shaped flowers 3–5 inches long and broad, the 5 petals changing color during the day from yellow with dark red base to orange and red; and (3) elliptic seed capsules 1–$1^1/_2$ inches long, densely yellow brown hairy, which split into 5 parts.

Evergreen medium–sized planted tree becoming 80 feet high with tall straight trunk 15 inches in diameter, or larger. The bark is gray, smooth to finely fissured. Inner bark is fibrous, whitish to whitish green, and slightly bitter. Twigs are green with star–shaped hairs when young and have rings at nodes.

The alternate leaves have slender round leaf–stalks $2^1/_2$–4 inches long and 2 oblong short–pointed slightly hairy light green stipules $1^1/_4$ inches long, which shed, leaving a ring scar around twig. Blades are about 5–7 inches long and broad, abruptly short– or long–pointed at apex and heart–shaped or notched at base, very slightly thickened, with straight or finely wavy edges, the upper surface is green and hairless.

One to 3 flowers each lasting a day are borne at leaf bases at the ends of twigs on stout green stalks $^1/_2$ inch long. Around the flower is a light green tube (involucre) $^1/_2$ inch long, ending in 9 pointed lobes. The calyx is $1^1/_2$–2 inches long, light green and finely hairy, composed of tube and 5 long spreading lobes. The corolla consists of 5 spreading narrowly elliptic petals $3^1/_2$–5 inches long, united at base, when opening in the morning yellow with large dark red spot inside at base, in late afternoon turning to range and red or red bronze. Many stamens are on a whitish or pinkish column or tube with corolla at base. The pistil inside the stamen tube has a hairy 5–celled ovary, long style, and 5 rounded stigmas. The seed capsules contain many hairy seeds. Flowering irregularly through the year.

The sapwood is light brown, the heartwood bluish, greenish, or streaked when freshly cut. The wood varies from soft, fine–textured, with silky luster to hard, of medium to coarse, texture, and rather dull. Where this species is native, wood with

**Fig. 182 : Mahoe (*Hibiscus elatus*)**
Flowering twig and fruits (right), two–thirds natural size.

attractive colors has been prized for cabinetwork, furniture, interior trim, and gunstocks. Other uses include construction, railway ties, and shingles.

Bark of young trees makes good ropes, but this use destroys the growing tree and is to be discouraged. Formerly the lacelike inner bark was used for tying bundles of Havana cigars and was called Cuba bark. An infusion of the mucilaginous leaves and young twigs has been used in home remedies.

This species is being tested locally in Assam in experimental forest plantations and is sparingly used for ornament. It grows rapidly, reaching a height of 60 feet in 10 years. There is a red-flowered form in Manipur.

Planted as a fast growing handsome shade tree in sourthern Florida, being resistant to salt spray.

**Public Forests :** Cambalache, Carite, Guajataca, Guilarte, Luquillo, Rio Abajo, Toro Negro, Vega.

**Range :** Cuba and Jamaica. Planted also in Puerto Rico and Hispaniola.

**Other Common Names :** Emajagua excelsa (Puerto Rico); majagua, majagua azul (Cuba); Cuban bast (Florida); blue mahoe, mountain mahoe, Cuba-bark (Jamaica).

**Botanical Synonym :** *Paritium elatum* (Sw.) G. Don.

## 183. Pavona, hibiscus, Chinese hibiscus (*Hibiscus rosa–sinensis* L.*)

Pavona or amapola, the well–known introduced ornamental shrub, sometimes becomes a small tree and is included here. It is recognized by : (1) ovate shiny dark green leaves with 5 main veins at base and coarsely toothed toward long–pointed apex; and (2) solitary large red, purple, or white trumpet–shaped flowers 3–5 inches long and broad, with 5 spreading rounded petals and a long column bearing many stamens and 5 style branches.

Evergreen cultivated shrub commonly 10 feet or less in height, sometimes a small tree to 15 feet high and 3 inches in trunk diameter, commonly with several stems from the base. Bark gray, smooth, the inner bark pale green and tasteless. The twigs are green to dark purple with light dots (lenticels), almost hairless.

The alternate leaves have long dark purple leafstalks $^1/_2$–2 inches long and paired hairlike stipules $^1/_4$ inch or more in length, finely hairy. Blades are 3–5 inches long and $1^1/_2$–$2^1/_2$ inches wide, sometimes somewhat larger, rounded or short–pointed at base, thin, hairless, the lower surface dull light green.

The flowers are borne one at a time at base of an upper leaf on a slender stalk 3 inches or more in length. A flower has at base 6–7 spreading narrow green scales; bell–shaped light green calyx $^7/_8$ inch long, with 5 spreading pointed lobes; corolla of 5 spreading red, purple, or white petals rounded and sometimes wavy–toothed, united at base; many stamens in a long column or tube 3–4 inches long, of the same color, united with corolla at base; and pistil with 5–celled ovary, long style 5–branched at apex, and 5 rounded stigmas. Seed capsules, not produced in the cultivated plants, are about 1 inch long and many–seeded. Flowering throughout the year.

Pavona is a popular ornamental, widely grown for blossoms and as a hedge. It roots readily from cuttings and can be trimmed into hedges of various heights including wind–breaks. There are many varieties and hybrids differing in flower color and other characters. A double–flower variety has additional petals. The petals turn black when crushed and have been used elsewhere as shoe polish, as suggested by the common name shoe–black.

Also, it is reported that the petals rubbed on paper produce a bluish purple tint that serves like litmus paper in chemical acidity tests. The plants are very susceptible to attack by an introduced scale insect (*Cerococcus deklei*).

Commonly planted in India.

**Fig. 183 : Pavona, hibiscus, Chinese hibiscus (*Hibiscus rosa-sinensis L.**)**

Natural size.

## 184. Virginia (*Sida eggersii*)

This species, now known from a single tree at Assam, is one of the very rarest native trees. It is recognized by : (1) broadly ovate wavy–toothed leaves with 7–9 main veins from base and with both surfaces gray green from the cover of minute star–shaped hairs; (2) the light gray, twigs covered with pale yellow scale hairs when young; and (3) yellow flowers about $^3/_4$ inch long at leaf bases.

Deciduous tree 20 feet high and 4 inches in trunk diameter, recorded to 26 feets tall and 6 inches in trunk diameter. The bark is gray and smooth, the inner bark yellow brown, fibrous, and tasteless. Branches and twigs are light gray and smooth. The stout twigs bear many raised half–rounded leaf scars.

The alternate leaves are borne several together near the ends of the twigs. There are 2 minute stipules at the base of a leaf. Petioles are $^1/_2$–$1^1/_2$ inches long. The blades are mostly $1^1/_2$–$2^1/_2$ inches long and $1^1/_4$–$1^3/_4$ inches wide, recorded to 6 inches long, long–pointed at apex, slightly notched or heart–shaped at base, finely wavy–toothed at edges, slightly thickened.

Flower clusters (panicles) at base of leaves bear a few crowded nearly stalkless flowers. The flower is composed of bell–shaped velvety hairy calyx $^1/_2$ inch long with 5 pointed lobes, 5 yellow petals, many stamens in a long column, and pistil with hairy 5–6–celled ovary. The fruit separates into 5–6 hairy parts $^1/_8$ inch long, each containing 1 hairy seed. Collected with flowers in March.

The wood is light brown and slightly hard.

Very rare is seasonal deciduous forest or dry scrub woodland at about 2800 feet altitude on Jorhat.

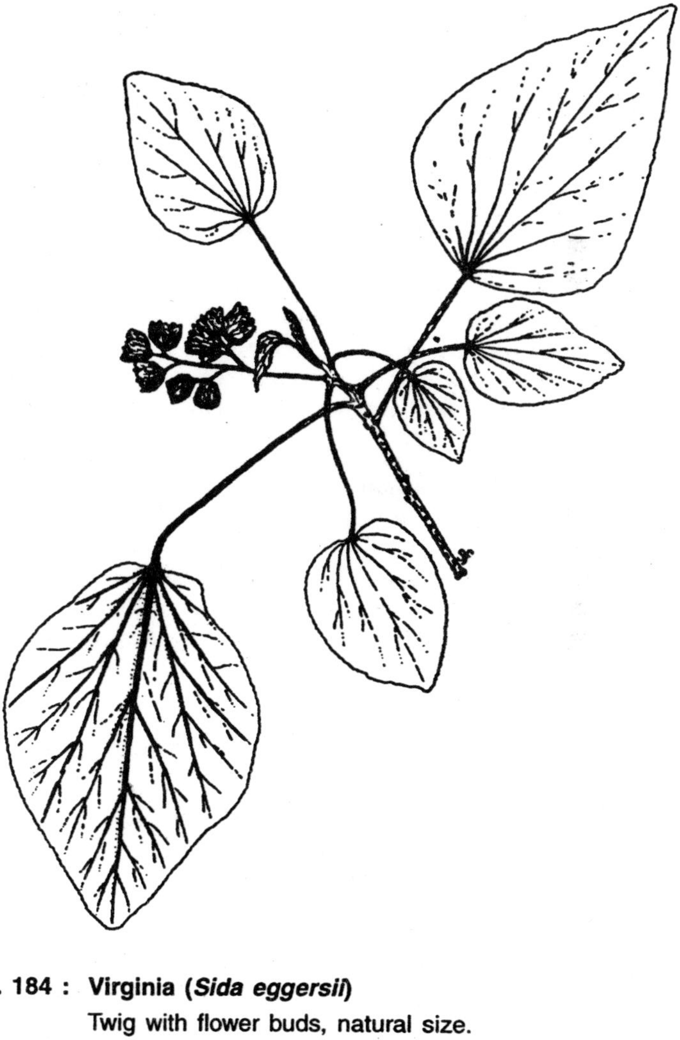

**Fig. 184 : Virginia (*Sida eggersii*)**
Twig with flower buds, natural size.

# BOMBAX FAMILY (BOMBACACEAE)

Trees often giants, rarely shrubs, known by : (1) trunks often stout or very broad, sometimes greenish and spiny, with mucilaginous sap; (2) leaves generally deciduous, alternate, mostly digitate, also simple and sometimes lobed, and palmate-veined, with star-shaped hairs or minute scales, with stipules; (3) flowers generally large and showy, often when leaves are absent, solitary or in lateral clusters, bisexual, usually regular, sometimes irregular, with calyx cuplike or 5-lobed, 5 petals sometimes hairy on outside, 5 to many stamens separate or united in tube, and pistil with superior ovary of 2–5 cells each with 2 to many ovules, style, and 1–5 stigmas; and (4) fruit a capsule or berry with many seeds sometimes enclosed in hairs.

**Key to Species**

A. Leaves palmately compound (digitate) with 5–9 leaflets.

B. Leaflets saw-toothed on edges.

C. Petioles short, less than $2^1/_2$ inches long.—*Ceiba aesculifolia.**

CC. Petiols longer, $2^1/_2$–5 inches long—*Chorisia speciosa.**

BB. Leaflets not toothed on edges.

D. Leaflets narrow, mostly lance-shaped—Ceiba, silk-cotton-tree, *Ceiba pentandra* (L.) Gaertn.

DD. Leaflets mostly broad, oblong or obovate to elliptic.

E. Leaflets oblong, hairy beneath—*Adansonia digitata.**

EE. Leaflets obovate to elliptic, hairless or nearly so beneath—*Pachira insignis.**

AA. Leaves simple.

F. Leaves with 5–9 main veins from base—Guano, balsa, *Ochroma pyramidale* (Cav.) Urban.

FF. Leaves with 1 main vein or midrib—Garrocho, *Quararibea turbinata* (Sw.) Poir.

## 185. Baobab (*Adansonia digitata*)

Baobab, an African tree noted for its enormous trunk, is a rare shade and ornamental tree in the Andaman Nicobar Islands. Distinguishing characters are : (1) the very thick stout but short tapering trunk; (2) palmately compound leaves (digitate) of 3–7 elliptic leaflets not toothed on edges, hairy beneath; (3) large drooping whitish flowers 5–6 inches across, borne on long stalks when trees are leafless; and (4) large oblong yellowish–brown gourdlike fruits 6–12 inches long and 2–3$^1/_2$ inches in diameter, bristly hairy on the outside and mealy within, hard–walled and not splitting open.

A medium–sized to large deciduous planted tree about 50 feet high and 6 feet or more in trunk diameter, somewhat larger in age where native. The crown is relatively broad, exceeding the height, but is leafless in the relatively dry winter season. The bark is smoothish and the inner bark is fibrous. Twigs stout, finely hairy when young.

The alternate leaves have finely hairy petioles 4–5 inches long. Leaflets are 3–6 inches long and 1–2$^1/_2$ inches wide, abruptly long–pointed at apex and gradually narrowed to long–pointed base. Leaves on young plants have only 3 leaflets.

The large flowers are borne singly on very long drooping stalks at leaf bases. The calyx is 2–3 inches long, deeply 5–lobed, leathery, and silky hairy within; 5 narrow petals becoming curved back; very many white threadlike stamens with purple dotlike anthers spreading 2 inches across from a tube 1$^1/_2$ inches high in lower half; and the pistil has a bristly hairy 7–10–celled ovary, slender style, and 7–10–lobed stigma. The hard woody fruits hang from long stalks and contain many kidney–shaped seeds in dry mealy pulp.

Baobab is a tree of great size and age. In trunk diameter, as much as 30 feet, it ranks among the world's largest trees, though its maximum height is only about 60 feet. Giant trees apparently are very old, but the hollow trunks of soft spongy lightweight wood cannot be dated precisely.

The spongy mealy fruits of baobab are sour but reported to be edible. These hard–walled fruits do not contain hairs or fibers as do the seed capsules of ceiba or silk–cotton–tree, balsa, and other relatives. Where the trees are native, the leaves have been cooked and eaten as a vegetable. Also, the seeds have been prepared into meal. Rope has been made from the fibrous bark.

Baobab is rare in Goa. It was recorded more than a century ago as naturalized Goa but would not be so classed now. Young trees were grown experimentally in Goa but apparently were not preserved. Planted in parks and gardens in tropical regions of the world, north to southern States.

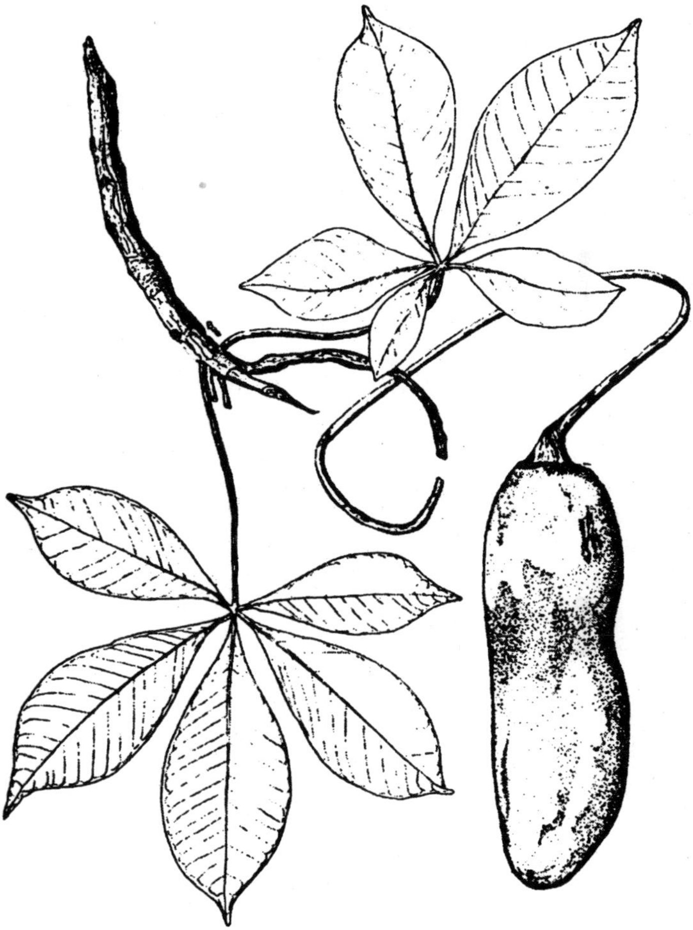

**Fig. 185 : Baobab (*Adansonia digitata*)**
Leafy twig and fruit (lower right), two–thirds natural size.

## 186. Pochote (*Ceiba aesculifolia*)

This ornamental and shade tree is related to the native species, ceiba or silk–cotton–tree, *Ceiba pentandra* (L.) Gaertn. Recognized by : (1) spines on trunk and twigs; (2) palmately compound leaves (digitate) with 5–8 elliptic leaflets saw–toothed on edges; (3) large flowers with 5 narrow yellowish hairy petals 4–6 inches long; and (4) large narrowly elliptic or oblong brown seed capsules 5–7 inches long, containing many seeds in cottony hairs.

A medium–sized to large deciduous planted tree with stout conic spines on the large smoothish trunk and with spreading crown. Twigs are stout and often spiny.

The alternate leaves have slender finely hairy petioles $1^1/_2$–$2^1/_2$ inches long. The leaflets are 2–6 inches long and $^3/_4$–2 inches wide, whitish green beneath, short–to long–pointed at apex, long–pointed at base with hairy stalk less than $^1/_4$ inch long.

The large flowers on short stout stalks have a bell–shaped calyx $^3/_4$–$1^1/_4$ inches long; corolla of 5 narrow yellowish hairy petals 4–6 inches long, spreading and curved; 5 slender stamens longer than petals; and pistil with 5–celled ovary and long slender style. The smooth seed capsules split open along 5 lines. Many small elliptic brown seeds $^3/_{16}$ inch long are borne in a mass of brownish or white cottony hairs.

Common as an ornamental and shade tree in Goa. In South the trees have been planed also for cottony fibers, which have been exported. These fibers like kapok from related species serve for stuffing pillows and mattresses, for insulation, and other uses.

**Fig. 186 : Pochote (*Ceiba aesculifolia*)**
Two–thirds natural size.

## 187. Chorisia, floss–silk–tree (*Chorisia speciosa*)

This relative of ceiba is one of the world's most beautiful flowering trees. It has been sparingly introduced for its showy flowers mostly in winter when leafless or nearly so. Characters for recognition are : (1) spines on trunk and branches; (2) palmately compound (digitate) leaves with 5–7 narrowly elliptic leaflets saw–toothed on edges; (3) large showy mostly pink flowers 3–4 inches long and 4 inches broad, covering the trees when leafless; and (4) fruit a large pear–shaped capsule about 8 inches long bearing many seeds in whitish silky hairs.

Medium–sized deciduous planted tree 35 feet high with spiny columnar greenish trunk 1 foot in diameter, probably becoming larger and swollen in age, and with widely spreading crown of coarse branches.

The alternate hairless leaves have slender petioles about $2^1/_2$–5 inches long. The leaflets are attached at end of petiole by short stalks $^1/_4$–$^1/_2$ inch long. Blades are 25 inches long and $^1/_2$–$1^1/_2$ inches broad, long–pointed at apex and short–to long–pointed at base, paler beneath.

The large flowers at leaf bases or in clusters (racemes) have a stout stalk $^1/_2$ inch long. The bell–shaped tubular calyx is about 1 inch long, irregularly 3–5–lobed at apex. There are 5 oblong or spoon–shaped spreading petals 3–4 inches long, mostly deep pink with whitish or yellowish base, spotted with purplish streaks, wavy margined, finely hairy on outside; the stamens are united into a double tube, the outer short with hairy sterile stamens, the inner $2^1/_2$–3 inches long and very narrow, with ring of anthers at apex; and within the stamen tube the pistil with 5–celled ovary containing many ovules, long slender style longer than stamen tube, and 5–lobed stigma. The fruit is a woody pear–shaped capsule 6–8 inches long. Many small seeds like peas are produced in a mass of silky or cottony hairs. Flowering in winter.

The wood is soft, lightweight, and not used.

The soft silky and cottony hair or fiber of the seed capsules, like that of ceiba and related species, has served for pillows and cushions.

Rarely planted as an ornamental and street tree in Goa. Cultivated elsewhere in the tropics including West Indies.

**Fig. 187 : Chorisia, floss–silk–tree (*Chorisia speciosa*)**
Two–thirds natural size.

The plants are propagated by seeds and cuttings and grow rapidly in good well drained soil to a maximum of 50–75 feet in height, flowering in about 8–10 years. Because of their large size, these trees are better for streets, parks, and gardens, than for small yards. The trees will tolerate occasional light frost. Flowers differ in size and in color, ranging in different varieties from deep or pale pink to white and purple.

## 188. Shaving–brush–tree (*Pachira insignis*)

An ornamental and shade tree, characterized by : (1) palmately compound leaves with mostly 7 (5–9) large obovate to elliptic leaflets almost stalkless; (2) very large handsome flowers 7–10 inches long with 5 narrow brownish–red petals and numerous spreading threadlike whitish stamens like a shaving brush; and (3) large brown seed capsules 6–10 inches long, rounded or elliptic.

A medium–sized evergreen planted tree with stout trunk becoming buttressed, stout branches in whorls, and spreading crown. The bark is greenish or gray, smooth or becoming slightly fissured, the inner bark light brown and slightly bitter. The twigs are dark green and very stout.

Leaves are alternate but grouped toward end of twig, measuring 10–15 inches long including round petiole 4–7 inches long and enlarged at both ends. Leaflets mostly 7, sometimes 5–9, stalkless or nearly so, 4–9 inches long and $1^1/_2$–$3^1/_2$ inches wide, broadest toward the rounded or notched apex and gradually narrowed toward base, not toothed on edges, thickened, hairless or nearly so, the upper surface slightly shiny green, and the lower surface dull light green with raised veins.

Flowers are borne 1 or 2 on very stout stalks 1–5 inches long from narrow brown finely hairy buds 7–9 inches long. The calyx is cup–shaped, about 1 inch long and broad, dark brown, finely hairy, and not lobed; the 5 petals are 7–9 inches long, finely hairy; the several hundred threadlike stamens are much branched ending in yellow anthers and united into a reddish tube below; and the pistil inside tube consists of a rounded 5–angled 5–celled ovary, very long style, and lobed stigma. The seed capsules are mostly longer than broad, blunt pointed, finely hairy, heavy thick–walled, and opening into 5 parts. There are many rounded and 4–angled brownish seeds about 1 inch across. Flowering probably through the year.

The woos is described as whitish, soft and lightweight, and not used.

It is reported that elsewhere the seeds are eaten after toasting or raw, having a flavor like that or chestnuts, also similar common names. However, the trees are not related to the true chestnuts. A honey plant.

Rare as a planted ornamental and shade tree in Goa and perhaps Lakhwasdeep Islands. Elsewhere in the tropics occasionally grown in parks.

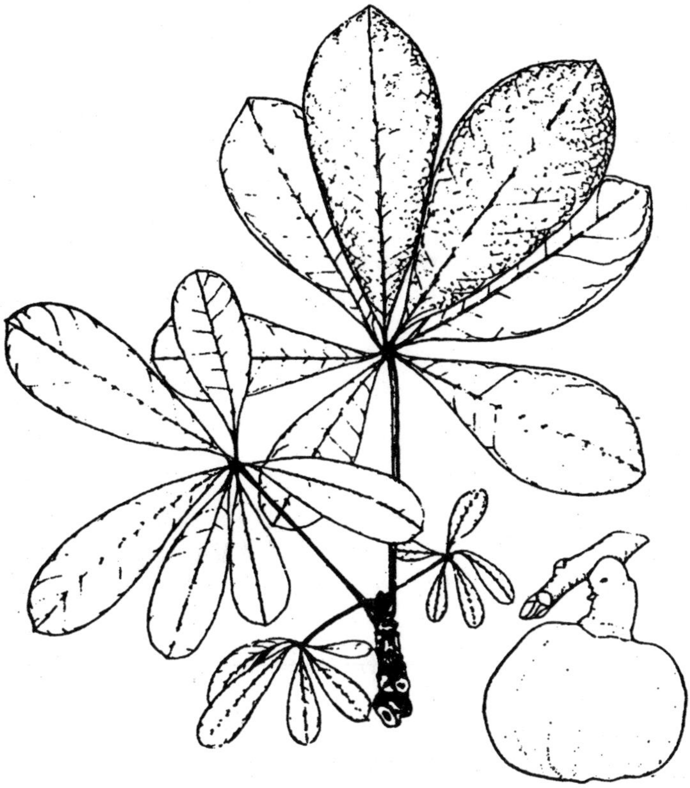

**Fig. 188 : Shaving–brush–tree (*Pachira insignis*)**

Leafy twig, one–half natural size; fruit (lower right), about one–fourth natural size.

# CHOCOLATE FAMILY (STERCULIACEAE)

Shrubs, herbs, vines, and trees, known by : (1) alternate simple leaves often palmate–veined and palmately lobed, sometimes digitate, with star–shaped hairs, the petiole often with enlargement at apex, with stipules; (2) flowers usually in lateral or terminal branched clusters (cymes) or sometimes along trunks, generally bisexual and regular with parts in 5's, calyx of 3–5 lobes, 5 petals or sometimes none, 5 stamens united in a tube or separate, sometimes with 5 staminodes, and pistil composed of superior ovary generally with 5 (1–4) cells with 2 to many ovules and 1–5 styles often lobed, sometimes the stamens and pistils on a long stalk; and (3) fruit a capsule or berry or 5 follicles.

**Key to species**

A. Leaves palmately compound (digitate) with 5–11 large elliptic leaflets—*Sterculia foctida.**

AA. Leaves simple

B. Leaves deeply 5–lobed, with 5 main veins from notched or heart–shaped base; petiole nearly as long as blade—*Anacagüita, panama–tree, Sterculia apetala* (Jacq.) Karst.*

BB. Leaves not lobed.

C. Leaves toothed on edges, mostly hairy, more or less unequal or oblique at base.

D. Leaves coarsely toothed, with 7 main veins from base, soft hairy—*Helicteres jamaicensis.*

DD. Leaves finally saw–toothed, with 3 or sometimes 5 main veins from base, hairy or nearly hairless—156. Guácima, jacocalalu, *Guazuma ulmifolia* Lam.

CC. Leaves not toothed on edges, mostly hairless, the base not notched and with equal sides.

E. Leaves with 5 or 7 main veins from base, broadly ovate—*Kleinhovia hospita.**

EE. Leaves with 1 main vein from base, elliptic or oblong.

F. Leaves 4–8 inches long, short–pointed at base—*Cola acuminata**

FF. Leaves 8–14 inches long, rounded at base—Cacao, chocolate–tree *Theobrom cacao* L.*

## 189. Cola–nut (*Cola acuminate (Beauv)*

Cola–nut from tropical Africa has been planted experimentally in India. Its characteristics for identification are: (1) oblong to ovate leathery leaves 4–8 inches long and $1^1/_2$–$3^1/_2$ inches wide; (2) yellowish or whitish star–shaped flowers clustered at leaf bases, $^3/_4$–$1^1/_4$ inches across the starlike spreading 5 calyx lobes; and (3) large fruit of 5 or fewer warty gray–green pods 4–5 inches long, each splitting open in 1 line.

A small evergreen planted tree becoming 15 feet or more in height and 6 inches in trunk diameter. The twigs sometimes have scattered hooked spines $^1/_8$ inch long.

The alternate leaves have slender petioles $^1/_2$–3 inches long. Leaf blades are abruptly long–pointed at apex, short–pointed at base, not toothed on edges, hairless.

Flower clusters (racemes) to 3 inches long bear at leaf bases several to many flowers partly of one sex, composed of a spreading 5–lobed star–shaped calyx yellowish or whitish with reddish or pulplish markings toward base, no corolla, about 10 stamens united into a column, and pistil with 5 or fewer cells and style. The 5 or fewer fruits (follicles) each contain 6–12 large white to reddish seeds 1–$1^1/_4$ inches long that become brown on drying. Collected with flowers from October to December.

The bitter seeds or cola–nuts (about 300–600 to a pound) contain caffeine and serve as a stimulant when chewed by persons living in Africa where the trees are native. The seeds have been exported in limited quantities for preparations such as wine and medicine. The trees, propagated from seeds, are grown in plantations for the nuts.

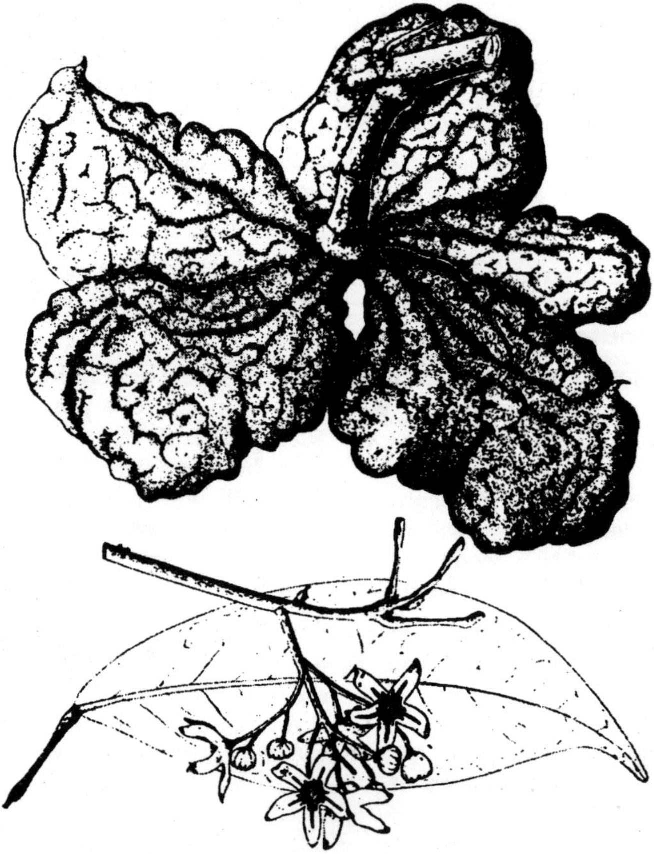

**Fig. 189 : Cola–nut (*Cola acuminate (Beauv)***

Fruit (above), flowers and leaf (below), two–thirds natural size.

## 190. Cuernecillo, cowbush (*Helicteres jamaicensis*)

This sparsely branched shrub or sometimes a small tree is easily recognized by : (1) dense soft covering or minute gray star–shaped hairs on foliage, flowers, and fruits; (2) coarsely toothed ovate leaves with unequal sides and 7 main veins from the notched base, spreading in 2 rows on long twigs; (3) few whitish irregular flowers with 5 slightly unequal petals $1^1/_4$ inches long and threadlike column $2^1/_2$–3 inches long bearing stamens and pistil; and (4) distinctive conelike cylindric, much twisted gray woolly hairy fruit $1^1/_2$–2 $^1/_2$ inches long and $^3/_4$ inch in diameter on a slender stalk 3 inches long above the calyx.

Evergreen shrub about 10 feet high, with few long spreading branches, sometimes a small tree 20 feet high and 3 inches in diameter, not forming a crown. Bark gray, slightly thickened, with conspicuous warts. Inner bark yellowish, turning orange on exposure, slightly bitter. Twigs stout, densely hairy, gray green, becoming gray, ending in hairy buds composed of stipules and young leaves.

Leaves are alternate and spreading in 2 rows and have paired narrow hairy stipules $^1/_4$–$^5/_8$ inch long and hairy leafstalks $^3/_8$ – $^3/_4$ inch long. Blades are 3–7 inches long and 1 $^1/_2$–4 inches wide, long–pointed, coarsely saw–toothed, often with few larger teethor 2 shallow lobes, soft hairy and slightly thickened, the upper surface green, and lower surface gray green, densely hairy, and with raised veins.

Flowers are few, 1–3 at leaf base or at end of twig, on hairy stalks $^1/_4$–$^3/_4$ inch long, only 1 or 2 opening at a time. The densely hairy gray–green calyx is bell–shaped, about 1 inch long and broad, slight 2–lipped, with 5 narrow unequal pointed teeth. The 5 white petals are narrow, with 2 lobes on sides, soon falling. A whitish hairy threadlike curved column from the base of the flower bears 10 stamens $^3/_8$ inch long and 5 smaller nonfunctional stamens (staminodes) and the pistil, composed of hairy 5–lobed, 5–celled ovary with many ovules and short bent threadlike style.

The peculiar seed pod, at the end of a slender stalk above the persistent calyx is pointed at both ends and fuzzy gray–green hairy, persistent and weathering to gray and blackish. It is composed of 5 hard parts (follicles) much twisted in a spiral, each opening along a line within. There are many oblong blackish seeds $^1/_8$ inch long. Flowering irregularly through the year.

The wood is yellowish, hard and heavy. The fibrous bark is strong and suitable for making rope.

Local at low altitudes in openings, clearings, and thickets in dry and moist coastal and limestone forests from sea level to 400 feet altitude in Kerala.

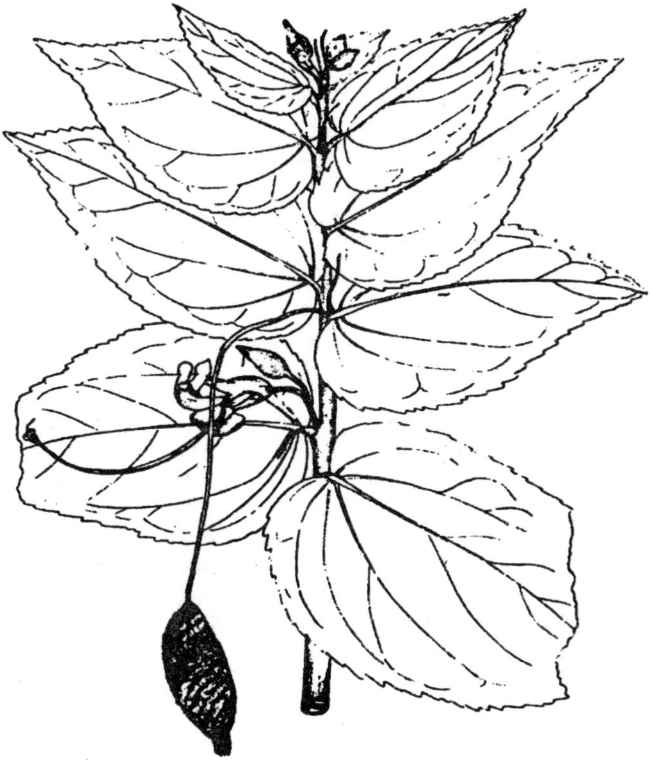

**Fig. 190 : Cuernecillo, cowbush (*Helicteres jamaicensis*)**
Two–thirds natural size.

## 191. Guest–tree (*Kleinhovia hospita* L.*)

The showy masses of many small bright pink flowers make this tree an attractive ornamental. Other characters for identification are : (1) broadly ovate or heart–shaped leaves with 5 or 7 main veins from slightly notched base; and (2) bladderlike fruits, greenish–brown papery or bladdery pods nearly 1 inch long and broad, 5–angled and broadest near apex.

A small evergreen planted tree to 25 feet high, the short often crooked trunk 6 inches in diameter or larger, with dense rounded crown. The bark is gray, fissured and rough. Twigs greenish and finely hairy when young, becoming light brown.

The alternate leaves have slender petioles 2–3$^1/_2$ inches long, slightly enlarged at both ends. The blades are 3–6 inches long and broad, abruptly long–pointed at apex, not toothed on edges, thin, dull green on upper surface, and paler with minute hairs beneath.

Flower clusters (panicles) are terminal and upright, 6–12 inches or more in length, with many finely hairy pinkish branches. The short stalked slightly irregular flowers about $^3/_8$ inch long and broad are composed of 5 narrow spreading pink finely hairy sepals $^3/_8$ inch long; shorter spurlike yellow–tipped lobe; and at end of a curved stalk 5 stamens and pistil with hairy 5–celled ovary and short style. The bladdery pods or capsules split into 5 parts, each with 1 or 2 rounded seeds about $^1/_8$ inch long. Flowering and fruiting irregularly through the year.

The wood is whitish.

Common as a planted ornamental in Sikkim flowering as a shrub. Elsewhere the trees are grown along roadsides.

**Fig. 191 : Guest–tree (*Kleinhovia hospita* L.*)**
Fruits (left), leaf, flowers (right), two–thirds natural size.

## 192. Anacaguita, hazel sterculia (*Sterculia foetida* L.*)

This handsome ornamental and shade tree is recognized by : (1) palmately compound leaves with 5–11 large elliptic long-pointed leaflets at end of long petiole; (2) large reddish, or purplish star–shaped flowers nearly 1 inch across the 5 narrow finely hairy calyx lobes, with unpleasant odor, several in narrow clusters with new leaves; and (3) fruit of 5 or fewer reddish egg-shaped hard woody pods 2–3 inches in diameter, with seeds attached to margins.

Large deciduous planted tree becoming 50 feet or more in height. Trunk stout, unbranched, often with buttresses at base, with smooth gray bark. Twigs hairless, with stout portions bearing many raised half–round leaf scars and long slender portions without leaf scars.

Leaves alternate, crowded at ends of twigs, hairless except when young, with long slender round petioles to 1 foot long. Leaflets 4–9 inches long, $1^1/_2$–$2^1/_2$ inches wide, slightly thickened, ending in long narrow point at apex, tapering to narrow stalkless base, not toothed on edges, with many parallel, slightly curved side veins.

Flower clusters (racemes and panicles) mostly unbranched, 4–6 inches long, several near ends of twigs with new leaves, each bearing several flowers on slender stalks $^1/_8$–$^1/_2$ inch long. Flowers partly male and female on the same tree (monoecious), with unpleasant odor, have reddish or puplish calyx divided almost to base into 5 narrow spreading lobes nearly $^1/_2$ inch long, the upper side densely hairy with star–shaped hairs, and no corolla. Male flowers bear 10–15 minute stamens at the end of a slender curved hairy stalk $^1/_4$ inch long. Female flowers have at the end of a stalk a pistil consisting of rounded 5–celled ovary, short style, and 5 short stigmas.

The fruit consists of 5 or fewer pods (follicles) from a flower, each with pointed stigma at apex, finely hairy, hard, thick–walled, turned back and splitting open widely along 1 line, the brownish inner wall without hairs. Seeds mostly 10–15, oblong, $^7/_8$ inch long, blackish.

The wood is reported to be soft and easily worked but subject to attack by insects and not durable.

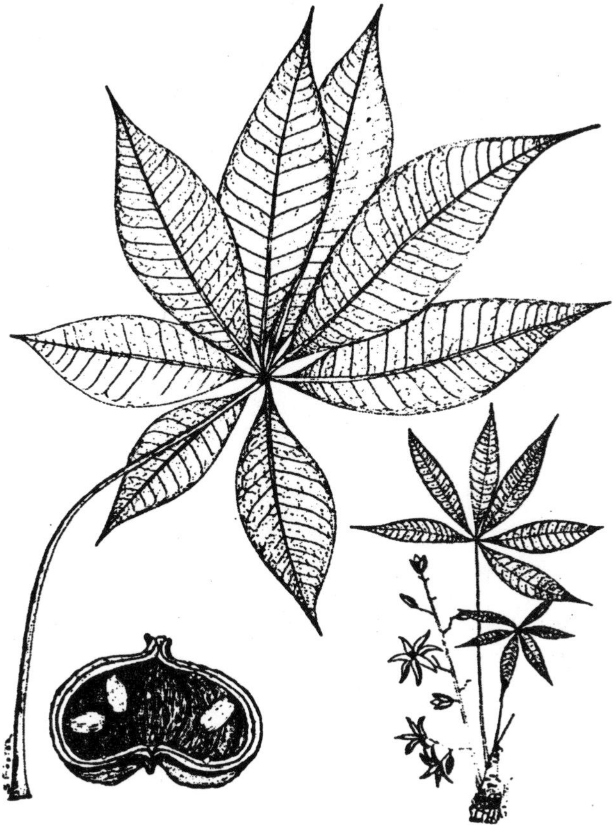

**Fig. 102 : Anacaguita, hazel sterculia (*Sterculia foetida* L.*)**
Leaf (above), opened fruit with seeds (lower left), and flowers and young leaves (lower right), all one–half natural.

Elsewhere the large seeds are eaten, especially after roasting. Also, they have served as a substitute for cacao or chocolate, though lacking the distinctive flavor.

Occasionally grown for ornament and shade in Puerto Rico but unpopular because of the bad odor during the short flowering season. A hand–some fast–growing ornamental for parks and large gardens. Introduced also through the West Indies, in India.

# OCHNA FAMILY (OCHNACEAE)

Small to large trees, shrubs, and rarely herbs, known by : (1) alternate simple leaves with border generally toothed, pinnate–veined with many lateral veins, often leathery, hairless, with stipules sometimes large and divided; (2) small to large flowers often showy in clusters (panicles, racemes, or cymes), bisexual, usually regular, with calyx of 4–5 sepals separate or united at base, 4–5 (10) petals often yellow and falling early, stamens 5, 10, or many with anthers often opening by pores, sometimes with staminodes, and pistil composed of superior ovary often deeply lobed on large base (receptacle), generally with 2–5 (rarely 10–15) cells with 1 to many ovules, 1 style, and 1–5 stigmas; and (3) fruit of 1–5 drupes or berries on the very large base (receptacle) or a capsule.

**Key to species**

A. Leaves thick and leathery, with several large spiny teeth on wavy edges—*Ouratea olicifolia.*

AA. Leaves only slightly thickened, with fine inconspicuous teeth or none.

B. Leaves broadest toward blunt or rounded apex, with fine inconspicuous teeth—*Ochna mossambicensis.**

BB. Leaves broadest toward base or near middle.

C. Leaves with fine inconspicuous teeth and many curved side veins becoming nearly parallel with edge; flowers many, $^3/_4$ inch across—*Ouratea littoralis.*

CC. Leaves without teeth, with few curved side veins that appear to cross many fine parallel veins; flowers few, $^1/_2$ inch across—*Ouratea striata.*

## 193. Mozambique ochna (*Ochna mossambicensis*)

An ornamental planted for the showy bright yellow flowers. Distinguished by : (1) oblanceolate or obovate finely saw–toothed leaves, broadest toward the blunt or rounded apex; (2) laterat clusters of large flowers $1^1/_4$ inches across the 5 spreading stalked yellow petals, on slender jointed stalks; and (3) fruit of 5 or fewer black elliptic fruits on enlarged base and bordered by 5 red sepals.

A planted evergreen shrub or small tree to 16 feet high and 4 inches in trunk diameter, reported to become a medium–sized tree where native. Flowering when a shrub of 4 feet.

Leaves alternate, hairless, 2–5 inches long and $^3/_4$–2 inches wide, gradually narrowed toward long–pointed base and short petiole less than $^1/_4$ inch long. The blades are slightly thickened, the shiny upper surface with many fine curved side veins all slightly sunken, and the lower surface slightly shiny.

The flowers at leaf bases consist of a slender jointed stalk of $^1/_2$–1 inch; 5 narrow spreading sepals; 5 bright yellow stalked petals, falling early; many stamens; and pistil with deeply 5–lobed ovary and slender style. From a flower 5 or fewer elliptic black 1–seeded fruits (drupes) $^3/_8$ inch long develop on an enlarged fleshy base with persistent style and red sepals.

Common as an ornamental in Goa.

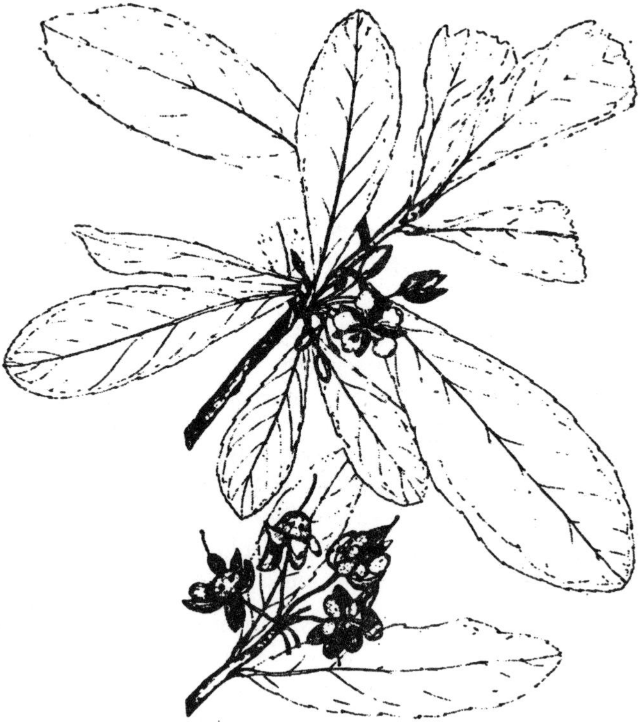

**Fig. 193 : Mozambique ochna (*Ochna mossambicensis*)**
Flowering twig (above), fruits (below), two–thirds natural size.

## 194. Arneau (*ouratea ilicifolia*)

This small tree is easily recognized by ; (1) elliptic leaves with several large spiny teeth on wavy margins; (2) several showy flowers about $^{5}/_{8}$ inch across the 5 rounded bright yellow petals, which shed early; and (3) fruits of 5 or fewer blackish or bluish elliptic drupes $^{3}/_{8}$ inch long from the enlarged rounded red base.

Evergreen small tree to 15 feet high and 3 inches in trunk diameter. Twigs hairless.

Leaves alternate, with petioles $^{1}/_{8}$ inch long. Blades $1^{1}/_{2}$–$2^{3}/_{4}$ inches long and $^{3}/_{4}$–2 inches wide, sharp–pointed at apex, straight or blunt at base, and with several large spiny teeth $^{1}/_{8}$–$^{1}/_{4}$ inch long along margin, thick and leathery, shiny green, hairless.

Flower clusters (panicles) 2–5 inches long at leaf bases. The flower is composed of 5 narrowly elliptic sepals; 5 rounded yellow petals $^{1}/_{4}$–$^{3}/_{16}$ inch long, narrowed at base, spreading, sheddy early; 10 short–stalked stamens, with large anthers opening by pores; and on the enlarged base the pistil with ovary deeply 5–lobed, 5–celled, and 5–ovuled, and erect style. From one flower the base enlarges and becomes $^{1}/_{4}$ inch across and red. There are 5 or fewer elliptic blackish or bluish 1–seeded fruits from a flower. Flowering and fruiting irregularly.

Local and uncommon in dry and moist coastal and lower Kerala forests near sea level along northern, northeastern, and southeastern coasts of Kerala.

**Fig. 194 : Arneau (*ouratea ilicifolia*)**
Flowering twig (left), fruiting twig (right), natural size.

## 195. Abey Amarillo (*Ouratea littoralis*)

This shrub or small tree known only from Goa and Islands is easily recognized by : (1) narrowly elliptic to ovate leaves finely and inconspicuously toothed, long–pointed, slightly thickened, with many fine, long, much curved side veins becoming nearly parallel with edge; (2) many fragrant yellow flowers $^3/_4$ inch across the 5 spreading petals, fan–shaped and notched, which shed early; and (3) distinctive fruits, 1–5 elliptic 1–seeded blue–black drupes $^5/_{16}$ inch long on an enlarged fleshy red base $^3/_8$ inch across.

Evergreen shrub or small tree to 20 feet high and 4 inches in trunk diameter, hairless. Bark gray, smooth, becoming slightly fissured. Inner bark pink or light yellow, with a green outer layer, slightly bitter. Twigs yellow green and slightly angled when young, becoming brown.

The alternate leaves have paired narrow long–pointed stipules $^3/_{16}$ inch long and short leafstalks $^1/_8$ inch long. Blades are $^1/_2$–4 inches long and $^3/_4$–$1^3/_4$ inches wide, short–pointed or rounded at base, finely and inconspicuously toothed, especially toward apex, slightly thick and stiff, the sides slightly turned up at midrib, upper surface green and slightly shiny, lower surface dull yellow green.

Flower clusters (panicles) 2–5 inches long and broad at ends and sides of twigs bear many fragrant flowers on spreading branches. Each flower is composed of calyx of 5 pointed yellow–green sepals that form the bud $^1/_4$ inch long, 5 spreading yellow petals, fan–shaped and notched, which shed early, 10 short stamens, and on a disk the pistil with 5–lobed ovary and slender style. In fruit the rounded base or disk becomes much enlarged and red and bears 5 or fewers elliptic blue–black drupes with thin flesh, large stone, and elliptic seed. Flowering irregularly during the year.

The wood is whitish and hard.

Common along eastern and northwestern coasts and other lowland open areas in moist coastal hills near sea level to about 100 feet altitude in Western Ghat.

**Fig. 195 : Abey Amarillo (*Ouratea littoralis*)**

Flowers (upper left), fruits (center left), twig with flower bud (right), natural size.

## 196. Guanabanilla (*Ouratea striata*)

This shrub or small tree rare and local in high mountains of Konkan Area is identified by : (1) the lance–shaped shiny leaves with a few long curved fine side veins over many fine parallel veins, slightly thickened and turned under slightly at edges; (2) few yellow flowers $^1/_2$ inch across the 5 bright yellow petals; and (3) distinctive fruits, 1–5 elliptic and deep blue, nearly $^1/_4$ inch long, on the enlarged fleshy red base of a flower.

Evergreen shrub or small tree to 15 feet high and 6 inches in trunk diameter, hairless throughout. The bark is brown and smoothish, the inner bark light brown and almost tasteless. Twigs are green and slender when young, turning brown.

The alternate leaves have petioles $^1/_8$–$^3/_8$ inch long. Blades are $1^1/_2$–4 inches long and $^5/_8$–$1^1/_4$ inches wide, long–or short–pointed at apex and rounded or short–pointed at base, the upper surface green to dark green and shiny, and the lower surface yellow green to green and slightly shiny.

Flower clusters (panicles) 1–2 inches long at near ends of twigs. Flowers few on stalks $^1/_8$–$^1/_4$ inch long, composed of 5 pointed sepals that form bud $^3/_{16}$ inch long; 5 yellow petals $^1/_4$ inch long, shedding early; 10 stamens; and pistil with 5–lobed 5 celled ovary, and curved style. The fruits from a flower are 1–5 blue drupes, each with 1 large seed, attached to the red enlarged rounded or fig–shaped base. Flowering intermittently.

The sapwood is whitish and hard.

Rare and local in forests and dwarf forest at 2,500–4,000 feet altitude in high mountains of Western Ghat.

**Fig. 196 : Guanabanilla (*Ouratea striata*)**
Fruiting twig (above), flowering twig (lower left), natural size.

# TEA FAMILY (THEACEAE)

Trees and shrubs, known by : (1) alternate simple leaves usually leathery, evergreen, sometimes with lines parallel to midvein, without stipules; (2) flowers often large and showy and aromatic, generally solitary or few and lateral, bisexual, regular, often with 2 scales (bracts) at base, with calyx of 5–7 sepals usually separate, overlapping, and persistent, corolla of 5 petals commonly white or pink, separate or united at base, overlapping, many stamens often united to corolla in 5 opposite groups, and pistil composed of 2–5–celled ovary generally superior with 2 to many ovules in each cell and 2–5 persistent styles, often united at base; and (3) frit usually a hard capsule with central persistent column, berry, or drupe.

### Key to Species

A. Leaves only slightly thickened, with finely wavy stalkless, almost 2 inches across; fruit a capsule splitting into 6–10 parts—Maricao verde, *Laplacca portoricensis* (Krug & Urban) Dyer.

AA. Leaves thick and lethery, with edges turned under; flowers stalked (except No. 516), less than 1 inch across; fruit a berry or opening irregularly.

B. Leaves with veins inconspicuous, often with minute teeth near apex; flowers with petals silky hairy on outside—*Cleyera albopunctta.*

BB. Leaves with veins mostly not visible, without teeth; flowers with petals hairless—*Ternstroemia.*

C. Flowers almost stalkless; leaves obovate, black–dotted beneath—*Ternstroemia subssessilis.*

CC. Flowers with stalks; leaves mostly elliptic, not black–dotted beneath.

D. Flower stalks less than $^5/_8$ inch long.

E. Leaves 1–2 inches long; flowers $^3/_8$ inch across—*Ternstroemia heptasepala.*

EE. Leave $1^1/_2$–4 inch long; flowers nearly $^3/_4$ inch across—*Ternstroemia stahlii.*

DD. Flower stalks more than $^3/_4$ inch long.

F. Leaves elliptic or obovate, about twice as long as wide, flowers about $^3/_4$ inch across—*Ternstroemia peduncularis.*

FF. Leaves elliptic or oblong, about 3 times as long as wide; flowers nearly 1 inch across—*Ternstroemia luquillensis.*

## 197. Copy Vera (*Cleyera albopunctata*)

This rare, usually small tree of high mountains is characterized by : (1) elliptic leaves 2–4 inches long and $1–1^3/_4$ inches wide, leathery, the upper surface dull dark green with edges turned under, and the lower surface light green; (2) narrowly bell–shaped flowers $^5/_8$ inch long, with 5 whitish petals, 1–3 borne at base of a leaf; and (3) the elliptic dark red berry $^1/_2$ inch long, ending in a long point, with greenish sepals at base.

An evergreen tree to 40 feet high and 1 foot in trunk diameter with compact dense crown, or smaller in dwarf forest of mountain summits. The bark is brown, smooth, covered with mosses and liverworts, the inner bark pinkish and slightly bitter. Twigs light green, hairless, becoming brown, ending in a narrow hairy pointed bud formed by a rolled young leaf without stipules.

Leaves alternate in 2 rows, becoming hairless, with stout light green petioles $^1/_4$–$^3/_8$ inch long, grooved above. Blades are blunt–pointed at apex, short–pointed at base, with edges turned under and often minutely toothed toward apex, with midrib sunken and veins inconspicuous.

The fragrant flowers are borne at leaf bases on separate light green stalks $^3/_8$–$^3/_4$ inch long, usually curved downward. The flower is composed of calyx of 5 overlapping dark green, thick, rounded or notched sepals $^1/_4$–$^3/_8$ inch long; corolla of 5 whitish or cream–colored elliptic petals $^1/_2$–$^5/_8$ inch long, slightly united at base, finely silky hairy on outside, and minutely toothed; about 30 stamens $^1/_4$–$^3/_8$ inch long, attached to base of corolla; and pistil $^3/_8$ inch long, with conic 3–celled ovary of many ovules, long style, and 3 short stigmas.

The berry ends in a long point formed by the style and is covered at base by the persistent overlapping sepals. It is thick–walled, slightly juicy, has few brown elliptic seeds $^1/_8$ inch long, and does not split open. Flowers and fruits are borne through the year.

Rare in upper forests and dwarf forest to ridges and peaks at 2,500–3,500 feet altitude. Ascends to summit of Nilgiri.

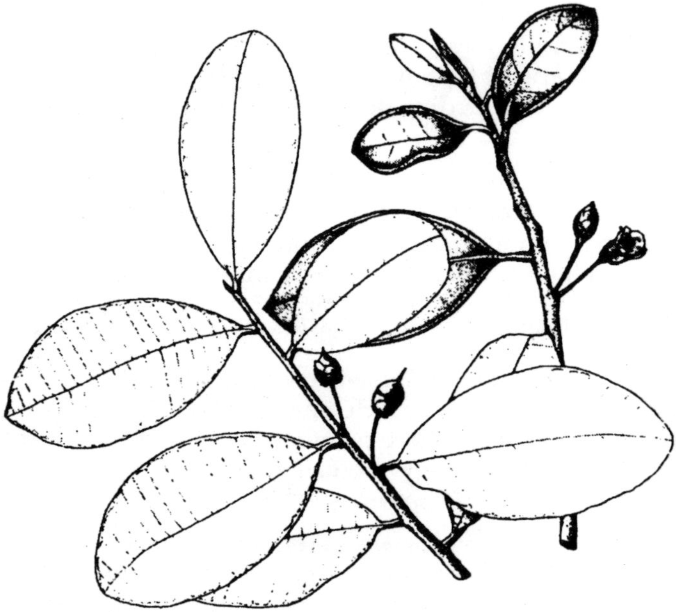

**Fig. 197 : Copy Vera (*Cleyera albopunctata*)**
Fruits (left), flowers (right), two–thirds natural size.

## 198. Heptasepala (*Ternstroemia heptasepala*)

This shrub or small tree of dwarf forests of summits, known only from Kerala Mountains, is distinguished from related species by : (1) the small thick elliptic leaves 1–2 inches long and $^5/_8$–1 inch wide, and (2) the small flowers about $^3/_8$ inch across the 5 whitish petals.

Usually a shrub to 10 feet high but becoming a small tree 25 feet high and 5 inches in trunk diameter, evergreen. The smooth gray bark is covered with mosses and liverworts, the inner bark pinkish and bitter. The twigs are brown or gray and hairless, relatively stout, ending in a minute pointed bud formed by a rolled young leaf, without stipules.

The leaves are alternate or sometimes opposite, hairless, petioles about $^1/_8$ inch long. Blades are elliptic, stiff and leathery, the apex rounded or blunt–pointed, the base short–pointed, the edges turned under, the upper surface dull dark green with midrib sunken and veins not visible, the lower surface light green.

Flowers are single or paired at leaf bases or back of leaves on curved light green stalks $^3/_8$–$^5/_8$ inch long. The fragrant flower is composed of calyx of 5 unequal rounded sepals $^1/_8$–$^3/_{16}$ inch long, greenish and tinged with brown; corolla of 5 obovate whitish petals $^3/_{16}$ inch long, united at base, minutely toothed on edges; about 25 stamens $^1/_8$ inch long attached to base of corolla; and pistil $^1/_8$ inch long, including rounded ovary, 2–celled with several ovules, and long stiff style. The egg–shaped fruits $^5/_{16}$–$^1/_2$ inch long, green when immature, has the narrow pointed style at apex and sepals at base. It is dry, 2–celled, few–seeded. Flowering intermittently.

The wood with light brown sapwood and brown heartwood is hard and heavy.

Rare in dwarf forest at 2,500–3,500 feet altitude on summits and ridges in Kerala.

**Fig. 198 : Heptasepala (*Ternstroemia heptasepala*)**
Natural size.

## 199. Palo Colorado (*Ternstroemia luquillensis*)

This medium-sized tree known only from Mountains reaches the largest size of the 5 native species of its genus, 4 of which are endemic to Goa. Distinguishing characteristics are : (1) elliptic or oblong leaves $2^1/_2$–$4^1/_2$ inches long and 1–$1^3/_4$ inches wide, thick and leathery; and (2) showy flowers nearly 1 inch across the 5 spreading white or cream–colored rounded concave petals, borne on long curved whitish stalks $1^1/_4$–$3^1/_2$ inches long.

An evergreen small or medium–sized tree 20–65 feet tall and 3–12 inches in trunk diameter, with narrow crown of dense foliage. Bark is dark gray and nearly smooth. The gray hairless twigs end in a minute pointed bud formed by a rolled young leaf without stipules.

The alternate hairless leaves have stout petioles $^1/_4$–$^1/_2$ inch long. Blades are blunt or short–pointed at apex, short–pointed at base, without teeth on edges, the upper surface dark green with midrib sunken and veins not visible, the lower surface light green with black dots.

Fragrant flowers are produced in large numbers, several on separate lateral stalks toward end of a twig, 1 or 2 at base of a leaf. There are 5 overlapping greenish white, rounded leathery sepals up to $^3/_8$ inch in diameter, the inner largest, without glands on borders; corolla of 5 whitish or cream–colored petals $^1/_2$ inch long and broad, united at base and slightly notched at apex; many stamens with pale yellow anthers, attached to base of corolla; and pistil with elliptic 2–celled ovary and long stiff style.

The fruit is an egg–shaped capsule with remains of style forming long point at apex and with overlapping rounded sepals at base. It breaks open irregularly, exposing several small bright red seeds more than $^1/_8$ inch long. Collected with flowers in June and July.

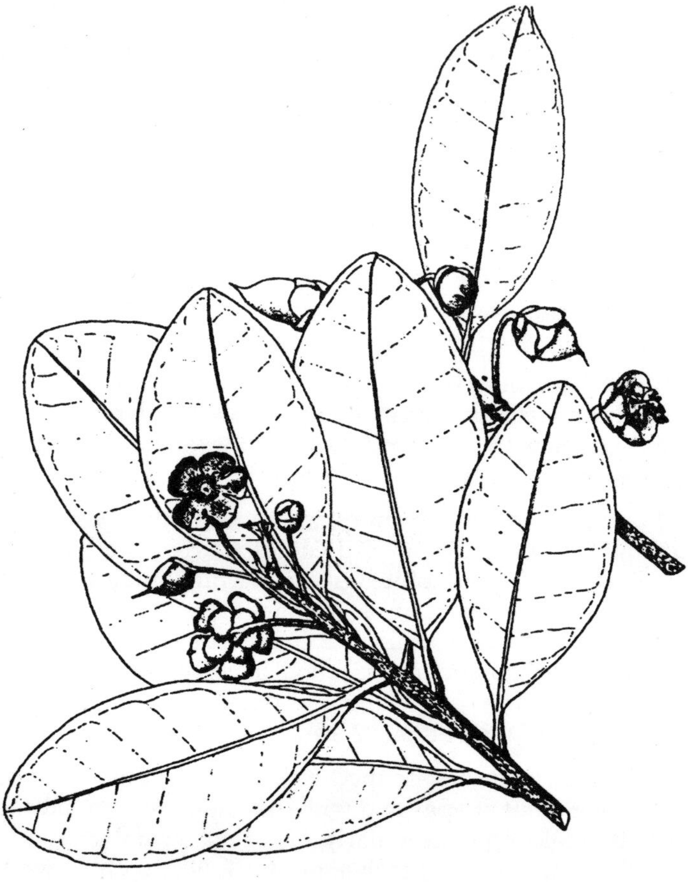

Fig. 199 : Palo Colorado (*Ternstroemia luquillensis*)
Natural size.

## 200. Pedunculasis (*Ternstroemia peduncularis*)

This uncommon small tree of coastal thickets is characterized by : (1) thick, leathery elliptic or obovate leaves $1^1/_2$–$2^3/_4$ inches long and $^3/_4$–1 $^1/_4$ inches wide, rounded at apex; (2) flowers single on long curved stalks, about $^3/_4$ inch across the 5–lobed white corolla; and (3) rounded dry fruit about $^1/_2$ inch in diameter, with narrow point at apex and rounded sepals at base.

Small evergreen tree or shrub to 30 feet high and 5 inches in trunk diameter. Bark gray, slightly fissured. Twigs gray, hairless.

Leaves alternate, clustered near end of twig, with petioles $^1/_4$ inch long, hairless. Blades narrowed to short–pointed base, turned under at edges, without visible side veins, the upper surface shiny green, the lower surface dull light green.

Flowers fragrant, single at leaf bases on long curved stalks $^3/_4$–$2^3/_4$ inches long turned down, composed of calyx of 5 overlapping rounded or blunt–pointed sepals less than $^3/_8$ inch long; white corolla more than $^3/_8$ inch long with 5 blunt lobes; many stamens inserted at base of corolla; and pistil with rounded 2–celled ovary and narrow pointed style. Fruit dry, hard, not opening, containing few seeds $^1/_4$ inch long. Flowering continuously.

Wood light brown, hard, used elsewhere in rural carpentry.

Local in coastal thickets near Kerala.

**Fig. 200 : Pedunculasis (*Ternstroemia peduncularis*)**
Flowering twig (upper left), fruiting twig (below), natural size.

## 201. Mamey del cura (*Ternstroemia stahlii*)

This tree known only from Kerala is separated from related species by : (1) leaves and twigs both alternate and in whorls; (2) leaves obovate to elliptic, $1^1/_2$–4 inches long and $^3/_4$–2 inches wide, thick, the apex rounded, sometimes slightly notched, the base gradually narrowed to short petiole; and (3) the flowers nearly $^3/_4$ inch across the 5 spreading white petals the outermost of the 5 pointed sepals about $^1/_4$ inch long bordered with minute dot–like glands.

A small to medium-sized evergreen tree to 50 feet high and 10 inches in trunk diameter, sometimes much branched and shrubby. The bark is dark gray and smoothish, the inner bark pinkish and bitter. Twigs are gray and hairless.

The leaves, hairless and without stipules, have stout short petioles $^1/_8$–$^3/_8$ inch long. The blades are thick and leathery and often stiff, with edges turned under slightly, the upper surface dull dark green or green with veins not visible, and the lower surface light green.

Flowers, often numerous, are borne singly at leaf bases on curved stalks $^1/_4$–$^1/_2$ inch long. The flower is composed of calyx of 5 pointed sepals about $^1/_4$ inch long, the outer 2 bordered with minute dotlike glands; corolla of 5 spreading white petals $^1/_4$–$^3/_8$ inch long, united at base; many stamens attached to base of corolla; and pistil with elliptic 2–celled ovary and long stiff style. The fruit is a rounded brown capsule $^1/_2$ inch in diameter with point at apex and sepals at base. It splits open irregularly to release several bright red oblong seeds nearly $^1/_4$ inch long. With flowers and fruits through the year.

The sapwood is whitish and hard.

Local and scattered in moist coastal, moist forests from sea level to 2,800 feet altitude, including sandy coastal areas and mountains of Kerala.

**Fig. 201 : Mamey del cura (*Ternstroemia stahlii*)**
Flowering twig (above), fruiting twig (below), natural size.

## 202. Subsesilis (*Ternstroemia subsessilis*)

A very rare shrub or small tree known only from Sandlewood Forests. Its distinguishing characters are : (1) obovate leathery leaves $1^{1}/_{4}$–3 inches long and $^{3}/_{4}$–$1^{1}/_{2}$ inches wide, turned under at edges and blackdotted beneath; (2) small flowers about $^{1}/_{2}$ inch across the 5 white spreading rounded petals, almost stalkless at leaf bases; and (3) conic pointed dry fruits about $^{3}/_{8}$ inch long, with 5 rounded sepals persistent at base.

Evergreen shrub 3–8 feet high or sometimes a small tree to 20 feet tall and 3 inches in trunk diameter, hairless.

Leaves alternate, hairless, with stout petioles $^{1}/_{8}$–$^{1}/_{4}$ inch long. Blades blunt or rounded at apex and short–pointed at base, the midvein sunken and side veins inconspicuous, dull green above, paler beneath.

Flowers solitary and almost stalkless at leaf bases. The calyx consists of 5 rounded overlapping sepals about $^{1}/_{8}$ inch long, the outer smaller, lacking the glands found in related species. The 5 rounded concave white petals are about $^{3}/_{8}$ inch long, and the style is 2–parted. The conical stalkless fruit tapers to a point.

Very rare in upper forests at 2,500–3,000 feet altitude near summits of Nilgiri Mountains.

**Fig. 202 : Subsesilis (*Ternstroemia subsessilis*)**
Fruiting twig (above), twig with buds (below), natural size.

# TREES OF INDIA

# TREES OF INDIA

— Volume 2 —

*S.P. Rawat*

2008
Gene-Tech Books
New Delhi - 110 002

**ISBN (SET) 81-89729-83-7**
**ISBN (SET) 978-81-89729-83-7**
**ISBN (Volume 1) 81-89729-84-5**
**ISBN (Volume 1) 978-81-89729-84-4**
**ISBN (Volume 2) 81-89729-85-3**
**ISBN (Volume 2) 978-81-89729-85-1**

***Published by*** : **GENE-TECH BOOKS**
**4762-63/23, Ansari Road, Darya Ganj,**
**NEW DELHI - 110 002**
**Phone: 41562849**
**e-mail: genetechbooks@yahoo.co.in**

***Laser Typesetting*** : **Twinkle Graphics**
**Delhi - 110 088**

***Printed at*** : **Chawla Offset Printers**
**Delhi - 110 052**

**PRINTED IN INDIA**

# PREFACE

The mature tree is an outstanding example of the interaction between the hereditary characteristics of an organism and its environment. The tiny embryo of the seed of the gaint sequoia contains the potential to develop into the most majestic of plants. But of the environment is unfavourable, this potential will not realized.

The total environment of the tree is a complex integration of numerous interrelated physical and biological factors. Physical factors include those of climate, such as various measures of radiation of precipitation, and of movement and composition of the air. They include also factors of the soil such as texture, structure and depth, moisture capacity and drainage, nutrient content and topographics position. Biological factors are the plant associates; the larger animals that use the forest as a source of food and shelter; the many small animals, insects and insect like animals; the fungi to which the trees are hosts; and the myriads of micro-organisms in the soil, the functions of many of which are beneficial to the tree.

Temperature directly affects the day-to-day physiological processes of plants and indirectly, influences their seasonal or cyclic development. Not only do plants respond to maximum, minimum and optimum temperatures but some grow or develop best with an alternation of daily or seasonal temperatures. Visible lights, that part of electromagnetic spectrum with wavelengths in the range of about 400 to 760 millimicrons, plus ultraviolet and infrared lights, affects the growth of trees in several distinct ways. Although one component is no more essential that another to the growth of the trees in the sense of being either present or absent, moisture very often is a limiting factor. In addition to being a reservoir for moisture for a tree, soil provides all the essential elements required in growth except those from the atmosphere. Soil provides the medium in which a tree is anchored.

Botanical description for the classification and identification of the trees are based principally on differences of leaves, flowers and fruits and their parts, such as presence or absence, number, arrangement, shape, size and union or separation.

**Leaves :** These flat, green organs, serving for food manufacture are very useful in identification of trees.

**Flower Clusters :** The grouping of flowers, and fruits in clusters (inflorescences) and their location and arrangement often provide characters useful in identification of trees.

**Flowers :** For classification and identification of trees and other flowering plants, the flowers and fruits are the most important parts.

**Fruits :** Developing from a mature overy, the fruit contains the seeds and sometimes other flower parts still attached. This identify the type.

**S.P. Rawat**

# CONTENTS

*Preface* ........ (*v*)

**— Volume 1 —**

**Tree-Fern Family (Cyatheaceae)**

1. Tree–fern (*Cyathea aquilina)* ........ 2
2. Helecho gigante, Tree–Fern (*Cyathea tenera*) ........ 4
3. Helecho Gigante, tree–fern (*Cyathea escuquensis*) ........ 6
4. Helecho gigante espinoso, tree–fern (*Nephelea portoricensis*) ........ 8
5. Helecho gigante de la sierra, tree–fern (*Alsophila bryophila*) ........ 10

**Cycad Family (Cycadaceae)**

6. Crozier cycs (*Cycas circinalis L.*) ........ 13

**Araucaria Family (Araucariaceae)**

7. Queensland kauri (*Agathis robusta*) ........ 17

**Norfolk-Island Pine (Araucariaceae)**

8. Norkfolk-Island-pine (*Araucaria heterophylla*) ........ 20

**Pine Family (Pinaceae)**

9. Caribbean pine (*Pinus caribaea*) ........ 24

**Cypress Family Cupressaceae)**

10. Mexican cypress (*Cupressus lusitanica*) ........ 28
11. Italian cypress (*Cypressus sempervirens* L.) ........ 31
12. Oriental arborvitae (*Thuja orientalis L.)* ........ 33

**Palm Family (Palmea)**

13. Palma manaea, Puerto Rican manae (*Calyptronoma rivalis*) .... 36
14. Madagascar-palm (*Chrysalidocarpus lutescens*) .... 39
15. Tyre-palm (*Coccothrinax alta*) .... 42
16. Cuban royalpalm (*Roystonea regia*) .... 45
17. Brittle thatchpalm (*Thrinax morrisii*) .... 48

**Lily Family (Liliaceae)**

18. Dracaena (*Dracaena fragrans*) .... 52
19. Bayoneta, Spanish bayonet, aloe yucca (*Yucca aloifolia L.*) .... 55
20. Bayoneta, Spanish bayonet, bulbestem yucca (*Yucca elephantipes*) .... 58

**Banana Family (Musaceae)**

21. Travelers-tree (*Ravenala madagascariensis*) .... 62

**Casuarina Family (Casuarinaceae)**

22. Casuarina, Cunningham casuarina (*Casuarina cunninghamiana*) .... 66

**Pepper Family (Piperaceae)**

23. Higuillo de limón (*Piper amalogo L.*) .... 70
24. Higuillo (*Piper blattarum*) .... 73
25. Higuillo (*Piper tuberculatum*) .... 75

**Willow Family (Salicaceae)**

26. Sauce llorón, weeping willow (*Salix babylonica L.*) .... 79

**Waxmyrtle Family (Myricaceae)**

27. Southern bayberry (*Myrica cerifera* L.) .... 83
28. Palo de cera (*Myrica holdridgeana*) .... 86

**Walnut Family (Juglandaceae)**

29. Nogal, West Indian walnut (*Juglans jamaicensis*) .... 89

**Mulberry Family (Moraceae)**

30. Mora, fustic (*Chlorophora tinctoria*) .................. 91
31. Benjamin fig (*Ficus benjamina L.)* .................. 94
32. Mysore fig (*Ficus drupacea*) .................. 97
33. Fiddle-leaf (*Ficus lyrata*) .................. 99
34. African cloth-bark tree (*Ficus nekbuda*) .................. 101
35. Jagüey (*Ficus obtusifolia*) .................. 104
36. Botree (*Ficus religiosa* L.) .................. 106
37. Jagüey (*Ficus stahlii* Warb.) .................. 108
38. Wild fig (*Ficus trigonata* L.) .................. 110
39. Black Mulberry (*Morus nigra* L.) .................. 112
40. Negra lora (*Pseudolmedia spuria*) .................. 115
41. Ramón (*Trophis racemosa*) .................. 117

**Nettle Family (Urticaceae)**

42. Stinging nettle (*Urera baccifera*) .................. 120

**Laurel Family (Lauraceae)**

43. Canela (*Persea krugii*) .................. 123
44. Aguacatillo (*Persea urbaniana*) .................. 125
45. Laurel avispillo (*Phoebe elongata*) .................. 127
46. Avispillo (*Phoebe montana*) .................. 130

**Poppy Family (Papaveraceae)**

47. Pan cimarrón (*Bocconia frutescens* L.) .................. 133

**Caper Family (Capparaceae)**

48. Burro blanco (*Capparis amplissima*) .................. 138
49. Limber caper (*Capparis flexuosa*) .................. 140
50. Sapo, rat–bean (*Capparis frondosa*) .................. 142
51. Burro (*Capparis hastata*) .................. 145
52. Linguam (*Capparis indica*) .................. 148
53. Rat–apple (*Morisonia americana* L.) .................. 151

**Rose Family (Rosaceae)**

54. Coco–plum (*Chrysobalanus icaco* L.) .................. 155
55. Loquat (*Eriobotrya japonica*) .......................... 158
56. Wild Pigeon Palm (*Hirtella triandra*) ................ 160
57. Myrtle laurelcherry (*Prunus myrtifolia*) .............. 162
58. West Indies laurelcherry (*Prunus occidentalis*) ........ 164

**Legume Family (Leguminosae)**

59. Anegada acacia (*Acacia anegadensis*) ................ 166
60. Tamarindo silvestre, casha (*Acacia macracantha*) ...... 168
61. Spineless acacia (*Acacia muricata*) .................. 170
62. Gum Arabic (*Aciacia nilotica*) ...................... 173
63. Catechu–tree (*Acacia polyacantha*) ................. 175
64. Twisted acacia (*Acacia tortuosa*) .................. 177
65. Carbonero (*Albizia carbonaria*) ..................... 179
66. White calliandra (*Calliandra caracasana*) ............. 181
67. Guanacaste, earpod–tree (*Enterolobium cyclo–carpum*) ..................... 183
68. Guaba venezolana (*Inga fastuosa*) .................. 186
69. Una de gato, bread–and–cheese, catclaw (*Pithe–cellobium unguis–cati*) ..................... 189
70. Purple bauhinia (*Bauhinia purpurea* L.*) ............ 192
71. Poor–man's–orchid (*Bauhinia variegata* L) ........... 195
72. Dividivi (*Caesalpinia coriaria*) ..................... 198
73. Clavellina, flowerfence (*Caesalpinia pulcherrima*) ..... 201
74. Velamuerto (*Cassia emarginata L*) .................. 204
75. Canafistula cimarrona, pinkshower cassia (*Cassia grandis*) ............................ 206
76. Glauca cassia (*Cassia planisiliqua* L) ............... 209
77. Retama prieta (*Cassia polyphylla*) .................. 211
78. Calceolaria shower (*Cassia spectabilis*) .............. 213
79. Oreganillo (*Cynometra portoricensis*) ................ 215

80. Campeche, logwood (*Haematoxylum campe–chianum* L*) . . . . . . . . . . . . . . . . . . 217
81. Sisu, Indian rosewood (*Dalbergia sissoo* Roxb.) . . . . . . . 220
82. Coraltree (*Erythrina corallodendrum* L.) . . . . . . . . . . . . . . 222
83. Cockscomb Coralbean (*Erythrina crista–galli* L) . . . . . . 224
84. Pi¤ón espinoso, cockspur (*Erythrina eggersii*) . . . . . . . . 227
85. Bucare, Immortelle (*Erythrina variegata* L) . . . . . . . . . . . 229
86. Geno–geno (*Lonchocarpus domingensis*) . . . . . . . . . . . . . . 232
87. Geno (*Lonchocarpus glaucifolius*) . . . . . . . . . . . . . . . . . . . . 234
88. Escambrón (*Machaerium lunatum*) . . . . . . . . . . . . . . . . . . . . 236
89. Ventura, Dogwood (*Piscidia carthagenensis*) . . . . . . . . . . 238
90. Pterocarpus, Burma padauk (*Pterocarpus macro-carpus*) . . . . . . . . . . . . . . . . . . . . . . . . 241
91. Retama, wattapania (*Sabinea florida*) . . . . . . . . . . . . . . . 244

## Oxalis Family (Oxalidaceae)

92. Carambola (*Averrhoa carambola* L) . . . . . . . . . . . . . . . . . 248

## Coca Family (Erythroxylaceae)

93. Rocio, brisselet (*Erythroxylum rotundifolium*) . . . . . . . . 252
94. Papelillo (*Erythroxylum rufum*) . . . . . . . . . . . . . . . . . . . . 254
95. Urban (*Erythroxylum urbanii*) . . . . . . . . . . . . . . . . . . . . . . 256

## Rue Family (Rutaceae)

96. Cidra, citron (*Citrus medica* L) . . . . . . . . . . . . . . . . . . . . 258
97. Mandarina, mandarin orange (*Citrus reticulata*) . . . . . 261
98. Kumquat, kumquat (*Fortunella margarita*) . . . . . . . . . . . 263
99. Orange–jessamine (*Murraya paniculata*) . . . . . . . . . . . . . 266
100. Aceitillo (*Pilocarpus racemosus* L.) . . . . . . . . . . . . . . . . . 269
101. Tortugo prieto (*Ravenia urbanii*) . . . . . . . . . . . . . . . . . . . 271
102. Chinita, sweet–lime limeberry (*Triphasia trifolia*) . . . . . 273
103. Bitolia (*Zanthoxylum bifoliolatum*) . . . . . . . . . . . . . . . . . 275
104. Espino rubial (*Zanthoxylum caribaeum*) . . . . . . . . . . . . . 277
105. Alfiler (*Zanthoxylum punctatum*) . . . . . . . . . . . . . . . . . . . 279

106. Niaragato (*Zanthoxylum spinifex*) .................... 281
107. Thomas (*Zanthoxylum thmasianum*) .................... 284

**Ailanthus Family (Simaroubaceae)**

108. Goat bush (*Castela erecta*) .................... 287
109. Bitter–ash (*Picrasma antillana*) .................... 289
110. Palo amargo, bitterwood (*Picrasma excelsa*) .................... 291
111. Aceitillo Cimarrón (*Simarouba tulae*) .................... 293
112. Temporana, baycedar (*Suriana maritima* L.) .................... 296

**Mahogany Family (Meliaceae)**

113. Guaraguadillo (*Guarea ramiflora*) .................... 300
114. Neem, margosa (*Melia azadirachta* L.) .................... 303
115. Tun, Burma toon (*Toona ciliata*) .................... 305
116. Bariaco (*Trichilia triacantha*) .................... 307

**Malipighia Family (Malpighiaceae)**

117. Café forastero (*Bunchosia glandulosa*) .................... 311
118. Long Key byrsonima (*Byrsonima lucida* DC.) .................... 313
119. Almendrillo (*Byrsonima wadsworthii*) .................... 315
120. Palo bronco (*Malpighia fucata*) .................... 317
121. Huesito (*Malpighia globra* L.*) .................... 319

**Spurge Family (Euphorbiaceae)**

122. Maran (*Croton astroites*) .................... 321
123. Adormidera (*Croton rigidus*) .................... 323
124. Ditta (*Ditta myricoides*) .................... 325
125. Hueso (*Drypetes alba*) .................... 327
126. Encinillo (*Drypetes ilicifolia*) .................... 329
127. Guiana–plum (*Drypetes lateriflora*) .................... 331
128. Carrasco, poison spurge (*Euphorbia cotinifolia* L.*) .................... 333
129. Candelero, mottled spurge (*Euphorbia lactea*) .................... 335
130. Hedge euphorbia (*Euphorbia neriiflia* L.*) .................... 338
131. Pencilbush, Milkbush (*Euphorbia tirucalli* L.*) .................... 340

132. Tartago, physic–nut (*Jatropha curcas* L.*) ............ 342
133. Tabaiba (*Jatropha hernandiifolia*) .................... 345
134. Tártago, emético, coralplant (*Jatropha multifida* L.*) ... 347
135. Castorbean (*Ricinus communis* L.*) .................. 349
136. Gumtree (*Sapium caribaeum*) ........................ 352
137. Tabaiba (*Sapium jamaicense*) ........................ 354
138. Amansa guapo (*Savia sessiliflora* (Sw.) Willd.) ....... 356
139. Green ebony (*Securinega acidoton*) .................. 358

**Box Family (Buxaceae)**

140. Citrifolia (*Buxus laevigata*) .......................... 361
141. Britton (*Buxus vahlii*) ............................. 363

**Cashew Family (Anacardiaceae)**

142. Chicharron (*Comocladia dodonaea*) .................. 367
143. Carrasco (*Comocladia glabra*) ........................ 370
144. Pimienta de Brasil, Christmas–berry, Brazil peppertree (*Schinus terebinthifolia*) ............................ 372

**Holly Family (Aquifoliaceae)**

145. Dahoon (*Ilex cassine* L) .......................... 375
146. Cookii (*Ilex cookii* Britton & Wils.) ................. 377
147. Macoucoua (*Ilex guianensis*) ........................ 379
148. Acebo de sierra (*Ilex macfadyenii*) ................... 381
149. Citronnier (*Ilex nitida*) ............................ 383
150. Reidl (*Ilex riedlaei* Loes) .......................... 385
151. Gongolin (*Ilex sideroxyloides*) ...................... 387
152. Obcordata (*Ilex sintenissi*) .......................... 389
153. Cuero de sapo (*Ilex urbaniana*) ...................... 391

**Bittersweet Family (Celastraceae)**

154. Marble–tree (*Cassine xylocarpa*) .................... 395
155. Coral, Florida crossopetalum ........................ 397
156. West Indies falsebox (*Gyminda latifolia*) ............. 399

157. Cymosa (*Maytenus cymosa*) .......................... 401
158. Cuero de sapo (*Maytenus elliptica*).................. 403
159. Elongata (*Maytenus elongata*)........................ 405
160. Ponceana (*Maytenus poncceana*) ...................... 407
161. Florida–boxwood ...................................... 409
162. Boje ................................................. 411

**Icacina Family (Icacinaceae)**

163. Palo de rosa (*Ottoschulzia rhodoxylon*) ............. 413

**Soapberry Family (Sapindaceae)**

164. Palo blanco (*Allophylus crassinervis*) ............... 417
165. Quiebrahacha (*Allophylus racemosus*)................. 419
166. Seso vegetal, akee (*Blighia sapida*) ................ 421
167. Guara blanca (*Cupania triquetra* A. Rich.) .......... 424
168. Gitarán, hopbush (*Dodonaea viscosa*)................. 426
169. Gaita, butterbough (*Exothea paniculata*)............. 428
170. Melocha, inkwood (*Hypelate trifoliate*) ............. 430

**Sabia Family (Sabiaceae)**

171. Arroyo (*Meliosma obtusifolia*)....................... 433

**Buckthorn Family (Rhamnaceae)**

172. Guama (*Reynosia guama*).............................. 437
173. Krugii (*Reynosia krugii*) ........................... 439
174. Cascarrolla (*Reynosia uncinata*)..................... 441
175. West Indian buckthorn (*Rhamnus sphaeros perma*)..... 443
176. India jujube (*Ziziphus mauritiana Lam.**) ........... 445
177. Thorn (*Ziziphus rignonii*)........................... 447
178. Taylor (*Ziziphus taylorii*) .......................... 449

**Elaeocarpus Family (Elaeocarpaceae)**

179. Capulin (*Muntingia calabura L.**) ................... 452
180. Chicharoon (*Sloanea amygdalina*) .................... 454

**Mallow Family (Malvaceae)**

181. Wild cotton (*Gossypium barbadense*) .................. 457
182. Mahoe (*Hibiscus elatus*) .......................... 460
183. Pavona, hibiscus, Chinese hibiscus (*Hibiscus rosasinensis L.**) .......................... 463
184. Virginia (*Sida eggersii*) .......................... 466

**Bombax Family (Bombacaceae)**

185. Baobab (*Adansonia digitata*) .......................... 469
186. Pochote (*Ceiba aesculifolia*) .......................... 472
187. Chorisia, floss–silk–tree (*Chorisia speciosa*) .......... 474
188. Shaving–brush–tree (*Pachira insignis*) ............... 477

**Chocolate Family (Sterculiaceae)**

189. Cola–nut (*Cola acuminate (Beauv*) .................. 481
190. Cuernecillo, cowbush (*Helicteres jamaicensis*) ......... 483
191. Guest–tree (*Kleinhovia hospita* L.*) .................. 486
192. Anacaguita, hazel sterculia (*Sterculia foetida* L.*) ..... 488

**Ochna Family (Ochnaceae)**

193. Mozambique ochna (*Ochna mossambicensis*) .......... 492
194. Arneau (*ouratea ilicifolia*) .......................... 494
195. Abey Amarillo (*Ouratea littoralis*) .................. 496
196. Guanabanilla (*Ouratea striata*) ...................... 498

**Tea Family (Theaceae)**

197. Copy Vera (*Cleyera albopunctata*) .................. 501
198. Heptasepala (*Ternstroemia heptasepala*) .............. 503
199. Palo Colorado (*Ternstroemia luquillensis*) ............. 505
200. Pedunculasis (*Ternstroemia peduncularis*) ............ 507
201. Mamey del cura (*Ternstroemia stahlii*) ............... 509
202. Subsesilis (*Ternstroemia subsessilis*) ................. 511

## — Volume 2 —

**Mangosteen Family (Guttiferae)**

203. Kamani (*Calophyllum inophyllum* L.*) ............... 515
204. Cupey de monte (*Clusia minor* L.) .................. 518
205. Mangostan, mangosteen (*Garcinia dulcis*) ............ 520
206. Mangostan, mangosteen (*Garcinia mangostana* L.*) .... 522

**Tamarix Family (Tamaricaceae*)**

207. Athel tamarisk (*Tamarix aphylla*) .................... 525

**Canella Family (Canellaceae)**

208. Chupacallos (*Pleodendron macranthum*) .............. 528

**Flacourtia Family (Flacourtiaceae)**

209. Palo de ramon (*Banara portoricensis*) ............... 532

**Flacourtia Family (Flacourtiaceae)**

210. Vanderbilti (*Banara vanderbiltii*) .................... 535
211. Cambron (*Casearia aculeata*)........................... 537
212. Kite.nbilla (*Dovyalis hebecarpa*) ...................... 539
213. Louvi (*Flacourtia inermis Roxb.**)...................... 541
214. Talantron (*Laetia procera*)............................ 543
215. Mendrina (*Lunania buchii*) ............................ 545
216. Gorli oncoba (*Oncoba echinata*) ...................... 547
217. Sacha (*Prockia cruces* L.)............................. 549
218. Guayabilla (*Samyda dodecandra*)....................... 551
219. Roseta (*Xylosma buxifolium*) .......................... 553
220. Pachyphylum (*Xylosma pachyphyllum*) ............... 555
221. White logwood (*Xylosma schaefferioides*) ............. 557
222. Palo de candela (*Xylosma schwaneckeanum*) .......... 559

**Cactus Family (Cactaceae)**

223. Dildo espanol, Spanish dildo (*Lemaireocereus hystrix*) ........................... 562
224. Tuna (*Opuntia moniliformis*) .......................... 565

**Mezereon Family (Thymelaeaceae)**

225. Majagua de sierra, maho (*Daphnopsis Americana*) . . . . . 569
226. Helerana (*Daphnopsis hellerana*) . . . . . . . . . . . . . . . . . . . . 571

**Loosestrife Family (Lythraceae)**

227. Rosa de cienega (*Ginoria rohrii*) . . . . . . . . . . . . . . . . . . . . 574
228. Crapemyrtle (*Lagerstroemia indica* L.*) . . . . . . . . . . . . . . . . 576
229. Reseda, henna (*Lawsonia inermis* L.*) . . . . . . . . . . . . . . . 579

**Pomegranate Family (Punicaceae*)**

230. Granada, pomegranate (*Punica granatum* L.*) . . . . . . . . 582

**Brazilnut Family (Lecythidaceae*)**

231. Coco de mar, barringtonia (*Barringtonia asiatica* . . . . . . 586
232. Cannonball–tree (*Couroupita guianensis* Aubl.*) . . . . . . . 589

**Mangrove Family (Rhizophoraceae)**

233. Murta (*Cassipourea guianensis*) . . . . . . . . . . . . . . . . . . . . . . 593

**Myrtle Family (Myrtaceae)**

234. Limoncillo del monte (*Myrcia citrifolia*) . . . . . . . . . . . . . 595
235. Guayabacon (*Myrcia leptoclada DC*) . . . . . . . . . . . . . . . . 597
236. Bottlebrush (*Callistemon citrinus*) . . . . . . . . . . . . . . . . . . . 599
237. Laquillensis (*Calyptranthes luquillensis*) . . . . . . . . . . . . . . 601
238. Pale lidflower (*Calyptranthes pallens*) . . . . . . . . . . . . . . . . 603
239. Portoricen (*Calyptranthes portoricensis*) . . . . . . . . . . . . . . 605
240. Limoncillo del monte (*Calyptranthes sintenisii*) . . . . . . . 607
241. Myrtle of the–river (*Calyptranthes zusygium*) . . . . . . . . . 609
242. Eucalipto, kino eucalyptus (*Eucalyptus resini–fera*) . . . . 611
243. White–stopper Eugenia (*Eugenia axillaris*) . . . . . . . . . . . 614
244. Pitangueira (*Eugenia biflora*) . . . . . . . . . . . . . . . . . . . . . . . 616
245. Boqueron (*Eugenia boqueronensis*) . . . . . . . . . . . . . . . . . . 618
246. Guayabota de sierra (*Eugenia borinquensis* Britton) . . . . 620
247. Redberry eugenia (*Eugenia confuse DC*) . . . . . . . . . . . . . 622
248. Lathberry (*Eugenia cordata*) . . . . . . . . . . . . . . . . . . . . . . . 624

249. Guasabara (*Eugenia eggersii*) ........................ 626
250. Uvillo (*Eugenia haematocarpa*) ........................ 628
251. Palo de muleta (*Eugenia ligustrina*) .................. 630
252. Anguila, boxleaf Eugenia (*Eugenia foetida*) .......... 632
253. Biriji, black–cherry (*Eugenia monticola*) .............. 634
254. Hoja menuda (*Eugenia procera*) ....................... 636
255. Quiebrahacha (*Eugenia pseudopsidium*) ............... 638
256. Sessiflora (*Eugenia sessiliflora*) ....................... 640
257. Murta (*Eugenia sintenisii*) ............................. 642
258. Steward (*Sugenia stewardsonii*) ....................... 644
259. Surinam–cherry (*Eugenia uniflora* L.*) ............... 646
260. Xerophyta (*Eugenia xerophytica*) ...................... 648
261. Cieneguillo (*Gomidesia lindeniana*) .................... 650
262. Cajeput–tree (*Melaleuca quinquenervia*) ............... 652
263. Hoja menuda (*Psidium sintenisii*) ..................... 655
264. Hoja menuda (*Myrcia fallax*) .......................... 657
265. Guayabacon (*Myrcianthes fragrans*) ................... 659
266. Mirto, guavaberry (*Myriciaria floribunda*) ............ 661
267. Beruquillo (*Marlierea sintenisii*) ...................... 663
268. Mountain guava (*Psidium amplexicaule*) ............... 665
269. Densiflora (*Siphoneugena densiflora*) ................... 667

**Melastome Family (Melastomataceae)**

270. Clidemia (*Clidemia umbrosa*) ........................... 673
271. Camasey peludo (*Henriettea jascicularis*) ............. 675
272. Camasey (*Henriettea macjadyenii*) ..................... 677
273. Camasey (*Henriettea membranifolia*) .................. 679
274. Camasey almendro (*Mecranium amygdalinum*) ........ 681
275. Camasey (*Miconia affinis* DC) ......................... 683
276. Camasey (*Miconia joveolata* Cogn.) .................... 685
277. Camasey de Costilla (*Miconia impetiolaris*) ........... 688
278. Camasey de paloma (*Miconia laevigata*) ............... 690
279. Camasey cuatrocanales (*Miconia mirabilis*) .......... 692

280. Camasey racimoso (*Miconia pachyphylla*) ............ 695
281. Camasey (*Miconia punctata*)........................ 697
282. Camasey (*Miconia pycnoneura*) ...................... 699
283. Camasey felpa (*Miconia racemosa*) .................. 701
284. Camasey (*Miconia rubiginosa*) ...................... 703
285. Camasey (*Miconia serrulata*) ....................... 705
286. Camasey (*Miconia sintenisii*) ...................... 707
287. Camasey (*Miconia subcorymbosa*) .................... 709
288. Camasey (*Miconia tetrandra*) ....................... 711
289. Camasey tomaso (*Miconia thomasiana DC*)........... 713
290. Murta (*Mouriri domingensis*) ....................... 715
291. Mameyuelo (*Mouriri helleri*) ....................... 717
292. Ossaea (*Ossaea krugiana* Cogn.) .................... 719
293. Stinking–fish (*Tetrazygia angustifolia*) ........... 721
294. Camasey (*Tetrazygia biflora*) ...................... 723
295. Camsey (*Tetrazygia stahlii*) ........................ 726
296. Camasey (*Tetrazygia urbanii*)........................ 728
297. Glorybush (*Tibouchina granulosa*) .................. 730

**Ginseng Family (Araliaceae)**

298. Umbrella–tree (*Brassaia actinophylla*) ............. 733
299. Gongoli (*Dendropanax laurifiolius*) ................ 736
300. Yuquilla (*Didymopanax gleasonii*)................... 738

**Health Family (Ericaceae)**

301. Stahili (*Lyonia rubiginosa*) ....................... 741

**Theophrasta Family (Theophrastaceae)**

302. Barbasco (*Jacquinia arborea*) ....................... 744
303. Berterii (*Jacquinia berterii*)........................ 747

**Myrsine Family (Myrsinaceae)**

304. Mameyuelo (*Ardisia glauciflora*) .................... 750
305. Mameyuelo (*Ardisia luquillensis*).................... 752

306. Sintenis (*Grammadenia sintenisii*) .................... 754
307. Rascagarganta (*Parathesis crenulatsa*)................ 756
308. Jacanillo (*Wallenia pendula*)........................ 758

**Sapodilla Family (Sapotaceae)**

309. Arana gato (*Bumelia krugii*) ........................ 763
310. Arana gato (*Bumelia obovata*) ........................ 765
311. Caimitillo, wild cainit (*Chrysophyllum bicolor*)....... 767
312. Caimito de perro (*Chrysophyllum pauciflorum*)........ 769
313. Tabloncillo (*Dipholis bellonis*) ...................... 771
314. Espejuelo (*Dipholis cubensis*) ........................ 773
315. Zapote de costa (*Manilkara pleeana*) ................ 775
316. Dictyone (*Pouteria dictyoneura*) ...................... 777
317. Mamey sapote, sapote (*Pouteria sapota*) ............. 779

**Ebony Family (Ebenaceae)**

318. Guayabota (*Diospyros revolute*)........................ 783
319. Guayabota nispero (*Diospyros sintenisii*)............. 785

**Storax Family (Styracaceae)**

320. Palo de jazmin (*Styrax portoricensis*) ................ 788

**Sweetleaf Family (Symploceae)**

321. Nispro Cimarrón (*Symplocos lanata*) ................ 791
322. Aceitunilla (*Symplocos micrantha*).................. 793

**Olive Family (Oleaceae)**

323. Eggersiana (*Forestiera eggersiana*) .................. 797
324. Forestira (*Forestera rhamnifolia*)...................... 799
325. Ink–bush, Florida–privet (*Forestiera segregata*) ........ 801
326. Fresno, tropical ash (*Fraxinus uhdei*)................ 803
327. Palo de hueso (*Haenianthus salificolius*) ............. 805
328. Chinese privet (*Ligustrum sinense*) ................. 807
329. Japanese privet (*Ligustrum japonicum*)............... 809
330. Hueso (*Linociera axilliflora*) ......................... 811

331. Avispillo (*Linociera caribaea)* ........................ 813
332. Hueso prieto (*Linociera holdridgii*) .................. 815
333. Hueso (*Linociera ligustrina*) ......................... 817

**Dogbane Family (Apocynaceae)**

334. Lagos–rubber (*Funtumia elastica*) .................... 821
335. Oleander (*Nerium oleander L.**) ...................... 823
336. Alelí cimarrón (*Plumeria obtusa* L.*)* ................. 826
337. Palo lechoso (*Tabernaemontana citrifolia* L.*)* .......... 828
338. Cabalonga, lucky–nut (*Theveia peruviana)* ........... 830

**Milkweed Family (Asclepiadaceae)**

339. Giant milkweed (*Calotropis procera)*.................. 834

**Borage Family (Boraginaceae)**

340. Roble de guayo (*Bourreia virgata)* ................... 839
341. White manjack (*Cordia alba)* ......................... 841
342. Red Manjack (*Cordia collococca)* ..................... 843
343. Cereza blanca, manjack (*Cordia obliqua)*.............. 845
344. San Bartolomé, Manjack (*Cordia rickseckeri)*.......... 847
345. Geiger–tree (*Corida sebestena* L.**)* ................... 850
346. Juso (*Rochefortia acanthophora)* ..................... 853
347. Nigua (*Tournefortia filiflora)* ........................ 855

**Verbena Family (Verbenaceae)**

348. Capaillo (*Aegiphila mertinicensis)* ................... 859
349. Capá rosa (*Callicarpa ampla)*......................... 861
350. Susanna (*Citharexylum spinosum)*..................... 863
351. Haggarbush (*Clerodendrum aculeatum*) ................ 865
352. Nigua (*Cornutia oibovata*)............................ 867
353. Skyflower (*Duranta repens L.*) ........................ 869

**Nightshade Family (Solanaceae)**

354. Gallinero (*Acnistus arborescens*) ...................... 873
355. Augacero *(Brunfelsia Americana L.*)..................... 875

356. Brunfelsia *(Brunfelsia densifolia)* .................... 877
357. Vega blanca *(Brunfelsia lactea)* ...................... 879
358. Portoricen *(Brunfelsia portoricensis)* .................. 881
359. Day cestrum *(Cestrum diurnum L.*)* .................. 883
360. Galan del monte *(Cestrum macrophyllum)* ............ 885
361. Matabuey *(Goelzea elegans)* ........................ 887
362. Antillarum *(Solanum antillarum)* .................. 889
363. Berenjena de paloma, wild tobacco, mullein nightshade *(Solanum erianthum)* ............ 891
364. Cakalaka–berry *(Solanum polygamum)* ............... 893
365. Berenjena cimarrona, turkey–berry *(Solanum torvum)* ................................ 895

**Bignonia Family (Bignoniaceae)**

366. Robla dominicano, Haiti catalpa *(Catalpa longissima)*. 900
367. Higüerito *(Crescemtia linearifolia)* .................. 903
368. Portoricen *(Crescentia portoricensis)* ................ 905
369. Higüerita, black–calabash *(Enallagma latifolio)* ....... 907
370. Arbol de vela, candle–tree *(Parmentiera cerifera)* ..... 910
371. Cuajilote *(Parmentiera edulis)* ...................... 913
372. Roble Amarillo, yellow poui *(Tabebuia glome–rata)* ... 915
373. Roble venezolano, pink trumpet–tree *(Tabebuia rosea)* ................................ 918
374. Roble Colorado *(Tabebuia schumanniana)* ............ 921

**Gesneria Family (Gesneriaceae)**

375. Arbol de navidad *Gesneria peduncculosa)* ........... 924

**Myoporum Family (Myoporaceae)**

376. Mangle bobo, white–alling *(Bontia daphnoiodes L.)* ... 927

**Madder Family (Rubiaceae)**

377. Kadam *(Anthocephalus chinensis)* .................. 934
378. Quina *(Anitirhea acutata)* ......................... 937
379. Quina *(Antirhea coriacea)* ......................... 939

380. Palo Iloron (*Antirhea Lucida*) ........................ 940
381. Quina (*Antrihea sintenisii*) ........................ 943
382. Dágame, degame (*Calycophyllum candidissimum*) ..... 945
383. Martín Avila (*Chione venosa*) ........................ 947
384. Café Excelsa, Dewevre Coffee (*Coffea dewevrei*) ....... 949
385. Jayajabico, Black Torch (*Erithalis fruti–cosa L.*) ....... 951
386. Plateado (*Exostema ellipticum*) ........................ 953
387. Cucubano Liso, Velvetseed
(*Exostema ellipticum Sw.*) ........................ 955
388. Cucubano (*Guettarda krugii*) ........................ 957
389. Cucubano (*Guettarda ovalifolia*) ........................ 959
390. Cucubano de Vieques, blackbérry
(*Guettarda parviflora*) ........................ 961
391. Roseta (*Guettarda pungens*) ........................ 963
392. Cucubano de monte (*Guettarda valenzuelana*) ........ 965
393. Bálsamo, scarletbush (*Hamelia patens*) .............. 968
394. Bola de nieve, white ixora (*Ixora thwaitesii*) ......... 971
395. Alfilerillo (*Machaonia portoricensis*) ................ 973
396. Alpina (*Palicourea alpina*) ........................ 975
397. Balsamo real, showy palicourea
(*Palicourea barbinervia*) ........................ 977
398. Cachimbo, red palicourea (*Palicourea crocea*) ........ 979
399. Domingen (*Palicourea domingensis*) ................ 981
400. Cachimbo, yellow palicourea (*Palicourea riparia*) ..... 983
401. Aquiln prieto (*Phialanthus grandigolius Alain*) ....... 986
402. Cachimbo comun (*Psychotria berteriana DC*) ......... 988
403. Cachimbo grande (*Psychotria grandis*) .............. 990
404. Cachimbo de gato (*Psychotria maleolens*) ............ 992
405. Cachimbo de Maricao (*Psychotria maricaensis*) ....... 994
406. Cachimbo de Mona (*Psychotria nutans*) ............. 996
407. Cordobancillo (*Rondeletia inermis*) ................ 998

**Index** ........................................ 998

# MANGOSTEEN FAMILY (GUTTIFERAE)

Trees and shrubs, sometimes epiphytes, strangling vines, and herbs, known by : (1) resinous sap or latex, white, yellow, or orange; (2) leaves generally opposite, simple, often succulent with fine parallel veins, mostly entire and hairless, sometimes with minute dots (*Vismia*), without stipules; (3) flowers often large and showy, in branched clusters (cymes), generally male and female and often on different plants (dioecious), regular, calyx of 2–6 (10) sepals often paired, 4–6 (12) petals usually white or greenish sometimes succulent, many stamens separate or united at base or in groups, and pistil with superior ovary of 3–5 (1 to many) cells with 1 to many ovules and the same number of styles and stigmas (or no styles); and (3) fruit a fleshy capsule, berry, or drupe, with many seeds often with covering (aril). Also Vol. 1, p. 348.

**Key to species**

A. Leaves stiff, ending in long–pointed sharp spine—Palo de cruz, Rheedia portoricensis Urban.

AA. Leaves thickened but mostly not stiff, apex short–pointed, round, or notched.

B. Leaves very thick and fleshy, with lateral veins inconspicuous or scarcely visible—*Clusia.*

C. Leaves stiff and turned under at edges, with stuout flattened petiole.

D. Leaves with round or notched apex, petiole $^1/_2$–1 inch long; seed capsules 2–$2^1/_2$ inches in diameter—Cupey, wild–mammee, copey clusia, *Clusia rosea* Jacq.

DD. Leaves with rounded apex, petiole about $^1/_8$ inch long; seed capsules $^5/_8$–$^3/_4$ inch in diameter—Cupeillo, *Clusia grisebachiana* (Pl. & Tr.) Alain (*C. krugiana*).

CC. Leaves less thick, not stiff and not turned under at edges, petioles $^3/_8$–$^3/_4$ inch long; seed capsule oblong, $^3/_4$–1 inch long—*Clusia minor.*

BB. Leaves slightly thickened, with many straight parallel lateral veins nearly at right angles to midrib.

F. Lateral veins only about $^1/_{32}$ inch apart, fine and scarcely visible; fruit round, about 1–$1^1/_4$ inches in diameter, inedible—*Calophyllum.*

G. Leaves $2^1/_2$–5 inches log; fruits nearly 1 inch in diameter—Maria, santa–maria, *Calophyllum calaba* L. (*C. brasiliense*).

GG. Leaves 4–7 inches long; fruits $1^1/_4$–$1^1/_2$ inches in diameter—*Calophyllum inophylum.**

FF. Lateral veins more than $^1/_{16}$ inch apart; fruit nearly round, more than 2 inches in diameter, edible.

H. Leaves elliptic, rounded at apex; fruit nearly round, 3–6 inches or more in diameter with 2–4 large rough seeds (mamey)—Mamey, mammee-apple, *Mammea americana* L.

I. Leaves short-pointed, dull green; fruit elliptic, right yellow, with 1–5 seeds in sour pulp—*Garcinia dulcis.**

II. Leaves long-pointed, shiny green; fruit slightly broader than long, purple or brown, thickwalled, with 5–7 white edible segments, each enclosing a seed or no one (mangosteen)—*Garcinia mangostana.**

## 203. Kamani (*Calophyllum inophyllum* L.*)

This handsome introduced ornamental tree with cream–colored resinous sap or latex is identified by : (1) paired elliptic shiny dark green leaves, stiff and thick, with numerous very fine, nearly straight parallel side veins very fine, nearly straight parallel side veins scarcely visible without a lens or until dried; (2) fragrant white flowers about 1 inch across, several in branched clusters at leaf base; and (3) fruits a few round balls $1^1/_4$–$1^1/_2$ inches in diameter, light green, becoming yellow or brown, containing 1 large round seed. Distinguished from its native relative, No. 161, maria, santa–maria, *Calophyllum calaba L.*, by the larger leaves and larger fruits.

An evergreen planted tree 45 feet high and 1 foot in trunk diameter, probably becoming somewhat larger at maturity, with spreading crown of irregular branches broader than high. The bark is light gray, smooth on branches, becoming slightly cracked into shallow broad furrows and long flat ridges. Inner bark is pink, fibrous, and bitter. Twigs stout, green, turning to brown, hairless, with cream–colored or light yellow resinous latex. Bud about $^1/_4$ inch long, narrow and pointed, dark brown, composed of minute leaves, without scales or stipules.

The hairless leaves are paired or opposite, in 4 vertical rows on a twig. The light green petioles are $^5/_8$–$^3/_4$ inch long, stout and flattened above. Leaf blades are 4–7 inches long, $2^1/_2$–$3^1/_2$ inches wide, slightly notched at the rounded apex, rounded or blunt at base, slightly turned up from the yellow–green midvein and concave, with a narrow whitish line along the straight border, the flower surface yellow green and slightly shiny.

Flower clusters (racemes) 4 inches or less in length are borne at base of a leaf, usually only 1 on a twig. Several flowers on long stout stalks to $1^1/_2$ inches long are composed of 4 rounded concave white sepals $^1/_4$ inch or more in length, in 2 pairs; 2–8 elliptic to oblong concave spreading petals $^1/_2$ inch long; many stamens about $^5/_{16}$ inch long, with orange anthers and white filaments slightly united at base; and pistil with round red 1–celled ovary containing 1 ovule, slightly curved style, and disk stigma. The fruit (drupe) has thin flesh and 1 large brown seed abcut $1^1/_4$ inches in diameter.

The wood, like that of the Puerto Rican relative, is utilized in the native home for general construction, cabinetmaking, boat-building, and similar purposes.

The latex or resin has served in home remedies, while the oil extracted from the seeds has been used in medicine and for light. Reportedly the seeds are poisonous. Regarded as sacred, this tree was grown around temples in the South Sea Islands.

A common street tree in Kerala. Occasionally planted here and elsewhere in the tropics as an ornamental or shade tree for the dense shiny dark green foliage and fragrant flowers. Slow growing.

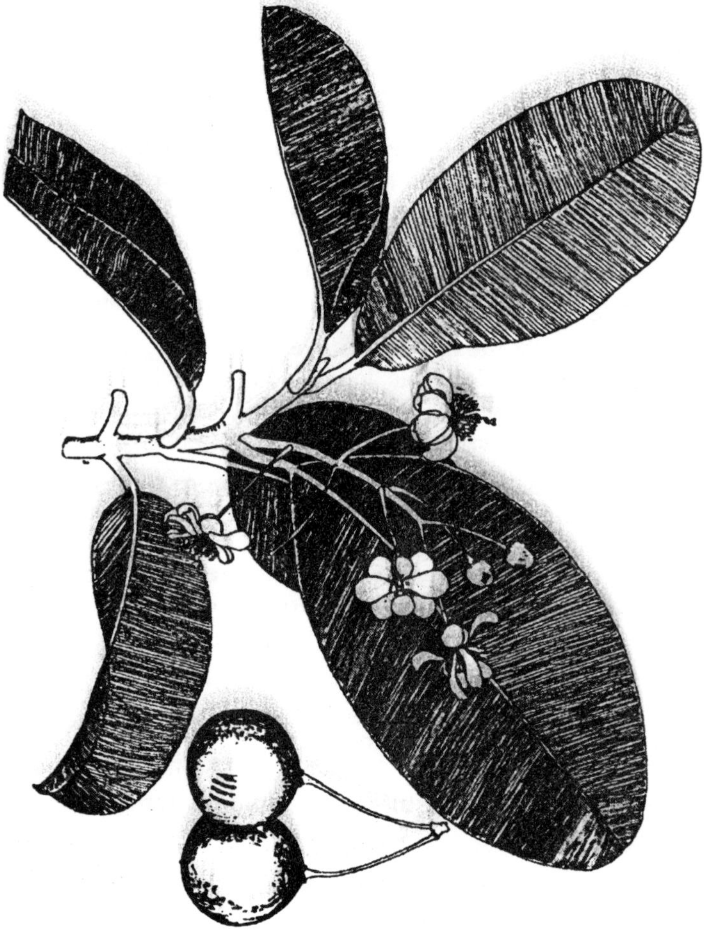

**Fig. 203 : Kamani (*Calophyllum inophyllum* L.*)**
Flowering twig (above), fruits (below), two–thirds natural size.

## 204. Cupey de monte (*Clusia minor* L.)

A vinelike shrub or sometimes a small tree with long spreading branches and with whitish latex, further distinguished by : (1) paired thick and fleshy obovate leaves rounded at apex and long–pointed at base; (2) flowers with 5 spreading pale pink petals about 1 inch across, few in short terminal clusters, male and female on different trees (dioecious); and (3) oblong fleshy green capsules $^3/_4$–1 inch long.

Evergreen vinelike shrub with long paired horizontal or slightly drooping branches climbing on trees or a small tree to 20 feet or more in height and 5 inches in trunk diameter. Hairless throughout. The gray bark is smoothish or slightly fissured, the inner bark pinkish and slightly bitter, with whitish or cream–colored slightly bitter latex. The brown twigs are stout.

Leafstalks of the opposite leaves are $^3/_8$–$^3/_4$ inch long. The thick blades are 2–4 inches long and 1–2 inches broad, with veins only faintly visible, green to dark green on upper surface and yellow green beneath.

The terminal flower cluster a little more than 1 inch long is much shorter than the leaves. There are 4 whitish sepals $^3/_{16}$–$^5/_{16}$ inch long and 5 widespreading large fleshy pale pink petals $^1/_2$–$^5/_8$ inch long. Female flowers have a greenish 5–celled ovary with 5 dark brown stigmas and a brownish sticky mass at base and are slightly fragrant.

Seed capsules are green and fleshy, oblong or broadest above middle, $^3/_4$–1 inch long and $^1/_2$–$^9/_{16}$ inch in diameter, with calyx at base and 5 stigmas at apex, splitting open into 5 parts to release the seeds $^3/_{16}$ inch long. Sometimes the ovary and capsule are 6–celled In flower and fruit through the year.

The whitish sapwood is soft.

Local in lower and upper forests at 600–4,000 feet altitude in mountains throughout Sivalik. Common locally on river banks, rocky areas, and peaks, becoming the dominant species in the upper part Forest.

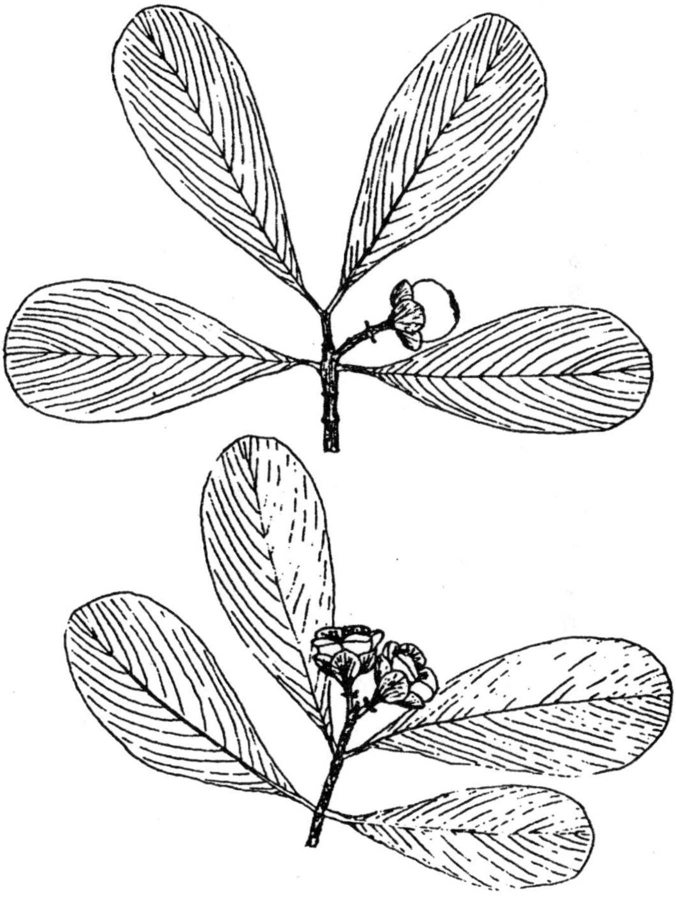

**Fig. 204 : Cupey de monte (*Clusia minor* L.)**
Fruiting twig (above), flowering twig (below), natural size.

## 205. Mangostan, mangosteen (*Garcinia dulcis*)

This introduced fruit tree related to the true mangosteen is identified by : (1) light yellow or whitish latex; (2) the slightly angled and flattened hairless green twigs with rings at nodes; (3) paired oblong leaves, stiff thick and leathery, 4–12 inches long and $1^1/_2$–$5^1/_2$ inches wide, dull green above and dull yellow green beneath; (4) few greenish white bell–shaped flowers about $^3/_8$ inch long and wide on short stalks at leaf bases and back of leaves; and (5) elliptic fruit 2–3 inches wide, pointed, light yellow, with 1–5 seeds in pale orange or yellow sour pulp.

Evergreen fruit tree 20 feet high and 3 inches in trunk diameter, hairless throughout. Bark brown, smoothish to slightly fissured. Inner bark pinkish, brittle and bitter. Twigs ending in a growing point covered by petioles of last pair of leaves.

Leaves opposite, with stout greenish petioles $^1/_4$–1 inch long, flattened above. Blades are short–pointed at both ends, not toothed at edges, slightly turned up from the prominent midvein, with many fine inconspicuous side veins.

Flowers few on stalks about $^1/_2$ inch long, composed of 5 unequal rounded light green sepals about $^1/_{16}$ inch long and 5 unequal elliptic greenish white petals about $^3/_8$ inch long, not spreading open. Male flowers have 5 stamens $^1/_4$ inch long. Female flowers have sterile stamens and on a disk the pistil with rounded green ovary $^1/_8$ inch wide and broad flattened whitish 5–lobed stigma. The thin–skinned fruits have sour juicy pulp. Seeds 1–5 large, brown. With flowers and fruits in summer.

The sapwood is light brown and hard.

This fruit and ornamental tree has been introduced experimentally into Chakrata and is rare. The fruits are too sour to be eaten raw but may be candied and male into jam. Gamboge paint, of reddish yellow hue, has been made from unripe fruit.

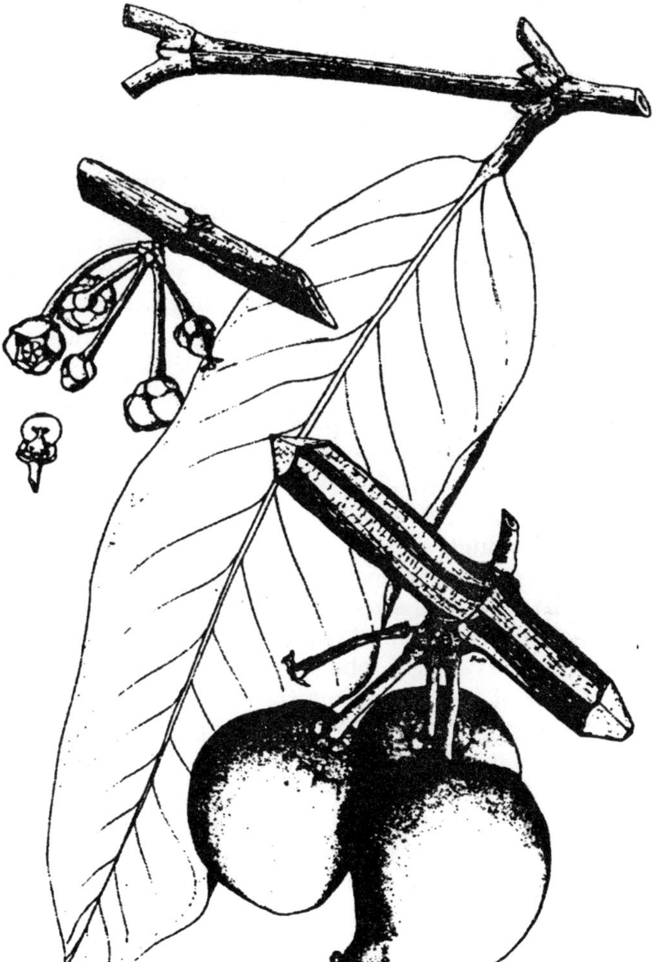

**Fig. 205 : Mangostan, mangosteen (*Garcinia dulcis*)**
Female flowers (upper left), leafy twig, and fruits (lower right), two–thirds natural size.

## 206. Mangostan, mangosteen (*Garcinia mangostana* L.*)

Mangosteen is a popular Old World fruit grown only rarely in India. It is characterized by : (1) light yellow latex; (2) large paired, elliptic oblong leaves, thick and leathery, 6–10 inches long and $2^1/_2$–$4^1/_2$ inches wide; (3) flowers large, $1^1/_2$–$1^3/_4$ inches across the 4 rose–pink fleshy petals; and (4) rounded reddish purple or brown berry 2–3 inches long with 5–7 white segments somewhat like tangerine, the juicy sweetish edible pulp coverings around the seeds.

Small evergreen planted tree to 30 feet high and 6 inches in trunk diameter, with compact crown of nearly horizontal branches and shiny green coarse foliage, hairless throughout. Bark brown, thick, rough and irregularly fissured into small long plates. The light brown bitter inner bark yields light yellow bitter latex. Twigs stout, slightly angled, yellow green, with rings at nodes.

Leaves opposite, with stout yellow–green petioles $^1/_2$–$^3/_4$ inch long, flattened above, those of end pair covering the bud. Blades are long–pointed at apex, short–pointed or rounded at base, not toothed on edges, with many fine parallel side veins, the upper surface shiny green to dark green, and the lower surface dull yellow green.

Flowers of 2 kinds (polygamous) at ends of twigs, the bisexual flowers single or paired, composed of 4 rounded pink to red sepals in pairs; 4 fleshy rose–pink petals $^3/_4$ inch long and wide, broadest toward the rounded apex; stamens many, small and nonfunctional in female flowers; rounded ovary of 5–7 cells and ovules and broad 5–7 lobe stigma. Also male flowers 3–9 together at end of twigs on same tree.

The berry slightly broader than long has a hard smoothish rind and pink bitter inedible wall about $^1/_4$ inch thick, which exudes small drops of yellow latex when cut. Four enlarged concave light green or brown sepals remain at base. The edible part consists of 5–7 white segments (arils) about $1^1/_4$ inches long and $^3/_4$ inch wide, each enclosing a seed or none. The rounded flattened seeds are about $^5/_8$ inch long. With flowers and fruits in summer.

Mangosteen is classed among the most delicious tropical fruits, having a flavor somewhat between that of a grape and a peach. Fruits are opened by cutting off the top half and exposing the segments. This fruit free is cultivated sparingly in the tropics, mainly Old World. Growth is slow, and fruits are seldom borne before 8–15 years.

**Fig. 206 : Mangostan, mangosteen (*Garcinia mangostana* L.*)**
Flowering twig, female flower (lower left), seed and fruit (lower right), two–thirds natural size.

## TAMARIX FAMILY (TAMARICACEAE*)

Trees mostly small and shrubs of dry and salty areas, native of the Old World, known by : (1) very slender often drooping twigs; (2) leaves minute, scalelike, alternate, pressed against twig, without stipules; (3) minute flowers mostly crowded and short-stalked in unbranched clusters (racemes), bisexual, regularly, composed of 4–5 sepals, 4–5 petals, stamens as many or twice as many as petals, and pistil with superior 1–celled ovary of 3–4 carpels and few to many ovules, 3–4 styles or none, and 3–4 stigmas, and (4) fruit a capsule with many hairy seeds.

## 207. Athel tamarisk (*Tamarix aphylla*)

Tamarisk, a rare introduced tree of dry areas, is easily recognized by : (1) drooping wiry gray–green twigs, jointed; (2) minute scale leaves $^1/_{16}$ inch long, circling twig and ending in minute point; (3) many small whitish pink flowers less than $^1/_8$ inch long in slender mostly branched clusters; and (4) seed capsules $^3/_{16}$ inch long, splitting in 3 parts, with many tiny hairy seeds.

Evergreen small to medium–sized planted tree to 50 feet high and 2 feet in trunk diameter, with many spreading branches and rounded to irregular gray–green crown. Bark light gray brown or reddish brown, becoming thick and deeply furrowed into long narrow hard ridges. Inner bark light brown, slight bitter. The branches are purplish brown and smooth. The wiry twigs less than $^1/_{16}$ inch in diameter are gray green and mostly deciduous, the older twigs greenish brown.

The minute scale leaves are alternate, hairless, each forming a joint along twig, shedding together.

Flower clusters (racemes and panicles) mostly at ends of twigs, slender and $1^1/_4$–$2^1/_2$ inches long or branched and to 6 inches, bearing many nearly stalkless flowers. There are 2 over–lapping rounded sepals, 5 petals less than $^1/_8$ inch long and shedding early, 5 stamens attached between lobes of disk, and pistil with 1–cell ovary and 3 stigmas. Many narrow pointed seed capsules split into 3 parts. Each seed has a tuft of whitish hairs.

The wood is light brown and hard. Tests elsewhere indicate that it would be suitable for furniture and turned articles.

Rarely planted around houses in dry areas of Rajasthan. The trees grow rapidly, are drought resistant, and tolerant of alkaline and saline soils. Elsewhere used for windbreaks and hedges as well as shade.

**Fig. 207 : Athel tamarisk (*Tamarix aphylla*)**
Flowering twig and fruits (lower right), natural size.

# CANELLA FAMILY (CANELLACEAE)

Small and medium–sized trees, sometimes shrubs, known by : (1) leaves aromatic (also other parts), alternate, simple, entire, leathery, hairless, with gland dots, without stipules; (2) flowers small often in branched clusters (cymes) or solitary, bisexual, regular, with 4–5 thick overlapping sepals, corolla generally of 4–5 separate petals or none, 20 or fewer short stamens united by filaments in a tube around ovary, and pistil with superior 1–celled ovary with 2–5 placentas with 2 to many ovules, short stout style, and 2–5 lobed stigma; and (3) fruit a berry.

**Key to Species**

A. Leaves obovate or spoon–shaped, rounded at apex, $1^1/_4$–$3^1/_2$ inches long, leathery with indistinct veins; flowers several to many in branched terminal clusters with 5 dark red petals $^1/_4$ inch across—Barbasco, canella, *Canella winterana* (L.) Gaertn.

AA. Leaves elliptic short–pointed, $3^1/_2$–5 inches long, thin with many fine parallel side veins; flowers single at leaf bases with 12 white petals $^3/_4$ inch across—*Pleodendron macronthum*.

## 208. Chupacallos (*Pleodendron macranthum*)

This distinctive aromatic tree is one of the very rarest in India. Its characters for recognition are : (1) elliptic leaves $3^1/_2$–5 inches long and $1^1/_2$–$2^1/_4$ inches wide, shortpointed at both ends, with many fine parallel side veins; (2) large yellowish white flowers single at leaf bases, $^3/_4$ inch across, with 3 sepals, 12 petals, and many united stamens; and (3) rounded purple to black edible fruits $^3/_4$–1 inch in diameter, many-seeded.

An evergreen small or medium-sized tree to 40 feet high and 10 inches in trunk diameter, apparently aromatic. The brownish twigs are slender and hairless.

The alternate hairless leaves have petioles about $^1/_4$ inch long. Blades are not toothed on edges, the upper surface shiny green with midrib slightly sunken, and the lower surface dull green with midrib raised.

Flower stalks are about $^1/_2$ inch long or to 1 inch on fruit. The flower has a yellowish white saucer-shaped fleshy calyx about $^1/_4$ inch across and $^1/_8$ inch high, becoming slightly 3-lobed and turned back, persistent on fruit. The corolla is composed of 12 ovate to oblong yellowish white slightly thickened petals $^1/_2$ inch long and $^3/_{16}$–$^5/_{16}$ inch wide, rounded at apex, in 4 rows of 3 each. Stamens many, united into a tube $^3/_8$ inch long. The narrow pistil $^3/_8$ inch long inside tube has a 1-celled ovary with many tiny ovules, slender style, and 6 stigmas.

The fruits have calyx at base and short style at apex. They are purple to black, green when immature, aromatic, with odor described as like oranges. Seeds many, rounded, about $^1/_8$ inch long, shiny brownish black. Collected with flowers in February and April–June and with fruits in June–August.

The wood is described as nearly white, hard, and heavy.

Very rare in forests in lower altitude.

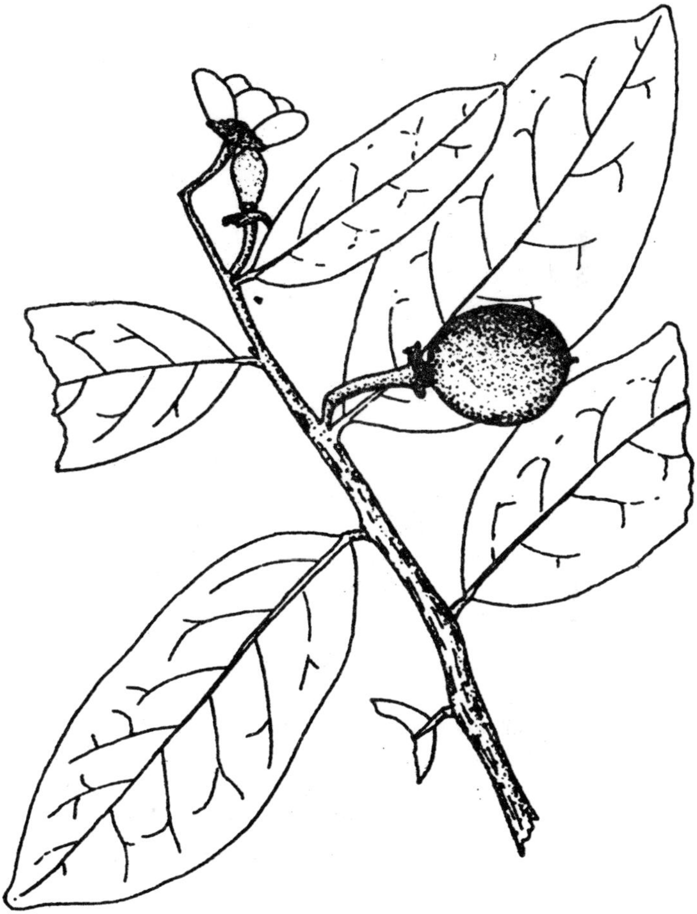

**Fig. 208 : Chupacallos (*Pleodendron macranthum*)**
Natural size.

# FLACOURTIA FAMILY (FLACOURTIACEAE)

Trees usually small or sometimes large and shrubs, known by : (1) leaves alternate in 2 rows, simple, usually toothed, sometimes with gland dots and lines (visible with lens against light); (2) flowers generally small or minute, green or white, many in branched clusters (cymes), usually bisexual, sometimes male and female, regular, with 2–15 sepals, 2–15 petals or none, many stamens, often with disk, and pistil composed of 1–celled ovary generally superior with 3–5 (2–10) parietal placentas with many ovules and 1–5 styles; and (3) fruit a capsule or berry (sometimes edible) with 1 to many seeds often with covering (aril).

**Key to Species**

A. Leaves with 3 or more main veins from base.

B. Leaves with edges straight or minutely wavy–toothed; flowers whitish; $^{3}/_{16}$ inch across, many nearly stalkless on narrow drooping axis—*Lunania buchii.*

BB. Leaves with edges toothed.

C. Leaves with 5 or 7 main veins from base; flowers yellowish, $^{3}/_{8}$ inch across, few with long stalks—*Prockia crucis.*

CC. Leaves with 3 main veins from base.

D. Leaves hairless, with finely saw–toothed edges; flowers greenish, $^{3}/_{8}$ inch broad, several on slender stalks—*Banara portoricensis.*

DD. Leaves soft hairy beneath, with wavy–toothed edges; flowers yellow, more than $^{1}/_{2}$ inch across, single at ends of side twigs—*Banara vanderbiltii.*

AA. Leaves with 1 main vein.

E. Flower in long stalked lateral clusters; seed capsules resembling dried flowers with 6 or 7 large brown sepals attached—Caracolillo, *Homalium racemosum* Jacq.

EE. Flowers at leaf bases, mostly short–stalked; fruit a seed capsule or berry, rounded or elliptic.

F. Flowers yellowish, mostly male and female on different plants; berries more than $^{1}/_{2}$ inch in diameter; uncommon introduced species.

G. Leaves oblong; flowers about 1 inch across; fruits 1–2 inches in diameter, densely spiny, not edible—*Oncoba echinata.**

GG. Leaves mostly ovate; flowers small, about $^{1}/_{4}$ inch across; fruits less than 1 inch in diameter, edible.

H. Leaves finely hairy on both surfaces, not toothed on edges—526 *Dovyalis hebecarpa.**

HH. Leaves hairless, wavy-toothed on edges—*Flacourti inermis.**

FF. Flowers greenish or whitish, bisexual; fruits smaller; native species.

I. Leaves soft hairy on both surfaces, gray green beneath; fruit dry elliptic, splitting open at top—*Samyda dodecandra.*

II. Leaves hairless or nearly so; fruit a fleshy capsule or berry.

J. Leaves thin, with gland dots and dashes visible with lens against light; fruit a fleshy capsule—*Casearia.*

K. Leaves few, elliptic; seed capsules more than $^1/_4$ inch in diameter.

L. Spines scattered on twigs; seed capsule $^1/_4$–$^5/_{16}$ inch in diameter, bright red—*Caseari aculeata.*

LL. Spines absent; seed capsules about $^3/_8$ inch in diameter.

M. Leaves less than 3 inches long, hairless; seed capsules round—Tostado wild honey-tree, *Casearia decandra* Jacq.

MM. Leaves $2^1/_2$–$5^1/_2$ inches long, hairy on veins; seed capsules elliptic—Palo blanco, wild-coffee, *Casearia guianensis* (Aubl.) Urban.

KK. Leaves many, evenly spaced on long slender twigs, appearing pinnate.

N. Leaves short-pointed at base, mostly lance-shaped.

O. Leaves with lower surface gray green, hairy; edges finely saw-toothed—Rabo raton, *Casearia arborea* (L. C. Rich.) Urban.

OO. Leaves green on both surfaces, hairless, edges wavy and appearing as without teeth—Cafeillo, *Casearia sylvestris* Sw.

NN. Leaves rounded at base, oblong, hairless, with finely wavy border; seed capsules light green, $^3/_8$–$^5/_8$ inch in diameter—*Leatia procera.*

JJ. Leaves thickened, without gland dots; fruit a berry mostly red—*Xylosma.*

P. Spines often much branched on branches and trunk; leaves elliptic.

Q. Spines also single and unbranched at nodes on twigs; leaves mostly ending in spine—*Xylosma buxifolium.*

QQ. Spines absent from twigs; leaves blunt at apex—*Xylosma Pachyphyllum.*

PP. Spines absent.

R. Leaves obovate or oblancelate, rounded or blunt at apex, slightly thickened—*Xylosma schaefferioides.*

RR. Leaves ovate, long-pointed, thick and leathery, with prominent network of small veins—*Xylosma schwaneckeanum.*

## 209. Palo de ramon (*Banara portoricensis*)

Shrub or small tree rare in central mountains, identified by: (1) lance–shaped to ovate, long–pointed leaves with 3 or 5 main veins from base and with edges finely saw–toothed; (2) several greenish flowers on slender stalks, $^3/_8$ inch broad, with 3–lobed calyx, 3 petals, and many stamens; and (3) fruit a rounded yellow or brown berry $^3/_8$–$^1/_2$ inch in diameter.

Evergreen shrub or small tree to 35 feet high and 8 inches in trunk diameter, with irregular spreading crown. Bark brown, slightly fissured. Inner bark light brown, bitter. Twigs are brown, very slender, hairless, with many dots (lenticels).

Leaves alternate in 2 rows, hairless. Petioles slender, $^1/_4$–$^1/_2$ inch long. Blades are 2–5 inches long, 1–2$^1/_2$ inches wide, rounded or short–pointed at base, thin or slightly thickened, slightly shiny, the upper surface green to dark green with veins slightly raised, and the lower surface lighter green with raised veins.

**Fig. 209 : Palo de ramon (*Banara portoricensis*)**

Fruiting twig (above), flowering twig (lower right), natural size.

# FLACOURTIA FAMILY (FLACOURTIACEAE)

Flower clusters (like racemes) at ends of twigs bear 5–10 flowers on long slender green stalks becoming $^1/_2$–1 inch long, enlarged toward upper end. The flower has 3 rounded calyx lobes $^3/_{16}$ inch long, 3 rounded green petals $^3/_{16}$ inch long, very many stamens, and pistil with 1–celled ovary and many ovules attached on 6 lines and long–pointed style. The berry has enlarged calyx and corolla at base and pointed style at apex. Many shiny seeds $^1/_{16}$ inch long are borne along 6 lines (placentas). Flowering and fruiting mainly in winter and spring, also in summer.

The wood is light brown and hard.

Rare in upper forest and dwarf forest at 2,500–3,500 feet altitude in Central India.

## 210. Vanderbilti (*Banara vanderbiltii*)

This very rare shrub or small tree known only from Bihar is characterized by : (1) wavy–toothed elliptic or lanceolate leaves $1^1/_2$–5 inches long and $^5/_8$–2 inches wide, with 3 main veins from base; and (2) yellow hairy flowers solitary at ends of short side twigs more than $^1/_2$ inch across the 3 pointed calyx lobes and 3 petals.

Evergreen shrub or small tree to 30 feet high and 5 inches in trunk diameter. Twigs slender, when young with dense spreading yellowish hairs.

Leaves alternate with small paired threadlike stipules and short hairy petioles about $^1/_4$ inch long. Leaf blades blunt or rounded and often slightly unequal at base, long–pointed at apex, wavy toothed, thin, with few curved side veins on each side and prominent beneath, the lowest pair parallel with margin, soft hairy on both surfaces, becoming rough and nearly hairless above.

Flowers solitary at ends of short side twigs on stalks $^5/_8$–$^3/_4$ inch long. Calyx about $^1/_4$ inch long, composed of 3 triangular hairy spreading lobes, persistent; petals 3, $^1/_4$–$^5/_{16}$ inch long, pointed, hairy, yellow, numerous spreading hairless stamens to $^5/_{16}$ inch long; and pistil composed of rounded hairless 1–celled ovary with many ovules attached on 4 lines, slender curved style, and 4 minute stigmas. Collected with flowers in May. The fruits have not been described, but those of this genus are many–seeded berries.

**Fig. 210 : Vanderbilti (*Banara vanderbiltii*)**
Natural size.

## 211. Cambron (*Casearia aculeata*)

This shrub or small tree of Chattisgarh is characterized by: (1) simple or branched spines $^1/_4$–$1^1/_4$ inches long usually scattered on branches; (2) elliptic leaves 2–4 inches long and $^7/_8$–$1^3/_4$ inches wide, short–pointed or blunt at both ends, inconspicuously wavy–toothed on edges, with minute dots and lines visible against the light with a lens, alternate in 2 rows and shedding in winter; (3) many small greenish–white flowers about $^5/_{16}$ inch across, clustered at leaf bases; and (4) round bright red seed capsules $^1/_4$–$^5/_{16}$ inch in diameter.

A deciduous shrub less than 12 feet, sometimes a small tree to 30 feet high and 4 inches in trunk diameter. The bark is light gray, smooth with warts (lenticels) and scattered spines. Inner bark is light brown and almost tasteless. The slender twigs are gray and finely hairy. Short twigs develop into spines.

The leaves have short finely hairy petioles $^1/_{16}$–$^1/_4$ inch long. Blades are thin, the upper surface green and hairless or nearly so, the lower surface pale green and hairy or nearly hairless.

Flowers several, fragrant, clustered at leaf bases on stalks less than $^1/_{16}$ inch long, consisting of calyx with 5 hairy lobes about $^3/_{16}$ inch long; 8 stamens alternate with 8 smaller sterile stamens (staminodes); and pistil with hairy 1–celled ovary, slender style, and rounded stigma. The seed capsules are slightly 3–angled and split into 3 parts. Seeds few. Flowering and fruiting intermittently through the year.

The sapwood is light brown and hard.

An uncommon understory shrub in moist and dry coastal and limestone forests from sea level to 300 feet altitude along Hazaribagh.

**Fig. 211 : Cambron (*Casearia aculeate*)**
Flowering twig (lower left) and fruiting twig, natural size.

## 212. Kitembilla (*Dovyalis hebecarpa*)

A shrub or small tree sometimes cultivated for its edible fruits. Identified by : (1) twigs often with long slender sharp gray spines to $1^1/_2$ inches long at leaf bases; (2) leaves elliptic to ovate, 2–4 inches long and $1–1^1/_2$ inches wide, often finely wavy-toothed on edges; (3) few greenish flowers about $^5/_8$ inch across at leaf bases, male and female on different plants (dioecious); and (4) round purplish velvety hairy berries $1–1^1/_4$ inches in diameter.

Evergreen planted shrub or small tree to 15 feet high. The bark is gray, slightly scaly, the inner bark whitish and bitter. Twigs are long and slender, greenish gray, finely hairy.

Leaves are alternate in 2 rows, with pinkish hairy petioles about $^1/_4$ inch long. Blades are long-pointed at apex and rounded at base, thin, curved up slightly from midvein, pinnate-veined with 2 long curved lateral veins. The upper surface is shiny green with minute hairs, and the lower surface yellow green with pinkish veins and soft hair.

Male flowers many on short stalks, composed of mostly 4–7 hairy sepals, no petals, and many stamens on a fleshy disk. Female flowers 1 or few, consisting of 5–9 hairy sepals persistent on fruit and pistil with ovary of several cells and several styles. The fruit has styles at apex, purplish sour edible pulp, and several elliptic hairy seeds $^1/_4$ inch long.

The wood is light brown and hardy.

Uncommonly planted in Bihar for the fruits, which have an acid flavor like goose-berry and which are made into jelly. Tested also in southern States.

**Fig. 212 : Kitembilla (*Dovyalis hebecarpa*)**
Fruiting twig (left), flowering twig (right), two–thirds natural size.

## 213. Louvi (*Flacourtia inermis Roxb.**)

This shrub or small tree sometimes grown for its edible sour cherrylike fruits is identified by : (1) ovate to elliptic wavy–toothed leaves; (2) small yellowish flowers about $^1/_4$ inch long and broad, in short clusters at leaf bases, male and female on different plants (dioecious); and (3) round shiny cherrylike fruits, red, purple, or black, $^1/_2$–1 inch in diameter.

A shrub or small tree to 30 feet or more in height, with light brown smoothish bark. The twigs sometimes have spines at leaf bases.

Leaves alternate, with petioles $^1/_4$–$^3/_8$ inch long. Blades are 2–8 inches long and 1–3 $^1/_2$ inches wide, long–pointed at apex and short–pointed at base, thin or slightly thickened, becoming hairless or nearly so, the long curved side veins slightly sunken.

Several male flowers are borne on stalks of $^1/_4$ inch in short clusters (racemes) 1–1$^1/_2$ inches long, composed of 4–5 rounded hairy sepals more than $^1/_{16}$ inch long, no petals, and many stamens. Female flowers few or single along twigs have pistil with 5–celled ovary and 5 short spreading styles. The fruits (drupes) are clustered along twigs mostly back of leaves, have remains of styles at apex, and contain 8–10 small seeds in juicy pulp.

Occasionally planted in U.P. and the fruits. Though very sour, the edible fruits make good jelly and preserves. The spiny variety has been used elsewhere in hedges.

**Fig. 213 : Louvi (*Flacourtia inermis Roxb.**)**

Flowering twig (above), fruits (below), two–thirds natural size.

## 214. Talantron (*Laetia procera*)

This rare tree in mountain forests is recognized by : (1) the leaves alternate in 2 rows on very long twigs, oblong, long–pointed, with finely wavy border and minute translucent dots and lines; (2) small greenish flowers, few on separate stalks at base of leaves; and (3) fruit a round fleshy capsule $^3/_8$–$^5/_8$ inch in diameter, which opens in 3 parts and contains many seeds.

A medium–sized evergreen tree to 75 feet high and 1 foot or more in trunk diameter. Bark gray or greenish gray, smooth, with warts (lenticels). The interior bark is orange, streaked with yellow, with pink dots at surface, and with gritty hairy. The buds less than $^1/_8$ inch long are covered with stipules.

The leaves are alternate in 2 rows on long twigs, with slender petioles $^3/_8$ inch long, grooved, light green. Leaf blades are $2^1/_2$–6 inches long and $1^1/_4$–$1^3/_4$ inches wide, thin, hairless, rounded at base, slightly turned up on both sides of midvein, the upper surface dark green and slightly shiny, the lower surface dull light green.

The flowers lateral at leaf bases have slender stalks $^3/_8$ inch long. The flower is composed of calyx of 5 elliptic greenish sepals $^3/_{16}$ inch long that become bent back; no corolla; many stamens more than $^1/_{16}$ inch long; and pistil with rounded 1–celled ovary, many ovules on 3 lines, and short style. The seed capsules are fleshy and berrylike, light green, 1–celled, with calyx at base. There are many rounded black seeds less than $^1/_8$ inch long. Collected with fruit in February, March and December.

The wood is light yellow to orange, not differentiated into sapwood and heartwood, soft, and moderately heavy (specific gravity 0.75). Tested elsewhere, the wood was found to have commercial possibilities, being suitable for construction, veneer, and furniture.

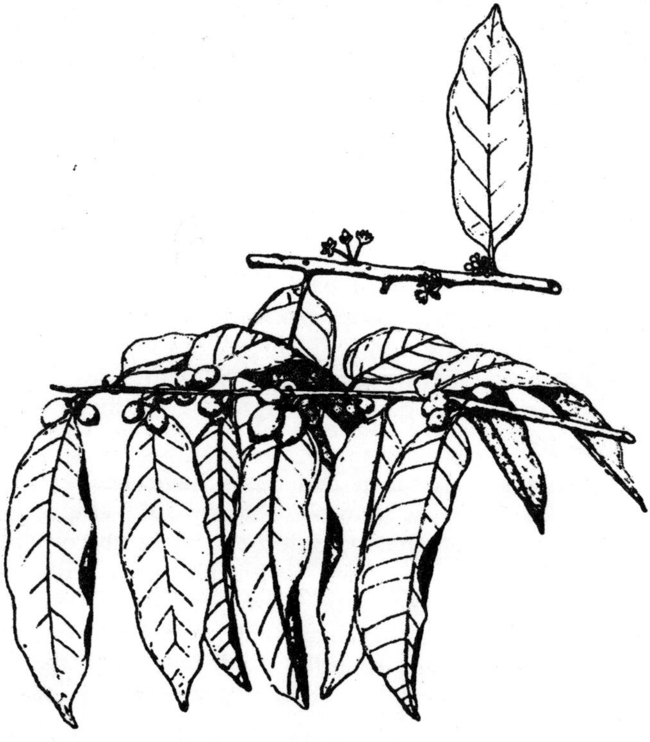

**Fig. 214 : Talantron (*Laetia procera*)**

Flowering twig (above), fruiting twig (below), two–thirds natural size.

## 215. Mendrina (*Lunania buchii*)

This small tree rare in upper mountain forests is recognized by : (1) leaves ovate with 3 main veins from base, the 2 side veins curved almost to apex; (2) many minute white flowers on narrow drooping axis 4–8 inches long; and (3) dark brown rounded seed capsules $^{5}/_{16}$ inch in diameter, splitting widely in 3 or 4 parts.

Evergreen small tree to 20 feet high and 5 inches in trunk diameter. Bark gray, smooth or finely fissured, the inner bark light orange, slightly bitter. Twigs are slender, light green, becoming gray.

The leaves are alternate in 2 rows on slightly zigzag twigs, hairless. Petioles are slender, $^{1}/_{4}$–$^{5}/_{8}$ inch long. The blades are 2 $^{1}/_{2}$–4 inches long and 1 $^{1}/_{4}$–1$^{3}/_{4}$ inches wide, long–pointed at apex, rounded at base, with straight or minutely wavy toothed edges, slightly thickened and stiff. The 2 prominent side veins extend from base almost to apex, and 2 smaller side veins extend from midvein along border of base. The upper surface is green and slightly shiny, the lower surface dull light green.

The flowers about $^{3}/_{16}$ inch long and broad have stalks less than $^{1}/_{8}$ inch long; calyx of 2–3 rounded concave white sepals $^{1}/_{8}$ inch long, no corolla; 7–8 stamens on a disk; and pistil with 1–celled ovary, many ovules, and short style. The seed capsules bear many seeds of less than $^{1}/_{16}$ inch along 3 or 4 lines (placentas) in a fleshy mass. With flowers and fruits in spring and summer.

Wood whitish, slightly hard.

Rare in upper Aravali forests at 1,500–2,800 feet altitude.

**Fig. 215 : Mendrina (*Lunania buchii*)**

Fruits (left), opened fruits (below), leafy twig (above), and flowers (right), natural size.

## 216. Gorli oncoba (*Oncoba echinata*)

This shrubby tree is sometimes planted as an ornamental. It is characterized by : (1) oblong leaves $4^1/_2$–7 inches long and $1^1/_2$–$2^1/_2$ inches wide; (2) few short–stalked flowers at leaf bases, about 1 inch across the spreading petals; and (3) fruit a rounded densely spiny greenish–yellow berry 1–2 inches in diameter, not edible.

An evergreen shrub or small tree, becoming hairless or nearly so. Leaves alternate in 2 rows, hanging down on petioles of less than $^1/_2$ inch from long slender twigs. The blades are abruptly long–pointed at apex and rounded at base, not toothed on edges, thin, with few curved lateral veins, paler beneath.

Flowers are clustered on short stalks along twigs at leaf bases and back of leaves, partly of one sex and bisexual (polygamous), about 1 inch across. There are 3–5 sepals and 5–10 widely spreading narrow petals, very many short stamens, and pistil with spiny 1–celled ovary and slender style. The thick–walled berry is covered with many long spines $^1/_4$–$^3/_8$ inch long and contains 1 or few seeds in the pulp. Collected with flowers and fruits in January.

Introduced experimentally in Himachal Pradesh an ornamental and medicinal plant. The seeds of this and related species contain chaulmoogric acid, used in the treatment of leprosy.

**Fig. 216 : Gorli oncoba (*Oncoba echinata*)**

Fruit (above), flowering twig (below), two–thirds natural size.

## 217. Sacha (*Prockia cruces* L.)

This shrub rarely treelike in size is identified by : (1) broadly ovate long–pointed saw–toothed leaves with 5 or 7 main veins from the notched base, with large long–pointed toothed stipules to $^3/_8$ inch long; (2) few long–stalked fragrant small yellowish flowers about $^3/_8$ inch across the 3 or 4 sepals and 3 or 4 petals; and (3) round black berries $^1/_4$–$^5/_{16}$ inch in diameter.

A much–branched shrub observed treelike to 15 feet high and $2^1/_2$ inches in trunk diameter, recorded (perhaps elsewhere) as sometimes a tree to 33 feet, evergreen or perhaps deciduous in dry areas. The slender twigs are hairless or finely hairy.

The alternate leaves have very slender petioles $^1/_4$–$^3/_4$ inch long. The thin blades are $1^1/_4$–$4^1/_2$ inches long and $^3/_4$–$2^1/_2$ inches wide, sometimes larger, hairless or finely hairy, paler beneath.

Flower clusters (racemes or corymbs) are terminal, 4 inches or less in length. The flowers on long stalks of $^3/_8$–$1^1/_4$ inches consist of 3 or 4 pointed hairy sepals nearly $^1/_4$ inch long, 3 or 4 short hairy yellowish petals that fall early, very many short threadlike stamens on a disk, and pistil with 3–5 celled ovary, many ovules, and slender style. The berries with calyx at base and style at apex are juicy and slightly sweet and contain many minute seeds. Collected with fruits in summer.

The wood is described as whitish, of medium hardness, and brittle.

Common at low altitudes near Punjab.

**Fig. 217 : Sacha (*Prockia cruces* L.)**
Fruiting twig, natural size.

## 218. Guayabilla (*Samyda dodecandra*)

This large shrub rarely treelike is identified by : (1) long slender finely rusty to yellowish hairy twigs; (2) oblong to elliptic leaves alternate in 2 rows, short–stalked, finely saw–toothed on edges, soft hairy on both surfaces, and with minute gland dots visible with lens against the light; (3) white bell–shaped flowers about $^{1}/_{2}$ inch long, nearly stalkless at leaf bases; and (4) elliptic seed capsules $^{3}/_{4}$–$1^{1}/_{4}$ inches long, bright red within.

A deciduous large shrub mostly 6–12 feet high, observed treelike to 15 feet high and 2 inches in trunk diameter, reported to reach 20 feet.

The leaves have short hairy petioles about $^{1}/_{8}$ inch long. The thin or slightly thickened blades are $1^{1}/_{2}$–4 inches long and $^{3}/_{4}$–2 inches wide, mostly short–pointed at apex and rounded at base, beneath gray green with network of raised veins.

The fragrant slightly irregular flowers have bell–shaped hairy calyx with 4–6 unequal lobes white above and green below, no corolla, 8–12 stamens on calyx and united into a tube, and pistil with hairy 1–celled ovary, many ovules, slender style, and dotlike stigma. The fruits with calyx at base and pointed style at apex split open at top along 3–5 lines and contain many seeds. Flowering from winter to summer.

Common and widely distributed at low altitudes in M.P.

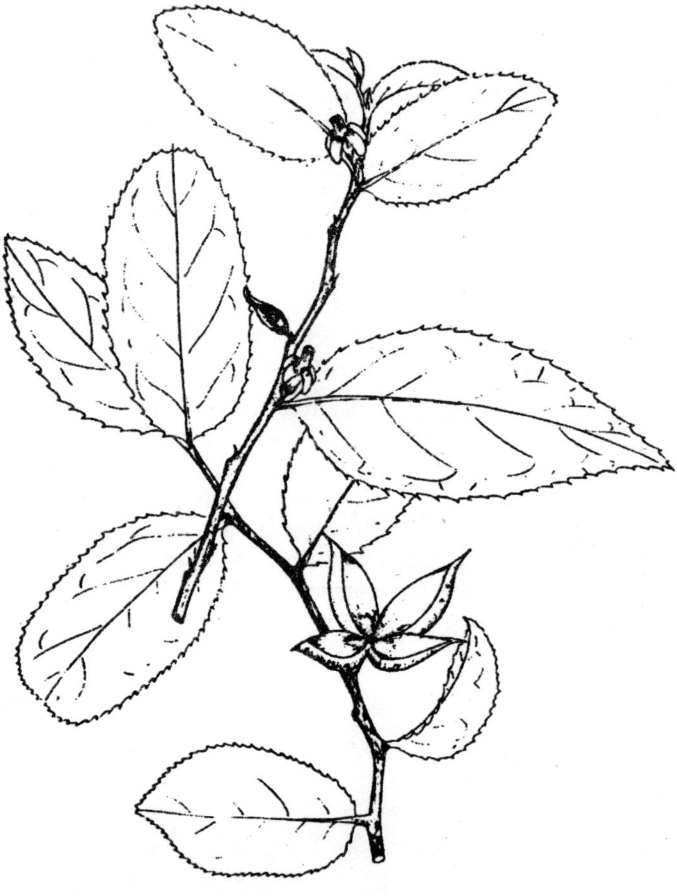

**Fig. 218 : Guayabilla (*Samyda dodecandra*)**
Flowering twig (above), fruiting twig (below), natural size.

## 219. Roseta (*Xylosma buxifolium*)

Roseta is recognized by : (1) single, unbranched, slender, gray spines $^{1}/_{4}$–$^{1}/_{2}$ inch long at nodes above leaf scars on twigs (sometimes absent) and the branched gray to red spines 1–3 inches long on trunk; (2) small, stiff, thick, elliptic leaves, usually spiny–pointed; and (3) lateral clusters of small yellowish flowers $^{1}/_{8}$ inch long (male and female on different trees); and (4) small red to black berries $^{3}/_{16}$ inch in diameter.

An evergreen tree to 45 feet high and 1 foot in trunk diameter, usually smaller. The gray bark is smoothish or slightly rough. Inner bark is light brown and bitter. The twigs are gray.

The alternate leaves have very short leaf–stalks $^{1}/_{16}$ inch long. Blades are 1–2$^{1}/_{4}$ inches long and $^{1}/_{2}$–1$^{1}/_{8}$ inches broad, short–pointed at base, the edges turned under and sometimes with 1 or more small teeth near the short spiny point, hairless, the side veins inconspicuous, shiny green on upper surface and paler beneath.

Male and female flowers are borne on different trees (dioecious), on slender stalks $^{1}/_{8}$–$^{3}/_{16}$ inch long, 4–8 at base of a leaf in a cluster (umbel) less than $^{1}/_{2}$ inch across. The calyx consists of 4 or 5 sepals less than $^{1}/_{16}$ inch long; petals absent; 8–30 stamens less than $^{1}/_{8}$ inch long are borne on a fleshy disk; and the pistil with 1–celled ovary and style 2–forked near apex. The elliptic in spring and collected with immature fruits in July.

The sapwood is light brown and hard.

Local and common in forests at 100–2,800 feet altitude in western mountains of A.P.

**Fig. 219 : Roseta (*Xylosma buxifolium*)**

Twig with female flowers (above), branched spine from trunk (lower right), natural size.

## 220. Pachyphylum (*Xylosma pachyphyllum*)

This uncommon shrub or small tree known only from western high mountains of Kerala is identified by : (1) much–branched spines $^1/_2$–1 inch long, dark brown but red when young, in scattered clusters on trunk and branches; (2) elliptic leaves $1^1/_2$–3 inches long and 1–2 inches broad, thick and leathery, toothed in upper half toward the blunt apex, and short–pointed at base; and (3) rounded bright red berries $^3/_{16}$ inch in diameter, single on short stalks at leaf bases.

An evergreen shrub or small tree to 20 feet high and 5 inches in trunk diameter. The bark is smoothish with warts (lenticels). Inner bark is pink and slightly bitter. Twigs are light brown and slender, hairless.

The alternate hairless leaves have short petioles less than $^1/_8$ inch long. Blades have edges turned under slightly, with inconspicuous wavy teeth in upper half, the upper surface dark green and slightly shiny, and the lower surface light green. There are few fine, long curved veins parallel with lower edges.

Flowers several, crowded at base of leaf from greenish buds less than $^1/_{16}$ inch long, not yet described. In this genus the small flowers are male and female on different trees (dioecious), without petals, with many stamens. The 4–5 broad hairy sepals $^1/_{16}$ inch long and short style with 3 short divided stigmas remain on fruit. The berries are borne singly at leaf bases on slender green stalks $^3/_{16}$–$^1/_4$ inch long. Seeds few, $^1/_{16}$ inch long. Flowering intermittently, collected with fruits in July and December.

The sapwood is hard and whitish.

Uncommon and local in serpentine forest at 2,000–3,000 feet altitude in western high mountains Kerala.

**Fig. 220 : Pachyphylum (*Xylosma pachyphyllum*)**

Twig with fruits (left and upper right), twig with flower buds (center right), and twig with old female flowers (lower right), natural size.

## 221. White logwood (*Xylosma schaefferioides*)

The shrub or small tree found in Kerala Forest is identified by : (1) small obovoate or oblanceolate leaves $^3/_4$–$1^3/_8$ inches long and $^1/_4$–$^1/_2$ inch wide, slightly thickened, rounded or blunt at apex, often finely wavy toothed on edges; (2) small yellow flowers $^1/_8$ inch across, clustered at leaf bases; and (3) round scarlet berries $^3/_{16}$ inch in diameter.

An evergreen shrub or small tree 10–13 feet high and 4 inches in trunk diameter, elsewhere recorded to 25 feet. Leaves alternate, with petioles less than $^1/_8$ inch long. Blades have short-pointed base and long fine side veins nearly parallel to edges.

Flowers male and female on different plants (dioecious), several at leaf bases spreading on equal stalks about $^1/_4$ inch long. The 4–5 yellow sepals are less than $^1/_{16}$ inch long, and there is no corolla. Male flowers have 10–20 stamens less than $^1/_8$ inch long. Female flowers have pistil with 1–celled ovary, very short style, and broad stigma. The berries have sepals at base and stigmas at apex.

Rare and local in serpentine forest at 2,000–2,500 feet altitude in western high mountains of Kerala.

**Fig. 221 : White logwood (*Xylosma schaefferioides*)**

Twig with male flowers (upper left), twig with immature fruits (upper right), and twig with female flowers (below), natural size.

## 222. Palo de candela (*Xylosma schwaneckeanum*)

This rare vinelike shrub or small tree is known only from mountain forests of eastern Sikkim. It is characterized by : (1) ovate leaves 2–4$^1/_2$ inches long and 1$^1/_4$–2 inches wide, thick and leathery, long-pointed at apex and with small teeth along edges, and with prominent network of small veins, the very young leaves reddish; (2) flowers small, white, about $^3/_{16}$ inch across, few clustered at leaf bases; and (3) red egg-shaped berries $^1/_2$–$^9/_{16}$ inch long at leaf bases.

An evergreen vinelike shrub or small tree 15–25 feet high and 3 inches in trunk diameter, with slender wandlike branches. The slender brown twigs are finely hairy when young, with raised dots (lenticels).

The alternate leaves have petioles $^1/_8$ inch long. Blades are hairless or nearly so, rounded or slightly notched at base, the upper surface shiny dark green with sides curved up from sunken midvein, the lower surface green and slightly shiny.

Flowers are male, female and bisexual on the same plant (polygamous), on short stalks $^1/_8$–$^1/_4$ inch long. There are 4–5 white rounded finely toothed sepals $^1/_8$ inch long; no corolla; 40–70 stamens less than $^1/_8$ inch long in male flowers, fewer in bisexual flowers; and pistil $^1/_8$ inch long with 1-celled ovary and short, thick, much divided, spreading styles. The berries clustered at leaf bases have sepals and styles persistent. Flowering and fruiting intermittently.

Rare in lower forests at 1,500–3,000 feet altitude in eastern mountains Sikkim.

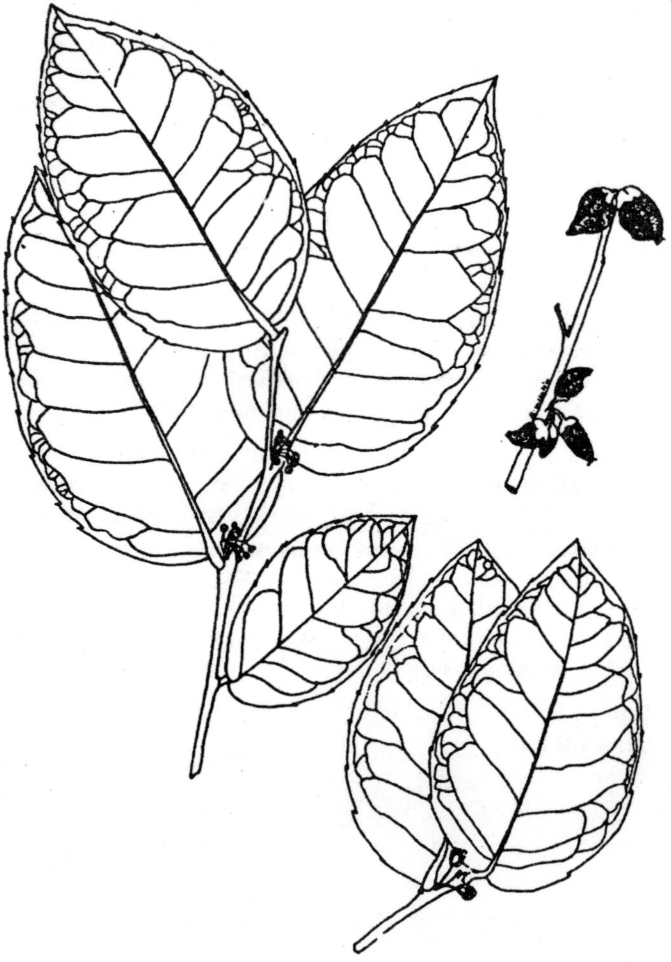

**Fig. 222 : Palo de candela (*Xylosma schwaneckeanum*)**

Twig with male flowers (left), fruits (upper right), and twig with female flowers (lower right), natural size.

# CACTUS FAMILY (CACTACEAE)

Succulent spiny plants of dry areas (xerophytes), including herbs, shrubs, few small trees, epiphytes, and vines, known by: (1) succulent spiny plants, herbaceous or woody, with enlarged cylindric or flattened stems often jointed, with many clustered spines and hairs spreading from a center; (2) leaves reduced to scales or none or rarely alternate, simple, flattened, and succulent (*Pereskia, Pereskiopsis*); (3) flowers generally solitary, stalkless, large, bisexual, generally regular, slightly fleshy, with tube (hypanthium) above inferior ovary, many colored sepals and petals and intermediates, and very many stamens inserted in tube, the petals generally yellow, white, or pink, and pistil with inferior 1–celled ovary with 3 to many parietal placentas and many ovules, style, and 2 to many stigmas; and (4) fruit a spiny berry often juicy and edible, with many black seeds.

**Key to Species**

A. Stems of many oblong flattened fleshy spiny pads or joints—*Opuntia*.

B. Flowers $^3/_4$ inch across the spreading petals—Tunna de petate, pricklypear, *Opuntia rubescens* Salm–Dyck.

BB. Flowers about 1 inch across the spreading petals—*Opuntia moniliformis*.

AA. Stems cylindric, with longitudinal ridges and grooves.

C. Stems and branches with mostly 6 longitudinal ridges; flowers 8 inches or more in length, white—*Cercus hexagonus.** .

CC. Stems and branches with 7–12 longitudinal ridges.

D. Stems mostly 1 trunk with several branches having 7–11 longitudinal ridges, with tufts of whitish hairs near apex; flowers 2 inches long, greenish, whitish, or purplish tinged; berries rounded but flattened, to 2 inches wide, red, spineless—Sebucán, dildo, *Cephalocereus royenii* (L.) Britton & Rose.

DD. Stems several nearly erect from base and mostly unbranched, with 9–12 longitudinal ridges, without tufts of whitish hairs; flowers 4–5 inches long, whitish and pinkish; berries round, about $1^1/_4$ inches wide, red, spiny—*lemaireocereus hystrix*.

## 223. Dildo espanol, Spanish dildo (*Lemaireocereus hystrix*)

Dildo espanol or Spanish dildo is one of the organpipe cacti, so named from the resemblance of the clustered nearly erect, mostly unbranched, columnar stems to a pipe organ. Found in southwestern Puerto Rico and small dry islands, this treelike cactus is characterized by : (1) several thick, gray–green cylindric stems or trunks unbranched from the base or with few branches mostly 4–6 inches in diameter, composed of 9–12 longitudinal ridges or ribs and grooves; (2) leaves absent but replaced by spreading slender sharp gray spines $^1/_2$–$1^1/_2$ inches long in clusters on ridges; (3) large funnel–shaped reddish flowers 4–5 inches long borne singly near ends of branches, with many spreading whitish and pinkish petals above the rounded spiny green ovary; and (4) fruit a round spiny red berry about $1^1/_4$ inches in diameter, juicy and edible.

Evergreen shrubby or treelike cactus mostly without a single trunk but with several nearly erect stout stems 10–15 feet high, sometimes to 25 feet, slightly spreading and curved up from base, often clustered and in colonies. These plants usually lack the single trunk and the crown of branches of trees but sometimes form a short trunk 1 foot high before branching. The stems, unjointed or with 1–3 constrictions, have 9–12 longitudinal ridges or ribs separated by grooves $^1/_2$–$^3/_4$ inch deep. The surface is smooth and gray green but becomes gray brown at base of older stems. The cut cross section consists of a dark green outer layer about $^1/_8$ inch wide bordering the yellow–green or yellowish tissue, which is soft, watery, and bitter. Within is a ring of light brown hard wood about 2 inches in diameter and $^1/_8$–$^1/_4$ inch thick, and a large soft central pith. The rounded growing point at apex of the stems is covered by minute whitish hairs. At intervals of about $^3/_4$ inch along the ridges are raised points (areoles) bearing clusters of about 10 spreading gray spines.

Flower buds develop on ridges, 1–3 near apex of stem. The flower is composed of rounded inferior ovary about $^1/_2$ inch in diameter, with scales and short spines; funnel–shaped tube 2–2 $^1/_2$ inches long, purplish to dark green; sepals many, fleshy, greenish; petals whitish and pinkish, about $^1/_2$ inch long; very many white stamens; and pistil with slender white style. The fruits have reddish pulp and break open irregularly to release

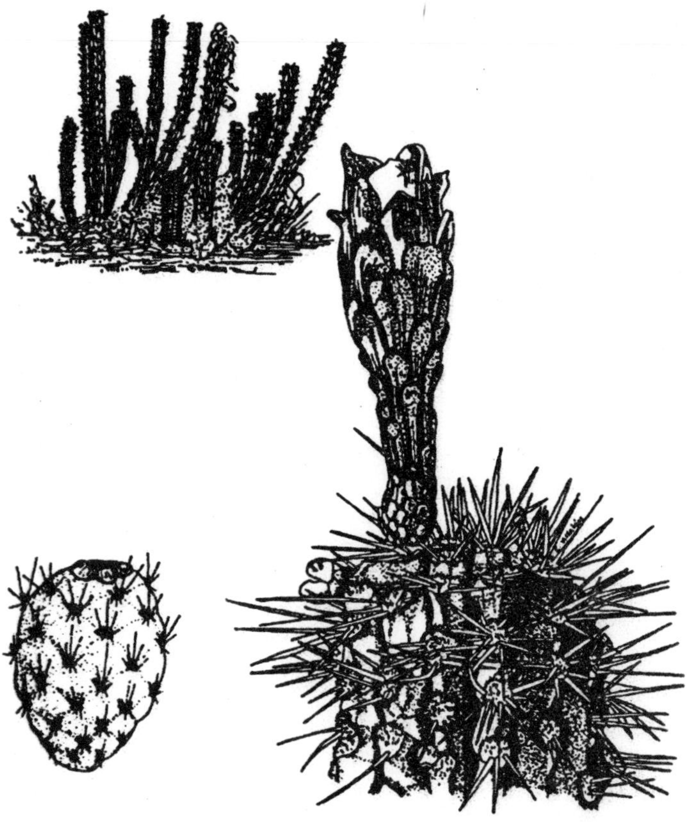

**Fig. 223 : Dildo espanol, Spanish dildo (*Lemaireocereus hystrix*)**
Fruit (lower left) and branch with flower (right), natural size.

many black seeds. Flowering in spring and summer and maturing fruits in summer.

Plants have been grown elsewhere as fences and hedges.

Locally common on dry limestone and coastal hills from sea level to 300 feet altitude in southwestern Rajasthan and small islands.

## 224. Tuna (*Opuntia moniliformis*)

This tree cactus of the group known as prickly pears or tunas is native of southwestern Rajasthan and the small islands. It is recognized by : (1) the single erect round spiny trunk on large plants; (2) many oblong flattened spiny gray–green fleshy pads or joints vertically oriented; (3) leaves minute and replaced by many clusters of sharp slender gray spines; (4) flowers few, mostly erect on edges of pads, $1^1/_2$–$1^3/_4$ inches long, with many petals orange red or sometimes yellow on narrow cylindric spiny green ovary; and (5) cylindric or pear–shaped spiny green to red berry fruit about $2^1/_2$ inches long.

Evergreen cactus, shrubby or becoming a small tree 15–20 feet high with single erect trunk, not jointed, 3–6 inches, sometimes 10 inches, in diameter and several spreading branches composed of oblong gray–green spiny pads. Trunks smoothish gray, covered with spines, but in age the surface and spines shedding and replaced by a thin layer of brown bark. The cut cross section consists of a very thin light green layer about $^1/32$ inch thick, a soft almost tasteless yellowish tissue about $^1/_4$ inch thick, and a yellowish hard woody cylinder $^1/_4$–$^1/_2$ inch thick with large oblong vertical holes, and central path.

The large oblong leaflike pads or joints are 6–12 inches long, 3–4 inches high, and $^1/_4$–$^3/_8$ inch thick, spreading horizontally or slightly curved upward, with the edge vertical. From many raised points (areoles) about $^3/_8$–$^1/_2$ inch apart and corresponding to nodes are borne singly the minute awl–shaped green fleshy leaves, which shed early, and afterwards clusters of spreading slender sharp gray spines mostly $^3/_8$–$^3/_4$ inch long (sometimes a few to $2^1/_2$ inches or more). Upon death, the soft succulent pads weather to reveal a skeleton or network of light brown woody strands.

Flowers are borne singly, stalkless and mostly erect on edges of pads. The narrowly cylindric green spiny inferior ovary $1^1/_4$ inches high and $^5/_8$ inch in diameter is fleshy, 1–celled and contains many ovules. It bears the other parts, many fleshy sepals, many petals about $^1/_2$ inch long and spreading 1 inch across, mostly orange red or sometimes yellow, very many stamens shorter than the petals, and a fleshy style with several spreading stigmas.

Green immature fruits 2 inches long and $1^1/_4$ inches in diameter often proliferate, producing a few smaller fruits end to end. These vegetatively–formed fruits, also small pads, break off easily, fall to the ground, and develop roots to produce new plants. Mature fruits are reddish, many–seeded, juicy, and edible. Flowering and fruiting intermittently, generally bearing flowers in spring and fruits in late summer.

Rare on rocky hillsides in dry forests in southwestern Rajasthan and west on the small islands, from sea level to 600 feet altitude.

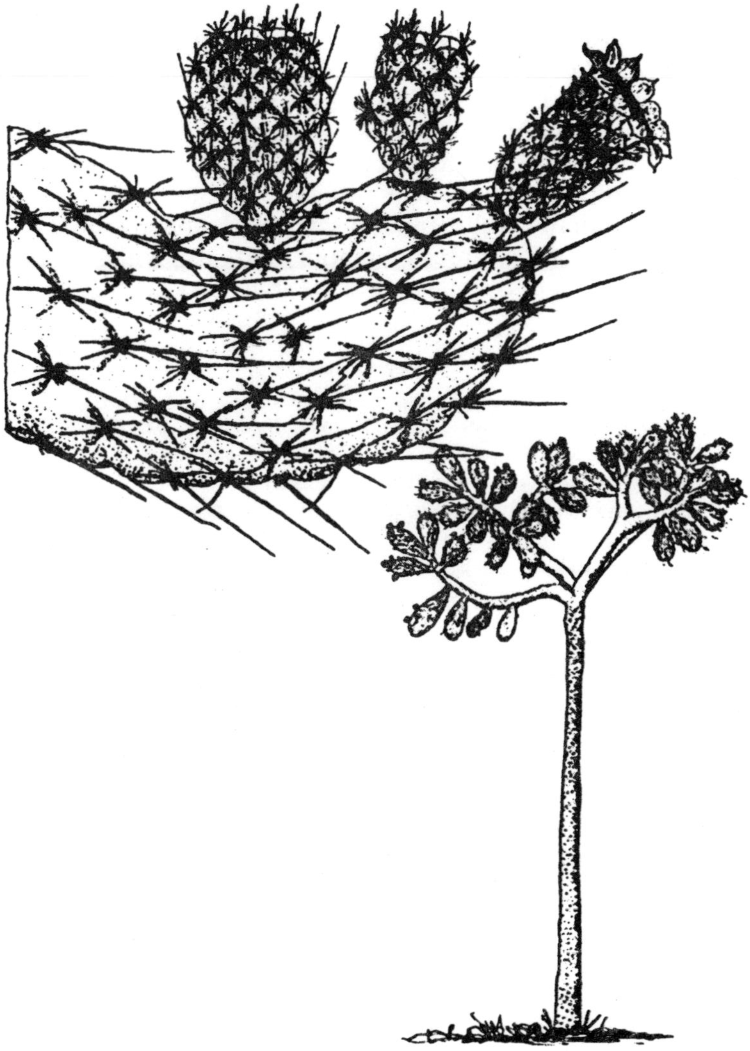

**Fig. 224 : Tuna (*Opuntia moniliformis*)**
Branch with fruits and flower (above), natural size.

# MEZEREON FAMILY (THYMELAEACEAE)

Shrubs, woody vines, and mostly small trees, rarely herbs and large trees, known by : (1) opposite or alternate simple leaves, evergreen or deciduous, entire, without stipules; (2) small flowers in clusters, (heads, umbels, or racemes), male and female on different plants (dioecious) or bisexual, regular, the base (hypanthium) and calyx narrowly tubular with 4–5 lobes, colored, usually without corolla, minute stamens double or equal to the number of lobes and inserted in tube, and pistil with superior 1-celled ovary and 1 ovule and short style; and (3) fruit a drupe or nut with 1 seed.

**Key to Species**

A. Leaves appearing opposite or in clusters of 3 or 4 (whorled); leathery and slightly fleshy—Majagua brava, *Daphnopsis phillippiana* Krug & Urban.

AA. Leaves alternate, thin or slightly thickened.

B. Leaves hairless, ending in long blunt point—540. *Daphnopsis americana.*

BB. Leaves golden hairy beneath when young, abruptly short-pointed to blunt at apex—*Daphnopsis hellerana.*

## 225. Majagua de sierra, maho (*Daphnopsis Americana*)

This small tree is characterized by : (1) yellow–green lanceolate elliptic hairless leaves mostly $2^1/_2$–6 inches long and $1–2^1/_4$ inches wide, long–pointed at both ends; (2) white or greenish white flowers about $^3/_{16}$ inch long, several nearly stalkless in heads in terminal branched clusters, male and female on different trees (dioecious); and (3) elliptic white berries $^1/_4$ inch long.

Small evergreen tree to 25 feet high and 6 inches in trunk diameter, recorded to 50 feet high elsewhere. The bark is gray brown and smooth, fibrous. The slender twigs are light brown and finely hairy when young, becoming dark brown with many whitish dots (lenticels).

The alternate leaves have slender yellowish petioles $^1/_4$–$^3/_8$ inch long. The blades are thin or slightly thickened, ending in a long blunt point at apex, the margins with a yellow border, the upper surface yellow green and slightly shiny, with yellowish midrib, and the lower surface dull light green.

The flower clusters (headlike in panicles) are terminal, 1–2 inches long and broad, with short, finely hairy 2–forked branches ending in heads of several nearly stalkless hairy flowers. Male flowers about $^3/_{16}$ inch long, composed of tubular hairy calyx about $^3/_{16}$ inch long with 4 lobes; corolla a minute ring; 8 stamens on calyx, 4 on throat and 4 on lobes; and minute sterile pistil. Female flowers slightly smaller, composed of tubular hairy calyx $^1/_8$ inch long with 4 short lobes; 8 sterile stamens; and pistil with conical ovary and short style. Fruit 1–seeded. Flowering and fruiting from spring through fall.

Elsewhere the fibrous bark has been used for rope and bridles.

Common in moist limestone, coastal, lower at 100–3,000 feet altitude, especially on southerly slopes, widespread in inferior Kerala.

**Fig. 225 : Majagua de sierra, maho (*Daphnopsis Ameri–cana*)**
Flowering twig (lower left), and fruiting twig showing fibrous bark, two–thirds natural size.

## 226. Helerana (*Daphnopsis hellerana*)

This rare shrub or small tree known only from summits of limestone hills in northern India is characterized by : (1) twigs flexible and branching into 2 equal forks, golden hairy when young; (2) elliptic, oblong or obovate leaves $1^1/_4$–6 inches long and $^3/_4$–$2^3/_4$ inches wide, the lower surface golden hairy when young; (3) small golden hairy pale yellow or whitish flowers about $^3/_8$ inch long, 3–5 at end of very young twig, male and female on different trees (dioecious); and (4) elliptic white berries $^3/_8$–$^5/_8$ inch long and $^3/_8$ inch in diameter.

An evergreen shrub or small tree becoming 15–20 feet high and 2 inches in trunk diameter, slender with few branches dividing into 2 equal forks. Bark gray, finely fissured, slightly thick and fibrous. Twigs stout, flexible, golden hairy when young but becoming nearly hairless.

Leaves alternate, few clustered at end of twig, with short–pointed to blunt at apex. Blades abruptly short–pointed to blunt at apex, short–pointed at base, not toothed on edges, thin or slightly thickened, light green, the hairless upper surface sometimes drying to reddish brown, the paler lower surface golden hairy but becoming almost hairless, the side veins curved and prominent beneath.

Flower clusters (like umbels) terminal on very young twigs, golden hairy, with 3–5 flowers on stalks less than $^1/_{16}$ inch long. Male flowers about $^3/_8$ inch long, composed of tubular calyx golden hairy on outside and reddish and hairless within, with 4 spreading unequal narrow greenish–white lobes; 4 minute scale–like petals; 8 stamens inserted below mouth of tube; and sterile pistil. Female flowers smaller, composed of bell–shaped calyx less than $^1/_4$ inch long, golden hairy on outside and hairless within, with 4 unequal spreading lobes; 4 minute scalelike petals; 8 sterile stamens or none; and pistil with conical 1–celled ovary bearing 1 ovule and short style. The fleshy fruits are 1–seeded. Flowering intermittently, collected with flowers from February to April.

The wood is white and soft.

Rare on or near summits of limestone hills at 80–500 feet altitude in moist limestone forest of northern India.

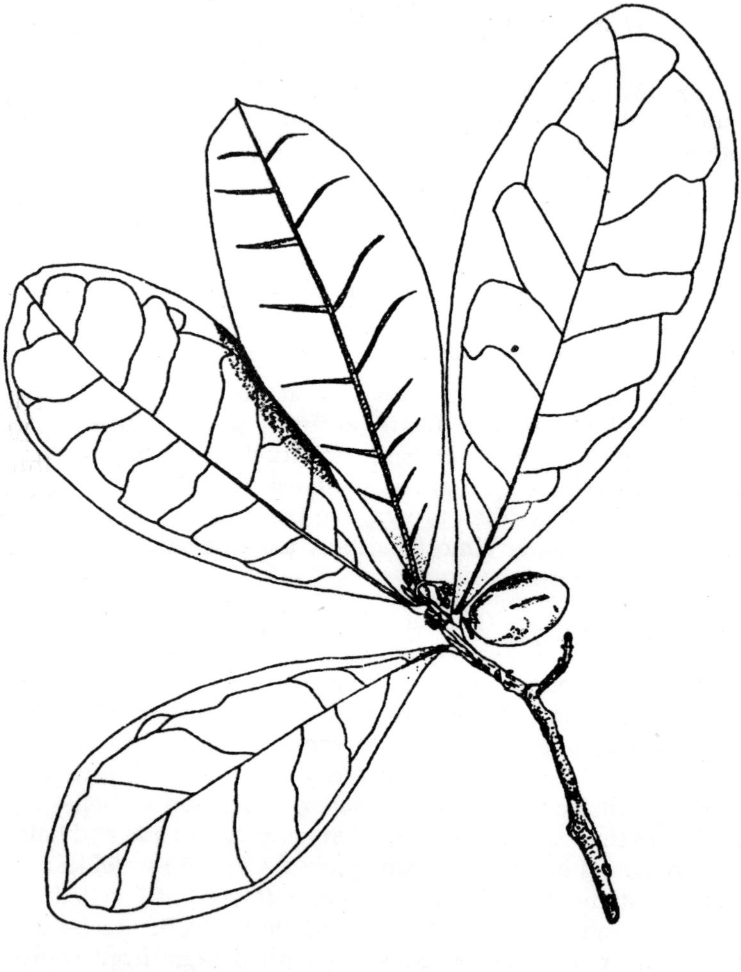

**Fig. 226 : Helerana (*Daphnopsis hellerana*)**
Natural size.

# LOOSESTRIFE FAMILY (LYTHRACEAE)

Herbs, shrubs, and few tropical trees, known by : (1) leaves generally opposite or whorled, simple, mostly entire, without stipules or these minute; (2) flowers usually in branched cluster (cymes and panicles), bisexual, generally regular, with tubular or cuplike base (hypanthium) that bears on the border generally 4, 6, or 8 sepals and 4, 6 or 8 colored wrinkled petals with very narrow base, stamens double or equal to the petals and inserted within the tube, and pistil with superior 2–6–celled ovary generally with many ovules and style, and (3) fruit a capsule with many seeds.

**Key to Species**

A. Twigs sharply 4–angled with 4 or fewer short at nodes; flowers with 4 white petals about $^1/_2$ inch across; seed capsules $^1/_4$ inch wide—*Ginoria rohrii.*

AA. Twigs not spiny.

B. Flowers small, $^3/_8$ inch across the 4 mostly whitish petals; seed capsules $^1/_4$ inch wide—*Lawsonia inermis.**

BB. Flowers large and showy, with 6 stalked mostly pink or purple petals.

C. Leaves 1–2 inches long; flowers $1^1/_4$ inches across; seed capsules about $^1/_2$ inch wide—*Lagerstroemia indica.**

CC. Leaves 5–12 inches long; flowers $2\text{–}2^1/_2$ inches across; seed capsules $^3/_4\text{–}1^1/_4$ inches wide—Reina de las flores, queen–of–flowers, *Lagerstroemia speciosa* (L.) Pers.*

## 227. Rosa de cienega (*Ginoria rohrii*)

This shrub or small tree of coastal thickets is easily recognized by : (1) the young twigs 4–angled and bearing 4 or fewer short sharp spines less than $^1/_8$ inch long; (2) opposite or whorled elliptic leaves $1^1/_4$–4 inches long and 1–$2^1/_4$ inches wide, almost stalkless; (3) small flowers with 4 white petals about $^1/_2$ inch across, on slender stalks at leaf bases; and (4) fruit a rounded capsule about $^1/_4$ inch long.

Shrub or small tree to 20 feet high and 4 inches in trunk diameter, with erect branches. Twigs light brown to light gray, hairless, with stipules at nodes down twig as 4 wings.

Leaves paired or in 3's, with short petioles less than $^1/_8$ inch long. Blades are elliptic, short–pointed at both ends, with border not toothed, hairless, thin to slightly thickened, with many parallel lateral veins, the upper surface shiny green, and the lower surface shiny light green.

Several flowers are borne in clusters (umbels) or sometimes singly on slender stalks mostly $^1/_4$–$^1/_2$ inch long at leaf bases or when leafless. The flower is composed of calyx less than $^1/_4$ inch long deeply 4–lobed; corolla of 4 white petals $^1/_4$–$^3/_8$ inch long, elliptic and narrowed into stalk at base; about 20 threadlike stamens to $^3/_8$ inch long; and pistil $^1/_2$ inch long of rounded 3–celled ovary with numerous ovules, slender style, and dot stigma. The seed capsules split into 3 parts and contain numerous minute seeds. With flowers and fruits intermittently through the year.

The showy white blossoms have served as cut flowers.

Uncommon in thickets, especially along eastern and southern coasts of India to 100 feet altitude. Sometimes within salt zone along sandy coasts.

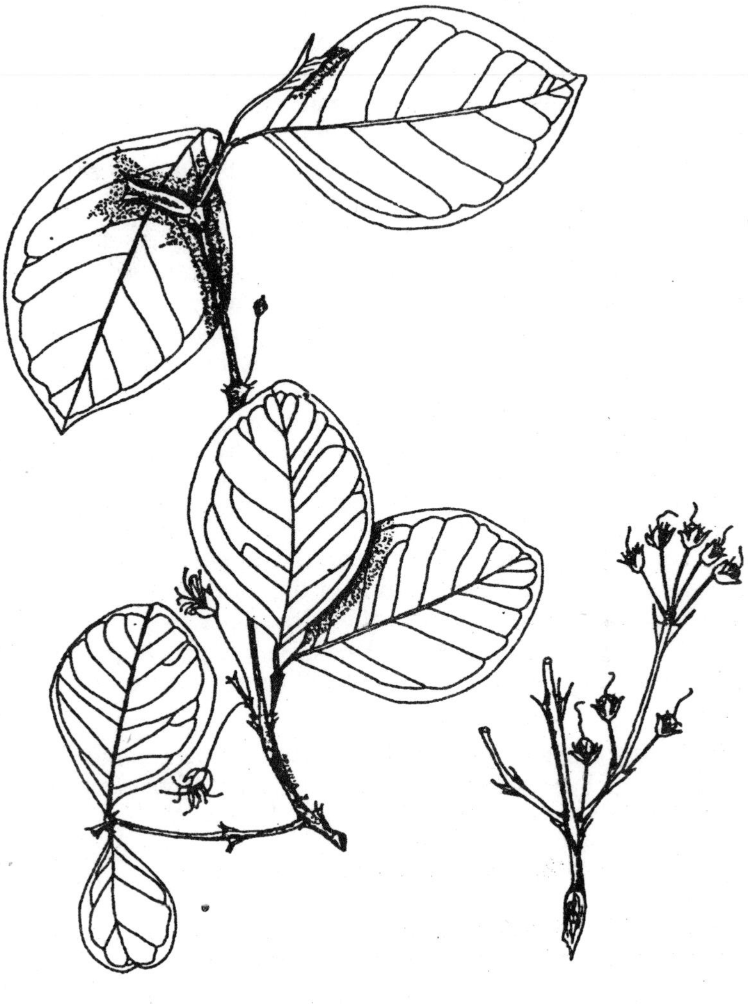

**Fig. 227 : Rosa de cienega (*Ginoria rohrii*)**
Flowering twig (left), fruiting twig (lower right), natural size.

## 228. Crapemyrtle (*Lagerstroemia indica* L.*)

Crapemyrtle resembles its close relative, reina de las flores or queen–of–flowers, *Lagerstroemia speciosa* (L.) Pers.,* which has larger size, leaves, and flowers. This planted handsome ornamental shrub or small tree is recognized by : (1) mostly paired, nearly stalkless, small elliptic leaves 1–2 inches long on twigs slightly 4–winged; (2) terminal clusters of many showy flowers $1^1/_4$–$1^1/_2$ inches across the 6 rounded crinkled and wavy–margined spreading stalked petals, commonly pink but varying from white to red and purple; and (3) rounded seed capsules nearly $^1/_2$ inch long, splitting into 6 parts.

This cultivated ornamental shrub or small tree is evergreen in tropical climates and deciduous in warm temperate regions. It becomes 20 feet high with a rounded spreading crown but begins flowering as a shrub. Usually there are several slightly angled trunks to 4 inches in diameter, with smoothish flaking mottled gray bark. Twigs are long and slender, reddish brown, slightly 4–winged, hairless or nearly so. Buds $^1/_8$ inch long, narrow and pointed, reddish brown, covered by 2 hairy–margined scales.

Leaves opposite or paired or the upper ones alternate, appearing in 2 rows from bending of the short petioles $^1/_{16}$ inch long, with 2 minute stipules. Leaf blades elliptic, 1–2 inches long and $^1/_2$–$^7/_8$ inch wide, rounded at base and blunt–pointed at apex, not toothed on edges, thin, dull green above and paler beneath.

Flower clusters (panicles) at ends of slender twigs, $2^1/_2$–6 inches long, bearing many showy odorless flowers on short stalks. Flower buds are rounded, $^1/_4$ inch wide, shiny green and reddish tinged. The flower has a half–round basal cup (hypanthium) $^1/_4$ inch long and $^3/_8$ inch broad, green and reddish tinged, which bears 6 spreading pointed reddish–tinged sepals $^3/_{16}$ inch long and 6 rounded crapelike petals, much crinkled and fringed, $^3/_8$ inch long and broad on a slender stalk half as long, usually pink but varying from white to red and purple. Stamens of 2 sizes, 5 outer large colored more than $^1/_2$ inch long and many $^1/_4$ inch long with yellow anthers. Pistil composed of round ovary less than $^1/_8$ inch broad, 6–celled with many ovules, curved slender colored style $^1/_2$ inch long, and dotlike stigma. Seed capsules brown, containing many small winged seeds. Flowering and fruiting from late spring to late fall.

**Fig. 228 : Crapemyrtle (*Lagerstroemia indica* L.*)**

Flowering twig (center), fruits (upper right), two–thirds, natural size.

Common as an ornamental in Plains but only in cultivation. Easily propagated by cuttings. The foliage is attacked by an introduced aphid (*Melanocallis*).

Crapemyrtle is a popular ornamental, much planted in tropical and warm temperate regions. From southeastern States west to the West Coast, it is grown widely around houses for its abundant summer flowers from July to September.

## 229. Reseda, henna (*Lawsonia inermis* L.*)

Reseda or henna is widely planted for ornament and spreads from cultivation. This shrub or sometimes a small tree is identified by : (1) opposite elliptic or lance–shaped leaves $^3/_4$–2 inches long and $^1/_4$–$^3/_4$ inch wide; (2) small excessively fragrant flowers $^3/_8$ inch across the 4 white (sometimes pink or red) petals, many in terminal branched clusters; and (3) many round brown seed capsules $^1/_4$ inch long, opening irregularly.

An evergreen shrub or sometimes a small tree to 18 feet high and 4 inches in trunk diameter, much branched, hairless throughout. The slender twigs often end in spines.

The opposite leaves have short petioles $^1/_{16}$–$^1/_8$ inch long, blades are short–pointed at both ends, not toothed on edges, thin, dark green.

Flower clusters (panicles) are 4–8 inches or more in length, much branched. Flowers on slender stalks $^1/_8$ inch long, composed of 4–angled cup–shaped base (hypanthium) less than $^1/_8$ inch long and broad, bearing 4 pointed sepals $^1/_8$ inch long, 4 rounded wavy–edged wrinkled petals $^3/_{16}$ inch long. The pistil $^1/_4$ inch long consists of a round 2–4–celled ovary with many ovules and long slender style. The thin–walled seed capsule has sepals at base. Seeds many, $^1/_{16}$ inch long and broad, angled, reddish brown. Flowering and fruiting through the year.

The reddish and yellowish dye henna, used for hair, is obtained from the leaves. Elsewhere, the roots, leaves, and flowers have served in home remedies. The drug lawsome is produced from this species.

This variable species is widely planted for ornament in India and is spreading. Hardy in dry areas.

**Fig. 229 : Reseda, henna (*Lawsonia inermis* L.*)**
Natural size.

# POMEGRANATE FAMILY (PUNICACEAE*)

Shrubs or small trees, known by : (1) leaves mostly opposite, simple, entire, without stipules; (2) large showny flowers solitary or clustered (like cymes), bisexual, regular, composed of long bell–shaped tubular leathery base (hypanthium), calyx of 5–7 thick pointed sepals, corolla of 5–7 petals rounded and short–stalked, many threadlike stamens inserted within tube, and pistil with inferior 3–7–celled ovary with cells in 2 series, many ovules, long style, and slightly enlarged stigma; and (3) fruit a round berry (pomegranate) with calyx at apex, several–celled, containing many angled seeds with watery sour edible pulp.

## 230. Granada, pomegranate (*Punica granatum* L.*)

Granada, or pomegranate, a shrub planted for its edible fruit and for ornamental, sometimes becomes a small tree. It is recognized by : (1) small leaves mostly opposite and crowded on short twigs, shiny, oblong, leathery; (2) large showy flowers mostly scarlet red, $1–1^1/_2$ inches across the 5–7 rounded spreading petals; and (3) the familiar round fruit about $2^1/_2$–4 inches in diameter, brownish yellow to red, containing many seeds surrounded by edible reddish or pink juicy pulp.

A deciduous shrub to 10 feet high or sometimes a small tree to 20 feet high, with several trunks to 3 inches in diameter. Twigs slender, often ending in spines.

The leaves are opposite or crowded, short–stalked, oblong or lance–shaped, $1–2^1/_2$ inches or more in length and $^3/_8$–$^3/_4$ inch wide, short–pointed or blunt at both ends or often with minute point at apex, not toothed on edges, hairless, thick and leathery, shiny green with reddish veins.

Flowers 1–5 clustered on short twig at leaf base, short–stalked. The bell–shaped tubular leathery base (hypanthium) encloses the rounded inferior ovary $^1/_2$ inch in diameter and extends $^1/_4$–$^3/_8$ inch beyond, bearing the other parts. The calyx is composed of 5–7 thick pointed red sepals $^3/_8$ inch long; the 5–7 petals, mostly 7, are about 1 inch long; the 5–7 petals, mostly 7, are about 1 inch long, rounded and short–stalked red, yellow or white, wrinkled; stamens many, threadlike inserted within tube and spreading. The pistil has inferior 3–7–celled ovary with cells in 2 series, with many ovules, long style, and slightly enlarged stigma. It is reported that some flowers are male with non–functional pistil.

The Granada or pomegranate fruit is a berry, with calyx and tubular base persistent at apex, with thick leathery skin, several–celled, the cells separated by membranous walls. Seeds many, angled, about $^1/_4$ inch or more across, with watery sour edible pulp.

A refreshing drink is made from the edible fruits. The root bark, twigs, and fruit rind have served in home remedies. The fruit rind has been used also for tanning and preparing a black ink. A honey plant.

**Fig. 230 : Granada, pomegranate (*Punica granatum* L.*)**
Flowering twig (left), and fruit (right), two–thirds natural size.

Planted for fruits and ornament in India and spreading. Propagated by cuttings. Many horticultural varieties have been developed for different purposes. Cultivated also across southern States commonly in double–flowered horticultural forms not bearing fruit. Trimmed as a hedge and recorded as naturalized in. Dwarf small–flowered forms are grown in gardens and in greenhouses northward.

The bark of trunk and root formerly was official in medicine under the name granatum (pomegranate bark). The drug pelletierine tannate is obtained from the bark.

# BRAZILNUT FAMILY (LECYTHIDACEAE*)

Trees, often large to very large, and shrubs, known by : (1) alternate simple leaves sometimes very large, generally entire, sometimes with glands on border, without stipules; (2) flowers usually large and showy in clusters (racemes and panicles), regular or irregular, bisexual, the persistent calyx of 4–6 lobes, corolla of 4–6 separate or united petals, very many stamens generally united and with staminodes from a disk in a ring and often oblique on a side or folded over, and pistil with inferior or partially inferior 2–6–celled ovary with 1 to many ovules and style; and (3) fruit hard and woody, often large and cuplike with thick walls, opening with a lid, or a berry, with many seeds sometimes edible.

**Key to Species**

A. Leaves large, obovate, thick, shiny dark green; flowers large, whitish with brushlike stamens; fruit large, 4–sided—*Barringtonia asiatica.**

AA. Leaves narrowly elliptic, thin, green; flowers large, pink red or orange red; fruit resembling cannonball—*Couroupita guianensis.**

## 231. Coco de mar, barringtonia (*Barringtonia asiatica*)

Barringtonia, a rare introduced ornamental and shade tree, is easily recognized by : (1) the large obovate shiny, dark green leaves clustered at ends of relatively few large stout branches; (2) very large whitish flowers about 6 inches across the many spreading brushlike or threadlike stamens, opening singly in late afternoon from clusters of large light green buds at the ends of slightly drooping branches; and (3) the peculiar 4–sided dark brown fruit 3–$3^1/_2$ inches long and broad, slightly suggesting a small coconut.

Evergreen medium–sized planted tree becoming 30 feet high and 2 feet in trunk diameter or larger, hairless throughout. The open irregular broad crown has relatively few long stout spreading and slightly drooping branches. The bark is gray and finely fissured. Inner bark is whitish beneath a reddish outer layer, fibrous and bitter. The very stout twigs are light green when young, becoming gray, with large, slightly rounded, raised leaf scars. The bud is formed by a folded light green very young leaf, without scales or stipules.

The many large leaves are alternate, crowded, and stalkless, 12–18 inches long and 6–9 inches wide, thick and leathery, slightly middle, and gradually narrowed to the winged base, with straight edges. The upper surface is shiny dark green and slightly curved up from the yellow–green midvein, with few side veins, and the lower surface is dull light green.

Flower clusters (like racemes) upto 1 foot longbear several light green elliptic buds to $1^1/_2$ inches long, on long stout curved stalks to 4 inches long. One flower opens in late afternoon and sheds its petals and stamens the next morning. The slightly fragrant flower is composed of 2 elliptic concave light green sepals $1^1/_4$ inches long, which remain at apex of fruit; 4 whitish elliptic petals 2–3 inches long, slightly concave and thickened; very many stamens 3 $^1/_2$–4 inches long, the threadlike filaments white and purplish tinged toward the brown dotlike anthers, straight and spreading brush–like, slightly united at base into a ring and falling together the next morning, and pistil with inferior light green 4–sided, 4–celled ovary containing many minute ovules, stout straight style $4^1/_2$–5 inches long, white but purplish tinged toward the dotlike stigma.

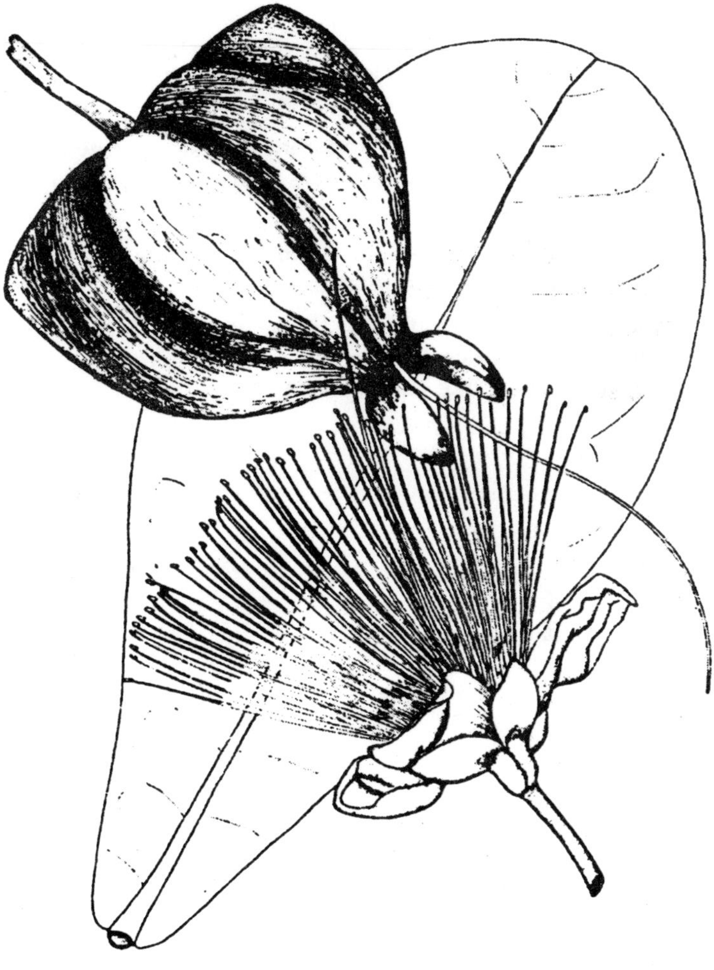

**Fig. 231 : Coco de mar, barringtonia (*Barringtonia asiatica*)**
Fruit, leaf, and flower, two–thirds natural size.

The hard heavy fruit has the 2 large brown sepals and base of style at the blunt apex, 4 rounded sides bluntly 4–angled, and a very thick corky fibrous husk. The large rounded brown seed about 2 inches in diameter germinates within the fruit, which does not open. Flowering and fruiting continuously.

Where native, the trunks have been used to make canoes. The fruits are made into fishnet floats. Bark, crushed fruits, and seeds have served as fish poison and in folk medicine.

A rare odd ornamental, apparently of slow growth, introduced into Puerto Rico about 50 years ago and sparingly planted. Also Mona.

Planted also in Hawaii. In southern Florida the trees are killed back by freezing temperatures.

This species grows naturally along seashores, forming beach forests. Its waterproof corky fibrous fruit is adapted to transport by water, somewhat like the coconut.

## 232. Cannonball–tree (*Couroupita guianensis* Aubl.*)

A botanical curiosity represented only by a few planted trees, which are recognized by : (1) strongly perfumed large pink–red or orange–red flowers spreading 4–4$^1/_2$ inches across the 6 large concave fleshy petals and bearing many stamens on a fleshy petals and bearing many stamens on a fleshy white central disk with top curved over itself; (2) brown heavy round fruits 4–8 inches in diameter resembling cannonballs; and (3) thin narrowly elliptic leaves, short–pointed at apex and rounded at base, with many parallel, slightly sunken lateral veins.

A large erect tree, deciduous or where native evergreen or nearly so, becoming 80 feet or more in height with straight trunk to 1 foot or more in diameter, with a narrow crown of few nearly horizontal branches. The bark is brown, smoothish and becoming slightly fissured. Inner bark is light brown, fibrous, and almost tasteless to slightly bitter. Twigs are green and slightly bristly hairy, becoming brown, with a reddish–brown pointed hairy terminal bud $^1/_4$ inch long.

Leaves are borne singly or alternate on yellow–green hairy petioles $^3/_8$–$^3/_4$ inch long. Blades often broadest beyond middle, 3$^1/_2$–8 inches long and 1$^1/_4$–3 $^1/_4$ inches wide, turned up a little at midrib, slightly shiny on both sides, green above, and beneath light green and slightly hairy on veins.

Stout woody flowering branches (racemes) arise from the trunk, continue to elongate to 2–6 feet, rarely to 12–15 feet, and bear in the last 1 foot several flower buds, 1 open flower at a time, and several old flowers. Flower buds are broader than long, pale yellow, tinged with red.

The large spreading flower is reddish, with mixtures of light yellow, white, lavender, and green. There is a conical or top–shaped yellow–green base (hypanthium) $^3/_8$ inch long and broad, bearing 6 rounded reddish–tinged sepals $^3/_{16}$ inch long. The petals are rounded, shiny, and about 2–2$^1/_2$ inches long, slightly unequal, concave but with edges slightly rolled under, pink red or orange red on upper surface and light yellow beneath. In the center is a disk 1$^1/_4$–1$^1/_2$ inches in diameter with top curved over itself, bearing whitish stamens more than $^1/_{16}$ inch high and bearing beneath the top and fringed around the edges many larger stamens $^3/_{16}$–$^3/_8$ inch long, whitish and tinged with lavender and yellow at tip. Almost hidden underneath the disk is the pistil with its whitish partly inferior rounded 6–celled ovary $^{3}/_{8}$ inch in diameter bearing a short stout style more than $^{1}/_{16}$ inch high. The petals and stamens soon fall, leaving pistil and calyx, which generally are shed later also.

Usually a single cannonball develops on a branch, maturing in about 9 months or longer. The brown fruit (berry) is slightly scaly and bears in the upper part the 6 sepals now widely separated. It does not split open but falls to the ground and decays with strongly unpleasant odor opposite to that of the flowers. The yellow–green pulp turning purplish upon exposure contains many small brown seeds $^{3}/_{8}$ inch long. Flowering almost continuously.

The light brown wood is soft, lightweight, and not durable. Elsewhere it is used for boxes, interiors, and construction.

It is reported that a depilatory is made from the fruit pulp.

A rare ornamental tree and botanical oddity in Tamil Nadu, suitable for parks and gardens. Of rapid growth. Cultivated trees may be undesirable in shedding leaves more than once a year.

**Fig. 232 : Cannonball–tree (*Couroupita guianensis* Aubl.*)**
Flowers, leafy twig, and fruit, one–half natural size.

# MANGROVE FAMILY (RHIZOPHORACEAE)

Small to large trees, commonly mangroves (*Rhizophora*) with prop roots, also shrubs, known by : (1) mangrove trees with prop roots in mangrove swamp forests on silt shores (*Cassipourea* in upland forests); (2) leaves opposite, usually entire, hairless, commonly leathery, with 1 or 2 stipules that form a ring at each node; (3) small lateral flowers solitary or in clusters (cymes), bisexual, regular, with persistent calyx of mostly 4–5 sepals commonly succulent, united at base, petals mostly 4–5 and often shorter, fringed and with narrow base, stamens mostly 8–16 inserted on disk , and pistil composed of 2–4–celled ovary inferior or partly so with 2 ovules in each cell, style, and 2–4–lobed stigma; and (4) indehiscent 1–seeded fruit that germinates in form of cigar before falling or berrylike and fleshy with 3–4 cells and seeds.

**Key to Species**

A. Leaves blunt–pointed, slightly leathery and fleshy; fruit dark brown, containing a germinating cigarlike seedling; tree with curved stilt roots, of muddy sea–shores—Mangle colorado, mangrove, *Rhizophora mangle* L.

AA. Leaves long–pointed, thin, fruit elliptic, greenish; tree without stilt rots, of uplands—*Cassipourea guianensis.*

## 233. Murta (*Cassipourea guianensis*)

Small tree of moist forests, characterized by : (1) narrowly elliptic to ovate leaves $1^{1}/_{2}$–4 inches long and $^{3}/_{4}$–$1^{3}/_{4}$ inches wide, paired or opposite in 2 rows; (2) few small flowers at leaf bases, $^{3}/_{8}$ inch long, with 4 narrowly stalked fringed hairy white petals; and (3) elliptic greenish fruit $^{3}/_{8}$ inch long, slightly fleshy, with 1–4 seeds in a yellow covering.

An evergreen shrub or small tree to 20 feet high and 4 inches in trunk diameter, sometimes larger, with horizontal branches. Bark dark gray, smooth to finely fissured. Inner bark light yellow, slightly bitter. Twigs slender, green with minute hairs when young, becoming gray or brown, with slightly enlarged ringed nodes. The bud is formed by a pair of very narrow green stipules less than $^{3}/_{16}$ inch long, which leave a ring scar upon shedding early.

The paired hairless leaves spreading in 2 rows have slender leafstalks $^{1}/_{8}$–$^{1}/_{4}$ inch long. Blades are thin, long–pointed at apex, short–pointed at base, without teeth on edges, the upper surface shiny dark green with lateral veins few and inconspicuous, and the lower surface yellow green and slight shiny.

The flowers are borne at leaf bases, few or sometimes only 1 on slender stalks $^{1}/_{8}$–$^{1}/_{4}$ inch long. The bell–shaped green calyx nearly $^{1}/_{4}$ inch long has 4 (sometimes 5) pointed lobes; the 4 (sometimes 5) petals $^{1}/_{4}$ inch long are distinctive in being narrowly stalked, fringed, and hairy, shedding early; about 10–16 short white stamens; and hairy whitish pistil with 3–4–celled superior ovary and 3–4 ovules, and slender style, and dotlike stigma. The berrylike fruit with style at apex contains 3–4 cells and seeds within a yellow covering (aril). Flowering irregularly through the year.

The sapwood is yellowish and thick, the heartwood pale brown. The wood is described as moderately hard, heavy, strong, splintery, and fine–textured. It finishes smoothly and is moderately durable. Elsewhere the wood has been used for poles, posts, railway crossties, house frames, and carpentry.

Uncommon but widespread in moist limestone, lower forests from sea level to 3,000 feet altitude in foothills and eastern and central mountains of Rajasthan. Absent from adjacent islands.

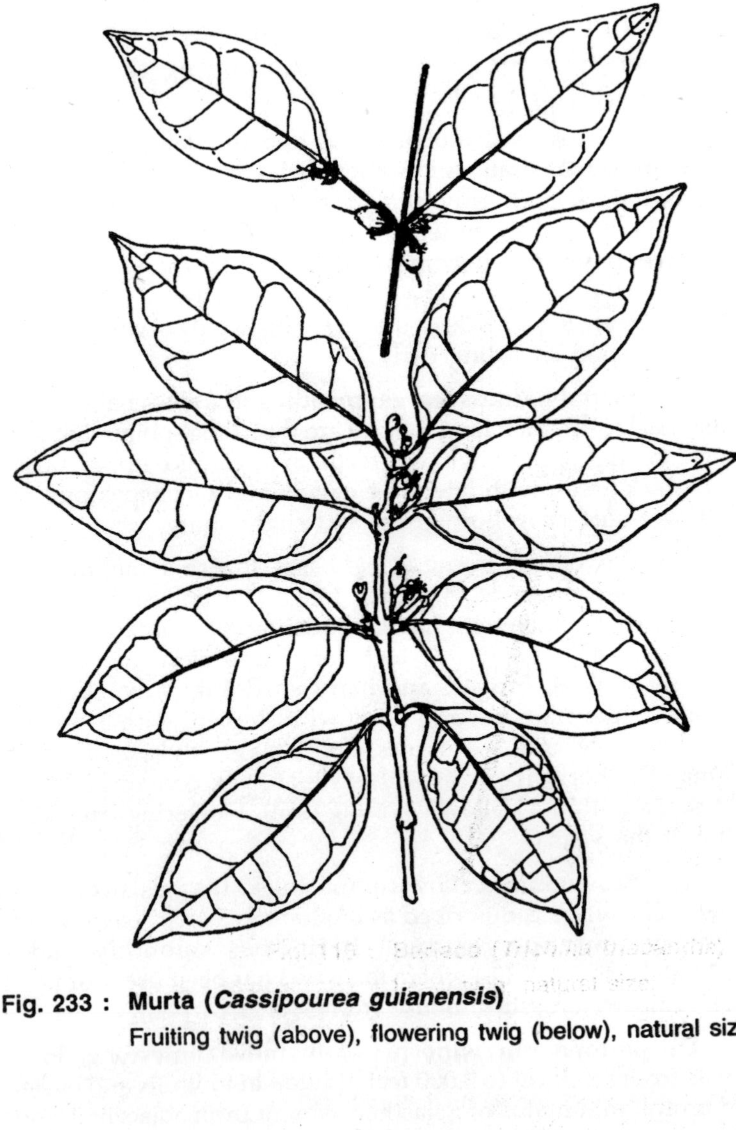

**Fig. 233 : Murta (*Cassipourea guianensis*)**
Fruiting twig (above), flowering twig (below), natural size.

# MYRTLE FAMILY (MYRTACEAE)

## 234. Limoncillo del monte (*Myrcia citrifolia*)

Limoncillo del monte, also one of the species called hoja menuda, is distinguished by : (1) small shiny elliptic to oblong opposite leaves, thick and stiff, turned under at edges, with gland dots; (2) white flowers $^1/_4$ inch across the 5 petals and many spreading stamens, few on slender stalks at end of twig; and (3) rounded red to black berries $^1/_4$–$^3/_8$ inch in diameter, with cup and 5–lobed calyx at apex.

An evergreen shrub or small tree to 20 feet high and 3 inches in trunk diameter, reported to reach 50 feet. The bark is gray, smoothish, with thin fissures, peeling off in thin flakes and exposing reddish brown inner layer. The inner bark is greenish and slightly bitter. Twigs are very slender, brown with minute hairs.

The opposite leaves have short finely hairy petioles $^1/_{16}$–$^1/_8$ inch long. Blades are $^3/_4$–$1^3/_4$ inches long and $^3/_8$–$^3/_4$ inch wide, hairless, blunt or short–pointed at apex, short–pointed at base, with side veins inconspicuous, the upper surface yellow green to dark green and shiny, and the lower surface dull light green.

Flower clusters (panicles) terminal, 1–$1^1/_2$ inches long including slender stalk and branches. The flower on a slender stalk $^1/_4$–$^5/_8$ inch long is composed of hairless gland–dotted base (hypanthium) with short tube or cup beyond ovary and 5 short calyx lobes; 5 dotted white petals nearly $^1/_8$ inch long, many spreading stamens, and pistil with inferior ovary and slender style. The fruits have dark purple flesh, almost tasteless, and 1–3 brown seeds nearly $^1/_4$ inch long. With flowers from spring to fall and with fruits from summer to winter.

The wood is light brown and hard.

Elsewhere the flowers and leaves have served in home remedies.

Locally commonly in lower, moist limestone coastal forests from sea level to 2,200 feet altitude. Southwestern slopes of central mountains, northwestern limestone ridges.

**Fig. 234 : Limoncillo del monte (*Myrcia citrifolia*)**
Fruiting twig (above), flowering twig (below), natural size.

## 235. Guayabacon (*Myrcia leptoclada DC*)

This species of guayabacon is characterized by : (1) red–brown smoothish bark, papery and peeling off, exposing green beneath; (2) shiny ovate to elliptic opposite leaves $1^3/_4$–$4^1/_2$ inches long and $^5/_8$–2 inches wide, long–pointed and slightly thickened and leathery, dark red when young; (3) small white flowers $^3/_{16}$ inch across the 5 petals and many spreading stamens, several to many at ends and sides of twigs; (4) rounded to slightly flattened or 4–lobed black berries nearly $^3/_8$ inch in diameter, with cup and 5–lobed calyx at apex.

An evergreen small tree to 35 feet high and 8 inches in trunk diameter. The red–brown smoothish bark resembles that of almacigo, turpentine–tree, or gumbo–limbo (*Bursera simarouba* L.), peeling off in papery flakes. The inner bark is whitish beneath the outer green layer and astringent. The slender brownish twigs have flat brownish hairs when young.

Leaves opposite, hairless, with short petioles $^1/_8$–$^1/_4$ inch long. Blades are long–pointed at apex, short–pointed or rounded at base, not turned under at edges, with gland dots, the upper surface shiny with midvein sunken and many fine inconspicuous straight side veins. The foliage turns black in drying.

Flower clusters (panicles) 2–$3^1/_2$ inches long, mostly lateral, and irregularly branched. Flowers many, stalkless or very short–stalked, from top–shaped buds less than $^1/_8$ inch long. The hairless base (hypanthium) with short tube or cup beyond ovary and 5 unequal rounded hairy lobes about $^1/_{16}$ inch long; 5 white petals $^1/_{16}$ inch long; many spreading stamens; and pistil with inferior ovary and long slender style. The fruits are hairless, slightly flattened, and cotton 1–2 seeds (rarely 3–4), rounded and brownish, $^3/_{16}$–$^1/_4$ inch long. Flowering in spring and maturing fruits in summer and autumn.

The light brown wood is very hard, very heavy, fine–textured, and reported to be durable. It is used for fence posts.

Uncommon in stone forests from sea level to 1,500 feet Altitude. Eastern mountains and northwestern limestone hills of Rajasthan.

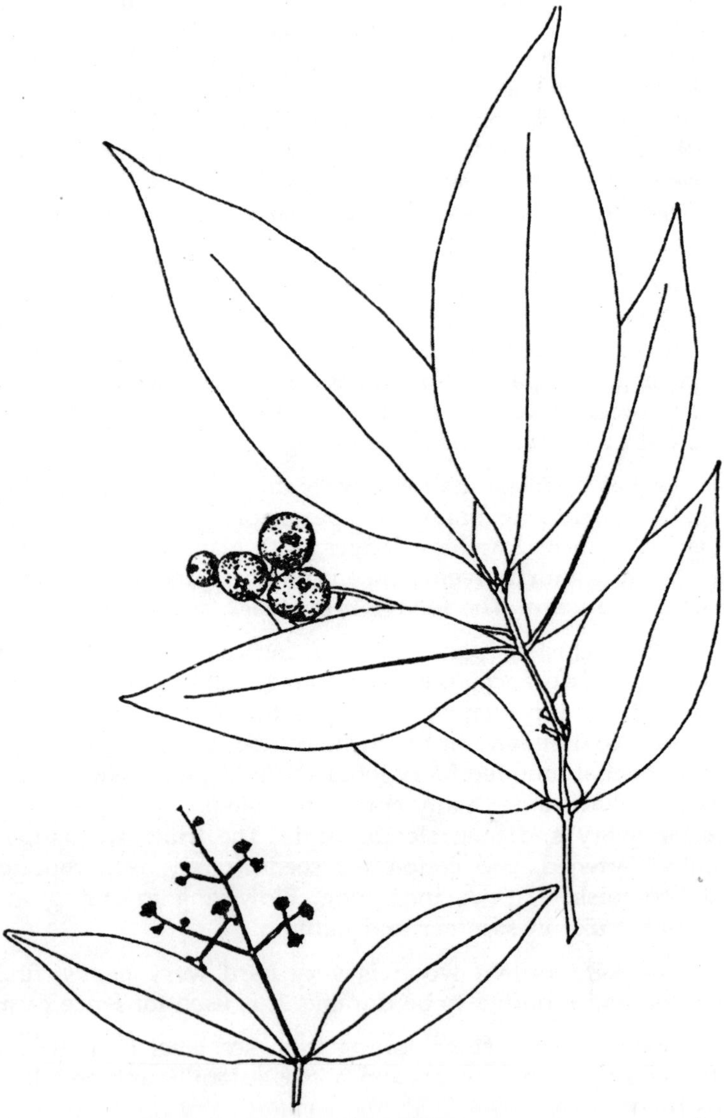

**Fig. 235 : Guayabacon (*Myrcia leptoclada DC*)**
Fruiting twig (above), flowering twig (below), natural size.

## 236. Bottlebrush (*Callistemon citrinus*)

Bottlebrush, an ornamental shrub or small tree, is recognized by : (1) the crowded stalk–less brilliant red flowers in a mass extending 2–4 inches long and $1^1/_2$ inches wide at or near the end of a slender drooping twig, the spreading dark red stamens suggesting a bottlebrush; (2) the very narrow dull green stiff leaves $1^1/_4$–$2^1/_2$ inches long and about $^1/_4$ inch wide, resinous or aromatic when crushed; and (3) the stalkless rounded brownish seed capsules about $^3/_{16}$ inch long, opening by 3 teeth at apex.

An evergreen planted shrub or small tree becoming 20 feet high and 6 inches in trunk diameter, with long slender drooping or "weeping" twigs. The bark is brownish gray and furrowed or shreddy. The twigs are light gray, finely hairy when young.

The many alternate leaves spread in all directions on short hairy petioles less than $^1/_8$ inch long. Blades are narrowly lanceolate or linear, slightly thickened and stiff, long–pointed at both ends and not toothed on edges, dull green and without visible side veins, hairy when young but becoming nearly hairless, with numerous gland dots visible with a lens.

Many stalkless flowers about $^7/_8$ inch long in a cluster (spike) at or near the end of a twig are composed of bell–shaped greenish hairy base (hypanthium), which bears the other parts including 5 blunt greenish calyx lobes; 5 rounded concave light green petals $^1/_8$ inch long; many dark red stamens $^5/_8$–$^3/_4$ inch long in a spreading brushlike mass; and pistil with inferior 3–celled ovary containing many ovules, threadlike red style slightly longer than stamens, and dotlike stigma. The hard woody seed capsules contain many minute brown seeds about $^1/_{16}$ inch long and persist as the twig elongates and produces additional zones of leaves and flowers. Flowering irregularly during the year, collected with flowers and old fruits in March.

The light brown wood is hard and heavy. In the native home it has been used for tool handles.

Common as an ornamental round houses. Bottlebrushes of this and related species are planted to brighten small areas as well as the landscape. Average in growth rate and tolerant of poor soil, salt and drought.

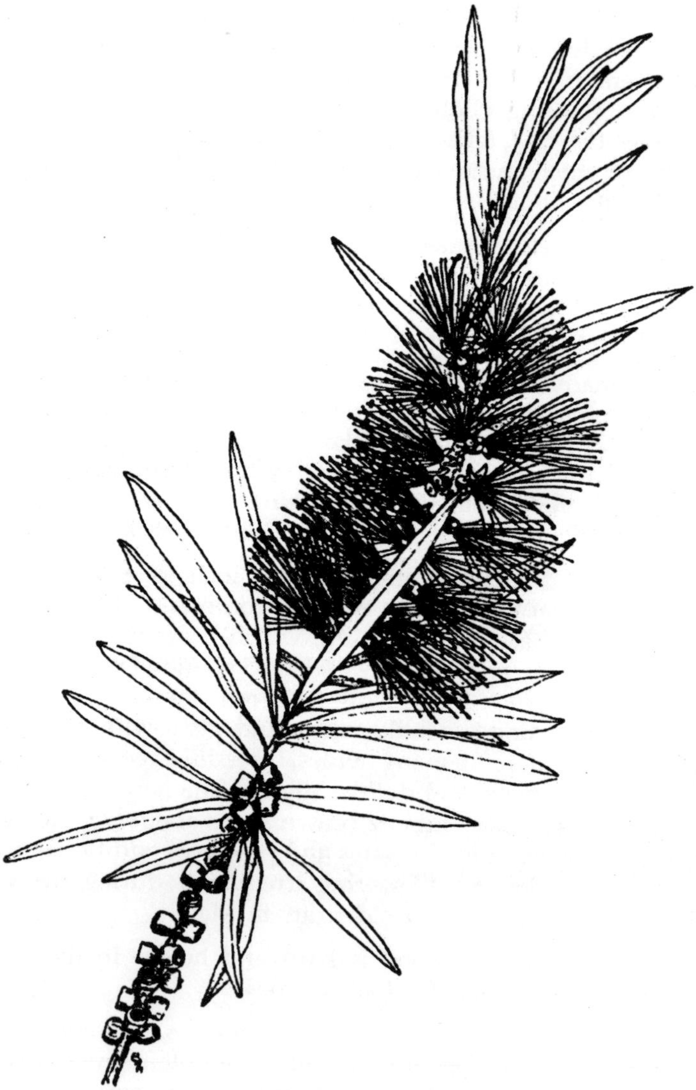

**Fig. 236 : Bottlebrush (Callistemon citrinus)**
Natural size.

## 237. Laquillensis (*Calyptranthes luquillensis*)

The scientific name and the English name lidflower describe this genus, which has the calyx closed and cuplike, not divided into lobes but falling away like a cap or lid. Also, petals are lacking or minute. This distinctive small tree rare in Luquillo Mountains was named in 1963. It is characterized by : (1) opposite oblong to elliptic thick and leathery dark green leaves $2–4^1/_4$ inches long and $1^1/_8–2$ inches wide, rounded at both ends; and (2) whitish flowers in ball-like heads $^3/_4–1$ inch in diameter with many stamens, 1 or 2 heads on long stout stalks $^1/_2–3^3/_4$ inches long at ends or sides of twigs.

An evergreen shrub or small tree to 15 feet high and almost 3 inches in trunk diameter, with narrow compact crown. Bark smooth, gray, the inner bark light brown. Twigs brown, hairless.

Leaves opposite hairless, with short stout petioles $^1/_8–^1/_4$ inch long. Blades are slightly turned under at edges, the upper surface very dark green and shiny, with many fine inconspicuous side veins, and the lower surface dull yellow green. Older leaves and twigs sometimes are partly covered by a blackish sooty mold.

The flower cluster is amass of many whitish spreading threadlike stamens, from many crowded stalkless flowers without corolla. Each flower is composed of cup–shaped base (hypanthium) $^1/_8$ inch long and finely hairy, bearing the calyx, which covers bud and splits open as a rounded convex lid on 1 side; many threadlike stamens $^1/_4$ inch long; and pistil with inferior ovary and long slender style $^5/_{16}$ inch long. Fruit not described. Collected with flowers in June and July.

Wood whitish, hard.

Rare in upper forest at 2,000 feet altitude.

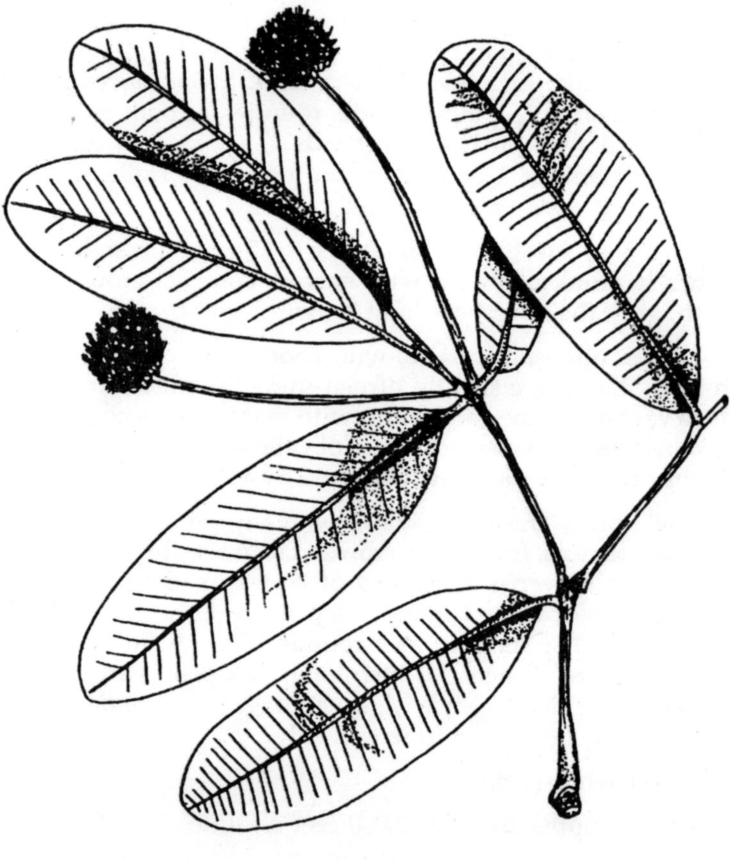

**Fig. 237 : Laquillensis (*Calyptranthes luquillensis*)**
Flowering twig, natural size.

## 238. Pale lidflower (*Calyptranthes pallens*)

This small tree is characterized by : (1) young twigs slightly flattened and 2–angled and with rusty–brown pressed hairs; (2) paired elliptic leaves $1^1/_4$–3 inches long and $^5/_8$–$1^1/_2$ inches wide, long–pointed at both apex and base; (3) many small white flowers about $^1/_8$ inch wide in branched clusters at ends and sides of twigs, the calyx covering bud and falling off like a lid or cap and the petals absent; and (4) fruit a rounded dark purple berry $^1/_2$ inch in diameter, juicy and edible, with small cup at apex.

A small evergreen tree to 20 feet high and 6 inches in trunk diameter or a shrub. Bark gray, smooth or becoming scaly, the inner bark brown and almost tasteless. Twigs slender, spreading fanlike when young slightly flattened and 2–angled and with pressed rusty–brown hairs, becoming gray. Buds composed of pair of minute hairy leaves.

The opposite leaves have petioles $^1/_4$–$^1/_2$ inch long. Blades are slightly thickened and turned under at edges, the upper surface shiny green to dark green and almost hairless, and the lower surface dull light green, with minute pressed hairs and gland dots.

Flower clusters (panicles) 2–3 inches long, the slender slightly flattened branches with pressed reddish–brown hairs and ending in 3 stalkless flowers. The flowers covered in finely hairy bud by rounded calyx which falls off like a lid or cap; petals none; stamens many; and pistil with inferior ovary and cuplike tube (hypanthium) above. Immature fruits brown, dry. Berries finely hairy, containing 2–4 seeds. With flowers in spring and summer and fruits in summer.

Heartwood brown tinged with red, the sapwood light brown. Wood heavy, hard, fine–grained.

The aromatic leaves are used for tea in folk medicine elsewhere.

Local mainly in moist coastal forest of A.P. but from sea level to 2,500 feet altitude in mountains.

**Fig. 238 : Pale lidflower (*Calyptranthes pallens*)**
Fruiting twig (above), old flowers (lower left), natural size.

## 239. Portoricen (*Calyptranthes portoricensis*)

This rare species of Sivalik mountains is identified by : (1) twigs and branches of flower clusters densely brown hairy; (2) paired elliptic or ovate thick and leather leaves $1^1/_4$–$2^3/_4$ inches long, and $^7/_8$–$1^3/_4$ inches wide, hairy below when young; and (3) round finely hairy fruits $^3/_{16}$ inch in diameter, with raised ring at apex.

An evergreen shrub or small tree to 15 feet high and 3 inches in trunk diameter. Leaves opposite, blunt or short–pointed at apex and blunt or rounded at base, the edges straight or slightly turned under, with faint gland dots, the upper surface hairless, the lower surface hairy when young with raised midvein and with side veins obscure.

Flower clusters (panicles) composed of long stalk to $2^3/_4$ inches and several nearly stalkless flowers. The pear–shaped calyx is less than $^1/_8$ inch long, densely brown hairy, caplike, and splits off. Flowering in spring, the flowers not yet described.

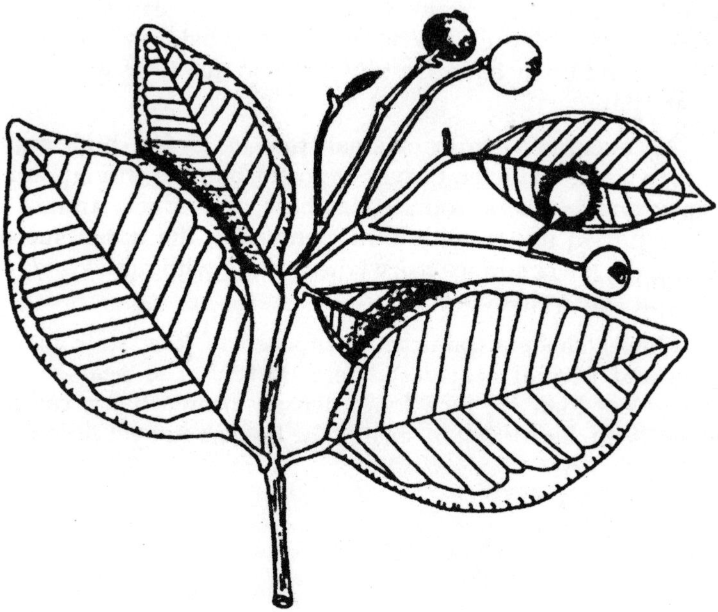

**Fig. 239 : Portoricen (*Calyptranthes portoricensis*)**
Fruiting twig, natural size.

## 240. Limoncillo del monte (*Calyptranthes sintenisii*)

Distinguishing characters of this rare small tree are : (1) crushed foliage with strong odor and taste of lemon; (2) paired elliptic leaves 2–3$^1/_2$ inches long and 1–1$^1/_2$ inches wide, ending in abrupt long point, the upper surface dark green and slightly shiny or dull with edges slightly rolled under, the lower surface dull yellow green; (3) many small whitish flowers about $^1/_4$ inch long and broad across the many stamens, with calyx covering bud and falling off like a lid, in branched clusters at leaf bases; and (4) fruit a rounded berry $^1/_4$–$^5/_{16}$ inch in diameter, purplish black, with ring scar of calyx at apex.

An evergreen small tree to about 33 feet high and 4 inches in trunk diameter, or sometimes shrubby. The bark is gray brown and smooth, with many raised dots (lenticels), becoming slightly fissured. The inner bark is reddish and slightly astringent. The twigs are slender and light brown, when young yellow green with minute brown pressed hairs, becoming slightly fissured, and ending in a minute hairy bud at base of last pair of leaves.

The opposite leaves have slender light green petioles $^1/_8$–$^1/_4$ inch long, with brown hairs when young. The blades are short–pointed at base, thin or slightly thickened, have minute gland dots visible with a lens, are hairless above, but have minute brown hairs along midrib beneath, the many lateral veins inconspicuous.

The branched flower clusters (panicles) at leaf bases are 1–2$^1/_2$ inches long. Flower buds are less than $^1/_8$ inch long, elliptic, greenish white or also pinkish tinged, with gland dots and slightly hairy. The flower is composed of basal tube (hypanthium) $^1/_{16}$ inch long, which extends cuplike beyond inferior ovary; rounded calyx like lid or cap which splits off and remains attached like 1 round sepal $^1/_{16}$ inch long; many spreading threadlike stamens about $^3/_{16}$ inch long; and pistil with minute inferior light green ovary, 2–celled with 2 ovules in each cell, and slender white style. Flowering and fruiting in spring.

The few berries are yellow green when immature, turning to puplish black, slightly shiny, the apex with ring scar of calyx $^1/_{16}$ inch in diameter. The pulp is thin, purple, aromatic, with spicy taste of lemon. There is 1 shiny light brown, rounded seed less than $^3/_{16}$ inch in diameter. Collected in flower in April and in fruit in June and July.

The wood has light brown sapwood and brown heartwood and is hard and heavy.

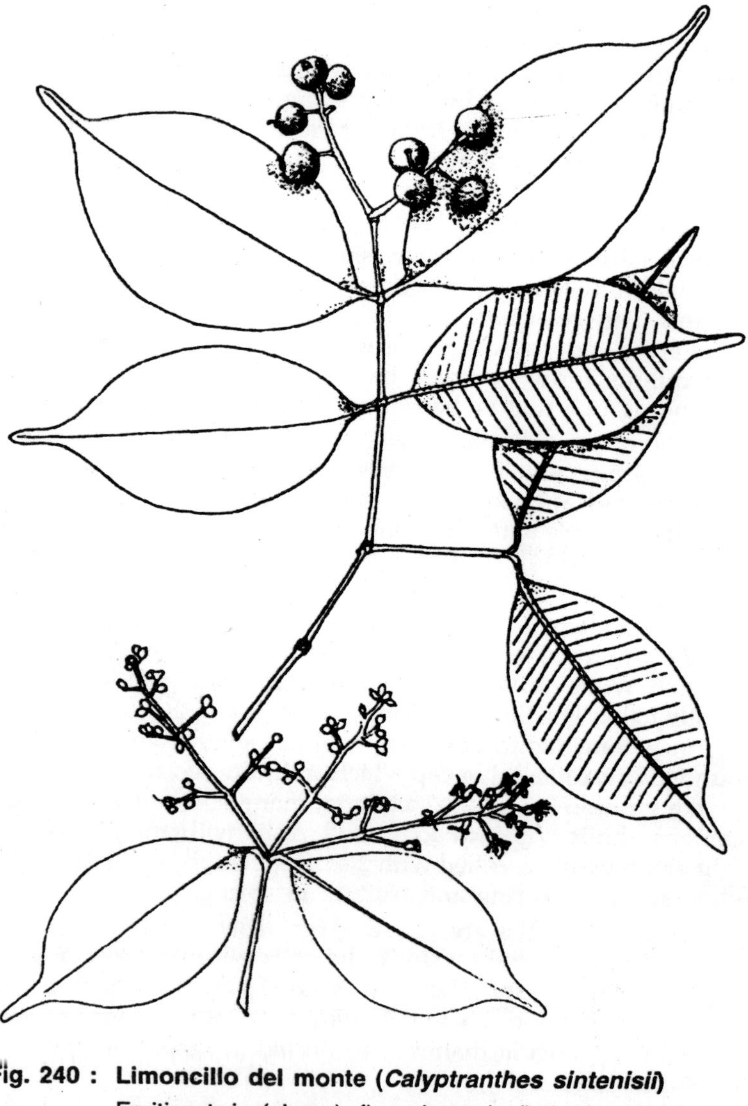

**Fig. 240 : Limoncillo del monte (*Calyptranthes sintenisii*)**
Fruiting twig (above), flowering twig (below), natural size.

## 241. Myrtle–of–the–river (*Calyptranthes zusygium*)

This rare tree first collection in Dehradun in 1950 is distinguished by : (1) paired elliptic leaves $1^1/_2$–$2^3/_4$ inches long and $^3/_4$–$1^1/_2$ inches wide, with raised mivein, hairless, with short petioles $^1/_8$–$^3/_{16}$ inch long; (2) few small whitish flowers $^1/_8$–$^3/_{16}$ inch across the many threadlike stamens, at ends of very slender hairless branches at leaf bases, the calyx covering bud and falling off like a lid or cap and the petals absent; and (3) fruit a rounded reddish berry $^1/_4$–$^5/_{16}$ inch in diameter, with small cup at apex.

A small evergreen tree to 25 feet high and 5 inches in trunk diameter, hairless throughout. Bark gray, smooth, the inner bark light brown and slightly bitter. Twigs very slender, branching in horizontal plane, when young slightly flattened and 2–angled and light green but becoming light gray, hairless, often forking widely. Buds minute, of paired young leaves.

Leaves opposite, short–pointed at both ends, turned under at edges, slightly thickened, the upper surface shiny green to dark green with raised midvein and faint side veins, and the lower surface dull yellow green with gland dots. New leaves are reddish.

Flower clusters (cymes) 2–3 inches long, the slender hairless branches ending in 1–3 flowers on short stalks of about $^3/_{16}$ inch. Flowers are covered in bud by the rounded greenish calyx which falls off like a lid or cap; petals none; stamens many; and pistil with inferior ovary, cuplike tube above, and slender style nearly $^1/_4$ inch long. Berries contain 1–4 shiny brown elliptic seeds $^3/_{16}$ inch long. Flowering in spring, with fruits in summer.

Wood light brown, hard.

Rare in moist forest to 300 feet altitude on north coast of Gujrat.

**Fig. 241 : Myrtle–of–the–river (*Calyptranthes zusygium*)**
Fruiting twig (above), flowering twig (below), natural size.

## 242. Eucalipto, kino eucalyptus (*Eucalyptus resinifera*)

This species of eucalyptus tested in forest plantations is characterized by : (1) reddish brown bark, rough, furrowed, thick and very fibrous; (2) the slender twigs, petioles, and old leaves pinkish tinged; (3) lance–shaped dark green leaves $3^1/_2$–6 inches long and $^3/_4$–$1^1/_4$ inches wide, slightly curved and unequal–sided from the short–pointed base, with long tapering point, stiff and leathery, with spicy resinous odor (when crushed) and taste; (4) cream–colored flowers with very many stamens in spreading mass nearly 1 inch across, several lateral at end of flattened stalk; and (5) cup–like green seed capsules $^3/_8$ inch long and $^5/_{16}$ inch broad.

A large evergreen tree becoming more than 50 feet tall and 1 foot in trunk diameter with straight axis and narrow to spreading dense crown of dark green foliage, hairless throughout. Bark on branches is light gray and smoothish. Inner bark is reddish brown, fibrous, and tasteless. The long slender twigs have brown dots (lenticels).

The alternate leaves droop from slender petioles $^3/_4$–1 inch long. Blades have light yellow midrib and many faint side veins, are not toothed on edges, are dull dark green above and slightly paler beneath, with many gland dots visible with lens against the light. The first few leaves of young plants are opposite and much smaller.

The clusters (umbels) of 6–10 short–stalked flowers are borne at the end of a slightly flattened stalk $^1/_2$–$^3/_4$ inch long at leaf bases. Flower buds are blunt $^5/_8$ inch long and $^1/_4$ inch wide, long–pointed at both ends, yellowish green, the paper half a long–pointed cap formed from calyx and corolla and splitting off. The cuplike base (hypanthium) bears on its rim many thread-like stamens more than $^3/_8$ inch long, which shed early. The pistil has an inferior 3–4–celled ovary with many ovules and greenish style. The persistent seed capsules contain many minute seeds that sift out through 3–4 pores below the rim. Flowering and fruiting in spring.

The dark red wood is hard, heavy, and fine–textured and has many uses.

One of the most promising of more than 30 species of *Eucalyptus* that have been tested experimentally in forest plantations in Dehradun. Apparently some trees are hybrids of this and other species. Elsewhere the wood of these fast–growing trees is used for paper pulp.

**Fig. 242 : Eucalipto, kino eucalyptus (*Eucalyptus resinifera*)**
Leafy twig (above), fruits and flowers (below), natural size.

## 243. White–stopper Eugenia (*Eugenia axillaris*)

Grajo, a small tree of dry forests, is characterized by : (1) opposite elliptic to ovate leaves $1^1/_2$–3 inches long and $^3/_4$–$2^3/_4$ inches wide, long–pointed at apex, the upper surface green to dark green and slightly shiny, with few inconspicuous veins, the petioles often purplish, and the foliage with slightly unpleasant odor; (2) small white 4–parted flowers $^3/_{16}$ inch across, few at leaf bases; and (3) round black fruits $^3/_8$ inch in diameter or often deformed and swollen to $^1/_2$ inch or more in diameter and yellowish.

Evergreen shrub or small tree 20 feet high and 5 inches in trunk diameter, reported to reach 40 feet in height and 1 foot in diameter, hairless. The bark is gray, becoming rough and slightly fissured or furrowed, the inner bark pink brown, hard, and bitter. Twigs are slender and light gray.

The opposite leaves have slender purplish or light green curved petioles $^1/_8$–$^1/_4$ inch long, without stipules. Blades are short–pointed at base, slightly thickened, slightly turned up at midvein and under at edges, with few inconspicuous side veins, with minute blackish gland dots visible with a lens on lower surface and against the light, the lower surface dull light green.

Flowers few or 1 at leaf base, on short stalks less than $^1/_4$ inch long or almost stalkless. The flower consists of a greenish cuplike base (hypanthium) less than $^1/_{16}$ inch long, which encloses the inferior ovary and bears 4 rounded light green, gland–dotted sepals $^1/32$ inch long, 4 round white gland–dotted petals $^1/_{16}$ inch long, many spreading stamens, and style $^1/_8$ inch long. The berry retains the 4 sepals at apex, changes color from green to red to black. It is edible, juicy and sweet but is often deformed by insect larvae within. There is 1 brown rounded seed $^1/_4$ inch in diameter. Flowering and fruiting in spring and summer, the fruits persisting.

The sapwood is light brown and the heart–wood brown, often tinged with red. The wood is described as heavy, hard, very fine–textured, and strong.

Common in moist limestone and coastal forests from sea level to 2,500 feet altitude in western coast and mountains of Dehradun also northeastern end.

**Fig. 243 : White–stopper Eugenia (*Eugenia axillaris*)**
Flowering twig (above), fruiting twig (below), natural size.

## 244. Pitangueira (*Eugenia biflora*)

Various species of small trees belonging to this and a related genus (*Myrcia*) are known as hoja menuda. They have the following characteristics in common : (1) opposite short–stalked small leaves usually hanging vertically, lance–shaped to elliptic, long or short–pointed at apex, with numerous minute gland dots; (2) small white flowers usually in short lateral clusters, with 4 or 5 calyx lobes, 4 or 5 white petals, numerous spreading white stamens, and inferior ovary; and (3) fruit a rounded or elliptic berry, usually blackish or sometimes reddish, with sepals (ring in *Myrcia*) at apex, and containing 1 to few seeds. This species is further distinguished by : (1) leaves $^3/_4$–$2^1/_2$ inches long and $^3/_8$–1 inch broad, mostly long–pointed at apex and short–pointed at base; (2) several stalked flowers about $^3/_8$ inch across; and (3) reddish berries about $^5/_{16}$ inch in diameter.

A small evergreen tree to 30 feet tall and 8 inches in trunk diameter, much branched and with dense bushy crown. The light brown or gray bark is smoothish. Inner bark is light brown or reddish and slightly bitter. The slender brown or light gray twigs are minutely hairy.

Leafstalks are $^1/_8$–$^1/_4$ inch long and finely hairy. Blades are turned up on both sides of midvein (not flat), slightly thickened and leathery, hairless or nearly so, above green to dark green and slightly shiny, and beneath lighter green.

The flower clusters (racemes) with slender, finely–hairy stalks are mostly $^1/_2$–1 inch long, many and showy. The flower has a hairy tubular base (hypanthium) $^1/_{16}$ inch long and broad, which encloses the inferior 2–celled ovary and bears 4 rounded hairy sepals $^1/_{16}$ inch or less in length, 4 rounded white petals less than $^1/_8$ inch long, numerous white stamens almost $^1/_4$ inch long, and curved style $^1/_4$ inch long. The reddish berries have the 4 sepals remaining at apex, thin edible flesh, and a large seed. In flower and fruit in spring and summer.

The sapwood is light brown and hard.

Sometimes planted as an ornamental for the showy white flowers.

Common in moist and dry limestone and lower forests from sea level to 2,200 feet altitude, middle elevations in mountains and on limestone Haridwar hills.

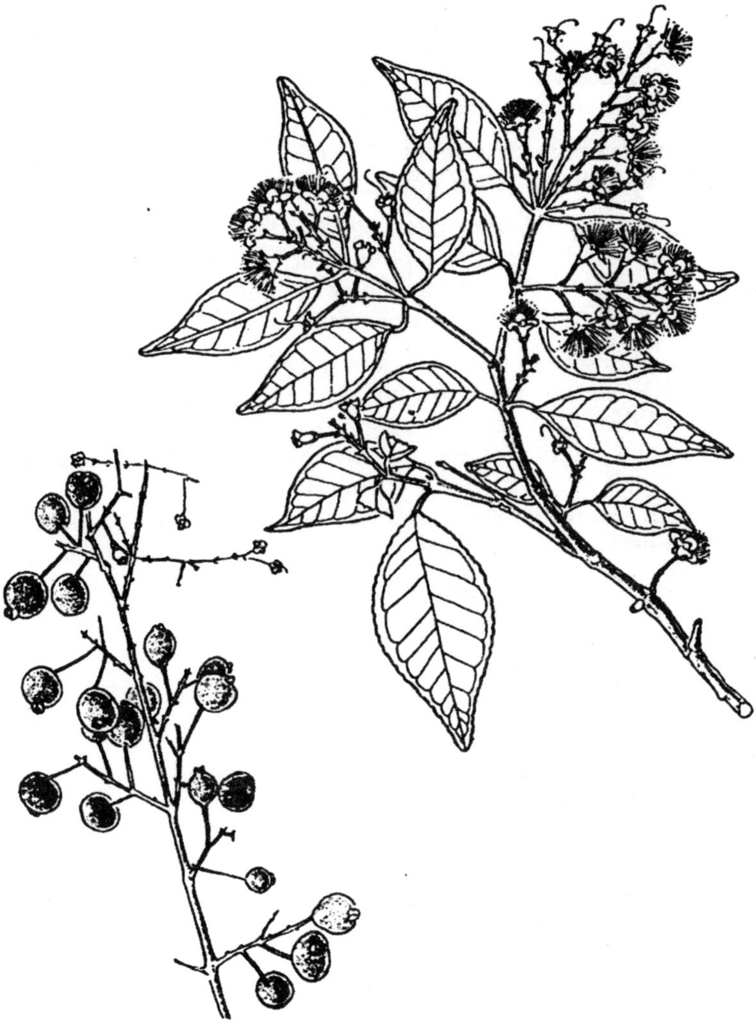

**Fig. 244 : Pitangueira (*Eugenia biflora*)**
Fruits (left), flowering twig (right), natural size.

## 245. Boqueron (*Eugenia boqueronensis*)

This shrub or small tree known only from southwestern Dehradun is characterized by : (1) opposite ovate leaves $^3/_4$–$1^1/_2$ inches long and $^3/_8$–$^1/_4$ inch wide, tapering to a blunt point at apex, shiny green with minute blackish gland dots; and (2) small flowers about $^3/_{16}$ inch across the 4 rounded white petals, few on short stalks at leaf bases.

Evergreen shrub or small tree to 25 feet high, wigs slender, gray, densely hairy when young.

The opposite leaves have petioles less than $^1/_8$ inch long, often hairy. Blades are short–pointed at base, slightly thickened, shiny green on both surfaces, hairy on midvein beneath when young but becoming hairless or nearly so, with fine side veins.

Flowers few in hairy clusters (racemes) at bases of uppermost leaves on slender stalks less than $^1/_4$ inch long. The flower has a cuplike hairy base (hypanthium) $^1/_{16}$ inch long, which encloses the inferior ovary and bears 4 rounded hairy gland–dotted sepals $^1/_{16}$ inch long, 4 rounded petals about $^1/_{16}$ inch long, many stamens, and pistil with inferior ovary and slender style. Collected with flowers in February. Fruit unknown.

Rare in dry limestone forest from sea level to 1,200 feet altitude in southwestern Dehradun.

**Fig. 245 : Boqueron (*Eugenia boqueronensis*)**
Flowering twig, natural size.

## 246. Guayabota de sierra (*Eugenia borinquensis* Britton)

Along summits of the eastern mountains of Nilgiri this distinctive small tree is rather common. Characterized by : (1) round stiff leathery leaves, $1^1/_4$–$3^1/_2$ inches in diameter with minute gland dots, opposite and stalkless; (2) large spreading white 4–petaled flowers $1^1/_2$–2 inches across the numerous white stamens, terminal and lateral on separate stalks; and (3) nearly round reddish green berries $^3/_4$ inch in diameter, with 4 unequal sepals remaining attached at apex. Distinguished from the closely related species No. 191, guayabota (*Eugenia stahlii* (Kiaersk.) Krug & Urban), by the leaves stalkless and round instead of short–stalked and elliptic.

An evergreen tree to 25 feet high and 8 inches in trunk diameter, or sometimes shrubby, with compact crown of dark green foliage. The bark is light brown, smoothish or becoming rough and fissured or scaly. Inner bark is reddish brown and slightly astringent. The stout twigs are light gray to reddish brown, flattened a little under leaves.

The leaves are mostly round at apex and slightly notched or heart–shaped at base, the edges turned under, the upper surface green or dark green and slightly shiny, and the lower surface paler green.

Flowers are a few together at ends of twigs or single above a leaf, on stout dark green stalks $^3/_8$–$1^1/_4$ inches long. The bell–shaped dark green base (hypanthium) $^1/_4$ inch long and bears 4 rounded concave calyx lobes tinged with dark red, 2 about $^3/_8$ inch long and 2 half as large; there are 4 spreading rounded white petals $^3/_4$–1 inchlong; numerous spreading white stamens $^3/_4$–1 inch long; and pistil with 2–celled inferior ovary and long white style $^7/_8$ inch long. The fleshy fruits have 1–4 large brown seeds. Flowering and maturing fruit from spring to late autumn.

The sapwood is light brown. The heartwood is hard, heavy, and durable. The wood is seldom used because of the small size and relative inaccessibility of the trees.

Locally common in dwarf forest and thickets on or near the ridges and peaks in upper Nilgiri forests at 2,500–3,500 feet altitude.

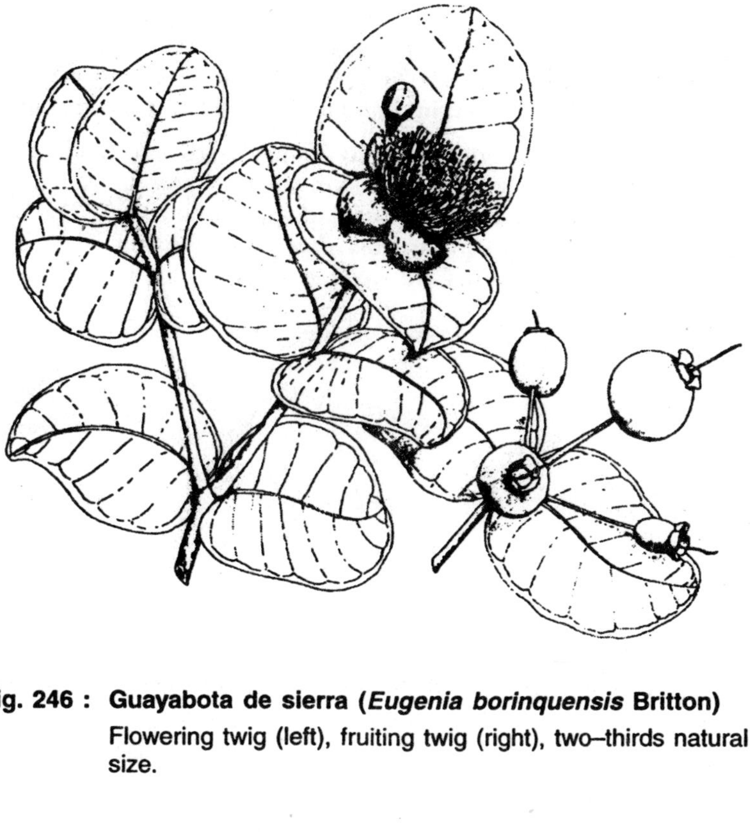

**Fig. 246 : Guayabota de sierra (*Eugenia borinquensis* Britton)**
Flowering twig (left), fruiting twig (right), two–thirds natural size.

## 247. Redberry eugenia (*Eugenia confuse DC*)

This species of western mountains Rajasthan is distinguished by : (1) opposite ovate to lanceolate shiny leaves slightly thick and stiff, ending in vary leaves slightly thick and stiff, ending in vary long point, often curved; (2) flowers about $^3/_8$ inch across the 4 spreading white petals, several or 1 at leaf bases on threadlike stalks $^1/_2$–1 inch long; and (3) rounded orange to scarlet berries $^1/_4$ inch in diameter, with 4 calyx lobes at apex.

An evergreen small tree to 20 feet high and 3 inches in trunk diameter or sometimes medium–sized tree to 40 feet tall and 1 foot in diameter, with rounded spreading crown. The bark is gray, rough and deeply furrowed or scaly, the inner bark light brown and bitter. Twigs are slender, hairless, light green when young, turning gray.

**Fig. 247 : Redberry eugenia (*Eugenia confuse DC*)**

Fruiting twig (above), twig with old flowers (lower right), natural size.

## 248. Lathberry (*Eugenia cordata*)

This shrub or small tree is easily recognized by its leaves. Distinguishing characters are : (1) paired elliptic leaves $^3/_4$–2 inches long and $^1/_2$–$1^1/_2$ inches broad, rounded at apex, stalkless and slightly notched or heart-shaped and clasping at base; (2) small white flowers less than $^1/_4$ inch across the many stamens, clustered and stalkless at leaf bases and at nodes back of leaves; and (3) stalkless fruits about $^1/_4$ inch long.

An evergreen shrub or small tree 30 feet in height and 5 inches in trunk diameter, with smooth gray bark. The slender twigs are light gray or brown, minutely hairy when young.

The opposite leaves commonly are $^3/_4$–1 inch long but sometimes twice as large, slightly thickened, hairless, shiny green above and dull light green beneath, the border yellow and slightly turned under, with inconspicuous veins and with minute gland dots.

The crowded clusters of stalkless flowers consist of calyx of 4 rounded sepals $^1/_{16}$ inch long, 4 rounded white petals less than $^1/_8$ inch long, many threadlike white stamens, and pistil with inferior ovary $^1/_{16}$ inch long and threadlike style. The fruit is rounded to elliptic, greenish, about $^1/_4$ inch long. With flowers and fruits in spring.

Common in eastern Assam and rare along south coast.

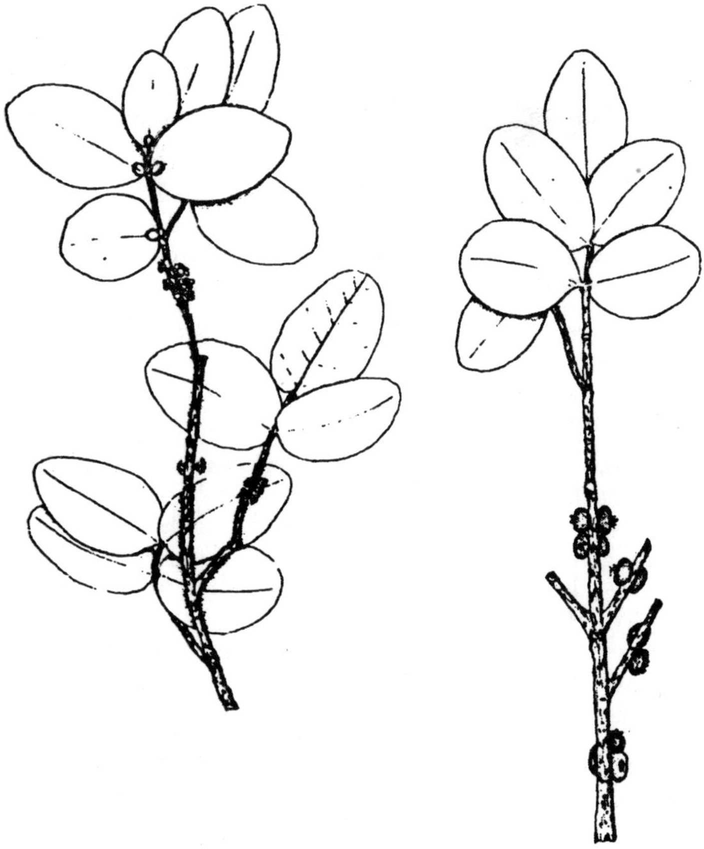

**Fig. 248. Lathberry (*Eugenia cordata*)**
Flowering twig (left), fruiting twig (right), natural size.

## 249. Guasabara (*Eugenia eggersii*)

Guasabara, a shrub or small tree restricted to mountain forests of Dehradun, is identified by : (1) opposite oblong to narrowly elliptic leaves 2–6 inches long and $^3/_4$–$2^1/_4$ inches wide, hairless, slightly thickened, leather, dark green and slightly thickened, leathery, dark green and slightly shiny on upper surface; (2) white flowers nearly $^1/_2$ inch across the 4 petals and many spreading stamens, several in short clusters at leaf bases; and (3) round warty fruit $^1/_2$ inch in diameter, with 4 rounded sepals at apex.

An evergreen shrub or small tree to 15 feet high and 3 inches in trunk diameter, reported to become a tree 65 feet tall. Bark gray, smooth to finely fissured, the inner bark greenish. Twigs slender, green, turning brown, hairless.

Leaves opposite, hairless, with petioles $^1/_8$–$^3/_8$ inch long. Blades mostly long–pointed at apex, short–pointed or blunt at base, not toothed on edges, the lower surface yellow green and slightly shiny.

Flower clusters (racemes) at leaf bases are $^1/_2$–$1^1/_2$ inches long. The fragrant flowers on stalks $^1/_8$–$^1/_4$ inch long have a cup–shaped green base (hypanthium) $^1/_{16}$ inch long, which bears 4 light green rounded sepals $^1/_8$ inch long; 4 rounded white petals $^3/_{16}$ inch long; many white spreading threadlike stamens $^3/_8$ inch long; and pistil with inferior 2–celled ovary containing many ovules and slender style. Flowering in spring and summer and fruiting in summer.

The wood is light brown and hard.

Rare as understory shrub or small tree in moist limestone, forests at 1,000–3,000 feet altitude in mountains of Dehradun.

**Fig. 249 : Guasabara (*Eugenia eggersii*)**
Flowering twig (left), fruiting twig (right), natural size

## 250. Uvillo (*Eugenia haematocarpa*)

This rare small tree known only from Luquillo Mountains was not named until 1963. Its distinguishing characters are : (1) paired, relatively large, oblong to elliptic leaves, thick and leathery, almost stalkless; (2) many small flowers about $^1/_4$ inch wide, borne short–stalked in clusters on trunk; and (3) round berries, $^3/_4$ inch in diameter, dark blood red, edible.

An evergreen small tree to 25 feet high and 6 inches in trunk diameter. Bark gray or whitish, smoothish shedding in plates. Twigs hairless, slightly 2–angled, rusty brown.

Leaves opposite, hairless, with short petioles less than $^1/_8$ inch long. Blades are mostly 4–7 inches long and 2–3$^1/_2$ inches wide, thin, long–pointed at apex, rounded or slightly notched at base, not toothed on edges, with many nearly slightly slender side veins slightly raised and forming prominent network, dull dark green on upper surface, and light green beneath.

Flowers many in clusters on trunk on slender nearly equal stalks of $^1/_8$–$^1/_2$ inch composed of bell–shaped tube (hypanthium) $^1/_{16}$ inch long, finely hairy, calyx of 4 rounded lobes $^1/_{16}$ inch long, 4 rounded light pink petals $^1/_8$ inch long, many stamens, and pistil with inferior ovary. Berries round or nearly so, dark blood red, containing 1 rounded light brown seed $^5/_8$ inch in diameter. Collected with flower buds and fruits in May.

Rare in lower slopes of Sivalik forest at 1,000–1,500 feet altitude.

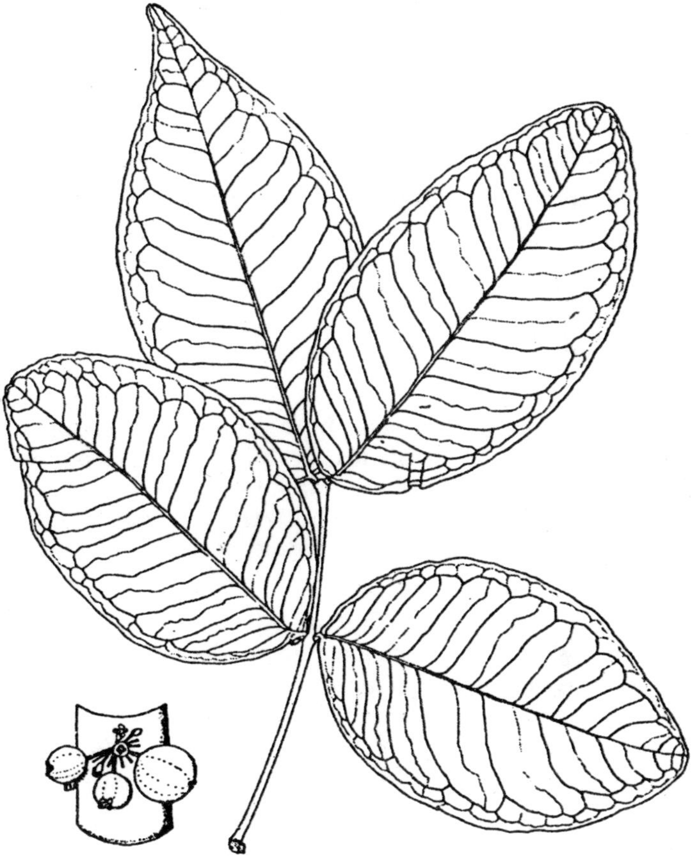

**Fig. 250 : Uvillo (*Eugenia haematocarpa*)**
Leafy twig and fruits (lower left), two–thirds natural size.

## 251. Palo de muleta (*Eugenia ligustrina*)

This uncommon shrub or elsewhere a small tree is recognized by : (1) paired, narrowly elliptic leaves $^3/_4$–$1^3/_4$ inches long and $^1/_4$–$^3/_4$ inch wide, very shiny green above; (2) smaller scalelike leaves at base of twigs; (3) flowers mostly single on slender stalks back of leaves, white, $^1/_2$–$^3/_4$ inch across the 4 white petals; and (4) round red berries $^1/_4$–$^5/_{16}$ inch in diameter, with 4 long narrow sepals at apex, sweetish and edible.

An evergreen shrub seldom taller than 8 feet but sometimes a tree 12 feet high and 3 inches in trunk diameter, reported to reach 15–20 feet elsewhere, hairless throughout. Bark gray, slightly fissured to peeling and slightly shaggy. The twigs are slender, light gray brown. The narrow pointed buds are up to $^3/_{16}$ inch long, 4–angled, and covered with overlapping pointed scales.

The opposite leaves have short slender yellowish petioles less than $^3/_{16}$ inch long. Blades are blunt or short–pointed at apex and short–pointed at base, turned under at edges, slightly thick and leathery with sunken midvein and many faint side veins, covered with minute gland dots, and dull yellow green beneath.

Flowers mostly single lack of leaves on slender stalks of $1^1/_2$ inches. Sepals 4, $^3/_{16}$ inch long, gland–dotted; petals 4, white, about $^3/_8$ inch long; stamens many; and pistil with inferior ovary containing many ovules and wit slender style. Berries on slender stalks are covered with gland dots and have few rounded seeds. Recorded with flowers in May.

The sweetish edible berries have been made into preserves elsewhere.

Uncommon and scattered mostly in dry and moist limestone forests of southern and western India, also scattered along north coast from sea level to 600 feet altitude.

**Fig. 251 : Palo de muleta (*Eugenia ligustrina*)**
Flowering twig (upper left), fruiting twig (right), twig with old fruit stalks (below) natural size.

## 252. Anguila, boxleaf Eugenia (*Eugenia foetida*)

This common shrub or small tree of the group of hoja menuda is distinguished by : (1) unpleasant, skunklike odor of foliage; (2) leaves paired, small, oblanceolate, narrow, rounded at apex, mostly broadest beyond middle and tapering to long–pointed base, almost stalkless; (3) few small white flowers $^3/_{16}$ inch across the many white stamens, clustered and almost stalkless at leaf bases and back of leaves; and (4) rounded blackish berries about $^1/_4$ inch in diameter.

A shrub or small tree to 20 feet high and 4 inches in trunk diameter, on Desecheo reaching 30 feet and 5 inches. Bark gray, smooth, becoming slightly fissured or mottled into large plates. Inner bark pink, slightly bitter. Twigs gray, slender, often finely hairy.

The opposite leaves $^3/_4$–$1^1/_2$ inches long and $^1/_4$–$^1/_2$ inch wide are hairless, slightly thickened and turned under at edges, dull or slightly shiny green above, and beneath light green with blackish gland dots. The leaves very greatly in shape and size.

Flowers few, slightly fragrant, consisting of 4 rounded calyx lobes more than $^1/_{16}$ inch long, 4 white gland–dotted petals less than $^1/_8$ inch long, many threadlike white stamens, and pistil with inferior ovary and long slender style. The fruits turn from yellow orange to black or dark brown at maturity and have calyx at apex. Seed usually 1, sometimes 2. Flowering and fruiting in spring and summer, the fruits persisting.

The heartwood is dark reddish brown and the sapwood light brown. The wood is hard, heavy, and fine–grained.

Common in moist and dry forests from sea level to 300 feet altitude in southwestern coast and foothills of A.P.

**Fig. 252 : Anguila, boxleaf Eugenia (*Eugenia foetida*)**
Flowering twig (upper left) fruiting twig (lower right), natural size.

## 253. Biriji, black–cherry (*Eugenia monticola*)

Biriji is common and widely distributed in Goa, mainly western part, and through the Virgin Islands. It is recognized by: (1) many twigs very slender and wandlike, finely hairy; (2) many crowded small leaves paired and in 2 rows, lance–shaped to narrowly ovate, pointed at both ends and often strongly turned under toward base; (3) many white flowers about $^3/_{16}$ inch wide, crowded on short stalks at leaf bases; and (4) fruit a round berry $^3/_{16}$ inch in diameter, turning from red to black at maturity.

An evergreen shrub or tree to 50 feet high and 6 inches in trunk diameter. Bark gray, smooth to slightly fissured, the inner bark brown and almost tasteless. Twigs very slender, light brown, finely hairy when young.

Leaves opposite in 2 rows, with slender finely hairy petioles less than $^1/_8$ inch long. Blades $^5/_8$–$1^1/_4$ inches long and $^3/_{16}$–$^1/_2$ inch wide (up to $^7/_8$ inch wide in abroad leaf form), the apex ending in a long blunt point, the base short–or long–pointed and may be strongly turned under, edges not toothed, thin or a little thickened, hairless or nearly so, with minute gland dots, dull green above, paler beneath. The foliage sometimes has an unpleasant, skunklike odor.

The flowers are composed of 4 rounded sepals $^1/_{16}$ inch long, 4 white petals less than $^1/_8$ inch long and gland dotted, many threadlike white stamens, and pistil with inferior ovary and slender style. The fruits, with sepals persistent at apex, are classed as edible. Flowering and fruiting irregularly from spring to fall.

Wood light brown, hard, and heavy.

Common and widely distributed in moist and dry forests from sea level to 2,500 feet altitude, mainly in Goa.

**Fig. 253 : Biriji, black–cherry (*Eugenia monticola*)**
Flowering twig (left), fruiting twig (right), natural size.

## 254. Hoja menuda (*Eugenia procera*)

This species of hoja menuda found in dry forests of south-western Puerto Rico and the islands eastward is characterized by : (1) opposite ovate to elliptic leaves 1–2 inches long and $^5/_8$–$1^1/_4$ inches wide, slight thickened; (2) white flowers about $^3/_8$ inch across the many stamens and 4 petals, several on very slender stalks $^1/_4$–$^5/_8$ inch long at nodes; and (3) rounded black sticky berries nearly $^1/_4$ inch in diameter.

An evergreen shrub or small tree to 15 feet high and 3 inches in trunk diameter. Bark gray, smooth, peeling off in rectangular plates. Twigs slender, finely hairy when young.

Leaves opposite, with petioles about $^1/_8$ inch long. Blades are bluntly long–pointed at apex, short–pointed at base, not toothed on edges, hairless, with minute gland dots, with inconspicuous veins forming a fine network, the upper surface dull green, and the lower surface paler.

The fragrant flowers are borne several at a node at base of a leaf on stalks of equal length (like an ambel). The cup–shaped base (hypanthium) bears 4 rounded calyx lobes less than $^1/_8$ inch long and, the other parts; 4 white petals more than $^1/_8$ inch long; many threadlike stamens about $^1/_8$ inch long; and pistil with inferior 2–celled ovary and style. Flowers and fruits produced from spring to autumn.

Rare in dry limestone forest of southwestern Goa.

**Fig. 254 : Hoja menuda (*Eugenia procera*)**

Flowering twig (upper left), fruiting twig (lower right), natural size.

## 255. Quiebrahacha (*Eugenia pseudopsidium*)

This shrub rarely reaching tree size is identified by : (1) opposite ovate to elliptic leaves $1^1/_2$–$4^1/_2$ inches long and $^3/_4$–2 inches wide, hairless, the upper surface green and slightly shiny; (2) whitish flowers about $^3/_8$ inch wide, 1 or few on long very slender stalks at leaf bases; and (3) the conspicuous rounded shiny red or yellow berries $^3/_8$–$^5/_8$ inch in diameter with calyx lobes at apex.

An evergreen shrub flowering when only 3 feet high, rarely a small tree at Vieques. Bark light fray, slightly fissured and scaly. Twigs slender, minutely hairy, green when young, turning brown.

The opposite hairless leaves have short petioles $^1/_8$–$^1/_4$ inch long. Blades are long–pointed at apex, short–pointed at base, not toothed on edges, slightly thickened, the lower surface dull light green.

Flowers are borne 1 or few on stalks $^3/_8$–1 inch long at leaf bases or sometimes at end of twig. The cup–shaped green base (hypanthium) $^1/_{16}$ inch long encloses the inferior ovary and bears the 4 rounded greenish calyx lobes about $^1/_8$ inch long and other parts; 4 rounded white petals $^1/_8$–$^1/_4$ inch long; many white stamens; and pistil with inferior 2–celled ovary containing many ovules and slender style. The rounded fruits have calyx persistent at apex and contain 1 rounded seed. Flowering and fruiting probably irregularly through the year.

Wood light or reddish brown, hard, reported to be very heavy (specific gravity 1.3).

Common in moist coastal forest at low and middle altitudes to 2,200 feet in Kerala.

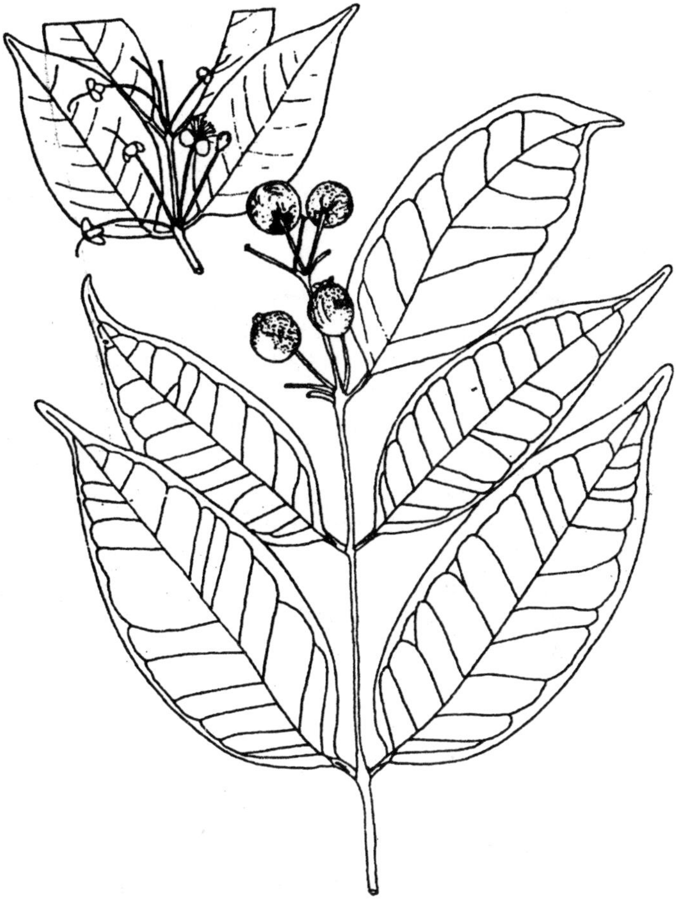

**Fig. 255 : Quiebrahacha (*Eugenia pseudopsidium*)**
Flowering twig (upper left) and fruiting twig, natural size.

## 256. Sessiflora (*Eugenia sessiliflora*)

This shrub or small tree is more widely distributed through the islands eastward than in Puerto Rico. Its characters for identification are: (1) opposite elliptic or nearly round leaves, thick and leathery, with edges much rolled under; (2) stalkless flowers clustered on twigs back of leaves, about $^1/_2$ inch across the many spreading threadlike stamens and 4 petals; and (3) round rose–colored berries about $^3/_4$ inch in diameter.

An evergreen shrub or rarely a small tree to 15 feet high and 3 inches in trunk diameter, hairless throughout. Bark gray, smooth. Twigs stout, light gray.

Leaves opposite, with stout yellow to brown petioles $^1/_8$–$^1/_4$ inch long. Blades are mostly $1^5/_8$–3 inches long and $1^1/_4$–$2^1/_4$ inches wide, sometimes to 4 by 3 inches, rounded or blunt at apex, short–pointed or blunt at base, with minute gland dots, the upper surface shiny yellow green with network of fine veins visible when dry, and the lower surface dull light green.

The flowers are composed of brownish hairy cup–shaped base (hypanthium) $^1/_8$ inch long, which bears the calyx with 4 yellow–green dotted lobes $^1/_8$–$^3/_{16}$ inch long and the other parts; 4 petals $^1/_4$ inch long; many threadlike stamens, and pistil with inferior 2–celled ovary and slender style. Flowering and fruiting in spring and early summer.

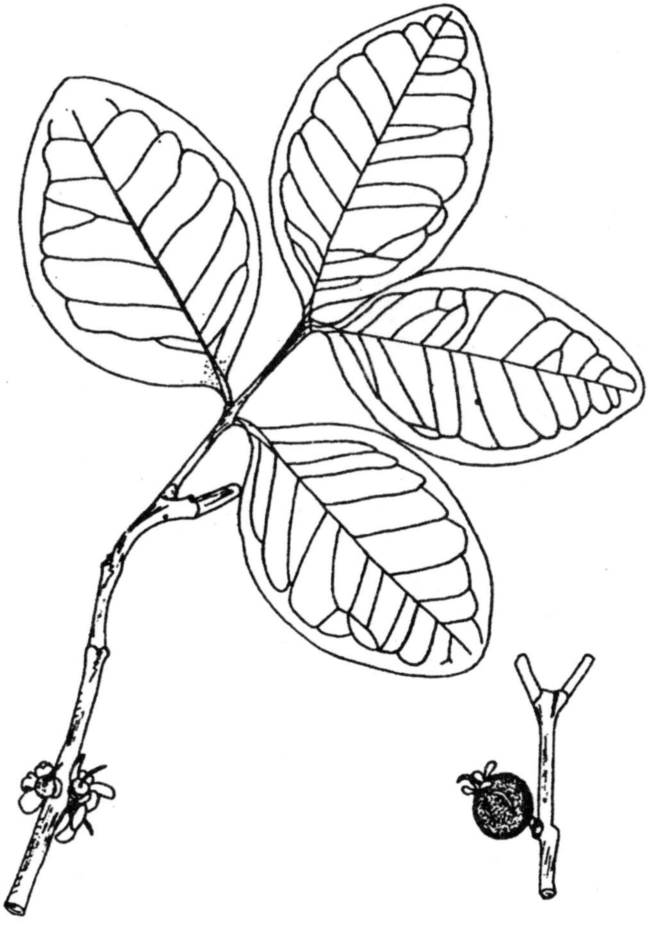

**Fig. 256 : Sessiflora (*Eugenia sessiliflora*)**
Flowering twig (left), fruiting twig (lower right), natural size.

## 257. Murta (*Eugenia sintenisii*)

This shrub or small tree is identified by: (1) paired, elliptic to obovate leaves narrowed at base, short–stalked; (2) small flowers about $^{3}/_{16}$ inch across the many stamens, stalkless in clusters back of leaves or at leaf bases; and (3) rounded red fruits, $^{3}/_{16}$ in diameter.

An evergreen shrub or small tree to 20 feet high and 4 inches in trunk diameter, hairless throughout. Leaves opposite with petioles less than $^{1}/_{8}$ inch long. Blades $1^{1}/_{2}$–3 inches long and $^{3}/_{4}$–$1^{1}/_{2}$ inches wide, blunt or rounded at apex and narrowed or short–pointed at base, dark green with whitish midvein, slightly thickened.

The flowers have calyx lobes less than $^{1}/_{16}$ inch long, rounded petals less than $^{1}/_{8}$ inch long, many threadlike stamens, and pistil with inferior ovary and long slender style. Flowering and fruiting in spring.

Rare in upper forest at 2,500–3,000 feet altitude in eastern and central mountains M.P.

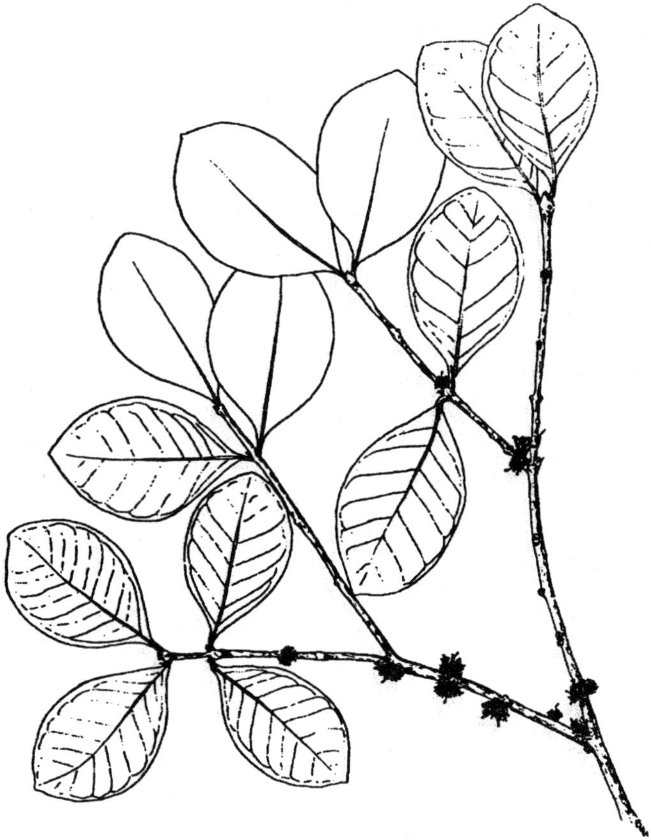

**Fig. 257 : Murta (*Eugenia sintenisii*)**
Flowering twig, natural size.

## 258. Steward (*Sugenia stewardsonii*)

This rare tree known only from Puerto Rico is described by: (1) paired ovate shiny green leaves almost stalkless, with prominent network of veins; (2) many stalkless white flowers nearly $^1/_4$ inch across the many stamens, crowded on twigs below leaves; and (3) rounded fruits $^1/_4$ inch in diameter.

A small evergreen tree 15–30 feet high and 4 inches in trunk diameter, hairless throughout Twigs slender, gray.

Leaves opposite, almost stalkless, $1^1/_4$–$2^1/_4$ inches long and $^3/_4$–$1^1/_2$ inches wide, slightly thickened, blunt or rounded at apex, rounded or slightly notched at base, not toothed on edges, ovate to elliptic, bright shiny green on both sides, with many straight fine side veins prominent in drying.

Flowers stalkless, crowded in clusters below leaves. The flower has 4 rounded greenish calyx lobes less than $^1/_8$ inch long, with gland dots; 4 white petals; many spreading threadlike white stamens; and pistil with inferior ovary and long style. The stalkless berries have calyx lobes at apex. Flowering and fruiting in spring.

Rare in upper at about 2,500–3,000 feet altitude in western mountains cf Kerala.

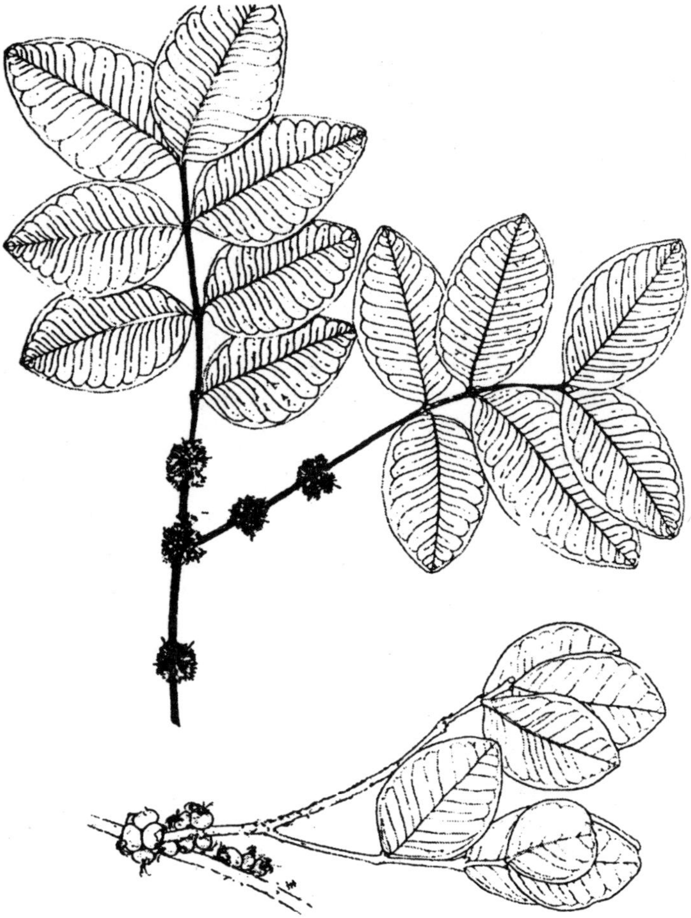

**Fig. 258 : Steward (*Sugenia stewardsonii*)**
Flowering twig (above), freiting twig (below), natural size.

## 259. Surinam–cherry (*Eugenia uniflora* L.*)

Surinam–cherry is an introduced shrub or small tree planted for its edible cherrylike fruits and for hedges. It is identified by: (1) opposite ovate leaves $1^1/_4$–$2^1/_4$ inches long and $^3/_4$–$1^1/_4$ inches wide, shiny dark green and slightly curved, with few side veins; (2) flowers about $^3/_8$ inch across the 3 spreading petals, white or tinged with pink, 1–4 at leaf bases on long slender stalks; and (3) rounded berries $^1/_2$–1 inch in diameter, red to deep purplish black, grooved or angled, resembling tiny tomatoes, juicy and slightly sour, 1–seeded.

A planted shrub or small tree to 15 feet high and 4 inches in trunk diameter, much branched, hairless. Twigs light brown, slender.

The opposite hairless leaves have light green petioles $^1/_4$ inch long. The blades are bluntly long–pointed at apex, rounded or slightly notched at base, not toothed on edges, slightly thickened, slightly curved and convex from base to apex, with minute gland dots, shiny dark green on upper surface, and dull light green beneath.

Flowers borne on slender stalks $^1/_4$–$1^1/_4$ inches long are composed of calyx with 4 light green lobes $^3/_{16}$ inch long, turned back and persistent; 4 spreading elliptic petals $^5/_{16}$ inch long, white or tinged with pink, soon falling; many threadlike white stamens $^3/_{16}$ inch long; and pistil with inferior rounded ovary $^1/_{16}$ inch long and threadlike white style. The fruits are broader than long, with 8 vertical ridges and grooves and with 1 large seed. A variety has purplish black fruits. With flowers and fruits in spring.

The cherrylike fruits are eaten raw and made into jellies and preserves. The shrubs can be pruned into attractive hedges with glossy dark green foliage. In temperate climates the plants in pots bear many showy fruits. It is reported that a home remedy like tea has been prepared from the leaves.

Rarely grown for the pleasantly acid fruits, also for hedges, in Uttaranchal.

**Fig. 259 : Surinam–cherry (*Eugenia uniflora* L.*)**
Flowering twig (left), twig with immature fruits (right), natural size.

## 260. Xerophyta (*Eugenia xerophytica*)

This rare shrub or small tree known only from southwestern Kerala characterized by: (1) opposite nearly round leaves, slightly thick and leathery, with prominent network of veins; (2) white flowers $^5/_8$ inch across the many spreading threadlike stamens and 4 petals, 1 to few on slender stalks $^1/_4$–$^1/_2$ inch long at nodes, mostly back of leaves; and (3) egg–shaped or rounded berries more than $^3/_8$ inch long, with 4 enlarged calyx lobes at tip.

An evergreen shrub or small tree to 15 feet high and 3 inches in trunk diameter, much branched. Bark brown, smooth, peeling off in plates and mottled, the inner bark whitish and tasteless. Twigs slender, brownish gray.

Leaves opposite, hairless, with short–petioles $^1/_8$–$^1/_4$ inch log. Blades are $^5/_8$–2 inches long and broad, rounded or blunt at apex, slightly notched or rounded at base, not toothed on edges, with minute gland dots, the upper surface shiny yellow green, often curved up on both sides, dull yellow green.

The flowers are composed of cup–shaped base (hypanthium) $^1/_{16}$ inch long, which bears 4 rounded or pointed gland–dotted calyx lobes $^1/_8$ inch long and the other parts; 4 white elliptic rounded petals more than $^1/_4$ inch long; many threadlike white stamens $^1/_4$–$^5/_{16}$ inch long; and pistil with inferior ovary containing many ovules and slender white style $^1/_4$ inch long. The fruits are green when immature and contain few seeds. Flowering and fruiting in spring, collected also with immature fruits in July and with flowers in October.

The wood is light brown and hard.

Rare in dry forest from sea level to 200 feet altitude in southwestern Kerala.

**Fig. 260 : Xerophyta (*Eugenia xerophytica*)**
Flowering twig (left), fruiting twig (upper right), natural size.

## 261. Cieneguillo (*Gomidesia lindeniana*)

Small tree of mountain forests and dwarf forests of mountain peaks, easily recognized by the paired very convex elliptic leaves, minutely gland–dotted, curved from the short–pointed base to the long–pointed apex and much turned under at edges. Other distinguished characters are : (1) twigs, leafstalks, buds, branches of flower clusters, and flowers all densely brown hairy; (2) many white 5–parted flowers $^3/_8$ inch across in branched clusters; and (3) round black berry $^1/_2$ inch or less in diameter with calyx at apex. Resembling the related species also known as cienguillo (No. 192, *Myrcia deflexa* (Poir.) DC.), which was brown hairy parts and similar shaped leaves but nearly flat, also larger flowers and smaller berries about $^1/_4$ inch wide.

A small evergreen tree 15–30 feet tall and 3–6 inches in trunk diameter, or sometimes shrubby. The crown is narrow or open, with few branches. The bark is gray and smooth, becoming slightly fissured. Inner bark is pink to brown and slightly astringent or bitter.

Leafstalks of the opposite leaves are $^1/_8$–$^3/_8$ inch long and stout. The blades are 2–6 inches long and 1–3 inches broad, thickened and leathery, above shiny green to dark green and hairless or nearly so, with lateral veins slightly sunken, and beneath yellow green or brownish green and soft hairy with raised, brown hairy veins.

Flower clusters (panicles) are wide, 2–6 inches long and broad, terminal and lateral near apex of twig, bearing many nearly stalkless flowers on the densely brown hairy branches. The slightly fragrant flowers have a broad brown hairy base (hypanthium) $^1/_{16}$ inch long, which encloses the inferior ovary and bears 5 pointed broad hairy brown sepals $^1/_{16}$ inch long, 5 rounded white gland–dotted petals $^1/_8$–$^1/_4$ inch long and hairy outside, numerous spreading white stamens $^3/_{16}$–$^5/_{16}$ inch long, and slender white hairy style $^1/_4$–$^3/_8$ inch long. The black berry is shiny and slightly brown hairy at maturity with astringent light brown flesh. Observed in flower and fruit from spring to early fall.

The light brown wood is hard.

Rare in upper and dwarf forests at higher altitudes of 2,000–3,000 feet including summits in central and western mountains of Chambal.

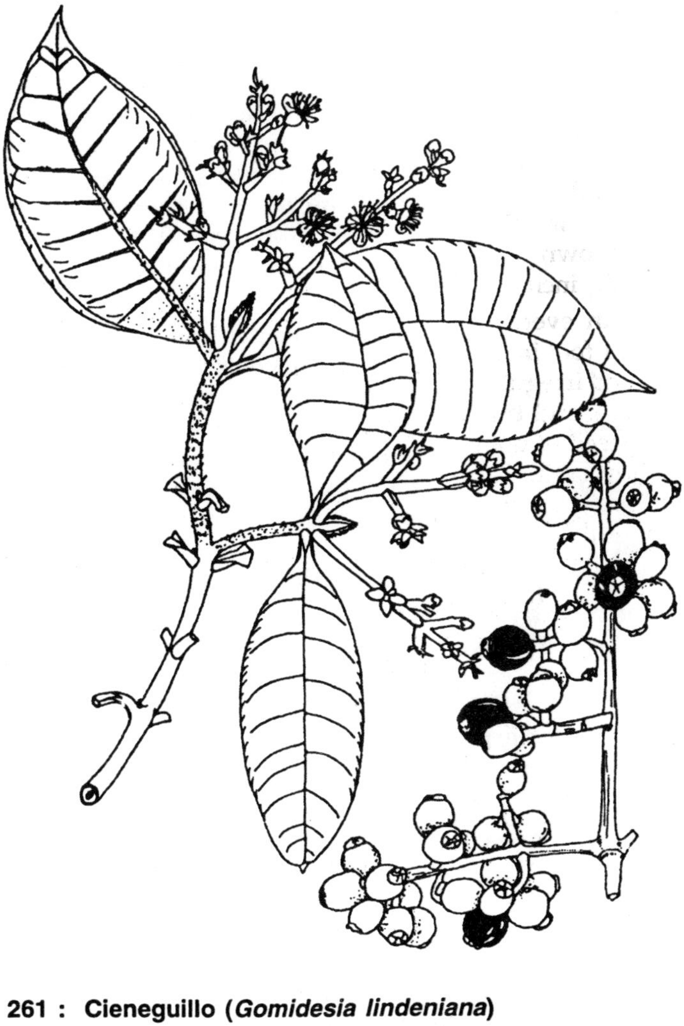

**Fig. 261 : Cieneguillo (*Gomidesia lindeniana*)**
Flowering twig (left) and fruits (right), natural size.

## 262. Cajeput–tree (*Melaleuca quinquenervia*)

The cajeput–tree, an introduced ornamental, is easily recognized by its odd whitish corky or spongy thick bark, which splits and peels in many papery layers, exposing light pink or brown beneath and becoming rough and shaggy. Other distinguishing characters are: (1) lance–shaped or narrowly elliptic leaves dull gray green with 5 (sometimes 7) veins (as specific name indicates) from base to apex, faint and nearly parallel, with resinous odor and taste when crushed; (2) many white flowers $^{5}/_{8}$ inch long, crowded and stalkless in masses 1–3 inches long and $1^{1}/_{2}$ inches across the mass of threadlike whitish stamens, which suggest a bottlebrush; and (3) many crowded gray–brown short hard cylindric seed capsules $^{1}/_{8}$ inch long and $^{3}/_{16}$ inch wide.

An evergreen resinous ornamental tree 50 feet high, the trunk 1 foot in diameter, slightly angled and grooved, with main axis and irregular branches forming narrow or open crown. Bark or trunk and branches is whitish and papery. The inner bark is composed of many light pink fibrous layers, light brown within, slightly sour. Twigs are slender, often drooping, light brown, and finely hairy when young, turning gray. End and side buds rounded to cylindric, $^{1}/_{8}$–$^{1}/_{4}$ inch long, greenish brown, composed of many rounded overlapping scales.

The alternate leaves have finely hairy light green petioles $^{1}/_{8}$ inch long. Blades are $1\,^{1}/_{2}$–$3\,^{1}/_{2}$ inches long and $^{1}/_{4}$–$^{5}/_{8}$ inch wide, long –pointed at both ends, not toothed on edges, slightly thickened and stiff, finely hairy and slightly paler on lower surface. Crushed leaves have a resinous odor and taste somewhat like that of eucalyptus.

The flowers are borne along twigs between groups of leaves or below. The calyx has 5 half–round lobes less than $^{1}/_{16}$ inch long; there are 5 concave whitish petals nearly $^{1}/_{8}$ inch long; about 30 threadlike white stamens nearly $^{5}/_{8}$ inch long, slightly united at base and to petals and falling together promptly; and pistil with inferior 2–4 celled ovary containing many ovules, long threadlike white style, and brown dot stigma. The seed capsules are crowded and stalkless in groups 1–3 inches long on gray twigs back of leaves or between groups of leaves. There are many minute, very narrow brown seeds less than $^{1}/_{16}$ inch long. Flowering from spring to fall, more than once a year, the fruits persistent.

**Fig. 262 : Cajeput–tree (*Melaleuca quinquenervia*)**
Natural size.

The sapwood is light brown and hard.

Common as an ornamental and shade tree near sea level in Gujrat. Fast growing and resistant to wind, drought, fires, and salt water. Suitable for windbreaks and beach planting. Elsewhere the bark has been used for packing fruits and for roofs and boats. Cajeput oil of medicine is obtained from the leaves and twigs of this and related species by steam distillation. Propagated from seeds.

## 263. Hoja menuda (*Psidium sintenisii*)

This species of hoja menuda, known only from eastern and western mountains of Puerto Rico, is characterized by: (1) opposite elliptic leaves $1^1/_4$–$2^1/_2$ inches long and $^5/_8$–2 inches wide, slightly thickened, with many minute gland dots; (2) slender twigs much branched and spreading in horizontal plane; (3) flowers from top–shaped bud nearly $^3/_8$ inch long, single on slender stalks at leaf bases; and (4) pear–shaped yellowish fruits about $^3/_8$ inch long, with calyx ring at apex.

An evergreen small tree to 30 feet high and 6 inches in trunk diameter. The trunk is slightly angled and grooved. Bark gray, slightly peeling, scaly and shaggy. The inner bark is pinky brown, gritty and slightly astringent. Twigs slender, hairless, with gland dots, light green when young, becoming light gray.

Leaves opposite, hairless, with petioles $^1/_8$–$^1/_4$ inch long. Blades are short–pointed or blunt at apex, short–pointed at base, and sometimes turned under at edges, the upper surface green to dark green and slightly shiny with midrib sunken but without visible side veins, and the lower surface yellow green and slightly shiny.

The fragrant flowers have slender stalks $^3/_8$–$1^1/_4$ inches long. The top–shaped hairless bud with gland dots is capped by the calyx, which splits and separates irregularly; the base (hypanthium) $^1/_8$ inch long encloses the inferior ovary and bears the other parts; corolla of 5 concave whitish petals $^3/_{16}$ inch long; many spreading stamens; and pistil with inferior 5–celled ovary containing many ovules, and slender style. The fruits change from light green to yellowish, have a hard thick wall, and contain several small seeds. Collected with flower buds in August, also with fruits from spring to autumn.

Wood whitish, hard.

Rare in upper forests at 2,500–3,000 feet altitude in eastern and western mountains M.P.

**Fig. 263 : Hoja menuda (*Psidium sintenisii*)**

Twig wigh flower buds (left), twig with old flowers (right), natural size.

## 264. Hoja menuda (*Myrcia fallax*)

This species of hoja menuda is uncommon in high mountains of eastern and central Puerto Rico. Its distinguishing characters are: (1) opposite elliptic leaves, long–or short–pointed at apex and short–pointed at base, slightly thickened, with minute gland dots, the side veins in a network but not numerous; (2) white flowers $^{1}/_{4}$ inch across the 5 spreading petals and many stamens, many in terminal and lateral clusters; and (3) cylindric pinkish fruits about $^{1}/_{2}$ inch long, with 5–calyx lobes at apex.

An evergreen small tree to 30 feet high and 6 inches in trunk diameter. The bark is gray, smoothish, becoming slightly fissured. Inner bark is light brown except for a dark red outer layer and is astringent. The slender twigs are light green and slightly hairy when young, becoming brown and fissured, ending in narrow bud of minute paired leaves.

The opposite leaves have brownish petioles $^{1}/_{8}$ inch long, slightly hairy when young. Blades are 1–2$^{1}/_{2}$ inches long and $^{5}/_{8}$–1$^{1}/_{4}$ inches wide, hairless, often slightly turned under on edges, the upper surface shiny green with sunken midrib, and the lower surface slightly shiny light green.

Flowers clusters (panicles) lateral and terminal, 2–3 inches long, with slender or sometimes stout branches and many small nearly stalkless fragrant white flowers. The greenish hairy base (hypanthium) $^{1}/_{16}$ inch long encloses the inferior ovary and bears other parts; calyx of 5 spreading rounded yellowish lobes; 5 rounded while petals $^{1}/_{16}$ inch long; many spreading white stamens; and pistil with inferior 2–celled ovary and curved hairy style. The fruit is about twice as long as wide and contains 1 shiny brown seed $^{3}/_{8}$ inch long. With flowers and fruits in spring and summer.

The wood is light brown and hard.

Common in forests and dwarf forest at 2,000–3,500 feet altitude in high mountains of eastern and central Bihar.

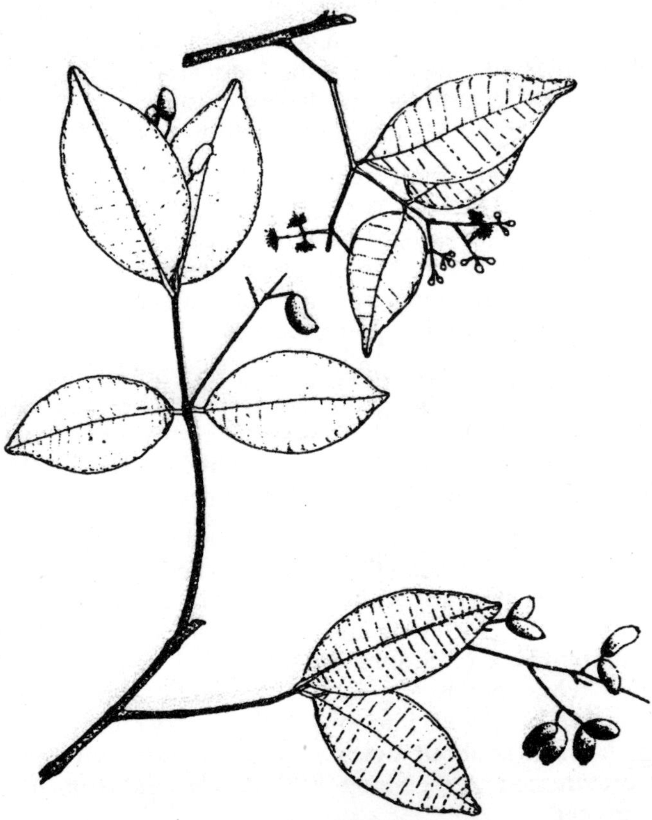

**Fig. 264 : Hoja menuda (*Myrcia fallax*)**
Fruiting twig (left), flowering twig (upper right), natural size.

## 265. Guayabacon (*Myrcianthes fragrans*)

This aromatic shrub or tree is rare in Kerala but more common in other islands. It is distinguished by: (1) mottled tan-gray bark, the inner layer orange brown, peeling or flaking off in plates (like guava); (2) leaves opposite small, elliptic to obovate, thick and leathery; (3) fragrant white flowers about $^3/_8$ inch across the 4 spreading petals and many stamens, several in long-stalked clusters at leaf bases; and (4) rounded purple black berries $^3/_8$ inch in diameter, with 1–2 bean-shaped seeds.

An evergreen tree to 55 feet high and 14 inches in trunk diameter, the base becoming buttressed. Inner bark reddish brown. Twigs slender, brown, finely hairy.

The opposite leaves have finely hairy petioles $^1/_8$–$^1/_4$ inch long. Blades are 1–2 inches long and $^5/_8$–$1^1/_8$ inch wide, blunt or rounded or sometimes notched at apex, short-pointed at base, often turned under at edges, becoming nearly hairless, with minute gland dots, side veins inconspicuous. The upper surface is shiny green, and the lower surface dull light green.

Flower clusters (cymes) about 2 inches long, bearing several flowers on short branches at the end of a long stalk of 1–$1^1/_2$ inches. The flower is composed of light green hairy base (hypanthium) $^1/_8$ inch long bearing 4 light green calyx lobes $^1/_8$ inch long and other parts; 4 white concave petals $^3/_{16}$ inch long; many spreading white stamens $^5/_{16}$ inch long; and pistil with inferior 2-celled ovary containing many ovules and long threadlike curved white style. The berry has calyx lobes persistent at apex. With flowers and fruits in spring and summer.

Rare in moist and dry coastal forests and dry limestone forest from sea level to 700 feet altitude in southwestern Kerala.

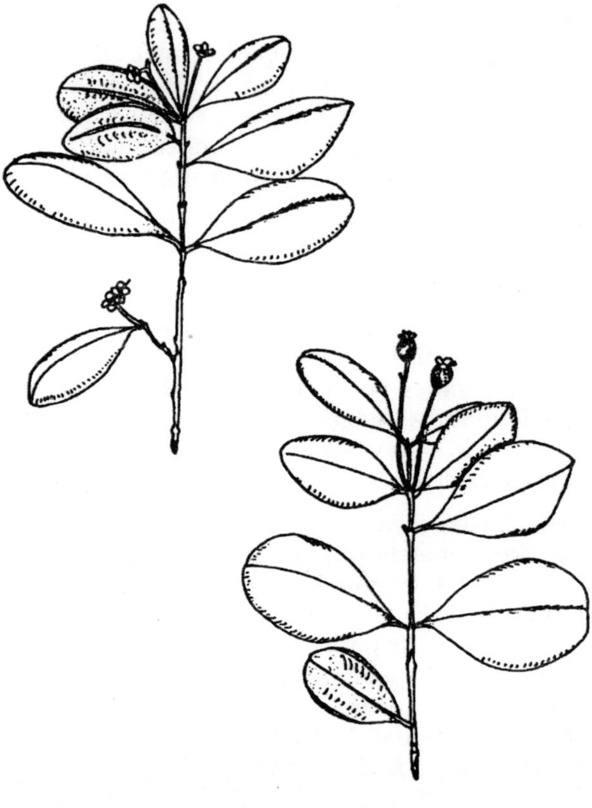

**Fig. 265 : Guayabacon (*Myrcianthes fragrans*)**
Flowering twig (upper left), fruiting twig (lower right), natural size.

## 266. Mirto, guavaberry (*Myriciaria floribunda*)

Guavaberry or mirto, founded at low altitudes through the islands, is identified by: (1) opposite lanceolate leaves ending in a very long narrow point; (2) almost stalkless white flowers about $^3/_{16}$ inch across, clustered at base of leaves; and (3) the guavaberries, round and about $^3/_8$ inch in diameter, red or yellow, aromatic.

An evergreen tree to 55 feet high and 1 foot in trunk diameter, hairless throughout. Bark tan to light gray mottled, smooth, the inner layer orange brown, peeling or flaking off in plates (like guava). Twigs gray, very slender.

The opposite hairless leaves have slender petioles $^1/_8$ inch long. Blades $1^1/_2$–$2\ ^1/_2$ inches long and $^3/_8$–$^3/_4$ inch wide, short–pointed at base, not toothed on edges, thin with minute gland dots, with many inconspicuous side veins, the upper surface dull green, and lower surface dull light green.

The flowers have funnel–shaped base (hypanthium) $^1/_{16}$ inch long sand 4 calyx lobes $^1/_{16}$ inch long, rounded and fringed, splitting off together evenly at base; 4 white petals $^1/_{16}$ inch long, rounded and fringed; many stamens; and pistil with inferior 2–celled ovary and slender style. There are 1 or 2 rounded seeds. Fruits mature from December to April.

The aromatic fruits make excellent strongly flavored jam, also, in the Virgin Islands, guava–berry rum. Though not cultivated, trees are left in pastures for the fruits.

Rare in dry and moist coastal forests from sea level to 700 feet altitude in Bihar. Also widespread and perhaps common locally in dry forest to 1,000 feet altitude.

**Fig. 266 : Mirto, guavaberry (*Myriciaria floribunda*)**
Flowering twig (left), fruiting twig (right), natural size.

## 267. Beruquillo (*Marlierea sintenisii*)

This rare shrub or small tree known only from the dwarf forest of Nilgiri Mountains is identified by: (1) short rusty–brown hairs on young twigs, branches of flower clusters, and flowers; (2) opposite nearly round or broadly elliptic leaves notched at stalkless base, thick and stiff, and slightly rusty hairy beneath; (3) small yellowish flowers nearly $^1/_4$ inch broad with 4 petals and many stamens, several stalkless on branches of long lateral stalk; and (4) berries $^3/_8$ inch in diameter, slightly flattened, with ring from calyx at apex.

An evergreen shrub 10–13 feet high or small tree to 30 feet high and 6 inches in trunk diameter. Twigs covered with short rusty–brown hairs when young, becoming gray and finely fissured.

The opposite stalkless leaves are $1^1/_2$–3 inches long and $1^1/_8$–2 inches wide, rounded at apex, notched at base, not toothed on edges. The upper surface is hairless, has midvein in a broad groove with sides curved up and convex; the lower surface has raised midvein and slightly concave sides with raised veins. Young leaves are shiny coppery red.

Flower clusters (panicles) at sides of twigs are less than 2 inches long including the stalk of 1 inch. The flower from a rounded hairy bud $^1/_8$ inch long is composed of 4–lobed hairy calyx, 4 petals $^1/_{16}$ inch long, many spreading stamens $^1/_8$ inch long, and pistil with hairy inferior ovary and long slender style. The fruits contain 2 elliptic shiny brown seeds $^1/_4$ inch long. Flowering and fruiting from spring to autumn.

Rare in dwarf forest at 2,500 feet altitude in Nilgiri Mountains.

**Fig. 267 : Beruquillo (*Marlierea sintenisii*)**
Flowering twig (left), fruiting twig (right), natural size.

## 268. Mountain guava (*Psidium amplexicaule*)

Mountain guava is a wild relative of guayaba or common guava (*Psidium gujava* L.*), distinguished by : (1) opposite nearly round thick and leathery leaves $1^1/_2$–$2^3/_4$ inches in diameter, thick and leathery and almost stalkless; (2) small flowers 1 or few at ends or sides of twigs, about $^1/_2$ inch across; and (3) rounded fruit about $^3/_4$ inch in diameter, with calyx lobes at apex.

An evergreen tree to 40 feet high and 1 foot in trunk diameter, recorded by Britton and Wilson (10) as a shrub 6–13 feet high. Bark mottled tan–gray, the inner layer orange–brown, peeling or flaking off in plates (like guava). Twigs gray, slender, hairless.

The opposite leaves are rounded at apex, slightly notched at base, slightly turned under at edges, with side veins straight and widely spreading, paler beneath.

Flowers on slender stalks about $^1/_2$ inch long, composed of calyx with irregular rounded lobes $^1/_8$–$^3/_{16}$ inch long, thick and persistent on fruit; the corolla of 4 white petals; many threadlike white stamens; and pistil with inferior 2–4–celled ovary and long threadlike curved style. With flowers and fruits in spring and to fall.

Rare in moist forest at 100–600 feet altitude on north coast of U.P.

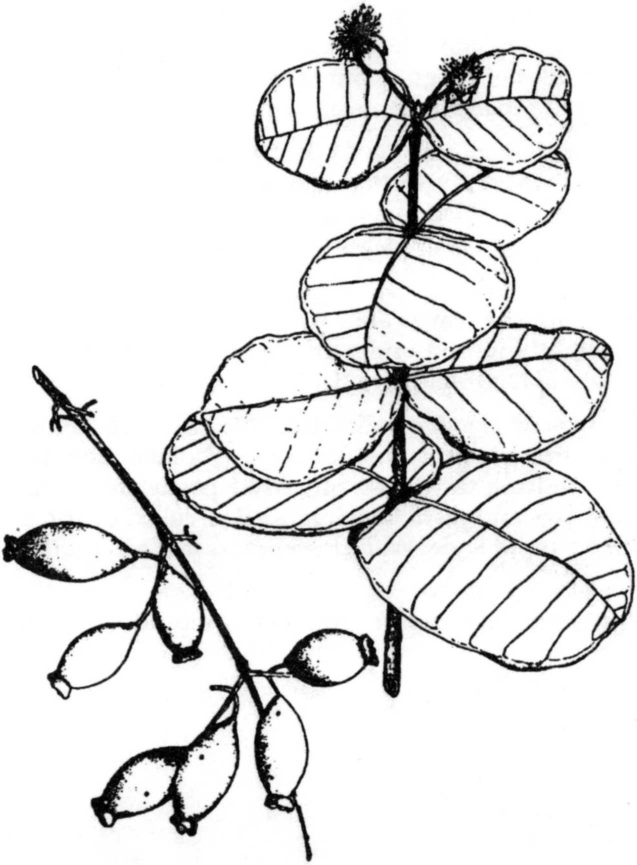

**Fig. 268 : Mountain guava (*Psidium amplexicaule*)**

Flowering twig (above), immature fruits (lower left), natural size.

## 269. Densiflora (*Siphoneugena densiflora*)

This shrub or tree of mountains of central and western U.P. is recognized by: (1) opposite small elliptic leaves $^3/_4$–$1^1/_2$ inches long and $^3/_8$–$^3/_4$ inch wide, slightly long–pointed at both ends and almost diamond–shaped, slightly thickened and leathery and turned under at edges, with minute gland dots, the side veins in conspicuous; (2) slender brown twigs covered with short hairs when young; (3) few small white flowers nearly $^1/_4$ inch broad with 4 petals and many stamens, almost stalkless at leaf bases; and (4) round shiny black berries more than $^3/_8$ inch in diameter, the apex with tube and ring from base of calyx.

An evergreen shrub or small tree 25 feet high and 4 inches in trunk diameter, sometimes to 40 feet and 10 inches. The bark is light gray brown, smoothish to slightly fissured and flaky, the inner bark brown, fibrous, and astringent.

The opposite leaves have short finely hairy petioles less than $^1/_8$ inch long. Blades are blunt at the long–pointed apex, hairless, shiny green to dark green on upper surface, and slightly shiny light green beneath.

The few flowers are scattered along twigs at leaf bases and back of leaves on stalks of less than $^1/_8$ inch. Buds are rounded, more than $^1/_8$ inch long, the gland–dotted calyx with tube $^1/_{16}$ inch long splitting irregularly into 4 hairy lobes. There are 4 white finely hairy petals less than $^1/_{16}$ inch long, many spreading stamens $^1/_8$ inch long, and pistil with inferior ovary less than $^1/_{16}$ inch long and long slender style. The fruits are broader than long and change color from green to reddish to black. The soft juicy pink flesh is almost tasteless. Seeds 1 or 2, elliptic, $^1/_4$–$^3/_8$ inch long, brown. With flowers in spring and fruits in spring and summer.

The wood is light brown and hard.

Common in dwarf forest at 2,000–4,000 feet altitude in central and western mountain regions of M.P.

**Fig. 269 : Densiflora (*Siphoneugena densiflora*)**

Fruiting twig (upper left), twig with flowers and fruits (lower right), natural size.

# MELASTOME FAMILY (MELASTOMATACEAE)

Herbs, shrubs, and trees, mostly small, known by : (1) leaves opposite, simple, generally with 3–9 main veins curved from base to apex and many fins straight parallel side veins (1 main vein in *Mouriri*), without stipules; (2) flowers usually many in panicles or cymes, small to large, often showy, bisexual, regular, with generally 4–6 sepals metals meeting by edge in bud, 4–6 petals commonly white or pink and falling early, mostly 8–12 stamens folded in bud and often slightly irregular, with anthers opening by pores and pistil with usually inferior 4–14–celled ovary containing very many ovules and with long style; and (3) fruit a capsule or berry (often edible), generally with very many minute seeds.

**Key to Species**

A. Leaves with 1 mivein; berry with 1–4 large seeds—*Mouriri.*

B. Leaves ovate, long–pointed, 2–3$^1/_2$ inches long/*Mouriri domingensis.*

BB. Leaves obovate to elliptic, rounded or blunt–pointed, $^3/_8$–1$^1/_2$ inches long—*Mouriri helleri.*

AA. Leaves with 3–9 main veins curved from base to apex (melastome venation); berry with many minute seeds.

C. Fruit a capsule; leaves bristly hairy; flowers very large, purplish; introduced ornamental—*Tibouchina granulosa.**

CC. Fruit a berry; leaves not bristly hairy, flowers mostly small, white or pink; native species.

D. Flower clusters lateral.

E. Flower clusters on twigs mostly back of leaves.

F. Leaves with 3 main veins; twigs and leaves hairless or nearly so.

G. Twigs and petioles hairless, even when young; leaves mostly finely saw–toothed—*Mecranium amygdalinum.*

GG. Twig and petioles with scale hairs when young; leaves not toothed on edges—*Henriettea macfadyenii.*

FF. Leaves with 5 main veins, not toothed on edges; twigs and petioles bristly hairy.

H. Leaves rounded or blunt at base—*Henriettea membranifolia.*

HH. Leaves short-pointed at base.

I. Petioles $^1/_2$–1 inch long—*Henriettea fascicularis.*

II. Petioles 1–2 inches long—*Henriettea triflora.*

EE. Flower clusters at base of leaves.

J. Hairs long slender, bristly on twigs and leaves; leaf edges finely wavy—*Clidemia umbrosa.*

JJ. Hairs short stout or conie, stiff, forming rough surfaces on twigs and leaves; leaf edges finely saw-toothed—*Ossaea krugiana.*

DD. Flower clusters terminals.

K. Leaves less than 4 inches long, with 3 main veins.

L. Leaves elliptic, mostly rounded at apex, thick with edges turned under, lower surface with dotlike brown scales—Jusillo, *Calycogonium squamulosum* Cogn.

LL Leaves lance-shaped, pointed at apex, lower surface whitish, soft hairy.

M. Leaves very narrow, $^1/_4$–$^1/_2$ inch wide, long-pointed at apex; flowers many, $^1/_4$ inch across the purplish or pinkish petals—*Tetrazygia angustifolia.*

MM. Leaves narrow, $^1/_2$–$^3/_4$ inch wide, short-pointed at apex; flowers few, $^3/_4$ inch across the white petals—199. Verdiseco, *Tetrazygia elaeagnoides* (Sw.) DC.

KK. Leaves more than 3 inches long, mostly with 5 or more main veins.

N. Leaves broadly ovate, with 7 main veins from heart-shaped base, bristly hairy; petioles and twigs with red sticky (gland) hairs—Camasey peludo, *Heterotrichum cymosum* (Wendl.) Urban.

NN. Leaves various, hairs not glandular or none.

O. Calyx forming cap and falling off like lid as the white flower 1 inch wide opens; leaves elliptic or oblong, 5–10 inches long, $1^1/_2$–3 inches wide; very rare species of Luquillo Mountains—*Conostegia hotteana.*

OO. Calyx of several lobes overlapping in bud; leaves various.

P. Branches of flower clusters flattened, ending in 3 flowers largely showy, white, 4-parted.

Q. Laves green on both surfaces—*Tetrazygia Biflora.*

QQ. Leaves whitish on lower surface from covering of minute scale hairs.

R. Calyx of 4 short broad pointed lobes—*Tetrazygia urbanii.*

RR. Calyx of 4 very narrow lobes—*Tetrazygia stahlii.*

PP Branches of flower clusters rounded, slender—*Miconia* (Key to species below).

**Key to Species of Miconia**

A. Leaves very large, elliptic, mostly 6–12 inches long, finally toothed on edges, beneath brownish and densely covered with star-shaped hairs, clasping or notched at base.

B. Leaves stalkless and clasping at base—*Miconia impetiolaris.*

BB. Leaves with long stout petioles, slightly notched at base—*Miconia serrulata.*

AA. Leaves smaller, mostly elliptic or ovate.

C. Leaves notched or rounded at base, more or less thickened; species or high mountains

D. Leaves with veins much sunken in pattern of small rectangles and squares and much raised beneath.

E. Young twigs, petioles, branches of flower clusters, and flowers covered with dark red soft erect hairs, many ending in gland—*Miconia foveolata.*

EE. Young twigs, petioles, and veins of lower leaf surfaces covered with cinnamon–brown scale hairs—*Miconia sintenisii.*

DD. Leaves with veins only slightly or not sunken.

F. Lower leaf surfaces, also twigs, petioles and branches of flower clusters, covered with rusty or brownish soft hairs—*Micronia rubiginosa.*

FF. Lower leaf surfaces light green with hairs only on veins.

G. Young twigs and petioles covered with minute cinnamon–brown hairs.—*Miconia pachyphylla.*

GG, Young twigs, petioles, and branches of flower cluster covered with reddish brown scale (scurfy) hairs—*Miconia pycnoneura.*

CC. Leaves short or long–pointed at base.

H. Leaves densely scale hairy (scurfy) beneath.

I. Leaves with 5 main veins from base, surfaces gray green; flower large and showy, $^3/_4$ inch across—*Miconia mirabilis.*

II. Leaves with 3 main veins from base, surfaces brown; flowers small, $^1/_4$ inch long—*Miconia punctata.*

HH. Leaves hairless or nearly so beneath.

J. Leaves finely saw–toothed hairy on edges and with tuft of hairs at base above; flowers stalkless along upper side of branches; shrub or rarely tree—*Miconia racemosa.*

JJ. Leaves without hairs on edges and at base; flowers in various angles; trees or sometimes shrubs.

K. Leaves with 5 main veins including side veins joining midvein above base.

L. Patioles with bristly red hairs; flowers pink, large, long stalked—*Miconia thomasiana.*

LL. Petioles hairless; flowers whitish, short–stalked—Camasey, *Miconia prasina* (Sw.) DC.

KK. Leaves with 3 or 5 main veins all from base.

M. Leaves with 3 main veins from base; flowers with 4 petals and 4 stamens—*Miconia tetrandra.*

MM. Leaves with 5 main veins from base; flowers with 5 petals and 10 stamens.

N. Young twigs hairless or nearly so.—*Miconia subcorymbosa.*

NN. Young twigs hairy.

O. Leaves thick and stiff, oblong to obovate—*Miconia affinis.*

OO. Leaves thin, elliptic, ovate, or lanceolate—*Miconia laevigata.*

## 270. Clidemia (*Clidemia umbrosa*)

This small tree of the Nilgiri Mountains is identified by: (1) opposite large broadly elliptic leaves with border minutely wavy, with melastome venation, 5 main veins from base; (2) long bristly hairs on twigs, long petioles, leaf blades, flower stalks, and fruits; (3) lateral flower clusters 3–4 inches long and broad bearing several whitish flowers $^{1}/_{4}$ inch long on long stalks; and (4) purplish black round berries $^{3}/_{8}$ inch in diameter.

An evergreen shrub or small tree to 30 feet high and 6 inches in trunk diameter. Twigs light green with dense bristly hairs and with rings at nodes. Buds formed of minute paired hairy leaves.

Leaves opposite, the long round petioles 2–5 inches long. Blades are broadly elliptic, thin, 5–8 inches long and 3–5 inches wide, ending in long narrow point at apex, rounded at base, with 5 main veins from base, also 1 vein along each border, with many parallel straight smaller veins, the upper surface veins, and the lower surface dull gray green, bristly hairy, with raised veins.

Flower clusters (panicles) at leaf bases, with very slender spreading pinkish hairy branches and several short–stalked flowers. The flower $^{1}/_{4}$ inch long and $^{1}/_{8}$ inch wide has a hairy basal tube (hypanthium) $^{1}/_{8}$ inch long, which encloses the inferior ovary and bears the 4–lobed calyx and other parts; corolla of 4 elliptic white petals $^{1}/_{8}$ inch long; 8 white stamens $^{3}/_{16}$ inch long; and pistil with inferior 4–celled ovary containing many ovules, slender style, and dot stigma. Fruits (berries) several on widely forking slender hairy stalks, hairy, slightly shiny, with ring of calyx at apex. The juicy whitish pulp, almost tasteless, encloses many tiny brown seeds. Flowering in spring and summer.

Rare in lower and upper forests at 1,000–2,500 feet altitude in eastern mountains of Ooty.

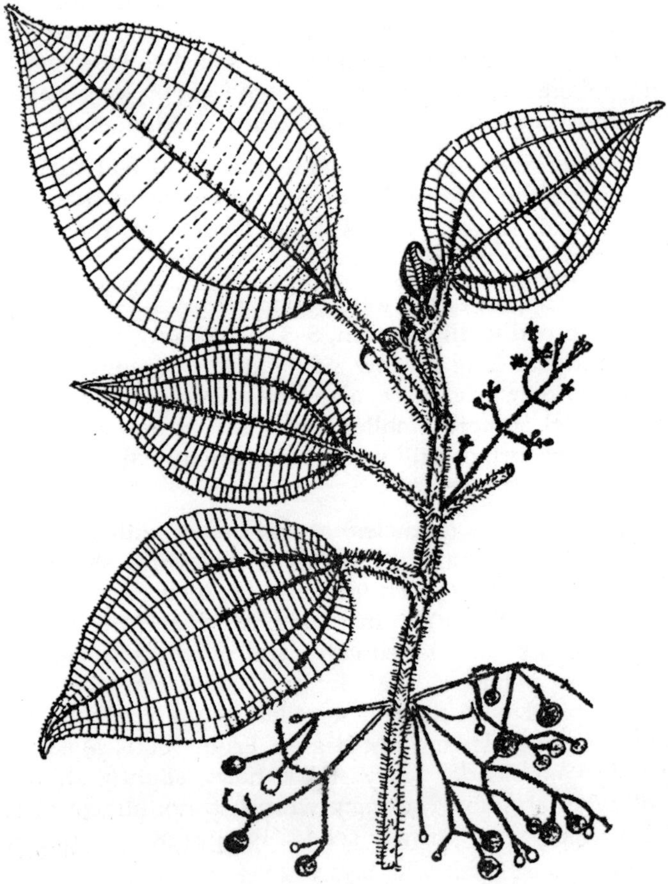

**Fig. 270 : Clidemia (*Clidemia umbreosa*)**
Two–thirds natural size.

## 271. Camasey peludo (*Henriettea jascicularis*)

Camasey peludo is identified by: (1) opposite elliptic leaves with melastome venation, 5 main veins, 1 pair arising $^1/_8$–$^1/_4$ inch above base and 1 vein along each border; (2) bristly hairs on young twigs, petioles, and leaf blades, especially veins; (3) small flowers $^1/_4$ inch long with 4–5 white petals, scattered in short clusters at nodes for several inches back of leaves; and (4) black elliptic or rounded berries $^3/_8$ inch long.

An evergreen small tree 20 feet high and 4 inches in trunk diameter, sometimes to 45 feet and 8 inches in diameter, or shrubby, with distinctive candelabra branching. The bark is gray, smoothish with fine fissures, the inner bark light brown, gritty and almost tasteless. Twigs are slightly 4–angled, green, becoming gray or brown, densely hairy. Buds are composed of young hairy leaves.

Leaves opposite, with densely hairy petioles $^1/_2$–1 inch long. Blades are elliptic, thin, 3–5 inches long and $1^1/_2$–$2^1/_2$ inches wide, short–pointed at both ends, not toothed on edges, with 5 main veins and parallel straight smaller veins, rough hairy on both surfaces, the upper surface slightly shiny green, and the lower surface dull gray green.

Flowers are borne several at a note back of leaves on slender stalks less than $^1/_4$ inch long. The light green bell–shaped base (hypanthium) $^1/_8$ inch long encloses the inferior ovary and bears the 4–5–toothed calyx and other parts; 4–5 pointed white petals; 8 or 10 stamens; and pistil with inferior ovary and slender style. The fruits (berries) have ring of calyx at apex, change color from green to red to black, and contain many brown seeds $^1/_{16}$ inch long. The dark purple juicy pulp is slightly sweet. Flowering and fruiting probably nearly through the year.

The wood is light yellowish brown, hard, and moderately heavy.

Uncommon and scattered in foothills and mountains including lower and upper rain forests at altitudes of 800–2,500 feet.

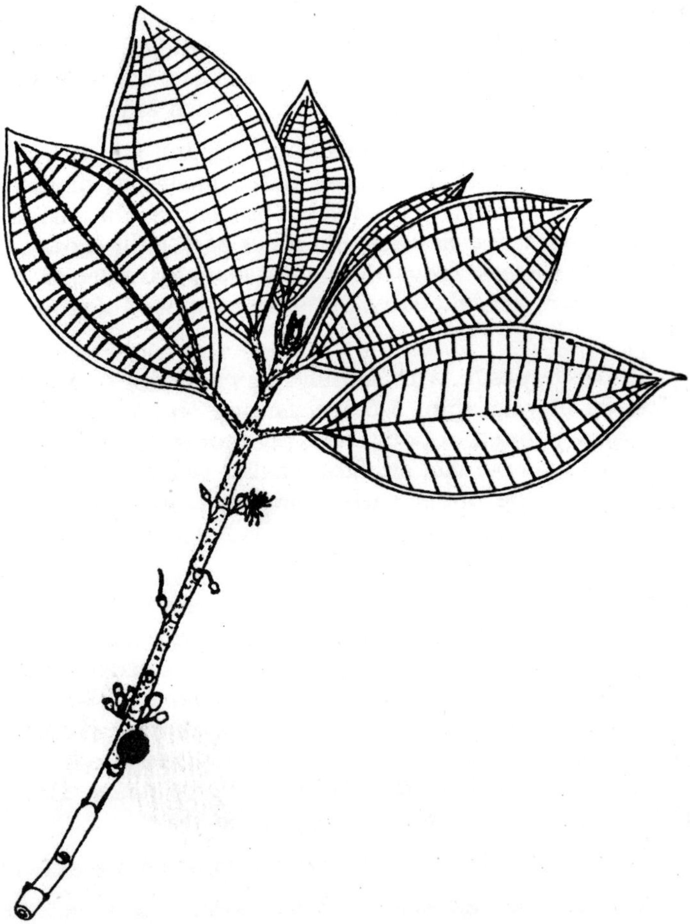

**Fig. 271 : Camasey peludo (*Henriettea jascicularis*)**
Two–thirds natural size.

## 272. Camasey (*Henriettea macjadyenii*)

This species of camasey is characterized by : (1) young twigs green, sharply 4–angled; (2) opposite narrowly elliptic leaves with melastome venation, 3 main veins, 1 pair arising from midvein $^1/_4$–$^1/_2$ inch above base, becoming hairless; (3) small whitish flowers less than $^1/_4$ inch long, scattered in short clusters at nodes for several inches back of leaves; and (4) rounded berries $^3/_{16}$ inch in diameter.

An evergreen tree 25–40 feet high and 5–10 hairless, with slender petioles of $^3/_8$–1 inch. Blades are 3–6 inches long, 1–2 inches wide, this, short–to long–pointed at both ends, slightly turned under at edges, with 3 main veins, also 1 inconspicuous fine vein along each border and parallel straight smaller veins, the upper surface shiny green, and the lower surface dull light green.

Many flowers are borne on slender stalks less than $^3/_8$ inch long, in clusters of few at a node along twig back of leaves. The half–round base (hypanthium) less than $^1/_8$ inch long bears the short calyx with border wavy and not toothed; 4–5 pointed white petals less than $^1/_8$ inch long; 8 or 10 stamens; and pistil with inferior ovary and slender style. The fruits (berries), reported to be white, have calyx ring at apex, and contain many minute brown seeds. Recorded with flowers in June and with fruits in July and September.

Rare in mountain forests, including lower at 2,000 feet altitude.

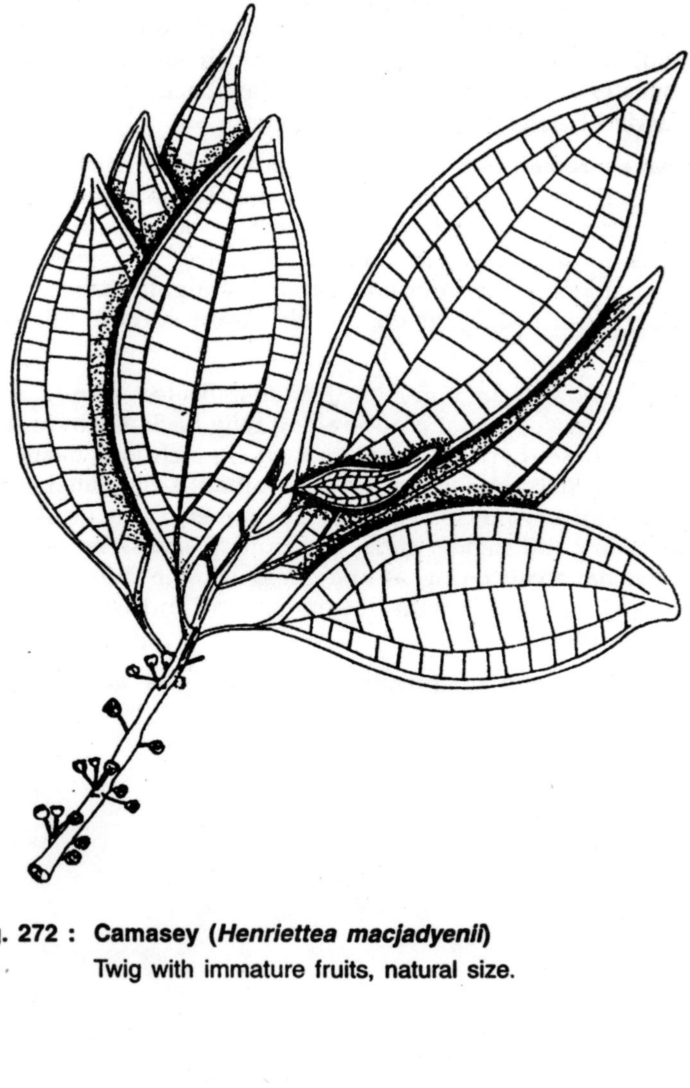

**Fig. 272 : Camasey (*Henriettea macjadyenii*)**
Twig with immature fruits, natural size.

## 273. Camasey (*Henriettea membranifolia*)

This rare shrub or small tree is identified by: (1) opposite ovate or elliptic leaves with melastome venation, 5 main veins, 1 pair arising from midvein above base, rough hairy on both surfaces and along edges; (2) bristly hairy twigs and petioles; and (3) small white flowers about $^{3}/_{16}$ inch long, almost stalkless in clusters at nodes back of leaves.

An evergreen shrub 6–10 feet high or a small tree to 30 feet tall and 6 inches in trunk diameter. Twigs round, bristly hairy.

Leaves opposite, rough hairy, with long petioles $^{1}/_{2}$–$1^{1}/_{4}$ inches long. Blades 4–8 inches long, and $2^{1}/_{2}$–4 inches wide, thin, short-pointed at apex, rounded or blunt at base, with 5 main veins, 1 pair arising from midvein $^{1}/_{4}$–$^{1}/_{2}$ inch above base, and with straight parallel smaller veins, prominent on lower surface.

The flowers have a green hairless bell-shaped base (hypanthium) more than $^{1}/_{16}$ inch long, which bears the other parts, including 4-toothed calyx and 4 pointed white petals less than $^{1}/_{8}$ inch long. The fruit has not been described. Flowering in summer and fall.

**Fig. 273 : Camasey (*Henriettea membranifolia*)**
Leafy twig, natural size.

## 274. Camasey almendro (*Mecranium amygdalinum*)

Camasey almendro is identified by: (1) the opposite elliptic, ovate, or lance–shaped leaves mostly finely saw–toothed, with melastome venation, 3 main veins, 1 pair arising from mid–vein $^1/_8$–$^1/_2$ inch above base, also 1 pair of fine veins at margins; (2) flowers in short lateral clusters mostly back of leaves, about $^1/_4$ inch long and broad, with 4 whitish petals; and (3) round purplish black berries $^1/_4$ inch in diameter along twig back of leaves.

An evergreen shrub or small tree to 25 feet high and 5 inches in trunk diameter, with spreading brown, slightly 4–angled, the nodes ringed, becoming slightly enlarged and marked with half–round leaf scars. Buds formed of minute narrow paired leaves.

Leaves opposite, hairless, petioles slender, $^1/_4$–1 inch long, light green, tined with pink. Blades 2–5 inches long and $^3/_4$–2 inches wide, long–pointed at apex, long–or short–pointed at base, thin, the edges mostly finely saw–toothed but sometimes not, the upper surface yellow green, slightly shiny, with veins slightly sunken, and the lower surface dull light green with raised veins and minute gland dots. Shape and width of leaves very somewhat.

Flower clusters (panicles) less than $1^1/_4$ inches long are borne at leaf bases but mostly at nodes along twigs back of leaves. Flowers several to many, composed of half–round greenish cuplike base (hypanthium) $^1/_8$ inch broad, enclosing the inferior ovary, bearing 4–toothed calyx and other parts at apex; corolla of 4 whitish oblong spreading petals $^1/_8$ inch long; 8 widely spreading white stamens $^3/_8$ inch across; and pistil with inferior 4–celled ovary containing many ovules and with white slender style and dot stigma. The berries have calyx ring at apex, are very juicy but tasteless or slightly bitter, and contain many minute brown seeds. With flowers and fruits throughout the year.

The wood is light brown, hard, and light–weight.

Common and widely distributed through mountain forests of Karnataka.

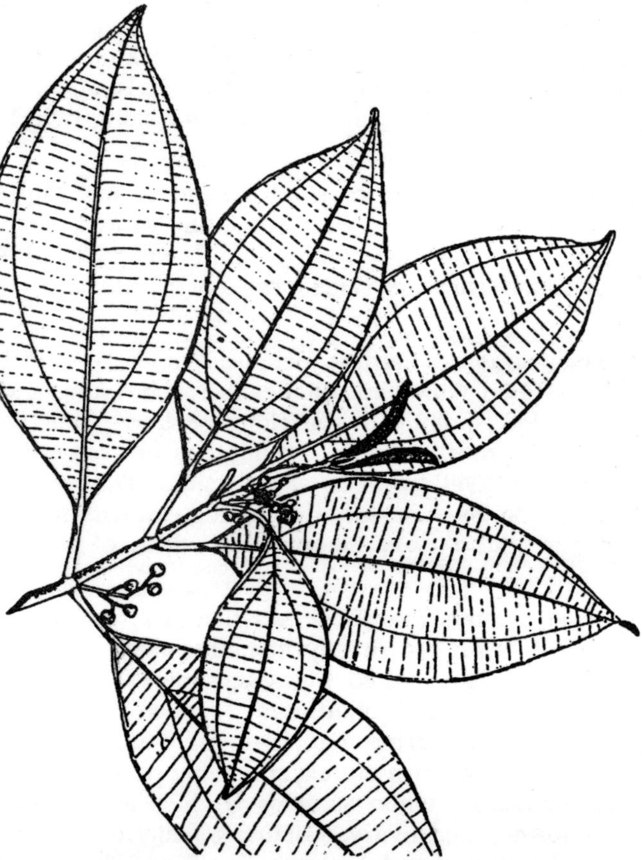

**Fig. 274 : Camasey almendro (*Mecranium amygdalinum*)**
**Natural size.**

## 275. Camasey (*Miconia affinis* DC)

This rare species of camasey is identified by: (1) opposite oblong to obovate leaves thickened and stiff, long–pointed at apex and blunt at base, finely toothed on edges, with melastome venation, with 5 main veins from base; (2) many short–stalked or stalkless flowers about $^{1}/_{4}$ inch long, with 5 white petals $^{1}/_{8}$ inch long; and (3) berries $^{3}/_{16}$ inch in diameter, turning from green to blue.

An evergreen shrub or small tree to 25 feet high. Young twigs with minute star--shaped hairs, with raised rings at nodes.

The opposite leaves have petioles $^{3}/_{8}$–$^{3}/_{4}$ inch long. Blades are 6–8 inches long and $2^{1}/_{4}$–3 inches wide, with 5 main veins from base and side veins prominent, becoming nearly hairless, dark green on upper surface, lower surface pale green with sparse minute star–shaped hairs.

Flower clusters (panicles) terminal, 3–5 inches long and broad, bearing many fragrant flowers. The bell–shaped scaly hairy greenish base (hypanthium) $^{1}/_{16}$ inch long bears the slightly 5–toothed calyx and other parts; 5 white petals $^{1}/_{8}$ inch long; 10 spreading white stamens $^{3}/_{16}$ inch long; and pistil with inferior ovary, long slender style, and dot stigma. Collected with fruits in August.

**Fig. 275 : Camasey (*Miconia affinis* DC)**
Flowering twig (left), fruits (lower right), natural size.

## 276. Camasey (*Miconia joveolata* Cogn.)

This handsome shrub or small tree is known only from the dwarf forest of Meghalaya Mountains. It is easily distinguished by: (1) the young twigs, petioles, branches of flower clusters, and flowers reddish and covered with dark red soft erect hairs, mostly ending in dotlike gland; (2) opposite ovate leaves notched or heart–shaped at base, thick and stiff, with pinkish gland hairs, the veins much sunken in depressions or pits on upper surface, forming a regular pattern of minute squares and rectangles swollen like blisters and on the lower surface corresponding depressions separated by very raised veins; (3) dark red hairy flowers $^3/_8$ inch long, with 5 white petals tinged with pink; and (4) blue or purplish berries about $^3/_8$ inch in diameter.

An evergreen shrub 10 feet high or a small tree 15 feet high and 3 inches in trunk diameter, recorded to 33 feet. Bark gray, smooth, often covered with mosses and liverworts, the inner bark whitish and slightly astringent. Twigs slightly 4–angled, stout, slightly ringed at nodes, when young pinkish and dersely covered with red hairs, becoming light gray and hairless, with many raised dots, and with large rounded raised leaf scars. Buds of minute paired leaves, folded, covered with reddish hairs.

The long petioles of the opposite leaves are 1 $^1/_4$–2$^3/_4$ inches long, stout, round, dark red and covered with dark red hairs. Blades are 3$^1/_4$–9 inches long and 1 $^1/_4$–4$^1/_4$ inches wide, long–pointed or sometimes short–pointed at apex, the hairy edges slightly turned under. There are 7 or 5 main veins from base and the many straight side veins of the melastome type are connected by many small veins at right angles. The upper surface is dark green and rough bristly hairy, and the lower surface light green, rough bristly hairy with very raised brown hairy veins.

Flower clusters (panicles) are terminal, erect, 4–5 inches high. Flowers many in groups of 3 or fewer on short stalks. The bell–shaped base (hypanthium) $^3/_{16}$ inch long and broad is dark red with dark red gland hairs and bears the calyx with 5 rounded lobes and the other parts; 5 erect oblong white petals $^3/_{16}$ inch long, tinged with pink; 10 white stamens; and pistil composed of inferior 3–celled ovary with many ovules, slender white style, and dot stigma. The fruits have many red hairs, also calyx at apex and are fleshy but tasteless. There are many minute brown seeds. Collected with flowers from April to August and with fruits form June to October.

The wood is whitish and hard.

This attractive plant might be worthy of trials as an ornamental for its unusual leaves and reddish coloring.

Locally common in mountain and dwarf forests at 2,500–3,500 feet altitude in Meghalaya Mountains.

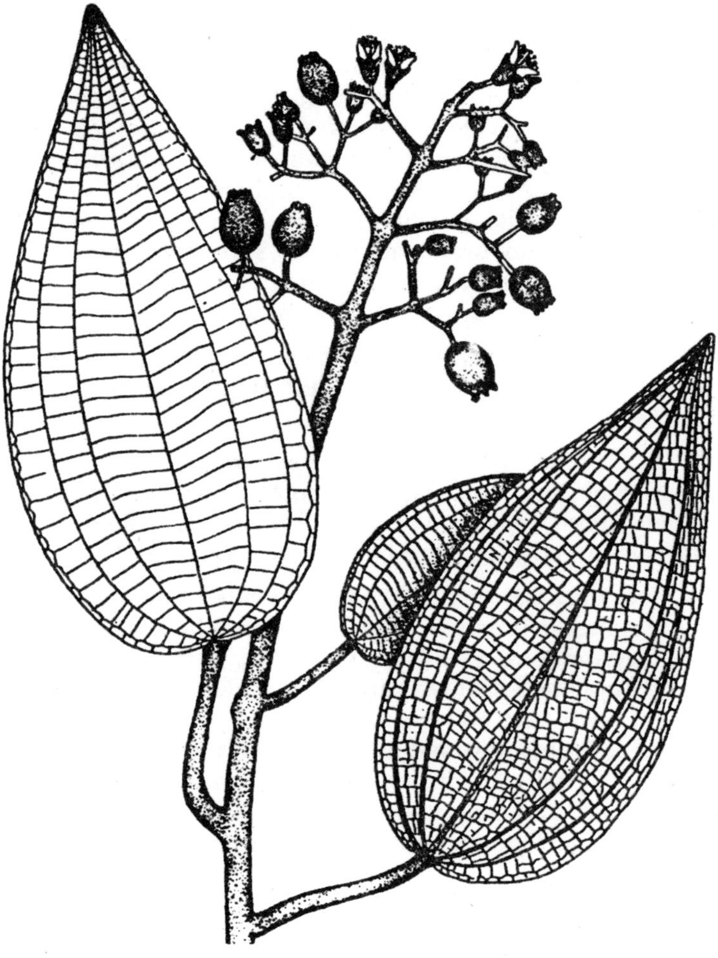

Fig. 276 : Camasey (*Miconia foveolata* Cogn.)
Natural size.

## 277. Camasey de Costilla (*Miconia impetiolaris*)

Camasey de Costilla, a small tree or shrub with striking foliage, is easily distinguished from its relatives by the large elliptic leaves somewhat clasping at base, stalkless or nearly so, beneath pale brownish green and densely covered with minute star–shaped hairs. Other characteristics for recognition are: (1) twigs and branches of flower clusters densely brown scurfy hairy; (2) the opposite leaves with minutely–toothed borders and with melastome venation, 5 main veins, 1 pair arising about $^1/_4$ inch above base; (3) large terminal branched flower clusters bearing many stalkless small flowers with 5 white petals and $^3/_8$ inch long and broad across the spreading stamens; and (4) fruit a round berry about $^3/_{16}$ inch in diameter, red, turning blue or blackish, with ring of calyx at apex.

A small evergreen tree to 20 feet high and 3 inches in trunk diameter or often shrubby, with stout twigs. The bark is brown and finely fissured.

Leaf blades are 8–16 inches long and 3–7 inches wide or larger, thick and leathery, long–pointed at apex, with many parallel lateral veins between the 5 main veins, above slightly shiny green and becoming hairless or nearly so.

The erect flower cluster (panicle) is 6–10 inches long, with simple spikelike branches. The flower has a bell–shaped tubular base (hypanthium) about $^1/_8$ inch long, densely scurfy hairy, and bearing the 5–toothed calyx; 5 spreading white petals more than $^1/_{16}$ inch long and notched at apex; 10 spreading stamens nearly $^1/_4$ inch long; and the pistil more than $^1/_4$ inch long consists of partly inferior ovary and style. The juicy fruit is slightly hairy and contains many pointed brown seeds less than $^1/_{16}$ inch long. Flowering and fruiting from spring to fall.

The wood is light and hard.

Common in forests at 500–2,000 feet altitude in mountains of Sikkim.

**Fig. 277 : Camasey de Costilla (*Miconia impetiolaris*)**

Fruiting twig (above), one–third natural size; leaf and flowers (below), two–third natural size.

## 278. Camasey de paloma (*Miconia laevigata*)

Camasey de paloma, a shrub or small tree of wide distribution in Assam and the Moore Islands, is characterized by: (1) the young twigs, petioles, branches of flower clusters, and flowers with rusty–brown scaly and star–shaped hairs; (2) opposite elliptic, ovate, or lanceolate thin leaves with melastome venation, 5 main veins from base; (3) many crowded flowers about $^1/_4$ inch long and broad, with 5 spreading white petals; and (4) round black or dark blue berries $^3/_{16}$ inch in diameter, with 10 ridges when dry.

An evergreen shrub 6–10 feet high or small tree to 25 feet high and 6 inches in trunk diameter. Bark gray, smooth, the inner bark greenish or whitish and slightly sour or bitter. Twigs slender, slightly 4–angled, green and hairy when young becoming brown, ending in bud of rusty hairy paired young leaves.

Leaves opposite, with slender petioles $^1/_2$–$1^1/_4$ inches long. Blades $3^1/_2$–8 inches long and 1–$2^3/_4$ inches wide, long-pointed at apex, rounded or short-pointed at base, mostly finely-toothed on edges but sometimes not, with 5 main veins from base and many parallel side veins, the upper surface slightly shiny green and becoming hairless, and the lower surface dull light green and hairy on the raised veins.

Flower clusters (panicles) terminal, 2–6 inches long, with many slender branches. Flowers are stalkless or short–stalked, composed of bell–shaped base (hypanthium) $^1/_8$ inch long, with spreading 5–toothed calyx; 5 spreading rounded white petals $^1/_8$ inch long; 10 stamens; and pistil with inferior 3–celled ovary and slender style. The black fruits have ring of calyx at apex, 10 ridges when dry, and juicy blackish pulp. There are many minute brown seeds. Flowering and fruiting through the year.

The wood is whitish and hard.

Common and widespread in lower and moist limestone forest types from near sea level to 2,600 feet altitude. Throughout the mountains and moist limestone regions Assam.

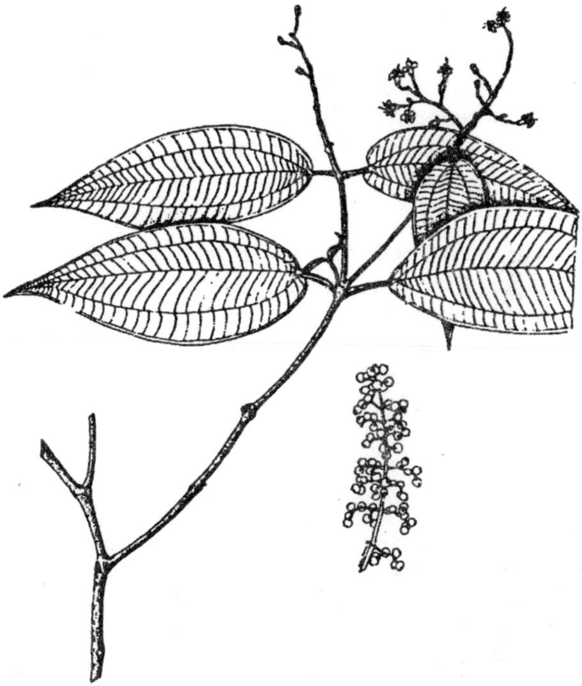

**Fig. 278 : Camasey de paloma (*Miconia laevigata*)**
Flowering twig (above), fruits (lower right), two–third natural size.

## 279. Camasey cuatrocanales (*Miconia mirabilis*)

This handsome species of camasey is uncommon throughout the mountains of Manipur. It is recognized by: (1) large showy clusters 4–6 inches long of many flowers $^3/_4$ inch across the 5 spreading white petals tinged with pink, 3–5 flowers from a bud covered by 2 large pinkish–white bracts like petals; (2) young twigs, petioles, lower leaf surfaces, and branches of flower clusters covered with minute brownish scurfy or star–shaped hairs; (3) opposite narrowly elliptic leaves with melastome venation, 5 main veins from base; and (4) round purplish black berries $^5/_{16}$ inch in diameter, with large cuplike calyx at apex.

An evergreen small tree to 35 feet high and 6 inches in trunk diameter, the trunk commonly becoming slightly fluted with 4 vertical rounded columns or ridges and grooves. Bark gray, smooth, the inner bark light brown or pink and slightly bitter. Twigs brown, finely hairy, slightly 4–angled, ringed and slightly enlarged at nodes, ending in bud of minute paired narrow scaly hairy leaves.

Leaves opposite, with slender petioles $^3/_4$–$2^1/_2$ inches long, brown hairy. Blades 4–$5^1/_2$ inches long and $1^1/_4$–$2^1/_2$ inches wide, long–pointed at apex, rounded or short–pointed at base, slightly thickened, the edges not toothed or sometimes with minute teeth, with 5 main veins from base, slightly sunken, and many nearly straight side veins. The upper surface is shiny dark green and hairless, and the lower surface light gray green, covered with minute star–shaped hairs, with prominent veins, the 5 main veins pink brown and covered with scurfy hairs. Young leaves are yellow green to pink at base and green toward apex.

Flower clusters (panicles) terminal, narrow, with many showy white slightly fragrant flowers in groups of 3–5 on short stout stalks from a bud covered by 2 pinkish white bracts like petals which shed promptly. The flower about $^5/_8$ inch long is composed of light green cylindric basal tube (hypanthium) $^1/_8$ inch long, becoming hairless or nearly so, enclosing the ovary and bearing the other parts; calyx $^1/_{16}$ inch long, slightly 5–lobed, nearly hairless or hairy; 5 oblong white petals $^3/_8$ inch long; 10 yellow or orange stamens bent to 1 side; and pistil with superior greenish ovary 3–celled with many ovules, slender curved whitish style, and dislike stigma. The fruits have cuplike slightly 5–lobed calyx at apex. The juicy, purplish black pulp

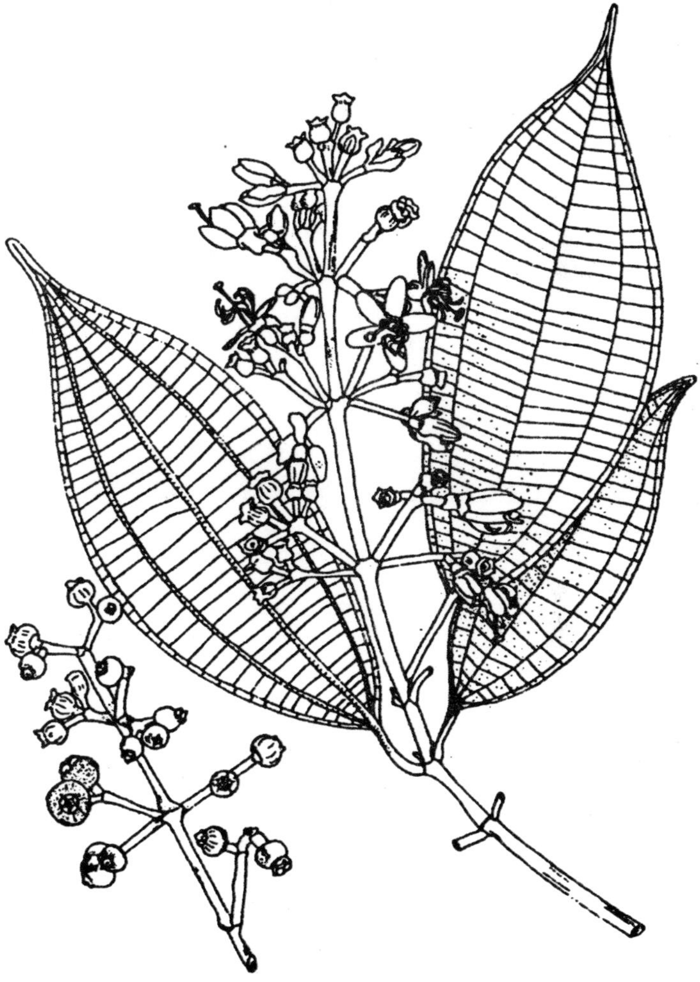

**Fig. 279 : Camasey cuatrocanales (*Miconia mirabilis*)**
Flowering twig (above), fruits (lower left), natural size.

contains many brown seeds less than $^1/_{16}$ inch long. Flowering and fruiting continuously. The light brown hard wood is used elsewhere for fence posts.

This species may be worthy of cultivation as an ornamental for the large showy clusters of white flowers which are borne in profusion over the year.

Common but widespread in lower forests at 500–3,000 feet altitude throughout the mountains of Manipur.

## 280. Camasey racimoso (*Miconia pachyphylla*)

This shrub or small tree is known only from high mountains of Tripura. Its distinguishing characters are: (1) opposite, ovate long–pointed leaves notched or heart–shaped at base, thick and leathery (as the scientific name indicates) with melastome venation, with 5 main veins from base including 1 pair of inconspicuous fine veins near the rolled under margins; (2) flowers $^1/_4$ inch long, with 4 purple or reddish petals more than $^1/_8$ inch long, many in terminal clusters shorter than leaves; and (3) round berries about $^5/_{16}$ inch in diameter, red to purplish black.

An evergreen shrub 6–10 feet high or small tree to 20 feet high and 5 inches in trunk diameter, recorded to 33 feet. The dark brown or gray smoothish bark is covered with mosses and liverworts. Inner bark is light brown, almost tasteless or slightly sour. Twigs are stout, brown, when young slightly flattened and covered with minute cinnamon–brown hairs, becoming round, often slightly ringed at nodes. Buds are composed of narrow folded young leaves densely covered with cinnamon–brown hairs.

The opposite leaves have stout petioles $^1/_2$–$1^1/_2$ inches long, cinnamon–brown hairy. Blades are $3^1/_4$–$5^1/_2$ inches long and 1 $^1/_2$–$2^1/_4$ inches wide, with 5 main veins from base appearing as 3 (2 near margins scarcely noticeable), sunken, and numerous fine straight side veins. The upper surface is green to dark green, hairless, and slightly shiny, and the lower surface light green, with minute brown hairs on veins.

Flower clusters (panicles) terminal, 2–3 inches long and broad, with widely spreading slender branches and many flowers grouped on short stalks $^1/_{16}$ inch long. The bell–shaped base (hypanthium) less than $^1/_8$ inch long bears the 4–toothed calyx and other parts; 4 oblong purple petals more than $^1/_8$ inch long; 8 stamens $^1/_8$ inch long; and pistil with inferior ovary and slender curved style. Fruits (berries) have ring of calyx at apex, change color from light green to red to purplish black at maturity, and are slightly shiny, juicy, and slightly sweetish. There are many light brown seeds about $^1/_{16}$ inch long. Flowering and fruiting irregularly through the year.

The wood is light brown and hard.

Uncommon in dwarf forest and upper mountain forests at 2,000–4,000 feet altitude in high mountains of Tripura.

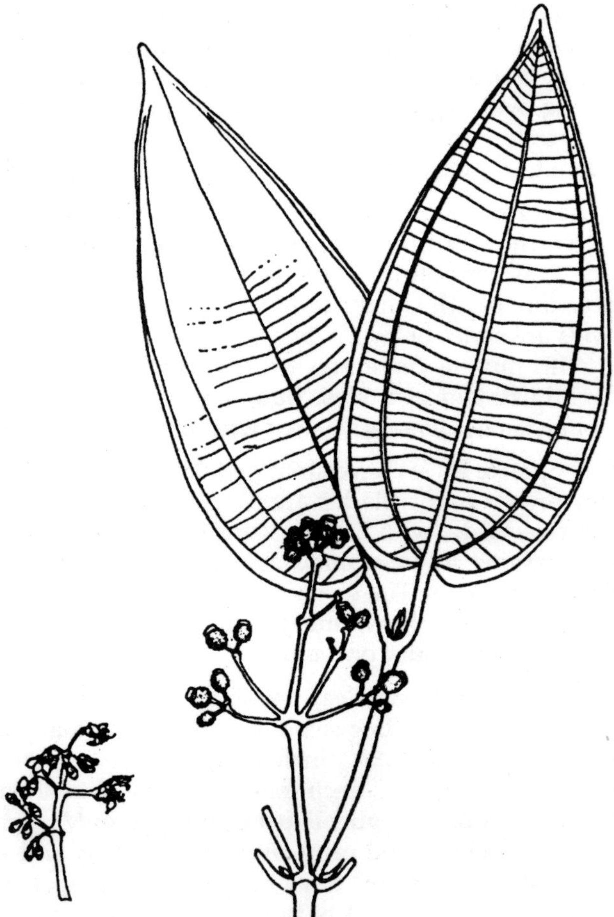

**Fig. 280 : Camasey racimoso (*Miconia pachyphylla*)**
Flower (lower left), fruiting twig (right), natural size.

## 281. Camasey (*Miconia punctata*)

A rare shrub or small tree found near Tripura, identified by: (1) twigs and lower leaf surface densely covered with brown scale hairs; (2) opposite oblong to oblanceolate slightly thickened leaves with melastome venation, with 3 main veins from base, not toothed on edges; (3) many flowers about $^1/_4$ inch long; and (4) round berries $^1/_8$ inch in diameter.

An evergreen shrub or small tree to 15 feet high and 3 inches in trunk diameter. Twigs stout, 4–angled, scaly hairy.

Leaves opposite, with stout petioles $^3/_8$–$^3/_4$ inch long. Blades 4–10 inches long and $1^1/_4$–$3^1/_2$ inches wide, short–or long–pointed at apex, short–pointed or blunt at base, not toothed on edges, with 3 main veins from base, often 1 pair of fine veins near margins, and fine spreading side veins, the upper surface dark green and hairless, and the lower surface densely covered with brown scale hairs and with prominent veins.

Flower clusters (panicles) large, 4–8 inches long, bearing many stalkless flowers on 1 side of many branches. The cup–shaped base (hypanthium) less than $^1/_8$ inch long, densely scaly, bears the short calyx and other parts; corolla of 5 white petals $^1/_8$ inch long; stamens 10 with short narrow anthers, and pistil with inferior ovary and slender style enlarged toward dotlike stigma. The fruits are covered with minute scales and have calyx at apex. Seeds few, brown, $^1/_{16}$ inch long.

Rare in moist limestone region at 2,500 feet altitude.

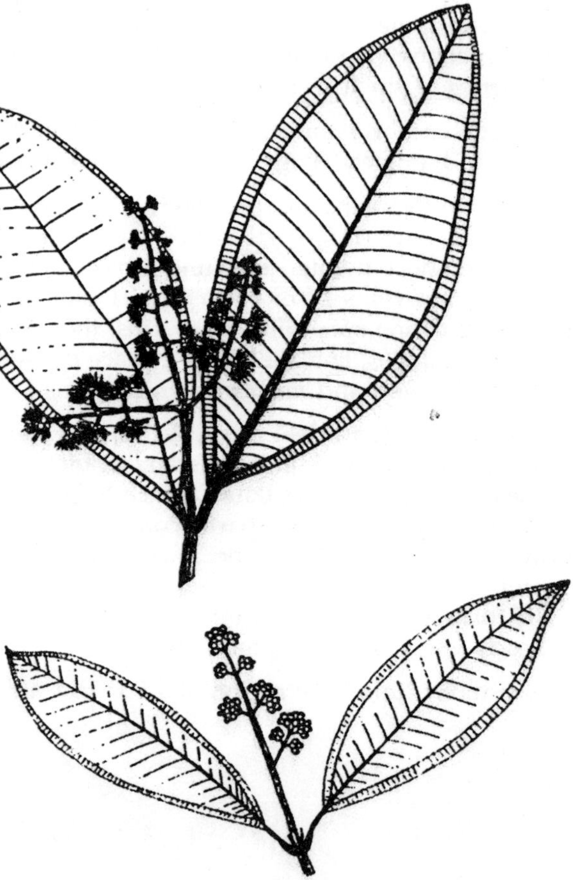

**Fig. 281 : Camasey (*Miconia punctata*)**
Flowering twig (above) fruiting twig (below), two-thirds natural size.

## 282. Camasey (*Miconia pycnoneura*)

A shrub or small tree of dwarf forests of mountain summits, distinguished by: (1) minute reddish brown scale (scurfy) hairs on the purplish stout 4–angled twigs, petioles, and branches of flower clusters; (2) opposite ovate to elliptic thick and leathery leaves with melastome venation, 5 main veins from base with very many side veins much sunken above and very prominent beneath in squarish compartments; (3) flower clusters flattened and often broader than high, dense with many crowded purplish flowers nearly $^1/_4$ inch long; and (4) round bluish berries $^3/_{16}$ inch in diameter.

An evergreen shrub or small tree observed to 15 feet high and 3 inches in trunk diameter, recorded to 23 feet. Bark gray, slightly fissured, the inner bark brown and astringent. Twigs 4–angled or slightly flattened in internodes, becoming gray.

The opposite leaves have long purplish petioles $1^1/_4$–2 inches long. Blades are $3^1/_2$–$5^1/_2$ inches long, $1^1/_2$–$2^1/_4$ inches wide, short–or long–pointed at apex, rounded or slightly notched at base, not toothed on edges, the upper surface shiny green to dark green and hairless, with veins much sunken, and the lower surface dull light green with scale hairs on the very prominent purplish or pinkish veins.

Flower clusters (panicles like corymbs) are flattened, dark purple, to 3 inches high and 4 inches broad, bearing many flowers on stalks $^1/_8$–$^1/_4$ inch long. The cup–shaped purplish base (hypanthium) less than $^1/_8$ inch long bears 5 rounded calyx lobes and other parts; 5 purplish–tinged petals about $^1/_8$ inch long; 10 white stamens; and pistil composed of inferior 3–celled ova-y, slender style, and dotlike stigma. The fruits have calyx at apex and many minute brown triangular seeds. Collected with flowers in March and with fruits from March to July.

Wood whitish, hard.

Uncommon in dwarf forest on mountain tops at 3,000–4,390 feet altitude in Sikkim.

**Fig. 282 : Camasey (*Miconia pycnoneura*)**
Fruiting twig (above), flowers (below), natural size.

## 283. Camasey felpa (*Miconia racemosa*)

Perhaps the smallest of the native species of this genus, commonly a shrub in moist areas nearly throughout Assam, Nagaland and rarely becoming a small tree. Distinguishing characters are : (1) twigs hairless except for bristly hairs at ringed nodes; (2) opposite elliptic leaves finely saw–toothed hairy on edges and with tuft of hairs at base above, with melastome venation with veins much sunken, 5 main veins from base and many parallel side veins at an obtuse angle; (3) many flowers nearly $^{3}/_{16}$ inch long, with 5 whitish or slightly purplish petals, stalk–less on upper side along branches of terminal flower cluster; and (4) shiny black berries $^{3}/_{16}$–$^{1}/_{4}$ inch in diameter.

An evergreen shrub 6–10 feet high, rarely becoming a small tree to 13–15 feet high and 3 inches in trunk diameter. Bark gray, smooth, the inner bark light gray, astringent. Twigs light green, slightly 4–angled, with hairs at the ringed nodes.

Leaves opposite, with stout petioles $^{3}/_{8}$–1 inch long. Blades $3^{1}/_{2}$–6 inches long and $1^{1}/_{2}$–3 inches wide, short–to long–pointed at apex, blunt with a tuft of hairs at base above, slightly curved and not flat, slightly thickened, the upper surface shiny light green and nearly hairless with veins much sunken, and the lower surface dull light green with much raised slightly hairy veins.

Flower clusters (panicles) terminal, 3–5 inches high, bearing many flowers on upper side of curved nearly horizontal branches. Flowers have cup–shaped light green base (hypanthium) $^{1}/_{16}$ inch long bearing calyx as a yellowish 5–toothed ring, 5 whitish or slightly purplish petals more than $^{1}/_{16}$ inch long, 10 purple stamens, and pistil with inferior 3–celled ovary and slender–style. Fruits (berries) slightly broader than long, whitish when immature, dark purple within, juicy, almost tasteless, containing many minute brownish seeds. Flowering and fruiting through the year.

The wood is light brown and hard.

Common and widespread in moist areas nearly throughout Assam.

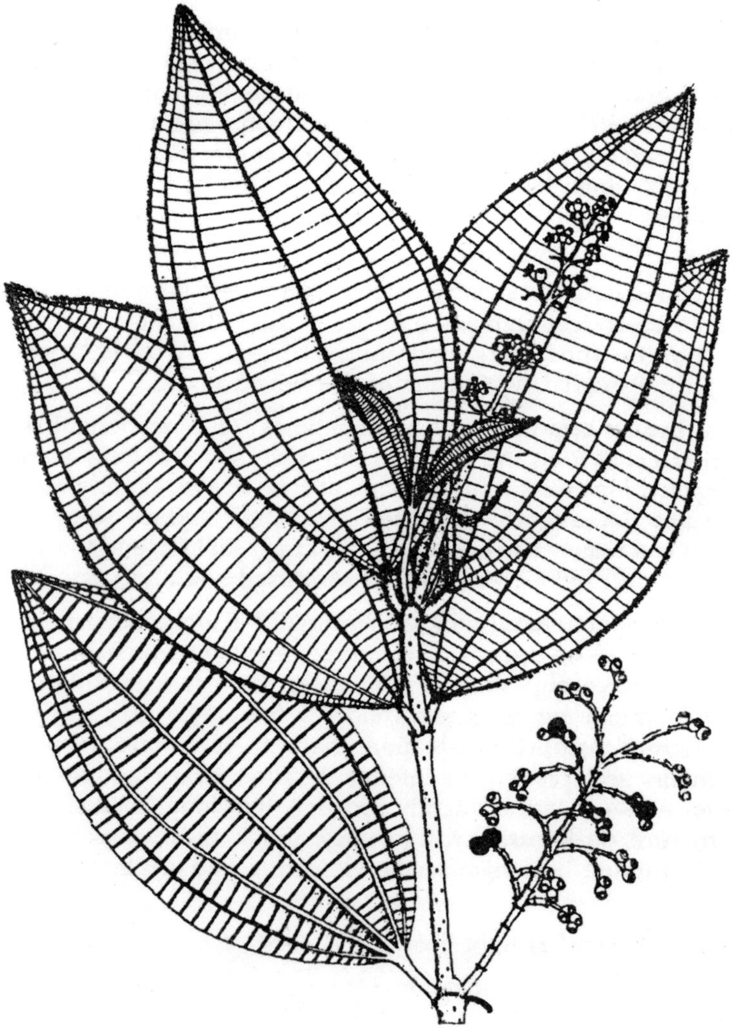

**Fig. 283 : Camasey felpa (*Miconia racemosa*)**
Natural size.

## 284. Camasey (*Miconia rubiginosa*)

This rare species of camasey in the western mountains is identified by : (1) rusty or brownish soft hairs covering twigs, petioles, lower leaf surfaces, branches of flower clusters, and flowers; (2) opposite ovate or elliptic leaves have melastome venation, 5 main veins from near base, with minute soft star–shaped hairs on flower surface, and with stout short petioles $^{1}/_{4}$–$^{3}/_{8}$ inch long; (3) many flowers stalkless in groups, about $^{1}/_{4}$ inch across the 5 elliptic petals, white and becoming light yellow; and (4) rounded blackish berries $^{3}/_{16}$ inch in diameter.

An evergreen shrub or small tree reported to become 25 feet high and 4 inches in trunk diameter. Bark gray, finely furrowed.

The blades of the opposite leaves are 3–5$^{1}/_{4}$ inches long and 1$^{1}/_{2}$–2$^{1}/_{4}$ inches wide, slightly apex, rounded or slightly notched at base, turned under along edges, with 5 main veins, including 1 pair of fine veins near margins and 1 pair joining midrib $^{1}/_{8}$ inch above base, and many parallel side veins. The upper surface is dark green and nearly hairless, with veins sunken, and the lower surface paler, with raised veins, and soft cover of star–shaped hairs.

Flower clusters (panicles) are terminal, cone–shaped, branched, to 4–6 inches long. Flowers many, stalkless in groups, composed of cup–shaped yellowish hairy base (hypanthium) nearly $^{1}/_{8}$ inch high, with 5–toothed hairy calyx; 5 elliptic petals more than $^{1}/_{16}$ inch long, white, becoming light yellow; 10 stamens, and pistil with inferior 3–celled ovary containing many ovules and slender style. The blackish fruits are slightly flattened, have minute cup at apex, scattered hairs, and have juicy purplish black flesh, almost tasteless. Seeds many, light brown, nearly $^{1}/_{16}$ inch long. Collected with flowers and fruits from January to August.

Rare in western mountains. Collected in clearing near Kerala in western mountains, altitude 2,000 feet.

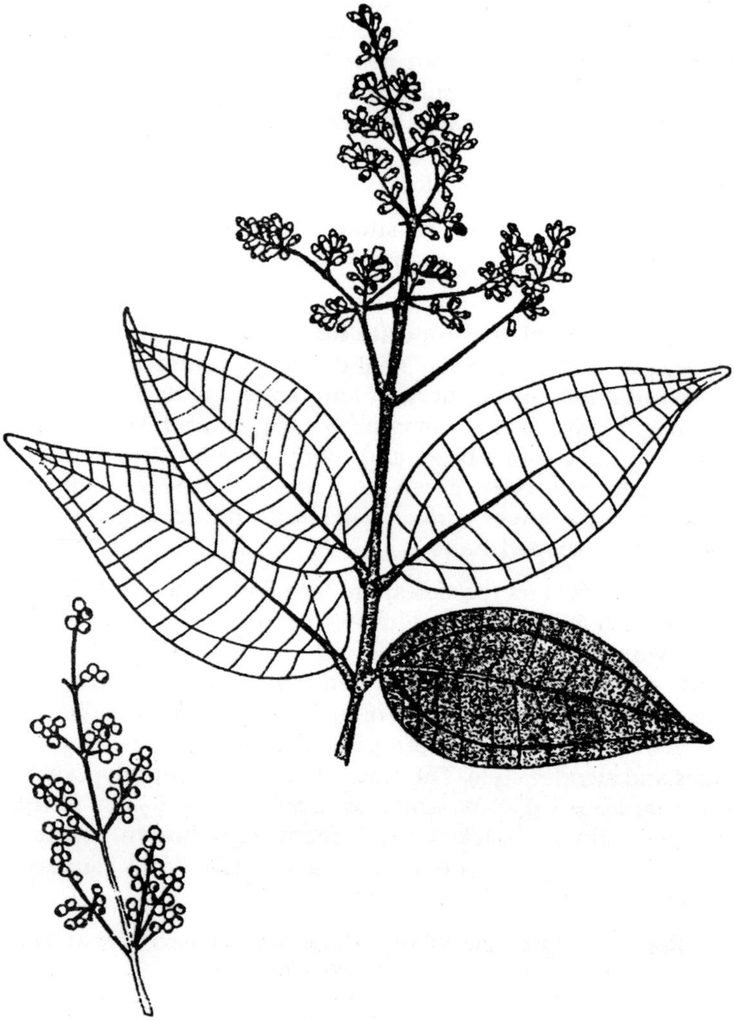

**Fig. 284 : Camasey (*Miconia rubiginosa*)**

Flowering twig (above), fruits (lower left), natural size.

## 285. Camasey (*Miconia serrulata*)

This distinctive species of camasey, a shrub or small tree, is recognized by : (1) the very large opposite ovate to elliptic, slightly thickened leaves 5–12 inches long and 3–$8^1/_2$ inches wide, with melastome venation, 5 or 7 main veins from base, distinctly and finely wavy toothed on edges; (2) twigs, petioles, raised veins on lower leaf surface, branches of flower clusters, and flowers densely covered with cinnamon–brown or rusty scale hairs; (3) flowers about $^3/_4$ inch long, with cylindric base, spreading calyx, and 6 white petals; and (4) rounded berries $^1/_4$ inch in diameter white to light blue.

An evergreen shrub 6–10 feet high or a small tree to 20 feet or more and 3 inches in trunk diameter. The bark is gray brown, finely fissured. Twigs stout, slightly flattened, covered with cinnamon–brown scaly hairs. Buds composed of rusy hairy paired minute leaves.

The opposite leaves have stout cinnamon–brown hairy petioles $1^1/_2$–$4^1/_2$ inches long. Blades are long–or short–pointed at apex, rounded or slightly notched at base, the upper surface dull dark green and becoming nearly hairless, with 5 or 7 main veins and many straight parallel side veins all slightly sunken, and the lower surface whitish green and soft with star–shaped hairs and network of small rectangles bordered by raised cinnamon–brown hairy veins.

Flower clusters (panicles) terminal, large, 5–10 inches long, with many flowers mostly stalkless at ends of finely hairy branches. The cylindric base (hypanthium) $^3/_{16}$ inch long is finely gray hairy and bears and slightly 6–toothed spreading calyx $^1/_8$ inch long and the other parts; 6 oblong whitish petals, reported also to be pink, $^1/_4$ inch or more in length, scaly hairy outside; purplish curved stamens 6 long and 6 short; and pistil with inferior ovary and long curved whitish style. The finely scaly hairy fruits have spreading ring of calyx at apex, are juicy but tasteless, and contain many dark brown seeds less than $^1/_{16}$ inch long. Flowering and fruiting mainly in spring and irregularly through the year.

Common in lower forests from near sea level to 1,500 feet altitude, as in Kerala.

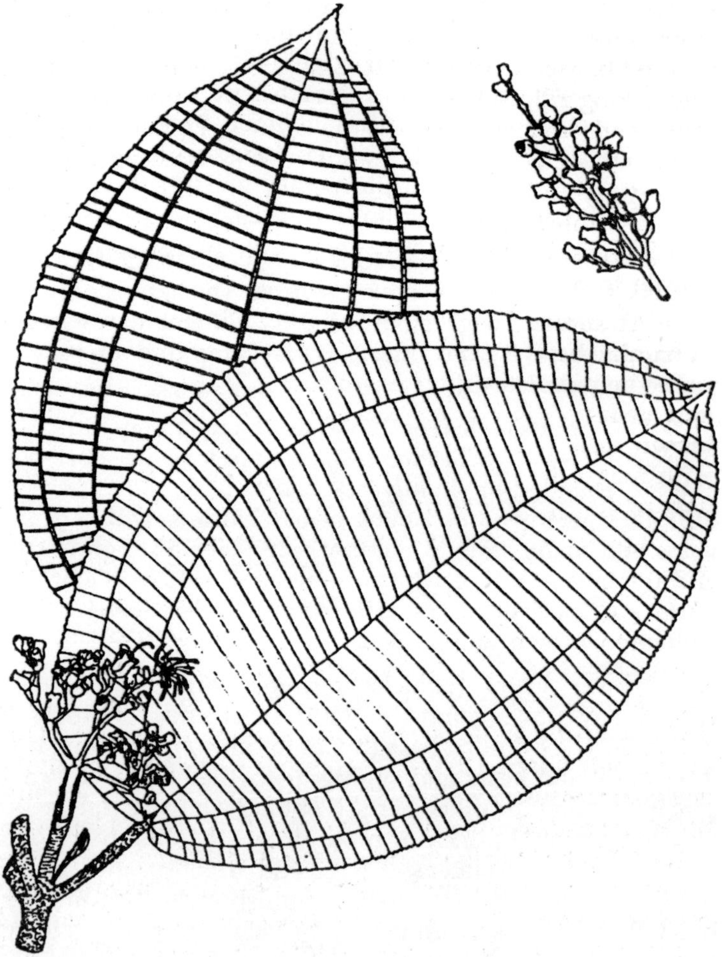

**Fig. 285 : Camasey (*Miconia serrulata*)**

Flowering twig (left) and fruits (upper right), two–thirds natural size.

## 286. Camasey (*Miconia sintenisii*)

This shrub or small tree known only from high mountains of Puerto Rico is recognized by: (1) opposite ovate to elliptic thickened and stiff leaves with melastome venation, 5 main veins from base, also 2 smaller veins near the turned under finely toothed borders and many straight cross veins, not flat but bent into 2 long ridges between midvein and 2 main side veins, all these and smaller veins deeply sunken in delicate pattern of small rectangles, the lower surface with raised cinnamon–brown hairy veins; (2) many flowers nearly $^1/_2$ inch long and broad, with 5 whitish petals in large long–stalked terminal clusters; and (3) rounded berries $^5/_{16}$–$^1/_2$ inch in diameter changing from purple to light blue.

An evergreen shrub or tree to 20–25 feet high and 4 inches in trunk diameter. The bark is gray and smooth, the inner bark light gray, whitish within, astringent. Young twigs, buds, petioles, and veins on lower leaf surface are covered with cinnamon–brown scale hairs. Twigs stout, with ringed nodes, light green with raised dots (lenticels), becoming nearly hairless, ending in narrow bud of folded young leaves.

Leaves opposite, with stout round petioles $^3/_4$–2 inches long, greenish to pinkish or purplish. Blades $3^1/_2$–$7^1/_2$ (10 inches long and $1^3/_4$–$3^3/_4$ inches wide, short–to long–pointed at apex, slightly notched or rounded at base, the hairless upper surface slightly shiny green to dark green, and the lower surface dull light green.

Flower clusters (panicles) 3–10 inches long bear many flowers mostly in 3's on stout stalks of $^1/_4$ inch or less. The light green bell–shaped base (hypanthium) nearly $^1/_4$ inch long encloses the inferior ovary and bears the other parts; calyx of 5 broad lobes less than $^1/_{16}$ inch long; 5 whitish oblong concave petals $^5/_{16}$ inch long, turning back and falling early; 10 white stamens; and pistil with inferior 5–6–celled ovary containing many ovules and slender white style. The rounded mature fruits become $^5/_{16}$–$^1/_2$ inch long and wide, have 5 calyx lobes at apex, juicy slightly sweet flesh, and numerous light brown seeds less than $^1/_{32}$ inch long. Flowering and fruiting probably through the year.

The wood is light brown, hard, and lightweight.

Common in dwarf forest at 2,500–4,000 feet altitude in high mountains of Gangtok.

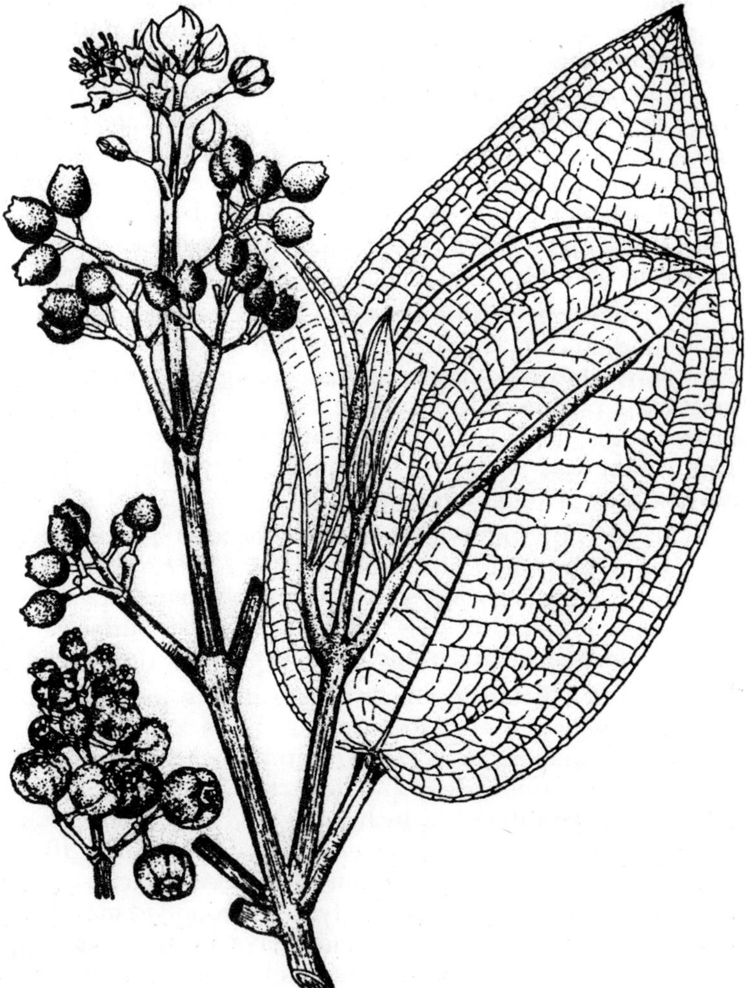

**Fig. 286 : Camasey (*Miconia sintenisii*)**

Twig with flowers and immature fruits (above), fruits (lower left), natural size.

## 287. Camasey (*Miconia subcorymbosa*)

This shrub or small tree rare in the central mountains is identified by: (1) opposite oblong to oblong–lanceolate leaves hairless or nearly so, with melastome venation, 5 main veins from base but 2 near margins; (2) many flowers $^5/_{16}$ inch long, with 5 white petals; and (3) round berries, $^1/_4$ inch in diameter, blue or turning white.

An evergreen shrub or small tree to 18 feet high and 4 inches in trunk diameter, recorded to 33 feet, becoming hairless or nearly so.

Leaves opposite, with petioles $^3/_4$–$1^1/_2$ inches long. Blades 3–$6^1/_2$ inches long and $^3/_4$–2 inches wide, long–pointed at apex, short–pointed or rounded at base, not toothed on margins, hairless or nearly so, with many fine cross veins, the upper surface shiny green, and the lower surface paler.

Flower clusters (panicles) 1–4 inches long, broad, slightly flattened, about as long as leaves, bearing many flowers on slender stalks $^1/_8$–$^5/_8$ inch long. The bell–shaped base (hypanthium) more than $^1/_8$ inch long bears 5 pointed calyx lobes and the other parts; 5 white petals more than $^1/_8$ inch long; 10 stamens; and pistil with inferior 5–celled ovary and slender style. The fruits reported to be reddish when immature, becoming blue or turning white, with calyx at apex. Seeds many, minute. Flowering and fruiting mostly in spring and summer but through the year.

Rare in lower forest at 2,500 feet altitude in central mountains of M.P.

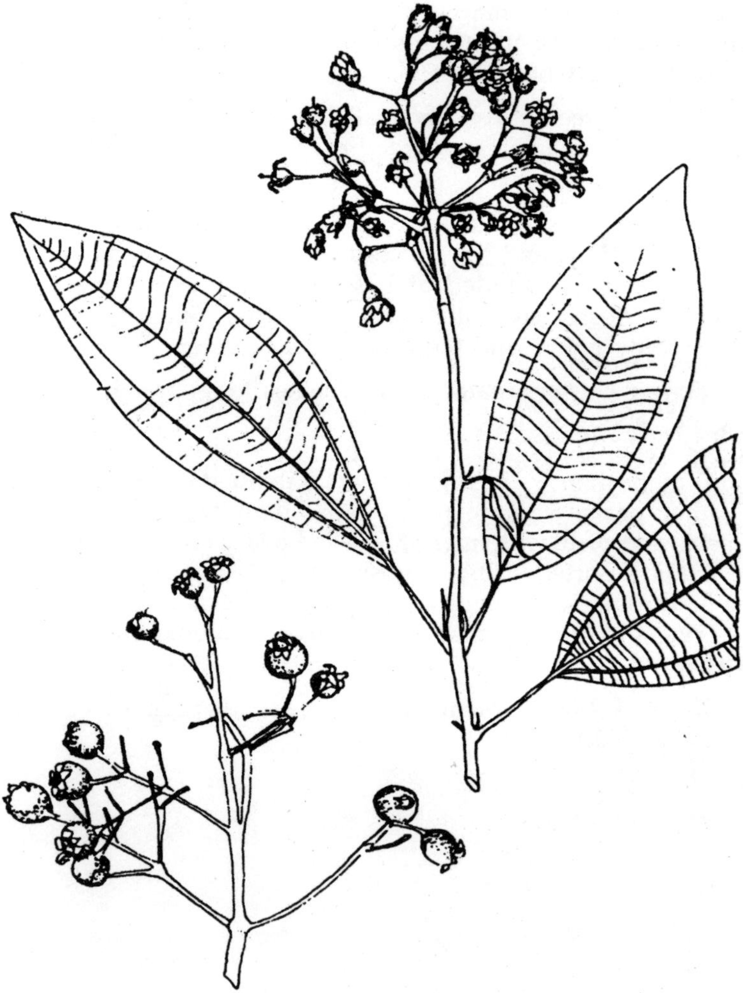

**Fig. 287 : Camasey (*Miconia subcorymbosa*)**
Flowering twig (above), fruits (lower left), natural size.

## 288. Camasey (*Miconia tetrandra*)

This species of camsey is a small to medium–sized tree common in mountain forests. It is identified by: (1) minute brownish scaly hairs on young twigs, petioles, branches of flower clusters, and flowers; (2) opposite elliptic to lanceolate leaves with melastome venation, 3 main veins from base, slightly thickened and rolled under at margins; (3) flowers about $^1/_4$ inch long, with 4 white (or greenish) petals $^1/_{16}$ inch long and 4 stamens (as the scientific name emphasizes) mostly stalkless, very numerous in terminal clusters; and (4) purplish or black round berries nearly $^3/_{16}$ inch in diameter.

An evergreen tree 20–60 feet high and 4–12 inches in trunk diameter. Bark light gray, smoothish or slightly fissured. The inner bark is light orange yellow, streaked, astringent. Twigs 4–angled, gray or brown, slightly ringed at nodes. Buds composed of paired narrowed folded gray–green young leaves.

Leaves opposite, with light brown petioles $^5/_8$–1 inch long. Blades 3–7$^1/_2$ inches long and 1$^1/_4$–2 inches wide, long–or short–pointed at apex, rounded or short–pointed at base, with the 3 main veins slightly sunken, often 1 pair of fine veins near margins, and many fine straight side veins, appearing hairless but with minute scattered scales, the upper surface slightly shiny green, and the lower surface dull light green.

Flower clusters (panicles) terminal, 3–5 inches long, bearing very numerous flowers mostly stalkless in groups of 3 at ends of short branches. The bell–shaped base (hypanthium) $^1/_{16}$ inch long, brownish green and covered with minute scales, encloses the inferior ovary and bears and the other parts; calyx slightly 4–toothed; corolla of 4 spreading elliptic white (or greenish) petals $^1/_{16}$ inch long; 4 white stamens $^1/_8$ inch long; and pistil with inferior 3–celled ovary containing many tiny ovules and slender white style. The fruits with ring calyx at apex are juicy and slightly sweet and contain many light brown seeds less than $^1/_{16}$ inch long. Recorded with flowers from January to July and with fruits from March to August.

The wood is light brown, hard, and moderately heavy.

Common in lower forests at 1,500–3,000 feet altitude in Puna.

**Fig. 288 : Camasey (*Miconia tetrandra*)**
Fruiting twig (above), flowers (lower left), natural size.

## 289. Camasey tomaso (*Miconia thomasiana DC*)

This shrubby species of camasey, which sometimes is a small tree, is identified by: (1) opposite elliptic leaves with melastome venation, 5 main veins including 1 pair of fine veins near margin and 1 pair joining midvein $^1/_8$–$^1/_2$ inch above base, petioles with red bristly hairs; (2) flowers more than $^5/_8$ inch across the 5 pink rounded petals, on very slender stalks $^1/_2$–$^3/_4$ inch long in terminal clusters; and (3) purplish black berries $^3/_{16}$–$^1/_4$ inch in diameter, with long dark red hairs.

An evergreen shrub 6–10 feet high or sometimes a small tree to 15 feet high and 3 inches in trunk diameter. Bark gray, smooth. Young twigs pink, with scattered minute light brown scale hairs. Buds composed of dark red paired young leaves.

Petioles of the opposite leaves are about $^1/_2$ inch long, pinkish, and have 2 rows of red bristly hairs above. Blades are 2–6 inches long and 1–3 inches wide, slightly thickened, hairless, short–or long–pointed at apex, rounded or slightly notched at base, the edges slightly turned under and sometimes with few minute teeth, not flat but slightly curved up from midrib. The upper surface is shiny green and has the 3 largest veins much sunken, and the lower surface is dull light green with prominent veins.

Flowers clusters (panicles) are terminal, as long as leaves or longer, bearing several to many flowers on very slender widely forking branches. The bell–shaped base (hypanthium) $^1/_8$ inch long is green, pinkish tinged, with long red hairs and bears the pink calyx with 5 rounded spreading lobes $^1/_{16}$ inch long and the other parts; 5 pink nearly round petals $^1/_4$–$^5/_{16}$ inch long; 10 very long and narrow pink stamens; and pistil with inferior 4–3–celled ovary and slender curved whitish style. Berries green to pink to purplish black at maturity, with pink 5–lobed calyx spreading at apex. Seeds several, $^1/_{16}$ inch long, shiny brown. Collected with flowers and fruits from March to September.

Locally common in moist coast and lower forests from sea level to 2,000 feet altitude on eastern and northern coast and western mountains of West Bengal.

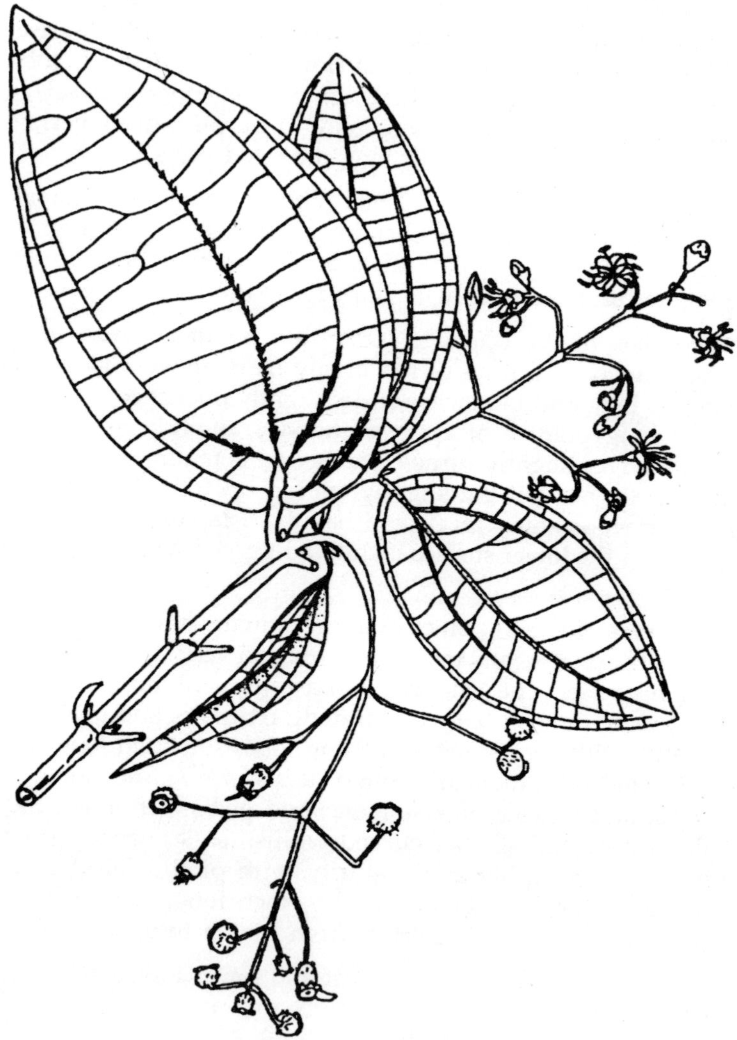

**Fig. 289 : Camasey tomaso (*Miconia thomasiana DC*)**
Natural size.

## 290. Murta (*Mouriri domingensis*)

This tree of the moist coastal forest is identified by: (1) opposite ovate leaves 2–3$^1/_2$ inches long and $^7/_8$–2 inches wide, appearing flat in 2 rows; (2) flowers about $^1/_2$ inch across the 5 spreading pink petals, several at base of a leaf; and (3) fruit an orange rounded berry $^5/_8$–1$^1/_4$ inches in diameter, edible.

An evergreen tree to 55 feet high and 10 inches in trunk diameter. Bark light brown gray, finely fissured, the inner bark red and bitter. Twigs slender, brown, hairless, with rings at nodes.

Leaves opposite, hairless appearing flat in 2 rows from twisting of twig and petioles, which are $^1/_8$ inch long. The blades are long–pointed at apex, rounded at base, not toothed on edges, thin, the side veins few, very fine, and scarcely visible, the upper surface shiny dark green, and the lower surface green and slightly shiny.

Flowers are clustered several at base of a leaf on stalks less than $^1/_4$ inch long. The top–shaped base (hypanthium) and 5–lobed calyx are about $^1/_8$ inch long; the 5 narrow spreading pink petals are $^3/_{16}$ inch long; stamens 10, longer than petals; and pistil with inferior ovary and long slender style. The fruit is an orange rounded berry wider than long, with ring of calyx on side, juicy, slightly sweet, and edible. Seeds 1–3, elliptic, $^3/_8$ inch long, brown. Flowering and fruiting in spring and summer.

The wood, described as yellowish, hard, and fine–textured, has been used elsewhere in cabinetmaking.

This species is sometimes cultivated as an ornamental for the shiny leaves and orange fruits.

Rare in moist coastal forest to 300 feet altitude in Kerala.

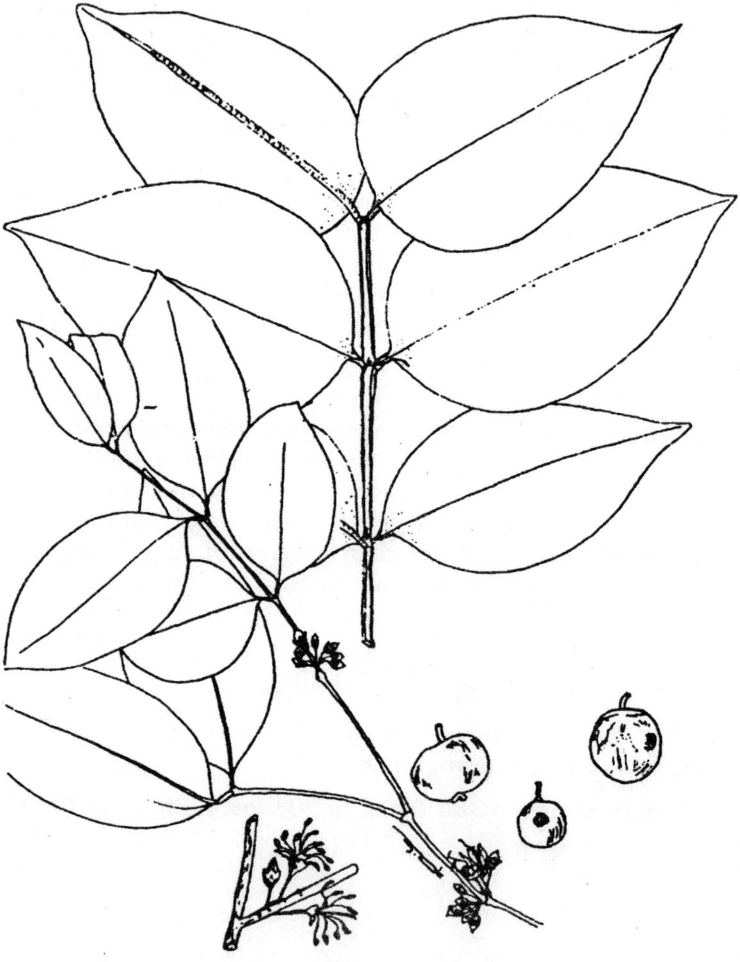

**Fig. 290 : Murta (*Mouriri domingensis*)**

Leafy twig (above), flowering (lower left), fruits (lower right), natural size.

## 291. Mameyuelo (*Mouriri helleri*)

This shrub or sometimes a medium–sized tree is characterized by: (1) opposite obovate or elliptic dark green leaves $^3/_4$–$1^1/_2$ inches long and $^3/_8$–$^3/_4$ inch wide, with edges turned under and without visible side veins; (2) star–shaped flowers $^1/_2$ inch across the 5 spreading pinkish petals 1–3 on slender stalks at base of a leaf; and (3) fruit an orange rounded berry $^3/_8$–$^7/_{16}$ inch in diameter.

An evergreen shrub 6–10 feet high or sometimes a tree to 60 feet high and 10 inches in trunk diameter, much branched. The bark is gray, thin, rough with many fissures, the inner bark pinkish or brownish and tasteless or slightly 4–angled when young.

The opposite hairless leaves have very short petioles about $^1/_{16}$ inch long. Blades are rounded or blunt at apex, short–pointed at base, slightly thick and leathery, with midvein but without visible side veins, the upper surface dark dull green, and the lower surface light green.

Flowers mostly few at leaf bases or, in a form at Susua Forest, 20 or more in small terminal clusters (panicles). The flowers have slender stalks $^1/_8$–$^3/_8$ inch long, jointed in middle. The tubular funnel–shaped base (hypanthium) and 5–lobed calyx are $^3/_{16}$ inch long; there are 5 narrow spreading petals $^3/_{16}$ inch long, pinkish or whitish tinged with pink; 10 stamens about $^1/_4$ inch long; and pistil with inferior 2–celled ovary and long slender style $^3/_8$ inch long. The fruit is a berry, rounded but slightly broader than long, with calyx at apex and with 1 or more shiny brown seeds. With flowers and fruits from spring to fall.

The wood is light brown and hard.

Scattered in moist limestone and lower forests from sea level to 2,500 feet altitude, north coast and western and northwestern mountains.

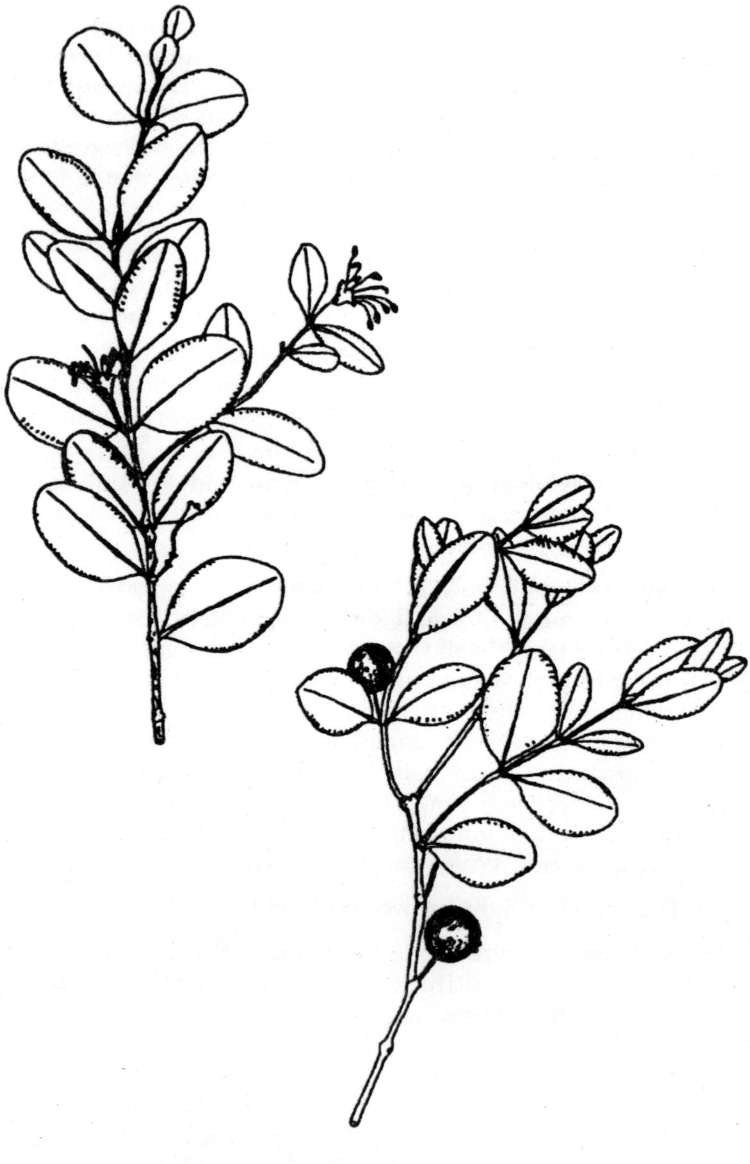

**Fig. 291 : Mameyuelo (*Mouriri helleri*)**
Flowering (upper left), fruiting twig (lower right), natural size.

## 292. Ossaea (*Ossaea krugiana* Cogn.)

A rare shrub or small tree known only from mountain forests of central India. Distinguished by: (1) opposite leaves with melastome venation, ovate to elliptic, with 5 main veins from base and with finely saw–toothed edges; (2) short stout stiff scalelike hairs forming rough surfaces on twigs, petioles, leaf blades, and branches of flower clusters; and (3) few flowers about $^1/_4$ inch long, with 5–6 pink or red petals, in lateral clusters.

An evergreen shrub 6–10 feet high or a small tree to 15 feet high and 3 inches in trunk diameter. Twigs slender, rough with pressed scalelike hairs.

Leaves opposite, with petioles $^1/_4$–$1^1/_4$ inches long. Blades $1^1/_2$–4 inches long and 1–2 inches wide, long–pointed at apex, rounded at base, with finely saw–toothed edges, thin, both surfaces rough with stiff hairs, the upper surface green with conic hairs, and the lower surface light green with many raised parallel side veins.

Flowers in small short–stalked clusters (cymes) at base of leaves. The basal tube (hypanthium) nearly $^1/_8$ inch long bears the 5–6–toothed, calyx, 5–6 rounded pink or red petals $^1/_8$ inch long, 10 or 12 stamens with yellow anthers, and slender style, and encloses the inferior ovary. Immature fruit a small rounded greenish berry $^3/_{16}$ inch in diameter, covered with scalelike hairs, with calyx at apex. Collected with flowers in June and July.

Rare in upper forest at 3,000–4,000 feet altitude in central mountains of India.

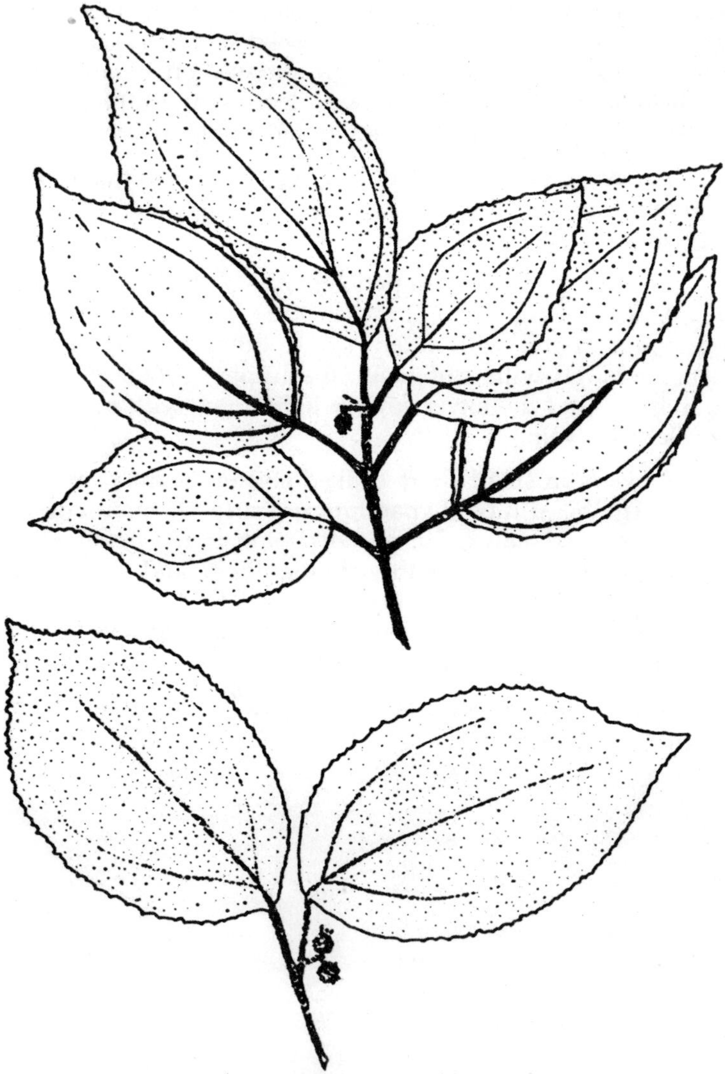

**Fig. 292 : Ossaea (*Ossaea krugiana* Cogn.)**
Twigs with immature flowers, natural size.

## 293. Stinking–fish (*Tetrazygia angustifolia*)

This small tree of the Andaman Nicobar Islands, also rare in Plains, is characterized by: (1) opposite very narrow, lanceolate leaves with melastome venation, 3 main veins from base and parallel cross veins like a ladder, the lower surface whitish hairy with raised brownish veins; (2) twigs, petioles, lower leaf surfaces, branches of flower clusters, and flowers covered with minute gray star–shaped hairs; (3) many crowded flowers in flat–topped terminal clusters, $^1/_4$ inch long and broad, the petals purplish or pinkish tinged; and (4) round blue–black berries $^3/_{16}$ inch in diameter.

An evergreen small tree commonly 20 feet high and 4 inches in trunk diameter, often shrubby, and recorded to 40 feet. Bark light gray, rough with many narrow furrows, the inner bark light brown, slight astringent. Twigs much branched, slender, covered with minute brown star–shaped hairs.

Leaves opposite, with slender petioles $^3/_{16}$–$^3/_8$ inch long. Blades $1^1/_2$–2 inches long and $^1/_4$–$^1/_2$ inch wide, very long–pointed at apex, short–pointed at base, not toothed on edges, thin, the upper surface dull light green and becoming almost hairless, with 3 long veins sunken.

Flower clusters (panicles) much branched, flattened, 1–2 inches long and broad. The flowers have stalks of about $^1/_{16}$ inch. The light brown tubular base (hypanthium) more than $^1/_{16}$ inch long bears the 4–lobed calyx $^1/_{16}$ inch long, 4 purplish or pinkish petals less than $^1/_8$ inch long, 8 purplish stamens, and long slender style and encloses the inferior ovary of pistil. The fleshy berry has spreading ring of calyx at apex and contains many minute seeds. Flowering and fruiting in summer and fall.

The wood is light brown and hard.

Local and scattered in moist coastal forest and foothills of Island from sea level to 1,000 feet altitude in northeastern, southeastern and northwestern Andaman Nicobar.

**Fig. 293 : Stinking–fish (*Tetrazygia angustifolia*)**
Fruiting twig (above), flowering twig (below), natural size.

## 294. Camasey (*Tetrazygia biflora*)

Three species, this and the next 2, known only from mountains of Nilgiri, have also been placed together in a separate genus (*Menendezia*). They are distinguished by many large showy white 4–parted flowers in terminal clusters, 3 or fewer stalkless at the end of a flattened branch, which is broadest and to $^1/_8$ inch wide at end. This species is further characterized by: (1) opposite narrowly elliptic leaves ending in an abrupt point, with melastome venation, 5 main veins from base, the lower surface light green with minute scale hairs, not whitish gray as in related species; and (2) large white flowers 1–$^1/_2$ inch across the 4 (or 5 spreading petals).

An evergreen tree to 50 feet high and 1 foot in trunk diameter. Bark gray, finely fissured and becoming slightly shreddy, the inner bark light brown and slightly sour. Twigs are slightly 4–angled, ringed at nodes, gray green and minutely hairy when young, becoming brown.

Leaves opposite, with finely hairy gray–green petioles $^3/_8$–1 inch long. Blades 2–5 inches long and $^3/_4$–2 inches wide, slightly thickened, long–pointed at apex, short–pointed or rounded at base, not toothed on edges, with 5 main veins from base and many straight parallel smaller veins. The upper surface is slightly shiny, green, becoming hairless, with veins slightly sunken, and the lower surface light green with raised brown veins and minute scale hairs.

Flower clusters (panicles) terminal, 2–4 inches long, bearing many fragrant flowers 3 or fewer stalkless at the end of a flattened finely hairy branch $^1/_2$–1 inch long. The flower about $^3/_4$ inch long is composed of bell–shaped base (hypanthium) $^1/_4$ inch long, yellow green and scaly hairy, slightly narrowed above inferior ovary, bearing calyx with 4 (sometimes 5) spreading pointed lobes $^3/_{16}$ inch long and the other parts; petals 4 (sometimes 5, rarely 6), white, $^1/_2$ inch long and broad but slightly unequal, reverse heart–shaped and notched at apex; stamens twice as many as petals, with long orange or yellow anthers curved to 1 side; and pistil with inferior ovary 4–celled (sometimes 5–celled) containing many ovules, and slender curved style.

Immature fruits $^1/_4$ inch or more in diameter, broader than long, yellow green, slightly 4–lobed (or 5–lobed) with calyx at apex, containing many narrow seeds less than $^1/_{16}$ inch long. Collected with flowers from May to September and with fruits in August–September.

The wood is light brown and hard.

The trees of the three species of this genus are very showy when covered by the abundant large white flowers and perhaps merit cultivation as ornamentals.

Local and scattered in middle to upper forests at 1,500–3,000 feet altitude.

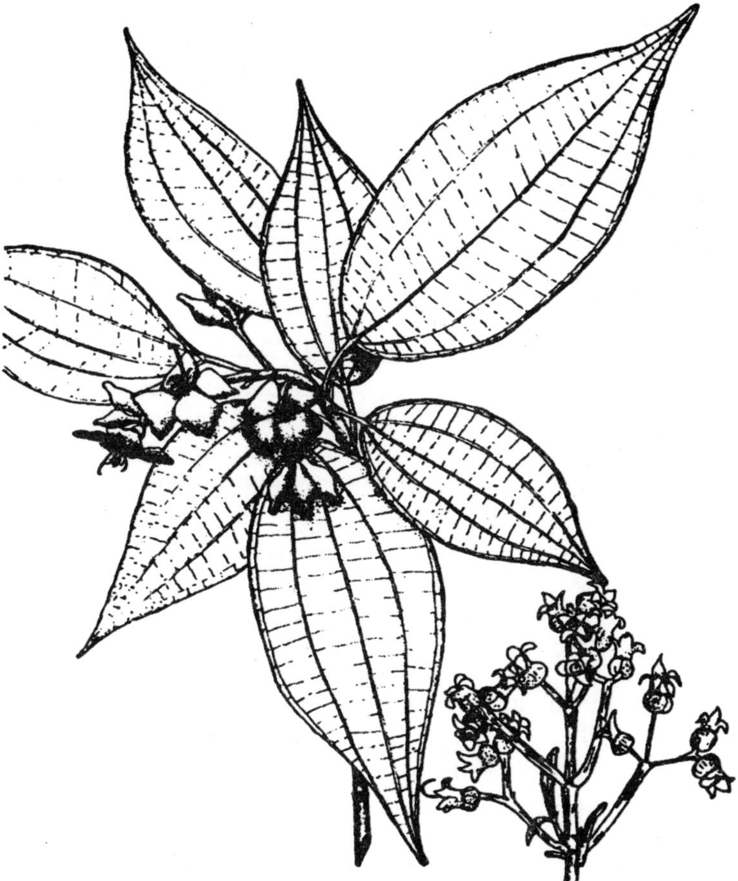

**Fig. 294 : Camasey (*Tetrazygia biflora*)**
Flowering twig (above), fruits (lower right), natural size.

## 295. Camsey (*Tetrazygia stahlii*)

This species of camasey is closely related to *Tetrazygia urbanii* Cogn., and possibly not distinct, differing in the very narrow longer outer calyx lobes. Other characters are : (1) 3 or fewer stalkless flowers at end of flattened branch of terminal cluster, the calyx of 4 very narrow lobes; (2) lanceolate to ovate leaves with very long narrow point at apex, with melastome venation, 5 main veins from base; and (3) minute whitish–gray scale hairs on young twigs, petioles, lower leaf surfaces, flower stalks, and flowers.

An evergreen tree recorded as 30–65 feet high and to 1 foot in trunk diameter. Twigs slender, slightly flattened when young, whitish gray hairy, ringed at nodes.

Leaves opposite, with petioles $^3/_8$–$^3/_4$ inch long. Blades 2–5$^1/_2$ inches long, $^3/_4$–1$^3/_4$ inches wide, slightly thickened, short– or long–pointed at apex, blunt or rounded at base, not toothed on edges, with 5 main veins from base and many straight parallel side veins, the upper surface dull green to dark green, and the lower surface whitish gray and covered with minute scale hairs and with prominent veins.

Flower clusters (panicles) terminal, narrow. The flowers have a bell–shaped base (hypanthium) $^3/_{16}$ inch long, gray hairy, which encloses the 4–angled inferior ovary and bears the other parts; calyx of 4 very narrow lobes $^1/_4$ inch long; 4 obovate white petals $^3/_8$ inch long; and 8 stamens with long yellow anthers. Fruits about $^1/_4$ inch broad, deeply 4–lobed, with scale hairs, and with calyx at apex. Collected with flowers in May and June and with fruits in June.

The showy white flowers suggest that this species has possibilities as an ornamental.

Common in lower forests at 1,000–3,000 feet altitude.

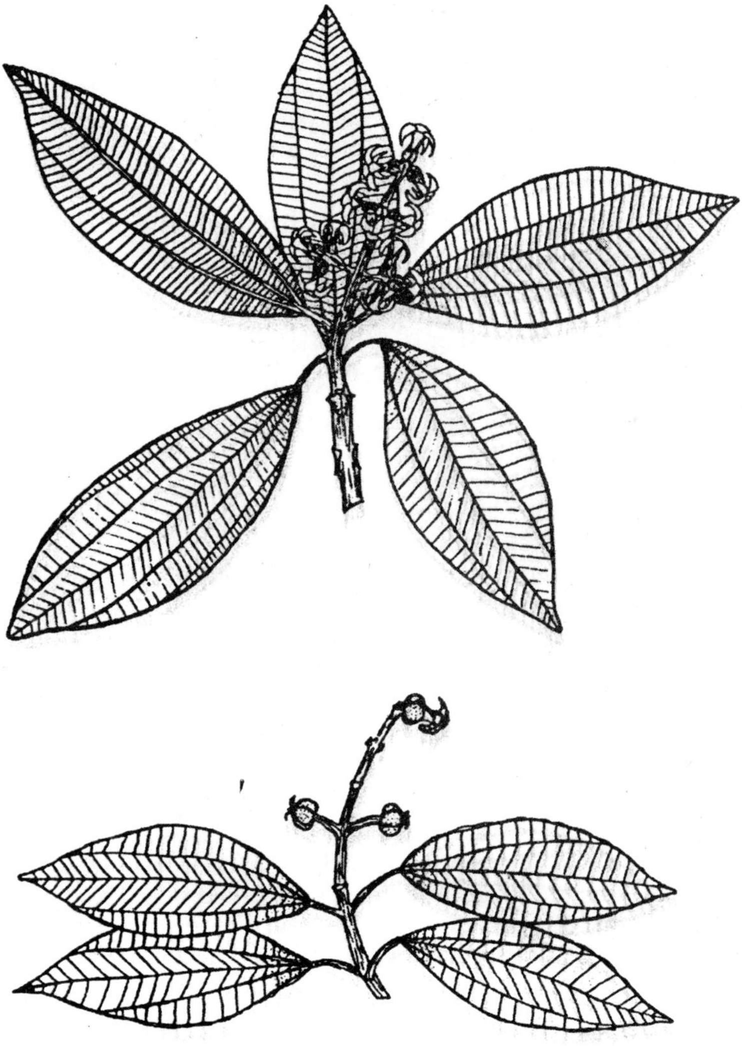

**Fig. 295 : Camsey (*Tetrazygia stahlii*)**
Flowering twig (above), fruits (below), natural size.

## 296. Camasey (*Tetrazygia urbanii*)

This species of camasey is distinguished by: (1) many showy white flowers $^{5}/_{8}$ inch across the 4 spreading petals, 3 or fewer stalkless at end of flattened branch of terminal cluster, the calyx of 4 short broad pointed lobes; (2) opposite narrowly elliptic or lanceolate leaves tapering to very long point at apex, with melastome venation, 5 main veins from base, the lower surface whitish gray hairy; and (3) minute pale whitish or light brown scale hairs on young twigs, petioles, lower leaf surfaces, flower stalks, and flowers.

An evergreen small tree to 33 feet high and 6 inches in trunk diameter or shrubby. Bark gray, slightly fissured, the inner bark orange brown and astringent. Twig 4–angled, covered with minute scale hairs, whitish gray, turning to brown, the nodes ringed and becoming slightly enlarged, with large raised half–round leaf scars. Buds composed of minute scaly paired leaves.

Leaves opposite, with petioles $^{1}/_{4}$–$1^{1}/_{4}$ inches long. Blades $2^{1}/_{2}$–$5^{1}/_{4}$ inches long and $^{3}/_{4}$–$1^{3}/_{4}$ inches broad, slightly thickened, rounded or short–pointed at base, not toothed on edges, with 5 main veins from base including 1 vein along each border and many straight parallel side veins, the upper surface dull green and hairless, and the lower surface whitish gray with raised veins and covered with minute scale hairs.

Flower clusters (panicles) terminal, 2–3 inches long bearing many fragrant flowers, 3 or fewer stalkless at end of flattened branch. The bell–shaped base (hypanthium) $^{3}/_{16}$ inch long, whitish green and covered with scale hairs, encloses the 4–lobed inferior ovary and as a narrow tube bears the other parts; calyx of 4 broad pointed lobes $^{1}/_{8}$–$^{3}/_{16}$ inch long, which covered bud and turn backward; 4 elliptic white petals $^{1}/_{4}$–$^{5}/_{16}$ inch long, slightly unequal; 8 stamens with long narrow yellow anthers bent to 1 side; and pistil with inferior 4–celled ovary containing many ovules, slender curved style, and dot stigma. Immature fruits nearly $^{1}/_{4}$ inch broad, deeply 4–lobed, hairy, with calyx at apex. Flowering in late spring and summer.

The wood is very light brown, hard, and moderately heavy.

This species with its showy white flowers has possibilities as an ornamental.

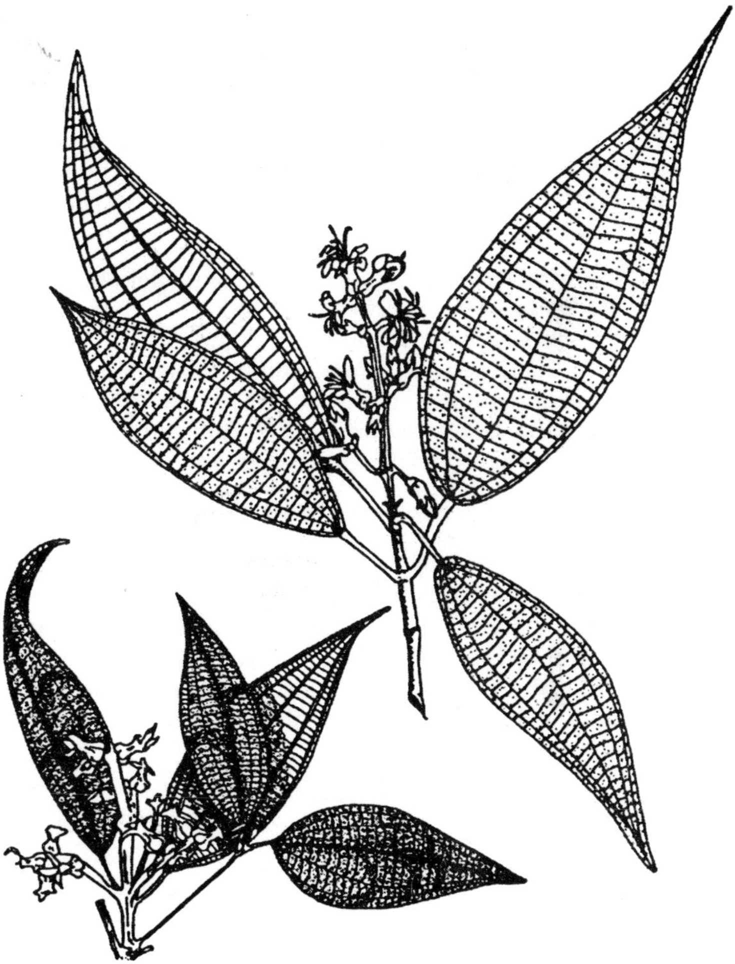

**Fig. 296 : Camasey (*Tetrazygia urbanii*)**
Flowering twig (above), fruiting twig (below), natural size.

## 297. Glorybush (*Tibouchina granulosa*)

This handsome ornamental with large purple to violet flowers is distinguished by: (1) hairy, broadly 4–winged twigs with ring of bristly hairs at nodes; (2) paired lanceolate or narrowly elliptic leaves with 5 main veins from base; (3) many flowers 2–2$^1/_2$ inches across the 5 spreading petals; and (4) fruit an elliptic pinkish hairy capsule $^3/_8$ inch long.

An evergreen shrub or small tree to 40 feet high and 8 inches in diameter. Bark gray, smoothish to fissured or furrowed. Twigs greenish and hairy, the 4 wings bordered by short stiff hairs.

The opposite leaves have brownish hairy petioles $^1/_2$–$^3/_4$ inch long. Blades are 3$^1/_2$–6 inches long and 1–1$^3/_4$ inches wide, long–pointed at apex and blunt at base, slightly turned under at edges, slightly thickened, with melastome venation of 5 main veins from base and sunken, the dark shiny green upper surface with rough stiff hairs, the lower surface dull light green with sparse pressed hairs, especially on the raised veins.

Flower clusters (panicles) are erect, terminal, branched, to 7 inches long. Flowers on short pinkish hairy stalks, composed of pinkish which bears short calyx with 5 narrow purplish lobes, 5 purple to violet petals 1–1$^1/_2$ inches long and broadest near apex, 10 threadlike, very hairy stamens bent in middle, and pistil with inferior 5–celled ovary, many ovules, and threadlike curved style. Capsules 5–celled, with many minute brown seeds. Flowering and fruiting nearly through the year.

A relatively recent introduction, becoming common around homes in moist areas of Kodaicanal.

**Fig. 297 : Glorybush (*Tibouchina granulosa*)**
Flowering twig and fruits (below), natural size.

# GINSENG FAMILY (ARALIACEAE)

Shrubs and trees, sometimes large or unbranched, also woody vines and epiphytes and few herbs, known by: (1) leaves alternate, large, crowded at apex of few stout twigs, palmately compound (digitate) or simple and mostly entire or palmately lobed, or rarely bi–or tripinnate, with usually thickened blades, often with star–shaped hairs, with petioles mostly long but often also short, and with stipules often forming a sheath; (2) flowers minute, greenish, yellowish, or whitish, in heads or umbels, generally compound, bisexual or unisexual (dioecious), regular, with 5–toothed calyx, 5 (–10) petals from a disk, 5 (–10) stamens alternate with petals, and pistil with inferior ovary of 2–5 (1–15) cells each with 1 ovule, and 2–5 styles sometimes united; and (3) fruit a berry.

**Key to Species**

A. Leaves simple—*Dendropanax*.

B. Leaves with additional fine veins parallel to side veins, dull green beneath; flower clusters (raceme of ubels) ball–like on branches along main axis—Pallo, *Dendropanax arboreus* (L.) Decne. & Planch.

BB. Leaves with network of raised veins on shiny green lower surface; flower clusters (compound umbels) branched on branches at end of axis—*Dendropanax laurifolius*.

AA. Leaves compound.

C. Leaves pinnate, with large ovate to elliptic or nearly round leaflets, mostly with short pointed teeth and white margined or blotched—*Polyscias guilfoylei*.*

CC. Leaves palmately compound (digitate).

D. Leaves with 7–12 oblong or elliptic hairless leaflets 6–12 inches long; flowers dark red, crowded in heads along many dark purple axes—*Brassaia actinophylla*.*

DD. Leaves with oblong leaflets hairy beneath at least when young; flowers brownish or yellowish green in much–brahcned clusters—*Didympanax*.

E. Leaflets mostly 10 or 11, 10–17 inches long, long–pointed, beneath brownish with satiny hairs—201. Yagrumo macho, matchwood, *Didymopanax morototoni* (Aubl.) Decne & Planch.

EE. Leaflets 3–8, $1^1/_2$–$3^1/_2$ inches long, slightly notched at rounded apex, beneath gray hairy when young–625. *Didymopanax gleasonii*.

## 298. Umbrella–tree (*Brassaia actinophylla*)

This distinctive ornamental has been introduced into Delhi in recent years. It is easily recognized by: (1) several trunks from base, mostly unbranched; (2) the few very large palmately compound leaves with mostly 7–12 oblong or elliptic shiny leaflets slightly dropping in a circle at the end of a very long slender petiole; (3) the large showy terminal flower clusters, consisting of 10–20 widely spreading dark purple axes $1^1/_2$–2 feet long, suggesting arms of an octopus; (4) the many showy dark red or crimson ball–like flowers $^3/_8$ inch wide, crowded in heads along each axis; and (5) round or top–shaped blackish fruits $^1/_4$ inch in diameter.

Small evergreen ornamental tree 20–35 feet high, with several trunks from base 2–8 inches in diameter, unbranched or with few stout branches, and with flattened crown of coarse foliage suggesting an umbrella, hairless throughout. Bark light gray, smoothish or becoming slightly fissured. Twigs few, very stout, $^3/_4$–2 inches in diameter, green, with light brown lines (lenticels).

Leaves alternate, palmate compound (digitate), about 2–3 feet long. Petioles very long, 1–2 feet, relatively slender, enlarged at both ends, round, light green. In the angle above the petiole, also forming bud at end of twig is a light green. In the angle above the petiole, also forming bud at end of twig is a light green stipule $1^1/_2$–2 inches long, very narrow and ending in long point, persistent. Leaflets mostly 7–12 (5–18), spreading in circle at end of petiole on slender spokelike stalks 2–$4^1/_2$ inches long. Leaflet blades are mostly 6–12 inches long and 3–5 inches broad, rounded and abruptly short–pointed at apex, rounded or short–pointed at base, slightly turned under at edges, slightly thickened and leathery. The upper surface is shiny dark green with grooved light green midrib and inconspicuous side veins, and the lower surface dull light green with slightly raised veins.

The large terminal flower clusters (panicles) consist of 10–20 widely spreading stout axes. Flowers are borne 10–12 stalkless in rounded heads $^3/_4$ inch across on dark purple stalks of $^3/_8$–$^1/_2$ inch long the axis. The top half of the rounded dark red bud nearly $^1/_4$ inch in diameter is composed of 10–12 narrow pointed thick petals $^3/_{16}$ inch long, dark red on outer surface and whitish on inner surface, shedding early as a half–round cap. Other flower parts are of the same number as petals, the calyx represented by a narrow rim with minute teeth. Stamens 10–12, $^{3}/_{16}$ inch long, red, with stout filaments and large anthers, erect and slightly spreading $^{3}/_{8}$–$^{1}/_{2}$ inch across, soon shedding. The half-round pistil, turning from light to dark red, consists of an inferior ovary whitish within and slightly resinous and aromatic, with 10–12 narrow cells each containing 1 ovule, and the same number of dot stigmas in a ring.

In fruit the head is composed of 10–12 berries, each bordered by 4 brownish black scales in form of a cup, which is persistent after shedding. The berry has a ring of stigmas at apex, a ring slightly above middle, also vertical ridges corresponding to the 10–12 cells. Seeds 1 in each cell, elliptic flattened, brown, $^{1}/_{8}$ inch long. Flowering and fruiting in summer.

The wood is soft and not durable.

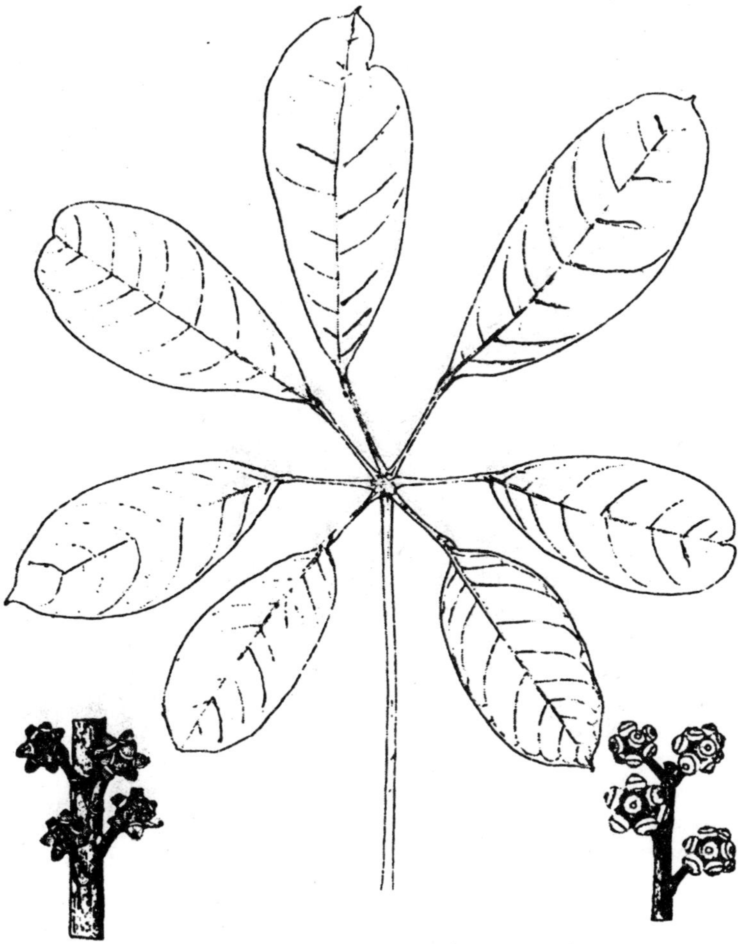

**Fig. 298 : Umbrella–tree (*Brassaia actinophylla*)**

Leaf, one–third natural size; flowers (lower left), fruits (lower right), two–thirds natural size.

## 299. Gongoli (*Dendropanax lauriflolius*)

Gongoli, a species restricted to mountain forests of Kerala, is identified by: (1) dark green obovate to elliptic leaves with unequal, short but mostly long, green leafstalks $^1/_2$–5 inches long, slender but often slightly enlarged at both ends; (2) branched flower clusters (compound umbels), with many small yellow–green 5–parted flowers $^1/_4$ inch across, spreading in rounded masses on stalks of equal length; and (3) many rounded fleshy fruits $^1/_4$ inch in diameter, with 5 short styles at apex.

Evergreen small to medium–sized tree to 30 feet high and 6 inches in trunk diameter, reported to reach 65 feet and 10 inches, hairless throughout. Bark on small trunks gray or brown and smooth. Inner bark whitish with a green outer layer, tasteless. Twigs are green when young, becoming gray, with both long and short internodes.

The alternate leaves are mostly grouped near ends of twigs. Blades are $2^1/_2$–$8^1/_2$ inches long and 1–$4^1/_2$ inches wide, slightly thickened and leathery, short–to long–pointed at apex, short–pointed to rounded, at base, edges straight or slightly wavy, the upper surface dark green and dull or slightly shiny, with faint veins, the lower surface shiny green with network of raised veins.

The branched flower clusters (compound) umbels), borne at sides and ends of twigs are 2–3 inches across and heave spreading stalks about 1 inch long with 3–4 scales near middle. Flowers are male and female on different branches on stalks $^1/_8$–$^1/_4$ inch long (monoecious). Male flowers have a 5–toothed calyx, 5 yellow–green petals more than $^1/_{16}$ inch long, and 5 alternate stamens. Female flowers have 5–toothed calyx, 5 petals, and pistil with inferior 5–celled ovary and 5 short styles. Fruits (berries) are fleshy, bear at apex the calyx and styles, becoming 5–grooved on drying, 5–seeded. Flowering in spring and maturing fruits from late spring to fall.

The wood is light brown and soft.

Common in lower forests at 1,500–3,000 feet altitude. Middle altitudes in mountains of Kerala.

**Fig. 299 : Gongoli (*Dendropanax lauriflolius*)**
Fruiting twig (left) and flowers (center right), natural size.

## 300. Yuquilla (*Didymopanax gleasonii*)

This distinctive small tree of high mountains is easily recognized by its leaves resembling those of tapioca or yucca, from which the common name was derived. Characters for identification include: (1) palmately compound leaves with 3–8 oblong–obovate leaflets spreading at the end of a long leafstalk; (2) many small yellowish green–4–5–parted flowers about $^1/_4$ inch wide, spreading from stalks of equal length (compound umbels); and (3) rounded and flattened black berries $^5/_{16}$ inch long and broad.

Evergreen shrub or small tree commonly to 15 feet high and 3 inches in trunk diameter, rarely to 35 feet tall and 10 inches in diameter. The bark is gray, smoothish to slightly fissured and warty. Inner bark is light brown and slightly bitter. The stout twigs are green, turning brown, with large leaf scars.

The alternate leaves are crowded near ends of twigs. Leafstalks are 2–4 inches long, slender but enlarged at both ends, light green. The leaflets have stalks $^1/_4$–$^1/_2$ inch long and blades $1^1/_2$–$3^1/_2$ inches long and $^5/_8$–$1^1/_8$ inches wide, slightly notched at the rounded apex, widest near apex and tapering to long–pointed base, slightly thickened, edges slightly turned under, with inconspicuous side veins, the upper surface shiny green to dark green and hairless, the lower surface dull pale green, densely gray hairy when young but becoming nearly hairless.

Flower clusters (compound umbels) are erect at or near ends of the twigs, 4–9 inches across, consisting of a long main stalk ending in many spreading shorter stalks of nearly equal length, each bearing many flowers on slender hairy stalks $^1/_4$–$^3/_8$ inch long in a rounded mass (umbel). The flower is composed of minute finely hairy basal tube (hypanthium) enclosing the inferior 2–celled ovary and bearing the minute 4–5–celled ovary and bearing the minute 4–5–toothed calyx, 4–5 yellowish green petals, 4–5 long spreading stamens, and 2 styles. The fleshy fruits (berries) topped by the 2 styles are finely hairy, green when immature, and bear 2 flattened seeds. Flowering and fruiting in spring and summer.

The wood is light brown, soft, and brittle.

Common in upper forests, also dwarf forest, at 1,500–4,390 feet altitude in upper mountains of Manipur.

**Fig. 300 : Yuquilla (*Didymopanax gleasonii*)**
Two–third natural size.

# HEALTH FAMILY (ERICACEAE)

Shrubs sometimes small, epiphytes, and trees generally small, known by : (1) leaves usually alternate, simple, elliptic, and entire, often thick and evergreen, without stipules; (2) flowers small or large and showy, bisexual, regular or slightly irregular, with calyx 4–7 lobed and generally persistent, corolla of 4–7 lobes or petals, often bell–or funnel–shaped, 8–10 stamens from a disk, and pistil with superior or inferior ovary of usually 5 cells and many ovules and styles; and (3) fruit a capsule, berry, or drupe.

## 301. Stahili (*Lyonia rubiginosa*)

The only tree species of the health family native in Moore Islands is this rare shrub or small tree of mountain forests. It is identified by : (1) elliptic leaves $1^1/_4$–$2^1/_2$ inches long and $^1/_2$–$1^1/_4$ inches wide, slightly thickened and leathery, the upper surface shiny green with midvein sunken and edges slightly turned under; (2) small flowers $^1/_4$ inch long with whitish urn–shaped corolla, 5–toothed at the narrow apex, several on slender stalks at leaf bases; and (3) elliptic seed capsules $^3/_{16}$ inch long, brown, 5–angled, splitting into 5 parts.

Evergreen shrub or small tree to 20 feet high and 1 foot in trunk diameter. Bark gray to brown, finely fissured and slightly shreddy, the inner bark dark brown and tasteless. The twigs are light green and angled and with dotlike scales when young, becoming reddish brown, with slightly raised leaf scars. End and side buds are elliptic, $^1/_{16}$ inch long, light green to light brown.

The alternate leaves have light green leafstalks $^1/_8$–$^3/_8$ inch long. Blades are blunt or short–pointed at apex, short–pointed at base, with inconspicuous side veins, the lower surface dull light green, with many minute rusty or gray dotlike scales when young.

Several flowers are borne spreading at base of a leaf on slender stalks about $^1/_4$ inch long. The flower, with parts bearing minute dotlike scales, is composed of short 5–lobed calyx, whitish urn–shaped corolla with 5 teeth at narrowed apex, 5 stamens, and pistil with 5–angled hairy ovary and slender style. The finely hairy seed capsules persist on twigs with calyx remaining at base and often the style at apex. There are many very small, narrow seeds. Flowering and fruiting irregularly from spring to fall.

The wood is light yellow.

Rare and local in dwarf at 1,800–4,390 feet altitude in mountains of Moore Island.

**Fig. 301 : Stahili (*Lyonia rubiginosa*)**

Flowering twig (upper left), fruiting twig (lower right), natural size.

# THEOPHRASTA FAMILY (THEOPHRASTACEAE)

Shrubs and trees mostly small, known by: (1) leaves generally alternate simple, without stipules; (2) flowers small, bisexual or unisexual (dioecious), regular, with 5 sepals, corolla 5–lobed and yellow or orange, 5 stamens opposite the lobes and 5 staminodes or teeth, and pistil with superior 1–celled ovary containing many ovules on central placenta; and (3) fruit a yellow berry or drupe with many to few seeds.

**Key to Species**

A. Leaves small, $^1/_2$–$1^1/_4$ inches long; flowering few on stalks curved down at leaf bases, with light yellow corolla—*Jacquinia berterii.*

AA. Leaves larger, more than $1^1/_4$ inches long; flowers several to many on erect stalks in terminal clusters, with white corolla.

B. Leaves with short–pointed base of blade and very short petiole $^1/_{16}$ inch long—*Jacquinia revoluta.*

BB. Leaves with gradually narrowed long–pointed base of blade and petiole $^1/_8$–$^1/_4$ inch long—*Jacquinia arborea.*

## 302. Barbasco (*Jacquinia arborea*)

This shrub or small tree common along the coasts is easily recognized by: (1) the dull yellow–green spoon–shaped (spatulate) leaves, which are thick, stiff, and slightly fleshy, with edges much turned under; (2) twigs much forking, stout, whitish gray, covered with minute mealy scales; (3) many small white bell–shaped 5–parted flowers $^1/_4$ inch long and broad on erect stalks in terminal clusters; and (4) many orange–red, round berries about $^5/_{16}$ inch in diameter and pointed at end, probably poisonous.

Generally a shrub but sometimes a small tree to 15 feet high and 6 inches in trunk diameter, with evergreen, compact, much branched rounded crown. The bark is smoothish or finely fissured and dark brown, and the light brown or yellowish inner bark is bitter.

The leaves are mostly in clusters of 3 or more together near ends of twigs but partly alternate. Leafstalks are $^1/_8$–$^1/_4$ inch long, yellowish and minutely scaly. Leaf blades are $1^1/_2$–$3^1/_2$ inches long and $^5/_8$–$1^3/_4$ inches broad, widest near the rounded or slightly notched apex and gradually narrowed toward the long–pointed base, hairless but with many tiny gland dots on both sides, paler beneath.

Many fragrant flowers are borne in terminal clusters (racemes) 2–4 inches long on stout erect scaly stalks $^3/_8$–$^3/_4$ inch long, which are thickened toward apex. There are 5 overlapping rounded green sepals about $^1/_8$ inch long, hairless, remaining at base of fruit; the white corolla $^1/_4$ inch long is tubular and bell–shaped with 5 rounded spreading lobes; 5 stamens on base of corolla tube opposite the lobes and alternating with 5 petallike sterile stamens (staminodes); and pistil with 1–celled ovary and short style.

The berries are sometimes so numerous that the clusters are curved downward from their weight. Each has a thin shell and 1–4 rounded brown seeds $^3/_{16}$ inch long in the bitter and probably poisonous, orange–red pulp. Flowering from winter to summer and maturing fruits from spring to falk.

The hard wood is light brown or yellowish.

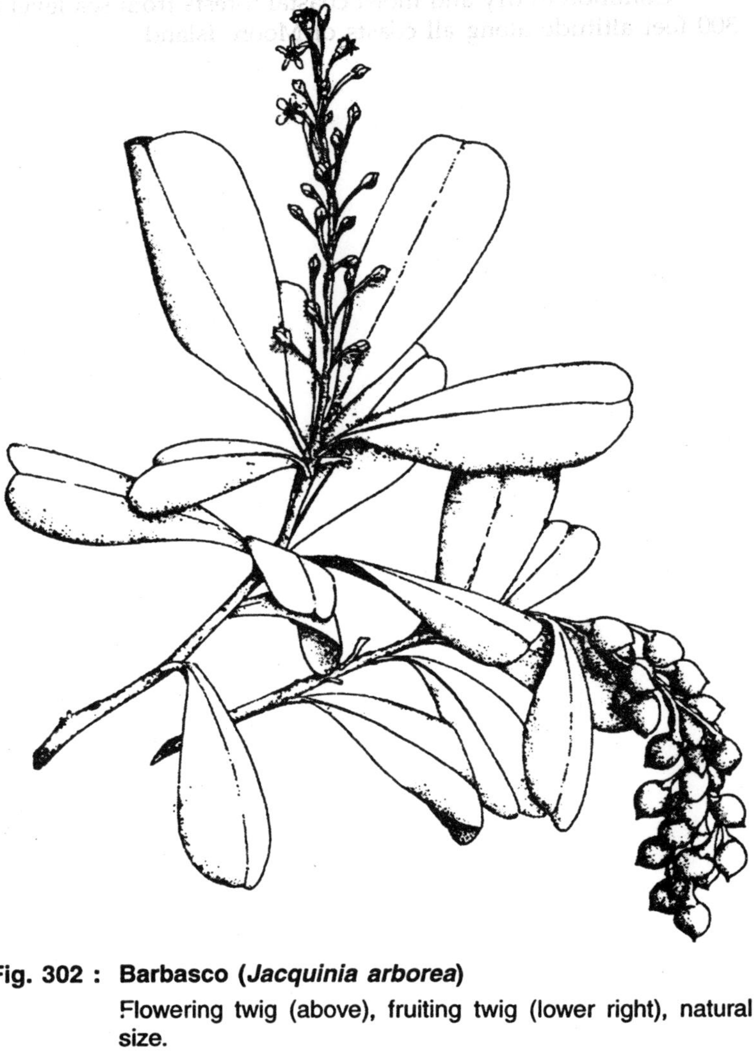

**Fig. 302 : Barbasco (*Jacquinia arborea*)**

Flowering twig (above), fruiting twig (lower right), natural size.

This species has possibilities as an ornamental, because of its compact yellow–green foliage, fragrant white flowers, and yellow fruits. Crushed fruits of this and related species have been used to poison or stupefy fish.

Common in dry and moist coastal forests from sea level to 300 feet altitude along all coasts of Moore Island.

## 303. Berterii (*Jacquinia berterii*)

This shrub or small tree of dry and moist coastal forests is recognized by: (1) small spoon–shaped dull yellow–green thickened leaves in whorls or sometimes opposite; (2) small light yellow 5–parted flowers $^{3}/_{16}$ inch long and broad, on stalks curved down at leaf bases; and (3) egg–shaped or rounded orange or yellow fruit $^{1}/_{4}$ inch in diameter, probably poisonous.

Evergreen shrub or a small tree 20 feet high, recorded to 25 feet, and 6 inches in trunk diameter, much branched and spreading. Bark dark gray to blackish, smooth to finely fissured, the inner bark light yellow and bitter. Twigs are light brown, covered with scale hairs, ending in minute bud.

The leaves are crowded, mostly 3 or more at a node or whorled, also paired or opposite, without stipules, almost stalkless, the leafstalk about $^{1}/_{16}$ inch long. The blades are mostly spoon–shaped (spatulate), $^{1}/_{2}$–$1^{1}/_{4}$ inches long and $^{1}/_{4}$–$^{1}/_{2}$ inch wide, varying in shape from narrow to broad, rounded or slightly notched at apex, gradually tapering to long–pointed base, thick and stiff with edges slightly curved down, without visible midvein or veins, hairless, with minute dots and dull yellow green on both surfaces.

One to several fragrant flowers are borne at leaf bases on slender stalks less than $^{1}/_{4}$ inch long and enlarged at apex, nodding or curved down. The flower is composed of calyx with 5 rounded overlapping lobes; light yellow bell–shaped waxy corolla with 5 rounded spreading lobes; 5 short stamens on base of corolla and opposite the lobes; and pistil with 1–celled ovary containing many ovules, tapering style, and dotlike stigma. The berrylike fruit is pointed at apex, hard–walled, and many–seeded. Flowering and fruiting in summer.

The wood is light brown and hard.

Uncommon and scattered in dry and moist coastal forests from sea level to 200 feet altitude on eastern and southern coasts of A.P.

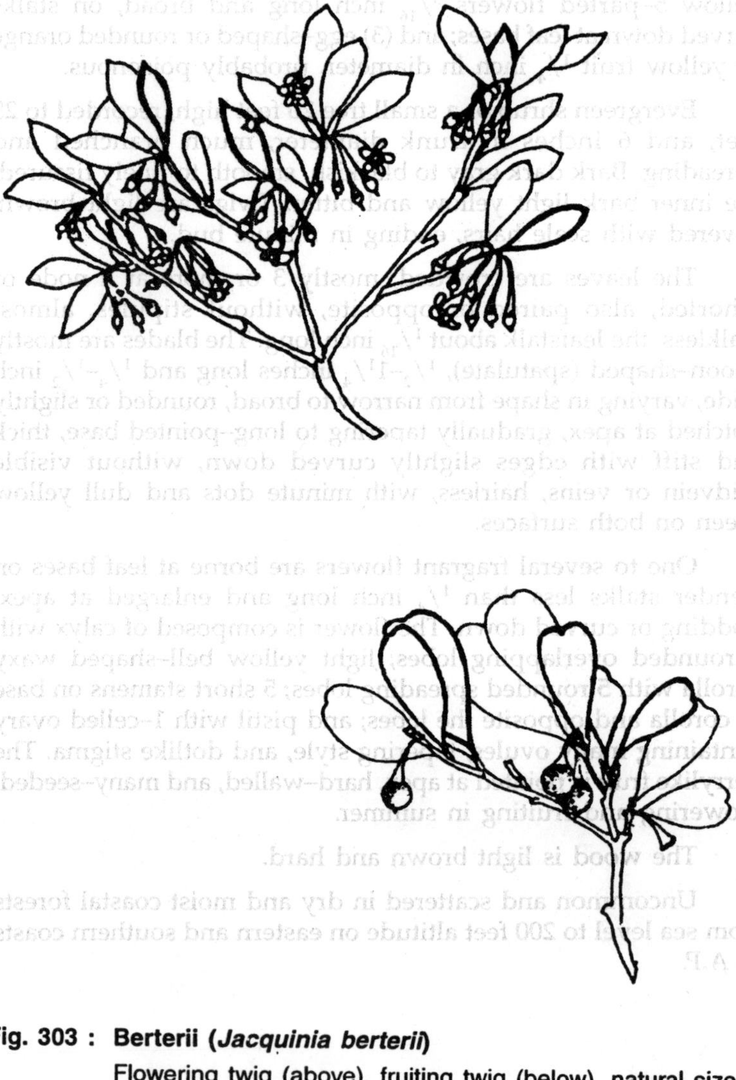

**Fig. 303 : Berterii (*Jacquinia berterii*)**
**Flowering twig (above), fruiting twig (below), natural size.**

# MYRSINE FAMILY (MYRSINACEAE)

Shrubs and small trees, rarely large, known by: (1) leaves mostly alternate, simple, thickened, with gland dots and lines, without stipules; (2) flowers small, often crowded and many, bisexual or sometimes unisexual (dioecious), regular, with calyx of 4–5 sepals or lobes and persistent, corolla of 4–5 spreading lobes mostly white or pink and with gland dots, 4–5 stamens inserted on the corolla opposite the lobes, and pistil with superior 1–celled ovary of 1 to several ovules and style; and (3) fruit a small rounded drupe, black or purplish, with 1 to few seeds.

### Key to Species

A. Flowers and fruits many in almost stalkless clusters along twigs mostly back of leaves—*Rapanea.*

B. Twig rusty–brown hairy; leaves slightly hairy beneath, lance–shaped, short–pointed—Manteequero, *Rapanea coriacea* (Sw.) Mez (*R. ferruginea*).

BB. Twigs hairless; leaves hairless, elliptic, rounded or blunt–pointed—Bádula, Guiana rapanea, *Rapanea guianensis* Aubl.

AA. Flowers and fruits in stalked clusters.

C. Flowers many in terminal erect much–branched clusters (panicles).

D. Leaves thin, mostly finely wavy–toothed on edges, long–pointed at apex—*Ardisia.*

E. Leaves 4–10 inches long, thick and succulent, with edges turned under; flowers nearly $^1/_2$ inch across—*Ardisia glauciflora.*

EE. Leaves mostly smaller, slightly thick and succulent, with edges slightly turned under.

F. Leaves $2^1/_2$–$5^1/_2$ inches long; flower clusters 2–3 inches long and broad, flowers $^3/_8$ inch across—Mameyuelo, *Ardisia obovata* Desv.

FF. Leaves 2–3 inches long; flowers clusters 2–3 inches long and broad, flowers $^5/_{16}$ inch across—*Ardisia luquillensis.*

CC. Flowers in unbranched clusters.

G. Leaves obovate, 3–10 inches long, crowded at ends of stout twigs; flowers stalked along very slender drooping axis (raceme)—*Wallenia pendula.*

GG. Leaves obovate to oblanceolate, 1–$2^1/_2$ inches long, crowded near ends of short side twigs; flowers several in slender stalked clusters at leaf bases—*Grammadenia sintenisii.*

## 304. Mameyuelo (*Ardisia glauciflora*)

A small tree known only from mountains of eastern Assam, recognized by: (1) elliptic leaves 4–10 inches long and 2–4$^1/_2$ inches wide, mostly large, thick and leathery and with edges turned under; (2) terminal much branched flower clusters bearing very many spreading 5–parted slightly succulent flowers nearly $^1/_2$ inch across; and (3) many crowded round berry fruits about $^1/_4$ inch in diameter.

A small evergreen tree recorded to 26 feet high and 4 inches in trunk diameter, with crown of few stout branches. The bark is gray and smooth, the inner bark pink and tasteless. Twigs few, stout, dark brown, finely hairy.

Leaves alternate, with short stout petioles less than $^3/_8$ inch long, hairless. Blades are blunt at both ends, with prominent network of veins when dry, with minute gland dots, green and slightly shiny on upper surface, and pale green beneath.

The flower clusters (panicles) 3–6 inches wide bear very many flowers, the branches brown and finely hairy, the flower stalks less than $^1/_4$ inch long. The calyx is composed of 5 rounded sepals $^3/_{16}$ inch long, covered with black dots and with minute hairs on edges; the corolla with short tube $^1/_8$ inch long and 5 narrow spreading lobes $^3/_{16}$ inch long, gray with black dots; stamens 5, inserted on corolla tube and opposite the lobes; and pistil with conical 1–celled ovary and slender style. Flowering and fruiting in spring and summer.

The wood is whitish to light brown, hard, and heavy. It has been used locally for furniture.

**Fig. 304 : Mameyuelo (*Ardisia glauciflora*)**

Fruiting twig (above), flowers (lower right), two–third natural size.

## 305. Mameyuelo (*Ardisia luquillensis*)

This rare small tree known only from Bihar Mountains is distinguished by : (1) obovate leaves 2–3$^1/_2$ inches long and $^7/_8$–1 $^1/_2$ inches wide, slightly thick and succulent and turned under at edges, with minute gland dots; (2) erect terminal, much branched flower clusters with very small flowers $^5/_{16}$ inch across, pale green tinged with pink; and (3) many round 1–seed fruits $^1/_4$–$^5/_{16}$ inch in diameter, black and slightly fleshy.

A small evergreen tree to 15 feet high and 4 inches in trunk diameter, hairless. Bark gray or brown, smoothes to slightly fissured, covered with mosses, the inner bark light brown or pink and bitter. Twigs are brown and slightly angled.

The alternate leaves are blunt–pointed at apex, broadest beyond middle, and narrowed into long–pointed base and petiole less than $^1/_4$ inch long. The upper surface is green to dark green, slightly shiny, with yellowish midvein and side veins indistinct, and the lower surface dull light green.

The flower clusters (panicles) 2–3 inches long and broad bear very many crowded short–stalked flowers at the ends of reddish branches. The calyx is composed of 5 rounded gland–dotted lobes $^1/_{16}$ inch long; corolla has short tube and 5 spreading gland–dotted lobes more than $^1/_8$ inch long; 5 stamens inserted near base of corolla tube and opposite the lobes; and pistil with elliptic 1–celled ovary and short style. The gland–dotted berries have short point of style at apex, calyx at base, change color from green to red to black at maturity, and contain 1 large round seed. Collected with flowers and fruits in August and with immature fruits in March.

The wood is light brown and hard.

Rare in dwarf forest at about 3,000–3,400 feet altitude on summits of peaks of Bihar Mountains.

**Fig. 305 : Mameyuelo (*Ardisia luquillensis*)**
Fruiting twig (left), flowering twig (upper right), natural size.

## 306. Sintenis (*Grammadenia sintenisii*)

Shrub or small tree of dwarf forest at high altitudes in Bihar Mountains characterized by: (1) obovate to oblanceolate leaves crowded near ends of short side twigs; (2) small whitish or greenish yellow 5–parted flowers $^1/_4$ inch across, several in clusters at leaf bases; and (3) elliptic berrylike fruit $^5/_{16}$ inch long, with minute point at apex, 1–seeded.

Evergreen shrub 10 feet or less in height, sometimes a small tree to 15 feet high and 4 inches in trunk diameter, hairless throughout. Bark smooth, gray, the inner bark brown and slightly bitter. Twigs brownish green, both slender and stout, becoming light brown. Buds of minute pointed young leaves without scales or stipules.

Leaves alternate along main twigs and crowded at enlarged ends of slender side twigs. Leafstalks $^1/_8$–$^1/_4$ inch long. Blades obovate to oblanceolate, 1–2$^1/_2$ inches long, $^1/_2$–$^3/_4$ inch wide, thick, stiff, and slightly succulent, with minute gland dots visible with lens. The apex is blunt–pointed, the base long tapering, the border slightly turned under, the upper surface green to dark green and slightly shiny, the sides bent up from the sunken midvein and without visible veins, the lower surface dull light green.

Flower clusters (like umbels) at base of leaves bear at the end of a slender stalk about $^3/_4$ inch long several flowers on slender stalks each about $^1/_2$ inch. Flowers bisexual and male, composed of light green calyx with 5 narrow pointed lobes, greenish yellow or whitish corolla of 5 rounded spreading lobes, 5 minute stamens borne near base of corolla and opposite the lobes; and minute pistil with rounded 1–celled ovary containing 3–4 ovules and very short style. Flowering continuously.

Known only from dwarf forest at altitudes of 2,500–3,500 feet in Hazaribagh Mountains of eastern Bihar.

**Fig. 306 : Sintenis (*Grammadenia sintenisii*)**
Flowering twig, natural size.

## 307. Rascagarganta (*Parathesis crenulatsa*)

This shrub, rarely small tree, of moist forests in Bihar, is identified by: (1) oblanceolate or narrowly elliptic thin leaves, long–pointed at both ends, with finely wavy–toothed or straight edges, with minute gland dots visible with a lens against the light; (2) many small pink flowers $^{1}/_{4}$–$^{5}/_{16}$ inch across in terminal branched clusters; and (3) many black round fleshy fruits $^{1}/_{4}$–$^{5}/_{16}$ inch in diameter, 1–seeded.

An evergreen shrub or rarely small tree to 18 feet high and 3 inches in trunk diameter. The young twigs are densely covered with minute brownish hairs and scales.

The alternate leaves have blades 3–6 inches long and $^{3}/_{4}$–$1^{3}/_{4}$ inches wide, the base tapering into the slightly winged slender petiole $^{3}/_{8}$–$^{3}/_{4}$ inch long. The upper surface is shiny green and hairless, and the lower surface dull and paler with minute hairs or nearly hairless.

Flower clusters (panicles) are terminal and 2–4 inches long, the slender branches covered with minute hairs and scales. The finely hairy flowers on stalks of $^{1}/_{8}$ inch or less are composed of calyx with 5 narrow pointed lobes, pink corolla $^{3}/_{16}$ inch long with 5 narrow pointed lobes, 5 stamens at base of lobes and opposite them, and pistil with conic ovary and long threadlike style. The berries have calyx at base and pointed base of style at apex and are covered with minute gland dots. The round seed is $^{1}/_{8}$ inch in diameter. Collected with flowers from May to July and with fruits from July to December.

Common in moist forests from sea level to 2,500 feet altitude in Bihar.

**Fig. 307 : Rascagarganta (*Parathesis crenulatsa*)**
Fruiting twig (above), flowering twig (below), two–thirds natural size.

## 308. Jacanillo (*Wallenia pendula*)

A shrub or sometimes small tree known only from high mountains of Bihar. Distinguishing characters are: (1) obovate leaves 3–10 inches long and $1^1/_4$–3 inches wide, slightly thickened and leathery, mostly clustered at ends of stout twigs; (2) flower clusters consisting of a very slender drooping axis and many short–stalked greenish white flowers about $^1/_4$ inch long, male and female on different trees (dioecious); and (3) rounded reddish berries less than $^1/_4$ inch diameter, 1–seeded.

An evergreen shrub 6–10 feet high or small tree to 20 feet high and 3 inches in trunk diameter, recorded as sometimes to 33 feet. Twigs gray or brownish gray, stout, hairless.

The leaves are alternate but clustered at ends of twigs, which have some long internodes. Petioles $^3/_8$–1 inch long. Blades are short–pointed or blunt at apex, broadest beyond middle, and gradually narrowed to short–pointed or blunt base, with straight or slightly wavy edges, with pointed network of fine veins, hairless, paler beneath, and with minute gland dots visible against the light with a lens.

Flower clusters (racemes) are unbranched, 5–8 inches long. Male flowers have stalked about $^1/_{16}$ inch long; calyx of 5 blunt sepals more than $^1/_{16}$ inch long; corolla tubular, $^3/_{16}$ inch long, with 5 rounded lobes; 5 slender stamens nearly twice as long as corolla and opposite the lobes; and small sterile pistil. Female flowers have calyx, corolla, stamens, and pistil with round ovary, 2 ovules, and narrow style. The berries are slightly broader than long, have calyx at base and slender point at apex. Flowering and fruiting irregularly over the year.

Rare in dwarf forest at 2,000–3,500 feet altitude in high mountains through Bihar.

**Fig. 308 : Jacanillo (*Wallenia pendula*)**
Fruiting twig (left), flowers (lower right), natural size.

# SAPODILLA FAMILY (SAPOTACEAE)

Tree and few shrubs, known by : (1) white latex or milky sap; (2) leaves alternate simple, generally entire, thickened, usually without stipules; (3) flowers small, generally, white, green, or light brown, crowded or single at base of leaves or below at nodes, bisexual, regular, with hairy calyx of 4–8 overlapping lobes, corolla with short tube and 4–8 short lobes, stamens very short, generally 4–8 (to many) inserted on corolla opposite the lobes, often with staminodes alternate, and pistil with superior ovary containing generally 4–5 (1–14) cells each with 1 ovule and with short style; and (4) fruit a berry with 1 to few large elliptic shiny seeds with large scar and milky pulp (sometimes edible). Also vol. 1, p. 436.

**Key to Species**

A. Leaves less than $2^1/_2$ inches long, obovate, rounded at apex.

B. Leaves with few long fine side veins nearly parallel with edges—*Bumelia.*

C. Leaves $^1/_4$–$^5/_8$ inch long; many short twigs as sharp slender spines—*Buymelia Krugii.*

CC. Leaves $^3/_4$–$1^1/_2$ inches long; not spiny—*Bumelia obovata.*

BB. Leaves $^3/_4$–$1^1/_2$ inches long, with several short side veins—*Dipholis cubensis.*

AA. Leaves larger.

D. Leaves with many straight parallel side veins nearly at right angle to midvein.

E. Leaves with side veins very close together and indistinct.

F. Leaves less than $3^1/_2$ inches long; fruits elliptic, $^3/_4$–$1^3/_4$ inches long–*Micropholis.*

G. Leaves abruptly short–pointed; the lower surface reddish brown, finely silky hairy—Caimitillo, *Micropholis chrysophylloides* Pierre.

GG. Leaves rounded or slightly notchd at apex, green on both surfaces, becoming hairless or nearly so—Caimitillo verde, *Micropholis garciniifolia* Pierre.

FF. Leaves 3–$4^1/_2$ inches long, short–pointed, very shiny; fruits $1^1/_2$–3 inches in diameter (sapodilla)–Nispero, sapodilla, *Manilkara zapota* (L.) v. Royen.*

EE. Leaves with side veins less close, distinct, mostly larger.

H. Leaves thin, elliptic, short–pointed at both ends, mostly silky hairy beneath, less than 5 inches long—*Chrysophyllum.*

I. Leaves with lower surfaces reddish brown or golden silky.

J. Fruit round, several–seeded, 2–3 inches in diameter from 6–11 celled ovary (star–apple).

JJ. Fruit oblong or elliptic, mostly 1–seeded, less than $1^1/_4$ inches long.

K. Fruit oblong, $^3/_4$–$1^1/_4$ inches long, from 5–9–celled ovary—*Chrysophyllum bicolor.*

KK. Fruit elliptic, $^3/_4$ inch long, from 5–celled ovary—Caimitillo de perro, satin–leaf, *Chrysophyllum oliviforme* L.

II. Leaves with lower surface finely hairy or becoming nearly hairless.

L. Leaves with lower surface grayish or silvery green and often silky hairy; fruit elliptic or round, dark blue, $^3/_8$–$^3/_4$ inch long, from 6–8 celled ovary—Lechecillo, *Chrysophyllum argenteum* Jacq.

LL. Leaves with lower surface becoming nearly hairless; fruit oblong, $^1/_2$–$^3/_4$ inch long, sharp–pointed, from 5 celled ovary–639. *Chrysophyllum pauciflorum.*

HH. Leaves thick and leathery, mostly elliptic, shiny, hairless—*Manilkara.*

M. Leaves often large, to 8 inches or more in length; fruits rounded, more than 1 inch in diameter.

N. Leaves elliptic, widest near middle; fruits round or elliptic, 1–$1^1/_4$ inches long, edible—Ausubo, balata, *Manilkara bidentata* (A. DC.) Chev.

NN. Leaves oblong to obovate, mostly widest beyond middle often crowded at end of long internode; fruits rounded but flattened, 2 inches in diameter—*Manikara pleeana.*

MM. Leaves smaller, elliptic or oblong.

O. Leaves slightly notched or rounded at apex; fruit $^3/_4$–$1^1/_4$ inches in diameter, slightly flattened–Manilkara jaimiqui.

OO. Leaves rounded at apex; fruit round, $^1/_2$–$^5/_8$ inch in diameter–642. *Manilkara albescens.*

DD. Leaves with several side veins at acute angle to midrib, mostly curved at end.

P. Leaves widest beyond middle.

Q. Leaves large, to 14 inches or more in length, mostly short–pointed at apex; fruits large, edible.

R. Leaves oblanceolate or narrowly obovate; fruits egg–shaped, about 4 inches long, brown (sapote)–*Pouteria sapota.**

RR. Leaves narrowly elliptic; fruits round or elliptic, $1^1/_2$–2 inches long, orange colored–Jácana, *Pouteria multiflora* (A. DC.) Eyma.

QQ. Leaves 3–6 inches long, broadly obovate, rounded at apex, fruits rounded at apex; fruits round $1^1/_4$–$2^1/_4$ inches in diameter, brown, warty–*Pouteria hotteana.*

PP. Leaves widest at or below middle.

S. Leaves mostly elliptic, petioles mostly long, slender, $^1/_2$–$1^1/_4$ inches long.

T. Leaves blunt, round, or slightly notched at apex, with yellow midrib and yellow–green petiole Tortugo amarillo, false–mastic, *Sideroxylon foetidissimum* Jacq.

TT. Leaves short–pointed at apex–*Dipholis bellonis.*

SS. Leaves mostly lanceolate, petioles short, mostly less than $^1/_2$ inch long.

U. Leaves with long whitish hairs beneath, short–pointed or blunt at apex—*Pouteria dictyneura.*

UU. Leaves hairless, long–pointed at apex—Singuinaria, wild mespel, willow bustic, *Dipholis salicifolia* (L.) A. DC.

## 309. Arana gato (*Bumelia krugii*)

Known only from dry areas of southwestern Tamil Nadu and Vieques, this spiny shrub or small tree is identified by: (1) twigs much branched, many short twigs $^1/_4$–$^1/_2$ inch long modified as sharp slender gray spines extending straight at right angle; (2) very small, obovate to nearly rounded leaves $^1/_4$–$^5/_8$ inch long and $^1/_8$–$^3/_8$ inch wide, with few long fine side veins nearly parallel with edges, slightly thick and stiff; (3) small light yellow or whitish fragrant flowers $^1/_8$ inch long clustered at leaf bases on very short stalks; and (4) rounded fruits $^1/_4$ inch in diameter, green turning dark red, containing 1 large seed.

An evergreen shrub 10 feet high or a small tree to 20 feet high and 6 inches in trunk diameter, much branched. The bark is gray, rough, furrowed into thick plates, the inner bark light brown and bitter. Twigs are slender, much branched finely rusty–brown hairy when young. Spines arise as short twigs 1 at a node above a leaf, while long spines bear leaves on sides.

The leaves are alternate or sometimes clustered at nodes on very short petioles less than $^1/_{16}$ inch long. The blades are broadest near the rounded apex, taper to short–pointed base, not toothed on edges, minutely hairy or nearly hairless, green above and light green beneath.

Several flowers are borne in a cluster $^1/_4$ inch across at the base of a leaf on brown hairy stalks $^1/_{16}$ inch long. The calyx about $^1/_{16}$ inch long has 5 hairy overlapping sepals; the light yellow or whitish corolla $^1/_8$ inch long has a short tube and 5 lobes with 2 small projections at each notch; 5 short stamens inserted near base of tube opposite the lobes and 5 sterile stamens (staminodes); and pistil with round hairy 5–celled and slender style.

The berries are borne singly on stalks $^1/_{16}$ inch long, have calyx at base and narrow point from style at apex, are slightly longer than broad, and change color from green to dark red at maturity. Inside the thin pulp is 1 large rounded shiny light brown seed. Collected with flowers from January to March and with fruits in May and July.

The wood is light brown and hard.

Rare and local in dry limestone and lower forests of southwestern Tamil Nadu.

**Fig. 309 : Arana gato (*Bumelia krugii*)**
Flowering twig (above), fruiting twig (below), natural size.

## 310. Arana gato (*Bumelia obovata*)

A shrub or small tree common and widespread in dry areas of southern Puerto Rico and other islands. Distinguishing characters are: (1) obovate or rounded leaves $^{3}/_{4}$–$1^{1}/_{2}$ inches long and $^{3}/_{8}$–$1^{1}/_{4}$ inches wide, with few long fine side veins nearly parallel with edges, slightly thick and leathery; (2) small whitish flowers $^{1}/_{8}$ inch long, few at leaf bases on very slender stalks about $^{1}/_{4}$ inch long; and (3) rounded fruits $^{5}/_{16}$ inch in diameter, green turning dark brown or black, containing 1 seed.

An evergreen small to medium–sized tree to 40 feet high and 1 foot in trunk diameter, sometimes larger, with dense compact crown. The bark is gray, thick, becoming fissured or furrowed and slightly rough. Inner bark is light brown and slightly bitter. The twigs are gray, slender, with minute pressed brown hairs when young.

The alternate leaves have short petioles $^{1}/_{8}$–$^{3}/_{16}$ inch long. Blades are rounded, blunt, or notched at apex, long–pointed to blunt at base, slightly turned under at edges, hairless, when dry with many fine lines parallel to side veins, green and slightly shiny on both surfaces.

The flowers are composed of calyx about $^{1}/_{16}$ inch long with 5 hairy overlapping sepals; whitish corolla $^{1}/_{8}$ inch long with short tube and 5 lobes with 2 small projections at each notch; 5 short stamens inserted near base of tube opposite the lobes and 5 sterile stamens (staminodes); and pistil with round hairy 5–celled ovary and slender style. The berries are borne singly at leaf bases, slightly longer than broad, and have calyx and threadlike style persistent. The large rounded seed is shiny brown. Flowering and fruiting probably irregularly through the year.

The wood is light brown and hard.

Uncommon but widely distributed at low altitudes in dry limestone forest and dry coastal hills and beaches to 800 feet altitude in southern and southwestern Chennai.

**Fig. 310 : Arana gato (*Bumelia obovata*)**

Twig with flowers and immature fruits (left), fruiting twig (right), natural size.

## 311. Caimitillo, wild cainit (*Chrysophyllum bicolor*)

Caimitillo is characterized by: (1) twigs golden or reddish brown silky hairy, with milky sap; (2) elliptic leaves dark green on upper surface and golden or brown silky hairy beneath; (3) several light green bell–shaped flowers $^3/_{16}$ inch long at leaf bases; and (4) oblong or rounded berries $^3/_4$–$1^1/_4$ inches long, usually 1–seeded.

An evergreen tree recorded as 25–50 feet high and 1 foot in trunk diameter, with milky sap.

The alternate leaves have petioles $^1/_4$–$^1/_2$ inch long. Blades are 2–5 inches long, 1–$2^1/_4$ inches wide, abruptly long–pointed at apex, blunt or rounded at base, not toothed on edges, thin, the upper surface dark green, and the lower surface golden or reddish brown silky hairy or becoming nearly hairless.

Several flowers are borne on slender reddish–brown hairy stalks $^1/_4$–$^1/_2$ inch long at base of a leaf. There are usually 5 overlapping reddish–brown hairy sepals $^1/_{16}$ inch long; the bell–shaped light green hairy corolla $^3/_{16}$ inch long has 5 spreading lobes mostly shorter than the tube; stamens 5 on corolla tube opposite the lobes; and pistil with hairy round 5–9–celled ovary, short style, and 5–9 stigmas. The fruits have 1 seed, sometimes 2 or more. Flowering and fruiting from spring to fall.

The trees have been cultivated for the edible fruits.

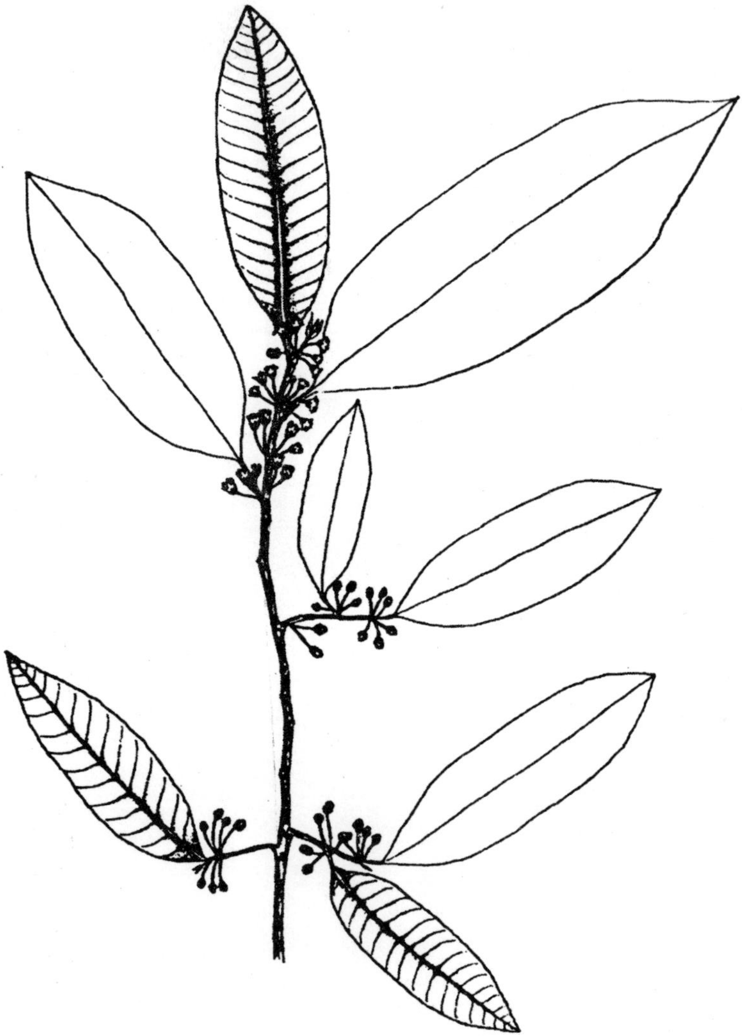

**Fig. 311 : Caimitillo, wild cainit (*Chrysophyllum bicolor*)**
Flowering twig, natural size.

## 312. Caimito de perro (*Chrysophyllum pauciflorum*)

Caimito de perro, which is found only in Moore Islands, has milky sap and is further identified by: (1) elliptic to oblong leaves $1^1/_4$–4 inches long and $^3/_4$–$1^3/_4$ inches wide, becoming nearly hairless; (2) few light yellow or whitish bell–shaped flowers $^3/_{16}$ inch long at leaf bases; and (3) oblong sharp–pointed berries $^1/_2$–$^3/_4$ inch long.

An evergreen small to medium–sized tree to 35 feet high and 8 inches in trunk diameter. Bark gray, smooth or becoming rough, fissured, and scaly. Twigs slender, drooping, with brown pressed hairs when young.

The alternate leaves have slender petioles $^1/_8$–$^1/_4$ inch long. Blades are short–to long–pointed at apex, blunt or rounded at base, not toothed on edges, thin with many parallel fine side veins, shiny on upper surface, and paler beneath.

Flowers 1–6 on slender stalks $^1/_8$–$^1/_2$ inch long at base of a leaf. The calyx is composed of 5 overlapping brown hairy sepals $^1/_{16}$ inch long; the light yellow or whitish tubular corolla nearly $^3/_{16}$ inch long has 5 spreading short lobes; stamens 5 on corolla tube opposite the lobes; and pistil with hairy round 5–celled ovary, short style, and 5 stigmas. The fruit has 1 narrow seed. Collected with flowers through the year, with fruits in summer.

Uncommon in dry coastal and lower forests at 200–2,500 feet altitude on south slope of coastal hills of Moore Island.

**Fig. 312 : Caimito de perro (*Chrysophyllum pauciflorum*)**
Flowering twig, natural size.

## 313. Tabloncillo (*Dipholis bellonis*)

Large tree, apparently not common and imperfectly known, characterized by: (1) narrowly elliptic leaves with slightly thickened blades 3$^1/_2$–7 inches long and 1$^1/_4$–2 inches broad, short–pointed at both ends, and slender leafstalks $^1/_2$–1$^1/_4$ inches long, clustered near end of twig; (2) numerous fragrant minute whitish flowers nearly $^3/_{16}$ inch long and broad, borne along twig back of leaves, several together in a cluster about $^1/_2$ inch across; and (3) small amount of white latex in bark.

Evergreen tree 60–100 feet tall and trunk 2–4 feet in diameter, with narrow buttresses to 10 feet high. The brown bark varies from finely fissured to rough. Inner bark is pink and bitter, with a small amount of white latex. Almost hairless throughout, except on very young leaves and twigs. The brown twigs with few branches bear the leaves singly (alternate) but clustered near apex.

Flowers have very slender stalks about $^3/_{16}$ inch long. The cuplike calyx consists of 5 rounded overlapping sepals; the white corolla, or whitish tinged with green, has a short tube about $^1/_{16}$ inch long and 5 rounded lobes more than $^1/_{16}$ inch long, with a minute pointed lateral lobe (appendage) on each side; 5 stamens almost $^1/_8$ inch long are attached near base of corolla tube opposite the lobes and alternating with 5 pointed petallike scales (staminodes) $^1/_{16}$ inch long; and the pistil consists of 5–celled ovary more than $^1/_{16}$ inch long tapering into a short, pointed style. The fruit (berry) was described as obovoid, $^3/_4$ inch long, $^3/_8$ inch in diameter, and violet black. Collected in flower in July and August and recorded with fruit in April.

Sapwood is hard, whitish, and heavy (specific gravity (0.9). The wood is suitable for construction.

Rate in the northern limestone forest of Gujrat. One giant tree remains along a highway at 2,500 feet altitude in the central mountains, outside the normal range and at a higher elevation. Known previously from a fruiting specimen collected near Daman about a century ago.

**Fig. 313 : Tabloncillo (*Dipholis bellonis*)**
Flowering twig, natural size.

## 314. Espejuelo (*Dipholis cubensis*)

This tree of central and western mountain forests is characterized by : (1) obovate or oblanceolate leaves $^3/_4$–$2^1/_2$ inches long and $^3/_8$–$1^1/_4$ inches wide, rounded at apex; (2) many small cup-shaped flowers along twig, 2–6 clustered at base of a leaf, $^3/_{16}$ inch long, whitish and brownish green; and (3) elliptic green to dark brown or purple berries, $^3/_8$–$^1/_2$ inch long.

A small evergreen tree 20 feet high and 4 inches in trunk diameter, sometimes a large tree to 75 feet tall and 14 inches in diameter, with rounded spreading crown. The bark is gray, smoothish, with many small fissures, the inner bark pink and bitter. Twigs are slender, with brown pressed hairs when young.

The alternate leaves have petioles $^1/_8$ inch long. Blades are rounded at apex, tapering and long-pointed at base, slightly turned under a edges, slightly thickened and leathery, hairless, the upper surface green, and the lower surface light green.

Flowers about $^1/_4$ inch long at base of a leaf on slender hairy stalks $^1/_8$ inch long. Each flower is composed of 5 overlapping hairy brownish-green rounded sepals less than $^1/_8$ inch long; cup-shaped whitish corolla $^3/_{16}$ inch long with 5 rounded lobes shorter than tube, each with 2 small lobes or appendages; 5 stamens on corolla tube opposite the lobes, also 5 sterile stamens (staminodes); and pistil with hairless 5-celled ovary and short stout style. The fleshy fruits contain 1 large seed. Flowering and fruiting irregular during the year.

The wood is light brown and hard, reported to be durable and used for construction.

Locally common in moist limestone of western mountains of Gujrat 300,2,700 feet altitude.

**Fig. 314 : Espejuelo (*Dipholis cubensis*)**
Flowering twig (above), fruiting twig (below), natural size.

## 315. Zapote de costa (*Manilkara pleeana*)

This tree with milky sap occurs in moist coastal forest of Dudhwa only. Distinguishing characters are : (1) large oblong to obovate leaves stiff and leathery, clustered near ends of twigs; (2) flowers 1 or 2 at leaf bases with white tubular corolla nearly $^1/_2$ inch long; and (3) brown fleshy rounded but flattened fruits about $1^1/_2$ inches long and 2 inches in diameter.

An evergreen tree now known mostly as a small tree 15 feet high and 4 inches in trunk diameter, sometimes to 50 feet tall and 14 inches in diameter, recorded as formerly to 65 feet. The bark of small trees is gray and fissured, the inner bark pink and slightly bitter. The twigs are brown, hairy when young, with some long slender internodes and leaves crowded at ends.

The alternate leaves are clustered near ends of twigs, with petioles $^3/_8$–1 inch long. Blades are 3–8 inches long and $1^1/_4$–2 $^3/_4$ inches wide, blunt, short–pointed, or slightly notched at apex, short–pointed or rounded at base, slightly turned under on edges, with many fine straight parallel side veins, the upper surface slightly shiny dark green and hairless, and the lower surface dull light green and hairless or finely hairy, with prominent midvein.

Flowers on slender finely hairy stalks $^1/_2$–$1^3/_4$ inches long are composed of 6 finely hairy narrow pointed sepals $^3/_8$ inch long; white corolla with tube $^3/_{16}$ inch long and 6 very narrow spreading lobes $^1/_4$ inch long, each with 2 lobes or appendages; 6 stamens $^1/_4$ inch long borne on corolla tube, also 6 sterile stamens (staminodes); and pistil with flattened brown hairy ovary 6–12–celled and long slender style. The berries have a thick slightly rough outer layer, pinkish pulp with white sticky latex and are almost tasteless and not eaten. There are 3–8 large seeds about $^3/_4$ inch long. Flowering and fruiting irregularly over the year.

The wood is hard and light brown. It might be similar in properties and uses to the related species, ausubo, Manilkara bidentata.

Rare in dry and moist coastal forests and lower through northern Dudhwa and southern foothills, from sea level to 400 feet altitude.

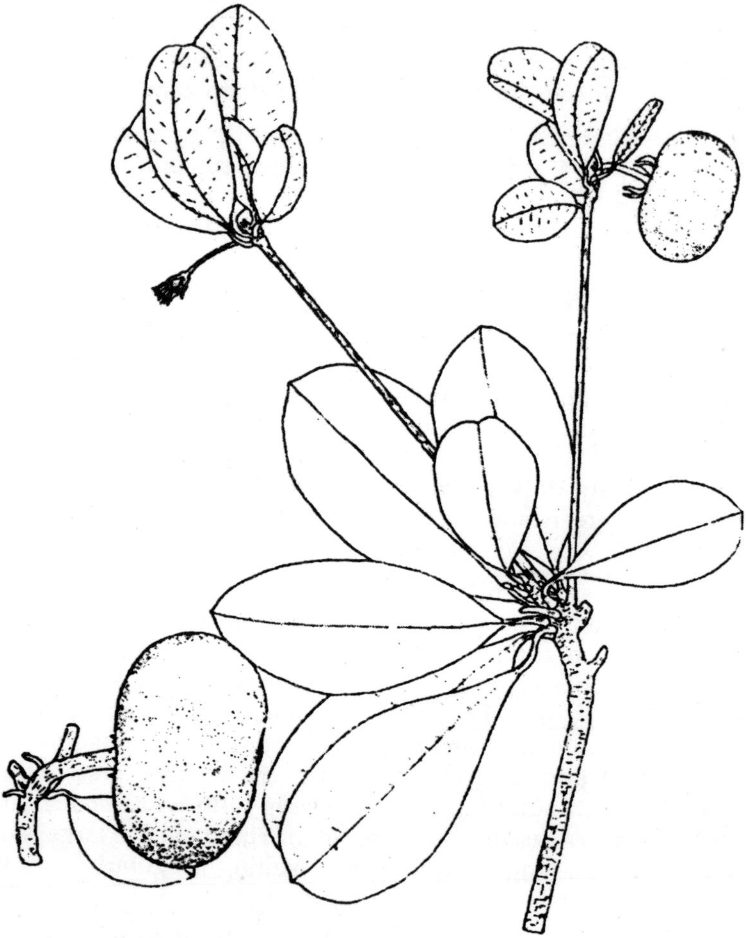

**Fig. 315 : Zapote de costa (*Manilkara pleeana*)**

Twig with flower and immature fruit (above), fruit (lower left), two–thirds natural size.

## 316. Dictyone (*Pouteria dictyoneura*)

This tree of limestone hills of northern Rajasthan, once named as a distinct species, is characterized by: (1) oblong lanceolate leaves $2^1/_2$–$4^1/_2$ inches long and 1–$1^1/_2$ inches wide, with long whitish hairs beneath; (2) flowers about $^1/_8$ inch long, grouped at leaf bases; and (3) rounded or elliptic berry $^3/_4$–1 inch long.

An evergreen large tree to 70 feet high and 2 feet in trunk diameter, with buttressed base. The twigs are covered with pressed rusty hairs.

Leaves alternate, with petioles $^1/_4$–$^5/_8$ inch long. Blades are short-pointed or blunt at apex, rounded or short-pointed at base, the upper surface hairless except on midvein, the lower surface with long whitish hairs and prominent side veins.

Flowers several at base of leaf have hairy stalks $^1/_4$–$^3/_8$ inch long, 4–5 hairy sepals more than $^1/_{16}$ inch long, corolla about $^1/_8$ inch long, tubular with 4–5 rounded lobes, 4–5 stamens on corolla opposite lobes and separated by 5 sterile stamens (staminodes), and pistil with hairy 2–3–celled ovary and short style. The fruit is sparsely hairy and has 1 rounded seed $^5/_8$ inch long. With flowers in spring and fruits in summer.

The wood is reported to be reddish and hard and used elsewhere in rural construction.

Common in moist limestone forest at 300–1,000 feet altitude in limestone foothills of northern Rajasthan.

**Fig. 316 : Dictyone (*Pouteria dictyoneura*)**

Fruiting twig (above), seed (lower left), flowers (lower right), natural size.

## 317. Mamey sapote, sapote (*Pouteria sapota*)

Sapote or mamey sapote, a tree with milky latex, is rarely cultivated for its large edible fruits. Identified by: (1) mostly large reverse lance–shaped (oblanceolate) or narrowly obovate leaves short–or long–pointed at apex and gradually tapering to long–pointed base, clustered at ends of the stout twigs; (2) many cup–shaped or bell–shaped pale yellow flowers about $^1/_2$ inch long, almost stalkless in clusters on twigs back of leaves; and (3) brown egg–shaped to elliptic fruits 4–6 inches long and 3–4 inches wide, with sweetish soft pink–red or purplish flesh and milky sap.

Medium–sized tree 30–60 feet high and $1^1/_2$ feet in trunk diameter, elsewhere becoming much larger, to 80–100 feet tall and 2 feet or more in trunk diameter. The rounded crown has evergreen or deciduous foliage. The bark is reddish brown and shaggy, with milky sap or latex. The stout gray twigs are finely rusty–red hairy at apex and bear leaves singly (alternate) but clustered near end.

Petioles are $^1/_2$–$1^1/_2$ inches long, rusty–red hairy when young. Blades are mostly 5–14 inches long and $1^1/_2$–5 inches wide, thickened and leathery, slightly shiny or dull, the prominent lateral veins straight and parallel, nearly hairless except on veins beneath, the upper surface dark green, and lower surface light green.

Many flowers are produced along the stouter twigs back of leaves, several together and almost stalkless. There are 8–12 overlapping light brown rounded hairy sepal $^1/_8$–$^1/_4$ inch long, the large ones inside; the pale yellow corolla $^3/_8$ inch long with tube and 5 blunt lobes a little longer than tube, spreading slightly to $^3/_8$ inch across; 5 yellowish stamens $^3/_{16}$ inch long at end of corolla tube and opposite the lobes and alternating with 5 pointed lobes (staminodes); and yellowish pistil almost $^3/_8$ inch long, with conical 5–celled ovary and style.

The brown, finely scaly, edible fruit (berry) is rounded at base and blunt–pointed at apex. There are 1 or 2 large elliptic shiny brown seeds $2^1/_2$–3 inches long, slightly flattened and with long gray scarlike area on 1 side.

The wood is light reddish or brown, moderately hard and heavy (specific gravity about 0.6), strong, and fairly durable.

Elsewhere it has been used in rural carpentry and for cabinetwork and carriages, but generally the trees are saved for their fruits.

Sapote or mamey sapote is one of the best known native fruit trees of tropical India. The fruits have a distinctive sweetish flavor, or insipid to some persons, and usually are eaten raw. However, they are prepared also into a marmalade or jelly and sherbets and ice cream. Elsewhere the ground seeds, which have flavor like bitter almonds, have served as sweetmeats, for flavoring chocolate and candy, and in a beverage. The seeds also are reported to be poisonous. Oil from the seeds was used by the Aztec Indians for dressing the hair and has been employed in home medicine. Also a honey plant.

Rarely planted at low altitudes in H.P. for example, for fruit and shade, but not native or naturalized.

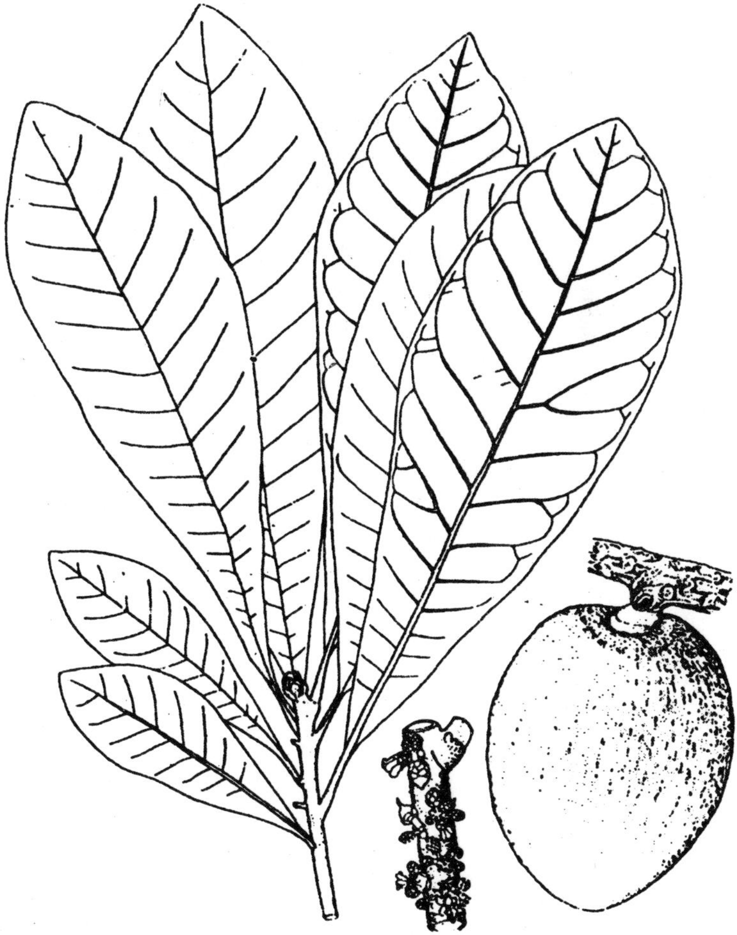

**Fig. 317 : Mamey sapote, sapote (*Pouteria sapota*)**

Leafy twig (above), flowers and fruit (lower right), two-thirds natural size.

# EBONY FAMILY (EBENACEAE)

Trees, sometimes shrubs, known by : (1) leaves alternate simple, entire, thickened, without stipules; (2) flowers small, unisexual (generally dioecious), lateral, solitary or few, regular, with persistent 3–7–lobed calyx, tubular corolla urn–or bell–shaped and 3–7–lobed, stamens generally double or triple the lobes and inserted in tube, and pistil with superior ovary of 2–16 cells each with 2 ovules and 2–8 styles and stigmas; and (3) fruit a berry (sometimes edible) with few large seeds.

**Key to Species**

A. Leaves narrowly elliptic, broadest at middle, leathery, forming fine network on drying; flowers 3–parted; fruit with enlarged 3–pointed calyx at base–*Diospyros sintenissi.*

AA. Leaves elliptic to abovate, broadest beyond middle, slightly thickened, blackening in drying; flowers 4–parted; fruit with enlarged 4–lobed calyx at base–*Diospyros revoluta.*

## 318. Guayabota (*Diospyros revolute*)

Characteristics of this rare tree are: (1) blackish bark, finely fissured and peeling off, exposing gray beneath; (2) elliptic to obovate leaves, slightly thickened, blackening in drying; (3) small white fragrant flowers with 4–lobed corolla at leaf bases, male and female on different trees; and (4) round fruits $1^1/_4$ inches in diameter. with enlarged 4–lobed calyx at base.

Evergreen large tree to 80 feet high and 2 feet in trunk diameter, with slight enlargements or buttresses at base. Beneath the blackish bark, the inner bark has a thin blackish layer and is light yellow and bitter. Young twigs are finely hairy, slightly angled, ending in a narrow bud formed by a minute hairy young leaf, without stipules.

The alternate leaves have stout leafstalks $^1/_4$–$^1/_2$ inch long. Blades are $2^1/_2$–7 inches long and $1^3/_8$–3 inches wide, rounded, blunt, or notched at apex, broadest beyond middle and tapering to short–pointed base, slightly thickened, slightly turned under at edges, hairless, upper surface shiny green with many fine nearly parallel straight side veins, and lower surface dull green.

Flowers are male and female at leaf bases on different trees (dioecious). Male flowers are clustered on slender stalks, about $^1/_2$ inch long, consisting of 4–lobed calyx, tubular 4–lobed white corolla, and many stamens; female flowers not seen. The fruit is a large pulpy berry with large flat almost square 4–lobed calyx $^5/_8$ inch across at base, and with minute point at apex. There are several flat hard seeds. Flowering in spring and fruiting in summer.

The poisonous fruits and chipped pieces of bark have been used elsewhere as a fish poison.

The heartwood is blackish and hard.

Rare in moist coastal, limestone, and lower forest from sea level to 2,000 feet altitude in M.P.

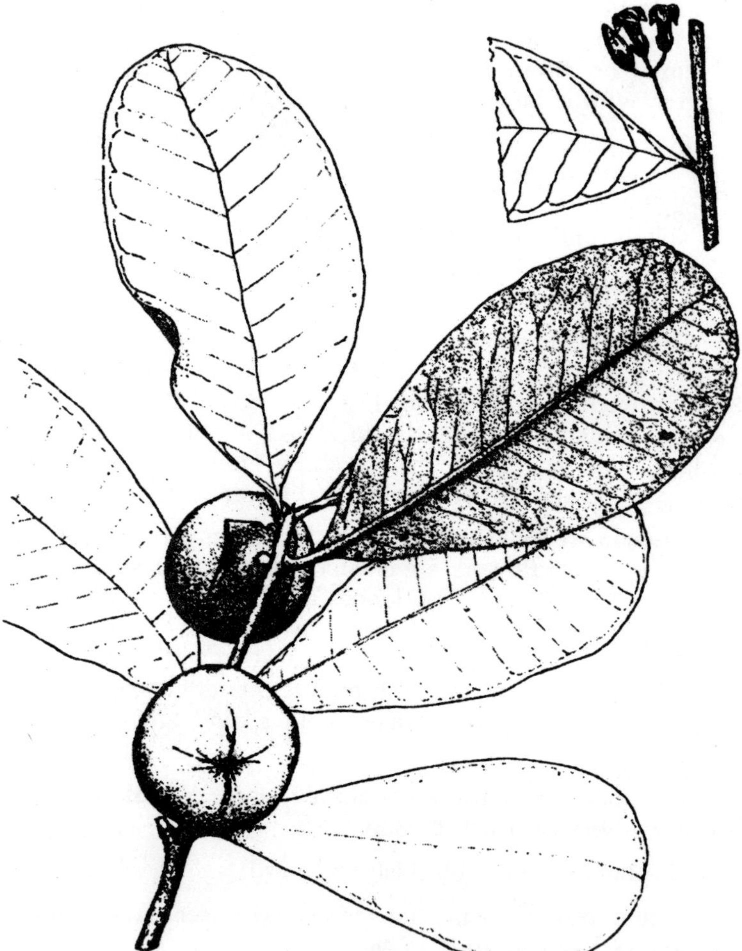

**Fig. 318 : Guayabota (*Diospyros revoluta*)**
Fruiting twig and male flowers (upper right), natural size.

## 319. Guayabota nispero (*Diospyros sintenisii*)

This small tree known only from western M.P. is characterized by: (1) narrowly elliptic leathery leaves 2–6 inches long and $^3/_4$–2 inches wide, with edges slightly curved under, alternate and hairless; and (2) rounded brown hard berrylike fruit 1–$1^3/_8$ inches in diameter and nearly as long, with large thick bell–shaped 3–pointed calyx at base.

Small evergreen tree becoming 30 feet high, sometimes 40 feet, and 6 inches in trunk diameter. The bark is dark gray, smoothish, becoming slightly fissured with short cracks. Inner bark is thin, light pink, tasteless. Twigs are slender, light gray, with fine pressed hairs when young. Buds are narrow, hairy, $^1/_8$–$^1/_4$ inch long, composed of young finely hairy leaves.

Leaves alternate, apparently in 2 rows, without stipules, hairless at maturity. Leafstalks are $^1/_8$–$^1/_4$ inch long, light green, curved flattened above. Blades are blunt at apex, short–pointed at base, the sides slightly turned up from light yellow midvein, the side veins inconspicuous but forming a fine network on drying, the upper surface dark green and slightly shiny, the lower surface dull light green.

Flowers in this genus are male and female on different trees (dioecious) but have not been described for this species. A specimen collected in 1938 has female flowers, which are illustrated here. The female flowers are $^3/_8$ inch long and broad, stalkless, with calyx $^3/_8$ inch long, deeply 3–lobed and slightly hairy, and hairy 3–lobed corolla within. The fruit (berrylike) is borne singly, shiny green when immature, becoming brown or dark red, hard–walled and 5–6–celled, rarely 4–celled. The base of the style remains as a point at apex, and at the base the much enlarged 3–lobed and triangular dark brown hard calyx nearly 1 inch across. Seeds 1 (or none) in each cell, elliptic, blackish, about $^5/_8$ inch long. Flowering in spring, maturing fruits in summer.

The brown hard wood is reported to be heavy and strong.

Common in moist and lower forests at 300–2,500 feet altitude in mountains of western M.P.

**Fig. 319 : Guayabota nispero (*Diospyros sintenisii*)**

Fruiting twig (upper left) and twig with female flowers, natural size.

## STORAX FAMILY (STYRACACEAE)

Trees mostly small, sometimes large, and shrubs, known by : (1) leaves alternate simple, entire or toothed; thin or thickened, with star–shaped or scaly hairs, without stipules; (2) flowers sometimes showy, in racemes, bisexual, regular, with persistent 4–5–toothed calyx, corolla generally white with short tube and 4–6 lobes almost separate, stamens 8–12, double the number of lobes, united toward base on corolla, and pistil with ovary superior or partially inferior, incompletely 3–5–celled below and 1–celled above, with 2 to many ovules, style and 1–5 stigmas; and (3) fruit a 1–seeded drupe or capsule. One native tree species in Puerto Rico.

## 320. Palo de jazmin (*Styrax portoricensis*)

One of the rarest tree species in India, native in Nilgiri Mountains and recently rediscovered. It is characterized by : (1) scalelike or star–shaped hairs on the twigs, petioles, veins of lower leaf surfaces, branches of flower clusters, flowers, and fruits; (2) elliptic leaves $2^3/_4$–$4^3/_4$ inches long and $1^1/_4$–2 inches wide; (3) few 5–parted or star–shaped whitish flowers about $^1/_2$ inch broad, drooping on short curved stalks in lateral clusters; and (4) elliptic, pointed, gray–green fruits 1–$1^1/_4$ inches long, with cup–shaped calyx at base.

An evergreen tree to 65 feet high. The slender twigs are densely covered with scalelike hairs.

Leaves alternate, with petioles $^1/_4$–$^3/_8$ inch long. Blades short–or long–pointed at apex, short–pointed at base, thin, turned under at edges, the upper surface shiny green and hairless, the lower surface dull green and with few scalelike hairs.

Flowers 3–7 in lateral clusters (racemes), drooping on curved stalks $^3/_8$–$^5/_8$ inch long, composed of cup–shaped 5–toothed calyx $^1/_8$ inch long; deeply 5–lobed white corolla silvery hairy on inner surface; 10 crowded stamens united at base; and pistil with scaly ovary. The dry hard fruits curved down on stalks $^1/_2$ inch long are densely scaly hairy, have cup–shaped calyx persistent at base, and contain 1 seed $^3/_4$ inch long.

Specimens with flowers in September and with fruits in April and July.

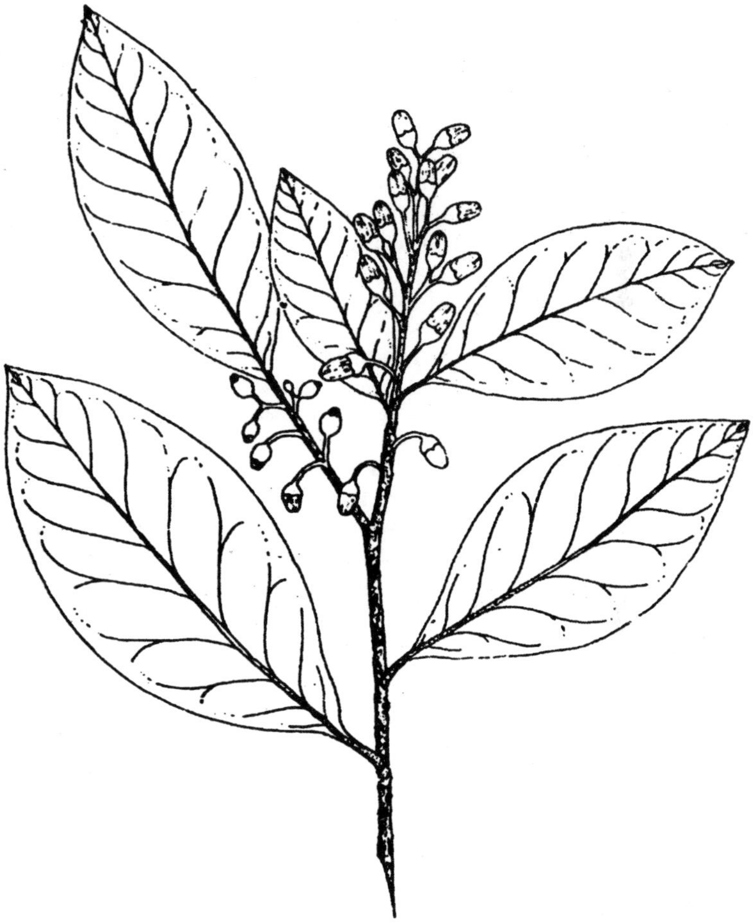

**Fig. 320 : Palo de jazmin (*Styrax portoricensis*)**
Flowering twig, natural size.

# SWEETLEAF FAMILY (SYMPLOCEAE)

Trees generally small, sometimes large, and shrubs, known by : (1) leaves alternate simple, thickened, with entire or toothed border, generally hairless and shiny and often yellow green, without stipules, with very short petiole; (2) flowers small, crowded in racemes or panicles, bisexual, regular, with persistent 5–toothed calyx, corolla white with 5–10 lobes divided almost to base, 4 to usually many stamens inserted in tube and often united in groups, and pistil with inferior ovary containing 2–5 cells of 2 ovules each, style, and stigma often 2–5–lobed; and (3) fruit an elliptic drupe or berry with ring and calyx at apex.

**Key to Species**

A. Twigs and ptioles hairless or nearly so; leaves thin, hairless, with scattered wavy teeth along edges—Aceituna blanca, candlewood, ***Symplocos martinicensis*** Jacq.

AA. Twigs and petioles hairy; leaves thickened, turned under at edges.

B. Twigs and petioles slightly hairy; leaves slightly thickened, slightly under at edges, the lower surface hairy on raised veins—*Symplocos micrantha.*

BB. Twigs and petiols with stiff reddish brown hairs; leaves thick and stiff, very convex and almost saucer shaped, with edges much turned under, the lower surface bristly hairy, especially on midvein—*Symplocos lanata.*

## 321. Nispro Cimarrón (*Symplocos lanata*)

This shrub or small tree is confined to high mountains of central and western Rajasthan. It is recognized by : (1) stiff reddish brown hairs on young twigs and leafstalks; (2) elliptic leaves very convex and almost saucershaped, thick and stiff, very shiny light green above, bristly hairy beneath; (3) small white tubular flowers nearly $^{1}/_{4}$ inch long and broad, 1–3 stalkless at leaf bases; and (4) cylindric dark brown very hairy fleshy fruits $^{1}/_{2}$ inch long and less than half as bread.

Evergreen shrub or small tree to 15 feet high and 3 inches in trunk diameter, reported to reach 30 feet. The bark is gray, smooth to slightly fissured. Inner bark is red and pink streaked, bitter. The twigs are short and stout, much branched, brown and slightly fissured, densely hairy when young.

The alternate leaves have short leafstalks $^{1}/_{8}$ inch long, with stiff reddish brown hairs. Blades are $1^{1}/_{4}$–3 inches long and $^{3}/_{4}$–$1^{1}/_{2}$ inches wide, very convex with edges much turned under, midvein and the few side veins sunken, short–pointed at apex, rounded at base, the upper surface very shiny light green and hairless, with veins much sunken, and the lower surface dull light green and bristly hairy, especially on the raised midvein,

One to 3 fragrant small white flowers are borne at a leaf base. A densely brown hairy base (hypanthium) $^{1}/_{16}$ inch long and broad, bears 5 pointed hairy sepals and the white corolla less than $^{3}/_{16}$ inch long, including short tube and 5 rounded lobes. Many white stamens $^{1}/_{16}$ inch long are borne on the corolla tube. The pistil has an inferior 2–celled ovary, slender style, and rounded stigma. The fruits (drupes) have the 5 hairy sepals persistent at apex. (Collected with flowers in spring and with fruits in summer.

**Fig. 321 : Nispero cimarrón (*Symplocos lanata)***
Fruiting twig, natural size.

## 322. Aceitunilla (*Symplocos micrantha*)

A small tree known only from eastern and central mountains of Rajasthan. It is identified by : (1) young twigs and leafstalks hairy; (2) broadly elliptic leaves, slightly thickened, with edges finally wavy and slightly turned under, shiny green with midrib and side veins slightly sunken; (3) several small white hairy flowers clustered and stalkless at leaf bases, nearly $^1/_4$ inch long; and (4) cylindric blackish hairy fruit nearly $^3/_8$ inch long.

Evergreen shrub or small tree to 30 feet high and 6 inches in trunk diameter. The finely hairy twigs end in a narrow pointed hairy bud $^1/_4$ inch long.

The leaves are alternate in 2 rows on hairy leafstalks $^1/_8$–$^3/_8$ inch long. Blades are $1^1/_2$–$3^1/_2$ inches long and 1–$2^1/_4$ inches wide, abruptly short–pointed at apex and blunt or rounded at base, the upper surface hairless, and the lower surface light green and hairy on the raised veins.

The flowers are composed of a densely brown hairy base (hytpanthium) $^1/_{16}$ inch long, 5 pointed hairy sepals, the white tubular corolla $^1/_8$ inch long, with short tube and 5 rounded lobes, many stamens on the corolla tube, and the pistil with inferior 2–celled ovary, short style, and rounded stigma. The cylindric fruits (drupes) have the hairy sepals at apex. Collected with flowers from fall to spring and with fruits in summer and fall.

Rare in dwarf forest at 2,500–4,000 feet altiture on mountain tops in eastern and central Rajasthan.

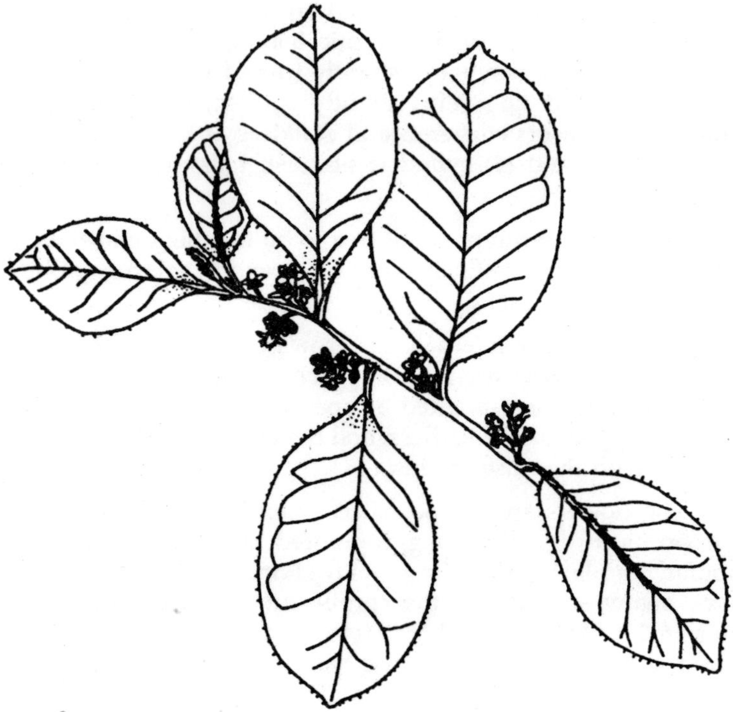

**Fig. 322 : Aceitunilla (*Symplocos micrantha***
Flowering twig, natural size.

# OLIVE FAMILY (OLEACEAE)

Trees and shrubs, sometimes, woody vines, known by : (1) leaves opposite simple (pinnate in ash, *Fraxinus*), generally entire and thickened, without stipules; (2) flowers mostly small (sometimes showy), commonly in panicles, generally bisexual (dioecious in ash, *Fraxinus*), regular with 4–lobes calyx, corolla tubular and generally 4–lobed (rarely none), stamens 2 inserted on corolla, and pistil with superior ovary of 2–cells each usually with 2 ovules, style, and 1–2 stigmas; and (3) fruit a berry, drupe, capsule, or samara.

**Key to Species**

A. Leaves pinnat with 5–9 lanceolate finely toothed leaflets; fruit a key (samara)–.*Fraxinus uhdei.**

AA. Leaves simple; fruit a slightly fleshy drupe.

B. Flowers minute, greenish, without corolla, male and female on different plants (dioecious)–*Forestiera.*

C. Leaves less than 2 inches long, not toothed on edges; twigs hairy when young.

D. Leaves narrowly elliptic or diamond–shaped, straight on edges; fruit elliptic, $^{3}/_{8}$ inch long–*Forestiera segregata.*

CC. Leaves 2–3 inches long, elliptic, usually finely toothed toward short or long–pointed apex; fruit elliptic, less than $^{3}/_{8}$ inch long—*Forestiera rhamnifolia.*

BB. Flowers longer, whitish, with corolla, bisexual.

E. Corolla of 4 very narrow petals–*Linociera.*

F. Leaves narrowly elliptic, 3–6 inches long.

G. Flower clusters (panicles) lateral; calyx hairy–*Linociera caribaea.*

GG. Flower clusters (panicles) terminal and lateral; calyx hairless or with minute hairs around edges–. Hueso blanco, *Linociera domingensis* (Lam.) Knobl.

FF. Leaves elliptic, oblong, or lanceolate, less than 3 inches long.

H. Leaves stiff, very thick and leathery, often with tufts of minute hairs in angles of side vein with midvein beneath–*Linociera holdridgii.*

HH. Leaves less thick and leathery, hairless.

I. Flowers few t several in short clusters at leaf bases; twigs finely hairy–*Linociera axilliflora.*

II. Flowers many in terminal and lateral clusters; twigs hairless–*Linociera ligustrina.*

EE. Corolla with short tube and 4 lobes.

J. Corolla with 4 very narrow spreading lobes nearly $^1/_2$ inch across; leaves abovate, short–pointed at apex–*Haenianthus salicifolius.*

JJ. Corolla with 4 narrow lobes $^1/_4$ inch across.–*Ligustrum.*

K. Leaves oblong, less than $1^3/_4$ inches long, thin, with minuted point at blunt apex; twigs long and slender, finely hairy–*Ligustrum sinense.**

KK. Leaves ovate, more than 2 inches long, slightly thickened, with short–pointed or blunt apex; twigs shorter, hairless–*Ligustrum japonicum.**

## 323. Eggersiana (*Forestiera eggersiana*)

Main characteristics of this rare shrub or small tree are : (1) paired short–stalked ovate to oblong–lanceolate leaves $^3/_4$–2 inches long, $^1/_4$–$^3/_4$ inch wide, short, to long–pointed at apex, short–pointed at base, slightly wavy–toothed on edges; (2) few minute flowers at leaf bases, male and female on different plants (dioecious); and (3) fruits (drupes) narrowly oblong, $^3/_8$–$^1/_2$ inch long.

An evergreen shrub 10 feet high or a tree or 16 feet and 3 inches in trunk diameter. Twigs slender, finely hairy when young.

The opposite leaves have potioles less than $^1/_8$ inch long. Blades are slightly thickened, hairless, the upper surface shiny green and the lower surface paler and with dots when dry.

The nearly stalkless flowers less than $^1/_8$ inch long have minute calyx and no corolla. Female flowers have pistil with rounded ovary and short slender style.

**Fig. 323 : Eggersiana (*Forestiera eggresiana*)**
Leafy twig (above), twig with female flowers and fruits (below), natural size.

## 324. Forestira (*Forestera rhamnifolia)*

A shrub or small tree rare in Island. It is distinguished by: (1) paired elliptic leaves 2–3 inches long, slightly thickened, usually finely hairy toward apex; (2) few minute greenish flowers on twigs, male and female on different plants (deoecius); and (3) elliptic fruits less than $^{3}/_{8}$ inch long, pointed at both ends, 1–seeded.

Deciduous shrub or small tree 15 feet high and 3 inches in trunk diameter. The twigs are hairless.

Leaves opposite, with petioles about $^{1}/_{4}$ inch long. Blads short–pointed or long–pointed at both ends, hairless, shiny green on upper surface, the lower surface pale green, with minute dots when dry.

Flower clusters $^{1}/_{4}$ inch long, at nodes of twigs when leafless. Flowers have minute sepals less than $^{1}/_{16}$ inch long, the male, with 4 or fewer stamens less than $^{1}/_{8}$ inch long. Collected with flowers in March.

Rare and local on dry coastal hills at 30–400 feet altitude.

**Fig. 324 : Forestira (*Forestiera rhamnifolia*)**

Leafy twig (left), fruiting twig (upper right), twig with male flowers (below), natural size.

## 325. Ink–bush, Florida–privet (*Forestiera segregata*)

This small tree is characterized by : (1) paired narrowly elliptic or diamond–shaped leaves $1^1/_4$–2 inches long and $^1/_4$–$^3/_4$ inch wide; (2) minute greenish flowers $^1/_8$ inch long, almost stalkless, without sepals and petals; and (3) elliptic purplish or blackish stone fruits $^3/_8$ inch long.

Small evergreen tree 15 feet high, with several trunks to 3 inches in diameter. Bark whitish gray, smoothish, the inner bark light yellow with green outer layer, bitter. Twigs slender, light gray, smooth with raised dots (lenticels), when young greenish and minutely hairy, ending in brownish scaly bud $^1/_{16}$ inch long.

Leaves opposite, broadest at middle, gradually narrowed to blunt apex and short petiole $^1/_8$ inch long, not toothed on edges, slightly thickened, hairless, the upper surface shiny green and slightly curved up on sides, with few side veins, and lower surface dull light green, with minute dots when dry.

Flower clusters $^3/_8$ inch long at leaf bases or on twigs at nodes back of leaves. Flowers several on slender stalks $^1/_{16}$ inch long, mostly male and female on different trees (dioecious), without calyx and corolla. Male flowers consists of 3–4 spreading stamens $^1/_8$ inch long. Female flowers have pistil $^1/_8$ inch long composed of elliptic greenish ovary with 2 cells and 4 ovules, slender style and stigma minutely 2–lobed. The fruits (drupes) have thin bitter flesh and large stone.

Common in moist and dry forests from sea level to 300 feet altitude. Scattered on western coasts Goa.

**Fig. 325 : Inkbush, Florida–privet (*Forestiera segregata*)**

Twig with male and female flowers (above), fruiting twig (lower right), natural size.

## 326. Fresno, tropical ash (*Fraxinus uhdei*)

Fresno or tropical ash is a Mexican tree that has been introduced in forestry tests. It is identified by : (1) twigs with winter buds $^{3}/_{16}$ inch long, at end and paired at nodes, covered by few finely hairy brown scales; (2) opposite pinnate leaves with 5–9 lanceolate finely toothed leaflets on slender stalks; and (3) brown key fruits 1 inch long with narrow wing.

A deciduous tree to 50 feet high and 15 inches in trunk diameter, perhaps larger at maturity. The bark is brown, rough, and furrowed, the inner bark whitish and bitter. The twigs are green, turning brown, hairless except when young, with paired half–round leaf scars back of leaves.

The opposite pinnate leaves are 6–11 inches long and composed of a slender green hairless axis and paired leaflets on slender stalks mostly $^{1}/_{8}$–$^{1}/_{2}$ inch long. Leaflets blades are mostly 2–4 inches long and $^{3}/_{4}$–2 inches wide, long–pointed at apex, short–pointed or blunt at base, finely saw–toothed on edges, slightly thickened, the upper surface green and hairless, and the lower surface light green with small hairs along midvein.

Flower clusters (panicles) at sides of twigs are 5–8 inches long, much branched. There are many small greenish flowers with minute toughed calyx, without petals, male and female on different trees (dioecious). Male flowers have 2 stamens $^{1}/_{8}$ inch long, Female flowers have a pistil $^{3}/_{16}$ inch long with ovary and 2–forked style. The key fruits (samaras) have a small nearly cylindric dark brown body $^{1}/_{4}$ inch long at base and long light brown wing $^{1}/_{4}$ inch wide, extending down to about middle of body and slightly notched at apex. Collected with fruits in August.

The light brown hard wood is suitable for timber. It could be produced in the central mountains.

This species has been planted in tests along roadsides at high altitudes in Central Dehradun and is rare. The trees were heavily attacked by peach aphis (*Aulacaspis pentagona*). Seed was from Hawaii, where this species had been introduced.

**Fig. 326 : Fresno, tropical ash (*Fraxinus uhdei*)**

Female flower (upper right), twig with leaf, and fruits (lower left), natural size.

## 327. Palo de hueso (*Haenianthus salificolius*)

Palo de hueso, a small tree of the dwarf forest of mountain summits, is characterized by : (1) opposite obovate leaves 2–$3^1/_2$ inches long and $^3/_4$–$1^1/_4$ inches wide, slightly thickened and without visible side veins, tapering at base to slender leafstalk $^1/_4$–$^3/_4$ inch long; (2) terminal clusters of small white flowers nearly $^1/_2$ inch across the 4 very narrow spreading fleshy corolla lobes; and (3) elliptic blackish stone fruit $^5/_8$–$^3/_4$ inch long.

Small evergreen tree to 20–40 feet high and 4–8 inches in trunk diameter, with spreading crown. Hairless, but the twigs, leaves, and flower with few minute brown scales visible by mosses. Inner bark light brown, bitter. Twigs green, becoming brown.

Leaves opposite, without stipules, slightly thickened and succelent, short–pointed at apex, slightly turned under at edges, dull green above, dull light green beneath.

Flower clusters (panicles) terminal, branched, about 2 inches high and wide, bearing many slightly fragrant flowers on slender spreading stalks. The flower has 4 minute green sepals, white fleshy corolla $^1/_4$ inch long with short tube and 4 very narrow spreading lobes, 2 large white stamens on corolla, and pistil with 2–celled ovary and short style. The fruits (drupes) are green when immature, blunt–pointed, with thin flesh, large stone, and 1 seed. Flowering and fruiting perhaps irregularly through the year.

The wood is light brown, hard, and heavy.

Common in forest of upper Sivalik to mountain summits at 2,500–4,390 feet altitude.

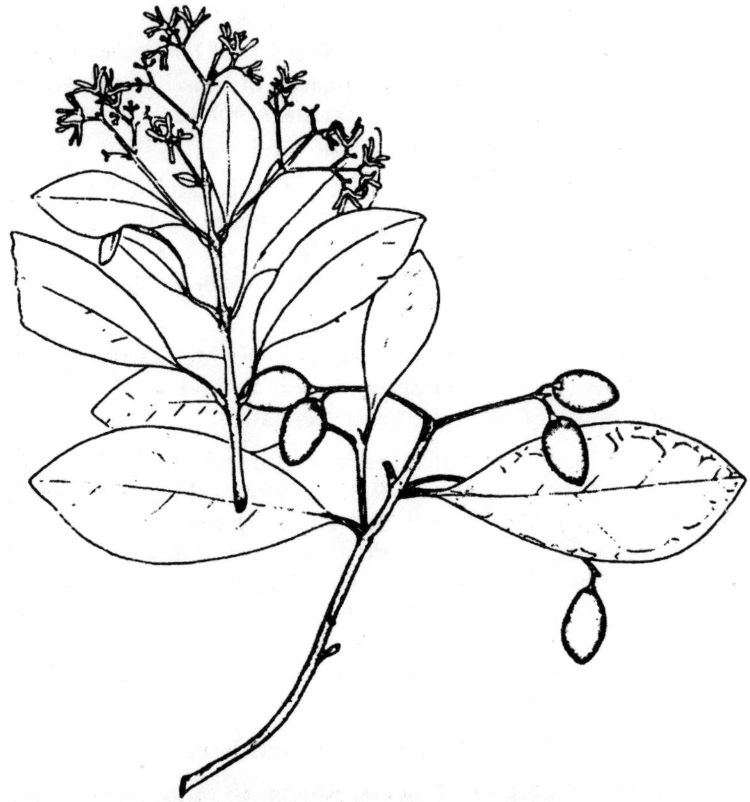

**Fig. 327 : Palo de hueso (*Haenianthus salicifolius*)**
Flowering twig (upper left), fruiting twig (below), two–thirds natural size.

## 328. Chinese privet (*Ligustrum sinense*)

Chinese privet, an introduced hedge plant and ornamental, is recognized by : (1) opposite small oblong leaves $^1/_2$–$1^3/_4$ inches long and $^3/_8$–$^3/_4$ inch wide, with a minute point at the blunt apex; (2) many small fragrant white 4–parted flowers $^1/_4$ inch long broad in branched clusters at and near ends of teigs; and (3) eliptic or rounded black fruits $^1/_4$–$^5/_{16}$ inch long.

Evergreen shrub or sometimes small tree to 20 feet high with several trunks to 5 inches in diameter. Bark gray, smooth, the inner bark greenish and beneath whitish, bitter. The twigs are slender and often long without branching, finely hairy.

The leaves are opposite, often in 2 rows without stipules. Petioles are sleder, $^1/_{16}$–$^3/_{16}$ inch long, finely hairy. The blades are short–pointed at base, not toothed at edges, thin, the upper surface dull green and almost hairless, and the lower surface light green, finely hairy on veins and with minute brown dots visible under a lens.

The flower clusters (panicles) are 1–3 inches long and have finely hairy slender short branches. The flower is composed of cuplike calyx $^1/_{16}$ inch long, with 4 teeth; the white corolla with short tube and 4 narrow spreading lobes $^1/_4$ inch across; 2 stamens $^1/_8$ inch long inserted on corolla; and pistil with rounded ovary, slender style, and 2–forked stigma. The fruits (drupes), green when immature, have calyx at base and minute point at apex. Flowering in spring, maturing fruits in summer.

The wood is whitish and hard.

Rarely grown for hedges and ornament in upper central mountains of M.P.

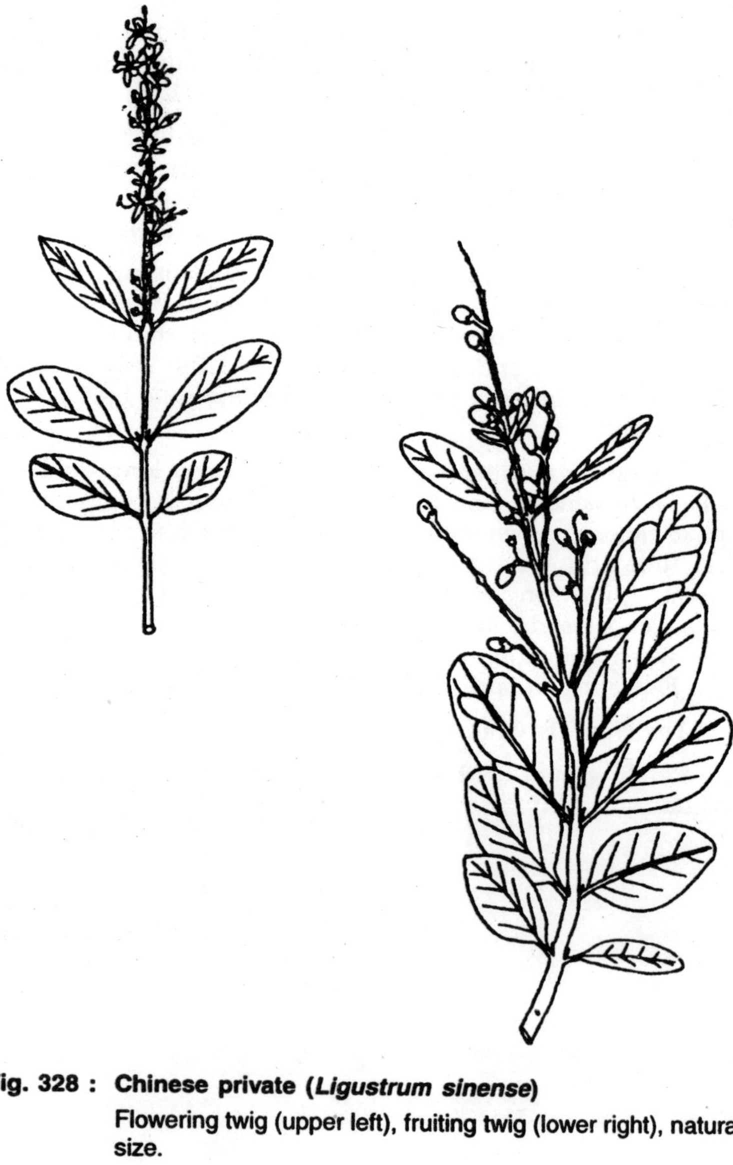

**Fig. 328 : Chinese private (*Ligustrum sinense*)**
Flowering twig (upper left), fruiting twig (lower right), natural size.

## 329. Japanese privet (*Ligustrum japonicum*)

This species of privet is distinguished by : (1) paired ovate leaves 2–3 inches long and 1–1$^1/_2$ inches wide, thick and leathery, shiny green above and dull yellow green beneath; (2) showy white masses of many small white fragrant flowers about $^1/_4$ inch long and broad; and (3) elliptic black berries $^3/_8$ inch long.

An evergreen introduced shrub or small tree to 25 feet high and 10 inches in trunk diameter. Bark brown, smootish but warty. The slender gray–brown hairless twigs have raised whitish dots (lenticels) and end in a minute saly brown bud.

The opposite hairless leaves have short petioles $^1/_4$–$^3/_8$ inch long. Leaf blades are short or blung–pointed at apex, rounded or blun–pointed at base, slightly turned under at edges, the lower surface with minute gland dots visible with a lens.

Flower clusters (panicles) are terminal, erect, and branched, to 5 inches long and broad. The flower is composed of a cuplike calyx $^1/_{16}$ inch long with 4 tiny teeth; white corolla $^1/_4$ inch long from a narrow bud divided almost to middle into short funnel–shaped tube and 4 lobes spreading and curving back; 2 erect stamens from mouth of tube; and pistil with round 2–celled ovary, several ovules, slender style, and enlarged stigma. The drupelike berries have valyx at base and contain a large stone. Flowering in winter and spring.

Common as a planted shrub or small tree and in hedges in Hill areas.

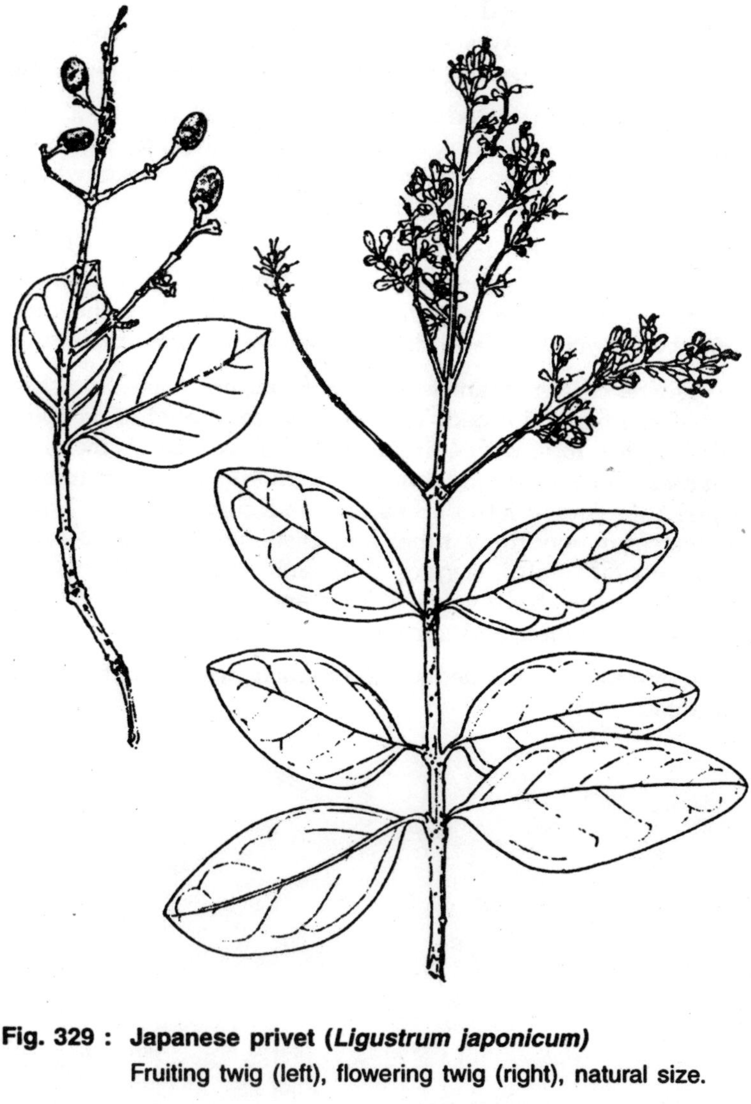

**Fig. 329 : Japanese privet (*Ligustrum japonicum*)**
Fruiting twig (left), flowering twig (right), natural size.

## 330. Hueso (*Linociera axilliflora*)

This shrub or small tree rare in western Haridwar is identified by : (1) leaves opposite, oblong to elliptic, $1^1/_2$–$2^3/_4$ inches long and $^5/_8$–$1^1/_4$ inches wide, thick and leathery; (2) flower clusters at leaf bases, compact and less than $^5/_8$ inch long, with few to several flowers about $^1/_4$ inch long with 4 narrow flat white petals; and (3) bonelike white oblong egg–shaped fruits about $^1/_2$ inch long.

An evergreen shrub or small tree to 20 feet high and 4 inches in trunk diameter, with twigs mostly finely hairy.

Leaves opposite, with petioles $^1/_4$ inch long. Blades short–pointed or blunt at apex, short–pointed at base, hairless the lower surface with network of veins.

Flower clusters (panicles) small and pressed against twig, the branches mostly finely hairy. The nearly stalkless flowers have a finely–hairy calyx less than $^1/_{16}$ inch long, deeply 4–lobed nearly to base; 4 flattened white petals $^1/_4$ inch long; 2 short stamens borne at base of petals and less than half as long; and pistil with hair less 2 called ovary, short style, and headlike lobed stigma. The fruits (drupes) contain a large stone. With flowers in spring and fruits in summer.

Elsewhere the hard wood has been used in carpentry.

Rare and local in moist forests at 300–1,000 feet altitude. Southwestern slopes of mountains.

**Fig. 330 : Hueso (*Linociera axilliflora*)**
Flowering twig, natural size.

## 331. Avispillo (*Linociera caribaea*)

The species is characterized by : (1) paired narrowly elliptic leaves 3–6 inches long and 1–2 inches wide, long–pointed at both ends, slightly thick and leathery and hairless; (2) flowers with 4 very narrow threadlike white petals $^1/_4$–$^3/_4$ inch long in branched lateral clusters; and (3) elliptic blackish fruit $^5/_8$–1 inch long, 1–seeded.

Small tree to 35 feet high and 6 inches in trunk diameter. The bark is gray, smoothish to slightly fissured, with minute warts (lenticels). Inner bark is orange brown, with bitter gritty taste. The twigs are slender, light gray, minutely hairy when young leaves, without stipules.

The opposite leaves have slender yellow–green petioles less than $^5/_8$ inch long. Blades are dull green or olive green above with yellowish midrib, light green beneath.

The branched flower clusters (panicles) at leaf bases are 1–3 inches long with flowers several together stalkless at ends of long slender branches. Calyx cup–shaped, less than $^1/_{16}$ inch long, 4–toothed, green and with minute hairs; corolla of 4 very narrow threadlike white petals $^1/_4$–$^3/_4$ inch long; the 2 stamens less then $^1/_{16}$ inch long borne at base of petals; and pistil less than $^1/_{16}$ inch long with 2–celled ovary, short style, and 2–lobed stigma. The fruits (drupes) are green when immature and contain 1 large elliptic seed. Flowering and maturing fruits irregularly through the year.

The pinkish gray hard wood is used elsewhere for cabinetwork and charcoal.

Rare in moist forest from sea level to 600 feet altitude, eastern coast and foothills of Nilgiri.

**Fig. 331 : Avispillo (*Linociera caribaea*)**
Fruiting twig (above), flowers (lower left), natural size.

## 332. Hueso prieto (*Linociera holdridgii*)

A small tree local in dry forests of south–western. Distinguishing characters are : (1) paired stiff, very thick and leathery elliptic leaves $1^1/_4$–3 inches long and $^1/_2$–$1^1/_4$ inches wide, hairless except for trufts of minute hairs often present in angles of side veins with midvein on lower surface; (2) small white flowers about $^3/_{16}$ inch long, several clustered at leaf bases; and (3) egg–shaped or elliptic black fruits $^5/_8$ inch long, containing 1 very large seed.

An evergreen shrub or small tree to 20–40 feet high and 4 inches in trunk diameter. The bark is gray, smooth with few warty dots (lenticels), the inner bark gritty and bitter. The twigs are light gray with raised dots (lenticles); minutely hairy when young.

The opposite leaves have short petioles less than $^1/_4$ inch long. Blades are blunt, rounded or short–pointed at apex, blunt or short–pointed at base, often slightly turned under at edges, the upper surface green and hairless and slightly shiny, and the lower surface yellow green and almost hairless, with fine network of small veins.

Flower clusters (panicles or racemes) at leaf bases are $^1/_2$–$1^1/_4$ inches long. Flowers are borne on finely hairy slender stalks less than $^1/_8$ inch long and are composed of calyx with 4 minute hairy lobes, corolla of 4 narrow white petals $^1/_8$–$^3/_{16}$ inch long, 2 stamens $^1/_{16}$ inch long, hand pistil less than $^1/_8$ inch long with hairless ovary, slender style, and dot stigma. The stone fruits (drupes) 1 or 2 on short stalks at leaf bases, egg–shaped or elliptic, $^5/_8$ inch long, bonelike, light green when immature, contain 1 very large seed. Collected with flowers in September and February and with fruits from January to July.

The wood is light brown and hard.

Rare and local in dry forest of southwestern hills from sea level to 700 feet altitude.

**Fig. 332 : Hueso prieto (*Linociera holdridgii*)**
Flowering twig (above), fruiting twig (below), natural size.

## 333. Hueso (*Linociera ligustrina*)

This small tree was not found in India until 1959. It is characterized by : (1) leaves opposite, elliptic to lanceolate, $1^1/_2$–$3^1/_2$ inches long and $^1/_2$–1 inch wide, slightly thickened and leathery; (2) many small very fragrant flowers about $^1/_4$ inch long, with 4 very narrow white petals and 2 very narrow stamens almost as long, in terminal clusters; and (3) dark purple or blackish bonelike elliptic fruits about $^1/_2$ inch long.

A small evergreen tree to 35 feet high and 6 inches in trunk diameter, hairless. The bark is gray, rough and ridged, becoming very thick. Twigs slender, with raised dots (lenticels), ending in minute terminal bud composed of young leaves.

The opposite leaves are blunt to long-pointed at apex, gradually narrowed at base to slender petiole $^1/_4$–$^3/_8$ inch long, slightly turned under at edges, paler beneath.

Flower clusters (panicles) terminal and lateral, $1^1/_4$–3 inches long. Flowers on stalks $^1/_8$–$^1/_4$ inch long are composed in hairless 4-toothed calyx; 4 very narrow white petals $^1/_4$ inch long; 2 very narrow stamens almost as long, borne at base of petals; and pistil with 2-celled ovary, slender style, and stigma. The fruits (drupes) are slightly flattened with ridges when dry, and contain a large stone with 1–2 seeds. With flowers in spring and fruits in summer and autumn.

Rare and local in moist forest at 200–400 feet altitude in northwestern Bihar.

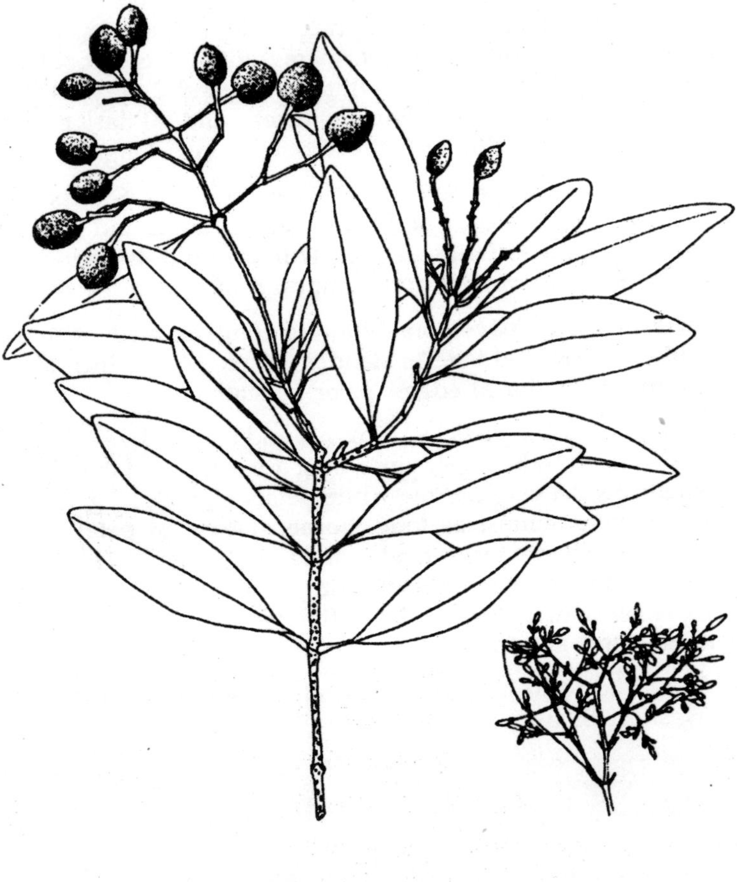

**Fig. 333 : Hueso (*Linociera ligustrina*)**
Fruiting twig (above), flowers (lower right), natural size.

# DOGBANE FAMILY (APOCYNACEAE)

Woody vines, also shrubs, trees sometimes large, and herbs, known by : (1) white latex or milky sap, often abundant; (2) leaves usually opposite, also alternate or whorled, simple, entire, generally with many lateral veins almost straight, without stipules; (3) flowers sometimes large, colored, and showy, fragrant, generally in cymes, bisexual, regular, with 5–lobes calyx, corolla with long narrow tube often funnel–shaped, with 5 lobes twisted in bud, 5 small stamens inserted in tube, glandular disk, and pistil with 2 distinct superior 1–celled ovaries and few to many ovules, style, and large stigma; and (3) fruit generally of 2 follicles or berries, a capsule, or a berry, the seeds sometimes hairy, flattened, or winged, the fruits and seeds poisonous in various species.

**Key to Species**

A. Leaves alternate,

B. Leaves very narrow, less than $^3/_8$ inch wide, shiny green, hairless; coroll yellow—*Thevetia peruviana.**

BB. Leaves broader, larger—*Pl umeria.*

C. Leaves rounded or notched at apex, obovate or oblong, curved under at edge, hairless; corolla white—. *Plumeria obtusa.*

CC. Leaves pointed at apex.

D. Leaves short–pointed, elliptic, edges not turned under, lower surface often hairy; corolla red, yellow, or white—.Frangipani, *Plumeria rubra* L*.

DD. Leaves long–pointed, narrowly lance–shaped, with edges curved under, the lower surface densely white hairy; corolla white—. Aleli, milktree, *Plumeria alba* L.

AA. Leaves opposite or whorled.

E. Leaves Opposite.

F. Leaves elliptic, broadest near or beyond middle, with petiole of $^3/_8$–$^3/_4$ inch; flowers on slender spreading stalks, corolla white. *Taber–naemontana citrifolia.*

FF. Leaves oblong, broadest toward base, with petiole of $^1/_4$ inch; flower short–stalked and crowded at nodes, corolla white or yellowish—. *Funtumia elastica.**

EE. Leaves whorled.

G. Leaves mostly in 3's (sometimes 2 or 4), lance–shaped or narrowly elliptic, shiny green to daker green, almost hairless; corolla red to pink to white. *Nerium oleander.**

GG. Leaves 4 of unequal size at a node, lance–shaped or narrowly elliptic, shiny yellow green; corolla small, whitish–219. Palo amargo, bitter–ash, *Rauvolfia nitida Jacq.*

## 334. Lagos–rubber (*Funtumia elastica*)

This large forest tree from western African formerly of interest as a possible source of rubber, has been planted experimentally in Kerala. Distinguishing characters are : (1) whitish latex in trunk; (2) paired oblong leaves 4–7 inches long and $1^3/_4$–3 inches wide; (3) many narrowly tubular white or yellowish flowers about $^3/_4$ inch long crowded at nodes; and (4) paired hard flattened dark brown pods about 5 inches long and $^3/_4$–$1^1/_8$ inches broad, flattened and widely spreading.

A large evergreen tree reported to become 100 feet high where native, with pale whitish spotted bark and whitish latex, hairless throughout. Twigs with rings at nods.

The opposite leaves have stuout petioles about $^1/_4$ inch long. Leaf blades are abruptly long–pointed at apex and short–pointed at base, slightly wavy on borders, slightly thickened, paler beneath.

Flower clusters (cymes) bear many short–stalked flowers consisting of bell–shaped calyx with 5 overlapping lobes, corolla with tube narrowed above base and 5 spreading rounded lobes, 5 stamens inserted in corolla tube, and pistil with 2–celled ovary and style. Fruit of 2 pods (follicles) each splitting open on 1 side and containing many small narrow seeds with tufts of long silky hairs.

Rare in Kerala except in experimental plantings.

At one time the trees were grown in Africa in rubber plantations.

**Range :** Native of western tropical Africa and sparingly introduced in other tropical regions including West Indies, for example botanical gardens.

**Other Common Names :** goma (Puerto Rico); caucho de Lagos (Cuba); ofruntum tree, silk–rubber (English); caoutchouc French.

**Botanical Synonym :** *Kickxia elastica* Preuss.

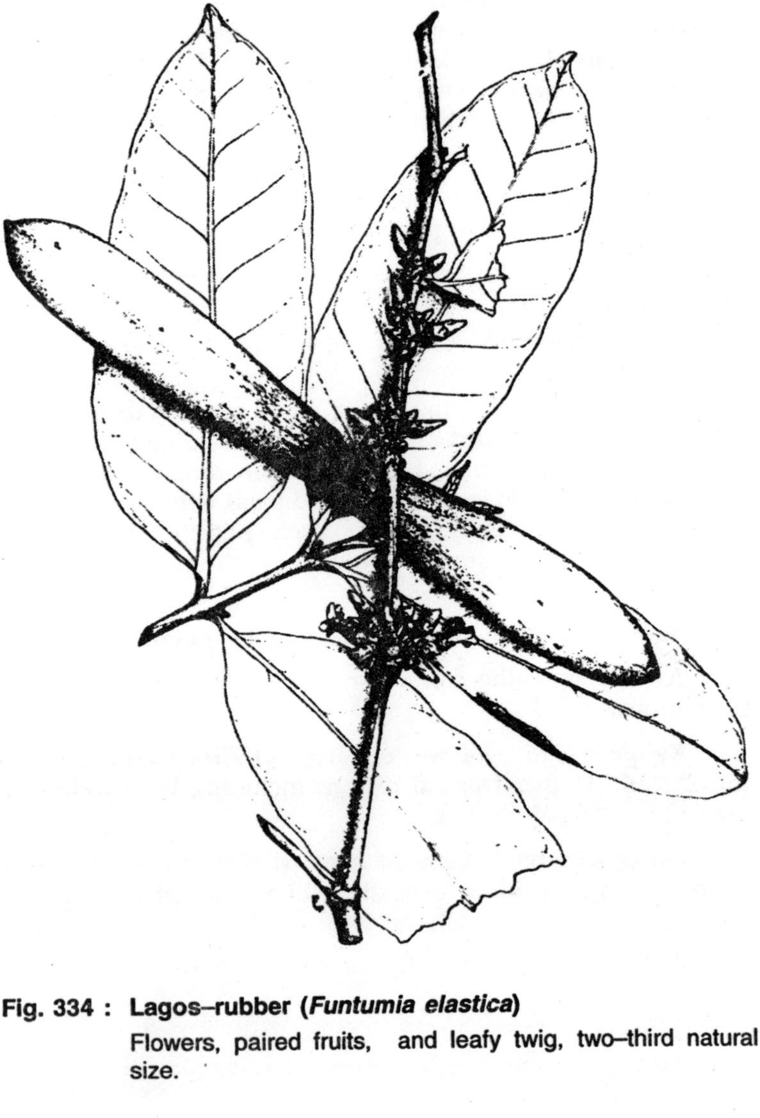

**Fig. 334 : Lagos–rubber (*Funtumia elastica*)**

Flowers, paired fruits, and leafy twig, two–third natural size.

## 335. Oleander (*Nerium oleander L.**)

Oleander is a familiar introduced ornamental shrub or sometimes small tree commonly planted for its clusters of large showy flowers. However, the plants including all parts and their whitish sap or latex are extremely poisonous to man and animals. Death has been caused by eating a few leaves or flowers. Distinguishing characters include : (1) narrow leathery shiny green leaves, long–pointed at both ends, mostly in 3's; (2) large tubular 5–lobed, flowers about 1 inch long and 2 inches broad, verying in color from red to pink to white, not scented; and (3) fruit of 2 long stut pods 3–6 inches long.

A handsome rounded spreading evergreen shrub commonly 6–15 feet high but sometimes a small tree to 20 feet high and 3 inches in trunk diameter. The twigs are greenish brown and minutely hairy. The sap or latex is whitish, turning gray, bitter, and poisonous.

The leaves are attached closely together along the twig in groups of 3 (sometimes 2–4) at a node, or whorled, on short petioles about $^1/_4$ inch long. Leaf blades are lance–shaped or narrowly elliptic, 3–5 inches long and $^1/_2$–1 inch wide, thick and stiff, with edges turned under slightly, almost hairless, with many parallel side veins, shiny green on lower surface.

Several to many short–stalked flowers are borne in terminal branched clusters (cymes) extending beyond the leaves. The calyx is about $^1/_4$ inch long, of 5 narrow–pointed green lobes, minutely hairy; corolla about 2 inch long, with a narrow tube and 5 broad rounded twisted spreading lobes about 2 inches across, with fringe of toothed scales at mouth of tube; 5 stamens almost $^1/_2$ inch long including the threadlike hairy tip, attached at mouth of corolla tube and united with stigma; and 2 separate ovaries with a slender style $^3/_8$ inch long. Frequently double–flowered, with many additional corolla lobes. The fruit of 2 stuout pods (follicles) from a flower contains many flattened hairy seeds with a tuft of hairs. Flowering and fruiting mostly in spring and summer but continuing through the year.

The wood is described as whitish yellow, slightly hard and brittle, moderately heavy (specific gravity 0.6).

Commonly planted for ornament, hedges, and borders and in gardens and parks in various parts of Puerto Rico from low to high altitudes and in Virgin Islands to Anegada. Apparently not escaping from cultivation here.

Oleander is hady, elsewhere withstanding both freezing temperatures of subtropical climates and also long droughts. Northward in temperate regions, it is a popular house plant grown in tubs. The shrubs are readily propagated by cuttings and can be heavily pruned and topped. Reported to have a deep root system, which does not affected underground pipes.

This species is extremely poisonous, all parts of the plant such as latex, flowers, leaves, bark, and roots being toxic. Persons have died from carelessly eating the flowers. Cattle have been killed by browsing the foliage. In spite of the deadly properties the plants have been employed in medicine as a heart stimulant and in various home remedies for assorted ailments. Insecticides and rat poisons have been made from the latex and other parts. The leaves contain varying amounts of rubbers latex. It is reported that smoke from the burning wood is toxic.

**Fig. 335 : Oleander (*Nerium oleander L.**)**
Fruits (left and lower right) and flowering twig, natural size.

## 336. Alelí cimarrón (*Plumeria obtusa* L.)

This species of alelí is readily distinguished from related species by its broader obovate or oblong leaves rounded or notched at apex. Other characters for identification include: (1) milky juice of white latex, which flows abundantly form cuts; (2) the from with few stout branches ending in clusters of leaves, without a crown of foliage; (3) the showy, very fragrant tubular, 5–lobed waxy, white flowers $1^3/_4$–2 inches long; and (4) paired cigarlike dark brown pods 3–6 inches long and about $^1/_2$ inch in diameter.

Small evergreen tree 15–40 feet high and to 10 inches in trunk diameter, hairless throughout. The bark is gray smoothish to slightly furrowed. Inner bark is light yellow beneath an outer green layer, bitter. The twigs are few and stout, green when young, turning gray, more than $^3/_8$ inch in diameter, with abundant, slightly bitter, white latex.

The alternate leaves have slender leafstalks $^1/_2$–2 inches long. Blades are mostly 2–5 inches long and 1–2 inches wide, sometimes larger, slightly thickened and leathery, under at edges, tapering to base, to base, slightly shiny, the upper surface green, the lower surface yellow green.

Flower clusters (cymes) are borne at the end of a long stout stalk $1^1/_2$–5 inches long that arises among the leaves. Crowded on short stalks are many developing narrow, twisted flower buds up to $1^3/_4$ inches long whitish and reddish tinged and a few opening flowers. These are composed of 5–lobed calyx $^1/_{16}$ inch long; waxy cotolla with narrow tube about 1 inch long and 5 narrow spreading lobes inside; 5 small stament inserted near base of tube and alternate with lones; and pistil about $^1/_8$ inch long, composed of 2 separate ovaries partly inferior and united with 1 stigma. From a flower develop 2 long pods (follicles), which split along 1 line release many flat winged seeds. Flowering and fruiting nearly through the year.

The wood is light brown and slightly hard.

This native species is a potential ornamontal like its relative frangipani, *Plumeria rubra* L.* The white latex is irritating to the skin of some persons.

Locally common in dry and moist at 200–3,000 feet altitude in mountains of western Dehradun.

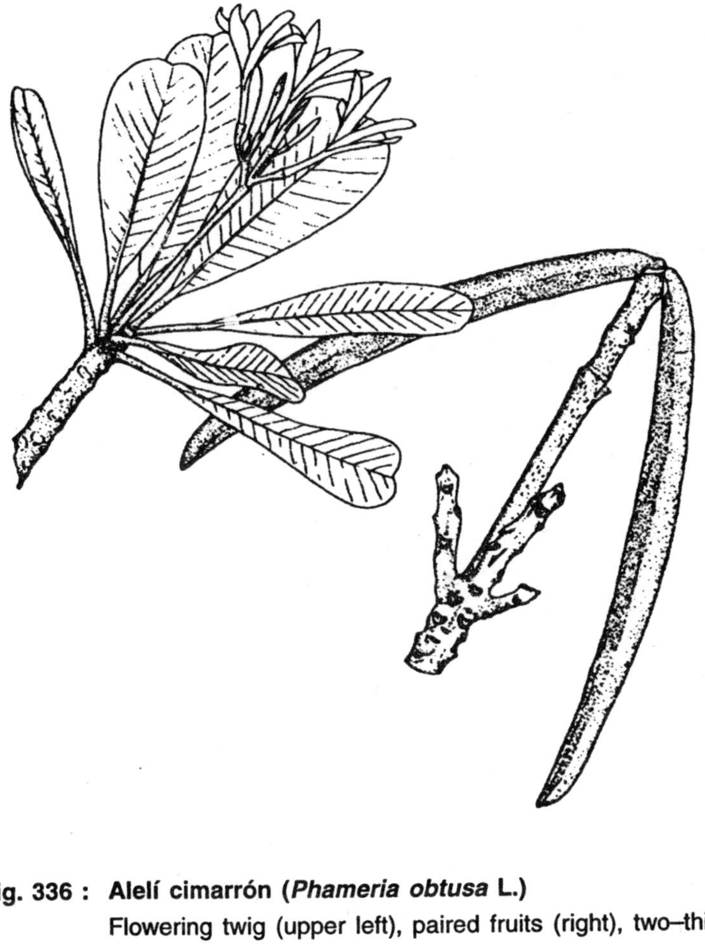

**Fig. 336 : Alelí cimarrón (*Phameria obtusa* L.)**
Flowering twig (upper left), paired fruits (right), two–thirds natural size.

## 337. Palo lechoso (*Tabernaemontana citrifolia* L.)

Tree or shrub with abundant white latex, characterized by: (1) paired elliptic leaves, long–pointed at base and abruptly long–pointed at apex slightly thickened and succulent, the lateral veins prominent, nearly at right angles to midrib, much sunken above and raised beneath; (2) showy fragrant tubular white flowers nearly $^3/_4$ inch long more than 1 inch across the 5 overlapping spreading lobes; and (3) green fruit of 2 elliptic pointed fleshy pods (follicles) $1^1/_2$–2 inches long and $^3/_8$–$^5/_8$ inch broad, which do not open.

Small evergreen tree 25 feet tall and 4 inches in trunk diameter, reported to reach 40 feet, or shrubby, hairless throughout. The light brown bark is smooth. Inner bark, also light brown, is bitter and contains much white latex which is slightly bitter. The green twigs ringed at nodes yield abundant white latex when cut.

Petioles of the opposite leaves are $^3/_8$–$^3/_4$ inch long, and the blades 3–7 inches long and $1^1/_2$–3 inches broad. Blades are broadest near or beyond middle, turned under at edges, above green or light and shiny or dull, and beneath pale whitish green.

The lateral branching flower clusters (cymes) near apex of twing are shorter than the leaves and few to several flowers on slender spreading green stalks. Pointed flower buds are about $^7/_8$ inch long. A flower has a green calyx $^3/_{16}$ inch long with 5 pointed overlapping lobes; white corolla with narrow cylindric tube $^9/_{16}$ inch long, and 5 broad overlap ping lobes $^5/_8$ inch long and rounded at apex; 5 stalkless green stamens $^1/_8$ inch long inserted inside corolla tube; and pistil composed of 2 light green ovaries $^3/_{16}$ inch long, 1 slender white style $^1/_4$ inch long, and enlarged stigma. The 2 separate fruits (follicles) developing from 1 flower are grooved and long–pointed and have the persistent calyx at base. Many seeds are imbedded in a flesh orange–red pulp. With flowers and fruits through the year.

The sapwood is whitish brown, medium hard, heavy, and fine–textured.

Because of showy white flowers, this species is suitable for planting as an ornamental. Elsewhere the caustic white sap or latex has served in home remedies.

Rare in moist forests at 200–2,500 feet altitude throughout the northern limestone hills and eastern and central mountains of Rajasthan.

**Fig. 337 : Palo lechoso (*Tabernaemontana citrifolia* L.)**
Flowering twig and fruiting twig (lower right), two–third natural size.

## 338. Cabalonga, lucky–nut (*Theveia peruviana*)

This introduced shrub or small tree with very poisonous milky juice is planted as an ornamental for its large yellow flowers. It is large yellow flowers. It is recognized easily by: (1) the funnel–shaped or bellshaped waxy flowers 2–2$^1/_2$ inches long and broad; (2) the very narrow, linear shiny leaves 3–6 inches long and only $^1/_4$ inches long and only $^1/_4$–$^3/_8$ inch wide, shiny green above and pale beneath; and (3) triangular–shaped, slightly flattened fruits, which are poisonous.

A spreading evergreen shrub or small tree 15 feet or sometimes to 25 feet high and 3 inches in trunk diameter, hairless throughout. Bark gray, slightly rough. The bark, stout twigs, and leaves abundantly exude white, poisonous latex when cut.

The very narrow leaves are borne single or alternate but numerous and crowded. They are stalkless or nearly so, long–pointed at both ends, thick and with edges slightly turned under, with prominent midrib but without lateral veins.

The handsome sweet–scented flowers are borne on slender stalks $^1/_2$–$_1{}^1/_2$ inches long, several together in short, branched terminal or lateral clusters (cymes). Calyx of 5 narrow, pointed sepals $^3/_8$ inch long; the yellow or sometimes pinkish corolla with a narrow tube at base and gradually expanding into 5 overlapping and spreading, very broad and rounded lobes, with 5 hairy scales at the end of the tube 5 small stamens on and within the corolla tube; ovary 2–lobed and 2–celled, with slender style and disk–shaped stigma.

The green yellow, or blackish fruit is triangular–shaped and slightly flattened about a $^3/_4$ inch long and 1$^1/_4$ inch broad and $^1/_2$ inch thick, with thin and large somewhat triangular stone almost as large as the fruit, containing only 2 large seeds. Flowering and fruiting throughout the year.

It is reported that the wood is brownish gray, hard, lightweight, fine textured, and easily worked.

The triangular poisonous fruits, which have a line around the edge and down the center, are carried or worn as pendants and good–luck charms, as the common name luck–nut indicates. A bright yellow oil of possible medicinal and illuminating uses can be pressed from the seeds.

**Fig. 338 : Cabalonga, lucky–nut (*Thevetia peruviana*)**
Flowering twig (above), fruiting twig (below), natural size.

Not only the juice and fruits but all parts of the plant are poisonous to humans and livestock, containing thevetin, a chemical like digitalis. The poison causes a slow and later an irregular pulse accompanied by vomiting and shock. The lucky-nut fruits have been shipped to the United States for preparation of certain heart medicines. The poisonous sap and bark have served also in folk remedies.

Common as a planted ornamental in India where it grows rapidly, has escaped from cultivation, and has become naturalized in coastal thickets. Not recommended for planting because of the dangerous poison.

## MILKWEED FAMILY (ASCLEPIADACEAE)

Herbs and shrubs, often vines, rarely trees (*Calotropis*), sometimes succulent, known by : (1) milky sap; (2) leaves opposite or whorled, simple, generally entire, with stipules often minute or none; (3) flowers in clusters (cymes, like racemes, or like umbels), bisexual, regular, consisting of calyx of 5 sepals often united at base, star–shaped corolla of 5 spreading or turned back lobes, 5–lobed crown (corona) between corolla and stamens and attached to either, 5 stamens attached to corolla near base and opposite lobes each bearing 2 flat yellowish masses of pollen (pollinia), and pistil within crown composed of 2 nearly separate nearly superior 1–celled ovaries with many ovules, 2 styles, and broad flat dislike stigma; and (4) fruit of 2 elliptic pods (follicles) with many seeds with tufts of silky hairs. One introduced tree species.

## 339. Giant milkweed (*Calotropis procera*)

This curious introduced giant milkweed is perhaps more like a large herb than a tree but rarely reaches treelike size. The only representative of the milkweed family included here, it is easily recognized by: (1) abundant white latex, slightly bitter, in all parts; (2) large paired leaves broadly–elliptic to nearly round, heart–shaped at base and nearly stalkless, slightly thick and succulent, pale yellow green above and whitish green with fine coat of soft hairs beneath; (3) many flowers more than $^3/_4$ inch across the 5 lobed slightly succulent corolla, whitish and tinged with purple; and (4) fruits 1 or 2 large swollen pods containing many seeds with tufts of long white silky hairs.

A naturalized evergreen shrub 3–6 feet high, rarely becoming treelike and 12–18 feet high, with a trunk 3–10 inches in diameter, with few stout spreading branches and crown. Bark whitish, thick and corky, the inner bark whitish, slightly bitter, with abundant latex. Twigs stout, $^3/_8$–$^1/_2$ inch in diameter, pale yellow green, minutely white hairy at tip.

The opposite leaves have stout light yellow petioles $^1/_8$–$^3/_8$ inch long. Blades are 3–7 inches long and 2–5 inches wide, abruptly short–pointed or blunt at apex and almost clasping at the heart–shaped base, without teeth on edges, slightly leathery. The upper surface is pale yellow green with broad light yellow midvein and hairless, and the lower surface is whitish green, with midvein and main veins broad and light yellow, with fine coat of soft hairs like felt that rub off.

Flower clusters (umbellike cymes) near or at ends of twigs have stalks about $2^1/_2$ inches long and bear many flowers on stalks 1 inch long in compound clusters about 3 inches across. The flower about $^5/_8$ inch high has calyx of 5 sepals $^1/_4$ inch long, whitish green with purplish point; star–shaped corolla with short cup $^1/_4$ inch long and 5 pointed spreading lobes $^3/_8$ inch long, whitish with purple tip and with central 5 lobed purplish crown (corona); 5 stamens attached to corolla near hase and opposite lobes, each bearing in anthers under stigma 2 flat yellowish masses of pollen (pollina); and pistil within crown, composed of 2 yellowish ovaries with many ovules, 2 slender styles, and broad flat 5–angled stigma.

The fruits are 1 or 2 large swollen elliptic pods (follicles) 3–$4^1/_2$ inches long and 2–$2^1/_2$ inches broad, slightly, fleshy, opening

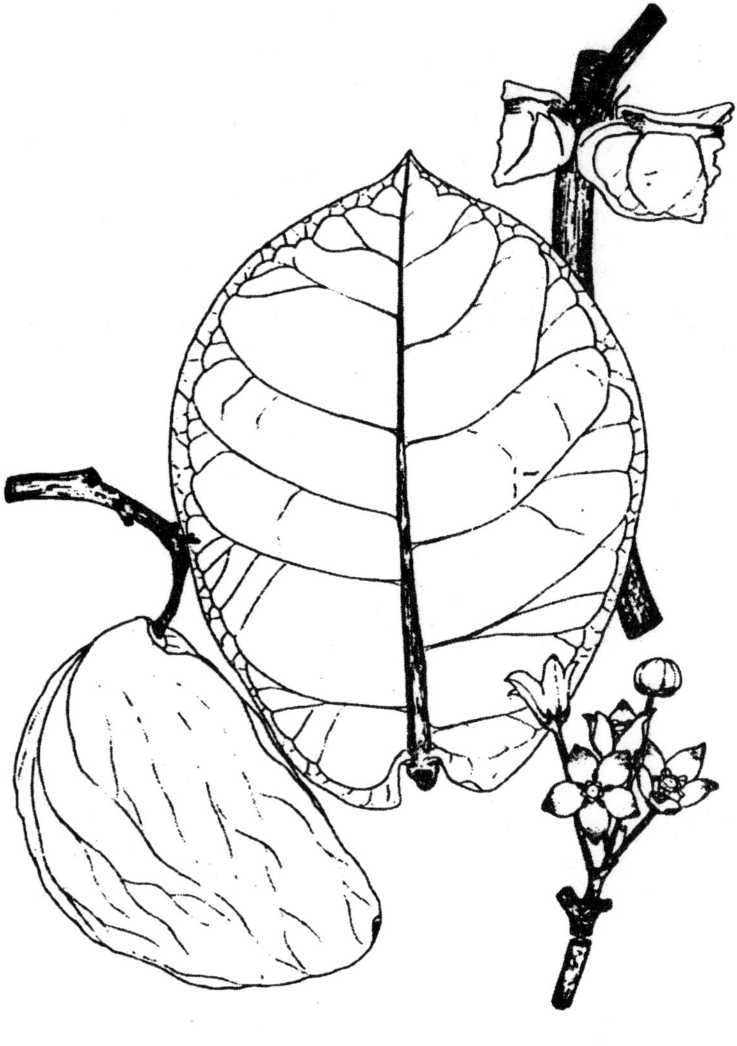

**Fig. 339 : Giant milkweed (*Calotropis procera*)**

Fruit (lower left), leaf, stem with leaf bases (upper right), and flowers (lower right), two–thirds natural size.

on 1 side, yellow green but becoming brown and dry. Seeds numerous flat, brown, $^5/_{16}$ inch long, with large tuft of white hairs $1^1/_4$ inches long and $2^1/_2$ inches in diameter. With flowers and fruits throughout the year.

The wood is whitish and soft.

The plants are often cultivated for ornament in dry areas of M.P and U.P. and elsewhere. Being unpalatable to sheep and cattle, they tend to spread and become common on heavily grazed pastors. In some places the silky hairs on the seeds serve for stuffing pillows. Also, the plants are used in home medicines. The foliage and fruit are covered, particularly in dry periods, by a large aphid (*Aphis nerii)*, yellow or orange–yellow with 2 black appendages.

Common and scattered in dry forest, waste place, heavily grazed pastures, and beaches to 600 feet altitude.

# BORAGE FAMILY (BORAGINACEAE)

Herbs, shrubs, and trees, know by : (1) leaves usually alternate simple, mostly entire, sometimes rough with stiff hairs, without stipules; (2) flowers in cymes often asymmetric in a spiral, sometimes large and showy, mostly bisexual, regular, with calyx of 5 sepals separate or united at base, corolla tubular often in form of funnel or bell, 5 stamens inserted in tube and alternate, and pistil with superior ovary 2–celled and becoming 4–celled with 4 ovules, style, and stigma sometimes with 2 or 4 lobes; and (3) fruit a drupe of 4–1 seeds or 4 nutiets.

**Key to Species**

A. Leaves small, less than $1^1/_2$ inches long.

B. Leaves less than $^1/_2$ inch long; slender straight spines $^1/_4$ inch or more in length at nodes; corolla light yellow–. *Rochefortia acanthophora.*

BB. Leaves mostly $^3/_4$–$1^1/_2$ inches long; spines none; corolla white–. *Bourreria virgata.*

AA. Leaves large.

C. Flowers large with funnel–shaped orange or corolla more than 1 inch long; fruit egg–shaped, covered by enlarged calyx.

D. Leaves ovate, slightly rough hairy above; flowers nearly $1^1/_2$ inches long and broad–. *Cordia sebestena.**

DD. Leaves elliptic or oblong, usually very rough above; flowers about $1^1/_4$ inches broad and 1 inch broad–. *Cordia rickseckeri.*

CC. Flowers smaller, less than 1 inch long, wide white or yellowish corolla.

E. Flowers tubular, narrow.

F. Flowers about $^1/_2$ inch long and broad with narrow tubular calyx, the white corolla persistent on nutlet and turning brown, crushed leaves with odoe of garlic–. Capá prieto, capa, *Cordia alliodora* (Ruiz & Pay). Oken.

FF. Flowers about $^3/_{16}$ inch long; fruits round white juicy, $^1/_4$ inch in diameter. *Tournefortia filiflora.*

EE. Flowers spreading or bell–shaped, with broad 5–lobed corolla; fruit a rounded drupe.

G. Leaves broadly elliptic to nearly round.

H. Leaves coarsely toothed, upper surface rough hairy, petioles less than 1 inch long; corolla whitish or pale yellow; fruits elliptic whitish–. *Cordia alba.*

HH. Leaves often with a few wavy teeth; petioles more than 1 inch long; corolla light yellow; fruits light pink, with sticky flesh–. *Cordia obliqua.*

GG. Leaves elliptic or ovate; corolla white.

I. Leaves large, mostly 9–12 inches long, rough hairy on upper surface and soft hairy beneath; flowers $^3/_{16}$ inch broad; fruit whitish, $^3/_8$ inch in diameter–. Moral, white manjack, *Cordia sulcata* DC.

II. Leaves smaller, mostly less than 6 inches long, hairless or nearly so (or slightly hairy beneath); fruit orange red or red.

J. Leaves thick and leathery, with prominent network of many veins; flowers about $^1/_4$ inch broad; fruit about $^3/_8$ inch in diameter, broader than long, oblique–. Muneco, *Cordia boringuensis* Urban.

JJ. Leaves thin or slightly thickened, with veins not conspicuous.

K. Flowers stalkless on forking branches, about $^1/_4$ inch wide, with hairy calyx–. *Cordia collococca.*

KK. Flowers stalked, about $^1/_2$ inch wide, with calyx hairless.

L. Flowers saucer–shaped, with spreading corolla, many in clusters 2–4 inches across and high, the branches forking by 2's; fruit bright red–. Capá colorado, red manjack, *Cordia nitida* Vahl.

LL. Flowers bell–shaped, with bell–shaped calyx, many in branching clusters 2–8 inches across and high; fruit orange re–. Palo de vaca, pigeon–berry, *Bourreria succulenta* Jacq.

## 340. Roble de guayo (*Bourreia virgata*)

The shrub or small tree of dry areas is distinguished by: (1) small elliptic or obovate leaves mostly $^3/_4$–$1^1/_2$ inches long and $^1/_4$–$^1/_2$ inch wide, sometimes larger, rough or smooth, slightly thick and stiff and with edges rolled under; (2) tubular flowers $^1/_2$ inch long and broad across the 5 widely spreading white corolla lobes, few in terminal clusters: and (3) fleshy orange or red rounded fruit about $^3/_8$ inch in diameter, wider than long.

An evergreen shrub or small tree to 30 feet high and 6 inches in trunk diameter, often with several stems from base, reported to become large. The bark is gray, smoothish to slightly fissured, the inner bark whitish and almost tasteless. The twigs are slender, often short and much branched, green when young and becoming gray. Twigs leaves, and branches of flower clusters vary from hairless to densely hairy.

The leaves are alternate but often crowded, the short petioles $^1/_{16}$–$^3/_{16}$ inch long. Blades are rounded or slightly notched at apex, gradually narrowed to the short–pointed base, the upper surface curved up slightly at sunken midrib, shiny green, rough with short stiff hairs or smooth, the lower with short stiff green with raised network of small veins and commonly hairy.

Flowers few in terminal clusters (cymes), almost stalkless, slightly fragrant. The bellshaped calyx is $^3/_{16}$ inch long, finely hairy, 5 lobed; the tubular white corolla with 5 widely spreading lobes; stamens 5 on corolla tube; and pistil. The fruits (drupes) with calyx at base contain 4 or fewer light brown nutlets. Flowering and fruiting through the year.

The wood is light brown and hard.

Locally common in moist and dry limestone forests 100–1,000 feet altitude in western foothills of Dehradun.

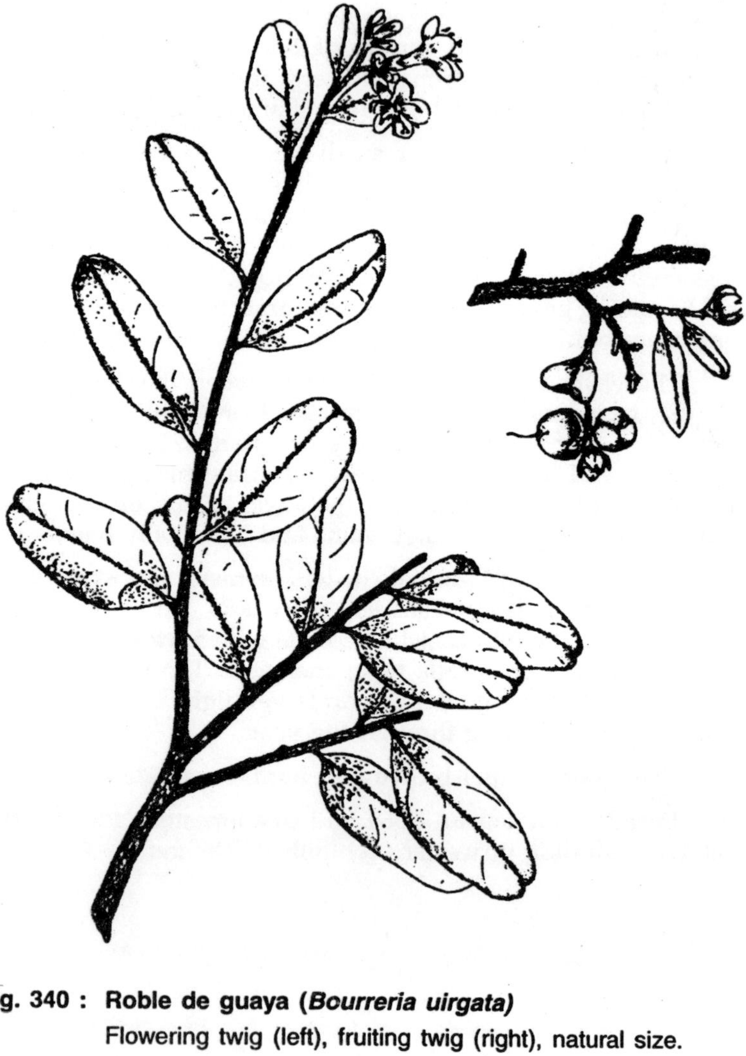

**Fig. 340 : Roble de guaya (*Bourreria uirgata*)**
Flowering twig (left), fruiting twig (right), natural size.

## 341. White manjack (*Cordia alba*)

This shrub or small tree of thickets and dry forests is distinguished by: (1) broadly elliptic leaves coarsely and irregularly toothed, with upper surface rough hairy; (2) many showy fragrant whitish or pale yellow flowers about $^5/_8$ inch across the spreading funnel–shaped 5–toothed corolla; and (3) elliptic whitish fleshy fruits $^3/_8$–$^5/_8$ inch long.

A deciduous shrub or small tree to 25 feet high and 8 inches in trunk diameter, with broad rounded crown, sometimes vinelike. Bark light gray, very thick, deeply furrowed into narrow ridges, the inner bark light brown and bitter. Twigs slender, brownish, hairy.

The alternate leaves have hairy petioles $^3/_8$–1 inch long. Blades are 2–5 inches long and $1^1/_4$–$2^1/_2$ inches wide, short–pointed at apex and blunt or rounded at base, slightly thickened, the upper surface green and rough, and the lower surface dull light green with hairy veins.

Flower clusters (panicles) terminal, erect, much branched, flattened above, often broader than bond and up to 10 inches wide. Flowers almost stalkless, consisting of tubular hairy calyx less than $^1/_4$ inch long, 10–ribbed, 3–5 toothed, forming cap in bud; whitish funnel–shaped corolla $^1/_2$ inch long; 5 stamens inserted near base of corolla; and pistil with ovary and 2 styles each 2–forked. The fruit (drupe) with calyx persistent at base contains a stone and whitish pulp, reported to be edible. Flowering and fruiting throughout the year.

The light brown wood is soft of medium weight, and not durable. Elsewhere it has served for interiors and construction.

Grown around houses for shade and in living fences. Elsewhere the flowers and leaves have served in home remedies. The mucilaginous pulp of the fruits has served as glue. The fruits are a good source of wildlife food, especially for doves and other birds. A honey plant.

Locally abundant in dry forest from sea level to 500 feet altitude.

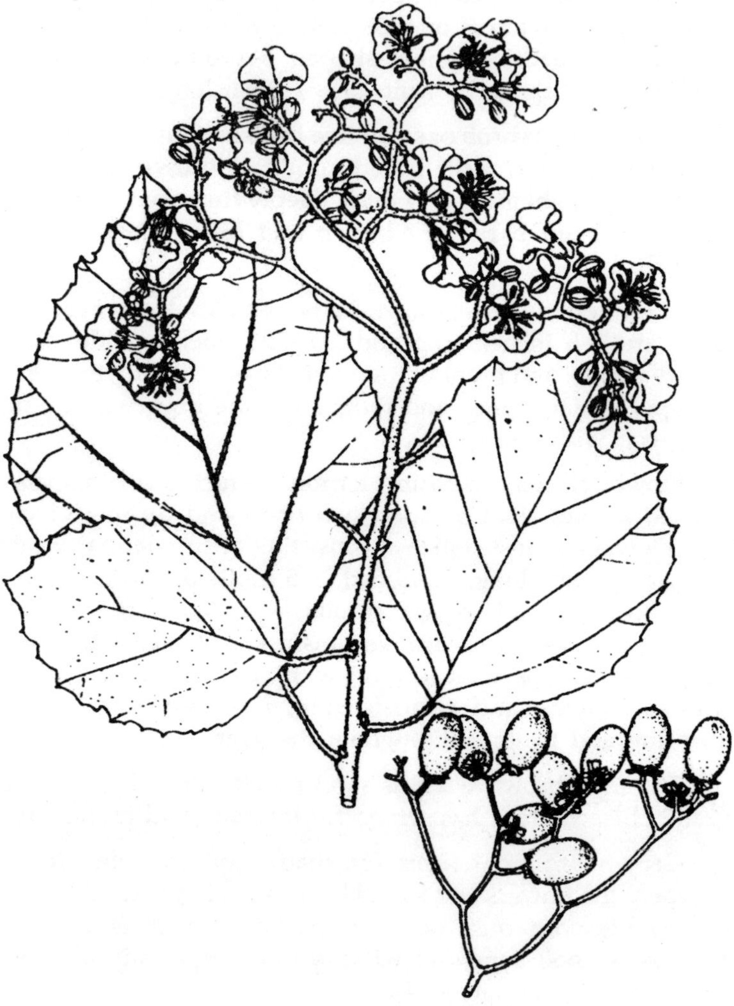

**Fig. 341 : White Majack (*Cordia alba*)**
Flowering twig and fruits (lower right), natural size.

## 342. Red Manjack (*Cordia collococca*)

This common tree at jow altitudes is recognized by: (1) shiny elliptic to obovate leaves with edges often wavy, often shed at time of flowering; (2) many small stalkless flowers with 5–lobed cup–shaped white corolla $^1/_4$ inch long and wide; and (3) round orange–red fleshy fruits $^3/_8$ inch in diameter.

Small to medium–sized widely spreading deciduous tree to 60 feet height and $1^1/_2$ feet in trunk diameter, often with horizontal and widely forking branches. Small plants have whorls or rings of horizontal branches. Bark gray, smoothish to finely fissured and rough, the inner bark light brown and almost tasteless. The twigs are green and finely hairy when young, becoming light gray and hairless.

The alternate leaves have leafstalks $^1/_4$–2 inches long, without stipules. Blades are mostly $2^1/_2$–6 inches long and $1^1/_4$–$2^3/_4$ inches wide slightly thickened and leathery, short–pointed or blunt at apex, pointed or rounded at base, the upper surface very shiny green and nearly hairless, the lower surface light green and slightly hairy with slender bristles along midvein.

The flower clusters (cymose panicles) erect at ends of twigs contain many fragrant stalkless flowers on slender widely forking branches, male and female on different plants (dioecious). The flower consists of cup–shaped hairy green calyx $^1/_8$ inch long, irregularly 3–5 toothed from round bud; cup–shaped corolla with short tube and 5 lobes spreading and bent down; 5 stamens inserted on tube and alternate with lobes, small and nonfunctional in female flowers; and pistil with 4–celled ovary and 2 stules each 2–forked. The muculaginous sweetish fruits (drupes) are edible but astringent and contain a large1–seeded stone. Flowering and fruiting in spring and summer.

The wood is whitish brown, soft, and light–weight.

In other areas the fruits are eaten by hogs and chickens and the roots are used in home remedies.

Locally common and widespread in dry coastal forest from sea level to 400 feet altitude in eastern and southern coasts Tamil–Nadu.

**Fig. 342 : Red Manjack (*Cordia collococca* L.)**

Leafy twig with old fruit stalks (left), fruits and flowering twig (lower right), natural size.

## 343. Cereza blanca, manjack (*Cordia obliqua*)

This introduced ornamental tree is identified by: (1) broadly ovate or elliptic to nearly round leaves 2–$3^1/_2$ inches long and $1^1/_2$–4 inches wide, with slender leafstalks 1–$1^3/_4$ inches long; (2) many flowers $^1/_4$ inch long and broad with light yellow 5–lobed corolla, on slender widely forking branches; and (3) many light pink round fruits $^1/_2$–$^5/_8$ inch in diameter, with slightly sweet, sticky flesh, planted evergreen spreading small tree 25–30 feet high and 8 inches in trunk diameter, perhaps becoming larger. The bark is gray smoothish to finely fissured or furrowed. The inner bark is light yellow, turning to brown on exposure, almost tasteless. Twigs are green, turning to brown, hairless, with raised light brown dots (lenticels) and raised half–round leaf scars.

The blades of the alternate leaves are slightly thickened and leathery, blunt or rounded at apex, rounded or notched are base, often with a few wavy teeth at border, the upper surface green to dark green, and dull or slightly shiny, and hairless, the lower surface light green with tufts of hairs in vein angles.

The flower clusters (corymbose cymes) at ends of twigs are 2–5 inches broad, forking regularly by 2's. Flowers are composed of cylindric green calyx $^3/_{16}$ inch long, minutely hairy, 5–toothed; light yellow corolla with short tube and 5 spreading lobes; 5 spreading stamens inserted in tube and alternate with lobes; and pistil with elliptic 4–celled ovary and 4–forked style.

Fruits (drupes) are borne in quantities, causing the weighted branches to droop, turning from whitish green to light pink at maturity. The fruits have enlarged light green cup–like calyx at base, point from style at apex pinkish, very mucilaginous, juicy and slightly sweet flesh, flattened light brown stone $^3/_8$ inch in diameter, and 1–2 seeds. Flowering and fruiting throughout the year.

Grown as an ornamental for the foliage, flowers, and fruits and as a fence row tree. Elsewhere, the sticky fruit pulp has served as mucilage.

Planted and locally common at law altitudes at the northeast end of Agartala.

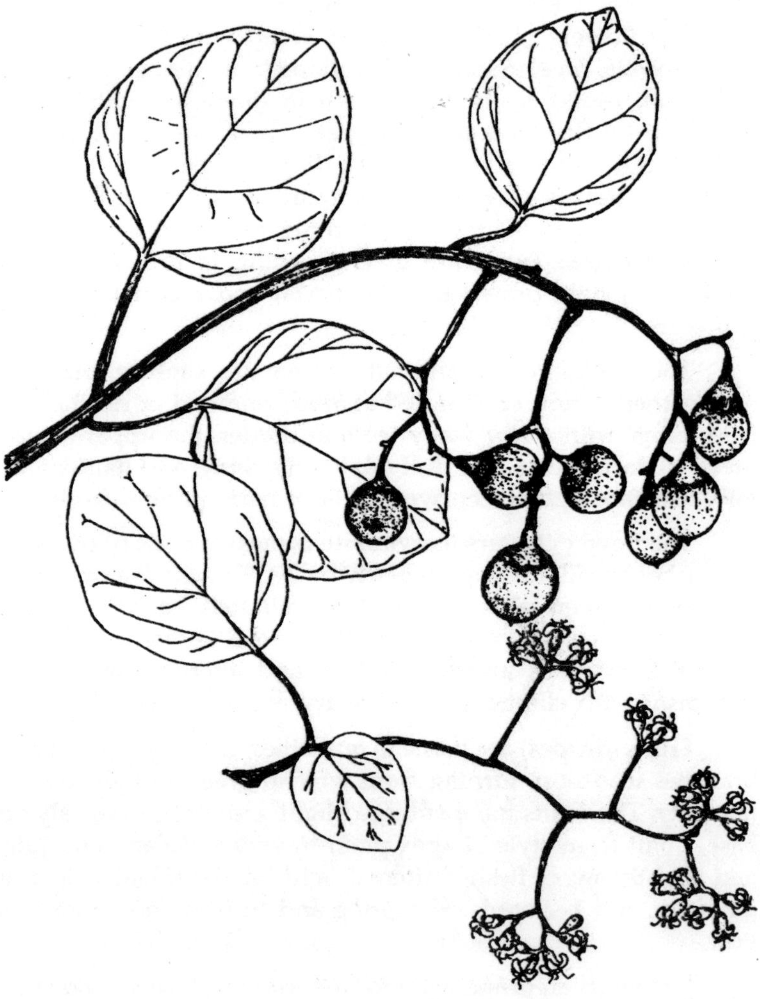

**Fig. 343 : Cereza blance, manjack (*Cordia obliqua*)**
Fruiting twig (above), flowering twig (below), natural size.

## 344. San Bartolomé, Manjack (*Cordia rickseckeri*)

This uncommon tree with bright orange Kerala flowers, native only in the dry forests of and the Virgin Islands, is characterized by: (1) elliptic or oblong leaves $3^1/_2$–9 inches long and 2–4 inches wide, usually very rough above like sandpaper with minute sharp flattened hairs; (2) showy, clustered orange flowers about $1^1/_4$ inches long and $^7/_8$–1 inch across the 5 spreading lobes of the funnel–shaped corolla; and (3) the egg–shaped fruits 1 inch or more in length and $^3/_4$ inch in diameter, covered by the enlarged calyx, brownish green, becoming yellow or orange. From the closely related introduced ornamental No. 675, vomitel colorado or Geiger–tree, *Cordia sebestena* L.,* this native tree is distinguished by the narrower leaves usually very rough above, with shorter petioles, the smaller flowers with shorter, often hairless calyx, and by the fruits not whitish.

A small deciduous (?) tree to 35 feet high and 14 inches in trunk diameter or sometimes a bushy shrub. The bark is gray, thick smoothish, slightly fissured, or with deep long furrows. The dead outer bark is gray, and the inner bark yellowish green or light brown and bitter. The twigs are stout and light gray, with stiff hairs when young.

The alternate leaves have stout petioles $^1/_4$–$1^1/_2$ inches long, with stiff hairs. Leaf blades are short–pointed or rounded at both ends, usually thickened and stiff with edges turned under, the upper surface dark green and usually very rough though sometimes almost smooth, with lateral veins often reddish thinged and sunken, and the lower surface light green, smoothish and nearly with lateral veins often raised.

Many or several are borne in short spreading branched terminal clusters (cymes) about $2^1/_2$–4 inches broad at the end of leafless twigs. The flower has an oblong tubular orange calyx nearly $^1/_2$ inch long, hairless, slightly fleshy, ending in 3–5 short teeth; corolla funnel–shaped, $1^1/_4$ inches long, orange or orange or orange red, finely hairy outside, with many long ridges, ending in 5 (sometimes 6) spreading rounded wavy–margined lobes; 5 staments $^3/_8$ inch long inserted in throat of corolla tube; and pistil about 1 inch long, composed of short conic ovary of 4 cells and 4 ovules and threadlike style 4 forked at end.

The fruit (drupe) is covered by the enlarged calyx, often has the threadlike remains of style at apex, has thin pulp and large, very thick–walled stone enclosing 1–4 whitish elliptic seeds nearly $^1/_2$ inch long. Flowering and fruiting through the year.

The sapwood is whitish or light brown and hard.

Locally common in dry and moist forests from sea level to 700 feet altitude on eastern and southern coasts of Kerala.

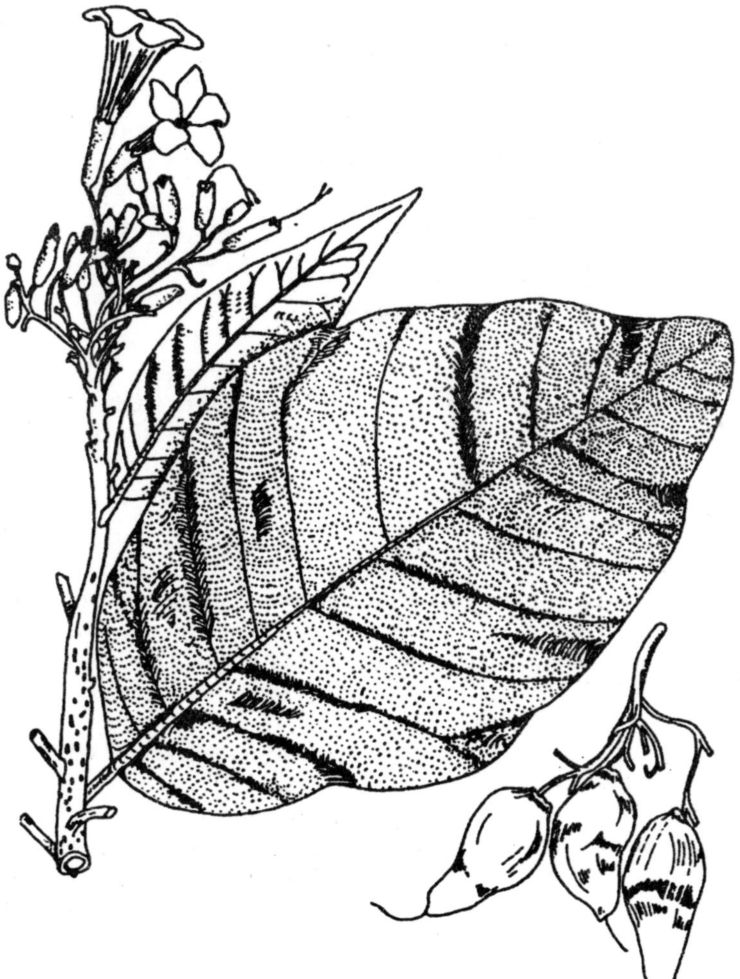

**Fig. 344 : Manjack (*Cordia rickseckeri*)**
Flowering twig (left), fruits (lower right), natural size.

## 345. Geiger–tree (*Corida sebestena* L.*)

This uncommon ornamental is easily recognized by the showy clusters of many funnel–shaped orange or red flowers nearly $1^1/_2$ inches long and broad, with usually 5 or 6 spreading rounded lobes. Other characters are: (1) the ovate leaves $3^1/_2$–8 inches long and $2^1/_2$–$5^1/_2$ inches wide, slightly rough hairy above; and (2) the egg–shaped hard fruits 1–$1^1/_2$ inches long and $^3/_4$ inch broad, covered by the enlarged whitish calyx.

A small deciduous or evergreen tree to 20 feet high and 6 inches in trunk diameter, with compact rounded crown. The bark is thick, dark brown and scaly, deeply furrowed into narrow scaly ridges. The stout twigs are green and covered with rusty hairs, becoming gray.

The stout petioles of the alternate leaves are 1–2 inches long. Blades are short–pointed at apex, rounded or slightly heart–shaped at base, sometimes with a few teeth, slightly thick and stiff, the upper surface dark green and slightly rough with stiff hairs and the lower surface paler and usually hairy, especially along the raised lateral veins.

The flower clusters (cymes) are terminal, branched, and flattened about 6 inches across the many crowded flowers. Each flower is composed of the narrow oblong tubular calyx more than $^5/_8$ inch long, densely hairy, ending in 3–5 short teeth; the funnel–shaped orange or red corolla finely hairy outside, with many long ridges, ending in usually 5 or 6 spreading rounded wavy–margined lobes 1–$1^1/_2$ inches across; stamens 5 or 6, $^1/_4$ inch or more in length inserted in throat of corolla tube and alternate with the lobes; and pistil about $1^1/_4$ inches long, composed of short conic ovary of 4 cells and 4 ovules and threadlike style 4–forked at end.

The fruit (drupe) is covered by the white enlarged calyx, often has the threadlike remains of style at apex, has thin pulp and large grooved, very thick–walled stone containing 1–4 white seeds $^1/_2$ inch long. Flowering and fruiting throughout the year.

The thick sapwood is light brown and the heartwood dark brown and the heartwood dark brown. The wood is hard heavy (specific gravity 0.7), fine–textured, and has been used elsewhere in carpentry.

**Fig. 345 : Geiger–tree (*Cordia sebestena* L.*)**
Flower (upper left), fruits (lower left), and leaf (right), natural size.

Common as a planted ornamental about homes and along roadsides, especially in the drier parts of Bihar. Propagated form seeds and suttings and of slow growth. Escaping from cultivation and naturalized on the south coast but not native here.

This is a popular ornamental in the Indian tropics, hardy in dry areas and poor soils and salt–tolerant. Recommended for south India. It is reported that the fruits are both medicinal and edible.

## 346. Juso (*Rochefortia acanthophora*)

This much–branched spiny shrub or small tree of dry areas is recognized by; (1) the zigzag gray twigs with a slender straight spine $^1/_4$ inch or more in length at each node; (2) the very small obovate leaves $^3/_8$–$^1/_2$ inch long and $^1/_8$–$^3/_{16}$ inch wide, clustered on very short lateral twigs; (3) few greenish flowers more than $^1/_{16}$ inch clustered at leaf bases; and (4) round, yellowish to red fruit less than $^1/_4$ inch in diameter.

A deciduous shrub 6–10 feet high or rarely small tree becoming 15 feet high and with several trunks to 3 inches in diameter, with thin crown of many branches and very small leaves. The bark is gray and smoothish. Twigs are gray, hairless, becoming fissured, with a slender straight sharp spine at each node and with a very short twig less than $^1/_8$ inch long above a spine.

The leaves are alternate but often several together on the very short twigs, almost stalkless. Blades are sometimes to $^3/_4$ inch long and $^1/_2$ inch wide, rounded at apex, gradually narrowed at base into slender petiole less than $^1/_{16}$ inch long, turned under at edges, slightly thick, stiff, and leathery, minutery, minutely hairy, the upper surface green and shiny, the lower surface dull light green.

A few stalkless flowers are clustered are leaf bases on short twigs. The flower is composed of calyx of 5 round green hairy sepals $^1/_{16}$ inch long; corolla of 5 light yellow rounded concave petals less than $^1/_{16}$ inch long, slightly united at base; 5 stamens attached near base of corolla and alternate with the lobes; and yellowish pistil with conical 2–celled ovary and 2 hairy styles. The fruit is a round yellowish drupe less than $^1/_4$ inch in diameter, with calyx at base fleshy and containing 4 hard nutlets. Flowering and fruiting in spring and summer.

The wood is hard, the sapwood light brown and the heartwood dark brown. Used elsewhere in cabinetmaking

Locally common in dry limestone, and lower forests from sea level to 600 feet altitude in southwestern Kerala.

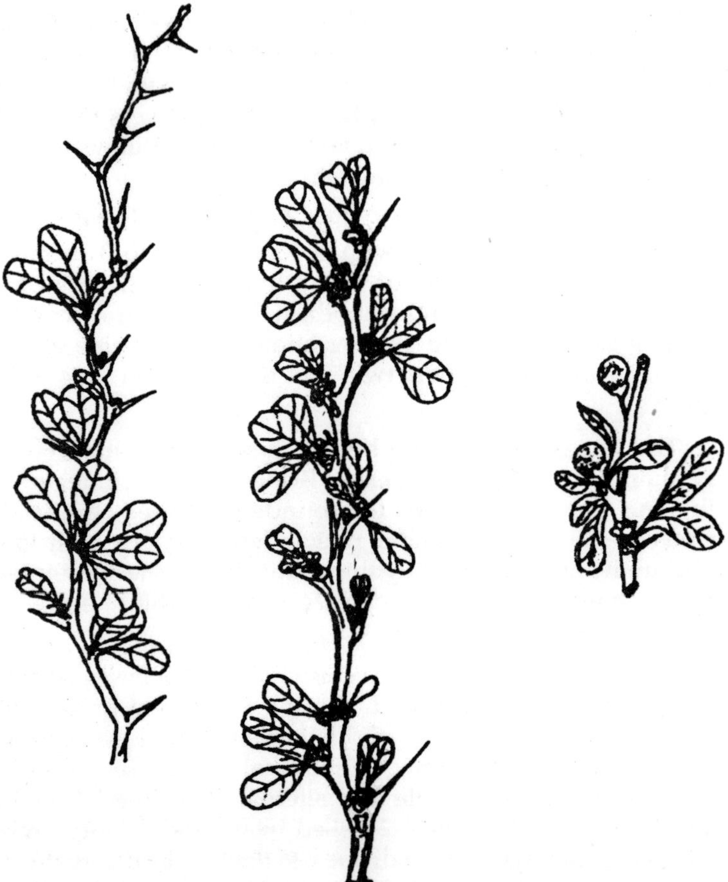

**Fig. 346 : Juso (*Rochefortia acanthophora*)**

Leafy twig (left), flowering twig (center), fruiting twig (right), natural size.

## 347. Nigua (*Tournefortia filiflora*)

This rare shrub, sometimes a small tree, is identified by: (1) twigs angled, green and stout; (2) large elliptic thin hairless leaves; (3) many small white 5–parted flowers about $^3/_{16}$ inch long, stalkless on top side of spreading nearly horizontal forks of long–stalked clusters; and (4) round white juicy fruits $^1/_4$ inch in diameter.

Evergreen shrub or small spreading tree to 20 feet high and 4 inches in trunk diameter. Break dark gray, finely fissured. Inner bark light brown, bitter, the twigs are finely hairy when young.

The alternate leaves have stout green leafstalks to 2 inches long. Blades are 4–17 inches long and $1^1/_2$–7 inches wide, short–pointed or long–pointed at both ends, not toothed on edges, the upper surface green and slightly shiny with side veins slightly sunken, the lower surface dull light green.

The large flower clusters (cymes) have several horizontal and curved slender branches 1–$2^1/_2$ inches long at the end of long stalk as much as 6 inches long, the branches finely hairy. The flowers have green 5–toothed hairy calyx about $^1/_{16}$ inch long; white hairy corolla $^3/_{16}$ inch long with very narrow tube and 5 lobes; 5 stamens inserted within tube and alternate with lobes; and pistil with 4–celled ovary and 2–lobed style. The fruits (drupes) have calyx at base and contain 2 brown nutlets. Collected with flowers in December and April, with fruits from March to August.

The wood is whitish, soft, and brittle.

Rare in moist forest at 100–800 feet altitude in northwestern of Bihar.

**Fig. 347 : Nigua (*Tournefortia filiflora*)**

Flowers (upper right) and fruiting twig, two–thirds natural size.

# VERBENA FAMILY (VERBENACEAE)

Herbs, shrubs, woody vines, and trees often large, known by: (1) twigs often 4–angled; (2) leaves opposite (sometimes whorled), generally simple (palmately compound or with 3 or fewer leaflets in *Vitex*), often deciduous without stipules; (3) flowers mostly small, often colored, commonly in cymes, racemes, or spikes, bisexual, usually irregular, with calyx mostly 5–lobed or 5–toothed, corolla with short tube and 5 short unequal spreading lobes or in 2 lips, stamens usually 4 in pairs and sometimes 1 staminode inserted in tube, and pistil with superior ovary of 2 or 4 cells each containing 1 ovule, long style, and 1–2 stigmas; and (4) fruit generally a drupe with 2–4 nutlets or only 2–4 nutlets (capsule in black–mangrove, *Avicennia*

**Key to Species**

A. Leaves compound.

B. Leaves digitate (plamately compound) with 5–7 narrowly lanceolate leaflets–*Vitex agnus–castus.**

BB. Leaves with 3 elliptic leaflets often only 1 or sometimes 2, the end one largest–. Higuerillo, white fiddlewood, *Vitex divaricata* Sw.

AA. Leaves simple.

C. Twigs often spiny, long, slender, and often vinelike.

D. Spines slender, straight, 1–2 at a node; flowers light blue; fruits yellowish.–*Duranta repens..*

DD. Spines stout, curved, 2–3 at a node; fruits blackish.–*Clerodendrum aculeatum.*

CC. Twigs not spiny, erect, not vinelike.

E. Leaves very large, elliptic, 12–15 inches or more in length, thick, leathery, and harsh,–Teca, teak, *Tectona grandis* L. f.*

EE. Leaves less than 8 inches long.

F. Leaves hairless or nearly so (velvety hairy beneath in a variety).

G. Leaves elliptic, long–pointed; flower branched (cymes in panicles). *Aegiphila martinicensis.*

GG. Leaves elliptic to oblong, short–to long–pointed; flower clusters not branched (racemes).–*Citharexylum.*

H. Leaves with few side veins, mostly blunt at apex, dull green, not forming prominent network when dry; flower stalks about $^{1}/_{16}$ inch long.–226. Pendula de sierra, *Citharexylum coudatum* L.

HH. Leaves with prominent network, of small veins when dry.

I. Leaves shiny yellow green, turning red before falling, sometimes hairy on veins beneath, petiole pink or orange; flowers almost stalkless.–Pendula, pasture fiddlewood, Florida fiddlewood *Citharexylum fruticousm* L.

II. Leaves green, with tufts of hairs in vein angles beneath; flowers with stalks $^1/_{16}$–$^1/_8$ inch long.–*Citharexylum spinosum.*

FF. Leaves hairy, white, gray, or yellow green on lower surface (often hairless).

J. Leaves lance–shaped or narrowly elliptic, 2–4$^1/_2$ inches long; tree of mangrove swamp forest. Mangle prieto, black–mangrove, *Avivennia germinans* (L). L. (*A. nitida).*

JJ. Leaves mostly elliptic, larger; upland trees.

K. Petioles long, 1–$^1/_2$ inches, long, leaf blades rough, lower surface pale yellow green, with minute hairs and scales and prominent veins–. Capa blanco, *Petitia domingensis* Jacq.

KK. Petioles mostly short, mostly less than 1 inch long.

L. Leaves shiny above, long–or short–pointed at both ends; flowers white.–*Callicarpa ampla.*

LL. Leaves dull above, not long–pointed; flowers purplish.

M. Leaves obovate, broadest toward blunt of rounded apex.–*Cornutia obouata.*

MM. Leaves ovate to elliptic, broadest below middle, short–pointed at apex.–*Cornutia pyramidata.**

## 348. Capaillo (*Aegiphila mertinicensis*)

Capaillo, a rare shrub but sometimes a small tree, is identified by: (1) opposite oblong to elliptic thin hairless leaves 3–8 inches long; (2) the mostly terminal flower clusters bearing many small regular flowers $^1/_4$–$^3/_8$ inch long with tubular 4-lobed white or pale yellow corolla; and (3) orange, yellow, or red rounded fleshy fruits nearly $^3/_8$ inch in diameter, with irregularly split enlarged calyx at base and containing 1–4 nutlets.

An evergreen shrub 3–10 feet high or sometimes a slender tree to 20 feet high and 3 inches in trunk diameter. Young twigs 4–angled, finely hairy.

Leaves opposite, with slender petioles $^1/_4$–$^5/_8$ inch long. Blades 3–8 inches long and 1–3 inches wide, long–pointed at apex and short–pointed or rounded at base, thin, hairless, paler beneath.

Flower clusters (cymes grouped in panicles) 2–6 inches long are mostly terminal but sometimes also lateral, the branches finely hairy. The flower has a very slender stalk about $^1/_8$ inch long; calyx green, bell–shaped or more than $^1/_{16}$ inch long, with nearly straight border; corolla about $^1/_4$–$^3/_8$ inch long, tubular, deeply 4–lobed; staments 4, equal, inserted in tube and extending $^1/_8$ inch; and pistil with ovary incompletely 4–celled and 4 ovules, slender style, and 2 threadlike stigma lobes. The fruits (drupes) have irregularly split enlarged calyx at base and contain 1–4 nutlets, each with 1 seed $^3/_{16}$ inch long. Flowering and fruiting irregularly through the year.

Rare in moist limestone and lower Luquillo forests at 200–800 feet altitude in northern and northeastern foothills of Puerto Rico. Also in St. Croix and St. Thomas.

**Public Forests :** Cambalache, Guajataca, Luquillo.

**Range :** Cuba, Jamaica, Cayman Islands, Puerto Rico and Virgin Islands, and through Lesser Antilles from St. Eustatius to Grenada, Barbados, and Trinided. Also rare on continent in Panama, Colombia, Venezuela, and French Guiana.

**Other Common Names :** Spiritweed (Barbados); bois cabrit (Guadeloupe, Martinique, Dominica); bois de fer, sureau gros (Guadeloupe).

**Fig. 348 : Capaillo (*Aegiphila martinivensis*)**
Fruiting twig (above), flowers (lower right), natural size.

## 349. Capá rosa (*Callicarpa ampla*)

Capá rosa is a rare shrub or medium–sized tree known only form wet forests of Assam. It is characterized by: (1) twigs, lower leaf surfaces, and branches of flower clusters densely white scurfy; (2) leaves opposite, elliptic to oblong–elliptic, 4–10 inches long and $1^1/_2$–3 inches wide; (3) many small flowers with white 4–lobed corolla about $^1/_8$ inch long, in lateral clusters; and (4) rounded fruits nearly $^1/_4$ inch in diameter, with calyx at base.

An evergreen shrub 10 feet high or recorded as a tree 20–50 feet high. Young twigs 4–angled. The opposite leaves have stout petioles $^1/_4$–$1^1/_4$ inches long. Blades are long–or short pointed at both ends, straight or finely wavytoothed on edges, slightly thickened, the upper surface shiny green and hairless, with long curved side veins, and the lower surface densely white scurfy with prominent network of raised veins.

Flower clusters (cymes) are large, long–stalked, spreading and flat–topped. The nearly stalkless flowers have a short bell–shaped whitish calyx about $^1/_{16}$ inch long, slightly 4–toothed; white short tube and 4 equal lobes; 4 stamens much longer than corolla, and pistil. The fruits (drupes) change color from white to pink to pale purple at maturity. Collected with flowers January to June and with fruits in September–October.

Rare in wet forests in mountains of Assam.

**Fig. 349 : Capá rosa (*Callicarpa ampla*)**
Fruiting twig (above), flowers (lower right), two–thirds natural size.

## 350. Susanna (*Citharexylum spinosum*)

This tree is recorded as wild and planted in F.R.I. It is distinguished by; (1) twigs 4–angled, hairless (not spiny); (2) opposite elliptic to oblong leaves mostly large and thin, densely net–veined when dry; (3) long narrow clusters of small white flowers nearly $^3/_8$ inch across the 5 slightly unequal corolla lobes, on stalks $^1/_{16}$–$^1/_8$ inch long; and (4) elliptic shiny black fruits $^3/_8$ inch long.

An evergreen medium–sized tree, reported to reach 65 feet in height, not spiny, as the scientific name erroneously suggests. The opposite leaves have slender petioles $^3/_8$–1 inch long. The blades are $2^1/_2$–8 inches long and $^3/_4$–4 inches wide, long–pointed, short–pointed, or blunt at apex and short–pointed at base, not toothed on edges (except rarely on young shoots), hairless on upper surface, the lower surface with tufts of hairs in vein angles.

Flower clusters (recemes) terminal or lateral, to 1 foot long, narrow and drooping. Flowers many, fragrant, the cup–shaped calyx more than $^1/_8$ inch long, hairless and slightly 5–toothed; corolla white, with tube $^1/_4$ inch long, hairy at throat; 4 stamens in pairs inserted in corolla tube; and pistil with 4–celled ovary, style, and 2–lobed stigma. The fruits (drupes) are red when immature, turning black, have enlarged cup–shaped calyx at base, and contain 2 nutlets each 2–seeded.

Recorded from forests in F.R.I. but very. Planted for ornament and shade there and elsewhere in the tropics and indoors northward.

**Fig. 350 : Susanna (*Citharexylum spinosum* L.)**
Flowering twig (above), fruiting twig (below), natural size.

## 351. Haggarbush (*Clerodendrum aculeatum*)

This vinelike spiny climbing shrub sometimes becomes treelike. It is distinguished by: (1) the long slender twigs with 2 or 3 curved spines at each node; (2) the elliptic leaves 1–2 inches long and $^3/_8$–1 inch broad, paired or 3 at a node; (3) the clustered flowers about $^3/_4$ inch long, with white corolla in very narrow tube with 5 short lobes; and (4) fruits blackish, juicy, $^1/_4$–$^3/_8$ inch in diameter.

Usually a shrub with vinelike branches climbing on trees, to 30 feet long and 1 inch in diameter, sometimes treelike and to 20 feet high and 6 inches in trunk diameter. Bark light gray, thick, rough, furrowed. The slender light brown twigs are covered with minute hairs and bear 2 or 3 stout sharp curved spines $^1/_8$–$^1/_4$ inch long at each node, the spine from the basal part of the petiole.

The leaves are opposite or sometimes whorled, attached on the upper side of a spine near the tip, and have slender petioles less than $^1/_4$ inch long. Blades are rounded or short–pointed at apex, short–pointed at base, not toothed on edges, thin, hairless, covered with minute translucent dots, the upper surface dull green, and the lower surface dull light green.

Flower clusters (cymes) are terminal and lateral, 1–2 inches across, consisting of several slightly fragrant flowers on slender branched stalks. The slightly irregular flower is composed of bell–shaped calyx $^1/_8$ inch long, green, hairy, 5–lobed; the white corolla with very narrow tube about $^5/_8$ inch long and 5 elliptic spreading lobes $^1/_4$ inch long; 4 threadlike whitish to purplish stamens about $^3/_4$ inch long in unequal pairs, attached inside tube and widely spreading; and pistil with minute ovary, threadlike purplish style about $1^1/_4$ inches long, and 2–forked stigma. The fruit (drupe) is shiny black, slightly broader than long, 4–grooved, slightly juicy, and contains 4 brown nutlets less $^1/_4$ inch long, united impairs. With flowers and fruits throughout the year.

Sometimes grown as a hedge plant and climbing on walls at St. Thomas and elsewhere. It is reported that a poultice in home remedies is made form the leaves. This species is host of one of the largest leafhopper insects in Punjab (*Gypona portoricensis* Caldwell).

Locally common in moist and dry thickets and forests Punjab.

Fig. 351 : Haggarbush (*Clerodendrum aculeatum*)
Natural size.

## 352. Nigua (*Cornutia oibovata*)

This rare small tree known only from Dehradun is recognized by: (1) the 4–angled twigs, branches of flower clusters, and lower leaf surfaces minutely hairy; (2) opposite obovate leaves blunt or rounded at apex, the lower surface light green with a coat of fine hairs; (3) many flowers $^5/_8$ inch long, with purplish corolla tubular and irregularly 4–lobed, in showy erect narrow clusters; and (4) rounded finely hairy fruits $^3/_{16}$ inch long, with calyx at base.

A small evergreen tree 25 feet high and 5 inches in trunk diameter, recorded to 33 feet in height. Bark light gray, rough with many long fissures, the inner bark light brown, slightly fibrous, and almost tasteless. Twigs 4–angled, light brown, minutely hairy.

The opposite leaves have petioles $^3/_8$–$^3/_4$ inch long. Blades are obovate, $2^1/_2$–6 inches long and $1^1/_2$–4 inches wide, blunt or rounded at apex, short–pointed at base, edges slightly turned under, slightly thickened, the upper surface green and becoming hairless, and the lower surface light green with a coat of fine hairs.

Flower clusters (panicled cymes) are erect, terminal, and narrow, 3–10 inches long, only $1^1/_2$–2 inches wide. Flowers have stalks about $^1/_{16}$ inch long; calyx bell–shaped, less than $^1/_2$ inch long, 4–lobed, finely hairy; corolla purplish, finely hairy, tubular and 4–lobed, the lowest lobe larges and with yellow spot at base inside; stamens 2 and 2 sterile (staminodes) on corolla; and pistil with 4–celled hairy ovary and 2–lobed style. Fruits (drupes) recorded as violet are finely hairy, have enlarged calyx at base, and contain a 3–4–celled stone with 4 or fewer seeds. Collected with flowers from June to August, with fruits in September–October.

The wood is light brown and hard.

These purple–flowered plants might make attractive ornamentals.

Rare and local in moist forests at 1,000–3,000 feet altitude in central mountains.

**Fig. 352 : Nigua (*Cornutia oibovata*)**
Fruits (upper left), flowering twig (right), natural size.

## 353. Skyflower (*duranta repens L.*)

This shrubby species, which sometimes is vinelike or a small tree, grows wild in dry or open areas and is planted as an ornamental for its showy light blue flowers and many yellowish fruits. Recognized by: (1) the long, slender, unbranched, gray twigs, often with slender straight gray spines paired or single at nodes; (2) paired small elliptic leaves mostly $^3/_8$–1 inch long and $^1/_4$–$^1/_2$ inch wide, often with few shallow teeth toward rounded or short–pointed apex; (3) many small light blue irregular tubular flowers $^3/_8$ inch long; and (4) many elliptic or rounded yellow fruits nearly $^1/_2$ inch long.

Evergreen shrub with many stems, or vinelike or sometimes a small tree to 15 feet high and 3 inches in trunk diameter. The bark is light gray, becoming rough and fissured. Inner bark is light gray and slightly bitter.

The opposite leaves have short finely hairy leafstalks $^1/_{16}$ inch long. Blades are sometimes $1^1/_2$ inches long, and $^7/_8$ inch wide, short–pointed at base, hairless, thin, with few veins, dull light green on both surfaces.

The flowers are short–stalked in narrow curving unbranched clusters (racemes) 2–6 inches long at ends or sides of twigs. The flower about $^3/_8$ inch long is composed of tubular 5–toothed calyx; light–blue corolla with cylindric tube and 5 unequal spreading lobes $^3/_8$ inch across; 4 stamens in pairs inside corolla tube; and pistil with ovary partly 8–celled, short style, and enlarged stigma.

The fruits (drupes) are covered by the yellow calyx with slightly 5–lobed point at apex. Inside the thin yellow bitter flesh is an elliptic yellow stone or 4 nutlets. Flowering and fruiting mostly in spring and summer.

The wood is light brown and hard.

Often planted as an ornamental shrub and spreading from cultivation, such as along roadsides. Propagated by seeds and cuttings. Elsewhere the leaves have served in home remedies. It is reported that the fruits are poisonous to humans.

Common and scattered in dry limestone forest from sea level to 400 feet altitude on coastal hills of southwestern Kerala and islands eastward, including Lakasdeep.

**Fig. 353 : Skyflower (*duranta repens L.*)**
Natural size.

# NIGHTSHADE FAMILY (SOLANACEAE)

Herbs, shrubs, vines, and sometimes trees mostly small, known by : (1) leaves alternate mostly simple, sometimes, lobed, often with star–shaped hairs, without stipules; (2) flowers in cymes, bisexual, regular, with 5–lobed persistent calyx, corolla tubular with 5 lobes often spreading and star–shaped, 5 stamens inserted on corolla alternate with lobes, disk generally present, and pistil with superior ovary usually 2–celled with axial placentation and many ovules, style, and 2–lobed stigma; and (3) fruit a berry or capsule with many seeds.

**Key to Species**

A. Flowers with corolla widely spreading and star–shaped, white or bluish.—*Solanum*.

B. Twig with spines or prickles.

C. Leaves with border lobed or wavy, ovate to elliptic—*Solenum torvum.*

CC. Leaves with border straight, not toothed or lobed.

D. Leaves oblong to ovate, $1^1/_4$–6 inches long, short–pointed–*Solanum polygamum.*

DD. Leaves narrowly oblong to lanceolate, 3–8 inches long, long–pointed.—*Solanum drymophilum.*

BB. Twigs not spiny.

E. Leaves opposite but very unequal, ellipic, hairless or nearly so—*Solanum antillarum.*

EE. Leaves alternate, with minute star–shaped hairs.

F. Leaves slightly rough, with hairs inconspicuous.—Tabacón, *Solanum rugosum* Dunal.

FF. Leaves soft hairy, the under surface whitish with dense star–shaped hairs.—697. *Solanum erianthum.*

AA. flowers with corolla in narrow tube, mostly funnel–shaped.

G. Flowers with 6–lobed orange corolla, nearly 1 inch long.—*Goetzea elegans.*

GG. Flowers with 5–lobes corolla, white or yellowish.

H. Flowers less than 1 inch long, several to many along sides of twigs.

I. Flowers with bell–shaped corolla $^3/_8$ inch long, mostly back of leaves—*Acnistus arborescens.*

II. Flowers with very narrow funnel–shaped or salverform corolla.

J. Flowers few, $^1/_2$ inch long, short–stalked at leaf bases and back of leaves.—*Cestrum macrophyllum.*

JJ. Flowers several to many at end of long stalk.

K. Flowers $^1/_2$–$^5/_8$ inch long in clusters longer than the long–pointed leaves; berries purplish black.—*Cestrum diurnum.**

KK. Flowers about 1 inch long inlarge clusters longer than the long–pointed leaves; berries white.—*Cestrum Nocturnum.**

HH. Flowers 2 or more inches long, single or sometimes few, mostly at ends of twigs.—*Brunfelsia.*

L. Leaves very narrow, mostly less than $^1/_4$ inch wide; flowers very long and narrow, 4–5$^1/_2$ inches long.—*Brunfelsia densifolia.*

LL. Leaves broad, more than $^3/_4$ inch wide; flowers 2–4 inches long.

M. Leaves broadest toward abruptly short–pointed apex, tapering to base; flowers 4 inches long.—*Brunfelsia portoricensis.*

MM. Leaves broadest near middle.

N. Leaves thick and leathery; flowers 3–4 inches long.—*Brunfelsia lactea.*

NN. Leaves only slightly thickened; flowers 2 inches long.—*Brunfelsia americana.*

## 354. Gallinero (*Acnistus arborescens*)

This shrub or small tree of openings in moist areas is characterized by: (1) elliptic leaves 3–6 inches long and $1^1/_4$–3 inches wide, sometimes larger, mostly long-pointed at both ends; (2) many whitish bell-shaped flowers about $^3/_8$ inch long, clustered along the twigs mostly back of leaves; and (3) many shiny yellow or orange berries $^1/_4$–$^5/16$ inch in diameter along the twigs.

An evergreen shrub or small tree becoming 20 feet high and 6 inches in trunk diameter, with crown of open branches. The light brown or gray bark is finely fissured, the inner bark whitish and bitter. The long stout twigs are light gray or brown and finely hairy.

The alternate leaves have light green petioles $^3/_4$–1 inch long. The blades are thin or slightly thickened, not toothed on edges, the upper surface dull light green and hairless, and the lower surface paler and sometimes hairy.

Flower clusters (like umbels) up to 2 inches across consist of many fragrant flowers spreading on slender nearly equal stalks $^1/_2$–1 inch long. Each flower has a whitish cup-shaped calyx $^1/_8$ inch long, 5-toothed at apex; whitish tubular bell-shaped corolla nearly $^3/_8$ inch long, with 5 spreading pointed lobes; 5 stamens borne near base of tube and slightly longer; and slender pistil with green 2-celled ovary and long slender style.

The fruits (berries) with calyx persistent at base are 2-celled and contain many light yellow seeds about $^1/_{16}$ inch long. The juicy pulp is almost tasteless or slightly bitter and not edible, though eaten by birds. With flowers in spring and with fruits in spring and summer.

The wood is light brown and hard.

Elsewhere the shrubs have served in hedges and living fence posts. Chickens eat the fruits. However, according to an old report, the fruits can cause death.

Uncommon and scattered in openings and clearings in upper Luquillo and upper Sivalik forests at 1,500–3,000 feet altitude in mountains of eastern and central Punjab.

**Fig. 354 : Gallinero (*Acnistus arborescens*)**
Flowering twig (left), fruiting twig (right), natural size.

## 355. Augacero (*Brunfelsia Americana L.*)

Aguacero, a shrub or small tree of eastern mountains Assam is often planted in gardens as an ornamental. It is distinguished by: (1) elliptic to obovate leaves $1^1/_2$–4 inches long, slightly thickened; (2) narrow flowers mostly single at twig ends, the corolla about 2 inches long with 5 spreading rounded lobes, white with purplish center, turning to yellow; and (3) round yellow fruits $^3/_8$–1 inch in diameter. From the 3 related tree species of this genus described and illustrated here, this species differs in its smaller flowers and fruits.

An evergreen shrub 3–10 feet high or a small tree to 15 feet. The twigs are hairless or often hairy.

The alternate leaves are crowded on short side twigs and have petioles mostly less than $^1/_4$ inch long. Blades are abruptly short–pointed, blunt, or notched at apex, short–pointed or blunt at base, slightly turned under at edges, the upper surface shiny and almost hairless, and the lower surface paler and often finely hairy.

The solitary fragrant flowers have stalks less than $^1/_4$ inch long; a bell–shaped calyx $^1/_4$ inch long, finely hairy with 5 rounded lobes; corolla about 2 inches long with narrow tube about $^3/_{16}$ inch wide and 5 rounded spreading lobes about $1^1/_2$ inches across. The berry fruits with calyx at base contain many elliptic brown seeds more than $^1/_2$ inch long. Flowering in spring and fruiting in summer.

Common locally in moist forest from sea level to 1,800 feet altitude in eastern and central M.P.

**Fig. 355 : Augacero *(Brunfelsia Americana L.)***
Fruit (upper left), flowering twig, and flower (lower left), natural size.

## 356. Brunfelsia (*Brunfelsia densifolia*)

This distinctive small tree known only from mountain forests of western Chakrata is identified by: (1) very narrow shiny leathery leaves clustered at ends of twigs; (2) very long narrow flowers single at twig ends with corolla 4–5$^1/_2$ inches long and about $^1/_2$ inch in diameter of tube, with 5 spreading lobes, white, turning to yellow; and (3) rounded ball–like fleshy fruits 1 inch or more in diameter, turning from green to yellow.

A small evergreen tree 15–30 feet high and 5 inches in trunk diameter, with several main stems and many branches, hairless. The gray bark is smoothish with few warts, becoming slightly furrowed, the light brown inner bark slightly bitter. Twigs are light brown, slender.

Leaves alternate but crowded mostly on short side twigs. The linear to linear–oblanceolate blades are 2–4 inches long and mostly less than $^1/_4$ inch wide, short–pointed or blunt at apex, mostly widest beyond middle, not toothed on edges, and tapering gradually to the short slender petiole mostly less than $^1/_4$ inch long. The upper surface is shiny green, with fine raised side veins, almost parallel with midvein, and the lower surface is yellow green with raised veins.

Flowers are borne singly on stalks of $^3/_8$ inch or less. The narrowly bell–shaped calyx about $^1/_4$ inch long has 5 short rounded lobes; the corolla 4–5$^1/_2$ inches long has a cylindric tube slightly more than $^1/_8$ inch in diameter, ending abruptly in 5 widely spreading rounded lobes almost $^1/_2$ inch long and nearly 1 inch across, white when fresh but turning to yellow; stamens within the tube, 2 fertile and 2 sterile; and pistil with 2–celled ovary containing many ovules, long threadlike style, and 2–lobed stigma. The berry fruits surrounded by longer leaves have enlarged split open. There are many brown elliptic seeds more than $^1/_8$ inch long. With flowers in spring and summer and fruits in summer and fall.

The wood is light brown and hard.

Common in upper forest at higher altitudes (2,000–2,700 feet) on serpentine bedrock in western Dehradun.

**Fig. 356 : Brunfelsia *(Brunfelsia densifolia)***
Flowering twig (left), fruiting twig (right), natural size.

## 357. Vega blanca (*Brunfelsia lactea*)

A shrub or small tree known only from high mountains of eastern Bihar, distinguished by: (1) elliptic leathery leaves 2–6 inches long and 1–$2^1/_4$ inches wide, with edges turned under; (2) large very fragrant flowers single at twigs ends, 3–4 inches long, with narrow tube and 5 rounded lobes 3 inches across, pale yellow or whitish or becoming pink tinged; and (3) round fleshy fruits $^3/_4$–1 inch in diameter, turning from green to orange brown.

Evergreen shrub 3--10 feet high, or a small tree to 20 feet high and 4 inches in trunk diameter, recorded to 40 feet, hairless throughout. Bark light brown, slightly fissured, covered with mosses and liverworts, the inner bark whitish and almost tasteless. Twigs light brown, with narrow ridges.

Leaves alternate, scattered or clustered at ends of short side twigs, on petioles $^3/_{16}$–$^5/_8$ inch long. Blades varying from long–pointed to blunt at both ends, leathery, green on both surfaces with network of many fine raised veins, slightly shiny above.

The large flowers are borne singly on stalks about $^3/_8$ inch long at ends of twigs but sometimes appearing lateral on short twigs. The bell–shaped calyx nearly $^5/_8$ inch long has 5 short rounded lobes; the corolla 3–4 inches long has a tube less than $^1/_4$ inch diameter and 5 rounded lobes about 3 inches across nearly as broad as length of tube, pale yellow or whitish or becoming pink tinged; stamens within the tube, 2 fertile and 2 sterile; and pistil with 2–celled ovary containing many ovules, long threadlike styles, and 2–lobed stigma. The berry fruits with calyx at base and hard walls do not open. Seeds many, elliptic, brown, about $^1/_2$ inch long. With flowers and fruits through the year.

The wood is hard and light brown.

This shrub with large very fragrant flowers might be suitable as an ornamental. The common name jazmin del monte was suggested.

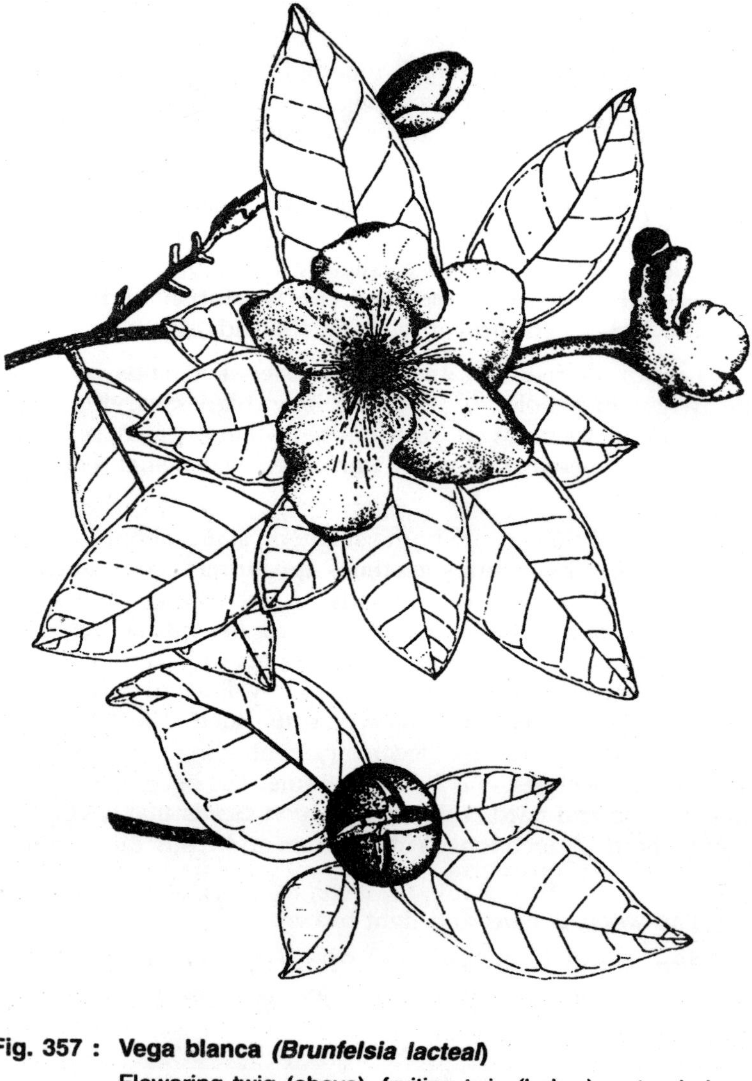

**Fig. 357 : Vega blanca *(Brunfelsia lacteal)***
Flowering twig (above), fruiting twig (below), natural size.

## 358. Portoricen (*Brunfelsia portoricensis*)

This handsome rare shrub of Luquillo Mountains seldom reaches tree size. It is easily recognized by: (1) obovate to oblanceolate leaves abruptly short–pointed at apex; (2) large flowers few or single, about 4 inches long, the white corolla with narrow cylindric tube and 5 large rounded lobes about $2^1/_2$ inches across; and (3) yellow rounded fruits 1–$1^1/_4$ inches in diameter.

An evergreen shrub 3–10 feet high, rarely a small tree to 15 feet high and 3 inches in trunk diameter, hairless throughout. Twigs stout, brown, fissured.

Leaves alternate, sometimes crowded, with stout petioles $^1/_4$ inch long. Blades are 3–6 inches long and $^1/_2$–2 inches wide, broadest beyond middle and tapering to long–pointed base, not toothed on edges, slightly thick and leathery, the upper surface shiny green with sunken midvein and many fine side veins at about $45^0$ angle, and the lower surface dull light green.

Flowers few or single, at ends or sides of twigs on short stalks of less than $^1/_2$ inch. The tubular cylindric calyx $1^1/_4$–$1^3/_4$ inches long splits irregularly to middle on 1 side or into 2 lobes; the corolla tube $2^1/_2$–$3^1/_2$ inches long and only $^1/_2$–$^3/_{16}$ inch in diameter, finely hairy with 5 spreading lobes about 1 inch long; stamens 4 in pairs inside corolla tube; and pistil with 2–celled ovary, long slender style, and stigma slightly 2–lobed. The berrylike fruit has calyx at base, and thick wall. There are many elliptic brown seed $^3/_{16}$ inch long. With flowers in spring and summer and with mature fruits in summer.

This species might an attractive ornamental in cultivation for its large white flowers and shiny foliage.

Rare in lower forest at 1,500–1,800 feet altitude in Sivalik Mountains only.

**Fig. 358 : Portoricen (*Brunfelsia portoricensis*)**
Flowering twig (above), fruit (below), natural size.

## 359. Day cestrum (*Cestrum diurnum L.**)

This common introduced shrub or rarely small tree of gardens, roadsides, and pastures is identified by: (1) oblong leaves $2^1/_2$–4 inches long and 1–$1^1/_2$ inches wide, shiny green to yellow green above and dull whitish green beneath; (2) several to many flowers clustered at bases of uppermost leaves, $^1/_2$–$^1/_4$ inch long, the white corolla with narrow tube and 5 short blunt lobes turned back; and (3) elliptic purplish black berries $^5/_{16}$ inch long.

An evergreen shrub or rarely small tree 15 feet high and 3 inches in trunk diameter. The bark is smoothish gray, the inner bark yellow brown and slightly bitter. Twigs are slender and greenish gray, often with fine pressed hairs.

The alternate leaves have light green petioles $^1/_4$–$^1/_2$ inch long. Leaf blades are short–pointed at apex, short–or blunt–pointed at base, not toothed on edges, slightly thickened, and hairless or hairy on veins beneath.

The slightly fragrant white flowers are in almost stalkless groups near the ends of slender stalks of $^1/_2$–$2^1/_2$ inches. The whitish green calyx is bell–shaped and 5–toothed; the white corolla has narrow tube and 5 short blunt lobes turned back; the 5 stamens with short straight stalks are within corolla tube near mouth and alternate with lobes; and the pistil has a rounded 2–celled ovary with several ovules, slender style, and exposed flat stigma. The juicy berries, whitish before maturity, have calyx at base and contain several brown angled seeds $^1/_8$ inch long. Flowering and fruiting probably irregularly through the year.

The wood is whitish and hard

Very common in gardens, roadsides, and along fences at low to middle altitudes in moist areas of M.P. Apparently introduced for ornament and spread by birds along fences and roadsides. Propagated by seeds and cuttings.

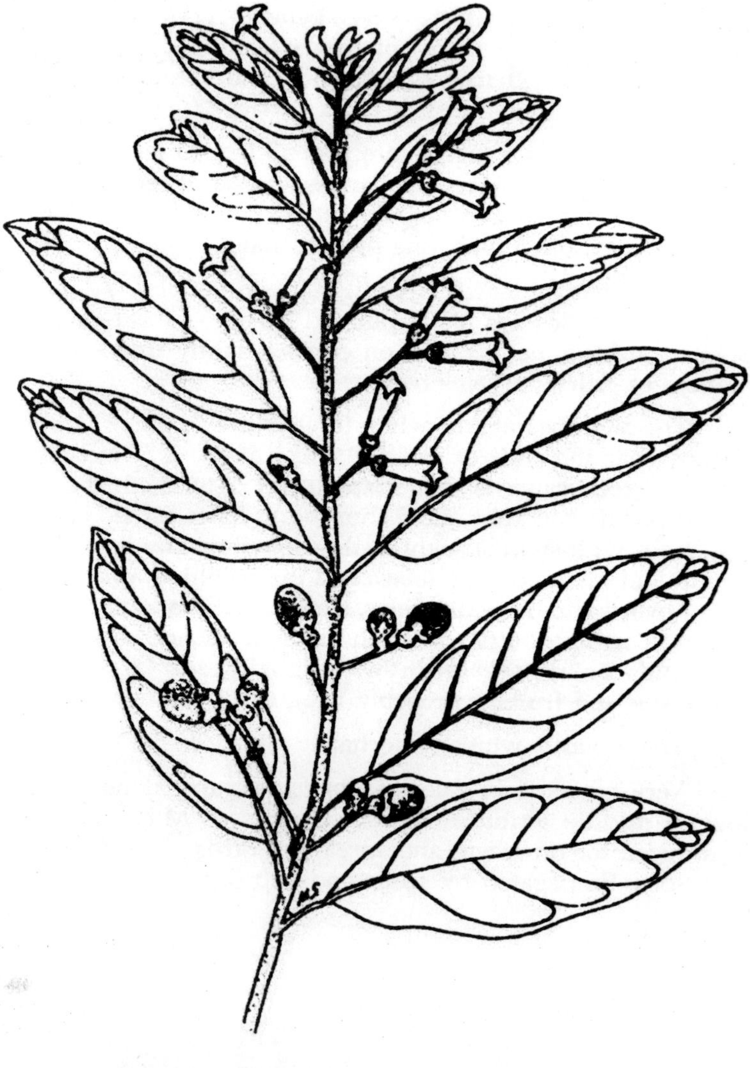

**Fig. 359 : Day cestrum *(Cestrum diurnum L.")***
Natural size.

## 360. Galan del monte (*Cestrum macrophyllum*)

Shrub or small tree of moist forests to high altitudes in mountains. Distinguishing characters are : (1) dull green broadly elliptic leaves $2^1/_2$–7 inches long and $1^1/_4$–4 inches wide, with stout leafstalks; (2) few narrowly tubular flowers $^1/_2$ inch long at leaf bases and back of leaves; and (3) shiny purplish black elliptic to rounded fleshy fruits more than $^3/_8$ inch long, like a small eggplant.

Evergreen shrub or vinelike small tree 15–20 feet high and 3 inches in trunk diameter, widely spreading, hairless. The bark is gray, smoothish. Inner bark light brown, streaked, with taste like raw potato. The light gray twigs have slightly raised half–round leaf scars.

The alternate leaves have stout leafstalks $^1/_2$–1 inch long. The blades are short–pointed at both ends, slightly turned under at edges, slightly thickened, with new curved side veins, the lower surface slight green.

The short–stalked slightly fragrant flowers have a bell–shaped 5–toothed green calyx $^1/_8$ inch long with narrow funnel–shaped tube and 5 rounded lobes, cream–colored, turning pale yellow; 5 alternate stamens on and within corolla tube; and pistil with 2–celled ovary, threadlike style and dot stigma. The berries with green calyx at base have green tasteless flesh and 6–3 narrow blackish brown seeds $^3/_{16}$ inch long. Flowering and fruiting throughout the year.

The wood is light brown or whitish and slightly soft.

Elsewhere the plants are grown as ornamentals.

Common in lower moist limestone forests, also dwarf forest, at 800–3,800 feet altitude throughout mountains of Sivalik.

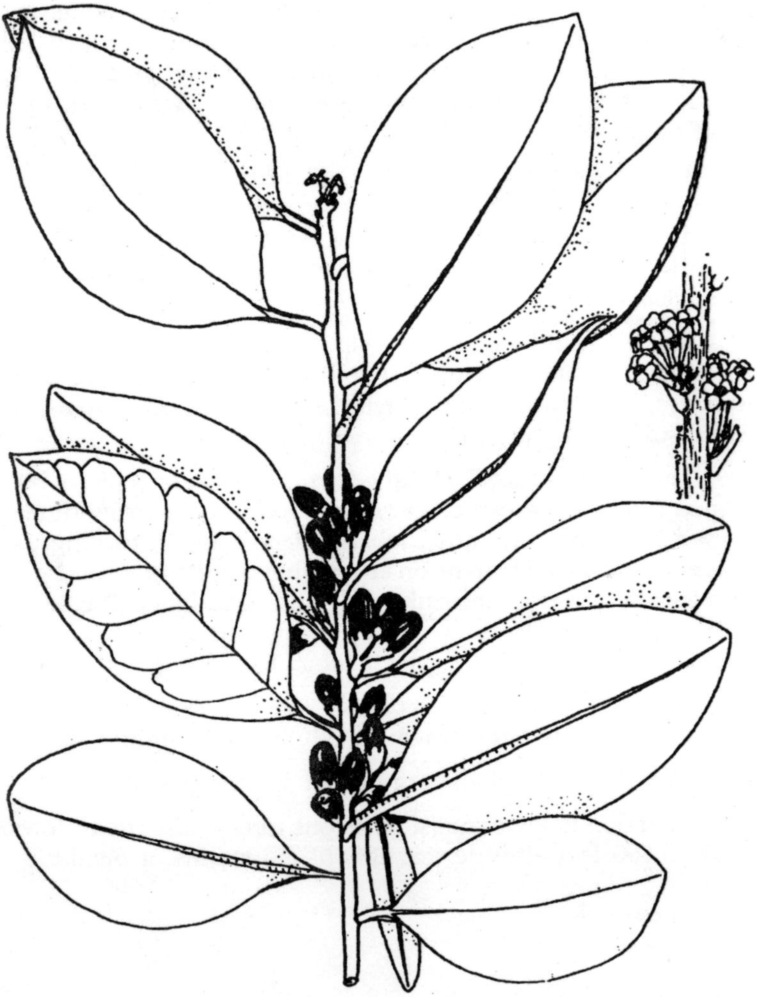

**Fig. 360 : Galan del monte *(Cestrum macrophyllum)***
Fruiting twig (center)), flowers (right), natural size.

## 361. Matabuey (*Goelzea elegans*)

Matabuey is one of the most distinctive of the rare endemic trees of Jharkhand, even having been made the type of a new small plant family. It is characterized by: (1) elliptic shiny leaves with many straight parallel raised side veins; (2) funnel–shaped orange flowers nearly 1 inch long, with 6–lobed corolla, usually borne singly at leaf bases; and (3) orange rounded or elliptic berries $^3/_4$ inch long.

An evergreen small tree 20–30 feet high and 5 inches in trunk diameter. Bark brown, becoming rough and much fissured, the inner bark whitish and almost tasteless. The slender twigs and petioles are brown and minutely hairy.

Leaves alternate, with petioles $^1/_2$–$^1/_4$ inch long. Blades elliptic, 1–4 inches long and $^1/_2$–2 inches wide, short–pointed at both ends, not toothed on edges, slightly thickened, with many straight parallel side veins raised on both surfaces, the upper surface shiny green and hairless, and the lower surface light green and minutely hairy.

Flowers are borne singly–at leaf bases on slender brown hairy stalks $^1/_4$–$^1/_2$ inch long, sometimes few in terminal clusters (cymes). The flower is composed of a bell–shaped calyx $^1/_4$ inch long, finely brown hairy, with 6 pointed lobes; funnel–shaped orange minutely hairy corolla $^3/_4$ inch long and $^1/_4$–$^3/_4$ inch across the 6 rolled back blunt lobes; 6 slender stamens borne near base of corolla and projecting beyond; and pistil with hairy 2–0celled ovary bearing few ovules, slender styles, and 2–lobed stigma. The berries are finely hairy, thick–walled, and bear persistent calyx at base. Seeds elliptic, $^1/_4$ inch long. Collected with flowers and fruits from May to August.

The wood is whitish and hard.

Worthy of cultivation as an ornamental with shiny leaves, showy orange flowers and orange fruits, as the name meaning elegant suggests. However, the plant, perhaps the fruit, was reported long ago to be poisonous.

Very rare and local, known only from a few places in the moist forests at 2090–600 feet altitude on north Bihar.

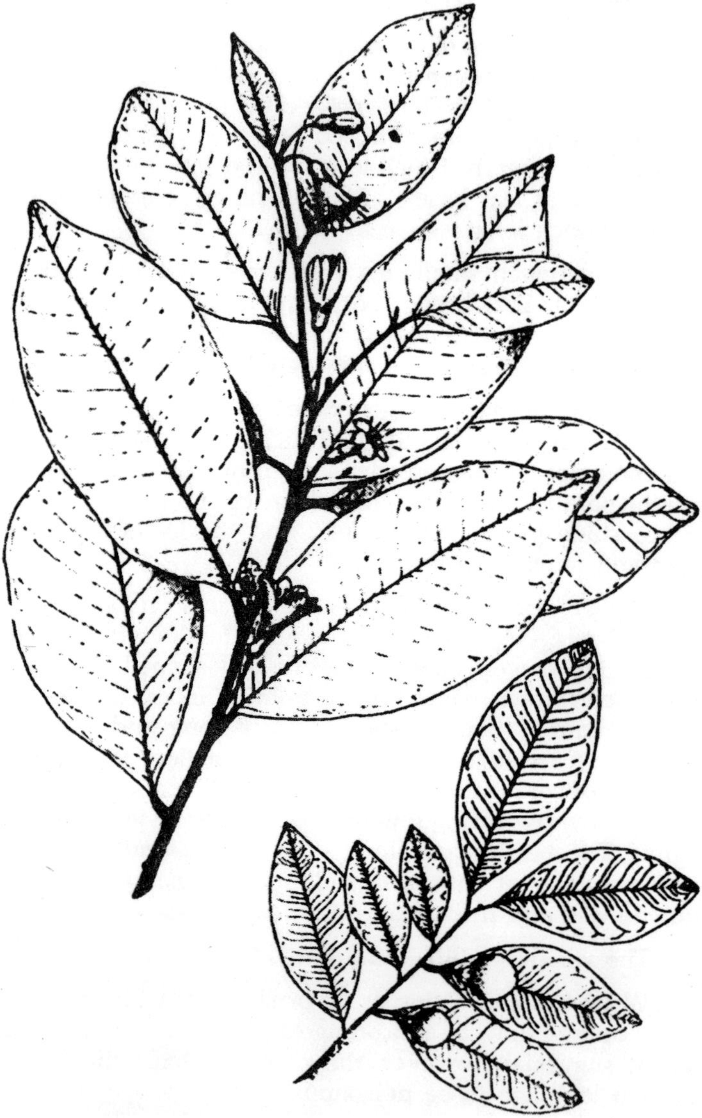

**Fig. 361 : Matabuey *(Goelzea elegans)***

Flowering twig (above), fruiting twig (lower right), two–thirds natural size.

## 362. Antillarum (*Solanum antillarum*)

This shrub or small tree is widely distributed in Bihar, especially higher mountains, but was not listed until recent years. It is easily distinguished from related species by: (1) spines absent and hairs almost lacking; (2) leaves in pairs but very unequal in size, elliptic, the larger 3–5 inches long and $1^1/_4$–2 inches wide or larger; (3) flowers several on slender stalks opposite the leaves, more than $^3/_8$ inch across the white 5–lobed star–shaped corolla; and (4) round orange berries $^5/_{16}$ inch in diameter.

Evergreen shrub or small tree to 20 feet high and 4 inches in trunk diameter, sometimes with several trunks, much branched, with spreading crown. The bark is smoothish, brown or gray, with many small warts (lenticels), the inner bark whitish and slightly bitter. Twigs are green, minutely hairy when young, sometimes zigzag.

The opposite leaves are long–pointed at apex short–pointed at base and tapering into the narrowly winged slender petiole $^1/_4$–$^1/_2$ inch long, with straight or slightly wavy margins, the upper surface dull green with veins slightly sunken, and the lower surface light green with raised veins and tufts of hairs in vein angles along midvein. The smaller leaf of the pair is 1–$2^1/_2$ inches long and $^1/_2$–$1^1/_4$ inches wide.

Flower clusters (short racemes like umbels) 1–$1^1/_2$ inches long bear few to several flowers on slender stalks $^1/_2$–$^3/_4$ inch long. The flower is composed of bell–shaped green calyx $^1/_8$ inch long with 5 short blunt lobes; star–shaped corolla with 5 widespreading turned back lobes more than $^3/_8$ inch across, white with purple streak on outside of each lobe; 5 yellow stamens $^1/_8$ inch long; and pistil with hairless ovary and slender style. Berries round, with enlarged calyx at base, turning from green to orange, with many seeds more than $^3/_{16}$ inch long. With flowers and fruits throughout the year.

The wood is whitish or light brown and hard.

Rare in upper mountain forests including understory, openings, and cutover areas. At altitudes mostly 1,500–3,500 feet in high mountains of eastern and central Bihar.

**Fig. 362 : Antillarum *(Solanum antillarum)***
Flowering twig (above), fruiting twig (lower right), natural size.

## 363. Berenjena de paloma, wild tobacco, mullein nightshade (*Solanum erianthum*)

Berenjena de paloma, a large shrub sometimes becoming a small tree , is recognized by: (1) whitish green star–shaped hairs on twigs, leaves, flowers and fruits; (2) ovate to elliptic leaves 4–8 inches long and $1^1/_2$–4 inches wide, sometimes larger, the lower surface whitish green and soft, covered flowers with 5–lobed spreading white corolla $^1/_2$–$^3/_4$ inch wide; and (4) rounded yellow berries $^1/_2$–$^3/_4$ inch in diameter. Further distinguished from some related species by the absence of spines.

An evergreen shrub 3–10 feet high, sometimes a small tree to 15 feet high and 3 inches in–trunk diameter. Bark greenish gray, smoothish with scattered raised dots (lenticels), the inner bark light gray and slightly bitter. The twigs are whitish with dense star–shaped hairs.

The alternate leaves have petioles $^1/_2$–2 inches long, sometimes larger, covered with star–shaped hairs. Blades mostly long–pointed at apex, rounded or short–pointed at base, the edges straight or very slightly wavy, thin, the upper surface green and finely hairy, and the lower surface whitish green and covered with dense star–shaped hairs.

The erect terminal branched flower clusters (cymes) 3–4 inches long bear several to many flowers on stout hairy stalks $^1/_4$–$^1/_2$ inch long. The flower consists of light green calyx about $^1/_4$ inch long, densely hairy, with 5 pointed lobes; white corolla with short tube and 5 widely spreading pointed lobes; 5 stamens with long yellow anthers inserted on throat of corolla; and pistil with hairy 2–celled ovary and slender style. The berries have calyx persistent at base, turn from green to yellow, and contain many small brownish seeds about $^1/_{16}$ inch in diameter. With flowers and fruits through the year.

The wood is light brown and hard.

Common but widespread in open areas, clearings, waste places, and thickets in dry and moist limestone forests from sea level to 300 feet altitude in western Rajasthan.

**Fig. 363 : Berenjena de paloma, wild tobacco, mullein nightshade (*Solanum erianthum*)**

Fruiting twig (above), flowering twig (lower left), two–thirds natural size.

## 364. Cakalaka–berry (*Solanum polygamum*)

This spiny shrub of dry areas rarely reaches tree size. Distinguishing characters are : (1) very slender sharp yellowish spines $^1/_2$–$^3/_8$ inch long on twigs and sometimes midveins of leaves; (2) minute grayish star–shaped hairs forming dense mostly yellowish–gray cover on twigs, petioles, and the oblong to ovate leaf blades; (3) flowers partly male and bisexual (polygamous), about $^3/_4$–1 inch across the 4–7 spreading purplish or whitish corolla lobes; and (4) round orange or red berries about $^3/_4$ inch in diameter, with spiny enlarged calyx.

A shrub 3–10 feet high, rarely a small tree to 15 feet and 3 inches in trunk diameter. The alternate leaves have petioles of $^1/_4$–$^1/_2$ inch (to $1^1/_4$ inches) in length. Leaf blades are mostly $1^1/_4$–4 inches (to $6^1/_2$ inches) long and $^3/_4$–$1^3/_4$ inches wide, short–pointed or blunt at apex, blunt, rounded, or unequal at base, straight or wavy along edges, slightly thickened, the upper surface green, and lower surface yellowish–gray hairy.

Male flowers are in clusters of 1–20, composed of hairy 4–5 lobed calyx, corolla with 4 or 5 narrow pointed spreading lobes, and 4–5 stamens alternate with lobes. Bisexual flowers borne singly are slightly larger and have parts mostly 5–7 and pistil with densely hairy rounded ovary, slender style, and enlarged stigma. The berries with enlarged spiny calyx at base are hairy and contain many brownish seeds $^1/_8$ inch in diameter. Flowering and fruiting from late winter (February) to summer.

Common in dry forests near sea level, collected in Orissa.

**Fig. 364 : Cakalaka–berry (*Solanum polygamum*)**
Flowering twig, natural size.

## 365. Berenjena cimarrona, turkey–berry (*Solanum torvum*)

A common shrub widespread through moist areas of Puerto Rico and the Virgin Islands sometimes becoming treelike, Its distinguishing characters are: (1) young twigs, petioles, lower leaf surfaces, and branches of flower clusters with gray–green cover of star–shaped hairs; (2) sharp green to brown straight spines or prickles to $^1/_4$ inch long scattered along twigs, sometimes also on petiole and midvein beneath; (3) mostly paired, ovate to elliptic leaves with border often wavy or with as many as 7 lobes the upper surface slightly rough hairy and lower surface soft hairy; (4) starshaped flowers than $^3/_4$ inch across the 5–lobed white corolla; and (5) round berries $^3/_8$–$^5/_8$ inch in diameter, green to yellow.

An evergreen shrub 3–10 feet high, sometimes a small tree 15 feet high and 3 inches in trunk diameter, spreading and much branched. Bark light gray, smooth with raised dots (lenticels). The inner bark is whitish within a light green outer layer, slightly bitter. The stout gray–green twigs are soft hairy, spiny, and widely forking.

Leaves opposite or sometimes 1 at a node, with stout round hairy petioles $^1/_2$–2 inches long. Blades are 3–8 inches long and $1^3/_4$–7 inches wide, short–pointed at apex, unequal and notched or rounded at base, the borders with as many as 7 short–pointed lobes or none, the upper surface dull yellow green, and the lower surface gray green with prominent network of veins.

Flower clusters (cymes) are lateral, 1–3 inches long, bearing several to many flowers on slender gland hairy stalks about $^1/_4$ inch long on 1 side of curving branches, 1 or 2 opening at time and soon shedding. The green gland hairy calyx about $^3/_{16}$ inch long has a short tube and 5 narrow pointed lobes; the white corolla has a very short tube and 5 long–pointed lobes about $^1/_2$ inch long, spreading in form of a star 5 stamens with large yellow anthers, inserted between the coroll lobes; and pistil with 2–celled ovary and slender style. The fruits, 1 or 2, hang down from stout stalks, have calyx at base, and are 2–celled. There are many rounded flat brown seeds $^1/_{16}$ inch long. With flowers and fruits through the year.

The wood is pale yellow or whitish and soft.

The roots and fruits have served in home remedies. Also the plants served as stocks for grafting cultivated egg plant or berenjena (*Solanum melongena* L.).

Abundant in open areas and thickets of all forest types except high peaks, a weed in waste places, from sea level to 3,000 feet altitude. Throughout moist areas of and Assam smaller islands eastward.

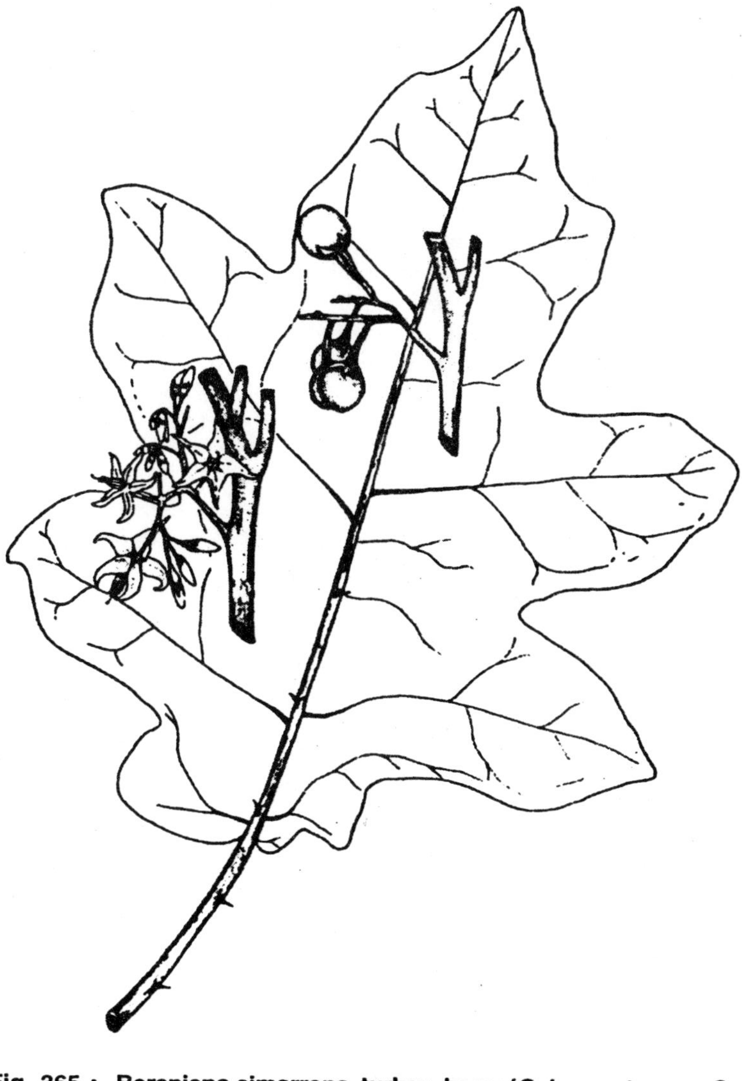

**Fig. 365 : Berenjena cimarrona, turkey–berry (*Solanum torvum Sw)***
Flowers, fruits, and leaf, two–thirds natural size.

# BIGNONIA FAMILY (BIGNONIACEAE)

Mostly woody vines, also shrubs and trees often large, rarely herbs, known by: (1) leaves mostly opposite, sometimes alternate, palmately or pinnately compound or bipinnate, sometimes simple, without stipules: (2) flowers generally large showy, in vymes and racemes, bisexual, slightly irregular, with tubular 5–toothed or 5–lobed calyx, large tubular corolla, colored and commonly yellow, pink, or whitish, funnel–or bell–shaped, with 5 unequal lobes sometimes in 2 lips, stamens usually 4 large in pairs and 1 stamindo inserted in tube, and on disk the pistil with superior 2–celled ovary containing many ovules, long thin style, and 2 stigmas; and (3) fruit usually a 2–parted capsule often long like a pod, with many winged seeds or a berry.

**Key to Species**

A. Leaves simple.

B. Leaves mostly crowded in clusters of 3–5 or more long stout twigs; rounded or elliptic.–*Crescentia.*

C. Leaves spoon–shaped, broadest beyond middle, blunt or short–pointed at apex.

D. Leaves thin; fruit round 4–12 inches in diameter.–Higuero, calabash–tree, common calabash–tree, *Crescentia cujete* L.

DD. Leaves thick and stiff; fruit cylindric, pointed, 3–4 inches long.–*Crescentia protoricensis.*

CC. Leaves very narrow, stiff, long–pointed and often spiny tipped; fruit or elliptic, $1^1/_4$–2 inches long.–*Crescentia linearifolia.*

BB. Leaves otherwise.

E. Leaves alternate, broadly elliptic to obovate, thick and leathery; fruit gourdlike, egg–shapped or elliptic, $2^1/_2$–4 inches long.–*Enallagma latifolia.*

EE. Leaves opposite or partly in 3's.

F. Leaves partly in 3's (whorled), oblong to lanceolate, with light green spot of blade above, slightly thickened, slightly wavy at edges; pod very long and narrow like wire.–*Catalpa longissima.**

FF. Leaves opposite, stiff and leathery, with edges turned under; fruit a cigarlike capsule.

G. Leaves elliptic to ovate, broadest near middle.–Roble de sierra, *Tabebuia rigida* Urban.

GG. Leaves obovate or oblanceolate, broadest beyond middle.–*Tabebuia schumanniana.*

AA. Leaves compound.

H. Leaves digitate (palmately compound) with mostly 5 or 3 leaflets.

L. Leaflets 3, with petiole narrowly winged; fruit cylindric, fleshy, not splitting open–*Parmentiera.*

J. Spines 2 (sometimes 1) at some nodes; fruit 4–6 inches long, angled, like a large pickle.–*Parmentiera edulis.**

JJ. Spines none; fruit 1–2 feet long, like a candle.–*Parmentiera cerifera.**

II. Leaflets mostly 5 or sometimes fewer; petiole not winged; fruit a cigarlike capsule.

K. Flowers with yellow corolla; introduced trees.

L. Leaflets 5 obovate to elliptic ending in long narrow point.–*Tabebuia glomerata.**

LL. Leaflets 7, elliptic, short–or long–pointed.–*Roseodendron donnell–smithii.**

KK. Flowers with pink or red corolla; native trees.

M. Corolla pink or sometimes purplish.

N. Leaflets 5 or fewer (sometimes only 1), elliptic, $1^1/_2$–6 inches long, blunt–pointed at both ends.–Roble blanco, "white–cedar", *Tabebuia heterophylla* (DC.) Britton.

NN. Leaflets 5, ovate to elliptic, 4–8 inches long, long–or short–pointed at apex.–tabebuia rosea.*

MM. Corolla dark red; leaflets 5 or 3, elliptic.–Roble cimarrón, *Tabebuia haemantha* (Bert.) DC.

HH. Leaves pinnate or bipinnate.

O. Leaves pinnate.

P. Leaflets 5–13, lance–shaped to elliptic, long–pointed; pod cigarlike but narrow, 4–8 inches long.–Roble amarillo, ginger–thomas, *tecoma stans* (L.) H.B.K.

PP. Leaflets mostly 11–17, elliptic, not toothed, abruptly short–pointed; pod lance–shaped or boatshaped, 5–10 inches long.–234. Tulipán african tuliptree, *Spathodea campanulata* Beauv.*

OO. Leaves bipinnate, fernlike, pod round flattened, 2 inches long.–Jacaranda, *Jacaranda mimosifolia* D. Don.*

## 366. Robla dominicano, Haiti catalpa (*Catalpa longissima*)

This uncommon ornamental and shade tree is distinguished by its very long narrow pods 1–1$^1/_2$ feet long but less than $^1/_8$ inch in diameter, which hang down in large numbers like wires. Other characters for identification are: (1) leaves 2–3 at a node, oblong to lanceolate, with a light green spot at the base of the blade above; and (2) the irregular flowers in spreading terminal clusters, opening 1 or 2 at a time, the corolla whitish with narrowly bell–shaped tube and 5 unequal spreading rounded wavy–margined lobes 1 inch across.

A large evergreen planted tree to 70 feet high and 2 feet in trunk diameter, perhaps larger in age. The trunk become slightly broad or but–tressed at base and slightly angled and continues as a straight axis. Horizontal to slightly drooping branches form a narrow crown. Bark gray, rough, furrowed into narrow ridges, the inner bark light brown and bitter. The twigs are light green, becoming greenish gray, with raised light brown corky dots (lenticels). Young twigs and leaves have minute whitish or brown dotlike scales visible with a leans. The bud is formed by minute young leaves lens. The bud is formed bug minute young leaves.

The opposite and whorled leaves have slender light green petioles 1–1$^1/_2$ inches long, slightly enlarged at base. Blades are 2$^1/_2$–5 inches long and 1–2 inches wide, blunt or short–pointed at apex, rounded or blunt at base with a light green spot above, slightly thickened, slightly wavy at edges, the upper surface shiny green to dark green with few slightly sunken veins, and the lower surface dull light with few prominent veins.

Flower clusters (panicles) are terminal, 3–4 inches long with slender spreading forks 2 inches across. Many flower are produced, 1 or 2 opening at a time, not fragrant, the corollas shedding promptly and giving a whitish tinge to the ground beneath. A flower is composed of 2 half–round concave minutely pointed with narrowly bell–shaped light yellow tube $^1/_2$ inch long, 2–lipped with 5 unequal whitish lobes; stamens 2 inside tube and inserted near base; and pistil with narrow 2–celled ovary, long style, and 2–lobed stigma.

One fruit is formed from a flower cluster. The wire–shaped pods (capsules) are dull green and straight; at maturity they become brown and in drying split longitudinally into 2 curving parts. There are many small brown seeds with 2 long narrow hairy wings totaling $^{3}/_{8}$ inch long, bordered by whitish hairs more than $^{1}/_{2}$ inch long. With flowers and fruits through the year.

The wood is light to pinkish brown with darker lines, heavy, easily worked, and durable. Local uses where the trees are native include carpentry, flooring furniture, boathbuilding, and general construction.

Rare as an introduced ornamental, shade, and forest tree in Puerto Rico and St. Thomas. In forestry, tested and promising on lower slopes of Luquillo Forest.

**Public Forest :** Luquillo.

**Range :** Native of Hispaniola and Jamaica. Introduced in Lesser Antilles, southern Floeida, and other tropical areas.

**Other Common Names :** roble, roble deolor (Dominican Republic); Haiti catalpa (United States); Jamaica–oak, French–oak, mastwood, yokewood (Jamaica); chêne noir, bois chêne (Haiti); radegonde, chêne d' Amerique, chêne noit, poix doux marron (Martinique).

**Botanical Synonym :** *Macrocatalpa longissima* (Jacq.) Britton.

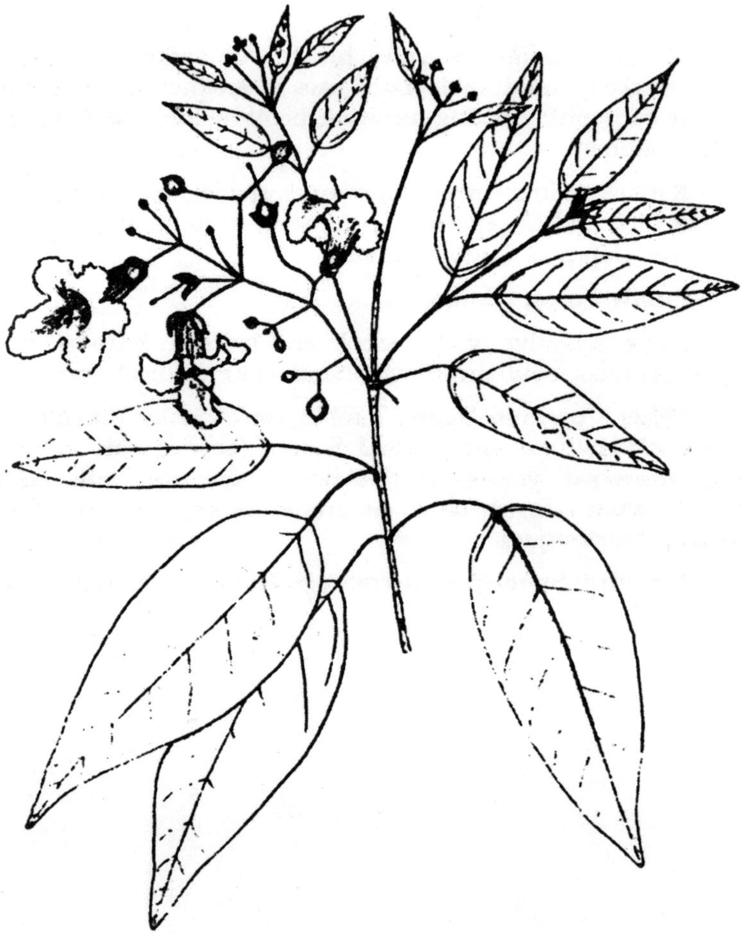

**Fig. 366 : Roble dominicano, Haiti catalpa (*Catalpa longissima*)**
Flowering twig and fruits, two–thirds natural size.

## 367. Higüerito (*Crescemtia linearifolia*)

Higüerito, a small tree of dry coastal areas, is recognized by: (1) many slender straight spreading wandlike branches; (2) leaves crowded at nodes, very narrow, shiny, stiff, long-pointed and often spiny tipped; (3) greenish irregularly bell-shaped flowers $1^1/_4$–2 inches long, the tubular corolla with 5 wavy lobes; and (4) brown rounded or elliptic gourdlike fruits $1^1/_4$–2 inches long.

A small deciduous tree or shrub to 23 feet high and 8 inches in trunk diameter, with spreading thin of many slender straight spreading wandlike branches, hairless throughout. Bark light gray, rough and fissured, the inner bark light brown, fibrous, and tasteless. Twigs light gray brown, slender when young, becoming $^1/_4$–$^1/_2$ inch in diameter, with enlarged nodes.

Leaves clustered at enlarged nodes, very narrow, linear to oblanceolate, $^3/_4$–4 inches long and mostly $^1/_8$–$^3/_8$ inch wide, sometime to 1 inch, gradually narrowed and long tapering to very short at base, not toothed on edges, slightly thickened, the upper surface green, and the lower surface light green.

Flowers single or few at upper nodes on short stalks less than $^3/_8$ inch long, the flower buds green and rounded, $^5/_8$ inch in diameter. The calyx is about $^3/_8$ inch long, deeply splitting into 2–3 lobes; corolla tubular, irregularly bellshaped, enlarged near base, $1^1/_2$–2 inches long and ending in 5 wavy lobes $1^1/_2$ inches across, green with narrow brown stripes; 4 stamens inserted in corolla in pairs; and pistil with 1–celled ovary, very slender style, and broad 2–lobed stigma. The flowers have an odor like rotten cabbage. The brown dry hard gourdlike fruits, which do not open, bear seeds in the pulp. Flowering and fruiting irregularly through the year.

Scattered and locally common in dry coastal forest to 200 feet altitude, mainly in southern coastal Gujrat.

**Fig. 367 : Higüerito (*Crescentia linearifolia*)**
Twig with flowers and fruit, natural size.

## 368. Portoricen (*Crescentia portoricensis*)

This rare vinelike shrub known only from Bihar western sometimes becomes a small tree. It is distinguished by: (1) oblanceolate to narrowly elliptic, shiny, stiff and leathery leaves mostly clustered at nodes; (2) tubular irregularly bell-shaped yellowish white flowers $1^1/_4$–$1^1/_2$ inches long, borne singly; and (3) fruit cylindric, pointed, 3–4 inches long and 1–$1^1/_2$ inches in diameter.

Evergreen vinelike shrub or sometimes a small tree to 20 feet high and 3 inches in trunk diameter, hairless. Bark light gray. Branches long and slender, light gray, with enlarged nodes, becoming fissured.

Leaves clustered at enlarged nodes or alternate, 2–6 inches long and $^3/_4$–3 inches wide, blunt to abruptly long-pointed at apex, tapering at petiole $^1/_4$–$^5/_8$ inch long, not toothed on edges, with prominent network of small veins. The upper surface is shiny green and the lower surface dull and paler.

Flowers single on slender stalks $^3/_8$–$^3/_4$ inch long. The calyx is $^3/_4$ inch long. Deeply 2-lobed, the corolla tubular, irregularly bell-shaped, net veined, with 5 broad entire lobes about $^1/_4$ as long as tube. The hard dry fruit is straight at base and does not split open.

**Fig. 368 : Portoricen (*Crescentia portoricensis*)**
Twig with fruit, natural size.

## 369. Higüerita, black–calabash (*Enallagma latifolio*)

This small tree of lowlands near coasts resembles higüero or calabash–tree, *Crescentia cujete*, L., in its flowers and fruits. It is characterized by: (1) leaves broadly elliptic to obovate, $3^1/_2$–6 inches long and $2^1/_2$–3 inches wide, thick and leathery, the apex rounded light green abrupt short point; (2) large tubular light green flowers $1^1/_2$–2 inches long and $^3/_4$ inch broad, borne singly; and (3) dark green fruits hard and egg–shaped or elliptic, $2^1/_2$ inches long and $1^1/_2$–$2^1/_2$ inches in diameter.

An evergreen small tree to 40 feet high and 1 foot in trunk diameter, hairless throughout. Bark gray, rough and furrowed the inner bark light brown and slightly bitter. Twigs greenish to light gray, stout.

The alternate leaves with short petioles $^1/_4$ inch or less in length are sometimes crowded on short side twigs. The blades are commonly broadest beyond the middle, short–pointed at base, not toothed on edges, the upper sunken, dark green slightly shiny with veins sunken, and the lower surface dull light green with raised veins.

The few flowers are solitary at solitary at bases on slender stalks about 1 inch long. The slightly fleshy green calyx covers the bud splits into a bell–shaped tube $^3/_4$ inch long and 2 or 3 irregular lobes $^5/_8$ inch long; the tubular bellshaped corolla $1^1/_2$–2 inches long, light green with 2–lipped wavy and irregularly toothed pink border, also a side fold; 4 stamens about 1 inch long in pairs and 1 shorter sterile stamen (staminode) inserted within the tube; and on a disk the whitish pistil with 1–celled ovary, long slender curved style, and 2–lobed stigma. The fruit (berry) resembles a small calabash or gourd, smooth, thin–walled, and fragile. The whitish, pulp contains many blackish heartshaped flattened seed about $^5/_8$ inch long, brownish within. Flowering irregularly during the year.

The wood is moderately hard and heavy, light brown or pinkish with orange tinge on exposure. Elsewhere it has been used for tool handles and plows.

Sometimes planted in Bangalore as an ornamental and for the seeds. The edible seeds are eaten and used like cacao or cocoa in a beverage, hence the local name cocoa. Elsewhere the pulp of the fruits has served in home remedies. The gourdlike fruits apparently are distributed along shores by ocean currents.

Uncommon in coastal swamps and moist forests at low altitudes near coasts back of mangroves and long streams at sea level around.

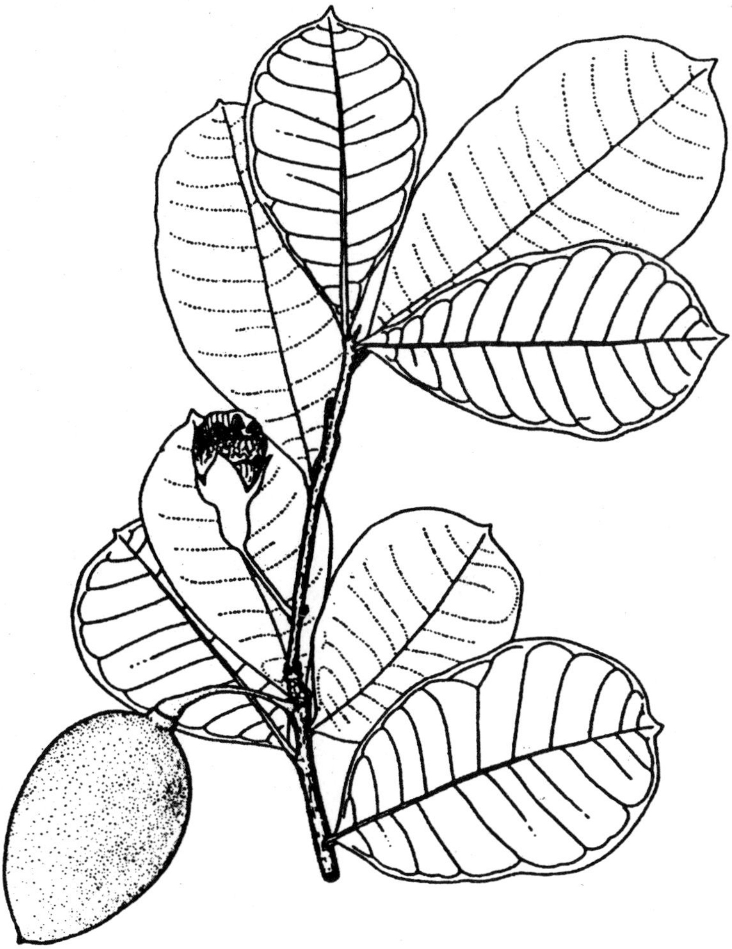

Fig. 369 : Higüerota, black–calabash (Enallagma latifolia)
Twig with flower and fruit, two–thirds natural size.

## 370. Arbol de vela, candle–tree (*Parmentiera cerifera*)

Arbol de vela, or candle–tree, is a rare introduced ornamental tree many cylindric white to yellow fruits resembling candles, hanging down from trunk and large branches. Other characteristics for identification are: (1) leaves opposite, with 3 leaflets and narrowly winged petiole; (2) large flowers irregularly bell–shaped, about $2^1/_2$ inches long, brone mostly along trunk and larger branches, the tubular corolla whitish with pink streaks. No. 705, the related cuajilote, *Parmentiera edulis* DC., has similar foliage and flowers but has 2 (sometimes 1) short spines at some nodes and flowers borne mostly are ends of twigs, also shorter and broader fruits yellowish green to pinkish brown.

Small to medium–sized deciduous cultivated tree, reaching 30 feet in height and often with several trunks from the base to 6 inches in diameter, hairless. Twigs light gray, slender, ending in bud $^1/_{16}$ inch long composed of several narrow scales. The bark is gray, smoothish to scaly and slightly fissured.

Leaves opposite or sometimes 3 or more on very short side twigs at nodes. Leaves 2–5 inches long, consisting of slender winged light green petiole $^1/_2$–2 inches long and 3 stalkless leaflets 1–3 inches long and $^1/_2$–1 inch wide, the central nearly twice as long as the 2 lateral. Leaflets obovate to elliptic, long–pointed at both ends, often toothed on edges, thin, dull light green, the lower surface paler and with tufts of hairs in vein angles along midvein.

The flowers are borne mostly along trunk and larger branches. The brownish calyx about 1 inch long splits to brownish calyx about 1 inch long splits to base on 1 side; the corolla tubular, irregularly bell–shaped, about $2^1/_2$ inches long, whitish with pink streaks, with 5 large rounded lobes; 4 stamens in pairs inserted near base of tube; and pistil with 2–celled ovary, many ovules, long style, and 2–lobed stigma.

The distinctive white to yellow candlelike fruits are 1–2 feet or more in length and $^3/_4$–1 inch in diameter, smooth fleshy, with odor like that of apples, berrylike and not splitting open seeds rounded, flat, less than $^1/_8$ inch in diameter. Flowering and fruiting through the year.

Fruits of wild trees are recorded as generally 2–3 feet, not infrequently 4 feet long. These fruits have been used to fatten cattle. It is reported that cattle strip and eat the bark also.

Rare as an ornamental tree in Maharashtra. Also planted in south and elsewhere in the tropics, India especially in botanical gardens.

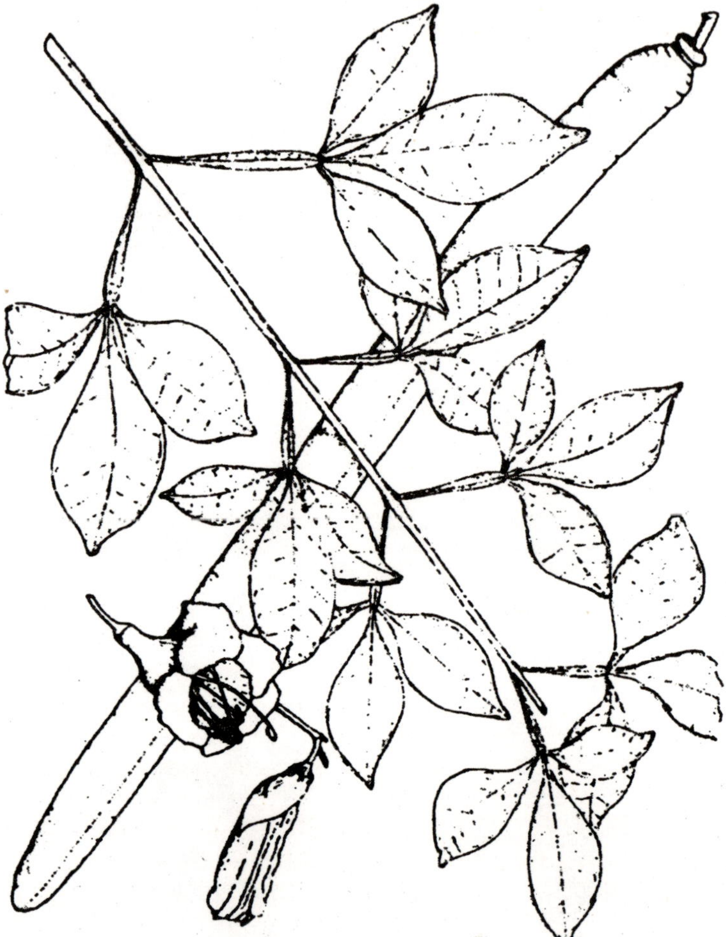

**Fig. 370 : Arbol de vela, candle–tree (*Parmentaiera cerifera*)**
Leafy twig, fruits, and flowers, two–thirds natural size.

## 371. Cuajilote (*Parmentiera edulis*)

Cuajilote, sometimes planted as an ornamental is easily recognized by: (1) 2 (sometimes 1) short spines at some nodes; (2) leaves opposite, with 3 leaflets and narrowly winged petiole; (3) large irregularly bell–shaped, flowers $1^3/_4$–$2^1/_2$ inches long, mostly single at ends of twigs, the tubular corolla greenish with purple streaks; and (4) fruit a cylindric berry 4–6 inches long and $1^1/_2$–2 inches broad, with many angles and grooves, yellowish green to pinkish brown, suggesting a large pickle.

Small deciduous cultivated tree 15 feet high and 3 inches in trunk diameter, elsewhere larger, with broad crown and many spreading branches, hairless. Trunk sometimes angled, with scaly, fissured light gray bark. Inner bark pink to brown, fibrous. Twigs light gray with raised dots (lenticels), enlarged below leaf bases, ending in bud $^1/_{16}$ inch long composed of several narrow scales. Some nodes have 2 (sometimes 1) slender straight sharp gray spines about $^1/_8$ inch long.

Leaves opposite or sometimes appearing clustered or whorled on very short side twigs at nodes, 2–5 inches long, composed of slender winged light green petiole $^1/_2$–2 inches long and 3 stalkless leaflets $^3/_4$–3 inches long and $^3/_8$–$1^1/_4$ inches wide, the central nearly twice as long as the 2 lateral. Leaflets obovate to elliptic, short–pointed at apex and short to long–pointed at base, not toothed on edges, thin, the upper surface dull light green, and the lower surface dull lighter green, with minute holes and tufts of hairs in vein angles along midvein.

Flowers mostly single at ends of twigs on stalks $^3/_4$–$1^1/_2$ inches long. The green calyx is $^3/_4$–$1^1/_4$ inches long, irregularly split to base on 1 side; corolla tubular, irregularly bell–shaped, $1^3/_4$–$2^1/_2$ inches long, greenish with purple streaks, with 5 large rounded lobes slightly unequal; 4 paired stamens and 1 sterile stamen inserted near base of tube; and on a disk the pistil with 2–celled ovary, many ovules, long style, and 2–lobed stigma. Berries hanging down, heavy, slightly flattened, 2–celled, with thick fibrous yellowish wall and many rounded flat yellowish seeds $^1/_8$ inch in diameter. Flowering irregularly through the year.

The scientific name indicates that the fruits are edible. Where the trees are native and become common in pastures, livestock eat the slightly bitter as well as fibrous. Elsewhere, they have been boiled or roasted or made into pickles or preserves.

The wood is described as whitish to pink and hard.

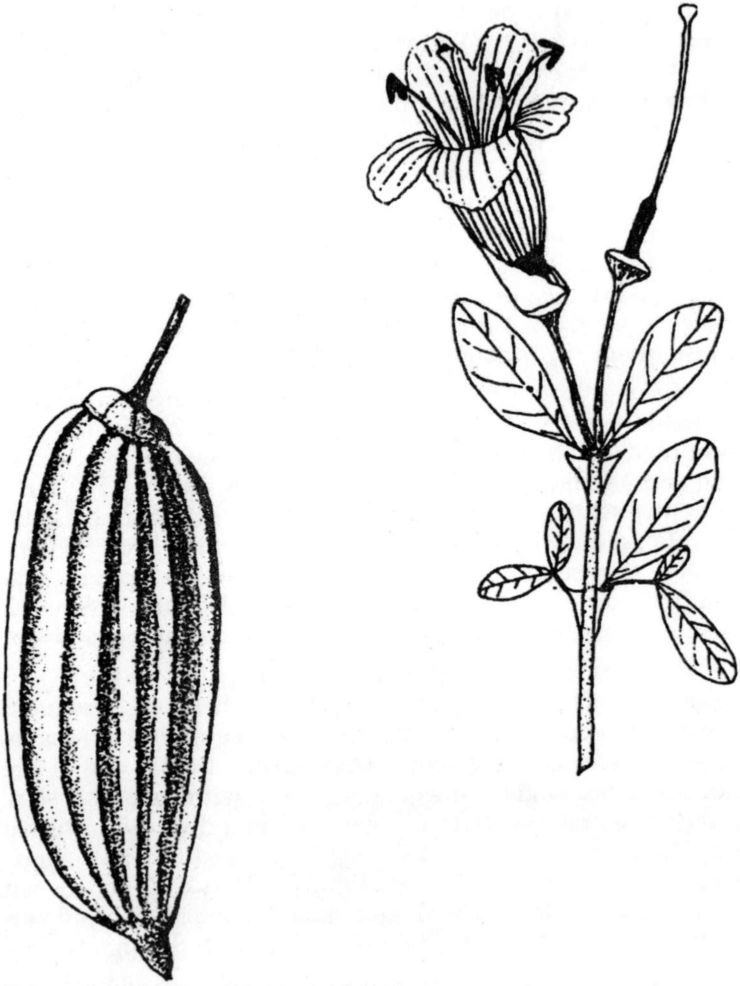

**Fig. 371 : Cuajilote (*Parmentiera edulis*)**
Fruit (left), flowering twig (right), natural size.

## 372. Roble Amarillo, yellow poui (*Tabebuia glomerata*)

This introduced ornamental tree is noted for its solid masses of yellow tubular 5–lobed flowers about 2 inches long and broad, borne in winter when leafless. Other characters for recognition are: (1) opposite palmately compound leaves with 5 obovate to elliptic leaflets ending in a long narrow point; and (2) the narrow dark brown pods about 1 foot long and $^3/_8$ inch wide, which split into 2 parts and curve irregularly, remaining on the tree.

Deciduous planted tree 30 feet high and 1 foot in trunk diameter, probably becoming larger, with spreading crown. The trunk has light brownish gray bark, with thin fissures, and the branches are gray and becoming fissured. Inner bark is light brown, fibrous, and bitter. Young twigs, petioles, and buds are covered with minute yellow–brown star–shaped hairs. The twigs are often flattened in last inter–node and become light gray and slightly fissured. Buds are formed by minute paired leaves, without scales or stipules. Some twigs end in a rounded cluster of many brown hairy flower buds $^1/_8$ inch high and $^3/_8$ inch across.

Leaves are opposite and palmately compound, without stipules. The slender rounded petioles are $2^1/_2$–6 inches long, rounded and enlarged at base. The 5 spreading leaflets have slender stalks $^3/_8$–$1^1/_4$ inches long, grooved above. Blades of the leaflets are 2–7 inches long and 1–3 inches wide, the 2 outer smallest and the middle one longest, the base varying from rounded to slightly notched, thin, and slightly wavy at edges. The upper surface is green, dull to slightly shiny, almost hairless, with midvein and the curved side veins sunken. The lower surface is dull light green, with the prominent light yellow veins bearing minute star–shaped hairs.

Flower clusters are headlike at ends of twigs. Flowers many on stalks less than $^1/_4$ inch long. Composed of cylindric rusty hairy calyx $^3/_8$–$^1/_2$ inch long, unequally 5–lobed; corolla yellow funnel–shaped, about 2 inches long, with 5 unequal spreading rounded wavy lobes 2 inches across; stamens 4 in pairs and 1 sterile within corolla tube near base; and on a disk the pistil with 2–celled ovary, slender style, and 2–lobed stigma. The short–stalked pods have minute star–shaped hairs and thin walls. There are many thin light brown seeds with 2 whitish papery wings nearly 1 inch across. Flowering in February and other winter months.

The wood is light brown and soft.

Uncommon and occasionally planted for ornament in residential areas throughout U.P. The trees are very showy covered with golden flowers while leafless or nearly so.

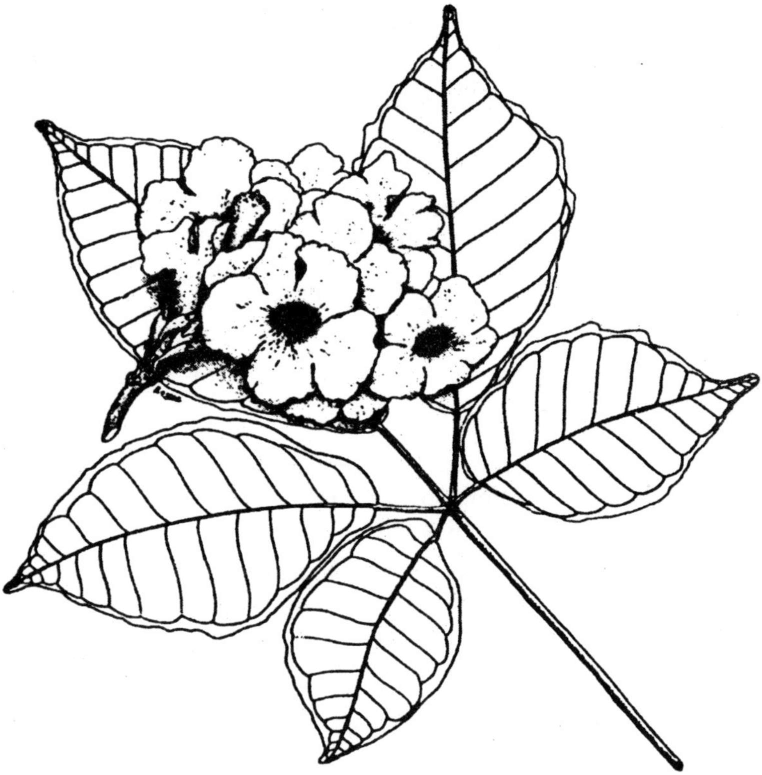

**Fig. 372 : Roble Amarillo, yellow poui *(Tabebuia glome–rata)***
Flowers and leaf, two–thirds natural size.

## 373. Roble venezolano, pink trumpet–tree (*Tabebuia rosea*)

This handsome ornamental tree native of continental America has been introduced into Puerto Rico in recent years for its abundant large showy tubular flowers more than 2 inches long, pale pink or reddish purple, produced usually in late spring when leaves are absent. Other characters are : (1) opposite palmately compound (digitate) leaves with long petiole and 5 elliptic to ovate leaflets 4–8 inches long and $1^1/_2$–$2^3/_4$ inches wide; and (2) narrow pod–like brown fruits 8–12 inches long and $^1/_2$ inch ion diameter.

A large deciduous planted tree 70 feet high and $1^1/_2$ feet in diameter, becoming somewhat larger in its native forests, with short erect trunk, slight buttresses, and rounded crown. The bark is gray or brown, thick, rough and deeply furrowed, the inner bark chocolate–colored. Twigs are stout and light gray.

The opposite leaves have petioles about 5 inches long, slightly flattened, light yellow green. The 5 slightly thickened leaflets with stalks $^1/_2$–2 inches long have short or long point at apex, the base rounded or short–pointed, the borders not toothed, and minute scales on both surfaces, the upper green and slightly shiny, the lower dull light green.

Flower clusters (corymbs) are terminal, branched, and spreading, and bear several to many flowers on short stalks. The calyx $^3/_4$ inch long is bell–shaped with 4 teeth or 2 lobes, brownish green and minutely scaly; the corolla with shape of funnel or bell is pale pink or reddish purple, the tube whitish and yellow within, with 5 slightly unequal rounded lobes widely spreading 2–$2^1/_2$ inches across; 4 stamens in pairs and 1 shorter sterile stamen (staminode) inserted within the tube; and on a disk the yellow–green pistil with 2–celled ovary, slender style, and 2–lobed stigma. The pods (capsules) split open along 2 lines. There are many flattened seeds about $1^1/_2$ inches across the 2 broad whitish wings. Flowering in late spring.

The yellowish or whitish sapwood turns pale brown on exposure. The heartwood is pale brown with fine dark lines. The wood is of medium weight (specific gravity 0.45–0.60), slightly soft, of medium texture to rough. It is easy to work, moderately durable, and takes a good polish. Locally it is used

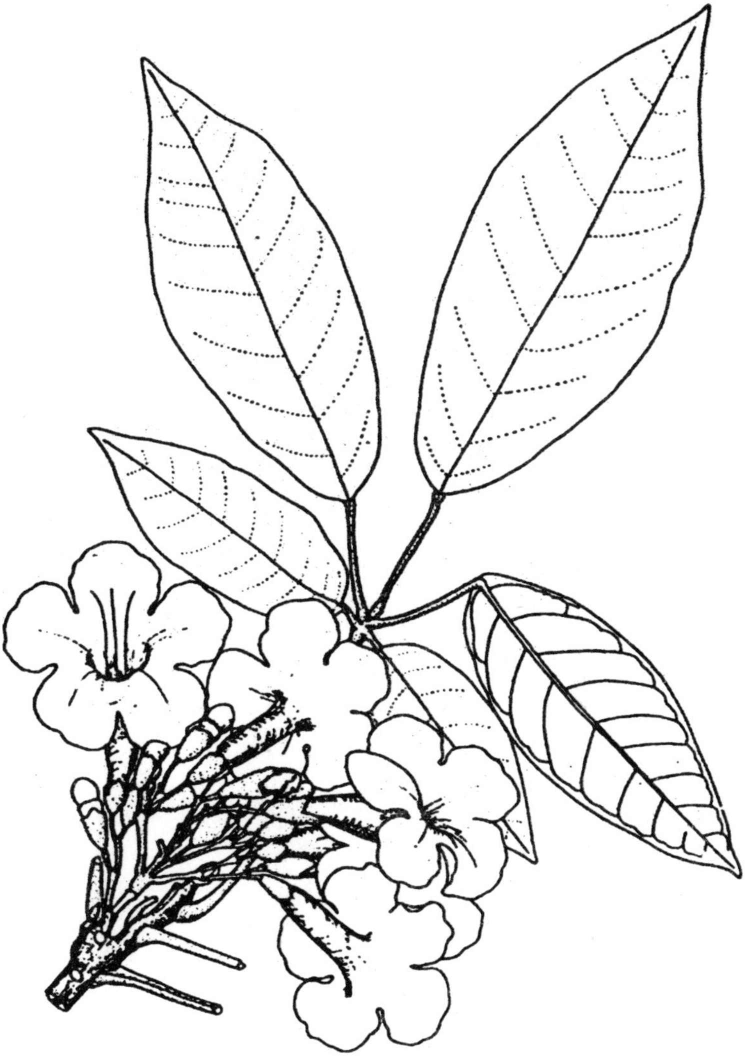

**Fig. 373 : Roble venezolano, pink trumpet–tree (*Tabebuia rosea*)**
Two–thirds natural size.

for posts and poles. Where this species is native, the wood is widely used for carpentry, furniture, interiors, flooring, doors and frames, construction, boat–building, tool handles, and boxes. It is reported that the bark is high in tannin content, also used in home remedies.

A popular ornamental and shade tree in cities of India.

## 374. Roble Colorado (*Tabebuia schumanniana*)

Roble Colorado, known only from high mountains of U.P. is distinguished by: (1) opposite simple leathery leaves 2–6$^1/_2$ inches long and $^3/_4$–2$^3/_4$ inches wide, broadest beyond middle (obovate or oblanceolate), with edges curved under, crowded on twigs; (2) few showy, slightly irregular flower 1$^1/_2$–2 inches long, the bright red tubular corolla with 5 slightly unequal lobes; and (3) dark brown cigarlike pods 3$^1/_2$–7 inches long and $^1/_4$–$^3/_8$ inch in diameter.

A small evergreen tree to 40 feet high and 1 foot in trunk diameter. The bark is gray, fissured and becoming rough, slightly corky and soft, the inner bark whitish and bitter. Twigs are stout, slightly angled, green when young but becoming light brown, with tiny dark brown dotlike scales.

The crowded opposite simple leaves have short stout petioles $^1/_4$–$^3/_4$ inch long. Blades are rounded or short–pointed at apex, gradually edges curved under, with minute dark brown dotlike scales, above dark green with inconspicuous veins, and beneath pale yellow green.

Few to several flowers, sometimes 1, are borne at end of twigs on slender stalks $^1/_2$–2 inches long. The tubular calyx is $^3/_8$–$^3/_4$ inch long, unequally 2–or 3–lobed dark red, densely covered with brown dotlike scales; the bright red tubular corolla 1$^1/_2$–2 inches long has 5 unequal rounded lobes spreading 1–1$^1/_2$ inches across; stamens 4 in 2 pairs of different length, attached near base of corolla; and pistil on disk composed of 2–celled ovary, slender style, and 2–lobed stigma. The pods (capsules) with many dotlike scales split lengthwise in 2 parts. There are numerous thin brown elliptic seeds 1–1$^1/_4$ inches long including the 2 whitish wings. Flowering and fruiting throughout the year.

The wood is light brown, hard, and heavy (specific gravity 0.9).

Common locally in dwarf and upper forests at 2,800–4,390 feet altitude on peaks of western and central mountains of U.P.

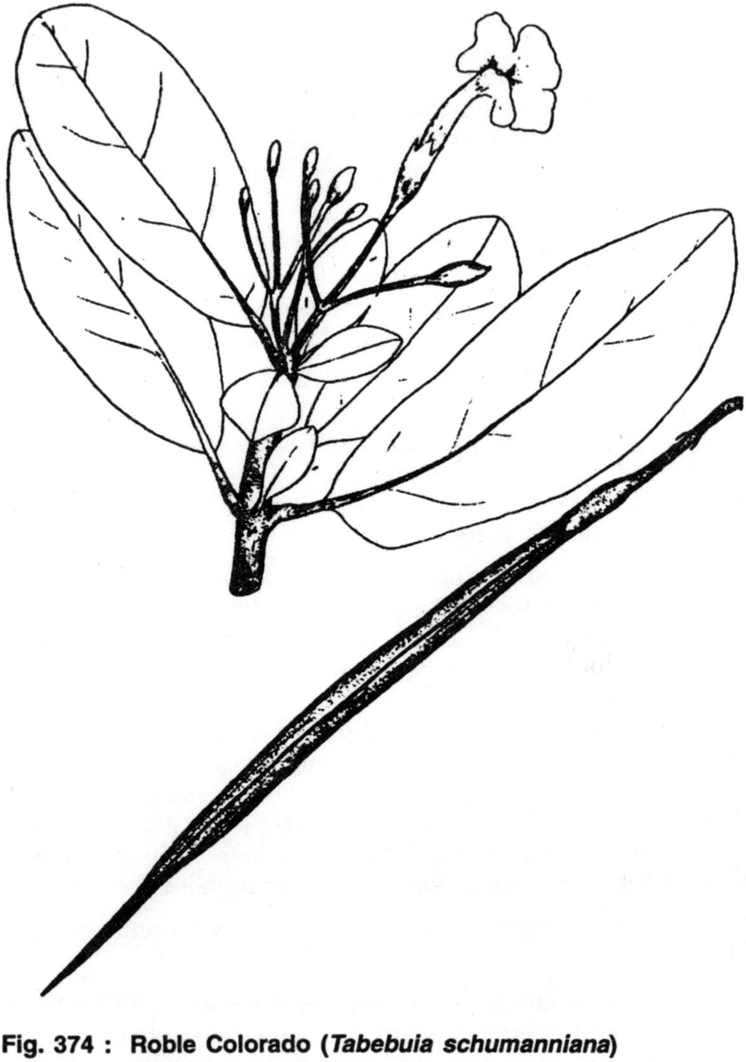

**Fig. 374 : Roble Colorado (*Tabebuia schumanniana*)**
Flowering twig (above), fruit (below), natural size.

# GESNERIA FAMILY (GESNERIACEAE)

Herbs and shrubs, rarely trees, known by (1) leaves alternate or opposite, often leathery or hairy, often toothed; (2) flowers showy, solitary or clustered at leaf bases or terminal, bisexual, irregular, composed of 5–lobed calyx, 5–lobed tubular corolla bell–shaped, funnel–shaped, or wheel–shaped, mostly 2–libbed, stamens 4 in pairs or 2 inserted within corolla, and pistil with superior or inferior ovary 1–celled with 2 parietal placentas and many ovules, slender curved style, and stigma often 2–lobed; and (3) fruit usually a capsule with many minute seeds. One native tree species.

## 375. Arbol de navidad *Gesneria peduncculosa*)

The gesneria family is a tropical group of herbs and shrubs, including several ornamentals, such as African–violet (*Saintpaulia*). The occurrence of a species of large shrub sometimes reaching the size of a small tree and confined to FRI Dehradun is unexpected. This unusual tree is recognized by : (1) shiny dark green elliptic leaves 2–4$^1/_2$ inches long and $^3/_4$–1$^3/_4$ inches wide, the edges often with minute wavy teeth; (2) flowers 1–4 more than 1 inch long at the end of a very long slender erect stalk of 3$^1/_2$–7 inches, the irregular bell–shaped whitish corolla with 5 unequal blunt lobes, the 2 upper largest, and with 4 long stamens; and (3) bell–shaped brown to gray seed capsules about $^1/_4$ inch long and broad, with 5 long narrow calyx lobes pointed upward.

An evergreen shrub mostly 10–15 feet high, with several spreading stems from base mostly 1 inch in diameter, sometimes a small tree. The light gray or brown bark is much furrowed and rough, the inner bark light brown, fibrous, and tasteless. Twigs are greenish to gray, slender, hairless.

The leaves are alternate but crowded at ends of twigs, hairless, with petioles $^1/_4$–$^5/_8$ inch long. Blades are thin or slightly thickened, short–or long–pointed at apex, short–pointed at base, the edges slightly turned under and often with minute wavy teeth, the upper surface dark green, and the lower surface light yellow green.

The flowers borne at the end of a very long stalk are composed of greenish conical base (hypanthium) $^3/16$ inch long which bears 5 narrow greenish calyx lobes $^1/_4$–$^3/_8$ inch long and the other parts; the irregular 2–lipped bell–shaped corolla $^3/_4$ inch long including 5 unequal blunt lobes, the 2 upper largest, whitish, yellowish, greenish, or mottled and purplish; 4 threadlike curved stamens about 1$^1/_2$ inches long, attached at base of corolla and extending beyond; and pistil with inferior 1–celled ovary, many ovules, long curved style, and enlarged blackish stigma. The capsules split along 4 lines and contain many very small reddish brown seeds. Flowering and maturing fruits nearly through the year. Pollinated probably by bats that feed upon the flowers. The wood is light brown and hard.

Scattered but widely distributed in moist and dry forests at low and middle altitudes in Dehradun.

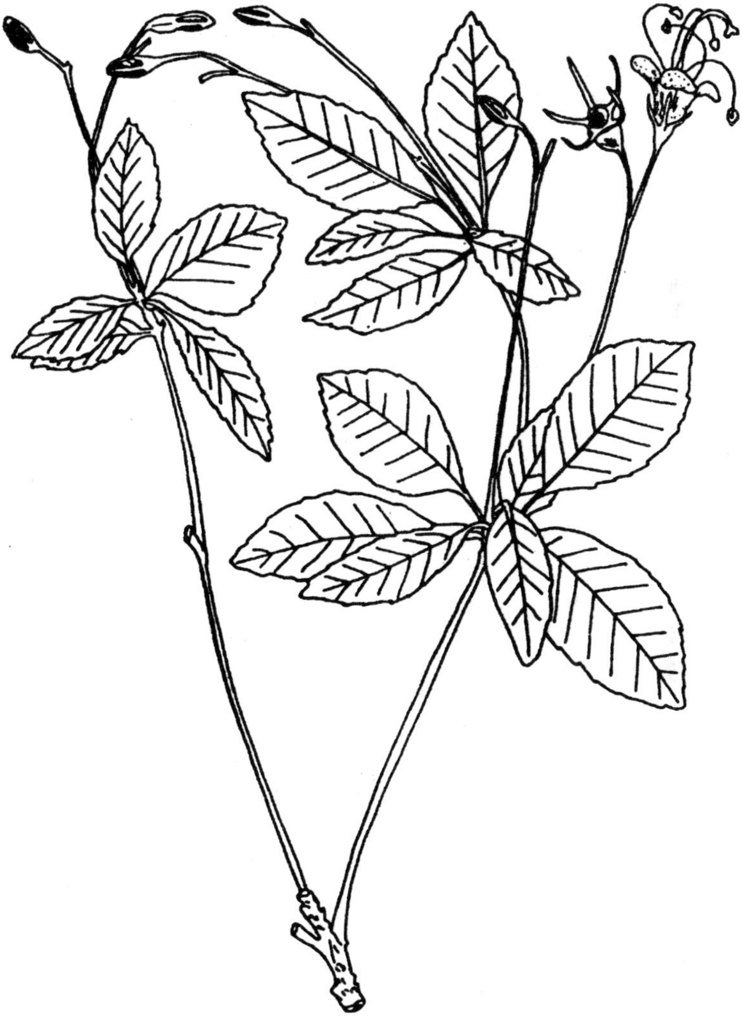

**Fig. 375 : Arbol de navidad *Gesneria peduncculosa*)**
Natural size.

## MYOPORUM FAMILY (MYOPORACEAE)

Shrubs or sometimes trees, mostly Old World, represented by *Bontia,* known by : (1) leaves alternate or opposite, simple, entire, without stipules; (2) flowering solitary or in clusters (cymes) at leaf bases, bisexual, irregular, consisting of 5–lobed persistent calyx, 5–lobed 2–lipped tubular corolla, stamens 4 in pairs inserted inside tube, and pistil composed of superior ovary 2–celled or sometimes 3–10 celled by false partition with 2–8 ovules in each cell slender style, and stigma; and (4) fruit a berry or drupe with few seeds. One tree species.

## 376. Mangle bobo, white–alling (*Bontia daphnoiodes* L.)

This shrub or small tree local on shores is sometimes found with mangroves and known by the corresponding Spanish name mangle. It is recognized by: (1) large dark green gland dots on young twigs, foliage, flowers, and fruits, giving and odor when crushed; (2) numerous crowded lance–shaped or narrowly oblong leaves slightly thick and fleshy, without visible side veins; (3) curved irregular tubular flowers $^3/_4$–$^7/_8$ inch long, the light yellow corolla with purplish spots; and (4) egg–shaped pointed stone fruit $^5/_8$–inch long and $^1/_2$ inch across, resembling an olive.

Evergreen shrub or small tree 10–20 feet high and to 6 inches in trunk diameter, nearly hairless throughout. The bark is light brown, thick, rough, and furrowed. Inner bark is light brown, bitter and pungent. The light green twigs, becoming brown, and in a very narrow pointed bud $^3/_4$ inch long, composed of rolled young leaves without stipules.

The numerous alternate crowded leaves have broad petioles $^1/_4$–$^1/_2$ inch long. Blades are 2–3$^1/_2$ inches long and $^5/_8$–$^7/_8$ inch wide, long–pointed at both ends, not toothed on edges, dull pale green on both surfaces, with many gland dots mostly small but few large, visible against the light with a lens.

Curved irregular flowers are borne singly at leaf bases on slender stalks $^1/_2$–$^3/_4$ inch long, enlarged at apex. The calyx $^1/_8$ inch long has 5 narrow–pointed hairy–edge teeth; the tubular corolla $^3/_4$–$^1/_2$ inch long has a cylindric tube and 2 unequal lips or lobes purplish hairy within, the lower lip shorter and much rolled backward; stamens 4, inserted in pairs near base of tube and extending beyond; and pistil consists of conic green ovary,. 2–celled with 4 ovules in each cell, and threadlike long style.

The stone fruits (drupes) turning from yellow green to light green at maturity, have calyx remaining at base and short point at apex from base of style, thin yellow slightly bitter flesh, a large whitish thick–walled stone, and few whitish seeds less than $^1/_8$ inch long. Flowering and fruiting through the year.

The wood is light gray brown, hard and heavy, fine–textured, with fairly straight grain, and with faint spicy odor, without special use.

Plants are grown for ornament, in hedges, and as windbreaks. Being tolerant of salt, they are common around houses on sandy shores of Guyana. Cultivated also at higher altitudes to 5,000 feet or more in the northern Andes of Venezuela. Also escaping from cultivation and becoming naturalized.

In forests, shores, and salty flats at sea level, mostly local, common, and scattered, but spread by planting. With mangroves uva de playa, seagrape, *Cocoloba wifera* (L.) L. In U.P. known from the dry southern and southwestern part.

**Fig. 376 : Mangle bobo, white–alling (*Bontia daphnoiodes L.*)**
Natural size.

# MADDER FAMILY (RUBIACEAE)

Shrub and trees sometimes large, also herbs and vines, known by: (1) leaves opposite, sometimes whorled, simple, entire with paired stipules that form bud and leave ring scars at nodes; (2) flowers generally many in cymes, small to large, bisexual, regular, with 5–4–lobed calyx often persistent, tubular corolla usually 5–4–lobed, generally colored and often showy, stamens 5–4 alternate and inserted in tube, and pistil with inferior ovary commonly 2–celled and containing many ovules (to 1), style, and 2 stigmas; and (3) fruit a capsule or berry, sometimes a drupe.

**Key to Species**

A. Plants spiny or sticky (viscid).

B. Plants spiny.

C. Twigs with spines, mostly paired.

D. Leaves and spines paired; leaves $^3/_8$–$1^3/_4$ inches long; side twigs short in 4 rows—Tintillo, box–briar, *Randia aculeata* L.

DD. Leaves and spines partly in 3's; leaves $^1/_4$–$^5/_8$ inch long; twigs long and slender—*Machaonia portoricensis.*

CC. Spines at end of stiff leaves—*Guettarda pungens.*

BB. Plants sticky (viscid) or resinous.

E. Buds not resinous—*Antirhea acutata*

EE. Buds in resin drop.

F. Flowers in terminal clusters—*Chione seminervis.*

FF. Flowers in lateral clusters.

G. Flowers crowded on 2 forks at end of long lateral stalk—Aquilón, *Terebraria resinosa* (Vahl) Sprague.

GG. Flowers crowded at leaf bases and back of leaves.

H. Leaves $^3/_4$–$1^3/_4$ inches long—*Phialanthus myrtilloides.*

HH. Leaves 1–$3^1/_2$ inches long.—*Phialanthus grandifolius.*

AA. Plants not spiny or sticky (viscid).

I. Flowers many in compact white terminal cluster 2–4 inches across, suggesting a snowball or with rounded stalked white lobes to $1^1/_4$ inches in diameter; introduced ornamentals.

J. Flower clusters suggesting a snowball—*Ixora thwaitesii.**

JJ. Flower clusters with rounded stalked lobes—*Calycophyllum candidissiumum.**

II. Flowers not suggesting a snowball, if white then in smaller clusters.

K. Fruit a capsule with many seeds, splitting open.

L. Flowers large, with long narrow tube and lobes from bud more than $1^3/_4$ inches long.

M. Leaves 4–$8^1/_2$ inches long—*Exostema sanctae–luciae.*

MM. Leaves less than $3^1/_2$ inches long.

N. Flowers many in terminal clusters—*Exostema ellipticum.*

NN. Flowers single at leaf bases—Albarillo, Caribbean princewood, *Exostema caribaeum* (Jacq.) Roem. & Schult.

LL. Flowers small, less than $^1/_2$ inch long.

O. Leaves 3–7 inches long; flowers several to many in branched lateral clusters—Juan tomás, *Rondeletia portoricensis* Krug & Urban.

OO. Leaves small, mostly less than 3 inches long; flowers few or 1 on short stalks at leaf base.

P. Twigs and lower leaf surfaces with long soft hairs—*Rondeletia pilosa.*

PP. Twigs and lower leaf surfaces slightly hairy or nearly hairless—*Rondeletia inermis.*

KK. Fruits dry or fleshy, not splitting open.

Q. Fruits (drupes) becoming hard and dry.

R. Fruits hairless, black, rounded but broader than long, 1–shaded; flowers on slender stalks; leaves becoming blackish upon drying—Cafeillo, false–coffee, *Faramea occidentalis* (L.) A. Rich.

RR. Fruits finely hairy, green., red, or black, rounded, with several seeds; flowers several clustered erect on forks at end of long stalk—*Guettarda.*

S. Leaves rough on upper surface—Palo de cucubano, "greenheart," roughleaf velvet–seed, *Guettarda scabra* (L.) Vent.

SS. Leaves smooth and hairless on upper surface.

T. Leaves hairy on lower surface.

U. Leaves thick, the lower surface soft hairy with network of raised veins—*Guettarda krugii.*

UU. Leaves thin, the surface with minute pressed hairs—*Guettarda elliptica.*

TT. Leaves hairless or nearly so on lower surface.

V. Leaves with deeply sunken side veins—*Guettarda ovalifolia.*

VV. Leaves with side veins not or slightly sunken.

W. Leaves $1^1/_2$–5 inches long, dull dark green—*Guettarda valenzuelana.*

WW. Leaves $^3/_4$–$2^1/_4$ inches long, shiny yellow green—*Guettarda parviflora.*

QQ. Fruits fleshy berries or drupes.

X. Fruits compound, from many flowers crowded in ball–like head; leaves large, elliptic, mostly 6–12 inches long.

Y. Flowers orange, in balls $1^1/_4$–$1^1/_2$ inches across—*Anthocephalus chinensis.**

YY. Flowers white, in balls 1 inch across—Morinda, pinkiller, *Morinda citrifolia* L.*

XX. Fruits simple, each from a flower; leaves mostly smaller.

Z. Seeds a few many.

(*a*) Fruits large, more than 1 inch wide.

(*b*) Fruit elliptic, $3^1/_2$–$4^1/_2$ inches long, with bluish dye, sour; leaves drying dark bluish green—Jagua, genipa, *Genipa americana* L.

(*bb*) Fruit round, $1^1/_4$ inches in diameter—*Vangueria madagascariensis.**

(*aa*) Fruits less than $^1/_2$ inch long.

(*c*) Berry elliptic, dark red to blackish; flowers orange red to red; leaves mostly in 3's—*Hamelia patens.*

(*cc*) Berry or drupe black; flowers whitish; leaves opposite.

(*d*) Flowers and fruits several to many, short–stalked at end of branching clusters; fruits round, $^3/_{16}$ inch in diameter, with 5–10 grooves and nutlets—*Erithalis fruticosa.*

(*dd*) Flowers and fruits clustered erect on forks at end of long stalk; fruits elliptic or oblong, about $^3/_8$ inch long, the stone with 2–6 cells and seeds—*Antirhea.*

(*e*) Leaves oblong to lanceolate, 3–4 times as long as wide—*Antirhea sintenisii.*

(*ee*) Leaves mostly elliptic or ovate, mostly 2–3 times as long as wide.

(*f*) Leaves shiny green above, the lower surface with several tiny holes in angle between side veins and midvein—*Antirhea lucida.*

(*ff*) Leaves dull green, slightly thickened.

(*g*) Leaves almost stalkless, rounded to heart-shaped at base—Quina, *Antirhea obtusifolia* Urban.

(*gg*) Leaves with short petiole, blunt or short–pointed at base.

(*h*) Petiole $^1/_4$–$^5/_8$ inch long; fruit with 2–seeded stone—*Antirhea coriacea.*

(*hh*) Petiole $^1/_8$–$^5/_{16}$ inch long, stout; fruit with 4–celled stone—716.*Antirhea portoricensis.*

ZZ. Seeds or nutlets 2, from 2–celled ovary.

(*i*) Flowers clustered at leaf bases and nodes back of leaves, short–stalked.

(*j*) Flowers more than 1 inch across the 5–7 spreading white corolla lobes; elliptic red berries with 2 large brown seeds (coffee beans)—Coffea.

(*k*) Leaves 3–7 inches long; corolla 5–lobed–240. Café, coffee, *Coffea arabica* L.*

(*kk*) Leaves 6–14 inches long; corolla 5–7 lobed—*Coffea dewevrei.**

(*jj*) Flowers $^1/_2$ inch long, the corolla with very narrow reddish or pink tube and 4–5 short white lobes; rounded red or pink berries—Cafeillo, *Ixora ferrea* (Jacq.) Benth.

(*ii*) Flowers in terminal branched clusters.

(*l*) Flower clusters with whitish axis; fruit (drupe) oblong, $^1/_2$–$^3/_4$ inch long, shiny purplish black with 2–celled stone; stipules shedding early—*Chione venosa.*

(*ll*) Flower clusters with greenish or reddish axis; fruit (drupe) rounded, about $^1/_4$ inch long, black, bluish, or red, with 2 nutlets; paired pointed stipules present.

(*m*) Corolla with long tube $^3/_8$–$^3/_4$ inch long, slightly curved and swollen at base, mostly yellow or red; axis orange or reddish; fruits black—*Palicourea.*

(*n*) Stipules deciduous; flowers nearly stalkless in clusters of 3, corolla white or pinkish—*Palicourea domingensis.*

(*nn*) Stipules persistent; flowers separate on slender stalks.

(*o*) Leaves less than 4 inches long; corolla yellow, turning to pale purple—*Palicourea alpina*.

(*oo*) Leaves mostly 4–10 inches long.

(*p*) Corolla white—*Palicourea barbinervia*.

(*pp*) Corolla yellow or red.

(*q*) Corolla yellow; flowers on orange–red branches of clusters about as broad as long, often more or less horizontal; leaves with small tufts of hairs where side veins join midvein beneath–738. *Palicourea riparia*.

(*qq*) Corolla red; flowers or yellow or orange branches of erect clusters longer than wide; leaves minutely hairy along midvein beneath—*Palicourea crocea*.

(*mm*) Corolla with short straight tube less than $^3/_8$ inch long, mostly white; axis greenish; fruits black or red—*Psuchotria*.

(*r*) Leaves thin, with straight edges.

(*s*) Leaves long–or short–pointed at apex, mostly more than 4–inches long.

(*t*) Leaves obovate or broadly oblanceolate, widest beyond middle, mostly 8–13 inches long—*Psychotria grandis*.

(*tt*) Leaves elliptic, widest at middle, 3–8 inches long—*Psychotria berteriana*.

(*ss*) Leaves blunt–or short–pointed at apex, less than 4 inches long; Mona Island only—*Psychotria nutans*.

(*rr*) Leaves slightly thickened and turned under at edges.

(*u*) Leaves elliptic to obovate, broadest near blunt–pointed apex and narrowed to long–pointed base; foliage without odor—*Psychotria maricaensis*.

(*uu*) Leaves oblong, widest near middle long–or short–pointed at both ends; foliage with bad odor—*Psychotria maleolens*.

## 377. Kadam (*Anthocephalus chinensis*)

Kadam, a rapidly growing introduced timber tree, is being tested experimentally in forestry plots in Puerto Rico. It is distinguished by: (1) opposite large elliptic yellow–green leaves with many sunken curved side veins; (2) orange flowers crowded in large balls or heads $1^1/_4$–$1^1/_2$ inches in diameter; and (3) fleshy fruits in balls about 2 inches in diameter.

A large evergreen introduced forest tree becoming at maturity 100 feet tall and $2^1/_2$ feet in trunk diameter, with straight axis and widely spreading nearly horizontal branches. Bark gray, smoothish to slightly fissured. The inner bark is light yellow with a thin green outer layer, fibrous, and bitter. Twigs stout, hairless, with ringed nodes, green when young, becoming brown. Buds very narrow and long–pointed, ]™ inch long, covered by paired stipules, which shed early.

Leaves opposite, hairless, with stout light green petioles $1^1/_4$–$1^3/_4$ inches long. Blades are 6–12 inches long and $3^1/_4$–8 inches wide, abruptly short–pointed at apex, rounded at base, not toothed on edges, slightly thickened and leathery, the upper surface yellow green and slightly shiny, the midvein and curved side veins light yellow and sunken, and the lower surface dull light green with raised veins.

Flowers numerous, fragrant, consisting of tubular base (hypanthium) $^1/_2$ inch long, which bears the 5–lobed calyx and other parts; corolla $^3/_8$ inch long, narrowly funnel–shaped with 5 elliptic lobes $^1/_4$ inch across; 5 stalkless stamens on throat of corolla and alternate with lobes; and pistil with inferior 2–celled ovary, long threadlike style, and enlarged stigma.

Fruits compound (multiple), composed of numerous crowded fruits pressed together in a fleshy ball around the central whitish base nearly 1 inch across. Each individual fruit about $^3/_8$ inch long and $^1/_2$ inch wide has 5 or sometimes 4 short brownish lobes $^1/_{16}$ inch long from a fleshy base and inferior ovary. There are 4 or fewer light brown seeds more than $^1/_{16}$ inch long. Flowering in spring and summer, maturing fruits in summer and autumn.

The wood is light in color and workable, suited especially for veneers and cabinetwork. It has not yet been used locally.

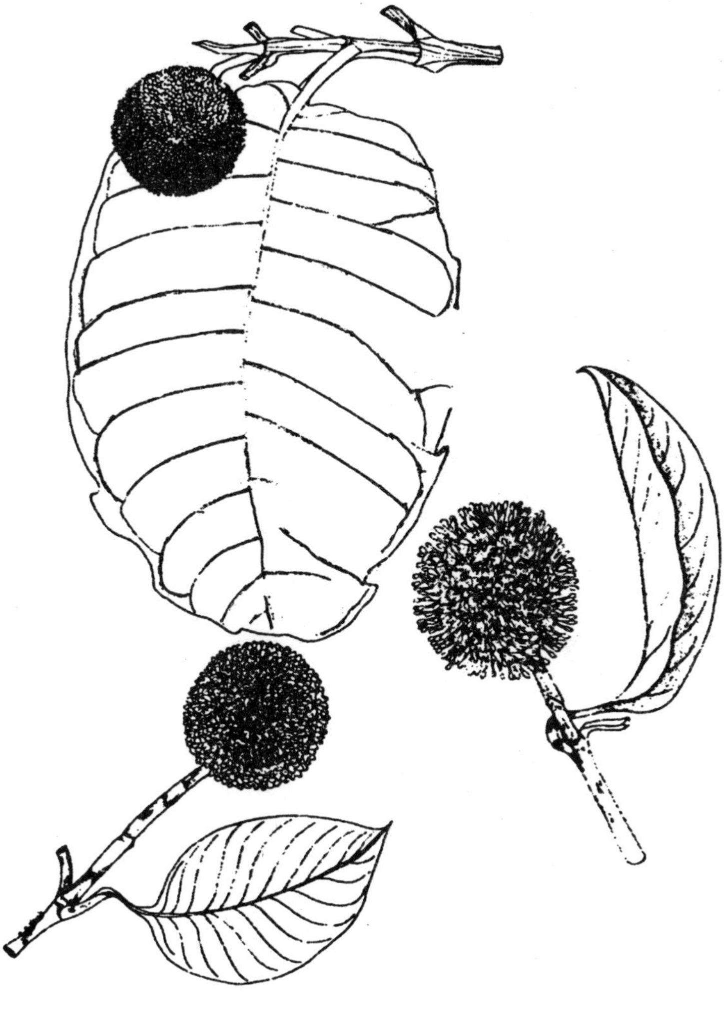

**Fig. 377 : Kadam (*Anthocephalus chinensis*)**
Flowering twigs (left), fruiting twig (right), two–thirds natural size.

Kadam is uncommon in Puerto Rico. This promising, very fast growing tree is planted mainly for timber but sometimes for ornament such as in plazas.

**Public Forests :** Luquillo, Maricao, Rico Abajo.

**Range :** Native of southeast Asia to East Indies and Philippines but planted as a forest tree and in botanical gardens in other tropical countries.

**Other Common Names :** Kadam (commerce); laran (Malay).

**Botanical Synonym :** *Anthocephalus cadaba* (Roxb.) Miq.

## 378. Quina (*Anitirhea acutata*)

This shrub or small tree is easily recognized by its small sticky or viscid, shiny yellow–green leaves. Other characters for identification include : (1) erect long slender wiry gray twings, mostly paired, with many crowed, slightly raised, half–round to round leaf scars; (2) leaves opposite and crowed at ends of twigs, elliptic to oblong, mostly $^{3}/_{4}$–$1^{1}/_{2}$ inches long and $^{3}/_{8}$–$^{5}/_{8}$ inch wide; (3) few very fragrant small white narrowly tubular flowers $^{3}/_{8}$–$^{5}/_{8}$ inch long borne erect in a line at end of short stalk from leaf base; and (4) small elliptic black fleshy fruits $^{3}/16$ inch long.

Evergreen much branched shrub or small tree to 15 feet high and 3 inches in trunk diameter, reported to reach 25 feet, often with several trunks from base. The bark is blackish gray, smoothish, becoming thick, rough and shaggy, with short scaly plates. Inner bark is pinkish brown and bitter. The twigs, finely hairy when young, have faint rings at nodes. Older twigs have corky outgrowths.

The leaves have light yellow finely hairy leafstalks $^{1}/_{10}$–$^{1}/_{8}$ inch long and blades sometimes to 2 inches long and $^{3}/_{4}$ inch wide. Paired narrow pointed brown stipulates $^{1}/_{8}$ inch or more in length leave a faint ring scar but persist sometimes on twigs after leaves fall. Blades are sticky, thin or slightly thickened, short–pointed at apex, short–pointed or rounded at base, slightly turned under at edges, hairless.

Flower clusters (cymes) bear few nearly stalkness fragrant flowers erect on one side of the end of a slender stalk $^{1}/_{2}$–1 inch long, which arises at base of a leaf. The flower is composed of short tubular yellow–green calyx $^{1}/_{16}$ inch long, with 5 long teeth; white corolla $^{3}/_{8}$–$^{5}/_{8}$ inch long, with very narrow tube and 5 spreading lobes $^{3}/_{16}$–$^{1}/_{4}$ inch across; 5 stamens inside throat of tube and alternate with lobes; and pistil with inferior 4–6 celled ovary and slender style. The fruits (drupes) bear the calyx at apex and contain few seeds within the 4–6 celled stone. With flowers and fruits through the year.

The wood is light brown, slightly hard.

Common in dry forests, also shrub thickets along beaches, from sea level to 800 feet altitude.

**Fig. 378 : Quina (*Antirhea acutata*)**
Natural size

## 379. Quina (*Antirhea coriacea*)

This tree, scattered in moist mountains and along the north Bihar, has : (1) opposite elliptic or ovate leaves 2–5 inches long and $1^1/_4$–$2^1/_2$ inches wide, slightly thickened, dull green; (2) several to many flowers $^3/_8$ inch long, the white corolla with slender tube and 4 short lobes, erect in few–branched clusters; and (3) oblong or elliptic black fruit $^3/_8$ inch long.

Evergreen tree to 40 feet tall and 8 inches in trunk diameter. Twigs hairless, with rings at nodes.

The opposite hairless leaves have leafstalks $^1/_4$–$^5/_8$ inch long. The paired oblong stipules are $^1/_4$–$^5/_{16}$ inch long. Blades are slightly thickened, the apex blunt or abruptly long–pointed, the base blunt or short–pointed, with $^5/_6$ pairs of side venins, prominent on under surface.

Flower stalks (cymes) arising from leaf base have slender stalk and few branches $1^1/_4$–2 inches, long, shorter than leaves. The fragrant flowers have a short tubular calyx $^1/_{16}$ inch long with 4 very short teeth; corolla $^5/_{16}$ inch long, the narrow tube much longer than the 4 lobes; 4 stamens inside throat of tube and alternate with lobes; and pistil with inferior 2–celled ovary and threadlike style. The fruit (drupe) has a 2–seeded stone. Bearing flowers in spring and fruits in summer.

The wood is described as yellowish, hard, heavy, and durable. Elsewhere it has served for posts and construction and the bitter bark in home remedies.

Scattered and common in moist coast, limestone, forests at 200–2,500 feet altitude on moist mountains and along the north of Bihar.

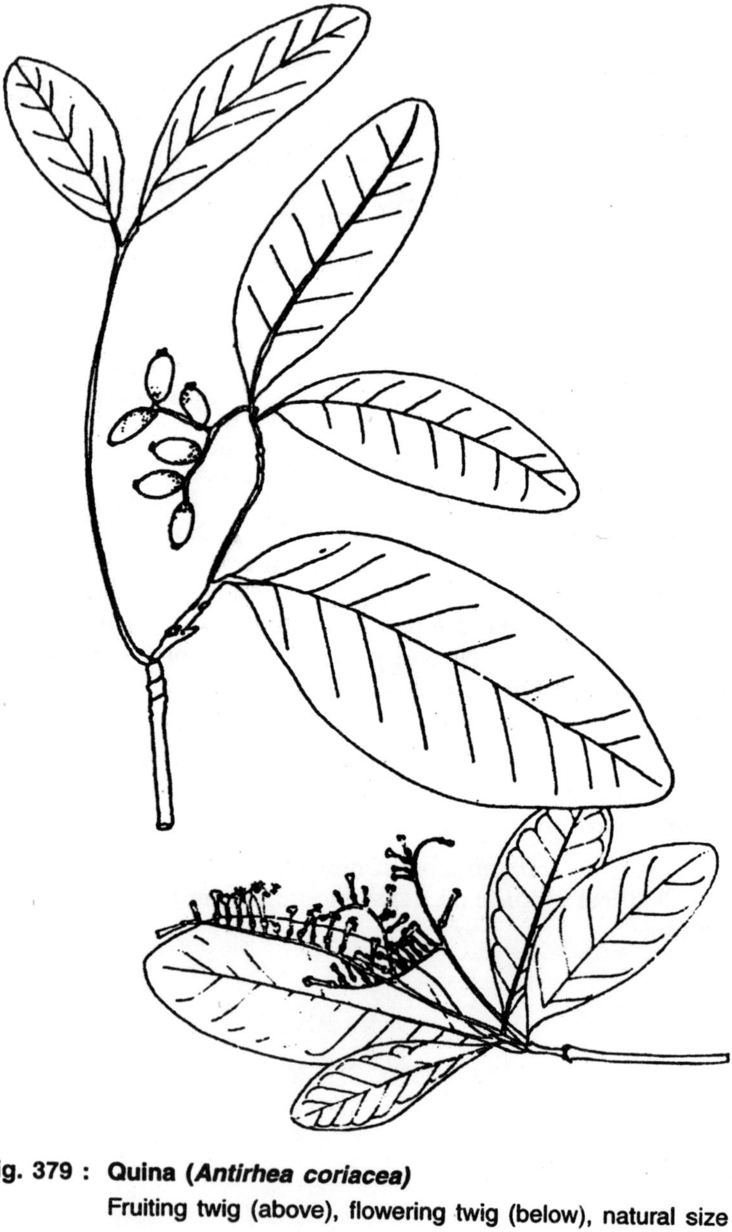

**Fig. 379 : Quina (*Antirhea coriacea*)**
Fruiting twig (above), flowering twig (below), natural size

## 380. Palo Iloron (*Antirhea Lucida*)

Palo Ilorón is distinguished by : (1) opposite leaves with faint lateral veins not distinct near margins, 1 or 2 pairs at end of light gray twig; (2) flower cluster (cyme) 3–5 inches long, consisting of 2 nearly horizontal, curved very slender, light green branches bearing on upper side many small erect stalkless light yellow and greenish tubular flowers nearly $^1/_4$ inch long; and (3) oblong fruit (drupe) $^1/_4$–$^3/_8$ inch in length, red to black.

A small deciduous tree 15–30 feet tall and 8 inches in trunk diameter, with spreading crown of smooth light gray branches, or a shrub, hairless almost throughout. The light bark is smooth with many dots (lenticels). Inner bark is light brown, turning darker on exposure, slightly bitter. Twigs are light gray, with faint rings at nodes. Young tips of twigs are green with a pointed bud about $^1/_4$ inch long formed by stipules, which are $^1/_4$ inch long, long–pointed, and shed very early.

The opposite leaves have leafstalks $^1/_4$–$^3/_8$ inch long and blades 3–6 inches long and $1^1/_2$–2 $^3/_4$ inches wide. Blades are short–pointed at base, short–pointed or rounded at apex, shiny green above and beneath light green and less shiny. On lower surface several tiny depressions or holes are formed in the angle between a lateral vein and midrib.

Each small flower has a green basal tube (hypanthium) $^1/_{16}$ inch long and broad, enclosing the inferior 2–celled ovary; green tubular calyx $^1/_{16}$ inch long, minutely 5–4 lobed at apex; light yellow corolla with tube about $^1/_8$ inch long and 5–4 spreading lobes nearly $^1/_4$ inch across, the tube bearing as many minute brown stamens as lobes; and light yellow style nearly $^3/_{16}$ inch long with 2–lobed stigma. The 2–seeded fruit retains calyx at apex. With flowers in spring and fruits in summer.

The wood is light brown and hard.

Common locally in dry limestone, moist coastal, and moist limestone forests from sea level to 800 feet altitude in southwestern Kerala.

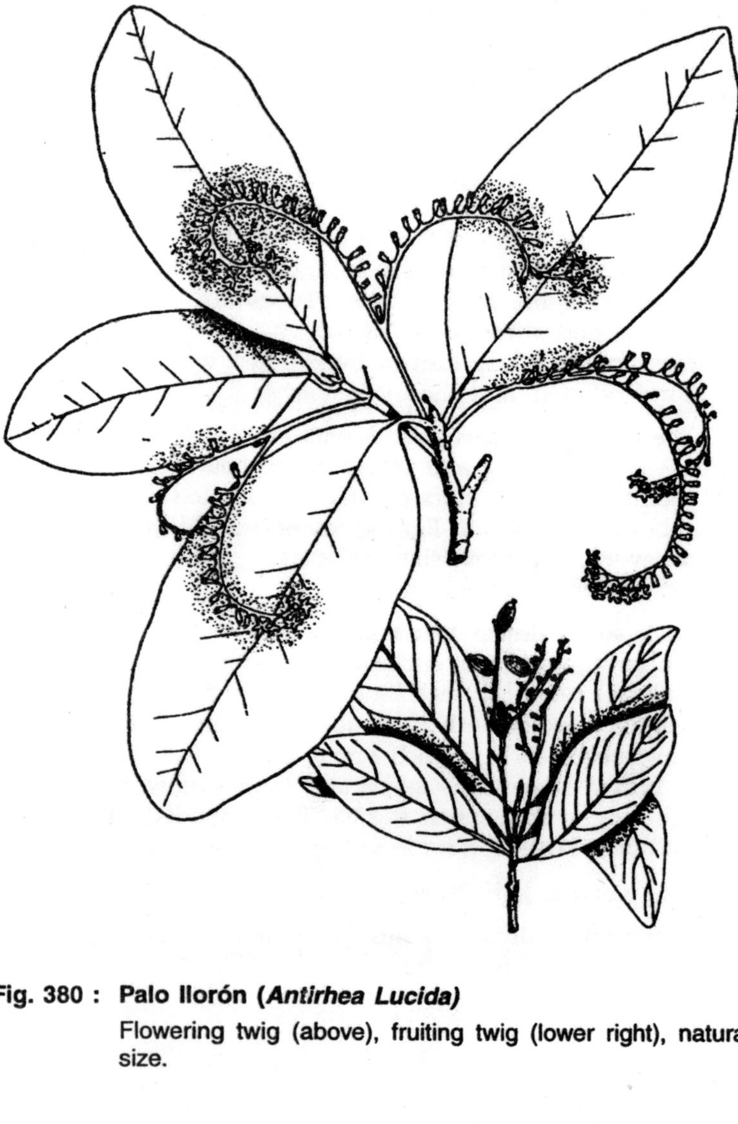

**Fig. 380 : Palo llorón (*Antirhea Lucida*)**

Flowering twig (above), fruiting twig (lower right), natural size.

## 381. Quina (*Antrihea sintenisii*)

A tree confined to moist parts of Assam and characterized by : (1) opposite oblong to lanceolate leaves 2–4$^1/_2$ inches long and $^5/_8$–1$^1/_4$ inches wide; (2) several flowers nearly $^3/_8$ inch long, the cream–colored corolla with narrow tube and 4 short lobes, stalkless along the end of a short stalk from base of leaves; and (3) elliptic fruit $^3/_8$ inch long.

Evergreen tree reported to reach 50 feet in height. Twigs slender, brown or gray brown, hairless, with rings at nodes.

The opposite hairless leaves have leafstalks $^3/_{16}$–$^3/_8$ inch long and paired ovate stipules $^1/_4$ inch long. Blades are slightly thickened, the apex blunt, rounded, or short–pointed, the base mostly short–pointed, the midvein prominent but side veins very obscure.

Flower clusters (cymes) from base of leaf have branches $^5/_8$–1 $^1/_4$ inches long bearing several stalkless flowers. The flowers are composed of short tubular calyx $^1/_{16}$ inch long with 4 short broad teeth; cream–colored hairy corolla $^1/_4$ inch long, the tube much longer than the 4 lobes; 4 stamens inside throat of tube and alternate with lobes; and pistil with inferior ovary, threadlike style, and 2–lobed stigma. The fruit (drupe) is black at maturity. Collected with flowers and fruits in January

The wood is described as yellowish.

Rare in moist forest at lower and middle altitudes in Assam.

**Fig. 381 : Quina (*Antirhea sintenisii*)**

Fruiting twig (above), flowering twig (below), natural size.

## 382. Dágame, degame (*Calycophyllum candidissimum*)

Dagame, a recent introduction, is a handsome ornamental and produces a useful wood. During the long flowering period, the trees are covered with white. Distinguishing characters are: (1) paired elliptic leaves mostly 2–4 inches long and $^3/_4$–$1^1/_2$ inches wide; (2) clusters of many small white flowers only about $^3/_8$ inch long and broad, a few with an enlarged rounded long–stalked white lobe or blade $^1/_2$–$1^1/_4$ inches in diameter formed from calyx; and (3) small cylindric slightly flattened blackish hairy capsules $^1/_4$–$^3/_8$ inch long.

An evergreen small planted tree 30 feet high and 5 inches in trunk diameter, attaining large size in age and in the forest. The bark is gray or reddish brown, smoothish, becoming fissured and scaly with thin plates. The twigs are brown and hairless or nearly so, ringed at nodes.

The opposite leaves have slender petioles $^1/_4$–$^3/_4$ inch long and at base a pair of scales (stipules) to $^3/_8$ inch long. Leaf blades are long or short–pointed at both ends, not toothed on edges, thin, becoming hairless or nearly so, shiny green on upper surface and paler beneath.

The flowers are in large branched, flattened clusters (corymbs) at ends of twigs, mostly stalkless in 3's from narrow pointed buds. The hairy cylindric base bears a minute calyx and white corolla with short tube and 4 spreading lobes; 4 short stamens attached on the hairy end of tube and alternate with lobes; and pistil with inferior 2–celled ovary, slender style and 2–forked stigma. A few flowers in a large cluster stigma. A few flowers in a large cluster produce from the calyx an enlarged rounded or broadly ovate long–stalked thin white lobe or blade, which makes the display of color. The 2–celled capsules contain many minute winged brown seeds. Flowering in winter, the whitish lobes long persistent into spring.

The heartwood is gray brown and the sapwood pale yellow. The wood is hard, heavy (specific gravity 0.65–0.70), with growth rings, fine–textured strong, and takes a good polish. Principal uses elsewhere include construction, carpentry, cabinetwork, agricultural implements, turned articles, and tool handles, Special uses are archery bows, fishing rods and fine–toothed combs.

Common as an ornamental in Bihar. Where native, classed as an important honey plant. The trees merit wider cultivation for the long–persistent showy white flowers and useful wood.

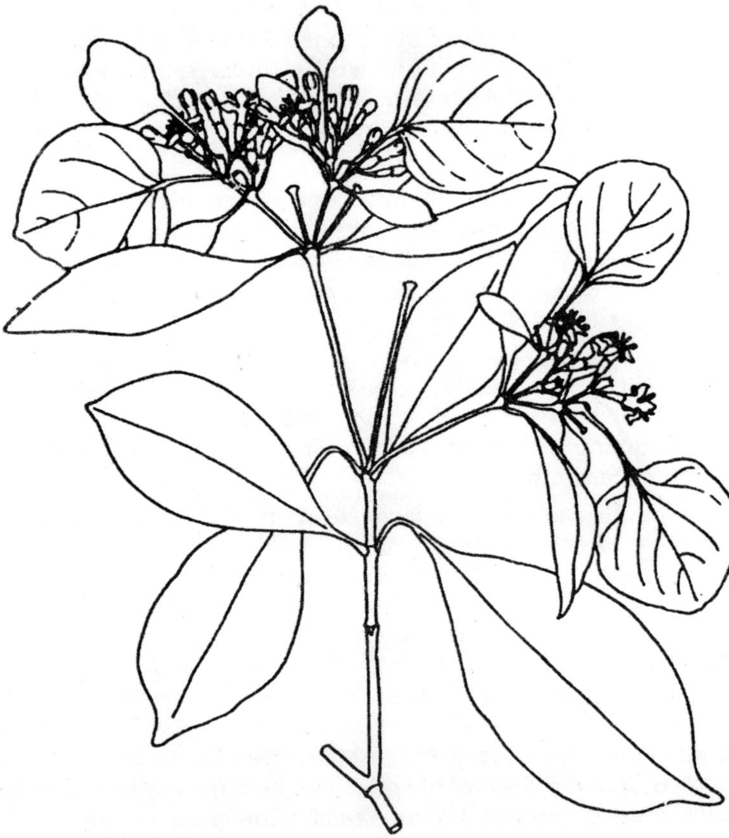

**Fig. 382 : Dágame, degame (*Calycophyllum candidissimum*)**
Flowering twig, natural size.

## 383. Martín Avila (*Chione venosa*)

This rare small tree is characterized by : (1) opposite narrowly elliptic leathery leaves, short–pointed at both ends and turned under at edges, with few side veins; (2) erect terminal flower cluster with whitish axis and branches bearing several to many fragrant tubular whitish flowers about $^1/_2$ inch long, with 5 rounded corolla lobes; and (3) shiny purplish black long fleshy fruits $^1/_2$–$^3/_4$ inch long.

Evergreen tree 20–60 feet in height, with smooth dark trunk 6–14 inches in diameter and with dense crown of foliage, hairless throughout. Twigs are gray, green when young, ringed at nodes.

The yellow–green slightly flattened leafstalks are $^1/_4$–$^3/_4$ inch long. Blades are $2^1/_2$–$4^1/_2$ inches long and $^3/_4$–$1^1/_2$ inches wide, above dark green and slightly shiny and beneath dull light green.

The erect whitish flower cluster (cyme) is 2–3 inches broad and 4–5 inches long including stalk, slightly fattened at top. The whitish basal tube (hypanthium) enclosing the inferior 2–celled ovary is $^3/_{16}$ inch long, bearing 5 minute calyx lobes at apex; the white corolla tube $^3/_{16}$ inch long and $^1/_8$ inch in diameter, faintly pink tinged in bud, has 5 rounded spreading lobes nearly $^1/_8$ inch long; 5 white stamens more than $^3/_8$ inch long, with narrow orange anthers, are attached near base of corolla tube; and the pistil has a white style $^3/_8$ inch long. The minute calyx lobes remain at apex of the fruit (drupe), which has slightly bitter flesh and a grooved stone containing 2 narrow seeds. With flowers and fruits in spring and summer.

Rare in moist forests from sea level to 1,500 feet altitude in northern limestone and foothills of Rajasthan.

**Fig. 383 : Martín Avlla (*Chione venosa*)**
Flowering twig and fruits (lower left), natural size.

## 384. Café Excelsa, Dewevre Coffee (*Coffea dewevrei*)

Café excelsa or Dewevre or Liberian coffee is uncommonly planted in mountains of Dehradun. It is distinguished from the common café or coffee, No. 240, *Coffea arabica* L.,* by : (1) larger size, a small tree 20–30 feet high unless pruned back, with a definite trunk 4–6 inches in diameter; (2) larger opposite elliptic leaves 6–14 inches long and 3–6 inches wide, slightly thick and leathery; (3) white fragrant flowers clustered at leaf bases and nodes back of leaves, the corolla with 5–7 (instead of 5 only) widespreading lobes $1^1/_4$ inches across; and (4) slightly larger elliptic dark red or reddish yellow berries $^5/_8$–$^3/_4$ inch long and $^1/_2$–$^5/_8$ inch broad.

An evergreen small cultivated tree, hairless throughout. The bark is gray and much fissured, the inner bark whitish and bitter. Twigs are stout, shiny, with enlarged ringed nodes and much raised large half–round leaf scars.

The paired leaves have 2 broad short–pointed stipules $^3/_{16}$ inch long, persistent at the base of the stout petioles $^1/_2$–$^3/_4$ inch long. The large thick, leathery blades are blunt at apex, short–pointed at base, and not toothed on edges, with relatively shiny green to dark green, and the lower surface pale yellow green.

The flowers almost stalkless at nodes have a calyx of hairy teeth on the base (hypanthium) less than $^1/_8$ inch long and broad; the white corolla has a narrow cylindric tube $^3/_8$ inch long and 5–7 widespreading lobes $^1/_2$ inch long, narrow and blunt–pointed; 5–7 stamens inserted in mouth of corolla tube; and pistil with 2–celled inferior ovary and long threadlike 2–forked style. The elliptic fleshy fruits on stalks $^1/_8$ inch long have thin pulp and usually 2 large elliptic seeds or beans, flattened on the side where they join. Flowering and fruiting throughout the year.

The wood is whitish and hard.

The seeds of the species of coffee (genus *Coffea*) contain caffein and when roasted and ground produce the well known beverage. The tree coffee is sometimes planted in the mountains of Dehradun but is nowhere common. It belongs to the Liberian group and is regarded also a variety of Liberian coffee (*Coffea liberica* Hiern.*) in a broad sense. This species is attacked by the coffee leafminer (*Leucoptera coffeella*), the worst insect pest of coffee in Dehradun.

**Fig. 384 : Café excelsa, Dewevre Coffee (*Coffea dewevrei*)**
One–half natural size.

## 385. Jayajabico, Black Torch (*Erithalis fruti–cosa L.*)

This shrub common near beaches and in openings of lowland forests sometimes becomes a small tree. It is recognized by : (1) compact rounded masses of dark green foliage, the paired leaves broadly elliptic or nearly round, mostly 2–$3^1/_2$ inches long and 1–2 inches wide, slightly leathery, the stipules forming a narrow ring around twig and leaving a ring scar; (2) terminal clusters of small tubular white fragrant flowers $^3/_8$ inch long and broad, with 5–7 corolla lobes; and (3) black rounded fleshy fruits $^3/16$ inch in diameter.

Evergreen compact shrub commonly less than 5 feet high, sometimes a large shrub or small tree to 30 feet high and 4 inches in trunk diameter, hairless throughout. Bark dark gray, smoothish, often warty, the inner bark light yellow, bitter. Twigs light green, with rings at nodes, becoming gray. Stipules forming slightly pointed bud $^1/_{16}$ inch long, afterwards a narrow ring $^1/_{16}$ inch high at each node.

Leaves opposite, with light green leafstalk $^1/_4$–1 inch long. In exposed areas, the leaves are reduced to only 1 inch long and $^1/_2$ inch wide. Blades blunt or rounded at apex, gradually narrowed to base, not toothed on edges, with few inconspicuous veins, the upper surface dark green and slightly shiny, the lower surface dull green.

Flower clusters (panicles) are terminal and lateral, erect, 2–3 inches long on a long stalk, with several to many short–stalked flowers. Each flower consists of tubular light green calyx 5–7 toothed; white corolla with short tube and 5–7 narrow spreading lobes; white stamens as many as corolla lobes, alternate and inserted near base; and pistil with inferior rounded light green ovary 5–10 celled with 1 ovule in each cell, slender white style, and lobed stigma. Fruits (drupes) light green, becoming black, have calyx persistent in ring at apex, as many longitudinal grooves as cells, and contain 5–10 narrow nutlets. Flowering and fruiting through the year.

The light brown wood has dark streaks in heartwood. It is very hard, heavy, and fine textured. It could be used for articles of turnery but is available only in small sizes. Elsewhere the durable wood has served for piles and posts. The resinous wood when split has been bound into torches.

Common near beaches and in openings of dry limestone, moist limestone, and lower forests from sea level to 400 feet altitude. Also in moist Bihar round.

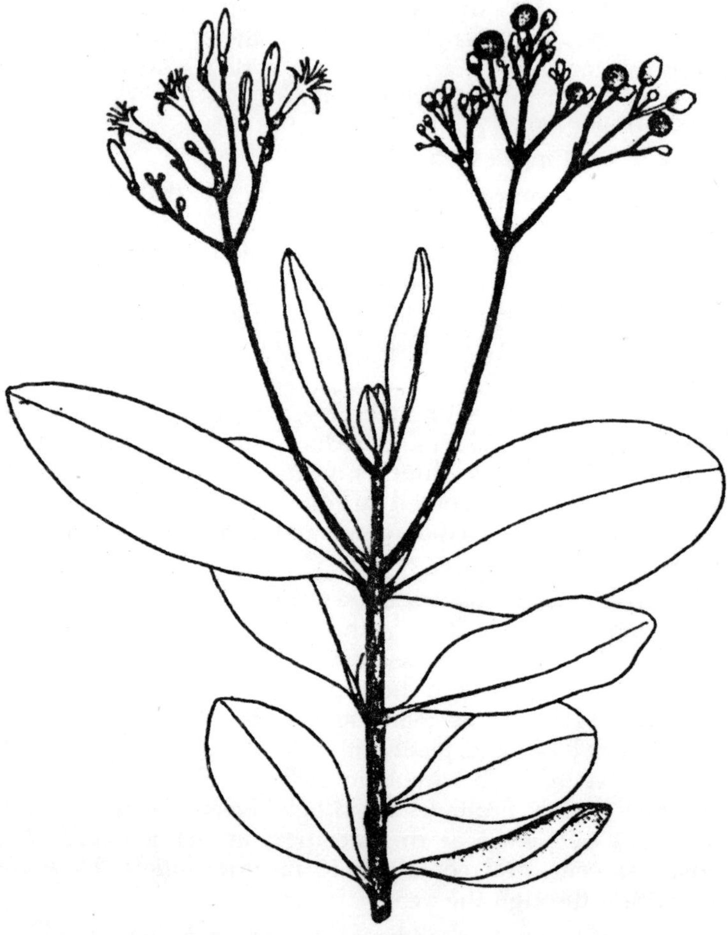

**Fig. 385 : Jayajabico, black torch (*Erithalis fruticosa L.*)**
Natural size.

## 386. Plateado (*Exostema ellipticum*)

This small tree is rare in central and western mountains of Rajasthan. Distinguished by : (1) opposite elliptic leaves $1^1/_2$–$3^1/_2$ inches long and $^3/_4$–$1^1/_4$ inches wide, with 2 blunt–pointed stipules $^1/_8$ inch long; (2) large flowers $2^1/_2$ inch long, the corolla with narrow tube and 5 very narrow lobes, changing color from white and pink to dark red; and (3) cylindric brown seed capsules $^1/_2$–$1^1/_4$ inches long.

An evergreen shrub or small tree to 20 feet high and 4 inches in trunk diameter. The bark is gray, smoothish, becoming fissured and rough, the inner bark pink and bitter. The hair less twigs are green when young, becoming light brown, ringed at nodes.

The opposite leaves have at base 2 blunt–pointed stipules $^1/_8$ inch long, green, turning brown, and short petioles $^1/_8$–$^1/_4$ inch long. Blades are mostly short–pointed at both ends, not toothed on edges, thick, leathery, and slightly succulent, the upper surface slightly shiny dark green and hairless, and the lower surface, green, sometimes with tufts of minute hairs in angles of side veins with midvein.

Flower clusters (panicles like corymbs) are terminal, slightly flattened, to 5 inches broad. The flowers borne on widely forking slender stalks $^1/_4$–$^3/_4$ inch long, the flower buds $1^1/_4$–$2\ ^3/_4$ inches long and less than $^1/_8$ inch wide, cream colored. The narrow tubular base (hypanthium) $^1/_4$ inch long bears the short calyx with 5 teeth $^1/_{16}$ inch long; the corolla has a narrow cylindric tube 1–$1^1/_2$ inches long and less than $^1/_8$ inch wide, and 5 very narrow spreading lobes $1^1/_2$ inches long and $^1/_{16}$ inch wide; 5 threadlike stamens united in a tube to base of corolla and extending $1^1/_4$ inches beyond; and pistil with 2–celled inferior ovary, very long threadlike style about $2^1/_4$ inches long, and enlarged stigma. The cylindric seed capsules have calyx teeth at apex, are hard–walled, and split into 2 parts. Flowering in spring and summer and with fruits from summer to winter.

The wood is light brown and hard. Elsewhere reported to be of good quality and used in rural carpentry.

Rare in lower forest at 1,500–3,000 feet altitude in central and western mountains of Rajasthan.

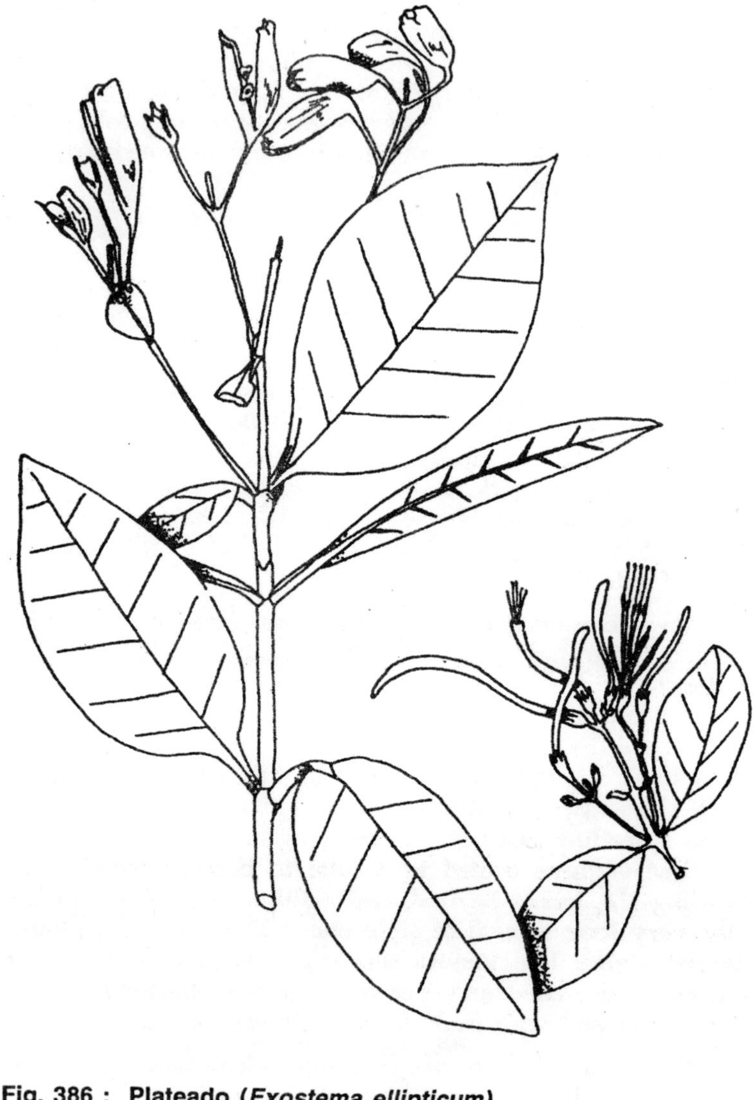

**Fig. 386 : Plateado (*Exostema ellipticum*)**

Fruiting twig (above), flowering twig (lower right), natural size.

## 387. Cucubano Liso, Velvetseed (*Exostema ellipticum Sw.*)

Characters for recognition of this shrub or small tree are : (1) paired short side twigs with crowded raised half–round leaf scars; (2) small elliptic to ovate leaves $^3/_4$–2 inches long and $^1/_2$–$1^1/_4$ inches wide, opposite and crowded at ends of short side twigs; (3) few flowers at the end of a stalk $^1/_2$–1 inch long at leaf base, about $^3/_8$ inch long, the yellowish or whitish corolla with narrow tube and 4 lobes; and (4) round fruit $^1/_4$ inch in diameter, turning to red and black.

Evergreen shrub or small tree to 200 feet high and 4 inches in trunk diameter, reportedly larger. The bark is slaty gray or gray brown, smooth, with minute dots or lines, sometimes mottled and peeling off in thin flakes. Inner bark is gray brown or pinkish and slightly bitter. The gray slender twigs are hairy when young, with whitish dots (lenticels) and rings at nodes.

The opposite leaves have slender hairy leafstalks $^1/_8$–$^1/_4$ inch long and paired long–pointed hiary stipules $^1/_8$ inch long, which form the bud and shed early. The thin blades silky hiary when young have a rounded apex with minute point, short–pointed or rounded base, edges not toothed, few curved side veins, upper surface dull green and almost hairless, the lower surface light green with fine, pressed silky hairs, especially on veins.

The flowers consist of a cup–shaped hairy calyx $^1/_8$ inch long; yellowish or whitish silky hairy corolla $^1/_4$ inch long with narrow tube and 4 lobes; 4 stamens in throat of tube alternate with lobes; and pistil with minute inferior 4–celled ovary, slender style, and dotlike stigma. Fruits (drupes) finely hairy, slightly fleshy, with calyx, large stone, and few seeds. Bearing flowers in spring and fruits in summer.

The wood is light brown, hard, heavy (specific gravity 0.83), and fine–textured.

Common in dry coast and limestone forests at 100–400 feet altitude along south coast of Tamil Nadu.

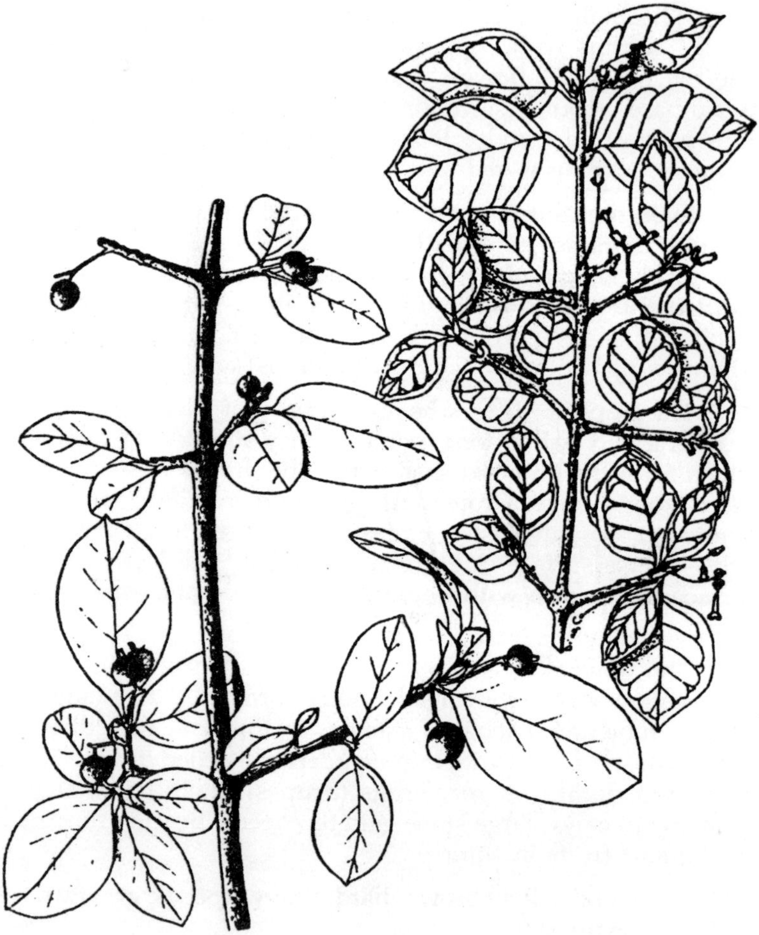

**Fig. 387 : Cucubano liso, velvetseed (*Guettarda ellipticum)***
Fruiting twig (left), flowering twig (right), natural size.

## 388. Cucubano (*Guettarda krugii*)

Shrub or small tree of south–western Bihar identified by : (1) paired stout gray twigs with raised half–round to round leaf scars; (2) opposite small mostly elliptic leaves 1–4 inches long and $^1/_2$–2 $^1/_2$ inches wide, with finely wavy edges turned under, thick and stiff, beneath soft hairy with prominent network of fine raised veins; (3) flowers 1 to few on short stalks at leaf base or bend of twig, $^3/_4$–1$^1/_4$ inches long, the white densely hairy corolla with narrow tube and 5–8 lobes; and (4) round or pear–shaped densely hairy yellow or reddish fleshy fruits $^1/_2$–1 inch long.

Evergreen shrub or small tree to 20 feet high and 3 inches in trunk diameter, reportedly to 30 feet tall, with few slender branches ending in clusters of leaves and without a definite crown. Bark gray, smooth, with few thin flakes peeling off and exposing green patches beneath. Inner bark light brown, tasteless. The twigs are densely brown hairy when young, becoming gray and hairless, stout with ringed nodes and both long and short internodes.

Leaves have densely hairy leafstalks $^1/_4$–$^5/_8$ inch long and paired brown hairy ovate pointed stipules $^1/_4$–$^1/_2$ inch long, which form the bud and shed early. Blades are mostly elliptic, also nearly round or obovate, blunt or short–pointed at apex, rounded or slightly notched at base, the upper surface almost hairless with several pairs of nearly straight side veins, the lower surface paler and yellowish green, with raised veins and soft hairy.

The fragrant flowers have a brown hairy tubular calyx $^1/_4$ inch long; densely white hairy corolla $^3/_4$–1 inch long with narrow tube and 5–8 narrow spreading lobes more than $^1/_2$ inch across; stamens in throat of corolla tube, as many as lobes and alternate; and pistil with minute inferior 4–celled ovary, long slender style, and dotlike stigma. Fruits (drupes) are greenish when immature, narrowed at base, have calyx at apex, almost tasteless pulp, large stone, and few seeds. Flowering and fruiting from spring to fall.

Wood light brown, hard and brittle.

Uncommon in dry and limestone forests at 100–500 feet altitude in southwestern Bihar.

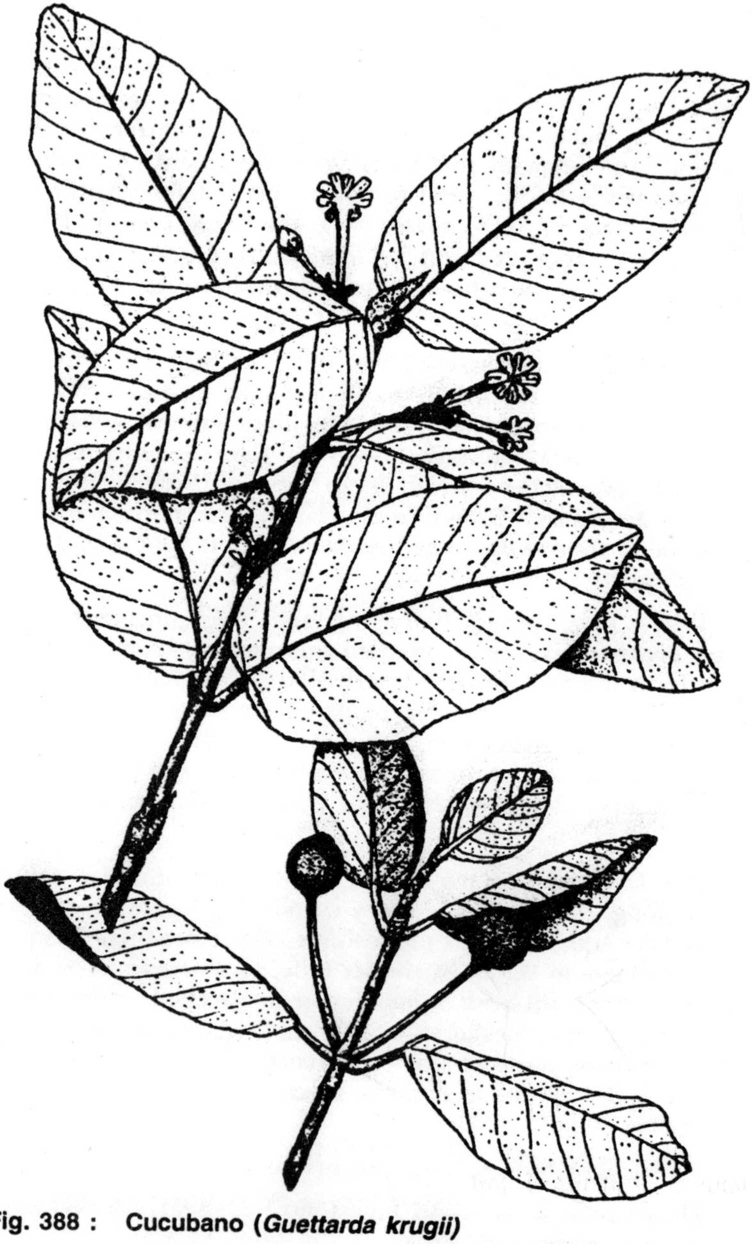

**Fig. 388 : Cucubano (*Guettarda krugii*)**
Flowering twig (above), fruiting twig (below), natural size.

## 389. Cucubano (*Guettarda ovalifolia*)

This small tree is recognized by : (1) its paired crowded elliptic leaves with finely wavy edges and deeply sunken curved side veins; (2) flowers $^3/_4$ inch long, several crowded at end of slender stalk 2–5 inches long, white, pink, or light purple, tubular with 5–6 lobes; and (3) rounded finely hairy fruits $^1/_4$ inch in diameter.

Evergreen small tree 30 feet high and 4 inches in trunk diameter. The bark is gray and smooth, slightly warty or becoming slightly fissured, the inner bark light pink and slightly astringent. The twigs are brownish, hairy when young, with rings at nodes.

The opposite leaves crowded near apex of twig have hairy leafstalks $^1/_4$–$^3/_4$ inch long and paired very narrow pointed hairy stipules $^1/_8$ inch long. The blades vary in size, $1^1/_2$–5 inches long and $^3/_4$–$2^1/_2$ inches wide, rounded with minute point at apex and short–pointed or rounded at base, thin or slightly thickened, the upper surface dull green and hairless, and the lower surface dull light green and finely hairy on the raised light yellow veins.

The flowers consist of cuplike hairy calyx $^1/_8$ inch long; white or light purple finely hairy corolla with narrow tube about $^5/_8$ inch long and 5–6 spreading lobes $^3/_8$ inch across; 5–6 stamens in throat of tube and alternate with lobes; and pistil with minute 4–6 celled inferior ovary, slender style, and dotlike stigma. The fruits (drupes) finely hairy, slightly fleshy, with large stone and few seeds. Flowering from spring to fall and with fruits in summer and fall.

Wood light brown and hard.

Uncommon in moist limestone and lower forests at 300–2,800 feet altitude in western mountains Mossurie.

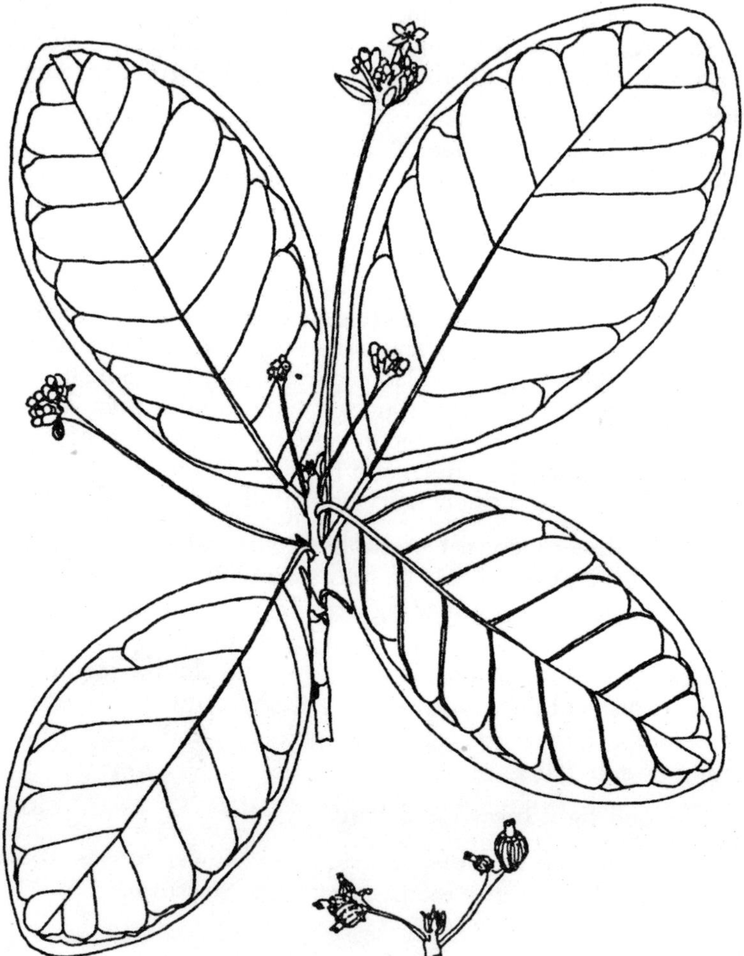

**Fig. 389 : Cucubano (*Guettarda ovalifolia*)**
Flowering twig (above), fruits (below), natural size.

## 390. Cucubano de Vieques, blackbérry (*Guettarda parviflora)*

Distinguishing characters of this shrub or small tree are : (1) paired slender twigs with ringed nodes formed by scalelike pointed stipules; (2) leaves opposite and crowded, oblong to elliptic or obovate, $^3/_4$–$2^1/_4$ inches long and $^3/_8$–1 inch wide, upper surface shiny yellow green; (3) flowers 1–3 on short stalk at leaf base, nearly $^3/_8$ inch long, the pinkish corolla with narrow tube and 4–6 unequal lobes; and (4) round black fruit $^1/_4$ inch in diameter.

Evergreen shrub or small tree to 25 feet in height, with 1 to several trunks to 5 inches in diameter. The bark is gray and smooth to scaly, the inner bark light brown and tasteless. Twigs are slender, gray or brown, finely hairy when young, with whitish dots (lenticels).

The leaves have hairy brown petioles about $^1/_8$ inch long and paired scalelike brown stipules less than 1/16 inch long, pointed and hairy, which form the bud and shed early. Blades are short–pointed or blunt at apex, rounded at base, not toothed on edges, thin upper surface shiny yellow green and hairless, lower surface dull light green and slightly hairy.

Flowers are composed of tubular hairy calyx 1/16 inch long on short 4–5 celled inferior ovary; pinkish hairy corolla $^1/_4$ inch long with narrow tube and 4–6 unequal spreading lobes more than $^1/_8$ inch across; stamens in throat of corolla tube, as many as lobes and alternate; and pistil with as many ovules as cells in ovary, slender style and dotlike stigma. Fruits (drups) finely hairy, slightly fleshy, with large stone and few seeds. Noted with flowers in spring and with fruits nearly through the year.

The wood is light brown and hard.

Scattered and uncommon in moist forest from sea level to 700 feet altitude in eastern and southeastern Bihar.

**Fig. 390 : Cucubano de Vieques, blackberry (*Guettarda parviflora*)** Twig with flower buds (above), fruiting twig (lowerleft), natural size.

## 391. Roseta (*Guettarda pungens*)

The paired small stiff elliptic leaves ending in a sharp slender yellowish spine $^1/_8$ inch long distinguish this shrub, sometimes a small tree, of mountains in western Puerto Rico. Other characters for identification are : (1) white flowers $^5/_8$–$^3/_4$ inch long, single at leaf bases, the white to pinkish corolla with narrow tube and 5–6 spreading lobes; and (2) round finely hairy fruit $^1/_4$ inch in diameter, blackish or violet

Evergreen shrub or small tree to 20 feet high and 4 inches in trunk diameter, often with several branches from base. The bark is gray and smooth or becoming scaly, the inner bark slightly bitter. Twigs are paired, gray, finely hairy when young, with ringed nodes.

The opposite, often crowded leaves have short finely–hairy leafstalks $^1/_8$ inch long and paired very narrow pointed stipules about $^1/_8$ inch long. Blades, are $^3/_4$–$1^1/_4$ inches long and $^3/_8$–$^7/_8$ inch wide, short–pointed at base, stiff and leathery, becoming hairless, the upper surface green, and the lower surface paler.

The flowers are single and almost stalkless at base of leaves. The calyx is less than $^1/_8$ inch long and slightly 2–lobed; the corolla is white to pinkish, finely hairy, with narrow tube more than $^1/_2$ inch long and 5–6 narrow lobes $^3/_8$–$^1/_2$ inch across; 5–6 stamens in throat of tube and alternate with oboes; and pistil with minute inferior 4–5 celled overy, slender style, and dot–like stigma. Fruits (drupes) finely hairy slightly fleshy, with large stone and few seeds. Flowering from spring to fall, with fruits in fall.

The wood is light brown and hard.

Locally common in lower forest at 1,000–2,700 feet altitude including serpentine slopes, in western mountains of Sivalik.

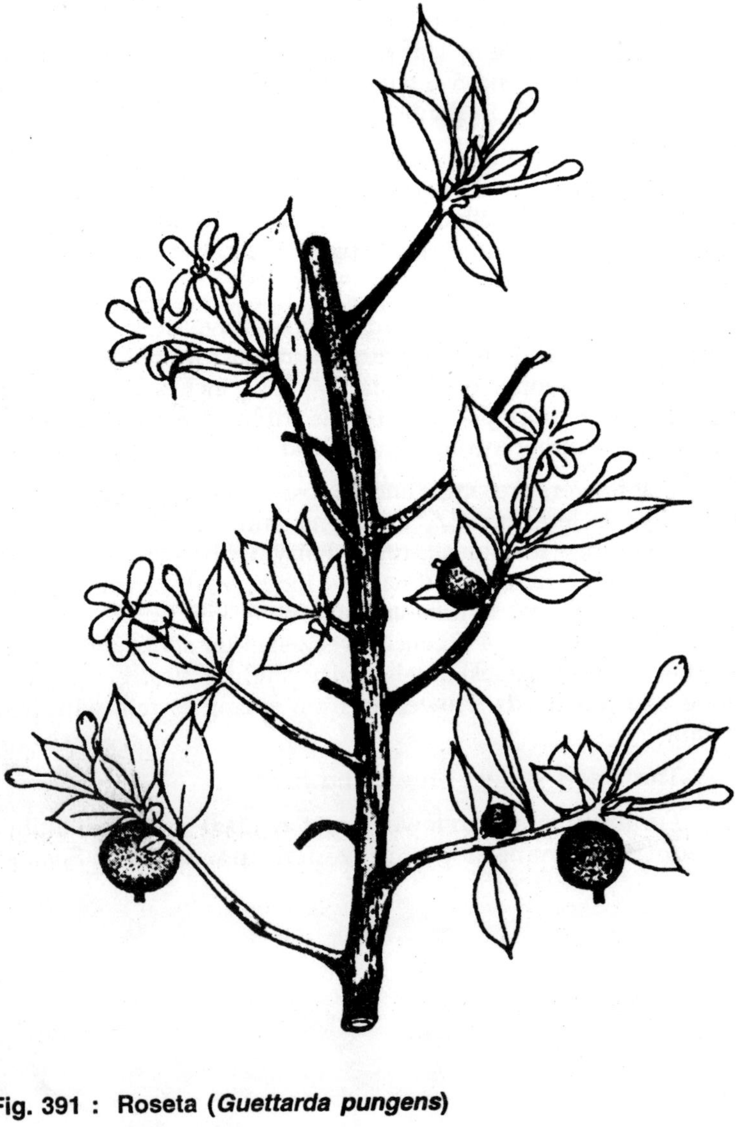

**Fig. 391 : Roseta (*Guettarda pungens*)**
Natural size.

## 392. Cucubano de monte (*Guettarda valenzuelana*)

This small to medium–sized tree of eastern mountains of Meghalaya is identified by : (1) opposite ovate to elliptic or obovate leaves $1^1/_2$–5 inches long and $^7/_8$–$2^3/_4$ inches wide, slightly thickened, with stout leafstalks $^3/_8$–$^3/_4$ inch long, opposite and crowded at ends of twigs; (2) few flowers clustered at end of slender stalk $1^1/_2$–$4^1/_2$ inches long at leaf base, $^3/_4$–1 inch long, the whitish and pinkish hairy corolla with narrow tube and 5–7 lobes; and (3) and round fruit $^3/_4$–$^5/16$ inch in diameter.

Evergreen small to medium–sized tree to 65 feet high and 1 foot in trunk diameter. Bark gray, smooth to slightly fissured. The twigs are brownish gray, hairy when young, with rings at nodes, with both long and short internodes, and with raised half–round leaf scars.

The leaves have paired hairy pointed green stipules $^3/_4$ inch long, which form the bud and shed early. Blades are rounded at apex, rounded or shot–pointed at base, curved under at edges, becoming hairless except for minute hairs on veins beneath, the upper surface dull dark green with several pairs of nearly straight slightly sunken side veins, and the lower surface light green.

The fragrant flowers have cuplike minutely hairy calyx $^1/_8$ inch long, whitish and pinkish hairy corolla nearly $^3/_4$ inch long with 5–7 spreading lobes $^1/_2$ inch across; 5–7 stamens in throat of tube and alternate with lobes; and pistil with minute 2–3 celled ovary, long slender style, and rounded stigma. Fruits (drups) are light green, finely hairy, with brown ring of calyx at apex. Flowering and fruiting from spring to fall.

The sapwood is cream to very light brown, and the heartwood golden brown, often with darker streaks. The attractive wood has good luster, fine texture, and generally straight grain. It is moderately hard, strong, heavy (specific gravity 0.65), and takes a good polish.

Tests of the wood have been made in Meghalaya. Cucubano wood seasons at a moderate rate with a moderate amount of degrade. It rated excellent in turning, good in planing, shaping and mortising, fair in boring and sanding, and poor in screw splitting. It is very susceptible to attack by dry–wood termites and probably is not durable in the ground.

Cucubano wood is seldom used except for fence posts and is available generally in small logs. It probably is suitable for tool handles, fancy boxes, brush backs, turney parts, interior trim, paneling, novelty items, construction, and bridges. Elsewhere it is used in rural construction.

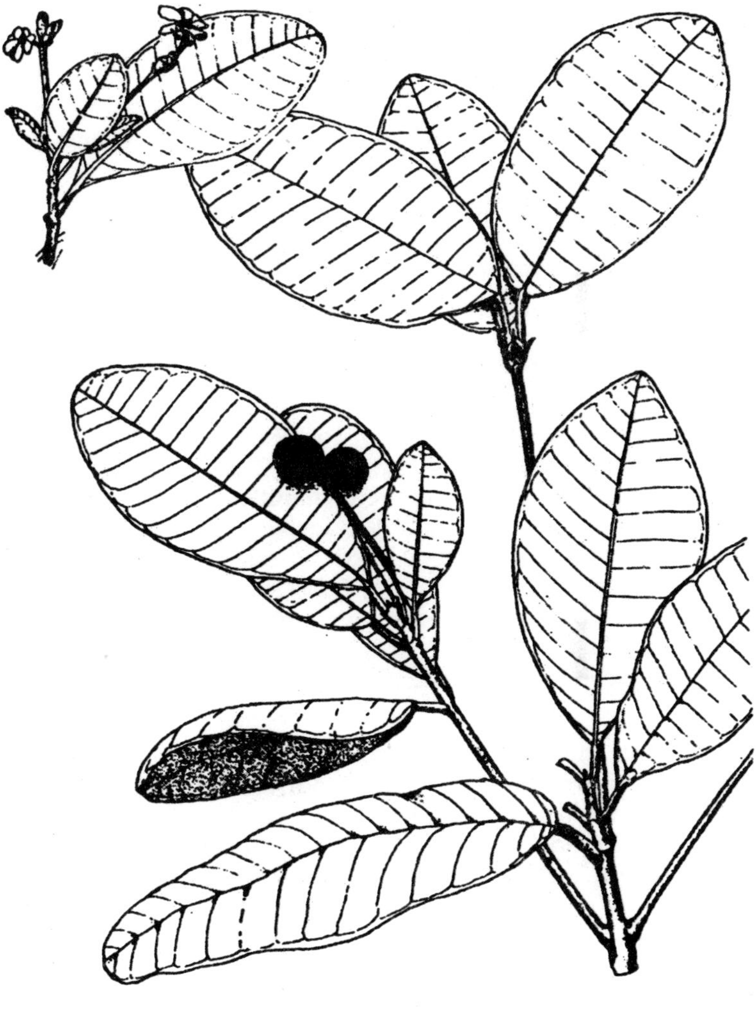

**Fig. 392 : Cucubano de monte (*Guettarda valeuzuelana*)**
Flowering twig (upper left) and fruiting twig, natural size.

## 393. Bálsamo, scarletbush (*Hamelia patens*)

Bálsamo is a handsome shrub of open areas sometimes attaining the size of small tree. It is easily recognized by : (1) reddish or pinkish tinge on various parts, including twigs, petioles, midvein, branches of flower clusters, flowers, and fruits; (2) elliptic leaves mostly in 3's, with long curved side veins; (3) narrow tubular orange–red or scarlet flowers $^3/_4$–$^7/_8$ inch long, erect and nearly stalkless on top side of widely forking branches, with successively smaller flower buds toward end; and (4) fruit an elliptic juicy berry, dark red to purple or blackish.

Commonly a shrub 6–10 feet high, often flowering when smaller, sometimes becoming a small tree 15 feet high and 3 inches in trunk diameter, evergreen. Bark gray, smooth, the inner bark light green and almost tasteless. Twigs purplish tinged, angled, minutely hairy, ending in bud of narrow pointed stipules $^1/_8$ inch long and becoming ringed from scars of leaves and stipules.

Leaves mostly 3 at a node, sometimes 2 or 4 (whorled), joined at base by stipules. Petioles $^1/_2$–2 inches long, slender, slightly grooved above, dark red, minutely hairy. Leaf blades elliptic, 2–6 inches long and 1–3 inches wide, mostly short–pointed at both ends, not toothed on edges, thin. The upper surface is slightly hairy or nearly hairless, dull green with dark red midvein and the side veins long curved and slightly sunken, and the lower surface finely hairy (densely hairy in dry areas), dull light green with prominent pink midvein and raised pinkish side veins.

Flower clusters (cymes) terminal, 2–4 inches long and broad, flattened, composed of widely–forking reddish finely–hairy branches, with fruits and oldest flowers at center and youngest buds towards sides. Flower buds narrowly tubular, orange red. The minute 5–toothed calyx above inferior ovary $^3/_4$ inch long and $^1/_8$ inch wide changes color from orange red to scarlet, is finely hairy, slightly fleshy and stiff, 5–angled and ending in 5 lobes $^1/_{16}$ inch long, and sheds soon after opening. Stamens 5, $^3/_8$ inch long, very narrow, within and inserted on corolla tube. The pistil consists of inferior rounded red ovary $^1/_8$ inch long, 5–celled, with several ovules in each cell, and threadlike reddish style $^5/_8$ inch long. The elliptic berries $^3/_4$–$^3/_8$ inch long are erect,

**Fig. 393 : Bálsamo, scarletbush (*Hamelia patens*)**
Flowering twig (above), fruits (lower left), natural size.

nearly stalkless, juicy, 5–celled, and bear ring of calyx at apex. There are many minute brown seeds. Flowering and fruiting throughout the year.

The wood is light brown and hard.

Elsewhere the plants are grown as ornamentals for the showy flowers, often borne when less than 2 feet high. The fruits and leaves have served for home remedies in some countries. It is reported that the sour fruits are edible as well as medicinal. The stems and leaves have been used in tanning leather. Propagated by seeds and cuttings.

Bálsamo is uncommon in open areas and thickets in the moist Sivalik and lower foothills from sea level to 2,000 feet altitude throughout moist areas of Punjab.

## 394. Bola de nieve, white ixora *(Ixora thwaitesii)*

Bola de nieve (snowball) is a shrub or sometimes small tree planted in gardens for its large clusters of white flowers. Distinguishing characters include: (1) opposite large narrowly elliptic to oblong leathery leaves; and (2) showy white flowers $1^1/_4$–$1^1/_2$ inches long with very narrow tube and 4 (or 5) spreading lobes more than $^1/_2$ inch across, in many–flowered compact terminal clusters 2–4 inches across, suggesting a snowball.

A tall evergreen shrub or small tree to 20 feet high and 4 inches in trunk diameter, hairless. Bark gray, smoothish, with corky warts (lenticels), becoming divided into small scales. The inner bark is whitish with orange streaks, slightly bitter. Twigs finely hairy and green when young, becoming gray, ringed at nodes.

Leaves opposite, with petioles $^1/_8$–$^1/_2$ inch long. Stipules paired, $^1/_8$ inch long, pointed and sheathing twig at node. Blades are 2–7 inches long and $^3/_4$–2 inches wide, blunt to long–pointed at apex, short–to long–pointed at base, not toothed on edges, the upper surface slightly shiny or dull green with few side veins, and the lower surface dull light green.

Flower clusters (corymbs) bear many fragrant white flowers nearly stalkless and crowded at ends of branches. Flowers are composed of red–brown base (hypanthium) bearing 4–5 light green calyx lobes nearly $^3/_{16}$ inch long; corolla with very narrow greenish–white tube $1^1/_4$–$1^1/_2$ inches long and less than $^1/_{16}$ inch in diameter and 4 (or 5) widely spreading lobes more than $^1/_2$ inch across; 4 (or 5) short stamens in throat of corolla and bending down between the lobes; and pistil with inferior 2–celled ovary with 1 ovule in each cell, very long slender style, and 2–forked stigma. Fruits (berries) not observed, described as rounded and $^1/_2$ inch in diameter, containing 2 nutlets. Flowering throughout the year.

The wood is light brown and hard.

Common in cultivation for ornament in residential areas and gardens of Chandigarh.

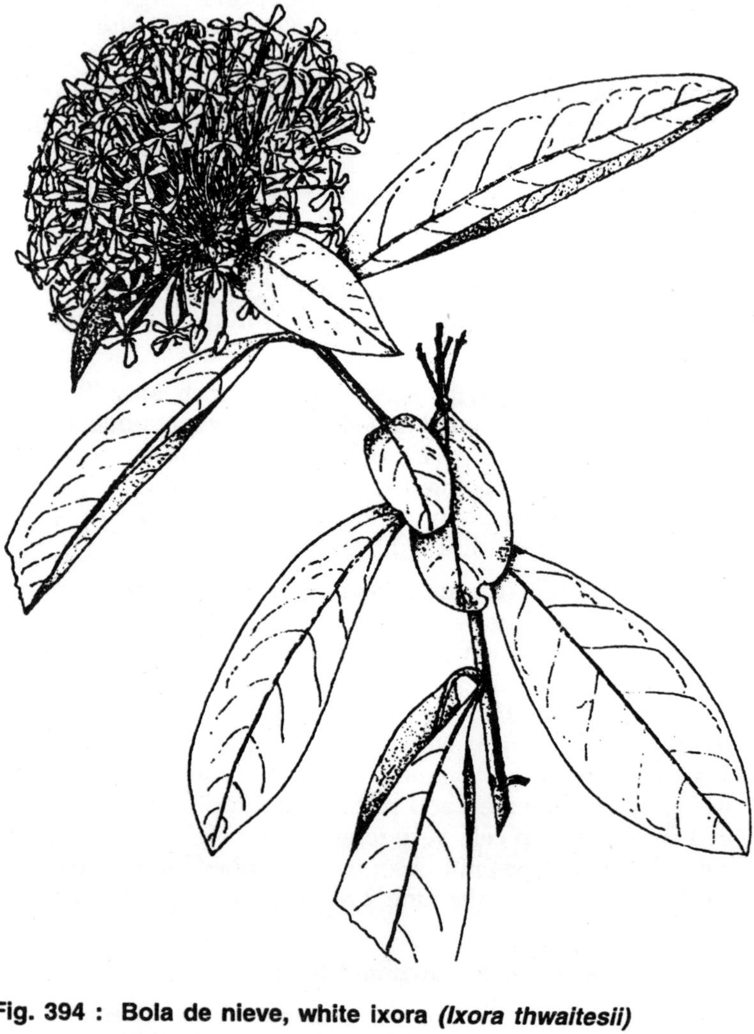

**Fig. 394 : Bola de nieve, white ixora *(Ixora thwaitesii)***
Flowering twig, natural size.

## 395. Alfilerillo *(Machaonia portoricensis)*

Alfilerillo is a much–branched spiny shrub or sometimes small tree known only from south–India. Its distinguishing characters are : (1) 2 or 3 slender spreading spines $^1/_4$–1 inch long at most nodes; (2) small elliptic or nearly round leaves $^1/_4$–$^5/_8$ inch long and $^3/_{16}$–$^3/_8$ inch wide, 2 or 3 almost stalkless at a node; (3) many small hairy flowers $^1/_8$ inch long and broad with tubular 4–lobed white corolla, crowded in small terminal clusters; and (4) top–shaped brown seed capsules $^3/_{16}$ inch long.

A deciduous, large, much–branched shrub with several stems from base, sometimes a small tree to 20 feet high and 3 inches in trunk diameter. The bark is gray, smooth, becoming fissured, with a few spines remaining, the inner bark whitish or light green and slightly bitter. Twigs brown, long and slender, finely hairy, bearing at the ringed nodes 2 or 3 spines each above a leaf and representing a branch and sometimes bearing minute leaves.

Leaves opposite or whorled, with minute pointed hairy stipules and very short hairy petioles about $^1/_{32}$ inch long, blunt or rounded at apex, short–pointed or blunt at base, not toothed on edges, thin, with side veins inconspicuous, the upper surface green and hairless, and the lower surface light green with midvein often slightly hairy.

Flower clusters (panicles) are terminal, $^3/_4$–2 inches long and broad, erect and short–stalked at ends of slender twigs. The crowded, almost stalkless flowers are composed of cylindric hairy base (hypanthium) $^1/_{16}$ inch long, which encloses the inferior ovary and bears the 4 shorter hairy calyx lobes and other parts; the white corolla has a short tube $^1/_{16}$–$^1/_8$ inch long and 4 spreading rounded lobes $^1/_{16}$ inch long; 4 short stamens on throat of corolla; and pistil with inferior 2–celled ovary and slender style. The seed capsules are top–shaped (obovoid), flattened, brown, and hairy, bear the calyx lobes at apex, split into 2 parts from base upward, and contain 2 brown seeds $^1/_8$ inch long. Flowering and fruiting irregularly through the year.

The wood is light brown and hard.

Common locally, often forming thickets, in dry forests at 200–1,500 feet altitude in southwestern foothills of Kerala.

**Fig. 395 : Alfilerillo *(Machaonia portoricensis)***
Flowering twig (above), fruiting twig (lower right), natural size.

## 396. Alpina *(Palicourea alpina)*

This shrub or small tree of dwarf forests of summits of mountains in Sivalik is identified by: (1) opposite elliptic leaves $1^1/_2$–4 inches long and $^5/_8$–$1^1/_2$ inches wide, slightly thickened and leathery, with narrow pointed 2–lobed stipules $^1/_8$–$^1/_4$ inch long persistent at a node; (2) flowers about $^3/_4$ inch long, with tubular 5–lobed corolla pale yellow, turning to pale purple, many on slender red to orange–red or pink branches in terminal clusters; and (3) black berries nearly $^3/_8$ inch long, egg–shaped and slightly flattened.

An evergreen shrub 12 feet high or small tree to 20 feet high and 5 inches in trunk diameter. The bark is gray and smoothish or slightly fissured, the inner bark whitish and almost tasteless. The twigs are green, hairless with persistent stipules at the enlarged ringed, 2–angled nodes.

The opposite leaves have petioles $^1/_4$–$^1/_2$ inch long. Blades are long–pointed at apex, long–or short–pointed at base, not toothed on edges, the upper surface green to dark green with curved sunken side veins, slightly shiny, hairless, and the lower surface light green and often hairy on the midvein and raised side veins.

Flower clusters (panicles) 2–4 inches long have a dark red axis and red to orange–red or pink branches. Flowers many, opening few at a time and spreading horizontally, composed of pale yellow base (hypanthium) $^1/_8$ inch long bearing 5–toothed calyx; corolla with cylindric tube $^5/_8$ inch long and 5 spreading lobes $^3/_8$ inch across, slightly succulent and swollen at base, changing color from pale yellow to purple; stamens 5, inserted within corolla tube; and pistil with inferior ovary and long style. The fruits are green with lighter ridges when immature, turning black, slightly angled when dry, with calyx persistent at apex, 2–celled with 2 nutlets. Flowering and fruiting throughout the year.

The wood is light brown and hard.

Common locally in forest at 3,000–4,390 feet altitude on high peaks and ridges of the central mountains of Sivalik.

**Fig. 396 : Alpina *(Palicourea alpina)***
Flowering twig (above), fruiting twig (lower left), natural size.

## 397. Balsamo real, showy palicourea *(Palicourea barbinervia)*

This rare shrub or small tree is identified by : (1) opposite elliptic large leaves 4–10 inches long and 2–5 inches wide, short–petioled, with persistent 2–lobed stipules; (2) flowers about $^3/_8$ inch long, with white tubular 5–lobed corolla, finely hairy, many on slender yellow or reddish branches in large long–stalked erect clusters; and (3) egg–shaped berry $^3/_{16}$ inch long.

An evergreen shrub or small tree to 20 feet high and 3 inches in trunk diameter. Bark brown gray, smooth to slightly fissured. The inner bark is light yellow with a green outer layer, bitter. Twigs stout, slightly flattened when young, green, ringed at nodes, hairless, ending in rounded flat green bud of $^3/_{16}$ inch formed by stipules.

Leaves opposite, with paired 2–lobed membranous stipules to $^5/_{16}$ inch long and short petioloes $^3/_8$–$^3/_4$ inch long. Blades are long–pointed at both ends, not toothed on edges, slightly thickened, slightly shiny, the upper surface green and hairless with many curved much sunken side veins, and the lower surface light green with very raised veins and minutely hairy on midvein.

Flower clusters (panicles) 4–6 inches long, terminal on long stalked of 3–4 inches with many yellow or reddish branches. Flowers many, with green base (hypanthium) $^1/_{16}$ inch long bearing 5 short rounded hairy–margined calyx lobes; corolla with cylindric tube $^5/_{16}$ inch long swollen at base and with 5 short lobes, finely hairy; 5 stamens inserted within corolla tube; and pistil with inferior 2–celled ovary, long slender style, and 2–lobed stigma. Flowering and fruiting throughout the year.

The wood is light brown and hard.

Rare on moist limestone, moist coastal, and lower forests at 300–2,000 feet altitude. Scattered throughout moist areas of River Bank of Ganga.

**Fig. 397 : Balsamo real, showy palicourea *(Palicourea barbinervia)***
Fruiting twig (above), flowers (lower right), natural size.

## 398. Cachimbo, red palicourea (*Palicourea crocea*)

This large shrub sometimes treelike is common in understory of moist forests. Its distinguishing characters are : (1) paired elliptic leaves $2^1/_2$–8 inches long and 1–$2^1/_2$ inches wide, with paired very slender scales (stipules) forming sheath at base and curved back; (2) many tubular red flowers $^3/_8$ inch long from slender yellow or orange stalks and branches of much–branched terminal clusters longer than wide; and (3) elliptic red to dark brown or black drupes less than $^1/_4$ inch long.

An evergreen shrub sometimes treelike and 15 feet high with slender trunk to 2 inches in diameter. The twigs are hairless and rounded or slightly angled, ringed at nodes.

Leaves are opposite, connected at base by paired 2–forked persistent scales (stipules) $^3/_{16}$ inch long and have slender petioles of $^1/_4$–1 inch. The blades are mostly long–pointed at both ends, thin, with many fine curved side veins, hairless except for minute hairs along midvein beneath.

The flower clusters (panicles) erect and longer than wide, with many branches mostly on i side. Flowers are short–stalked, composed of short yellow calyx, corolla with cylindric red tube and 5 short lobes, 5 stamens within tube, and pistil with inferior overy, style, and 2 forked stigma. The drupes contain nutlets with points at apex. Probably flowering and fruiting irregularly through the year.

Common in understory of moist forest including lower and upper forests from sea level to 3,300 feet altitude in Sivalik Hills.

**Fig. 398 : Cachimbo, red palicourea (Palicourea crocea)**
Flowing twing (above), fruits (right), natural size.

## 399. Domingen (*Palicourea domingensis*)

This understory shrub or sometimes small tree of moist forests is identified by: (1) twigs with ringed nodes formed by scars of minute scales (stipules) that shed early: (2) opposite narrowly elliptic leaves 2–8 inches long, and $1–2^1/_4$ inches wides, long–pointed at both ends, thin and hairless; (3) flowers in 3's nearly stalkless in terminal branched clusters, about $^1/_2$ inch long, the white or pinkish corolla with narrow tube and 5 narrow lobes; and (4) fruit (drupe) rounded, shiny black, about $^1/_4$ inch long.

Evergreen shrub or small tree to 16 feet high, with trunk to 3 inches in diameter, hairless throughout. Twigs slender, green, ringed and enlarged at nodes.

The opposite leaves have petioles $^1/_4–^3/_4$ inch long. Blades are not toothed on edges and have curved side veins, the upper surface dark green, and the lower surface paler.

Flower clusters (panicles) terminal and erect, about 2–3 inches long and broad, with several flowers in groups of 3, opening 1 or 2 at a time. Flowers consist of greenish basal cup (hypanthium), greenish 5–toothed calyx, corolla narrowly tubular with 5 short narrow lobes, 5 stamens within tube and pistil with 2 celled ovary slender style, and 2–forked stigma. The juicy fruity contains a stone with 5 points toward apex. Flowering in spring and summer, fruiting in summer and fall.

It has been suggested that this species might become an ornamental for its white flowers and wandlike appearance.

Uncommon in understory of moist forests at low and middle altitudes in Mussori.

**Fig. 399 : Domingen (*Palicourea domingensis*)**
Flowering twing (above), fruits (lower right), natural size.

## 400. Cachimbo, yellow palicourea (*Palicourea riparia*)

Abundant in the understory of moist forests, this shrub or small tree is recognized by: (1) the opposite dark green, narrowly elliptic leaves long–pointed at both ends, thin or only slightly thickened, with small tufts of hairs where side veins join midrib beneath and 2 pairs of erect bristlelike scales (stipules) at base; (2) the many light yellow narrowly tubular flowers about $^1/_2$ inch long in terminal flower clusters about as broad as long, with bright orange–red branches turning to purple, often more or less horizontal; and (3) the blackish, nearly round, fleshy fruits $^1/_4$–$^3/_8$ inch long and broad on branches which have turned red or purple.

Evergreen shrub 10 feet high or sometimes a small tree to 15 feet and 3 inches in trunk diameter. The bark is gray and smoothish to finely fissured. Inner bark is light green, slightly bitter and astringent. The twies are dark green and hairless, with slightly enlarged ringed nodes. Paired very narrow pointed stipules $^3/_{16}$ inch long form the bud.

The 2 leaves a node are connected by 2 pairs of bristlelike scales (stipules) $^1/_8$–$^3/_{16}$ inch long, which remain attached to twig after the leaves fall. Petioles are $^1/_4$–$^3/_4$ inch long. Leaf blades are 3–8 inches long and $1^1/_4$–$2\,^1/_2$ inches broad, the edges not toothed, the upper surface dark green, slightly shiny, with the curved lateral veins sunken, and the lower surface pale green, slightly shiny. The leaves darken on drying.

Flower clusters (panicles) with spreading orange–red branches are 1–3 inches long and broad beyond the purplish red main stalk $1^1/_2$–5 inches long. The many short–stalked flowers open few at a time. The yellowish tubular base (hypanthium) about $^1/_{16}$ inch long and broad bears the reddish–tinged 5–lobed calyx nearly as long; the light yellow tubular corolla $^3/_8$–$^1/_2$ inch long and $^1/_8$ inch broad, slightly swollen at base, has 5 short–pointed, slightly hairy lobes $^1/_4$ inch across; 5 stamens are attached inside the corolla tube; and the pistil is composed of inferior 2–celled overy, slender style, and 2–forked stigma.

The rounded, slightly flattened fruits (drupes) have juicy, tasteless, dark purplish flesh. There are 2 brown nutlets $^3/_8$–$^1/_4$ inch fruit throughout the year.

**Fig. 400 : Cachimbo, yellow palicourea (palicourea riparia)–** Fruiting twing (above), flowering twing (below), natural size.

The wood is light brown, hard, and light–weight.

Abundant in understory of moist coastal, moist limestone, Luquillo, and Cordillera forests from sea level to 2,500 feet altitude throughout moist areas of Dehradun

## 401. Aquiln prieto (*Phialanthus grandigolius Alain*)

This resinous shrub or small tree rare in western mountains is identified by: (1) foliage resinous or sticky the opposite leaves narrowly elliptic, 1–3$^1/_2$ inch long and $^3/_8$ inch wide, slightly thickened and leathery; (2) buds and young leaves in a drop of whitish wax or resin; and (3) very small white and almost stalkless flowers crowded at leaf bases and noded back of leaves, $^1/_8$ inch long, with 4 long spin–shaped papery calyx lobes, persistent on the top–shaped fruit.

An evergreen resinous shrub 8–12 feet high, rarely a small tree 15–25 feet and 3 inches in trunk diameter, with paired nearly horizontal branches. The bark is gray and smooth, the inner bark whitish and slightly bitter. The twings are green, with minute hairs, and have stipules in from of sheath $^1/_{16}$ inch high at the enlarged ringed nodes, branching in pairs, the side twigs short with nodes only $^1/_8$–$^3/_8$ inch apart. Buds develop within a round drop of whitish or cream–colored wax or resin $^1/_{16}$–$^1/_8$ inch long, and young leaves to $^1/_2$ inch long are within a larger wax or resin covering.

The opposite leaves have very petioles $^1/_{10}$–$^1/_8$ inch long, with minute hairs. The blades are blunt at apex, short–pointed at base, turned under at edges, hairless, with midrib but without visible side veins, the upper surface green and slightly shiny, the lower surface yellow green with prominent midvein.

The flowers have a small funnel–shaped tubular base (hypanthium) $^1/_{16}$ inch long, which bears the calyx with 4 long spoon–shaped papery lobes more than $^1/_{16}$ inch long and other parts; the 4–lobed corolla shorter than calyx; 4 short stamens inserted on corolla; and pistil with inferior 2–celled overy and slender style. The fruit (drupelike) has at apex the a enlarged papery light brown calyx lobes $^1/_8$ inch long. Flowering and fruiting probably through the year.

The wood is light brown and hard.

The drops of wax or resin in which the buds and young leaves develop can be chewed but are tasteless.

Rare in upper montane shrub on dry ridges at 1,000–2,600 feet in western Sivalik.

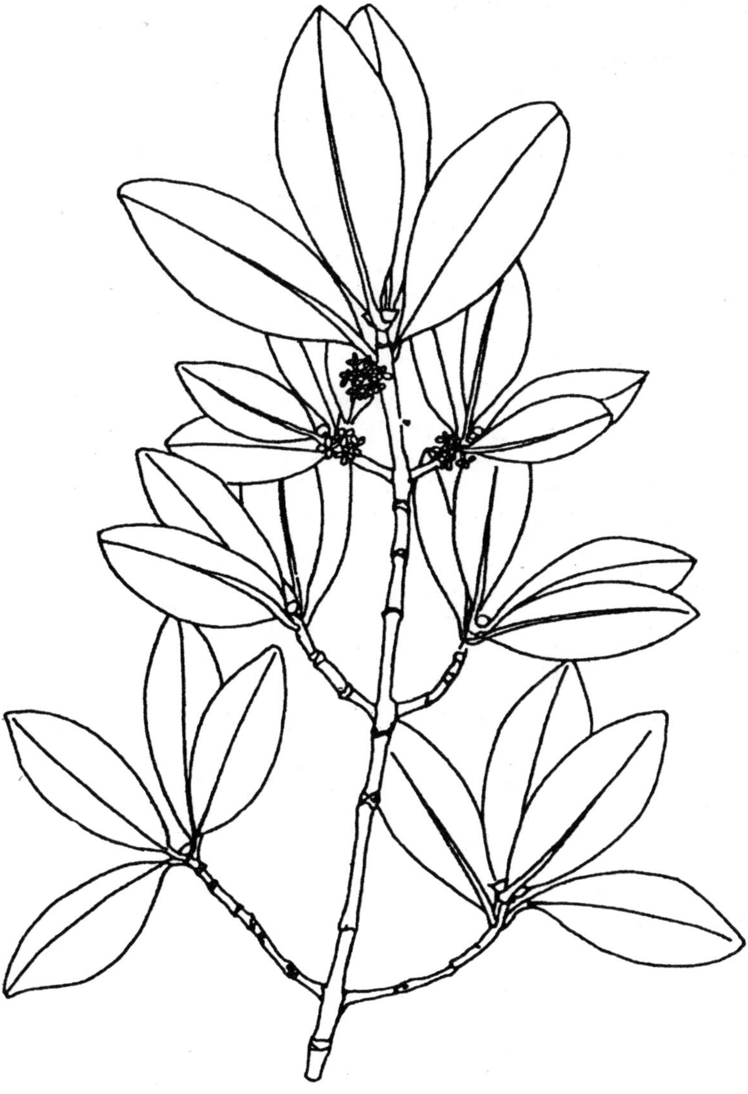

**Fig. 401 : Aquilon prieto–(Phialanthus grandifolius Alain)**
Natural size.

## 402. Cachimbo comun (*Psychotria berteriana* DC)

The most common of the shrubby species of cachimbo reaching tree size also one of the largest. It is distinguished by: (1) opposite elliptic thin leaves long and 3–8 inches long and $1^1/_4$–$3^1/_2$ inches wide, long–pointed at both ends and with many curved side veins; (2) flowers about $^1/_4$ inch long and broad, with 5–lobed pale yellow to corolla, many in large erect terminal clusters; and (3) many round black berries nearly $^1/_4$ inch in diameter.

An evergreen understory shrub or small tree to 20 feet high and 4 inches in trunk diameter. The bark is gray and smooth, the inner bark whitish and almost tasteless. Twigs are green, sometimes finely hairy often slightly 4–angled, becoming enlarged at the ringed nodes.

The opposite leaves have paired light green 2–pointed stipules $^1/_8$ inch long and slender petioles $^1/_2$–2 inches long. Blades are not toothed on edges, green and hairless or nearly so on upper surface, light green and sometimes slightly hairy on the raised veins on lower surface.

Flower clusters (panicles) 3–7 inches long, long–stalked, with greenish branches. Flowers many, opening few at a time, short–stalked or stalkless, consisting of very short hairy yellow–green base (hypanthium) less than $^1/_{16}$ inch long with 5 calyx lobes; tubular corolla $^3/_{16}$ inch long with 5 short lobes, which sheds promptly; 5 short stamens in notches of corolla; and pistil with inferior ovary and threadlike 2–forked style.

Fruits many in grapelike clusters, slightly broader than long with calyx at apex, green when immature, very juicy dark purple within, slightly bitter. The 2 nutlets are $^1/_8$ inch long, angled and with raised points at top. With flowers and fruits throughout the year.

The wood in understory of mountain forests including dwarf forest at 1,500–4,000 altitude in eastern, central, and western mountains.

**Fig. 402 : Cachimbo comun (*Psychoria berteriana* DC)**
Flowering twig and fruits (upper right), natural size.

## 403. Cachimbo grande (*Psychotria grandis*)

This rare understory tree or shrub is distinguished by: (1) opposite large leaves mostly 8–13 inches long and $2^1/_2$–5 inches wide, obovate or broadly oblanceolate, thin almost stalkless; (2) flowers less than $^1/_4$ inch long, with tubular 5–lobed white corolla, many, crowded, almost stalkless on short side branches of long–stalked clusters; and (3) round red berries nearly $^1/_4$ inch in diameter.

An evergreen small tree or shrub 20 feet high, rarely to 35 feet and 5 inches in trunk diameter. The brown or gray bark is smoothish, slightly fissured, the inner bark whitish. The twigs are green, slightly, angled, with ringed nodes, hairless.

The opposite hairless leaves have paired large united keeled long–pointed stipules about $^3/_4$ inch long, which shed before the leaves. Leaf blades are long–or short–pointed at apex, not toothed on edges, widest beyond middle, and tapering gradually to the stalkless base, blackening upon drying. The upper surface is shiny green with many curved side veins, and the lower surface dull green with prominent veins.

Flower clusters (panicles) terminal, large, 4–6 inches long on a long stalk. Flowers are composed of short hairy tubular base (hypanthium) less than $^1/_{16}$ inch long with 5 toothed calyx; tubular white corolla $^1/_8$–$^3/_{16}$ inch long with 5 short lobes; 5 short stamens in notches of corolla; and pistil with inferior 2–celled ovary, slender style, and 2–forked stigma. The fleshy fruits (drupes) are slightly grooved, containing 2 rounded light brown nutlets flattened on 1 side. Flowering and fruiting through the year.

The whitish wood is hard and upper forests at 300–3,000 feet altitude in moist mountains and wet lime–stone hilles of Mussore.

**Fig. 403 : Cachimbo (*Psychotria grandis*)**
Fruits as left, and flowering twig, two–thirds natural size.

## 404. Cachimbo de gato (*Psychotria maleolens*)

A shrub, rarely becoming a small tree in Mussore Forest, easily recognized by the bad odor of the folige, especially strong and skunk–like when crushed. Other characters are: (1) opposite oblong leaves 1–3$^1/_2$ inches long and $^3/_8$–1$^1/_4$ inches wide, slightly thickened and stiff, with edges under; (2) few small flowers $^1/_4$ inch long and broad with tubular 5–lobed white corolla; and (3) elliptic red berries $^3/_8$ inch long.

An evergreen shrub 3–10 feet high rarely in Maricao Forest becoming a small tree to 15 feet high and 3 inches in trunk diameter; however, reported to reach 33 feet. The bark is gray and smooth. The twigs are green and finely hairy when young, becoming gray, with ringed nodes formed by the pointed sheathing stipules which shed early.

The opposite hairless leaves have short petioles $^1/_8$–$^1/_4$ inch long. Blades are long–or short pointed at both ends, the upper surface shiny dark green, and the lower surface dull light green.

Flower clusters (panicles) terminal with few flowers on short stalks of less than $^1/_8$ inch. The minute base (hypanthium) $^1/_{16}$ inch long bears the 5 toothed calyx, white corolla about $^3/_{16}$ inch long, 5 minute stament in notches of corolla, and pistil with inferior ovary, slender style, and 2–forked stigma. The fleshy fruits have calyx at apex and are 10–ribbed when dry. There is 1 brown nutlet $^5/_{16}$ inch long. Flowering and fruiting through the year.

Wood light brown, hard.

Locally common as an understory shrub in lower and upper Luquillo and Cordillera forests and dwarf forest at 1,800–4.

**Fig. 404 : Cachimbo de gato (*Psychotria maleolens*)**
Natural size.

## 405. Cachimbo de Maricao (*Psychotria maricaensis*)

Thus shrub or small tree named for Mussori is found only in mountains of western Dehradun. It is distinguished by : (1) paired elliptic to obovate leaves $1^1/_4$–$2^3/_4$ inches long and $^3/_4$–$1^3/_4$ inches wide, slightly thickened and turned under at edges, with paired scales (stipules) forming sheath at base; (2) several stalkless small white flowers $^1/_4$ inch long in erect terminal clusters; and (3) fruit an elliptic reddish drupe $^1/_4$ inch long.

An evergreen shrub or sometimes a small tree to 20 feet high and 4 inches in trunk diameter, with many slender gray branches, hairless throughout. The bark is smooth and gray, the inner bark whitish and almost tasteless. Twigs are green, becoming gray.

The opposite leaves have at base paired scales (stipules) to $^3/_8$ inch high, which shed early, leaving ring scar around twig. Blades are blunt at apex and narrowed to long–pointed base and petiole of $^1/_8$–$^1/_2$ inch, slightly thickened, shiny green above and paler beneath.

The flowers stalkless at ends of branched clusters (panicles) 1–2 inches long consist of cuplike base with wavy border of calyx, white tubular corolla with 5 spreading lobes, 5 stamens within tube, and pistil with inferior 2–celled ovary, slender style, and 2–forked stigma. The berries with rim of calyx at apex contain a nutlet nearly $^3/_{16}$ inch long with ridges and grooves. Flowering and fruiting irregularly through the year.

The wood is whitish and hard.

Uncommon in understory at middle and higher altitudes of about 1,000–4,500 feet in upper forest, especially on serpentine bedrock, in western Mussori.

**Fig. 405 : Cachimbo de Maricao (*Psychotria maricaensis*)**

Twig with flowers and fruits (left), natural size; fruiting twig (lower right), twice natural size.

## 406. Cachimbo de Mona (*Psychotria nutans*)

This small tree rare at Noore Islnd is recognized by : (1) opposite oblanceolate or narrowly elliptic thin leaves blunt or short–pointed at apex and tapering to long–pointed base; (2) small flowers less than $^3/_8$ inch long with white tabular 5–lobed corolla; and (3) red elliptic berries about $^1/_4$ inch long.

A small evergreen tree to 15 feet high and 6 inches in trunk diameter, with spreading crown. The bark is very thick, with light brown corky ridges and deep furrows. The twigs have minute hairs when young, paired stipules $^1/_{16}$ inch long, which form the ringed nodes, and become light gray.

The opposite leaves have slender petioles $^1/_4$–$^3/_8$ inch long. Blades are $1^1/_4$–4 inches long and $^1/_2$–$1^1/_4$ inches wide, not toothed on edges, the upper surface dull dark green and hairless, with midvein and many curved side veins slightly sunken, and the lower surface light green with raised finely hairy brownish veins.

The flower clusters (panicles) terminal, less than 2 inches long, with few slender branches. Flowers few, composed of minute base (hypanthium), 5–toothed calyx, white tubular corolla $^1/_4$ inch long with 5 lobes; 5 minute stamens in notched of corolla; and pistil with inferior ovary and slender style. The berries have calyx teeth at apex and are often ribbed. Collected with flowers and fruits in August.

The wood is light yellow and hard.

Rare at Bajura de los Cerezos in dry forest at 150 feet altitude on Mood Island.

**Fig. 406 : Cachimbo de Mona (*Psychotria nutans*)**

Flowering twig (upper left), fruiting twig (lower right), natural size.

## 407. Cordobancillo (*Rondeletia inermis*)

Cordobancillo is a shrub small tree widely distributed through Himachal, with races differing in leaf shape and in hairiness. Distinguishing characters are : (1) paired, slender finely hairy twigs with ringed modes nodes formed by short–pointed stipules and with both long and short internodes; (2) opposite small, mostly elliptic leaves ending in abrupt point, with few curved side veins slightly sunken; (3) flowers $^5/_{16}$–$^3/_8$ inch long, 1 to few on short stalks at leaf bases, the hairy corolla with narrow tube and 4 rounded spreading lobes, whitish to yellow, pinkish, purple, or brown; and (4) seed capsule rounded, $^1/_8$ inch in diameter, gray, finely hairy, with 4 long calyx lobes persistent at apex.

Evergreen shrub or sometimes a small tree to 15 feet high and 3 inches in trunk diameter, infrequently vinelike. Bark gray, smoothish the inner bark light yellow, almost tasteless. The twigs are light green when young, becoming gray or light brown, usually finely hairy buit often hairless. Paired pointed hairy stipules $^1/_{16}$ inch long, green but turning brown, form a pointed bud and ringed nodes.

The opposite leaves have finely hairy petioles $^1/_8$–$^1/_4$ inch long. Blades are mostly elliptic but vary greately in shape and size from linear to oblong and obovate, mostly $^3/_4$–3 inches long and $^3/_8$–$1^1/_2$ inches wide, thin or slightly thickened, nearly hairless, short–pointed at base, not toothed on edges. The upper surface is green and slightly shiny, and the lower surface dull light green.

The fragrant flowers have an inferior rounded 2–celled hairy ovary $^1/_{16}$ inch long, which bears the other parts, including calyx with 4 whitish green hairy lobes nearly $^1/_8$ inch long; hairy corolla about $^1/_4$ inch long, pinkish in bud, with very narrow tube and 4 rounded spreading lobes of varying and changing color; 4 stamens in throat of corolla tube; and pistil with slender style and 2–lobed stigma. The fruit is a rounded seed capsule with calyx lobes bent back at apex, containing many minute seeds and persisting on old twigs. Flowering and fruiting through the year.

The wood is light brown and hard.

Uncommon but widely distributed through H.P. in moist limestone forest, both north and south, and in central mountains at 20006,500 feet altitude.

**Fig. 407 : Cordobancillo (*Rondeletia inermis*)**
Flowering twig (left), fruiting twig (lower right), natural size.

# INDEX


## A

*Abey amarillo*, 496
*Acacia anegadensis*, 166, 167
*Acacia macracantha*, 168, 169
*Acacia muricata*, 170, 172
*Acacia nilotica*, 173, 174
*Acacia polyacantha*, 175, 176
*Acacia tortuosa*, 177, 178
*Acebo de sierra*, 381
Aceitillo, 269, 293
Aceitunilla, 793
*Acnistus arborescens*, 873
*Adansonia digitata*, 469
Adormidera, 323
*Aegiphila mertinicensis*, 859
*Agathis robusta*, 17
*Aguacatillo*, 125, 126
Ailanthus family, 286-297
*Albizia carbonaria*, 179, 180
Aleli, 826
Alfiler, 279
Alfilerillo, 973
*Allophylus crassinervis*, 417
*Allophylus racemosus*, 419
Almendrillo, 315
Alpina, 975
*Alsophila bryophila*, 10
Amansa, 356
Anacaguita, 488
*Anegada acacia*, 166, 167
Anguila, 632
*Anitirhea acutata*, 937
*Anthocephalus chinensis*, 934
Antillarum, 889
*Antirhea coriacea*, 939
*Antirhea lucida*, 941
*Antrihea sintenisii*, 943
*Aquiln prieto*, 986
*Arana gato*, 763, 765
Araucaria family, 16-18
*Araucaria heterophylla*, 20
Arbol de navidad, 924
Arbol de vela, 910
*Ardisia glauciflora*, 750
*Ardisia luquillensis*, 752
Arneau, 494
Arroyo, 433
*Athel tamarisk*, 525
Augacero, 875
*Averrhoa carambola*, 248
Avispillo, 130, 131. 813

## B

Balsamo, 968, 977
Banana family, 61-64
*Banara portoricensis*, 532
*Banara vanderbiltii*, 535
Baobab, 469
Barbasco, 744
Bariaco, 307
*Barringtonia asiatica*, 586

*Bauhinia purpurea*, 192, 193
*Bauhinia variegata*, 195, 197
Bayoneta, 55, 57, 58, 59
Benjamin, 94, 95
*Berenjena cimarrona*, 895
Berenjena, 891
Berterii, 747
Beruquillo, 663
Bignonia family, 898-922
Biriji, 634
Bitolia, 275
Bitter-ash, 289
Bittersweet family, 393-412
Black mulberry, 112, 113
*Blighia sapida*, 421
*Bocconia frutescens*, 133, 135
Boje, 411
Bola de nieve, 971
Bombax family, 468-479
*Bontia daphnoides*, 927
Boqueron, 618
Borage family, 837-856
Botree, 106, 107
Bottlebrush, 599
*Bourreia virgata*, 839
Box family, 360-364
*Brassaia actinophylla*, 733
Brazilnut family, 587-591
*Brittle thatchpalm*, 48, 49
Britton, 363
*Brunfelsia*, 877
*Brunfelsia Americana*, 875
*Brunfelsia densifolia*, 877
*Brunfelsia lactea*, 879
*Brunfelsia portoricensis*, 881
Bucare, 229, 231
Buckthorn family, 435-450
*Bumelia krugii*, 763
*Bumelia obovata*, 765
*Bunchosia glandulosa*, 311
Burro, 138, 139, 145, 147
*Buxus laevigata*, 361
*Buxus vahlii*, 363
*Byrsonima lucida*, 313
*Byrsonima wadsworthii*, 315

## C

Cabalonga, 830
Cachimbo, 979, 983
*Cachimbo comun*, 988
*Cachimbo de gato*, 992
*Cachimbo de Maricao*, 994
*Cachimbo de Mona*, 996
*Cachimbo grande*, 990
Cactus family, 561-567
*Caesalpinia coriaria*, 198, 199
*Caesalpinia pulcherrina*, 201, 203
*Café forastero*, 311
Cafe, 949
Caimitillo, 767
Caimito de perro, 769
Cajeput, 652
Cakalaka, 893
Calceolaria shower, 213, 214
*Calliandra caracasana*, 181, 182
*Callicarpa ampla*, 861
*Callistemon citrinus*, 599
*Calophyllum inophyllum*, 515
*Calotropis procera*, 834
*Calycophyllum candidissimum*, 945
*Calyptranthes luquillensis*, 601
*Calyptranthes pallens*, 603

*Calyptranthes portoricensis*, 605
*Calyptranthes sintenisii*, 607
*Calyptranthes zusygium*, 609
*Calyptronoma rivalis*, 36, 37
Camasey, 675, 677, 679, 683, 685, 697, 699, 703, 705, 707, 709, 711, 723, 726, 728
*Camasey almendro*, 681
*Camasey cuatrocanales*, 692
*Camasey de costilla*, 688
*Camasey de paloma*, 690
*Camasey felpa*, 701
*Camasey racimoso*, 695
*Camasey tomaso*, 713
Cambron, 537
Campeche, 217, 219
Canafistula, 206, 207
Candelero, 335
Canella family, 123, 124, 527-529
Cannonball, 589
*Capa rosa*, 861
Capaillo, 859
Caper family, 136-153
Capparaceae, 136
*Capparis amplissima*, 138, 139
*Capparis flexuosa*, 140, 141
*Capparis frondosa*, 142, 143
*Capparis hastata*, 145, 147
*Capparis indica*, 148, 149
Capulin, 452
Carambola, 248
Carbonero, 179, 180
Caribbean pine, 24
Carrasco, 333
Carrasco, 370
Cascarrolla, 441
*Casearia aculeata*, 537
Cashew family, 365-373
*Cassia emarginata*, 204, 205
*Cassia grandis*, 206, 207
*Cassia planisiliqua*, 209, 210
*Cassia polphylla*, 211, 212
*Cassia spectabilis*, 213, 214
*Cassine xylocarpa*, 395
*Cassipourea guianensis*, 593
*Castela erecta*, 287
Castorbean, 349
*Casuarina cunninghamiana*, 66, 67
Casuarina family, 65-68
Casuarinaceae, 65
*Catalpa longissima*, 900
Catechu tree, 175, 176
*Ceiba aesculifolia*, 472
Cereza, 845
*Cestrum diurnum*, 883
*Cestrum macrophyllum*, 885
Chicharoon, 367, 454
Chinese privet, 807
Chinita, 273
*Chione venosa*, 947
*Chlorophora tinctoria*, 91, 93
Chocolate family, 480-490
*Chorisia*, 474
*Chorisia speciosa*, 474
*Chrysalidocarpus lutescens*, 39, 41
*Chrysobalanus icaco*, 155, 157
*Chrysophyllum bicolor*, 767
*Chrysophyllum pauciflorum*, 769
Chupacallos, 528
Cidra, 258
Cieneguillo, 650
*Citharexylum spinosum*, 863

Citrifolia, 361
Citronnier, 383
*Citrus medica*, 258
*Citrus reticulata*, 261
Clavellina, 201, 203
*Clerodendrum aculeatum*, 865
*Cleyera albopunctata*, 501
*Clidemia umbrosa*, 673
Clidemia, 673
*Clusia minor*, 518
Coca family, 251-257
*Coccothrinax alta*, 42, 43
Cockscomb, 224, 225
Coco de mar, 586
Coco palm, 155, 157
*Coffea dewevrei*, 949
*Cola acuminate*, 481
Cola-nut, 481
*Comocladia dodonaea*, 367
*Comocladia glabra*, 370
Cookii, 377
*Copy vera*, 501
Coral, 397
Coraltree, 222, 223
*Cordia alba*, 841
*Cordia collococca*, 843
*Cordia oblique*, 845
*Cordia rickseckeri*, 847
*Cordia sebestena*, 850
Cordobancillo, 998
*Cornutia oibovata*, 867
*Couroupita guianensis*, 589
Crapemyrtle, 576
*Crescemtia linearifolia*, 903
*Crescentia portoricensis*, 905
*Croton astroites*, 321
*Croton rigidus*, 323
*Crozier cycs*, 13
Cuajilote, 913
*Cuban royalpalm*, 45, 47
*Cucubano*, 955, 957, 959
*Cucubano de monte*, 965
*Cucubano de vieques*, 961
Cuernecillo, 483
Cuero de sapo, 391, 403
*Cupania triquetra*, 424
Cupey de monte, 518
Cupressaceae, 27
*Cupressus lusitanica*, 28, 29
*Cupressus sempervirens*, 31, 32
*Cyathea aquilina*, 2
*Cyathea escuquensis*, 6
*Cyathea tenera*, 4
Cycad family, 12-15
*Cycas circinalis*, 13
Cymosa, 401
*Cynometra portoricensis*, 215, 216
Cypress family, 27-25


## D


Dagame, 945
Dahoon, 375
*Dalbergia sissoo*, 220, 221
*Daphnopsis americana*, 569
*Daphnopsis hellerana*, 571
Day cestrum, 883
*Dendropanax lauriflolius*, 736
Densiflora, 667
Dictyone, 777
*Didymopanax gleasonii*, 738
*Dildo espanol*, 562
*Diospyros revolute*, 783
*Diospyros sintenisii*, 785

*Dipholis bellonis*, 771
*Dipholis cubensis*, 773
*Ditta myricoides*, 325
Ditta, 325
Dividivi, 198, 199
*Dodonaea viscosa*, 426
Dogbane family, 819-832
Domingen, 981
*Dovyalis hebecarpa*, 539
*Dracaena fragrans*, 52, 53
*Drypetes alba*, 327
*Drypetes ilicifolia*, 329
*Drypetes lateriflora*, 331
*Duranta repens*, 869

**E**

Ebony family, 782-786
Eggersiana, 797
Elaeocarpus family, 451-455
Elongata, 405
*Enallagma latifolio*, 907
Encinillo, 329
*Enterolobium cyclocarpum*, 183, 185
Eriobotrya japonica, 158, 159
*Erithalis fruti-cosa*, 951
*Erythrina corallodendrum*, 222, 223
*Erythrina crista-galli*, 224, 225
*Erythrina eggersii*, 227, 228
*Erythrina variegata*, 229, 231
*Erythroxylum rotundifolium*, 252
*Erythroxylum rufum*, 254
*Erythroxylum urbanii*, 256
Escambron, 236, 237
Espejuelo, 773
Espino rubial, 277
Eucalipto, 611
*Eucalyptus resinifera*, 611
*Eugenia axillaries*, 614
*Eugenia biflora*, 616
*Eugenia boqueronensis*, 618
*Eugenia borinquensis*, 620
*Eugenia confuse*, 622
*Eugenia cordata*, 624
*Eugenia eggersii*, 626
*Eugenia foetida*, 632
*Eugenia haematocarpa*, 628
*Eugenia ligustrina*, 630
*Eugenia monticola*, 634
*Eugenia procera*, 636
*Eugenia pseudopsidium*, 638
*Eugenia sessiliflora*, 640
*Eugenia sintenisii*, 642
*Eugenia stewardsonii*, 644
*Eugenia uniflora*, 646
*Eugenia xerophytica*, 648
*Euphorbia cotinifolia*, 333
*Euphorbia lactea*, 335
*Euphorbia neriiflia*, 338
*Euphorbia tirucalli*, 340
*Exostema ellipticum*, 953
*Exostema ellipticum*,. 955
*Exothea paniculata*, 428

**F**

*Ficus benjamina*, 94, 95
*Ficus drupacea*, 97, 98
*Ficus lyrata*, 99, 100
*Ficus nekbuda*, 101, 103
*Ficus obtusifolia*, 104, 105
*Ficus religiosa*, 106, 107
*Ficus stahlii*, 108, 109
*Ficus trigonata*, 110, 111
Fiddle-leaf, 99, 100

Flacourtia family, 529-533, 534-560
*Flacourtia inermis*, 541
Florida, 409
*Forestera rhamnifolia*, 799
*Forestiera eggersiana*, 797
*Forestiera segregata*, 801
Forestira, 799
*Fortunella margarita*, 263
*Fraxinus uhdei*, 803
Fresno, 803
*Funtumia elastica*, 821

**G**

Gaita, 428
*Galan del monte*, 885
Gallinero, 873
*Garcinia dulcis*, 520
*Garcinia mangostana*, 522
Geiger-tree, 850
Geno, 234, 235
Geno-geno, 232, 233
*Gesneria* family, 923-925
*Gesneria peduncculosa*, 924
Giant, 834
*Ginoria rohrii*, 574
Ginseng family, 732-739
Gitaran, 426
*Glauca cassia*, 209, 210
Glorybush, 730
Goat bush, 287
*Goelzea elegans*, 887
*Gomidesia lindeniana*, 650
Gongoli, 736
Gongolin, 387
*Gorli oncoba*, 547
*Gossypium barbadense*, 457
*Grammadenia sintenisii*, 754
Granada, 582
Green ebony, 358
Guaba, 186, 187
Guama, 437
Guanabanilla, 498
Guanacaste, 183, 185
Guara blanca, 424
Guaraguadillo, 300
*Guarea ramiflora*, 300
Guasabara, 626
Guayabacon, 597, 659
Guayabilla, 551
*Guayabota de sierra*, 620
Guayabota, 783
*Guayabota nispero*, 785
Guest-tree, 486
*Guettarda krugii*, 957
*Guettarda ovalifolia*, 959
*Guettarda parviflora*, 961
*Guettarda pungens*, 963
*Guettarda valenzuelana*, 965
Guiana-plum, 331
Gum Arabic, 173, 174
Gumtree, 352
*Gyminda latifolia*, 399

**H**

*Haematoxylum campechianum*, 217, 219
*Haenianthus salificolius*, 805
Haggarbush, 865
*Hamelia patens*, 968
Health family, 740-742
*Helecho gigante*, 4, 6
*Hedge euphorbia*, 338
*Helecho gigante da la sierra*, 10
*Helecho gigante espinoso*, 8

Helerana, 571
*Helicteres jamaicensis*, 483
*Henriettea jascicularis*, 675
*Henriettea macjadyenii*, 677
*Henriettea membranifolia*, 679
Heptasepala, 503
*Hibiscus elatus*, 460
*Hibiscus rosa*, 463
Higiierita, 907
Higuillo, 70, 71, 73, 74, 75, 77
*Hirtella triandra*, 160, 161
*Hoja menuda*, 655, 657
Hoja, 636
Holly family, 374-392
Huesito, 319
Hueso prieto, 815
Hueso, 327
Hueso, 811, 817
*Hypelate trifoliata*, 430

## I

*Icacina* family, 413-414
*Ilex cassine*, 375
*Ilex cookii*, 377
*Ilex guianensis*, 379
*Ilex macfadyenii*, 381
*Ilex nitida*, 383
*Ilex riedlaei*, 385
*Ilex sideroxyloides*, 387
*Ilex sintenissi*, 389
*Ilex urbaniana*, 391
*India jujube*, 445
*Inga fastuosa*, 186, 187
Ink-bush, 801
Italian cypress, 31, 32
*Ixora thwaitesii*, 971

## J

Jacanillo, 758
*Jacquinia arborea*, 744
*Jacquinia berterii*, 747
Jagiiey, 104, 105, 108, 109
Japanese privet, 809
*Jatropha curcas*, 342
*Jatropha hernandiifolia*, 345
*Jatropha multifida*, 347
Jayajabico, 951
Juglandaceae, 88
*Juglans jamaicensis*, 89, 90
Juso, 853

## K

Kadam, 934
Kamani, 515
Kitembilla, 539
*Kleinhavia hospita*, 486
*Krugiana*, 719
Krugii, 439
Kumquat, 263

## L

*Laetia procera*, 543
*Lagerstroemia indica*, 576
Lagos, 821
*Laquillensis*, 601
Lathberry, 624
Lauraceae, 123
*Laurel avispillo*, 127, 129
Laurel family, 123
Laurel family, 123-131
*Lawsonia inermis*, 579
Legume family, 166-246
Leguminosae, 166
*Lemaireocereus*, 562

*Ligustrum japonicum*, 809
*Ligustrum sinense*, 807
Liliaceae, 51
Lily family, 51-60
Limber, 140, 141
Limoncillo del monte, 595, 607
Linguam, 148, 149
*Linociera axilliflora*, 811
*Linociera caribaea*, 813
*Linociera holdridgii*, 815
*Linociera ligustrina*, 817
*Lonchocarpus domingensis*, 232, 233
*Lonchocarpus glaucifolius*, 234, 235
Long key *byrsonima*, 313
Loosestrife family, 573-580
Loquat, 158, 159
Louvi, 541
*Lunania buchii*, 545
*Lyonia rubiginosa*, 741

## M

*Machaerium lunatum*, 236, 237
*Machaonia portoricensis*, 973
Macoucoua, 379
Madagascar palm, 39, 41
Madder family, 930-999
Mahoe, 460
Mahogany family, 298-308
*Majagua de sierra*, 569
Mallow family, 456-467
Malpighia family, 309-320
*Malpighia fucata*, 317
*Malpighia globra*, 319
Mamey del cura, 509
*Mamey sapote*, 779
Mameyuelo, 717, 750, 752
Mandarina, 261
*Mangle bobo*, 927
Mangostan, 520
Mangostan, 522
Mangosteen family, 513-523
Mangrove family, 592-594
*Manilkara pleeana*, 775
Maran, 321
Marble tree, 395
*Marlierea sintenisii*, 663
*Martin avila*, 947
Matabuey, 887
*Maytenus cymosa*, 401
*Maytenus elliptica*, 403
*Maytenus elongata*, 405
*Maytenus poncceana*, 407
*Mecranium amygdalinum*, 681
*Melaleuca quinquenervia*, 652
Melastome family, 669-730
*Melia azadirachta*, 303
*Meliosma obtusifolia*, 433
Melocha, 430
Mendrina, 545
Mexican cypress, 28
Mezereon family, 568-572
*Miconia affinis*, 683
*Miconia impetiolaris*, 688
*Miconia joveolata*, 685
*Miconia laevigata*, 690
*Miconia mirabilis*, 692
*Miconia pachyphylla*, 695
*Miconia punctata*, 697
*Miconia pycnoneura*, 699
*Miconia racemosa*, 701
*Miconia rubiginosa*, 703
*Miconia serrulata*, 705
*Miconia sintenisii*, 707

*Miconia subcorymbosa*, 709
*Miconia tetrandra*, 711
*Miconia thomasiana*, 713
Milkweed family, 833-836
Mirto, 661
Mora, 91, 93
Moraceae, 91
*Morisonia americana*, 151, 153
*Morus nigra*, 112, 113
*Mountain guava*, 665
*Mouriri domingensis*, 715
*Mouriri helleri*, 717
*Mozambique ochna*, 492
Mulberry family, 91-118
*Muntingia calabura*, 452
*Murraya paniculata*, 266
Murta, 593, 642, 715
Musaceae, 61
Myoporum family, 926-929
*Myrcia citrifolia*, 595
*Myrcia leptoclada*, 597
*Myrica carifera*, 83, 85
*Myrica fallax*, 657
*Myrica holdridgeana*, 86, 87
Myricaceae, 82
*Myricanthes fragrans*, 659
*Myriciaria floribunda*, 661
Myrsine family, 749-759
Myrtle family, 162, 163, 595- 668
Mysore, 97, 98

**N**

Neem, 303
Negra lora, 115, 116
*Nephelea portoricensis*, 8
*Nerium oleander*, 823
Nettle family, 119-122
Niaragato, 281
Nightshade family, 871-897
Nigua, 855
Nigua, 867
Nispro, 791
Nogal, 89, 90
Norfolk-island pine, 19-22

**O**

Obcordata, 389
Ochna family, 491-499
*Ochna mossambicensis*, 492
Oleander, 823
Olive family, 795-818
*Oncoba echinata*, 547
*Opuntia moniliformis*, 565
Orange-jessamine, 266
Oreganillo, 215, 216
*Oriental arborvitae*, 33, 34
Ossaea, 719
*Ottoschulzia rhodoxylon*, 413
*Ouratea ilicifolia*, 494
*Ouratea littoralis*, 496
*Ouratea striata*, 498
Oxalis family, 247-250

**P**

*Pachira insignis*, 477
Pachyphylum, 555
Pale lidflower, 603
*Palicourea alpina*, 975
*Palicourea barbinervia*, 977
*Palicourea crocea*, 979
*Palicourea domingensis*, 981
*Palicourea riparia*, 983
Palm family, 36-50

*Palma manaea*, 36
Palmea, 36
*Palo amargo*, 291
*Palo blanco*, 417
*Palo bronco*, 317
*Palo colorado*, 505
*Palo de candela*, 559
*Palo de cera*, 86, 87
*Palo de hueso*, 805
*Palo de jazmin*, 788
*Palo de muleta*, 630
*Palo de ramon*, 532
*Palo de rosa*, 413
*Palo iloron*, 941
*Palo lechoso*, 828
Pan, 133, 135
Papaveraceae, 132
Papelillo, 254
*Parathesis crenulatsa*, 756
*Parmentiera cerifera*, 910
*Parmentiera edulis*, 913
Pavona, 463
Pedunculasis, 507
Pencilbush, 340
Pepper family, 69-77
*Persea krugii*, 123, 124
*Persea urbaniana*, 125
*Persea urbaniana*, 126
*Phialanthus grandigolius*, 986
*Phoebe elongata*, 127, 129
*Phoebe montana*, 130, 131
*Picrasma antillana*, 289
*Picrasma excelsa*, 291
*Pilocarpus racemosus*, 269
*Pimienta de Brasil*, 372
Pine family, 23-26
Pinon, 227, 228
*Pinus caribaea*, 24
*Piper amalogo*, 70, 71
*Piper blattarum*, 73, 74
*Piper tuberculatum*, 75, 77
Piperaceae, 69
*Piscidia carthagenesis*, 238, 240
Pitangueira, 616
*Pithecellobium unguis-cati*, 189, 191
Plateado, 953
*Pleodendron macranthum*, 528
*Plumeria obtuse*, 826
Pochote, 472
Pomegranate family, 581-584
Ponceana, 407
Poor-man-orchid, 195, 197
Poppy family, 132-135
Portoricen, 605, 881, 905
*Pouteria dictyoneura*, 777
*Pouteria sapota*, 779
*Prockia cruces*, 549
*Prunus myrtifolia*, 162, 163
*Prunus occidentalis*, 164, 165
*Pseudolmedia spuria*, 115, 116
*Psidium amplexicaule*, 665
*Psidium sintenisii*, 655
*Psychotria berteriana*, 988
*Psychotria grandis*, 990
*Psychotria maleolens*, 992
*Psychotria maricaensis*, 994
*Psychotria nutans*, 996
*Pterocarpus macrocarpus*, 241, 243
Pterocarpus, 241, 243
Puerto Rican, 37
*Punica granatum*, 582
*Purple bauhinia*, 192, 193

## Q

Queensland kauri, 17
Quiebrahacha, 419, 638
Quina, 937, 939, 943

## R

Ramon, 117, 118
*Rascagarganta*, 756
Rat apple, 151, 153
*Ravenala madagascariensis*, 62, 63
*Ravenia urbanii*, 271
Red manjack, 843
Redberry, 622
Reidl, 385
Reseda, 579
*Retama prieta*, 211, 212
Retama, 244, 245
*Reynosia guama*, 437
*Reynosia krugii*, 439
*Reynosia uncinata*, 441
*Rhamnus sphaerosperma*, 443
*Ricinus communis*, 349
*Roble amarillo*, 915
*Roble colorado*, 921
*Roble venezolano*, 918
*Roble*, 839, 900
*Rochefortia acanthophora*, 853
Rocio, 252
*Rondeletia inermis*, 998
Rosa de cienega, 574
Rosaceae, 154
Rose family, 154-165
Roseta, 553, 963
*Roystonea regia*, 45, 47
Rue family, 258-285

## S

Sabia family, 432-434
*Sabinea florida*, 244, 245
Sacha, 549
Salicaceae, 78
*Salix babylonica*, 79, 81
*Samyda dodecandra*, 551
San bartolome, 847
*Sapium caribaeum*, 352
*Sapium jamaicense*, 354
Sapo, 142, 143
Sapodilla family, 760-781
*Sauce horon*, 79, 81
*Savia sessiliflora*, 356
*Schinus terebinthifolia*, 372
*Securinega acidoton*, 358
Seso, 421
Sessiflora, 640
Shaving-brush-tree, 477
*Sida eggersii*, 466
*Simarouba tulae*, 293
Sintensis, 754
*Siphoneugena densiflora*, 667
Sisu, 220, 221
Skyflower, 869
*Sloanea amygdalina*, 454
Soapberry family, 415-431
*Solanum antillarum*, 889
*Solanum erianthum*, 891
*Solanum polygamum*, 893
*Solanum torvum*, 895
Southern bayberry, 83, 85
Spineless acacia, 170, 172
Spurge family, 321-359
Stahili, 741
*Sterculia foetida*, 488

## W

*Wallenia pendula*, 758
Walnut family, 88-90
*Waxmyrtle* family, 82-87
West Indian, 443
West Indies, 164, 165, 399
White calliandra, 181, 182
White logwood, 557
White stopper, 614
Wild cotton, 457
Wild fig, 110, 111
Wild pigeon palm, 160, 161
Willow family, 78
Willow family, 78-81

## X

Xerophyta, 648
*Xylosma buxifolium*, 553
*Xylosma pachyphyllum*, 555
*Xylosma schaefferiooides*, 557
*Xylosma schwaneckeanum*, 559

## Y

*Yucca aloifolia*, 55, 57
*Yucca elephantipes*, 58, 59
Yuquilla, 738

## Z

*Zanthoxylum bifoliolatum*, 275
*Zanthoxylum caribaeum*, 277
*Zanthoxylum punctatum*, 279
*Zanthoxylum spinifex*, 281
*Zanthoxylum thmasianum*, 284
Zapote de costa, 775
*Ziziphus mauritiana*, 445
*Ziziphus rignonii*, 447
*Ziziphus taylorii*, 449